# Names and Formulas of Some Common Ions

| POSITIVE IONS | | NEGATIVE IONS | |
|---|---|---|---|
| *Single Charge*(+) | | *Single Charge*(−) | |
| Ammonium ion | $NH_4^+$ | Acetate ion | $C_2H_3O_2^-$ |
| Copper(I) ion | $Cu^+$ | Bromide ion | $Br^-$ |
| (cuprous ion) | | Chlorate ion | $ClO_3^-$ |
| Hydronium ion | $H_3O^+$ | Chloride ion | $Cl^-$ |
| (This ion exists only in water | | Cyanide ion | $CN^-$ |
| solutions and does not | | Fluoride ion | $F^-$ |
| occur in compounds.) | | Hydrogen carbonate ion | $HCO_3^-$ |
| Potassium ion | $K^+$ | (bicarbonate ion) | |
| Silver ion | $Ag^+$ | Hydrogen sulfate ion | $HSO_4^-$ |
| Sodium ion | $Na^+$ | Hydroxide ion   , | $OH^-$ |
| | | Iodide ion | $I^-$ |
| *Double Charge*(2+) | | Nitrate ion | $NO_3^-$ |
| | | Nitrite ion | $NO_2^-$ |
| Barium ion | $Ba^{2+}$ | Permanganate ion | $MnO_4^-$ |
| Calcium ion | $Ca^{2+}$ | | |
| Cobalt(II) ion | $Co^{2+}$ | *Double Charge*(2−) | |
| Copper(II) ion | $Cu^{2+}$ | Carbonate ion | $CO_3^{2-}$ |
| (cupric ion) | | Chromate ion | $CrO_4^{2-}$ |
| Iron(II) ion | $Fe^{2+}$ | Dichromate ion | $Cr_2O_7^{2-}$ |
| (ferrous ion) | | Oxalate ion | $C_2O_4^{2-}$ |
| Lead(II) ion | $Pb^{2+}$ | Oxide ion | $O^{2-}$ |
| Magnesium ion | $Mg^{2+}$ | Sulfate ion | $SO_4^{2-}$ |
| Manganese(II) ion | $Mn^{2+}$ | Sulfide ion | $S^{2-}$ |
| Mercury(I) ion | $Hg_2^{2+}$ | Sulfite ion | $SO_3^{2-}$ |
| (mercurous ion) | | | |
| Mercury(II) ion | $Hg^{2+}$ | *Triple Charge*(3−) | |
| (mercuric ion) | | Phosphate ion | $PO_4^{3-}$ |
| Nickel(II) ion | $Ni^{2+}$ | | |
| Tin(II) ion | $Sn^{2+}$ | | |
| Zinc ion | $Zn^{2+}$ | | |
| | | | |
| *Triple Charge*(3+) | | | |
| Aluminum ion | $Al^{3+}$ | | |
| Chromium(III) ion | $Cr^{3+}$ | | |
| Iron(III) ion | $Fe^{3+}$ | | |
| (ferric ion) | | | |

# INTRODUCTION TO CHEMISTRY

# INTRODUCTION

## TO

## CHEMISTRY

EDITION 7

## T. R. DICKSON

Cabrillo College

## JOHN WILEY & SONS, INC.

New York   Chichester   Brisbane

Toronto   Singapore

| | |
|---|---|
| *Acquisitions Editor* | Joan Kalkut/Nedah Rose |
| *Marketing Manager* | Catherine Faduska |
| *Production Editor* | Deborah Herbert |
| *Cover Designer* | Madelyn Lesure |
| *Text Designer* | Carolyn Joseph |
| *Manufacturing Manager* | Susan Stetzer |
| *Illustration* | Rosa Bryant |
| *Cover Image* | Ed Foster/Superstock |

This book was set in 10/12 New Baskerville by General Graphic Services and printed and bound by Malloy Lithographing.
The cover was printed by New England Book Components, Inc.

Figure 3-12    photograph Courtesy IBM

*Library of Congress Cataloging in Publication Data:*

Dickson, T. R. (Thomas R.), 1938–
  Introduction to chemistry / T. R. Dickson.—7th ed.
    p.   cm.
  Includes index.
  ISBN 0-471-04390-7 (pbk.)
  1. Chemistry.  I. Title.
QD31.2.D53  1995
540—dc20
                              94-33544
                              CIP

Printed in the United States of America

10 9 8 7 6 5 4 3 2 1

Printed and bound by Malloy Lithographing, Inc.

This Book is Dedicated to
All Curious and Enthusiastic
Chemistry Students

# PREFACE

This book is designed for a one-term course in chemistry. It presents the foundations of chemistry to students who need preparation for further study, as well as to those who wish to take only an introductory course. The story of chemistry is introduced at a beginning level and no previous background in chemistry is assumed.

A preparatory course should provide students with an honest view of chemistry as a central science and should give them a solid foundation and preparation for the future. This book is a product of more than thirty years of teaching, and it reflects my view that the introductory course is one of the most important chemistry courses. The purpose of the book is to teach chemistry by providing a dynamic, interesting, and relevant view of chemistry and its importance to society and our daily lives. This is accomplished through the interweaving of three main themes:

1. *Chemical thinking and chemical vision are emphasized.* Images and examples are used to clarify chemical concepts. Students are encouraged and reminded to use these images to develop a chemical imagination. A major goal is to provide them with chemical images that help them visualize chemical processes in the laboratory, their own bodies, and the environment. I hope that, long after they have forgotten many of the inert facts of chemistry, they will retain a chemical vision.

2. *Examples of chemical ideas and processes are emphasized by using activities and hands-on exercises.* These Activities and "Chemistry in Action" sections use safe and simple household chemicals to experience chemistry. These exercises are intended to engage students and to bring life and action to chemical concepts.

3. *Problem-solving methods of beginning chemistry are introduced, explained, and illustrated throughout the text.* Approaches to solving chemical problems by reasoning and the unit-equation, factor-label, or dimensional-analysis methods are explained in detail. A variety of worked-out examples are carefully explained. Most examples are followed by suggested practice activities. I have tried to use examples and questions that illustrate topics and applications of interest to students. A major goal is to instill problem-solving skills that will serve them in the future.

I have worked hard to retain the readability that has made this book useful as an introductory text. To supplement the interest level, occasional vignettes are dispersed throughout various chapters. These vignettes include topics related to health, technology, the environment, and society. Others refer to specific tools of direct use to students. Some examples are "Using a Calculator," "Seeing Atoms," "Food and Calories," "Hole in the Ozone Layer," and "Calcium in the Body." At the suggestion of many students a glossary of terms has been added at the end of the text for reference. In addition, several appendices are provided as optional reading and reference.

 **ORGANIZATION**

This edition has been completely revised based on comments from teachers and students who have used previous editions. A major change has been made in the presentation of topics. A new introductory chapter has been added to give students an overview of coming attractions. In this chapter, students are introduced to scientific thinking. The Kinetic Molecular Theory is presented to illustrate chemical thinking and to show the dynamic nature of chemistry. The fact that chemistry involves interesting and relevant topics is emphasized.

Chemistry is based on atoms and their behavior. Chapter 3 introduces the Atomic Theory and establishes a foundation for developing a chemical imagination. Chemistry is an experimental science, and ideas of chemical formulas and reactions are fundamental. Of course, the concept of the mole is at the heart of chemistry. These topics are presented in Chapter 4. Since problem solving involving stoichiometry is such an important skill, stoichiometry is presented in Chapter 5. The early introduction of this topic serves students well. If they learn anything about chemical problem solving, it is related to stoichiometric thinking.

With a basis of

1. an atomic and molecular vision,
2. a dynamic view of reactions and the Kinetic Molecular Theory, and
3. calculation and stoichiometry skills

students are well prepared to continue exploring the chemistry presented in later chapters.

The remaining chapters have been rewritten and rearranged to vitalize the topics and to include interesting examples, analogies, and images. I resisted the temptation to expand the material. The text retains its reasonable length. The topics covered in the chapters follows a purposeful order, but the topic and chapter order can be varied if desired. For instance, Chemical Nomenclature (Chapter 8) could be covered earlier. Furthermore, the last few chapters contain optional material.

 **SUPPLEMENTS**

A comprehensive package of supplements has been created to assist both the teacher and the student.

- **Study Guide by T. R. Dickson.** Each chapter contains objectives, glossaries, summaries, as well as further mathematical explanations as needed for text concepts. This valuable student guide also includes self tests with solutions.
- **Laboratory Experiments by T. R. Dickson.** This manual contains a choice of 33 experiments designed for students who have had no laboratory experience. Along with vitalizing important chemical concepts, the manual emphasizes laboratory techniques, procedures, and safety.
- **Instructor's Manual for Text and Laboratory Experiments by T. R. Dickson.** The Instructor's Manual includes solutions to all end-of-chapter problems and exercises in the text. Instructor's notes on materials and equipment as well as sample report sheets for the laboratory experiments are also provided.
- **Test Bank by T. R. Dickson.** The Test Bank contains approximately 1000 questions.
- **Computerized Test Bank.** IBM and Macintosh version of the entire Test Bank are available with full editing features to help you customize tests.

**ACKNOWLEDGMENTS**

I am very grateful to all of the students, teachers, and other individuals who have helped me in the preparation of this seventh edition. My special thanks to Francine Genta and Jennifer Barzee for help with the end-of-chapter problems. I appreciate the useful comments and suggestions of the following reviewers: Robert Bohn, University of Connecticut; James Coke, University of North Carolina—Chapel Hill; Roger Ernst, Southwest Missouri State University; Berner Gorden, Saginaw Valley State College; William Jensen, University of Cincinnati; Donald Kleinfelter, University of Tennessee—Knoxville; Julia Ellefson Kuehn, William Rainey Harper College; Evelyn Libertelli, Fairleigh Dickenson University; Ann Loeb, College of Lake County; Douglas Magnus, St. Cloud State University; Mark Marshall, Amherst College; Jerry Mills, Texas Tech University; Stephen Ruis, American River College; Elsa Santos, Colorado State University; Phil Silberman, Scottsdale Community College; Walter Volland, Bellevue Community College; John Whitman, Western Washingtion University, Cary Willard, Grossmont College.

In addition, I thank the editorial staff at John Wiley & Sons for their many years of professional support, advice, and suggestions.

*Aptos, California, 1994*                                        T. R. DICKSON

# CONTENTS

# CHEMISTRY

## 1-1 A TALE OF TWO TIMES

In 1991 the freeze-dried body of a Stone Age man was discovered preserved in the ice of a mountain high in the Italian Alps. The man apparently met a natural death about 5300 years ago. His is the oldest body every discovered fully dressed in the clothing of the times. He wore well-tailored deerskin pants and top complemented with leather boots and a parka-like outer garment made of reeds. Among his belongings was a pouch containing small tools and some fungus that may have had some medical use. His body had tattoos and he wore a small jewelry-like item that may have been of artistic or religious significance. A wood-framed leather backpack and a copper ax were found nearby.

Let's compare his clothing and possessions to a modern-day hiker. A typical hiker might wear well-tailored wool pants, an Orlon sweater and a parka made of Gore-Tex, two common synthetic fibers. She might have leather boots with synthetic rubber soles and carry a first aid kit containing synthetic medicines. Around her neck might be a chain with a piece of jewelry of artistic or religious significance and on her shoulder a tattoo of a butterfly. Her backpack might be nylon with an aluminum frame and she might have a stainless steel ice ax. Some things have not changed much in over 5000 years. What has changed dramatically is the availability of materials that humans can put to a variety of uses. Furthermore, the

Stone Age man appears to have been very resourceful and aware of his environment but he did not have the opportunity to learn chemistry.

### 1-2  USEFULNESS AND CURIOSITY CREATE CHEMISTRY

**Chemistry:**

*(kem ə strē), noun*
*chemistries, plural  1. the*
*study of substances which*
*compose the universe and*
*the processes by which they*
*act on one another.  2. the*
*science dealing with charac-*
*teristics of elements or sub-*
*stances, the changes that*
*take place when substances*
*combine to form other sub-*
*stances, and the laws of the*
*behavior of substances un-*
*der various conditions.*
*3. the application of the sci-*
*ence of chemistry to certain*
*subjects: the chemistry of*
*iron, the chemistry of life.*
*4. Slang, the existence of*
*some factor: the chemistry*
*between us is good, you*
*make my molecules move.*

**Chemistry** is the science of the properties, composition, and behavior of matter. Matter refers to the materials or substances that comprise the universe. Chemistry has evolved from the question: What are things made of and how can they be used? One human activity that has influenced chemistry is the search for and the isolation of useful materials. This is applied chemistry, or chemical technology, that today involves the isolation of hundreds of useful materials from the environment and the manufacture of thousands of different consumer products. Some examples of useful materials are iron ore and aluminum ore, sources of iron and aluminum metals; and crude oil, the raw material for petroleum products, plastics, and many medicines. Another part of chemistry is philosophical and involves human curiosity and speculation about the nature of materials. This is theoretical chemistry, which provides a chemical view of nature and explanations of varied natural processes. What are materials like iron ore, aluminum ore, and petroleum made of? What are the natures of metals, medicines, and plastics? Theoretical chemistry provides answers to these kinds of questions. Modern chemistry includes both the theoretical investigations and the practical applications of materials. In your study of chemistry you'll learn many interesting applications and you'll gain a vivid chemical vision of the world.

There are four basic branches of modern science and chemistry is one of the four:

**biology**—the science of living systems

**chemistry**—the science of the properties, composition, and behavior of matter

**geology**—the science of the earth and earth processes

**physics**—the science of matter, energy, forces, and motions

**Chem-, Chemi-, and Chemo:**

*prefixes referring to chemistry:*
    *chemotherapy: the treatment of a disease by use of specific chemicals.*
    *chemiluminescence: the production of light during a chemical process without a rise in temperature.*

Chemistry contributes ideas and information to the other sciences and, in turn, borrows from them. There is a close relation between the evolution of chemistry and the development of medical sciences. Modern medicine depends upon a chemical understanding of life processes and some important chemical ideas have come from the curiosities and imaginations of medical practitioners. Knowledge of chemistry is vital to the large variety of scientific disciplines ranging from agriculture to zoology.

Chemical terminology is part of modern, everyday communications:

Olive oil contains monounsaturated fats.

I'm on a low sodium diet.

Watch your calories.

You need more iron and calcium in your diet.

This is iodized salt.

Chlorofluorocarbons contribute to the destruction of the ozone layer.

Your study of chemistry will provide a comprehensible view of nature and help you understand chemical methods and terminology. What you learn about chemistry in this book will serve you well in future chemistry courses and other science courses. In addition, you'll develop a chemical understanding of everyday natural occurrences.

There is a large variety of materials that make up the environment. Matter is defined as anything that has mass and occupies space. Mass refers to the amount of material in a sample of a material. Mass is a property of matter that gives it inertia, a resistance to being set in motion or resistance to any change in motion. A baseball is much easier to throw than a more massive shotput ball, and obviously easier to catch. Matter occurs typically in the physical forms of solids, liquids, and gases. Substances or chemicals are alternate terms used to refer to material objects and matter is a collective term to refer to all materials. Chemicals are within the realm of chemistry. Chemists observe matter in the environment and they collect, separate, and isolate samples of interest. In the laboratory, chemicals are examined, tested, measured, analyzed, purified, combined, and thoroughly studied. Chemists have isolated and classified a wide variety of chemicals from the environment, and have developed methods to make a variety of "synthetic" chemicals. Some have practical use while others are analyzed and studied to satisfy theoretical curiosity and add to the scientific understanding of nature. One organization of chemists is the **International Union of Pure and Applied Chemistry (IUPAC).** The name refers to the two aspects of chemistry. Pure chemical research deals with **theoretical chemistry** in which materials are investigated because they are scientifically interesting. **Applied chemistry** concentrates on practical uses of materials. The applied and theoretical parts of chemistry often overlap. Some applied research reveals new ideas about nature; some pure research generates practical applications.

Write a few sentences using some chemical terms that you know.

## 1-3   SCIENTIFIC METHOD

Albert Einstein, one of the most famous 20th-century scientists, once said that the most incomprehensible thing about the world is that it is comprehensible. **Natural science** refers to the accumulation of knowledge about nature. However, science is more than that. It represents a way of thinking and provides a vision of nature and natural occurrences. An important part of science is the open communication of scientific knowledge. We sometimes think of modern science as a European creation. However, the evolution of science was influenced by Asian, African, Arabic, Hindu, and Hispanic cultures as well as European and American cultures.

Let's first consider the general pattern of scientific thinking. Scientific thinking starts with observations. Observations can be expressed as facts. A **hypothesis** is an educated guess that serves as a possible explanation of specific observations. An interesting or useful hypothesis is subjected to careful scientific scrutiny. Experiments are then designed to test hypotheses. Experimental observations may support a hypothesis or may require the development of a new hypothesis. Often scientists try to reproduce experimental observations related to the hypothesis. If the hypothesis is supported by experiments over time and, if the hypothesis is of general importance, it takes on the status of a **scientific law.** A law is an expression of some notable and consistent pattern in nature that is supported by experiment. A common example is the law of gravity: attractive forces exist between all material objects. You experimentally confirm the law every time you step on a bathroom scale to weigh yourself.

Another aspect of scientific thinking is the development of theories or scientific models to give general explanations of natural occurrences. A law reveals what to expect and a scientific theory provides a grand explanation. Let's take a closer look at the important parts of scientific thinking.

## Observations

Careful observations are crucial in science. Noting details and looking for patterns requires focused attention. Some observations are qualitative descriptions: the material is black and a solid. And some are quantitative: the solid weighs 2.3 pounds. A quantitative observation can be expressed as a measurement having a number, like 2.3, and a unit, like pounds. Facts are verifiable observations. Facts and measurements of materials are part of the story of chemistry.

## Hypotheses

A hypothesis is a possible explanation of certain observations. Such explanations should make scientific sense. A hypothesis is an educated guess or a reasonable explanation. Extensive testing and experimentation may suggest alterations to a hypothesis or even its complete rejection as a valid explanation.

Find a new and shiny penny. Inspect and observe the penny as closely as you can. Record all your observations about the penny. (You can make a dull penny shiny by rubbing it with toothpaste and rinsing it with water.)

## Experiments

The term experiment relates to the word experience. Experiments are carefully controlled observations. Often experiments are designed to try to confirm or support a hypothesis, or to disprove it. One valid experiment that disproves a hypothesis may be enough to reject it, or at least require its alteration. **Chemistry is an experimental science.** Often when new and significant experimental results are reported in scientific literature, other scientists repeat the experiments to see if they get results that are the same or at least very similar. In 1989 two scientists claimed that they could carry out a process called nuclear fusion using simple laboratory equipment at room temperature. If their hypothesis was correct, it had the potential of providing a vast source of cheap energy. Scientists around the world tried to repeat the cold fusion experiments. After months of effort, no scientist was able to confirm the results as originally reported. Because the original experimental results could not be reproduced, the hypothesis put forth by the two scientists was rejected. Nevertheless, research on the topic of cold fusion is continuing. Experiments are important in chemistry; that's why the chemistry laboratory is an integral part of chemistry. The laboratory serves as the crucible for putting hypotheses to the acid test.

## Laws

A hypothesis can be supported and confirmed by experimental observations carried out by many scientists over a period of time. In this way a hypothesis is accepted. Some hypotheses are of general scientific importance and some achieve the status of scientific or natural laws. It is important to realize that a scientific law can be changed if experimental evidence warrants a change. In other words, scientific laws are not necessarily fixed and unchangeable. However, many laws have survived the test of time and the skepticism of scientists; they are considered unlikely to change. Under certain circumstances, objects can carry positive or negative electrical charges. An example of a law that is important to chemistry is the law that describes forces between electrical charges which is given in the margin. The law is called Coulomb's law, since it was first stated by the French scientist, Charles **Coulomb,** in the late 1700s.

In later chapters you'll learn about other laws that are important to chemistry. Learning chemistry does not mean that you have to learn about numerous laws. Actually, there are only a few laws that are vital to your study of chemistry.

## Theories or Models

Theories that provide explanations of natural phenomena are elegant treasures of science. They provide a framework with which we envision

**Coulomb's Law:**
*Like charges repel and unlike charges attract. The force of attraction or repulsion that exists between two charged objects is proportional to the product of the sizes of the two charges, divided by the square of the distance between them.*

Express the first part of Coulomb's law by using pictures to illustrate the idea.

physical reality and make sense of mystifying natural occurrences. **Theories** provide general explanations of observations and allow the prediction of events yet to occur. Often a theory is stated as a grand generalization about nature. Some theories are called **scientific models.** An example is the scientific model of the solar system. This theory provides a beautiful and comprehensible view of the planets in motion about the sun; it explains their relative positions. The model is used to predict future positions of planets and other phenomena, such as solar and lunar eclipses or phases of the moon. Furthermore, the model provides us with a mental image of the solar system. The image can be drawn or represented by a three-dimensional "scale model." (Of course you have never seen the solar system but you probably can draw a sketch of it. Sketches or even scale models are simplified visions.) As you continue your study of chemistry, you'll learn about a few very important chemical theories that are the foundations of chemical visions.

The development of scientific theories does not always happen easily, quickly, or smoothly. Evolution of thought takes time. The modern view of the solar system, for example, took thousands of years to develop. At times, new ideas meet significant resistance. The famous Italian scientist, Galileo Galilei (1564–1642), was forced by church authorities to retract his views that the earth moved around the sun. Furthermore, he was put under house arrest for the last few years of his life so that he could not communicate his ideas to others. There are many cases in which scientists were suppressed by political, social, governmental, and moral pressures when their views conflicted with official views. Such suppression not only occurred in the medieval times of Galileo but also happens in modern times. In the 1950s, Linus Pauling, an American chemist, had his passport restricted and was not allowed to travel out of the United States. In the 1970s and 80s, Andrei Sakharov, a Russian physicist, was exiled to a small city and not allowed to communicate with other scientists. These two scientists were punished for speaking against the development of nuclear weapons. Despite his harassment by the government, Linus Pauling was awarded the 1963 Nobel Peace prize for his work to ban the testing of nuclear weapons in the atmosphere. Before his death, Sakharov was recognized for his scientific integrity and opposition to the government of the former USSR, and recently the Catholic Church admitted that Galileo was treated unfairly in the 1600s.

Scientists are only human and subject to personal value judgments, biases, prejudices, misinterpretations, self-serving behaviors, and peer pressures. Our discussion of the evolution of scientific ideas is based on hindsight. We will describe the results and leave out a description of the hard work, mistakes, erroneous experiments, rejected hypotheses, and personal foibles.

International communication of scientific ideas aids scientific progress. Results of scientific studies and research are reported in numerous scien-

"Freedom is the oxygen without which science cannot breathe."
David Sarnoff—1954

tific publications around the world. The **American Chemical Society (ACS)** is an organization of research chemists, industrial chemists, and chemists involved primarily in chemical education. The ACS sponsors periodic conferences and meetings, and publishes specialized books and various scientific journals. Similar organizations exist in other countries. The IUPAC is an international organization of chemists that holds meetings in various countries. Open and international communication is essential to the development of science.

---

### Fact or Fiction?

Beware of certain "facts." A **scientific fact** is a statement of a verifiable observation. A fact should be unambiguous, and it should be possible to reproduce it or at least confirm it. The use of facts can vary from quite reliable or quantitative to outright distortions or deceptions. Some examples are:

A fact: the normal freezing point of water is 0 degrees Celsius or 32 degrees Fahrenheit.

A vague or questionable fact: "low calorie dessert," a claim on some packaged desserts.

A distortion of fact: "contains no chemicals," a claim on some packaged foods and bottled water.

A nonfact: little green men have landed in Texas.

Sometimes apparent facts are used to mislead or deceive the reader. A scientist reads "facts" from a skeptical point of view. You've probably seen a product label that reads "new and improved." What does such a statement mean?

---

### 1-4 THE BEHAVIOR OF GASES

To illustrate scientific thinking let's consider the behavior of gases (see Chemistry in Action 1-1). **Air** is a mixture of gases; its main components are nitrogen and oxygen. Some other common gases that you have heard of are natural gas, propane, helium, and carbon dioxide. Gases are interesting forms of matter and they have intriguing properties. Many gases are colorless and are, therefore, invisible; we only know they are present by their other properties. This section illustrates scientific methods as related to some specific behaviors of gases. It is important to realize that development of scientific ideas occurs over long periods of time and involves the work and communication of many scientists. What is left out of our discussion is the historical perspective and the many years over which scientific ideas evolved. The point of the example is to illustrate the important aspects of the scientific method referred to in the previous section.

ACTIVITY 1-4

Analyze the "facts" in the box and explain why they are classified as indicated. What is the difference between the following statements?

Little green men have landed in Texas.

Little green men are reported to have landed in Texas.

## Observations

All gases are transparent. Many gases, because they have no color, are invisible. They are observed by noting their behaviors. One general observation is that hot air rises. We observe that a hot-air balloon rises in the cooler surrounding air. Another observation is that heating a gas causes it to expand and cooling a gas causes it to shrink in volume. The tires on a car heat up and expand as the car is driven. On cold days the tires decrease in volume and appear somewhat flat. To make closer, more careful experimental observations we use samples of gas kept in convenient, special containers. If we place a gas sample in a flexible container, it is described by the volume it occupies (represented as $V$) and the temperature it has (represented as $T$). In addition, a gas sample exerts a pressure. If we attach a pressure gauge to the container of the gas sample we can measure a specific pressure. When we heat such a sample we observe that the volume increases. When the sample cools we observe that the volume decreases.

## Hypothesis

A hypothesis that describes the behavior of all gases can be based upon the observed behavior of a few gas samples. Such a hypothesis is worded to describe the behavior in a simple but general way. One possible hypothesis is: the volume of a gas sample varies directly with the temperature; an increase in temperature increases the volume and a decrease in temperature decreases the volume.

## Experiments

Experiments are controlled observations. In designing experiments, it is important to try to consider all factors that influence the observations. When observing the relation between the volume and temperature of a gas sample, it is required that the pressure of the gas sample remains constant. This is necessary because the pressure of a sample can influence the volume of the sample. It is also desirable to design experiments that give quantitative observations or measurements. The measurements are the experimental data.

An apparatus, like that shown in Figure 1-1, is used to collect data on the relation between the volume and temperature of a gas sample. Use of such an apparatus allows the pressure of the sample to remain constant when volumes and temperatures are measured. The temperature is increased or decreased and the volume of the sample is observed at a specific temperature. Measurements have numbers and units. The typical unit used in science to measure temperatures is the degree Celsius, °C,

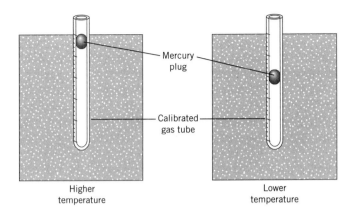

FIGURE 1-1
A simple apparatus used to measure the temperature and volume of a gas sample.

and a typical unit of volume is the liter, L. Suppose we ran an experiment on a gas sample and obtained the following data:

| VOLUME (L) | TEMPERATURE (°C) |
|---|---|
| 7.8 | 200 |
| 7.5 | 180 |
| 7.0 | 150 |
| 6.3 | 101 |
| 5.6 | 65 |
| 5.1 | 35 |
| 4.5 | 0 |
| 4.3 | −10 |
| 4.0 | −30 |

Note that as the temperature of the gas decreases, the volume decreases. Higher temperatures correspond to larger volumes and lower temperatures to smaller volumes. The data reflect this behavior as a set of numerical measurements, and they reveal that the volume of the sample varies directly with the temperature. Many similar experiments with different gas samples reveal the same general relation between volume and temperature. Scientific information is sometimes shown in a graphical form. A graph of the experimental data provides a nice visualization of the relation between volume and temperature. In other words, a plot of the data gives a picture of the variation of volume with temperature.

To graph the data, we note the ranges of the volumes and temperatures. The temperature range in the data is 230°C (from 200°C to –30°C) and the volume range is about 4 L (from 7.8 L to 4.0 L). The graph is marked to accommodate these ranges, then the axes are labeled.

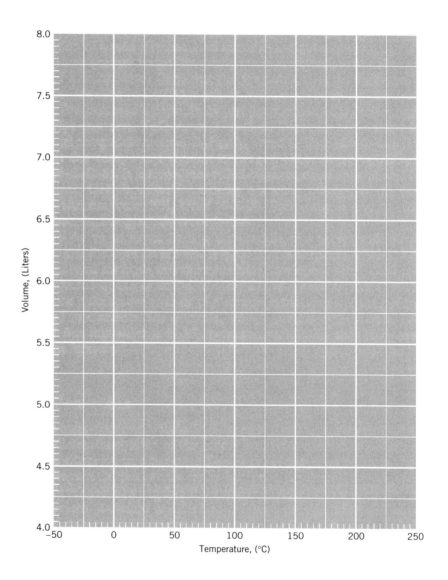

Once the axes are numbered and labeled the data points are placed on
the graph.

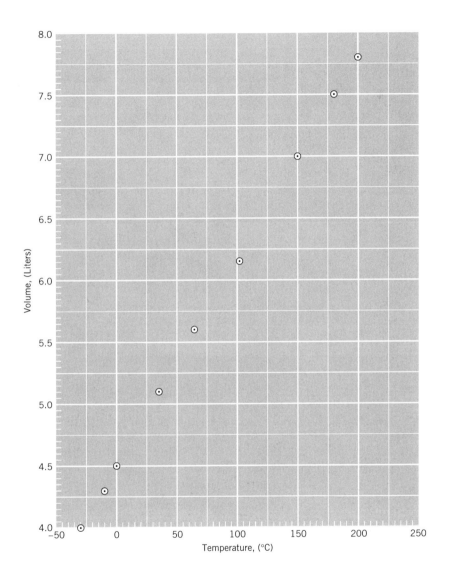

Since the volume appears to vary with the temperature in a direct straight-line manner, the data points are connected by a straight line.

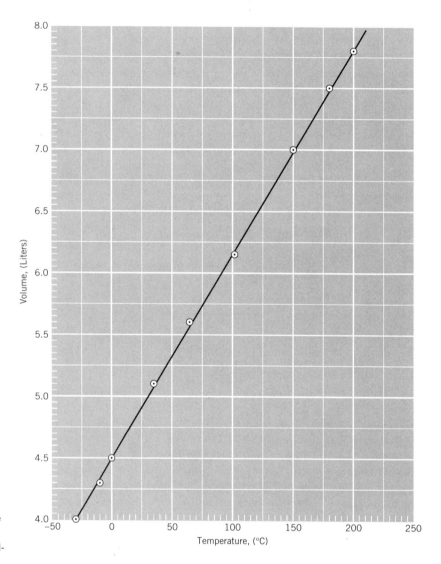

Temperature, (°C)

Information can easily be read from a graph. Practice reading the graph shown above by answering the following questions.

(a) What would the volume of the gas sample be at 175°C?

(b) At what temperature would the volume of the gas sample be 6.8 L?

The graph is important because it shows visually that the relation between volume and temperature follows a straight line or is linear. The volume and temperature do not vary in some complicated fashion; they follow a simple and direct relation. Repeated experimental observations of gases over various temperature ranges support the original hypothesis. The volume of the gas changes directly with the temperature.

### Law

The relation between the volume and temperature of a gas is important enough to be expressed as a law as given in the margin. The law is called **Charles' law** in honor of Jacques Charles, a French scientist and hot-air balloon enthusiast, who first stated it in 1787.

The term "varies directly" means that when the temperature increases the volume increases and when the temperature decreases the volume decreases. Charles' law is a formal statement of the simple idea that a gas, at constant pressure, expands when it is heated and contracts when it is cooled. It is a statement concerning the consistent behavior of gas samples and, therefore, is viewed as a scientific law.

**Charles' Law:**
*The volume of a gas sample at constant pressure varies directly with its temperature.*

### Theory or Model

A general theory of gases has been developed to explain the properties and behaviors of gases. The theory used to explain the dynamic behavior of gases is known as the kinetic molecular theory. The term "molecular" refers to an important chemical vision. This vision is that all matter is composed of minute particles much too small to be seen even with microscopes. The characteristic particles of many materials are molecules. Actually molecules are specific groups of atoms that we can view as tiny discrete particles. Samples of such molecular materials are viewed as collections of vast numbers of molecules. The term "kinetic" refers to motion or movement. A moving object has energy of motion, called **kinetic energy.** A moving ball, for instance, has kinetic energy and this energy is exchanged with objects that contact the ball. This is why a catcher's mitt gets warm when a ball hits it. Simply stated, the kinetic molecular theory provides a picture of a gas as a collection of molecules in motion. However, a theory is best stated in brief but precise terms or postulates. The **kinetic molecular theory** is stated as follows:

1. A gas consists of a collection of molecules that are on the average very far apart from one another.

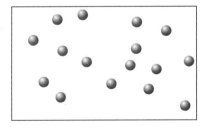

2. The molecules of a gas are in rapid, random, straight-line motion, and are constantly colliding with one another and any object in their envi-

ronment. The molecules of a gas travel at various speeds, and the motion of the molecules gives them kinetic energy.

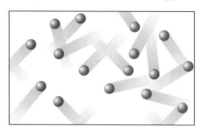

3. The attractive forces between molecules are negligible, so they do not stick together but instead rebound upon collision. Molecules can exchange kinetic energies when they collide.

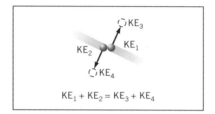

This theory provides a dynamic view of a gas as a collection of molecules rapidly darting about and constantly colliding with one another and other objects. The random motion of the molecules causes them to "fill" any container. The molecules of a gas sample are somewhat like minute balls on a three-dimensional pool table. A given molecule moves randomly and rapidly about until it collides with another molecule or the walls of the container. The molecules travel at various speeds, and the motions of molecules gives them kinetic energy. Thus, some molecules have relatively high kinetic energy and some have relatively low kinetic energy. Molecules can exchange kinetic energy when they collide. Most molecules have similar kinetic energies, which fall in the range of the average or typical kinetic energy of the collection of molecules. As an example of rapid motion consider that, in air, a typical oxygen molecule travels at about 1000 miles per hour. At such a speed it does not take long for a given molecule to collide with another molecule or some object. A typical molecule may average about four billion collisions each second. This means that the molecules travel relatively short distances between collisions. Upon collision they rebound, like pool balls, and continue their "cosmic dance." The difference between molecules and pool balls is that the molecules keep moving and do not roll to a stop.

It is not possible to see the molecules of a gas darting about, but the theory is very useful for explaining and making sense of the behavior of gases. The theory explains how gases exert pressures, and how the volume and temperature vary directly. In addition, the theory provides an important chemical vision, a mental image of gases.

## 1-5   USING THE KINETIC MOLECULAR THEORY

The kinetic molecular theory is a very useful model. It gives us a way to picture the behavior of matter, and a picture is worth a thousand words. It helps us to understand concepts, to explain observations, and to predict future observations. Often a scientific model provides new ideas or applications and, just as importantly, suggests new questions to answer. Consider some uses of the kinetic molecular theory.

### Gas Pressure

Even though you do not feel it, you are constantly being bombarded by the molecules of the air, as are all objects exposed to the atmosphere. The pressure exerted by a gas results from this constant bombardment. The various gases found in air create air pressure or **atmospheric pressure.**

To illustrate the idea of pressure, place the tip of a pen or pencil on a magazine or notebook and push down to make an indentation. Now jab the pen or pencil repeatedly into the paper. The repeated jabbing applies a pressure to the paper. This is similar to what happens when gas molecules collide with objects. The continuous collisions of gas molecules with objects in their environment cause the pressure. The molecules continually "jab" or "push" on the surface of objects. The pressure exerted by a gas depends upon how many collisions are occurring and how fast the molecules are moving.

### Temperature and Heat

The kinetic molecular theory provides a simple explanation of temperature and heat. Temperature and heat are related to the motions of the particles of matter (see Chemistry in Action 1-2). When a mercury thermometer is placed in a gas, the gas particles collide with the glass and, thus, indirectly with mercury particles composing the thermometer. This causes the motions of the particles in the thermometer to increase or decrease, since they exchange energy with the gas particles. As a result, the thermometer registers a specific temperature that reflects the motion of the gas particles (see Fig. 1-2). In a mercury or alcohol thermometer, as the liquid in the thermometer exchanges heat, it expands or contracts to reveal the temperature. In an electronic thermometer, a solid-state device called a thermistor changes its electrical properties depending upon the temperature.

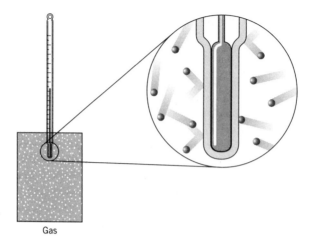

**FIGURE 1-2**
Collisions of gas molecules with a thermometer. (Not to scale. Molecules are very small.)

Gas

Explain the observations you made in Chemistry in Action 1-2 with food coloring in hot and cold water. Use your knowledge of temperature and heat in the explanation.

When a gas sample is heated, the average kinetic energy of the particles increases as the temperature of the gas increases. Cooling a sample decreases the average kinetic energy and, therefore, the temperature of the gas decreases. It is the motion of the particles that gives an object its measurable temperature, and **temperature** is a reflection of the average kinetic energy of the particles. Higher kinetic energy corresponds to higher temperature; lower kinetic energy corresponds to lower temperature. A direct relation exists between the average kinetic energy of the particles and the temperature.

Heat is also related to the motion of particles. Heat is thermal energy and is a medium of energy exchange between objects. When two objects at different temperatures are placed in contact, their particles collide and exchange energy, somewhat like pool balls that exchange energy when they collide. The hotter object cools down because the average kinetic energy of its particles decreases, while the cooler object warms up because the average kinetic energy of its particles increases. In time, the objects in contact come to "thermal equilibrium," which means that the average kinetic energies of their particles are the same, and therefore the objects have the same temperature. Note that heat and temperature are not the same thing. As an analogy, consider that two objects can have the same shade of red color, but one may be larger than the other. Two objects can have the same degree of hotness or coldness, that is, the same temperature, but one has a greater capacity to transfer heat than the other. For example, a burning match and a bonfire may have the same temperature, but the fire has a much greater capacity to transfer heat. Temperature does not depend upon the size of an object, but the ability to transfer heat does depend upon size. Temperature reflects the average kinetic energy of the particles, while the ability to transfer heat depends upon the number of particles and their motions.

### Chemical Vision

The kinetic molecular theory gives pictures of gases that allow us to understand their behaviors. We envision a gas sample as a collection of minute particles darting about at high speeds; it is this motion that contributes to the interesting properties of a gas. Our picture allows us to imagine the particles of a gas even if we can't see them. Most gases are colorless, so it is interesting that some are colored. Chlorine gas, for example, has a pale greenish-yellow color that fills the sample space. We can imagine that the molecules of chlorine gas have a greenish-yellow color and they "fill" the entire space of the sample by their continuous motion. The result is a gas sample that we can see through, but which has a distinct color. Another common colored gas is nitrogen dioxide, a reddish-brown gas that is one component of smog. This gas gives a reddish-brown color to a smoggy atmosphere.

How does the kinetic molecular theory explain Charles' law? Suppose we have a sample of a gas that has a certain volume at a specific temperature and is in a flexible container. Within the sample, the gas particles are moving about, filling the volume, and colliding with one another and the walls of the container. The sample has a measurable temperature, volume, and pressure. The pressure depends upon the number of collisions that are occurring with the walls of the container. The rapid and continuous motion of the molecules "pushes" on the surface of the container.

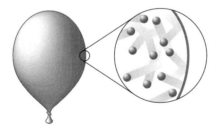

Heating a sample increases the average speed of the gas particles. This can increase the number of collisions with the container. A gas that behaves according to Charles' law has a constant pressure. To keep the pressure constant when a gas is heated, the gas sample expands and gives a larger volume to the flexible container. If the container is not flexible, there would be no way to keep the pressure constant. As a result, it makes sense that the volume changes directly with a change in temperature. If the volume does not increase when the speeds of the molecules increase, the pressure has to increase. This is what happens if the container is rigid. In a flexible container the gas expands upon heating, giving more surface area to the container; thus the pressure remains constant.

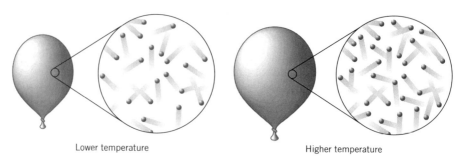

Lower temperature                    Higher temperature

## Celsius and Kelvin Temperature Scales

The invention of the Kelvin temperature scale illustrates how a scientific law generated a new idea. The most common temperature scale used in science is the **Celsius scale.** This scale is based upon the definition of the freezing point of water as zero degrees (0°C) and the boiling point of water as one hundred degrees (100°C). The Celsius degree, °C, is defined as one-hundredth of the temperature difference between the freezing point and the boiling point of water. Defining a temperature scale allows thermometers to be calibrated in appropriate units. Most laboratory and clinical thermometers are calibrated in degrees Celsius.

William Thomson was a Scottish scientist who worked in the late 1800s. When he was honored as an English lord he became Lord Kelvin, using the name of the Kelvin river that runs through the University of Glasgow. Kelvin suggested that it was useful to establish a temperature scale having a zero point corresponding to the lowest temperature possible, **absolute zero.** Temperature scales, like the Celsius scale, have arbitrary zero points. Zero on the Celsius scale, for instance, is defined as the temperature of the normal freezing point of water. On an absolute scale, temperature measurements would start at absolute zero. No temperature could lie below this value and there would be no negative temperatures. When gases are cooled to very low temperatures they become liquids. Kelvin reasoned that if the volume of a gas decreased with temperature according to Charles' law and if the gas did not become liquid, then it would theoretically have zero volume at the lowest temperature possible, absolute zero. A careful plot of the experimental data for the volume and the temperature of a gas sample is used to find the Celsius temperature corresponding to absolute zero. The plot gives a straight line, revealing the direct relation between volume and temperature. The resulting straight line on the graph is extended or extrapolated to the point at which the volume would theoretically be zero. Figure 1-3 shows such an extrapolation. The results of extrapolations of many graphs, using a variety of experimental data for

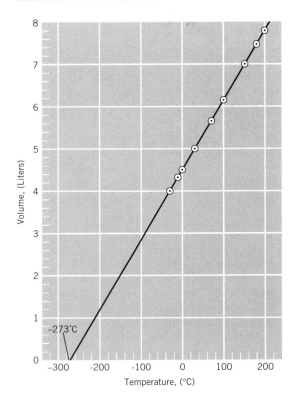

**FIGURE 1-3**
Extrapolation of a plot of the volume of a gas versus the temperature.

gases, reveal that absolute zero corresponds to −273°C (or more precisely −273.15°C). This seems to be the lowest temperature possible.

The **Kelvin scale** uses absolute zero as a starting point (0 K) and all other Kelvin temperatures have positive values. The Kelvin degree (abbreviated as K with no degree sign) is defined to be the same size as the Celsius degree. The difference between the two scales is that their zero points differ by 273 degrees. Typically, scientific thermometers are calibrated in degrees Celsius and temperatures are most often expressed in this scale. To change a **Celsius temperature to a Kelvin temperature we just add 273: $T(K) = T(°C) + 273.$** As you will see, when you study gases in more detail in Chapter 9, the Kelvin scale is useful when working with gas laws like Charles' law. Incidentally, Kelvin's name lives on in the brand name refrigerator, Kelvinator.

## Predicting Behavior

Using Charles' law to make qualitative predictions of gas behavior is quite simple. If a gas is in a flexible container, its volume increases with an increase in temperature and its volume decreases with a decrease in temper-

You need a balloon and some vanilla extract or cologne. Use an eyedropper or a drinking straw to place a few drops of vanilla extract or cologne into a balloon. Blow into the balloon to inflate it and tie it off. Place the balloon in a small room, like a bathroom or closet, and close the door. Open the door of the room after 10 minutes and note the odor. Explain your observations using the kinetic molecular theory of gases and draw pictures to illustrate your explanation.

Add a question to the list of questions.

ature. The law is also used to predict the quantitative behavior of gas samples; it is possible to predict the numerical increase in volume of a gas that occurs when it is heated. This kind of prediction is discussed in Chapter 9. The most important part of learning chemistry is the development of a chemical vision that you can use to make sense of many things that happen in nature. You now have a theory of gases that you can use to explain many observations to other people.

## New Questions

A theory provides explanations but also stimulates some interesting questions. A few questions suggested by the kinetic molecular theory are:

How are molecular motions related to other properties of gases?

How are molecular motions involved in the properties of liquid and solid materials?

How are molecules and molecular motions related to chemical changes and processes?

What is meant by saying molecules are minute particles? How minute are they?

What are the natures of molecules as minute particles?

These are valid questions. To answer them we need to continue the story of chemistry.

## Applications

A theory or law allows us to understand observations and might also suggest some novel applications. In addition, a theory allows us to understand natural phenomena of environmental interest. As an air mass or pocket of air near the earth's surface is warmed, it behaves as a flexible container and increases in volume. The increase in volume makes the pocket less dense than the air above it, so it moves up and mixes with the cooler air. This is how hot air rises. If we heat air in a lightweight balloon, the air expands and becomes less dense than the surrounding cooler air. The warmer air rises and takes the balloon with it. Hot-air balloons work nicely, but it is a good idea to take an air heater along on the ride. Why? Similarly, a fire heats the air around it; the hot air rises and carries the smoke with it. For this reason smoke detectors are placed on ceilings or high on walls. Furthermore, during a structural fire, the cooler, less smoky air is usually near the floor.

Air near the earth's surface is warmed by the water and rocks that have been heated by sunlight. The warmer air becomes less dense and rises. It is replaced by cooler air above. Figure 1-4 illustrates this normal vertical

**FIGURE 1-4**
The vertical circulation of air.

circulation or convection of air. Normally the temperature of air decreases with an increase in altitude, as shown in Figure 1-5a. In certain geographical regions, air masses get temporarily trapped in the atmosphere. This can happen when weather conditions allow a cooler air mass to move below the warmer air mass, as shown in Figure 1-5b. The warmer air mass becomes trapped between the cooler air below and above it. Normal vertical circulation of air does not happen under these conditions. This phenomenon is called a temperature inversion. In a **temperature inversion,** the temperature of the air decreases with altitude until the warm air mass is reached. The temperature of this air mass increases in altitude until the cooler air mass above is reached. In other words, the normal temperature

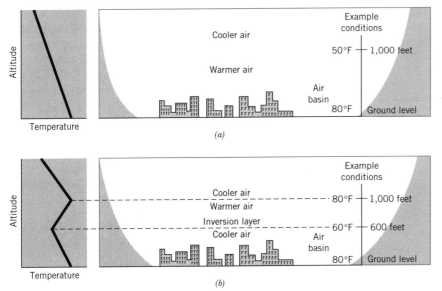

**FIGURE 1-5**
The formation of a temperature inversion in the atmosphere.

variation is inverted. The warmer air layer lies between the cooler air layers above and below.

A temperature inversion temporarily traps the warmer air mass. Environmental problems result when human-made air pollutants accumulate in the inversion layer and make smog. When this happens, you can often see a distinct hazy layer in the sky in the region where the warm layer contacts the cooler layer above it. The nature of air pollution and smog is discussed in Section 9-20.

### 1-6 LEARNING CHEMISTRY

This chapter is a preview of coming attractions and is designed to help you in your study. The ideas and examples presented are intended to give you a general view of scientific thinking and what you'll learn in this text. You are going to embark on an exploration of the basic ideas and techniques of chemistry. This text and your teacher are your guides.

This is your first course in chemistry so Chapter 1 serves as a general introduction. The next chapter introduces basic ideas about chemistry and gives you tools to use throughout the remainder of the book. We'll start with fundamentals and build on them as you explore the science of chemistry. You'll find the exploration to be interesting and useful to your future studies.

Chemistry has its own language. Many new words and terms are introduced along with the "grammar" for using them. Learning the language of chemistry gives you a new way to describe the world. In learning chemistry you'll gain a new vocabulary and new visions. Learning chemistry involves learning new ideas, new words, and new ways of looking at the world. You need to develop your own approach to studying and learning chemistry. The following sections will aid you in your adventure.

### Tools

Your textbook is the main source of information for your study of chemistry. Look at the table of contents and note that there are some useful appendixes at the end of the book along with a glossary of important terms. Your instructor will assign specific chapters in the text. Some students find it useful to scan a chapter first to see what topics and information are presented. Chapters contain focus sections that review or summarize some important ideas. As you carefully read assigned chapters, make notes on your reading. Underlining of important ideas can be useful. Many examples of problem solving are given in the text. Study them carefully for better understanding. Important new terms are printed in the text in boldface.

Each chapter has a set of end-of-chapter questions. Use them to test your knowledge and to practice problem solving. A variety of questions and problems are given to provide a range of choice. Material related to specific sections in the chapter are set off by section number. You are not expected to answer all of the end-of-chapter exercises, and your instructor may choose to assign selected questions. Those problems marked with an asterisk (*) have answers listed in the answer section at the back of the book.

Some students find the *Student Study Guide for an Introduction to Chemistry* to be a useful supplement. Your science department or school library has various chemistry reference books. Consult them for additional information about chemistry. A calculator is an invaluable tool to help you in calculations. A scientific calculator is the most useful type.

## Resources

Your teacher is your resource for guidance and information. Be creative and impress your teacher by asking good questions and participating in groups and discussions. Lecture notes are very useful, so practice taking careful notes and use them for study purposes. Homework questions and problems are assigned to allow you to practice what you are learning. Some students find study groups to be very useful. Groups can consist of two or more students. Group sessions are a good place to do homework, talk about ideas, or study for exams. Your school may have tutoring services available. Use the chemical visions you learn to explore the world. Practice telling others about what you are learning.

## Activities and Chemistry in Action

Each chapter has several activities designed to emphasize and illustrate the chemical ideas and concepts discussed in the text. Sophocles, an ancient Greek playwright, once said, "One must learn by doing the thing; though you think you know it, you have no certainty until you try." Try these activities individually or with study groups. You'll find activities on various pages. Read the material on a page before doing an activity. In addition, each chapter has one or more "Chemistry in Action" sections. These sections allow you to experience some interesting chemical processes using simple and safe household chemicals. You can do these hands-on exercises individually but it is recommended that you do them with study groups if possible. Your instructor may assign specific exercises. The exercises involve you in your exploration of chemistry and emphasize that chemistry is fascinating and fun. Appendix 6 gives an itemized list of materials and chemicals you need for specific exercises.

**ACTIVITY 1-9**

Give definitions for the following terms: chemistry, matter, experiment, theory, Charles' law, temperature, and heat.

## Terms and Vocabulary

Notice that chemistry has its own jargon and unique terminology, such as matter, molecule, and atom. Furthermore, some commonly used terms have special meanings in chemistry. The study of chemistry involves learning new terms. You'll have to memorize the definitions of some important terms. However, be selective about the terms you decide to memorize. If you are not sure about a term, ask your instructor or your study group. Many students use flash cards to aid memorization. Some students record tapes as study aids. Writing definitions and terms allows you to better picture them and tune into them.

## Problem Solving

Part of learning chemistry includes problem solving and calculations. Techniques of problem solving and calculation are introduced in Chapter 2 and emphasized throughout the remaining chapters. To learn proper methods of problem solving, carefully study examples given in the text and work some problems for practice. Some students find practicing problem solving with groups to be essential.

## Concepts and Visions

Two important goals of your study of chemistry are developing a chemical vision and developing an understanding of fundamental chemical concepts. Their accomplishment requires practice. Consider some examples.

1. A scientific fact should be logical. This means a fact should make sense, not just appear to make sense. Get used to thinking critically when you see a stated fact or a supposed fact. To practice, comment on the following "facts":

   Smoking causes cancer.

   This product contains 25% fewer calories.

   This product contains 75% less fat than the original kind.

2. Use mental images or draw pictures as aids for understanding concepts. Use your chemical vision. To practice, draw pictures to explain why bread rises as it is baked. (Hint: Charles' law is involved.)

3. Scientific thinking sometimes requires you to make a guess or simple hypothesis about some idea or observation. Work with another person and give a hypothesis for the following statement.

Refer to the discussion of the Kelvin scale in Section 1-5. Will the extrapolation of any set of volume-temperature data for gas samples give essentially the same extrapolated value for absolute zero? Suggest a method that can be used to test your hypothesis.

You are going to make two gases: carbon dioxide and oxygen. The chemicals used in these exercises are common household materials. Nevertheless, be careful when working with these chemicals and wear safety goggles when you do experiments. You need: white vinegar, baking soda, matches, household liquid bleach, 3% hydrogen peroxide (typical household medicinal variety), a small piece of plastic wrap, toothpicks, a paper towel or newspaper, and three small glasses (plastic if available). Dry baker's yeast can be used instead of bleach.

CAUTION: Although liquid bleach is a household chemical it is quite dangerous. Read the label on any bleach container that you use. Do not spill the bleach or let it splash on your skin or clothing. If you spill or splash it, quickly rinse the contaminated area with large amounts of water. Use paper towel or newspaper on your work area in case of spills. You may want to work in a laundry room, kitchen, or laboratory with a sink.

Record your observations. Describe the materials that are mixed and describe what happens when they are mixed. In addition, describe what happens when the gases are tested.

To make some carbon dioxide gas:

1.  Pour a small amount of white vinegar into a glass so that it fills the glass to a level of about 1 inch. Place the glass on a piece of paper.
2.  Obtain a heaping teaspoon of baking soda. Carefully stir the baking soda into the vinegar. Observe.
3.  Let the mixture bubble for about one minute. Strike a match and slowly immerse the burning match into the gas near the top of the glass. Instead of a match you can use the match to light a toothpick, then immerse the flaming toothpick into the gas. Observe. Repeat with another burning match or toothpick.

To make some oxygen gas:

1.  See Step 6 if you want to use baker's yeast. Pour a small amount of household liquid bleach into a glass so that it fills the glass to a level of about half an inch.
2.  Wrap the top of the glass with some plastic wrap.
3.  Add two tablespoons of 3% hydrogen peroxide to an empty glass.
4.  Lift an edge of the plastic wrap on the bleach glass and pour in the hydrogen peroxide. Quickly replace the plastic wrap.
5.  Light a toothpick and let it burn. Blow out the flame so that the toothpick is glowing hot. Lift an edge of the plastic wrap and immerse the glowing toothpick into the gas in the glass. When the toothpick ignites, blow it out and continue dipping the glowing toothpick into the glass. Repeat the process a few times using a new toothpick if needed. Be very careful to pour the contents of the glass into a drain followed by rinse water.
6.  Place one teaspoon of powdered baker's yeast in the glass instead of the bleach. Add two tablespoons of 3% hydrogen peroxide and stir the mixture. Oxygen gas will be trapped in the bubbles of

the foam that is produced. Immerse a flaming toothpick into the foam bubbles. If it stops burning, light another toothpick and try again.

Observations include descriptions of all that you see in an experiment. They serve to describe what happened. Part of the process of chemistry is to try to explain why things happen. Why were gases formed in these reactions and why do the gases have different properties? Answers to these questions are part of your study of chemistry. For now, however, we are just concerned with making observations. Just record your careful observations; do not try to explain them in chemical terms. Explanations or interpretations will come later. One very important observation involved in these exercises relates to gases. We observed that the gases have different properties related to their abilities to support burning. The curious observation was that the gases were not visible. The presence of these invisible forms of matter was observed by noting that they had properties that differed from the surrounding air.

Here is a simple exercise to compare hot and cold water. You need: red food coloring (or any other color), two glasses, hot water, and cold water.

1.  Fill a glass with cold water and set it on a flat surface.
2.  Boil some water or use very hot tap water.
3.  Fill the second glass with hot water and set it next to the first glass. **CAUTION:** If you are using glass instead of plastic, be sure to warm the glass first using hot tap water.
4.  Add two drops of red food color to each glass of water. Allow them to sit undisturbed for a few minutes then record your observations. What do you notice about the way in which the coloring disperses in the water samples?

## QUESTIONS

### Sections 1-1 to 1-3

1. Give a definition of chemistry that uses terms that are different from those used in the definition given in the text.

2. Do you think that the Stone Age man "knew" some applied chemistry? Why?

3. What two human activities lead to the development of the science of chemistry?

4. What is natural science?

5. What is the difference between a qualitative observation and a quantitative observation? Give some observations about yourself to illustrate the difference.

6. What are the general parts or aspects of scientific thinking?

7. Describe each of the following.

   (a) hypothesis
   (b) experiment
   (c) scientific law

8. Draw a picture to illustrate the first part of Coulomb's law given on page 5.

9. What is a theory or a model?

10. Why are theories useful in science?

11. Critically analyze the following "facts."

    (a) 50% less salt

    (b) eating sweets makes you fat

    (c) 95% fat free

    (d) contains no cholesterol

    (e) an apple a day keeps the doctor away

    (f) a monkey is reported to have given birth to twin humans

## Sections 1-4 and 1-5

12. Give a written statement of Charles' law.

13. Explain Charles' law using terms different from the formal statement given in the text.

14. Using graph paper, plot the following experimental data relating the volume and temperature of a gas sample. Plot the volume versus the temperature.

| Volume (L) | Temperature (°C) |
|---|---|
| 4.09 | 225 |
| 3.68 | 175 |
| 3.31 | 130 |
| 2.65 | 50 |
| 2.36 | 14 |
| 2.16 | −10 |
| 2.04 | −25 |

15. Describe the kinetic molecular theory and draw pictures to illustrate it.

16. State the essence of the kinetic molecular theory in a few sentences.

17. How do gases exert pressures?

18. What is temperature?

19. What is heat? How do heat and temperature differ?

20. How is heat exchanged between objects?

21. Explain why the volume of a gas sample in a flexible container decreases when the temperature decreases. Use pictures in your explanation.

22. What is the Celsius temperature scale and how is it defined?

23. What is the Kelvin temperature scale and how is it defined?

24. Prepare a piece of graph paper with a temperature axis extending from −300°C to 225°C and the volume axis extending from 0 L to 10 L. On the graph plot the two sets of volume–temperature data given in Section 1-4 and Question 14. After plotting two lines on the graph, extend or extrapolate the lines to find the temperature corresponding to zero volume.

25. Explain why hot air rises.

26. Explain how a hot-air balloon works. Instead of submitting a written answer to this question you can explain how a hot-air balloon works to a friend, relative, or acquaintance and get a note stating that you provided an adequate explanation.

27. Why should smoke detectors be placed on ceilings or high on walls?

28. Where would you expect to find cooler, less smoky air in a structural fire?

29. Why is a ceiling fan useful in the winter? Can a ceiling fan be useful in the summer?

30. Why are forced air heater vents usually placed at floor level?

31. What is a temperature inversion in the atmosphere?

32. What happens when a gas in a rigid container is heated? Draw a picture to illustrate your answer.

33. Suppose that you are camping and using an air mattress. You fill the mattress with air at 25°C. What happens to the volume of the mattress late at night when it cools to 0°C? Assume that the pressure is constant.

34. Why is a heater needed on a hot-air balloon?

35. Explain how a fireplace chimney works.

## Section 1-6

36. Here is a question involving scientific ethics. The tissue of the Stone Age man found frozen in the ice was essentially freeze-dried. Scientists have analyzed the DNA in the cells of a sample of his tissue. What are your opinions about such an analysis, or even the use of this DNA from an ethical perspective?

37. Here is a question asking you to use available tools and resources. Temperatures in science are most often expressed in degrees Celsius, °C. It is easy to change from degrees Celsius to kelvins since both have the

same degree size and their zero points differ by 273; $T(K) = T (°C) + 273$. In the United States, temperatures are often expressed in degrees Fahrenheit, °F. No simple relation exists between the Celsius and Fahrenheit scales. The two scales have different zero points and degree sizes ($1°C = 1.8°F$). For temperatures in the range of about 10°C to 60°C, it is possible to estimate the corresponding Fahrenheit temperature. To do this, double the Celsius temperature and add 25 to give the approximate Fahrenheit temperature. For example, 40°C is about 105°F ($40 \times 2 + 25 = 105$). Actually it's 104°F. To estimate Celsius temperatures that correspond to Fahrenheit temperatures in the range of about 40°F to 140°F reverse the process. That is, subtract 25 from the Fahrenheit temperature and divide by 2. For example, 85°F is about 30°C ($85 - 25 = 60$, $60/2 = 30$). It's actually 29°C. This estimation method only works within the ranges indicated.

Mathematical formulas are available to precisely convert Celsius temperatures to Fahrenheit and Fahrenheit temperatures to Celsius. Find these formulas, then apply them to the following temperature conversions.

(a) What is 98.6°F (typical body temperature) in degrees Celsius? (Express your answer to three digits.)

(b) Convert 10°C, 20°C, 30°C, 40°C, 50°C, and 60°C to degrees Fahrenheit. (Express your answers to two digits.)

Compare the precisely converted temperatures to those you find by using the estimation methods.

# MEASUREMENT

# IN CHEMISTRY

## 2-1 TAKING MEASURE

Chemicals are responsible for the shapes, textures, colors, and appearances of the things that surround you. Descriptions of materials are often based upon measurements, as shown by Chemistry in Action 2-1. Before we explore some fundamental concepts and theories of chemistry, we need to learn something about numbers and measurements. According to William Thomson (Lord Kelvin, mentioned in Chapter 1), "When you can measure what you are speaking about, and express it in numbers, you know something about it." In this chapter we'll learn about measurement and the significance of his statement.

Describe yourself. In response you might give some qualitative observations, for example, your hair color or the color of your eyes. More quantitative observations would include measures of your height and weight. Human societies have developed varieties of units of measurement and measuring devices. Modern supermarkets have electronic balances used to weigh produce, while merchants in rural China accomplish the same thing using hand-held balances. Historically, commerce, trade, and taxation have fostered the development of units of measure along with appropriate measuring instruments and calculating devices.

A few examples of the origins of units show how closely they relate to humans. The cubit, a biblical measure, was the length from the elbow to the tip of the middle finger. The foot originated from the length of the human foot, and the inch from the width of a human thumb. A mile was first defined by the Romans as 1000 double paces and was more precisely defined as 5280 feet by the English. The Latin word for weight is pound. Pound became the name of a commercial unit of weight first defined and used in France. A unit of weight still used in the pharmaceutical industry is the grain, which originated from the weight of a grain of wheat.

Until the beginning of the 1800s, various city-states, countries, and regions had their own systems of measuring and weighing. These systems often used units that were different from those used by others. This resulted in confusing commercial communication. Scientists of the time also used some of the common commercial units of measure. The lack of standardized units not only made commerce and trade difficult, it complicated communications between scientists. In 1791, French scientists began to establish a new system of measurement that has become the modern **metric system.** Normally it is quite difficult to convince the general populace to adopt a new system of measure. But the late 1700s and early 1800s were the years of the French revolution, a time of significant social change. Consequently, the metric system was widely adopted as a commercial system and a scientific system of measure. The **metric system** uses units like the meter and the gram instead of miles, yards, feet, inches, pounds, and ounces.

How is it possible to describe the objects and events that are the physical environment? From a qualitative view, poets and writers of literature have done this throughout the ages. In a sense, measurement is the poetry of science. To measure means to find the extent, size, quantity, or capacity of something. It is possible to describe the environment by measuring objects within it. To express measurements, we need fundamental units for length, mass, time, temperature, and volume.

## 2-2  THE METRIC SYSTEM

The metric system is the common system of reference units used in science. The metric system has evolved into the **International System of Units** sometimes called **SI units** from the French, "Système International d'Unités." Over the years, the metric system was refined; it currently consists of a set of specifically defined units that serve as an internationally accepted system of measurement. Nearly all countries of the world use metric units or SI units for business and commerce, and for scientific applications. These units provide for worldwide agreement in measurement.

The basic units or base units of the metric system are listed on the next page. Each unit has an abbreviated symbol that is used for convenience.

| OBSERVATION | UNIT | SYMBOL |
|---|---|---|
| mass | **gram** | **g** |
| length | **meter** | **m** |
| time | **second** | **s or sec** |
| temperature | **kelvin** | **K** |
| | or | |
| | **degree Celsius** | **°C** |
| volume | **liter** | **L** |

A variety of measuring instruments is available that allows precise measurements in these metric units. For example, balances are used to measure mass. The process of measuring the mass of an object on a balance is called weighing. Calibrated wheels or rulers are used to measure distances. Thermometers are used to measure temperature. Precise clocks or chronometers are used to measure time.

## 2-3  THE LARGE AND THE SMALL

Objects and the distances between them range from very large to very small. The invention of the telescope allowed scientists to look at the solar system and the stars. Telescope comes from the Greek *tele* 'distant' and *skopos* 'watcher'. The invention of the microscope allowed scientists to observe very small objects. Microscope comes from the Greek *mikros* 'small' and *skopein* 'watcher'. Those objects that are visible only with the aid of a microscope are said to be microscopic in size. Things that are too small to be seen with microscopes are said to be submicroscopic.

A **meter** is slightly longer than a yard, and it is a good unit to measure people-sized objects. The height of a typical person is between one and two meters. The diameter of the sun is about 1,400,000,000 meters and the diameter of the earth is about 13,000,000 meters. In contrast the average diameter of a red blood cell is 0.000009 meter or the length of an oxygen molecule is about 0.0000000001 meter.

In science large and small numbers are often written in a form called **exponential notation** or **scientific notation.** Exponential notation is a method by which numbers are expressed in terms of powers of ten rather than the normal form. Any multiple of ten can be expressed as 10 raised to a power.

$$100 = (10)(10) = 10^2 \qquad \text{one hundred}$$
$$1,000 = (10)(10)(10) = 10^3 \qquad \text{one thousand}$$
$$10,000 = (10)(10)(10)(10) = 10^4 \qquad \text{ten thousand}$$
$$100,000 = (10)(10)(10)(10)(10) = 10^5 \qquad \text{one hundred thousand}$$
$$1,000,000 = (10)(10)(10)(10)(10)(10) = 10^6 \qquad \text{one million}$$
$$1,000,000,000 = (10)(10)(10)(10)(10)(10)(10)(10)(10) = 10^9 \qquad \text{one billion}$$

A fraction of ten can be expressed as 10 raised to a negative power.

$$1/10 = 10^{-1} \qquad \text{one tenth}$$
$$1/100 = 1/(10)(10) = 10^{-2} \qquad \text{one hundredth}$$
$$1/1{,}000 = 1/(10)(10)(10) = 10^{-3} \qquad \text{one thousandth}$$
$$1/1{,}000{,}000 = 1/(10)(10)(10)(10)(10)(10) = 10^{-6} \qquad \text{one millionth}$$

To write a number that is greater than one in exponential notation form, move the decimal point to the left until it is between the last digit and the next to last digit on the left.

$$1\,400\,000\,000$$

Count the number of places the decimal was moved; use the count as an exponent of 10. Drop the extra zeros, then write the number as the remaining digits multiplied by 10 raised to this exponent.

$$1{,}400{,}000{,}000 \text{ becomes } 1.4 \times 10^{9}$$

To write a number that is less than one in exponential notation form, move the decimal point to the right until it is on the right of the first nonzero digit.

$$0.\,000\,009\,2$$

Count the number of places the decimal is moved; use the count as the negative exponent of 10. Drop the leading zeros, then write the number as the remaining digits multiplied by 10 raised to this negative exponent.

$$0.0000092 \text{ becomes } 9.2 \times 10^{-6}$$

Exponential notation is just a convenient way to write large and small numbers to avoid the use of numerous zeros. A few examples of measurements in exponential notation illustrate its convenience. The diameter of the sun is about $1.4 \times 10^{9}$ meters. Earth has a diameter of about $1.3 \times 10^{7}$ meters. The diameter of a typical red blood cell is about $9 \times 10^{-6}$ m. An oxygen molecule has a length of about $1 \times 10^{-10}$ m. Figure 2-1 shows the range of sizes of a variety of objects using exponential notation.

In the year 2000 the population of the earth is expected to be about 6.1 billion people. Write this number in exponential form and as a normal number using zeros.

**EXAMPLE 2-1**

The distance from our solar system to the nearest star is about 100,000,000,000,000,000 meters. What is this distance in exponential notation?

$$100{,}000{,}000{,}000{,}000{,}000 \text{ m becomes } 1 \times 10^{17} \text{ m}$$

**EXAMPLE 2-2**

A certain type of virus measures 0.0000008 meter in length. Express this
length in exponential form.

$$0.0000008 \text{ m becomes } 8 \times 10^{-7} \text{ m}$$

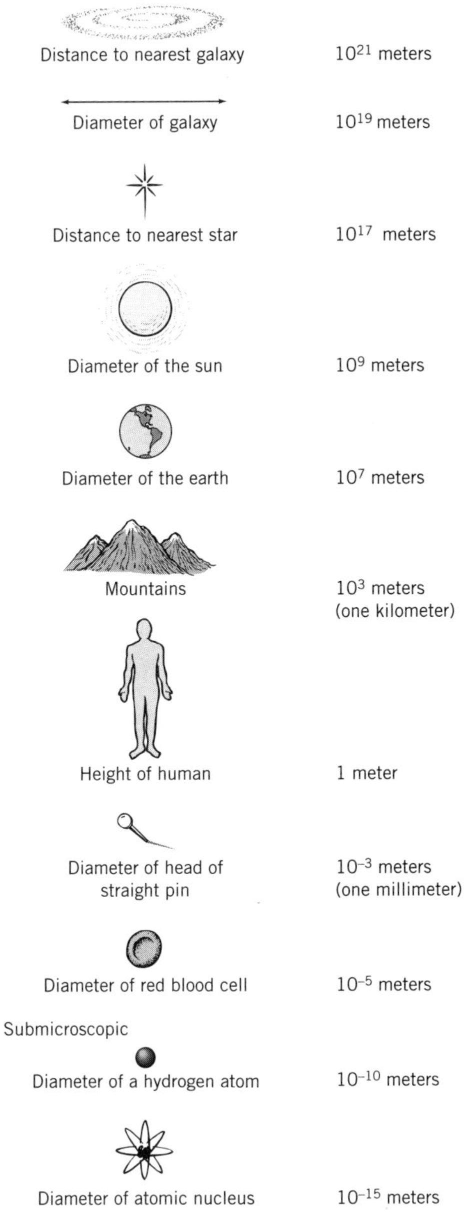

Distance to nearest galaxy          $10^{21}$ meters

Diameter of galaxy                  $10^{19}$ meters

Distance to nearest star            $10^{17}$ meters

Diameter of the sun                 $10^{9}$ meters

Diameter of the earth               $10^{7}$ meters

Mountains                           $10^{3}$ meters
                                    (one kilometer)

Height of human                     1 meter

Diameter of head of                 $10^{-3}$ meters
straight pin                        (one millimeter)

Diameter of red blood cell          $10^{-5}$ meters

Submicroscopic
Diameter of a hydrogen atom         $10^{-10}$ meters

Diameter of atomic nucleus          $10^{-15}$ meters

**FIGURE 2-1**
Variations in sizes and
distances.

Changing exponential numbers to normal form just requires moving the decimal point and adding zeros as needed. For example, to express the number $3.4 \times 10^6$ in normal form, we move the decimal six places to the right and add zeros; thus $3.4 \times 10^6 = 3,400,000$. When an exponential number has a negative exponent, convert to the normal form by moving the decimal point to the left. For example, to change $2.9 \times 10^{-9}$ to normal form, move the decimal nine places to the left; thus $2.9 \times 10^{-9} = 0.0000000029$.

## 2-4 METRIC PREFIXES

Prefixes are sometimes used in language to refer to size; a microcomputer is a small computer and megabuck is slang for a million dollars. It is easy to express very long and very short distances in the metric system. This is possible by using a set of prefixes that denote large and small multiples of the basic units. The prefixes are names for various multiples or fractions of 10.

To illustrate how prefixes work in the metric system, suppose we want to measure a distance somewhat less than a meter. By dividing a meter into 100 equal parts, we can measure lengths in units of 1/100 of a meter. This fraction is given a prefix name so length can be expressed in terms of the prefix. In the metric system, the prefix **centi-** means 1/100. Thus, a measurement of length can be expressed in units of centimeters (cm). We know that one centimeter is 1/100 of a meter and that there are 100 centimeters in one meter.

As another example, suppose you wanted to express your weight in metric units. You weigh several thousand grams, so the gram is not a convenient unit to express your weight. For example, 130 pounds is 59,000 grams. The metric prefix **kilo** means 1000, so you could express your weight in units of kilograms (kg). Doing this, we would know that one kilogram is 1000 grams. 130 pounds is 59 kilograms.

Prefixes are used to express multiples or fractions of any metric base unit. You'll find the three most commonly used prefixes listed below. (See the margin for some less common prefixes.)

### Metric Prefixes

| PREFIX | SYM-BOL | VALUE |
|---|---|---|
| tera | T | $10^{12}$ |
| giga | G | $10^9$ |
| mega | M | $10^6$ |
| kilo | k | $10^3$ |
| hecto | h | $10^2$ |
| deca | da | $10^1$ |
| deci | d | $10^{-1}$ |
| centi | c | $10^{-2}$ |
| milli | m | $10^{-3}$ |
| micro | μ | $10^{-6}$ |
| nano | n | $10^{-9}$ |
| pico | p | $10^{-12}$ |
| femto | f | $10^{-15}$ |
| atto | a | $10^{-18}$ |

| PREFIX | ABBREVIATION | MEANING |
|---|---|---|
| **kilo-** | **k** | 1000 or $10^3$ times unit |
| **centi-** | **c** | 1/100 or $10^{-2}$ times unit |
| **milli-** | **m** | 1/1000 or $10^{-3}$ times unit |

It is possible to use prefixes with any metric base unit. A prefix is placed in front of the base unit to represent the indicated multiple or fraction. The distance 200,000 m can be expressed as 200 km (read kilometers), and the mass 0.0023 g can be expressed as 2.3 mg (read milligrams). Note that the prefixes centi- and milli- have the same meaning they have in our monetary system. One cent is 1/100 of a dollar and one mill is 1/1000 of a dollar.

## 2-5  CONVERSION FACTORS

Experimental chemistry involves making measurements of the properties of matter. Suppose the mass of an object is found to be 288 g. The unit of grams reveals that this is a measure of mass. Take note that all measurements must include both a number and a unit. A measurement without a unit is meaningless. Computations with measurements are quite common in chemistry, so let's consider some typical computations using familiar examples.

If a wooden board is 78 inches long, how long is it in units of feet? The abbreviation in. is used for the inch. To change inches to feet we need the relation between them: (1 ft/12 in.). A slash (/) is read as per, so we say that there is 1 foot per 12 inches. Such a relation is a conversion factor because we use it to change or convert from one unit to the other. If we want to change a length in inches to feet, we just divide by 12. This process is important, so let's look closely at how a conversion factor works. Computation with the foot-to-inch conversion factor is represented as

$$78 \text{ in.} \times \frac{1 \text{ ft}}{12 \text{ in.}} = 6.5 \text{ ft}$$

A **conversion factor** of this type always has a numerator involving one unit and a denominator involving another unit. Notice how the units divide out or cancel as shown by the cancel marks. That is, the inches cancel and the desired units of feet remain. It is possible to cancel like units in the numerator and denominator by following the rules of algebra. The numerical part of the calculation involves dividing 78 by 12.

Since the use of conversion factors is important in many computations, let's see how their use is justified. Conversion factors express a relation between two units. For example, we know that the relation between inches and feet is

$$1 \text{ ft} = 12 \text{ in.}$$

From this relation we can obtain a conversion factor to convert from inches to feet, and a factor to convert from feet to inches.

$$\frac{1 \text{ ft}}{12 \text{ in.}} \quad \text{or} \quad \frac{12 \text{ in.}}{1 \text{ ft}}$$

Which factor is used depends on whether we want to convert from feet to inches or vice versa. Factors like these are called unit factors. They relate units of measure. As we shall soon see, unit factors are very useful in chemical computations. A few more examples illustrate the use of such factors.

---

**EXAMPLE 2-3**

A person is 5 ft 6 in. (5.5 ft) tall. What is this height in inches? You can do this calculation easily. However, consider the formal or logical "setup" needed for the calculation. The calculation involves multiplying the measurement by the unit factor relating inches to feet.

$$X \text{ in.} = 5.5 \text{ ft} \times \frac{12 \text{ in.}}{1 \text{ ft}} = 66 \text{ in.}$$

We start by deciding what unit is needed in the answer, then choose the factor that allows the unwanted units to cancel, thereby giving an answer in terms of the desired unit. The numerical parts of the factors are multiplied to give the numerical part of the answer.

---

**EXAMPLE 2-4**

Express 1 hr in units of seconds. In other words, find the number of seconds in one hour. One approach to this problem is to convert the hours to minutes, then the minutes to seconds. This requires the unit factor relating minutes to hours,

$$\frac{60 \text{ min}}{1 \text{ hr}}$$

and the factor relating seconds to minutes,

$$\frac{60 \text{ s}}{1 \text{ min}}$$

To solve the problem using these factors, first multiply 1 hr by the (60 min/ 1 hr) factor, then multiply this product by the (60 s/1 min) factor. Start by noting the unit needed in the answer.

$$X \text{ s} = 1 \text{ hr} \times \frac{60 \text{ min}}{1 \text{ hr}} \times \frac{60 \text{ s}}{1 \text{ min}} = 3600 \text{ s or } 3.6 \times 10^3 \text{ s}$$

Notice that the desired units result when we algebraically cancel the hour and minute units. If we had used the wrong factors, the units would not have worked out correctly; including the proper units is essential.

---

When you are traveling 55 miles per hour, how fast are you going in feet per second? Hint: The speed can be expressed as (55 mi/1 hr).

Sometimes a measurement or quantity needs to be multiplied by a series of factors to give a desired result. When a quantity is multipled by a series of factors, its units and numerical value change but it still represents the same measurement; 1.5 yards, 4.5 feet, and 54 inches are three ways of expressing the same distance.

## 2-6  METRIC UNIT CONVERSION

A quarter mile is 402 meters. How many kilometers is this? It is sometimes necessary to change from one base metric unit to a prefixed unit, or from a prefixed unit to the base unit. The conversion factors come from the definitions of the prefixes. Consider the relations between the meter as a base unit and the common prefixes used with the meter. (Similar relations can be expressed for any metric unit.)

$$1000 \text{ mm} = 1 \text{ m}, \quad 100 \text{ cm} = 1 \text{ m}, \quad \text{and} \quad 1000 \text{ m} = 1 \text{ km}$$

Conversion factors obtained from these relations are

$$\frac{1000 \text{ mm}}{1 \text{ m}} \quad \text{or} \quad \frac{1 \text{ m}}{1000 \text{ mm}}$$

$$\frac{100 \text{ cm}}{1 \text{ m}} \quad \text{or} \quad \frac{1 \text{ m}}{100 \text{ cm}}$$

and

$$\frac{1000 \text{ m}}{1 \text{ km}} \quad \text{or} \quad \frac{1 \text{ km}}{1000 \text{ m}}$$

To convert a measurement in a prefixed unit to the base unit, just multiply by the factor corresponding to the meaning of the prefix. The convenience of the metric system is that these computations involve simple multiplication or division by a multiple of ten.

---

### EXAMPLE 2-5

A quarter-mile running track is 402 meters around. What is this distance in kilometers? Multiplying by the factor (1 km/1000 m) gives the distance in kilometers.

$$X \text{ km} = 402 \text{ m} \times \frac{1 \text{ km}}{1000 \text{ m}} = 0.402 \text{ km} \text{ or } 4.02 \times 10^{-1} \text{ km}$$

Division by 1000 involves moving the decimal point three places to the left.

---

**EXAMPLE 2-6**

An inch is equivalent to 2.54 cm. How many meters is 2.54 cm? The factor (1 m/100 cm) is used in the conversion.

$$X\,\text{m} = 2.54\ \cancel{\text{cm}} \times \frac{1\ \text{m}}{100\ \cancel{\text{cm}}} = 0.0254\ \text{m}\ \text{or}\ 2.54 \times 10^{-2}\text{m}$$

Division by 100 involves moving the decimal point two places to the left.

---

A conversion factor is chosen so that the numerator has the desired unit and the denominator has the unit to be canceled. Consequently, it is absolutely necessary to include the correct units in a conversion factor and to make sure that the units cancel.

---

**EXAMPLE 2-7**

One pound is equivalent to 0.454 kg. How many grams is 0.454 kg? The factor relating grams to kilograms is

$$\frac{1000\ \text{g}}{1\ \text{kg}}$$

This factor is used to find the number of grams corresponding to 0.454 kg.

$$X\,\text{g} = 0.454\ \cancel{\text{kg}} \times \frac{1000\ \text{g}}{1\ \cancel{\text{kg}}} = 454\ \text{g}$$

Multiplication by 1000 involves moving the decimal point three places to the right.

---

**EXAMPLE 2-8**

A hand calculator can add two numbers in about 40 ms (milliseconds). How many seconds is this?

To obtain an answer in seconds and cancel milliseconds, we use the factor (1 s/1000 ms).

$$X\,\text{s} = 40\ \cancel{\text{ms}} \times \frac{1\ \text{s}}{1000\ \cancel{\text{ms}}} = 0.040\ \text{s}\ \text{or}\ 4.0 \times 10^{-2}\text{s}$$

---

**ACTIVITY 2-3**

Carry out the following metric conversions.

(a) One metric ton is 1000 kg. How many grams are there in 1000 kg? What is the common name for this number?

(b) A 10 K race is 10 kilometers long. How many meters is this?

(c) How many kiloseconds are there in 30 minutes?

## 2-7  SIGNIFICANT DIGITS

Typically we deal with three kinds of numbers: counted, defined, and measured. Counted items are expressed as exact whole numbers. If we count a number of people, for instance, the number is always an exact whole number, never a fraction. Defined relations also involve exact numbers but not always whole numbers. By definition there are exactly 12 inches per foot, there are exactly 60 seconds per minute, and one inch is exactly 2.54 centimeters. Such numbers come from definitions, not measurements. In contrast, measured numbers come from reading measuring devices; they are never exact. Imagine measuring a person's height, as shown in Figure 2-2. The measuring device shown in Figure 2-2 (*a*) is calibrated to measure lengths to the nearest tenth of a meter. The first digit of the measurement comes from the calibrations of the scale, and the final digit comes from interpolation. **Interpolation** means to read a measurement more closely than the calibration lines on a scale by estimating the distance between the lines. The measurement has two digits and a certainty of one tenth of a meter. If the measuring device is more precisely calibrated we can give more digits in the measurement, as shown in Figure 2-2 (*b*). In this case the height measurement has three digits and is certain to the nearest hundredth of a meter.

Some measuring devices, such as rulers, may require us to estimate the last digit by interpolation. Other devices, such as odometers on cars, are digital; they give the numbers of the measurement as a series of digits. The odometer in an automobile is a measuring device for distance. Imagine taking a trip in which the distance measured by the odometer is 370.2 miles. The odometer measures to the nearest 0.1 mile. This is its limit of precision or limit of certainty as a measuring device. The certainty means that the device measures precisely to the nearest tenth of a mile, but not to any smaller fraction of a mile. All measuring devices have limits of precision or certainty, which means they can be read to a specific number of digits.

Suppose each of five people carefully measures the length of a metal rod using a ruler calibrated to the nearest tenth of a centimeter. The group records the set of measurements shown on the next page.

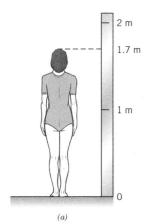
(*a*)

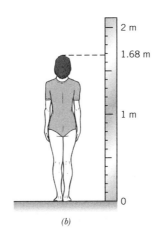
(*b*)

**FIGURE 2-2**
Measuring a height with rules of varying calibration.

$$25.3 \text{ cm} \quad 25.4 \text{ cm} \quad 25.3 \text{ cm} \quad 25.2 \text{ cm} \quad 25.3 \text{ cm}$$

Why are some of these measurements different from the others? A calibrated measuring instrument allows measurements to a specific certainty. In this case the ruler is calibrated to be read to a certainty of 0.1 cm. Such a certainty means that the ruler can be used to make measurements known to plus or minus one tenth of a centimeter. The average of the above measurements could be expressed as $25.3 \pm 0.1$ cm; all of the measurements agree within the 0.1 cm certainty of the measuring instrument.

Because measuring instruments have limits of certainty, measurements are never exact. Counting numbers and defined numbers are exact, but measurements are never exact numbers. All measuring instruments are calibrated to a specific certainty. They can be, and should normally be, read to that certainty. If calibrated correctly, instruments allow accurate measurements to be made to a specific number of digits. The ruler referred to above can be read to a certainty of plus or minus a tenth of a centimeter ($\pm 0.1$ cm) but not to the nearest hundredth of a centimeter. This would exceed its limit of certainty. A measurement made with an instrument of known certainty should be expressed to this certainty. For example, if you measure the mass of an object using a balance calibrated to be read to the nearest 0.01 grams you can express the measurement to $\pm 0.01$ g. Suppose you find the mass of an object to be 32.26 g. When a measurement is expressed this way you can assume that it is known to a certainty of plus or minus one digit in the last digit of the measurement (e.g., 32.26 g means $32.26 \pm 0.01$ g). Whenever you see a measurement in this book, you can assume it is known to a certainty of plus or minus one digit in the last digit. Normally it is not necessary to include the $\pm$ notation in a measurement.

A measurement is expressed in terms of a sequence of numbers containing a specific number of digits followed by a unit. Measuring instruments are designed to give specific numbers of digits. All the digits that are part of a measurement are called **significant digits** or **significant figures.** The measurement 370.2 mi has four significant digits; 25.3 cm has three significant digits. When dealing with a number obtained from a measurement, the following guidelines apply. Note that significant digits can be nonzero (1, 2, 3, 4, 5, 6, 7, 8, or 9) or zero and zeros can be significant digits or placeholders.

1. Count significant digits by starting with the first nonzero digit on the left and continue to the right. For example, 0.057 m has two significant digits; 12.67 cm has four significant digits.

2. Leading zeros occur at the beginning left-hand side of a number. They are never significant; they are merely decimal placeholders to show the position of the decimal point. For example, 0.0056 g has two significant digits. Notice that 0.0056 g and 5.6 mg are two ways of expressing the same measurement; both have two significant digits.

3. Zeros between nonzero digits are always significant. For example, 10,001 kg has five significant digits; 0.00003003 m has four significant digits.

4. Zeros that follow nonzero digits and are on the right of the decimal point are significant digits. For example, 125.0 m has four significant digits; 0.250 s has three significant digits. (The zero on the left is not significant but the zero on the right is.)

5. Trailing zeros are ambiguous; they may or may not be significant. Trailing zeros are zeros to the right of nonzero digits in numbers that contain no decimal point. For example, 25,000 m has five significant digits if it is measured to the nearest 1 m, or two significant digits if it is measured only to the nearest 1000 m. The ambiguity comes from the fact that zeros may be part of the measurement or they may be placeholders.

   Sometimes a trailing zero that is significant is written with a line above it. Following this pattern, 25$\bar{0}$ g has three significant digits and 3$\bar{0}$00 m has two significant digits. However, the use of lines is not a common practice. In this book you can assume zeros that follow nonzero digits are significant unless otherwise indicated by the context. For example, unless otherwise indicated, 20 g has two significant digits; 2500 miles has four significant digits. Furthermore, the context of a measurement sometimes suggests its certainty. Stating that the height of Mount McKinley is 20,300 ft suggests that the measurement is expressed to the nearest 100 ft, so we assume the height has three significant digits. Writing a number in exponential notation is the most effective way to show whether or not zeros are significant. If the height of Mount McKinley is written as $2.03 \times 10^4$ ft, we know the measurement is expressed to three significant digits. All significant digits are shown in the nonexponential part, so there is no ambiguity.

**EXAMPLE 2-9**

How many significant digits in each of the following measurements?

(a) 5.07 s (three significant digits; zeros between nonzero digits are significant)

(b) 0.0007 g (one significant digit; leading zeros are never significant)

(c) 203 km (three significant digits; zeros between nonzero digits are significant)

(d) 1.005 m (four significant digits; zeros between nonzero digits are significant)

(e) 200 g (three significant digits; zeros are significant unless otherwise indicated)

(f) 2500 m (four significant digits; zeros are significant unless otherwise indicated)

ACTIVITY 2-4

How much do you weigh? To one significant digit? To two significant digits? To three significant digits? (Scientific observation involves making and reporting honest measurements. However, you can use an imaginary weight in this exercise.)

(g) 10.0 s (three significant digits; otherwise there would be no reason to give the zero following the decimal point)

(h) 18,000,000°C (two significant digits likely, unless otherwise indicated)

(i) 2.80 × 10$^3$ m (three significant digits, as shown in the nonexponential part)

(j) 2346 (this is a trick question; significant digits only apply to measurements; this number has no unit and is not a measurement, so it has nothing to do with significant digits)

## Using a Calculator

An electronic calculator is quite convenient for most computations you need to do when solving problems. It doesn't solve problems for you, but it does find the numerical answer once a problem has been set up. The example given here is for typical algebraic calculators.

If a baseball pitch is timed at 92.1 miles per hour; what is this speed in units of feet per second? First the question can be summarized as

**GIVEN**

speed of 92.1 mi/hr

**GOAL**

change to units of ft/s or X ft/s = ?

To change the units, the given speed is sequentially multiplied by the factors relating hours to minutes, minutes to seconds, and feet to miles. The setup for the calculation is

$$X \text{ ft/s} = \frac{92.1 \text{ mi}}{1 \text{ hr}} \times \frac{1 \text{ hr}}{60 \text{ min}} \times \frac{1 \text{ min}}{60 \text{ s}} \times \frac{5280 \text{ ft}}{1 \text{ mi}}$$

To do the multiplications and divisions, the numbers are keyed into the calculator starting from the left and moving in sequence to the right. Each number is followed by the division or multiplication key according to the pattern given in the setup. You do not have to key the number one because multiplication or division by one does not change the result. For this example, the key sequence is

$$92.1 \div 60 \div 60 \times 5280 = 135.07982$$

Try it on your calculator. The answer rounded to three digits is 135 ft/s. A calculator can save you time and effort in most of the numerical calculations in this book. To practice, use your calculator to confirm answers given in examples throughout the book. See Appendix 2 for more about the use of calculators.

If you prefer, you can do the above calculation one step at a time, writing down the answer at each step, but be sure not to round off the intermediate answers. Keep more digits than you need in any intermediate answers; round off only when you have the final answer. Premature roundoff error can make your final answer incorrect.

## 2-8   SIGNIFICANT DIGITS IN CALCULATIONS

After driving a car from Mexico City to Guadalajara, suppose the odometer shows a driving distance of 558 km. Expressing this in terms of meters, the distance is 558,000 m. Does this mean that we know the distance in meters to six significant figures? No! The number of significant digits in a number is not changed by multiplying or dividing it by a conversion factor. When a measurement is used in a calculation, the number of significant digits in the measurement dictates the number of significant digits in the answer.

Significant digits can be important to everyday activities. Imagine buying 2.2 lb of something that costs $6.25/lb. How much should you pay? The price is a defined conversion factor and the amount in pounds is a measurement. The product of the pounds and the price per pound gives the cost.

$$2.2 \,\cancel{lb} \times \frac{\$6.25}{1 \,\cancel{lb}} = \$13.75$$

An important fact about significant digits in measurements is that the number of significant digits in the measurement cannot be increased or decreased by multiplying or dividing by a defined (exact) factor. It is not possible to change the certainty of the measurement by use of a factor. For the store to charge $13.75 based on the calculation above, their weighing device must allow weighing to the nearest thousandth of a pound. So if you buy 2.200 pounds, you pay $13.75; if you buy 2.2 pounds, you should pay $14. Stores assume that 2.2 pounds is an exact number, so significant digits don't apply and the cost is rounded to the nearest cent. However, the scales and balances used in stores need to be reasonably accurate, and many states require their periodic calibration.

### Significant Digits in Multiplication or Division

The rule for finding the number of significant digits in a multiplication or division calculation differs from that used in an addition or subtraction calculation. First, consider the case of multiplication or division. Suppose you drove a car 627.2 miles as measured by the odometer. Filling the tank before and after, you find that 14.1 gallons of gasoline were used. How many miles per gallon (gal) did you get? Simply divide the miles by the gallon to calculate the result. The calculator read out is

$$\frac{627.2 \text{ mi}}{14.1 \text{ gal}} = 44.48227 \text{ mi/gal}$$

Has the calculator created seven significant digits for us? No! Calculators cannot understand significant digits, but humans can. Our result can have no more significant digits than the measurement having the fewest significant digits. In other words, a calculated result can be no more certain than the data from which it is obtained. We know the gallons used to only three significant digits, so the mileage must be expressed to three significant digits. We can do the calculation to as many digits as we like, or to as many as the calculator produces, but we must round off the final answer to the correct number of significant digits. Thus, the answer for the mileage is rounded off to three significant digits to give 44.5 mi/gal. To round off numbers follow these simple rules.

1. If the number following the digits to be kept for the answer is less than 5, the digits kept are not altered. For example, 2.034 rounded to three digits is 2.03.
2. If the number following the digits to be kept for the answer is 5 or greater, add 1 to the last digit kept. For example, 1.999517 rounded to four digits is 2.000.

---

**EXAMPLE 2-10**

Round off the following numbers to the number of digits indicated.

(a)  253.507 to three digits   (add 1 to the last digit kept: 254)
(b)  0.0727 to two digits      (add 1 to the last digit kept: 0.073)
(c)  1.0039 to three digits    (the three digits kept are not changed: 1.00)
(d)  15.9994 to four digits    (add 1 to the last digit kept: 16.00)

---

When calculating with measurements, be aware of the number of significant digits allowed in the calculated result. The pattern is simple. When two or more measurements are used in multiplication and division calculations, the answer can have no more significant digits than the measurement with the fewest number of significant digits. So just count the digits in the measurements and use the least number of significant digits in the answer. Remember, factors that include exact numbers are not considered in the determination of significant digits in an answer. Simply put, in a multiplication or division calculation, the answer can have no more digits than the measurement that has the fewest number of digits. Remember, too, that in a multiplication or division calculation, the position of the decimal point has nothing to do with the number of significant digits in the answer. With practice you'll get used to expressing calculated answers to the correct number of digits. If you are not sure how many that is, just ask your instructor or study group for advice.

**EXAMPLE 2-11**

How many yards are in 237 inches?

**GIVEN**        **GOAL**

237 in.        how many yards or $X$ yd = ?

We need factors relating feet to inches and yards to feet.

$$X \text{ yd} = 237 \, \cancel{in.} \times \frac{1 \, \cancel{ft}}{12 \, \cancel{in.}} \times \frac{1 \, \text{yd}}{3 \, \cancel{ft}} = 6.5833333 \text{ yd rounds to } 6.58$$

The result should have three significant digits because the measurement has three significant digits.

You may question why the one and two digits in the conversion factors in the example above do not dictate the result. These exact factors come from defined relations and thus have no influence on the significant digits. That is, 1 ft = 12 in. is a defined exact equivalence and does not come from a measurement. Remember, whenever a factor is used that comes from a defined equivalence, it does not influence the number of significant digits. Significant digits only apply to measurements or factors derived from measurements.

**EXAMPLE 2-12**

If a car travels 15.2 miles in 21 minutes, what is the speed in miles per hour?

**GIVEN**          **GOAL**

15.2 miles        speed in miles per hour

21 minutes        $X$ mi/hr = ?

We need to change the ratio of miles to minutes to miles per hour using the factor relating minutes and hours.

$$X \text{ mi/hr} = \frac{15.2 \text{ mi}}{21 \, \cancel{min.}} \times \frac{60 \, \cancel{min}}{1 \text{ hr}} = \frac{43.428571 \text{ mi}}{1 \text{ hr}}$$

or rounded to two significant digits 43 mi/hr. The mile measurement has three significant digits, but the time measurement has only two significant digits, so the answer should have only two significant digits. Again, the defined conversion factor relating minutes and hours has no influence on the number of digits in the answer. An answer can be rounded off as you read from the calculator display.

A gasoline pump records volumes of gasoline. Check a pump at a gas station. How many significant digits are given in the digital read-out of volume?

## Significant Digits in Addition or Subtraction

When a calculation involves addition or subtraction, the number of significant digits is determined in a way that differs from multiplication and division calculations. The number of digits in a result calculated by addition or subtraction depends on the position of the decimal point. A calculated result can have no more significant digits with respect to the decimal point than the measurement with the fewest digits with respect to the decimal point. The reason for this is that the least precise quantity dictates the precision of the result. We cannot increase the precision with respect to the decimal point by just adding a more precise quantity. However, addition and subtraction of numbers can change the total number of digits in the result.

Suppose you get on an elevator with a group of people. First you want to add up the total weight of the passengers. Everyone knew their weight to the nearest pound except one person, who said their weight was about 190 pounds, which we'll assume is known to only two significant digits. When we add the weights, the quantity having the least certainty dictates the answer.

$$\begin{array}{r} 132 \text{ lb} \\ 122 \text{ lb} \\ 183 \text{ lb} \\ + \underline{190 \text{ lb}} \\ 627 \text{ lb} \end{array}$$

This calculation suggests we know the sum to the nearest 1 pound. But we cannot be that precise because the last measurement is known only to the nearest 10 pounds. The rule is that the sum or difference of measurements cannot be expressed more precisely than the measurement that is least precise with respect to the decimal point. Here the answer needs to be rounded to 630 lbs because the least certain measurement limits the answer to the nearest 10 pounds.

In most calculations, it is a good idea to carry along any extra digits until the final answer, then round off the result to the correct number of digits. Some examples are

| | |
|---|---|
| 52.50 m | (two digits past decimal) |
| + 2.017 m | (three digits past decimal) |
| 54.517 m | (answer is rounded to 52.52 m because the result may have only two digits past decimal) |
| | |
| 12.2 g | (one digit past decimal) |
| + 4.032 g | (three digits past decimal) |
| 16.232 g | (answer is rounded to 16.2 g because the result may have only one digit past decimal) |

**EXAMPLE 2-13**

Express the correct number of significant digits in the following cases.

(a)  63.4 g          (one digit past decimal)
   + 59.8 g          (one digit past decimal)
   ─────────
   123.2 g           (answer should have one digit past the decimal point,
                     but notice that this addition has increased the number
                     of significant digits to four)

(b)  63.4 g          (one digit past decimal)
   − 59.8 g          (one digit past decimal)
   ─────────
     3.6 g           (answer should have one digit past the decimal point,
                     but notice that this subtraction has decreased the
                     number of significant digits to two)

## 2-9  ENGLISH TO METRIC CONVERSIONS

Speed limits in Mexico are given in kilometers per hour, and distances in kilometers or meters. In Italy you buy grapes by the kilo (kilogram) and talk about the weather in °C. Nearly every major country of the world uses the metric system of measure for business and commerce. In the United States, we use a system of measure inherited from the past. This system is known as the English system of measure. However, the United Kingdom has changed to the metric system, so these units are now primarily used in the United States. Metric units do not have simple relations to English units. For rough estimations of how some of the metric units compare to our typical commercial units, consider the following. One inch is about 2.5 cm, so an inchworm would be a 2.5-cm worm in metric terms. One kilogram is about 2.2 lb, so a kilo of Italian grapes would be about 2.2 lb.

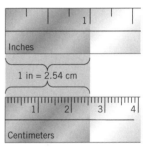

Inches

1 in = 2.54 cm

Centimeters

1 kilogram
||
2.205 pounds

1 gallon
||
3.785 liters

1 pound = 453.6 grams

Five miles is about 8 km, or there are about 0.6 mi/km; thus in Mexico, if you noticed a sign with a speed limit of 80 km/hr, this would be about 50 mi/hr, about 0.6 of 80. One meter is about 1.1 yd, so the 100-m dash is about 10 percent longer than the 100-yd dash.

The metric system is commonly used for technical and scientific purposes. Typically you'll see metric units of mass used on over-the-counter medicines and in nutrition data. It is sometimes necessary to convert precise measurements made in terms of English units to metric units, or from metric to English. Conversion factors that relate the two systems are used in such calculations. Appendix 1 lists the most common conversion factors, and a few comparisons of English and metric units are given on the previous page. For rough estimates between the two systems you can use the information given in the previous paragraph.

To convert a precise measurement made in one system to another, just look up the appropriate conversion factor. These factors are unit factors and should be used in forms having the wanted unit in the numerator and the unit to be canceled in the denominator. To get precise answers, do not round off a factor before using it.

---

**EXAMPLE 2-14**

A person is 60.0 inches tall. What is this height in centimeters?

$$X \text{ cm} = 60.0 \text{ in.} \times \frac{2.54 \text{ cm}}{1 \text{ in.}} = 152 \text{ cm}$$

---

**EXAMPLE 2-15**

If somebody weighs 80.7 kg, how much is this in pounds?

$$X \text{ lb} = 80.7 \text{ kg} \times \frac{2.205 \text{ lb}}{1 \text{ kg}} = 178 \text{ lb}$$

---

**EXAMPLE 2-16**

How many meters long is a 100-yard football field? (Assume that one hundred yards is known to three significant digits.)

$$X \text{ m} = 100 \text{ yd} \times \frac{0.9144 \text{ m}}{1 \text{ yd}} = 91.4 \text{ m}$$

**EXAMPLE 2-17**

A speed limit sign reads 80 km/hr. What is the speed limit in miles per hour? We need to convert the kilometers to miles.

$$X \text{ mi/hr} = \frac{80 \text{ km}}{1 \text{ hr}} \times \frac{0.6214 \text{ mi}}{1 \text{ km}} = 50 \text{ mi/hr}$$

**EXAMPLE 2-18**

A high jumper leaps 6.25 ft. How high is this in centimeters? The given measurement is in feet. We want the answer in centimeters. It is convenient to first convert the feet to inches, then the inches to centimeters. First, multiply the 6.25 feet by the feet-to-inches factor.

$$6.25 \text{ ft} \times \frac{12 \text{ in.}}{1 \text{ ft}}$$

Next, multiply this product by the inches-to-centimeters factor. The two steps can be included in a single setup as

$$X \text{ cm} = 6.25 \text{ ft} \times \frac{12 \text{ in.}}{1 \text{ ft}} \times \frac{2.54 \text{ cm}}{1 \text{ in.}} = 191 \text{ cm}$$

It is possible to use two or more factors in any conversion as needed.

**EXAMPLE 2-19**

One Olympic event is the 800-m run. How many miles is this? The calculation sequence can involve the conversion of the meters to kilometers, then the kilometers to miles. The setup includes multiplying by both the meters-to-kilometers factor and the kilometer-to-miles factor.

$$X \text{ mi} = 800 \text{ m} \times \frac{1 \text{ km}}{1000 \text{ m}} \times \frac{1 \text{ mi}}{1.609 \text{ km}} = 0.497 \text{ mi}$$

Notice the way the miles-to-kilometers factor is used. If the alternate form is needed, any conversion factor can be inverted or "turned over" by switching the numerator and denominator. In this case

$$\frac{1.609 \text{ km}}{1 \text{ mi}} \qquad \text{is inverted} \qquad \frac{1 \text{ mi}}{1.609 \text{ km}}$$

to give the factor needed to make the units cancel.

Do you think that the United States should change commercial measuring units to the metric system? Give reasons for your answer.

## *Problem Solving*

An important part of your study of chemistry involves learning problem solving methods. Each problem is unique, but there is a general pattern to follow in solving problems. Problems are stated in words, so you need to carefully read and interpret them. Let's consider the pattern by using an example problem.

A fast racehorse runs a distance of 7.0 furlongs in 80 seconds. What is the average speed of the horse in miles per hour?

1. Search the problem for what is to be solved for; the **unknown.** Look for a sentence with a question mark and look for key words such as "what," "what is," "how many," "how much," Express the sought unit or units of the unknown quantity. In this example, the unknown is the speed of the horse in units of miles per hour (mi/hr).

2. Search the question for any given information; the **givens.** Record the givens with their units. In this example, it is given that the horse runs 7.0 furlongs in 80 seconds. Make a summary of the givens.

   GIVEN     7.0 furlong     80 s

3. Decide what **concept** or **principle** is involved in the problem and, if possible, express the concept as a definition or a formula. In this example, the furlong is a unit of distance and the second is a unit of time. Thus, the concept is speed. Speed is the ratio of distance to time, $s = d/t$.

4. Use the givens as they relate to the concept. Write them down and follow them by the unknown, which is the goal of the problem. In this example, the given speed is 7.0 furlongs per 80 seconds.

<table>
<tr><td align="center">GIVEN</td><td align="center">GOAL SPEED</td></tr>
<tr><td align="center">$\dfrac{7.0 \text{ furlong}}{80 \text{ s}}$</td><td align="center">$\dfrac{? \text{ mi}}{\text{hr}}$ or X mi/hr = ?</td></tr>
</table>

5. Decide what conversion factors or other information are needed in the solution of the problem. In this example, you need to change seconds to hours and furlongs to miles. Since there are 60 seconds in 1 minute and 60 minutes in 1 hour, to convert seconds to hours use the factor

$$\frac{3600 \text{ s}}{1 \text{ hr}}$$

To change furlongs to miles we need to know how they relate. If you do not know a specific factor you need to look it up. The dictionary defines a furlong as a length used in horse racing that is equivalent to one-eighth of a mile. The factor, found by dividing 1 by 8, is

$$\frac{0.125 \text{ mi}}{1 \text{ furlong}}$$

6.  Use the concept, the givens, and the factors to set up the calculation sequence or do a sequence of intermediate calculations. Include units in the setup.

In the example, the setup is

$$\frac{7.0 \text{ furlong}}{80 \text{ s}} \times \frac{3600 \text{ s}}{1 \text{ hr}} \times \frac{0.125 \text{ mi}}{1 \text{ furlong}}$$

7.  Use the setup to calculate the numerical answer to the correct number of significant digits, and be sure that the units cancel to give the desired units for the answer. Canceling the units serves as a double check of the setup. In this example

$$\frac{7.0 \text{ furlong}}{80 \text{ s}} \times \frac{3600 \text{ s}}{1 \text{ hr}} \times \frac{0.125 \text{ mi}}{1 \text{ furlong}} = \frac{39 \text{ mi}}{1 \text{ hr}} \text{ or } 39 \text{ mi/hr}$$

## 2-10  VOLUME MEASUREMENTS

Matter is anything that has mass and occupies space. Objects occupy space. The term volume is used to describe the amount of space occupied; volumes represent amounts of three-dimensional space. In the English system, volumes are sometimes expressed in terms of cubic inches or cubic feet. In the metric system, cubic centimeters or cubic meters are used as units of volume.

Imagine a piece of clay in the form of a cube measuring 1 cm on an edge. The volume of the cube is said to be 1 cubic centimeter or 1 $cm^3$. Notice this volume unit corresponds to a cubic length (1 cm × 1 cm × 1 cm = 1 $cm^3$) and is read as cubic centimeter. Even if the cube of clay is rolled into a ball, flattened into a sheet, or distorted in another way, the volume remains 1 $cm^3$.

When working with volumes of liquids in the English system, we do not normally use units of cubic inches. Often, special units such as fluid ounces, pints, quarts, or gallons are used instead. A special metric unit is used in chemistry to measure and express volumes. This unit is the liter, L, and is defined as a volume equal to 1000 $cm^3$ (see Fig. 2-3). In this text L is used as the symbol for the liter.

Metric prefixes can be used with the liter. If 1 L of volume is divided into 1000 equal parts, each part is 1 milliliter (mL). There are 1000 mL in 1 L. One liter is also equivalent to 1000 $cm^3$, so the milliliter and cubic centimeter are the same amount of volume.

$$1 \text{ mL} = 1 \text{ cm}^3$$

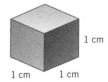

1 cm
1 cm     1 cm

**One Liter:**
*The volume equal to one thousand cubic centimeters.*

**1 L = 1000 cm³**

**FIGURE 2-3**   The liter. (The figure is not to scale.)

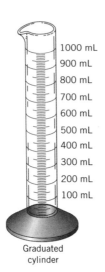

Graduated cylinder

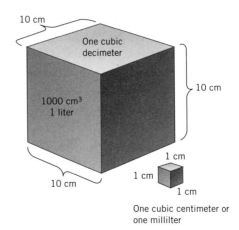

One cubic centimeter or one millilter

The liter (L) is equivalent to the volume occupied by a cube measuring 10 cm or 3.94 in. on an edge. One liter is equal to 1.057 quarts.

Since the milliliter and cubic centimeter have the same meaning, they are often used interchangeably. From the definition of the liter, we can write the unit factors

$$\frac{1\ L}{1000\ cm^3} \qquad \frac{1000\ cm^3}{1\ L} \qquad \frac{1\ L}{1000\ mL} \qquad \frac{1000\ mL}{1\ L}$$

These factors are used to change volumes from cubic centimeters or milliliters to liters or vice versa.

A beverage can lists its contents as 355 mL. What is this volume in liters? What is this volume in centiliters?

---

**EXAMPLE 2-20**

Express 0.250 L in units of milliliters. The volume is multiplied by the factor (1000 mL/1 L) to obtain the desired result.

$$0.250\ \cancel{L} \times \frac{1000\ mL}{1\ \cancel{L}} = 250\ mL$$

---

**2-11  DENSITY**

If you hold a rock in one hand and a piece of wood of comparable size in your other hand, the rock feels heavier. In fact, the piece of wood will probably float on water, while the rock will sink. How can we describe this difference between these two forms of matter? Remember, matter is anything that occupies space and has mass. In your immediate environment, substances might include wood, plastic, metals, rocks, various liquids, even gases. One way to compare samples of matter is to find the amount of mass contained in a specific unit of volume. To do this we carefully measure the mass and volume of a sample of a substance. The characteristic property of that substance is the ratio of its mass to its volume; this ratio is called the density. **Density** is defined as the mass per unit volume of a material.

Suppose a sample of iron metal has a mass of 18.8 g and occupies a volume of 2.39 cm³. We find the density by dividing the mass by the volume.

$$d = \frac{18.8 \text{ g}}{2.39 \text{ cm}^3} = \frac{7.87 \text{ g}}{1 \text{ cm}^3} \text{ or } 7.87 \text{ g/cm}^3$$

Notice that the numerical part of the calculation involves dividing 18.8 by 2.39. The units remain with grams in the numerator and cubic centimeters in the denominator. One cubic centimeter represents one unit of volume. Once a density is calculated, it is often expressed in terms of a certain number of grams per 1 cm³. Thus, the above density is expressed as 7.87 g/cm³ which indicates that the density is known to three significant digits. Since a density is found by experiment, the number of significant digits depends upon the number of digits in the mass and volume. Density tells us the mass of one cubic centimeter or a unit volume of a substance. Each cubic centimeter of iron has a mass of 7.87 g. A ratio, which expresses the amount of one quantity per unit amount of another, is quite common and useful in chemistry. Actually, we often use ratios like these every day (e.g., miles per hour, cents per pound, or dollars per gallon).

Since density is a ratio of mass to volume it can be represented by the algebraic expression

$$d = \frac{m}{V} = \frac{\textbf{mass in grams}}{\textbf{volume in cm}^3 \textbf{ or mL}} \qquad d = m/V$$

An algebraic expression of this type serves as a succinct definition of density, and suggests how the density of a substance can be found. Typical units for mass are grams; volumes are most often measured in cubic centimeters or milliliters; thus density is the number of grams of a substance per cubic centimeter or milliliter. Densities are found by experiment, then tabulated in reference books. For example, *The Handbook of Chemistry and Physics* is a reference book that contains a variety of information, including the densities of many chemicals. When densities are recorded, the temperature at which the density was determined is usually given. This is because volume and, therefore, density varies with temperature.

**Table 2-1** Densities of Some Common Chemicals

| Liquids | Density at 20°C (g/mL) |
|---|---|
| Water | 1.00 |
| Ethyl alcohol | 0.79 |
| Carbon tetrachloride | 1.59 |
| Mercury | 13.55 |

| Solids | Density at 20°C (g/mL) |
|---|---|
| Magnesium | 1.74 |
| Aluminum | 2.70 |
| Iron | 7.87 |
| Gold | 19.3 |
| Sulfur | 2.07 |
| Salt | 2.16 |
| Sugar | 1.59 |
| Ice (0°C) | 0.92 |

| Gases | Density at 0°C (g/L) |
|---|---|
| Air (dry) | 1.29 |
| Oxygen | 1.43 |
| Nitrogen | 1.25 |
| Carbon dioxide | 1.96 |
| Chlorine | 3.17 |
| Helium | 0.178 |

Table 2-1 lists the densities of a few representative substances (see Chemistry in Action 2-2). Gases typically have low densities, and some metals have very high densities. For example, the density of gold is 19.3 g/mL, whereas the density of helium gas is 0.000178 g/mL or 0.178 g/L. This means that gold is over 100,000 times more dense than helium. Notice that the relatively low densities of gases are typically expressed in terms of grams per liter, rather than grams per milliliter.

The gram, as a unit of mass, was originally defined as the mass of 1 cubic centimeter of water. Consequently, water has the easily remembered density of 1 $g/cm^3$. Since the density of a substance varies with temperature, water does not have a density that is exactly 1 $g/cm^3$ at all temperatures. However, the density of water is 1.00 $g/cm^3$ at room temperature and is very close to this value for liquid water between 0°C and 100°C.

Table 2-1 shows that helium gas has a density about seven times less than the density of air. A balloon or even a blimp filled with helium gas floats in the air. Helium has such a low density that it floats in air, and even allows its lightweight container to float. The average density of the helium and the container is less than the density of air. You know that the densities of gases decrease with increasing temperatures. Air at 40°C or 104°F, for instance, has a density of 1.12 g/L compared to air at 20°C or 68°F with a density of 1.20 g/L.

An object placed in water floats if it is less dense than the water. This is true as long as the object does not dissolve in water. When we pour two liquids together and they do not mix, the liquid of lower density forms a layer on top of the liquid of higher density. The oil in an oil and vinegar dressing floats on the vinegar. The oil must be less dense than the vinegar. Vegetable oils and petroleum oils are less dense than water, so they float on water. When an oil spill occurs in the ocean, the oil floats on the water.

There are some practical implications of the fact that oils and petroleum products float on water. One method used to obtain crude oil from nearly depleted wells involves the pumping of water into the well. The oil accumulates and floats on the water, so it can be more easily pumped out. Never try to put out a gasoline fire with water. The burning gasoline floats on the water and spreads the fire. A carbon dioxide fire extinguisher should be used on a gasoline fire, or the flames should be smothered using a blanket or dirt. For a similar reason, never try to put out a grease fire in the kitchen using water. Instead use a carbon dioxide fire extinguisher, dump baking soda on the flames, or cut off the air supply by covering the flames.

**ACTIVITY 2-8**

Why are gases less dense than liquids or solids? Draw pictures to illustrate your answer.

---

**EXAMPLE 2-21**

The density of water is 1.00 g/mL. Ice can float on liquid water, so it must be less dense. If a 3.3-g sample of ice has a volume of 3.6 mL, what is the density of ice?

GIVEN             GOAL                    CONCEPT

mass 3.3 g        density in g/mL         $d = m/V$
volume 3.6 mL

Dividing the mass by the volume gives the density.

$$d = \frac{3.3 \text{ g}}{3.6 \text{ mL}} = \frac{0.92 \text{ g}}{1 \text{ mL}} \text{ or } 0.92 \text{ g/mL}$$

### EXAMPLE 2-22

Ebony is a dark, hard wood. A rectangular piece of ebony has a mass of 522 g and a volume of 435 cm$^3$. Find the density of ebony and decide whether or not it will float on water. Dividing the mass by the volume gives the density.

$$d = \frac{522 \text{ g}}{435 \text{ cm}^3} = \frac{1.20 \text{ g}}{1 \text{ cm}^3} \text{ or } 1.20 \text{ g/cm}^3$$

A calculator may read 1.2 for this calculation. The extra zero is added because it is a significant digit. Again, calculators cannot understand significant digits, but humans can. The piece of ebony is more dense than water, so it won't float.

Since you know something about the behavior of gases, here is a question that you can answer. How and why does the density of a gas vary with temperature when the pressure is constant? Give examples or drawings to illustrate your answer.

A 325-mL sample of carbon dioxide gas has a mass of 0.637 g. What is the density of the gas sample in grams per liter? Would you expect a balloon filled with carbon dioxide gas to float in air? Why?

## Dense Humans

You might have some friends who are more dense than you are. If you place some butter or margarine in water it floats because fats are less dense than water. Body fat is also less dense than water. That is why a fatter person finds it easier to float on water than a leaner person. The most common chemicals in the human body are water, protein, bones, and fat. Proteins are the main components of muscles. The portion of the body consisting of bones, protein, and water is known as the lean body mass. Proteins and bones are more dense than fats, and do not tend to float in water. People differ in their amount of fat and lean body mass. This means that they have different densities. Overweight people are less dense than leaner people. Why?

One method used to measure the percentage of body fat is based upon how dense a person is. A large tank of water, into which a person can be totally immersed, is equipped with a seat attached to a weighing scale. This apparatus is used to weigh a person under water. See Figure 2-4. Actually, anyone will weigh less under water compared to their weight out of water. The greater the percentage of lean body mass a person has, the more they weigh under water relative to a person having a higher percentage of body fat. The more fat a person has, the more they tend to float and the less they weigh under water relative to a person having a higher percentage of lean body mass. By comparing the actual weight of a person out of water to their

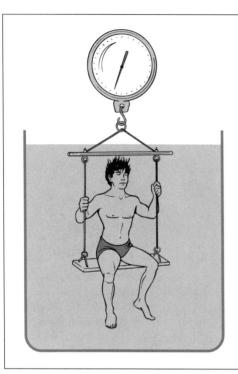

weight under water, it is possible to determine their lean body mass and percentage of body fat. If you want to win a competition with a muscular athlete, challenge them to a floating contest. Desirable body fat percentages for various persons are shown in the table.

| Desirable Body Fat Percentage | | |
|---|---|---|
| AGE | MEN | WOMEN |
| 20–29 | 10–15 | 18–22 |
| 30–39 | 12–16 | 20–24 |
| 40–49 | 14–18 | 21–25 |
| 50 and over | 15–19 | 22–26 |
| athletes | 5–8 | 12–18 |

**FIGURE 2-4** Determination of fat content by immersion.

**2-11**

Fill a glass with water. Add a small piece of butter or margarine to the glass. Add a small piece of hot dog, salami, turkey, and/or chicken meat to the glass. Give an explanation for your observations.

## 2-12 DENSITY AS A RATIO OF UNITS

Densities are useful in chemistry because they express the relation between the mass and volume of substances. The density as a ratio of units is used as a conversion factor to find the mass of a given volume, or the volume of a given mass. Let's see how this works. Gold has a density of 19.3 $g/cm^3$. A troy ounce of gold is 31.1 g. What volume would an ounce of gold occupy? The density is a ratio that can be expressed as a factor relating mass and volume, or inverted to relate volume and mass.

$$\frac{19.3 \text{ g}}{1 \text{ cm}^3} \quad \text{or} \quad \frac{1 \text{ cm}^3}{19.3 \text{ g}}$$

To find the volume, the mass can be multiplied by the second form of the factor, which is the reciprocal or inverse of the density.

$$X \text{ cm}^3 = 31.1 \text{ g} \times \frac{1 \text{ cm}^3}{19.3 \text{ g}} = 1.61 \text{ cm}^3$$

Density is an example of an experimentally derived ratio of units used to relate one property to another. In a sense, density is a property factor used to convert from one property to another. For any substance, the density relates mass and volume.

$$d = m/V$$

To use a factor like density, you have to be aware of the units of the desired answer. If you want units of mass, density is used directly as a factor. To find the mass corresponding to a given volume of a substance, multiply the volume by the density:

$$m = Vd$$

---

**EXAMPLE 2-23**

A gasoline has a density of 0.68 g/mL. What is the mass of 90.5 mL of this gasoline?

| GIVEN | GOAL | CONCEPT |
|---|---|---|
| volume 90.5 mL | find the mass | $d = m/V$ |
| density 0.68 g/mL | $X\,g = ?$ | |

Density is the factor. To get mass, the volume is multiplied by the density.

$$X\,g = 90.5 \; \cancel{mL} \times \frac{0.68 \; g}{1 \; \cancel{mL}} = 62 \; g$$

It is important to note that density is an experimental value based upon measurement. The number of significant digits in the density must therefore be considered when deducing the answer. In this example, the volume has three significant digits, but the density only has two, so the answer has only two significant digits.

---

If we want to obtain the volume of a substance from its mass, the inverse of the density is used as a factor. That is, the density is "turned over," switching the numerator and denominator. This gives a factor relating volume to mass.

$$V = m \times \frac{1}{d}$$

Carry the units in the factor. You know that you have the correct form of density if the unwanted units cancel and the desired unit remains.

**EXAMPLE 2-24**

A 10-carat diamond has a mass of 2.0 g. If the density of the diamond is 3.2 g/cm³, what is the volume of this diamond?

| GIVEN | GOAL | CONCEPT |
|---|---|---|
| mass 2.0 g | volume in cm³ | $d = m/V$ |
| density 3.2 g/cm³ | $X$ cm³ = ? | |

The density can be expressed as a factor.

$$\frac{3.2 \text{ g}}{1 \text{ cm}^3}$$

However, since we want volume from mass, we invert the density to give the factor.

$$\frac{1 \text{ cm}^3}{3.2 \text{ g}}$$

The mass is multiplied by the inverse of the density to give the volume.

$$X\text{cm}^3 = 2.0 \text{ g} \times \frac{1 \text{ cm}^3}{3.2 \text{ g}} = 0.63 \text{ cm}^3$$

**ACTIVITY 2-12**

The volume of a drop of mercury is 0.0500 cm³. If the density of mercury is 13.55 g/cm³, what is the mass of a drop of mercury? How does this mass compare to the mass of a drop of water having the same volume?

**CHEMISTRY IN ACTION 2-1**

Measuring provides information. A half-gallon milk carton has the form of a rectangular solid.

1. Pick up a full carton and guess how much it weighs in pounds.
2. Use a tape measure or a ruler to measure the height, length, and width of the carton in inches. Measure the height on the side having the highest height; disregard the tent-shaped top part. Just consider the container as a rectangular solid.
3. The weight of the filled carton can be estimated from its volume. Calculate the volume of the container in cubic inches.
4. To find the approximate weight in pounds, multiply the volume by 0.040. (Later you'll find out where this number comes from when you answer Question 1 in this chapter. For now just use the number given.)
5. You can check the estimated weight by weighing the carton on a bathroom scale, a kitchen scale, or an appropriate laboratory balance.

How is it possible to know the weight of an object by measuring its dimensions? The answer is related to the properties of materials and to geometry. (The volume of a rectangular solid is the product of its height, length, and width.) In this case, the property is the amount of weight per unit volume of milk.

**ACTIVITY**

A quart of milk weighs about 2.25 pounds. If the bottom dimensions of the carton are 2.75 inches by 2.75 inches, what is the height of the rectangular part of the carton? (Hint: Divide the weight by 0.040 to give the volume.) Measure the height of the rectangular part of a quart milk carton and compare it to your calculated result.

Here is an interesting way to estimate the density of ice. You need: two small glasses, water, vegetable oil (or mineral oil), and some pieces of ice.

1. Pour about one-fourth of a cup of water into a glass.
2. Pour about one-fourth of a cup of vegetable oil into the other glass.
3. Place a small piece of ice in each of the two glasses. Use pieces that do not have too many air bubbles trapped in them.

The density of water is 1 g/mL; the density of vegetable oil or mineral oil is about 0.8 g/mL.

Using these densities what do you conclude about the density of ice?

# QUESTIONS

## Section 2-1

1. The density of milk is about 1.1 g/cm$^3$. Calculate the density of milk in units of pounds per cubic inch. Compare your answer to the factor used in Chemistry in Action 2-1.

2. Imagine a system of measure based upon the following definitions for units of mass and length (the unit for time is the normal unit). The unit of mass is the stone (an old English unit corresponding to 14 pounds) and the unit of length is the league (an old English unit corresponding to 3 miles). How does a system of measure based on these units compare to the American and metric system. Express your height, weight, and a 55 miles per hour speed limit in terms of this new system of measure (use any prefix as needed).

3. Create your own system of measure by using references for units of mass and length (use the normal units for time). Determine how your system of measure compares to the typical units of the American system and the metric system. Express your height, weight, typical distances, and speed limits in terms of your units.

## Sections 2-2 to 2-4

4. List the names and abbreviations of the basic metric units.

5. List the three common metric prefixes along with their numerical meanings.

6. *Convert the following to exponential notation or standard form.

   (a) 695,950.0 km (the average radius of the sun)
   (b) $5.684 \times 10^{26}$ kg (the mass of Saturn)
   (c) $2.998 \times 10^{10}$ cm/s (the speed of light)
   (d) 0.0000000000000000000000001672 g (mass of a proton)
   (e) $3.1536 \times 10^{7}$ (the number of seconds in a non-leap year)
   (f) 0.000000028841 cm (the distance between the centers of two gold atoms in solid gold)
   (g) 125,200 cm/s (the average speed of a helium atom in the atmosphere)
   (h) $3.45 \times 10^{5}$ cm/s (the speed of sound at sea level)

7. Convert the following to exponential notation or standard form.

   (a) 36,000,000°F (the estimated temperature at the center of the sun)
   (b) $6.022 \times 10^{23}$ atoms (the number of atoms in 12 grams of carbon)
   (c) $1.67 \times 10^{-24}$ g (the mass of a typical hydrogen atom)
   (d) 0.0000000000000000000001675 g (the mass of a neutron)
   (e) $4.7 \times 10^{3}$ ft/s (the speed of sound in water)
   (f) 0.0000000186 cm (the radius of a typical sodium atom)
   (g) 299,800,000 m/s (the speed of light)
   (h) $1.86 \times 10^{5}$ mi/s (the speed of light)
   (i) $6 \times 10^{12}$ miles (the number of miles in a light-year to one digit)
   (j) 92,950,000 miles (the average distance of the earth from the sun)

## Sections 2-5 and 2-6

8. How many millimeters are there in 1 cm? How many centigrams are there in there in 1 mg?

9. *Which is a longer length, 0.017 m or 71 mm? Which is a greater mass, 75 mg or 0.5 g?

10. The RDA (recommended dietary allowance) for vitamin $B_{12}$ in adults is 0.0031 mg/kg of body weight. If a woman weighs 55 kg, how many grams of vitamin $B_{12}$ will she need each day?

11. A typical baby has a heart that weighs 30 g. What is this mass in milligrams? What is this mass in kilograms?

12. *The factor relating centimeters to meters is $\dfrac{100 \text{ cm}}{1 \text{ m}}$. Give the factors that relate the following units.

   (a) liters to milliliters
   (b) grams to kilograms
   (c) milliliters to deciliters
   (d) meters to centimeters
   (e) milligrams to grams
   (f) centimeters to millimeters
   (g) milliseconds to seconds
   (h) meters to kilometers

13. *Convert the following measurements to the units indicated.

   (a) 4.37 cm to m
   (b) 0.024 kg to mg
   (c) 2.8 cm to nm
   (d) 30 s to ms
   (e) 537 mm to m
   (f) 750 mL to L
   (g) 16,890 mg to kg
   (h) 0.0821 L to mL

14. Convert the following measurements to the units indicated.

   (a) 0.914 km to cm
   (b) 375 mg to g
   (c) 268 ms to s
   (d) 550 mL to L
   (e) 0.647 g to mg
   (f) 43.9 cm to mm
   (g) 0.0000001 L to μL
   (h) 219 mm to cm

## Section 2-7

15. *Give the number of significant digits in each of the following measurements.

   (a) 0.050 m
   (b) 28.30 g

(c) 0.046 L

(d) 0.0007 kg

(e) 974.32 g

(f) 5.080 s

(g) 33.6 L

(h) 103,000 cm

16. Give the number of significant digits in each of the following measurements.

(a) 20.8 km (the diameter of Phobos, a moon of Mars)

(b) 1500 g (the mass of a typical adult liver)

(c) 0.000072 cm (the average length of a human taste bud)

(d) 1769 mL (the measured volume of urine excreted in one day)

(e) 24,901.47 miles (the circumference of the earth at the equator)

(f) 0.860 s (the duration of a heartbeat)

(g) 103 m (the measured height of a redwood tree)

(h) 0.08 mm (the average thickness of human skin)

## Section 2-8

17. *Indicate the number of significant digits that should be expressed in the following calculations.

(a) $973 \text{ ft} \times \dfrac{12 \text{ in.}}{1 \text{ ft}} \times \dfrac{2.54 \text{ cm}}{1 \text{ in.}}$

(b) $26 \text{ mi} \times \dfrac{1760 \text{ yd}}{1 \text{ mi}} \times \dfrac{3 \text{ ft}}{1 \text{ yd}}$

(c) $0.500 \text{ m} \times \dfrac{1 \text{ km}}{1000 \text{ m}}$

(d) $4.70 \text{ mL} \times \dfrac{2.7 \text{ g}}{1 \text{ mL}}$

(e) $\begin{array}{r} 2.37 \text{ g} \\ +23.06 \text{ g} \\ \hline \end{array}$

(f) $\begin{array}{r} 0.776 \text{ cm} \\ +133.54 \text{ cm} \\ \hline \end{array}$

(g) $\begin{array}{r} 300.47 \text{ kg} \\ -189.290 \text{ kg} \\ \hline \end{array}$

(h) $\begin{array}{r} 7 \text{ s} \\ +4 \text{ s} \\ \hline \end{array}$

18. Indicate the number of significant digits that should be expressed in the following calculations.

(a) $3179 \text{ s} \times \dfrac{1 \text{ min}}{60 \text{ s}} \times \dfrac{1 \text{ hr}}{60 \text{ min}}$

(b) $6.28 \text{ hr} \times \dfrac{95 \text{ mi}}{1 \text{ hr}} \times \dfrac{1.609 \text{ km}}{1 \text{ mi}}$

(c) $25 \text{ mL} \times \dfrac{1 \text{ L}}{1000 \text{ mL}} \times \dfrac{1 \text{ gal}}{3.785 \text{ L}}$

(d) $5.0 \text{ g} \times \dfrac{1 \text{ oz}}{28.35 \text{ g}} \times \dfrac{1 \text{ lb}}{16 \text{ oz}}$

(e) $\begin{array}{r} 803 \text{ g} \\ +1647 \text{ g} \\ \hline \end{array}$

(f) $\begin{array}{r} 0.911 \text{ m} \\ +237.48 \text{ m} \\ \hline \end{array}$

(g) $\begin{array}{r} 562.1 \text{ kg} \\ -526.98 \text{ kg} \\ \hline \end{array}$

(h) $\begin{array}{r} 6 \text{ cm} \\ +8 \text{ cm} \\ \hline \end{array}$

(i) $\begin{array}{r} 750 \text{ m} \\ + \ 65 \text{ m} \\ \hline \end{array}$

19. A bag of nuts is labeled, "16 oz. (453.59 g)." Comment on the validity of the numbers used on the label in terms of the number of significant digits.

## Section 2-9

20. *To three significant digits how many millimeters are there in one yard?

21. A blood capillary has the same diameter as a red blood cell, 0.001 mm. How many inches is this?

22. *A home run hit over the back wall of a baseball field travels 403 ft. How many meters is this?

23. A 10-km race is how many miles?

24. Mary's little lamb weighs 127 lb. What is this in kg?

25. A thoroughbred racehorse has a mass of 549 kg. What is this mass in pounds?

26. Mount Olympus, Greece, is recorded as 9570 ft high. What is the height in meters?

27. *Express the speed 55 mi/hr in terms of kilometers per hour.

28. A baseball is pitched at 98 mi/hr. What is this speed in km/hr?

29. *An average bee carries 0.0018 oz of nectar per trip to the hive. What is this mass in milligrams? How many trips would it take to transport 1.00 kg of nectar?

30. The speed of light is $3.00 \times 10^8$ m/s. What is this speed in miles per hour?

### Section 2-10

31. What is a liter and how is it defined?

32. *Determine the number of gallons per 1.00 L, using 1 L = 1.057 qt, and 4 qt = 1 gal.

33. What is the volume of a cup of liquid in milliliters expressed to three digits? (4 cups = 1 qt)

34. Melons at their most active phase grow more than 5 cubic inches a day. What volume is this in liters? (Hint: Change cubic inches to cubic centimeters first.)

35. A winery ferments Cabernet Sauvignon in 250 gallon barrels. What is this volume in liters? How many 750-mL bottles can be filled with a barrel of wine?

36. The volume of blood flowing through an adult's liver is 25 mL per second. How many gallons per minute is this?

37. *If your economy car averages 41.5 mi/gal and the fuel tank holds 34.0 L of gasoline, how many miles can you drive using one tank of gasoline?

38. A can of soda contains 12 fluid oz. What is this volume in mL expressed to three digits?

### Sections 2-11 and 2-12

39. What is density and in what units is it usually expressed?

40. *Calculate the densities of the materials listed below using the experimental measurements given.

(a) a 34.1-mL sample of water with a mass of 34.2 g

(b) a 133-mL sample of gold with a mass of 2.568 kg

(c) a 47-mL sample of ice with a mass of 43.24 g

(d) a 61.9-mL sample of ethyl alcohol with a mass of 49.1 g

(e) a 58.7-$cm^3$ sample of aluminum with a mass of 158.5 g

(f) a 1.06-L sample of air with a mass of 1.35 g

41. Calculate the densities of the materials listed below using the experimental measurements given.

(a) a 42.43-$cm^3$ silver sample with a mass of 449.8 g

(b) a 95.5-mL sample of water with a mass of 96.0 g

(c) an 18.0-g sample of salt with a volume of 8.27 $cm^3$

(d) a 0.279-g sample of copper with a volume of 0.0314 $cm^3$

(e) a 63.1-L sample of ice at −10°C with a mass of 62.8 kg

(f) a 275-mL sample of ethane gas with a mass of 0.368 g.

42. Explain how it is possible for two substances to have the same density.

43. *Determine the mass or volume for each of the following, given the density and the measured mass or volume.

(a) What is the mass of 10.0 L of water if its density is 1.00 $g/cm^3$?

(b) What is the volume of a 6.72-g piece of sulfur if its density is 2.07 $g/cm^3$?

(c) What is the volume of a 31.3-g sample of aluminum if its density is 2.70 $g/cm^3$?

(d) What is the mass of a 456-$cm^3$ sample of iron if its density is 7.87 g/mL?

(e) A cup contains 237 mL of sugar. What is the mass of the sugar if its density is 1.59 g/mL?

(f) The total lung capacity of a typical male is $5.9 \times 10^3$ mL. What is the mass of air in the lungs if the density of air is 1.29 g/L?

(g) There are 3.785 L per gallon. What is the mass in grams of 15 gallons of gasoline if the density of gasoline is 0.70 g/mL. What is the mass in pounds?

44. Determine the mass or volume for each of the following, given the density and the measured mass or volume.

(a) What is the mass of 118 mL of milk if its density is 1.035 g/mL?

(b) What is the volume of 0.250 kg of water if its density is 1.00 $g/cm^3$?

(c) What is the mass of a piece of paper with a volume of 0.747 $cm^3$ if its density is 0.279 $g/cm^3$?

(d) A quintal is an esoteric unit of mass equal to 100 kg. What is the volume of 2.50 quintals of salt if its density is 2.16 $g/cm^3$?

(e) What is the mass of a 425-$cm^3$ sample of paraffin wax if its density is 0.88 g/cm³?

(f) What volume would a 0.639-g breath of air occupy if its density is 1.29 g/L?

(g) A croissant recipe calls for 700 g of flour. According to Jacque Pepin's data, the density of flour is 0.620 g/mL. How many mL of flour are needed for this recipe?

45. *The largest very pure gold nugget ever found had a volume of 3619 cm³. If the density of gold is 19.3 g/cm³, what was the mass of this nugget in grams? How many pounds is that?

46. The largest topaz ever found came from Brazil and had a volume of 11,026 cm³. If the density of topaz is 3.25 g/cm³, what is the mass of this topaz in grams? What is the mass in pounds?

47. *The density of water is 1.00 g/cm³. What is the density of water in pounds per cubic foot? (Hint: There are 28,317 cm³/1 ft³.)

48. If a very large hailstone weighs 2.0 lb and the density of ice is 0.92 g/cm³, what is the volume of the hailstone in units of quarts?

49. A bucket of molasses has a volume of 0.321 cubic feet and weighs 6.95 lb. Calculate the density of molasses in pounds per cubic feet and in grams per cubic centimeter. (See Question 47.)

50. When vegetable oil is poured into water, it forms a layer on top of the water. What can you deduce concerning the density of vegetable oil compared with the density of water?

51. When grenadine syrup is poured into orange juice, it sinks to the bottom. What can you deduce concerning the density of grenadine syrup relative to orange juice?

52. Ice will float on liquid water. What can you deduce concerning the density of ice compared with the density of liquid water? (See Question 40(a) and (c) to compare the actual densities.)

53. *A gold-colored ring has a mass of 31.6 g and a volume of 1.59 mL. Is the ring pure gold? The density of gold is 19.3 g/mL.

54. Why can a person float in salt water more easily than in fresh water?

55. Assuming a density of 1.0 g/cm³, what is the volume of a 139-lb person in cubic centimeters? What is this volume in liters?

56. *In certain types of stars, matter is thought to be highly compressed. If a handful of this matter, about 72 mL, has a mass of 1,440,000 kg, what is the density to two significant digits?

57. Which has a greater mass, 1 L of water or 1 L of gasoline? The density of water is 1.00 g/mL and gasoline is about 0.68 g/mL.

58. A small box is filled with liquid mercury. The box measures 3.44 cm wide, 6.88 cm long, and 1.56 cm high. If the density of mercury is 13.6 g/mL, what is the mass of mercury in the box?

59. *Pumice is a volcanic rock that contains many trapped air bubbles. A 550-g sample occupies 578.3 mL. What is the density of pumice? Will it float on water?

60. A reflecting pool is in the shape of a rectangular solid measuring 75 m by 22.5 m by 0.275 m. Using a density of 1.00 g/cm³ for water, determine the number of grams of water the pool contains. Also determine the number of pounds of water in the pool.

61. *A steel ball bearing in the form of a sphere has a mass of 0.216 g. The diameter of the bearing is 0.3742 cm. Calculate the density of the ball bearing. The volume of a sphere is given by the formula $V = 4/3(\pi r^3)$ or $(\pi d^3)/6$, where $r$ is the radius, $d$ is the diameter, and $\pi$ is 3.142.

62. A sample of aluminum metal is in the form of a cylinder of 0.17 cm radius and 1.18 cm height. If the mass of the sample is 0.289 g, calculate the density of aluminum. The volume of a cylinder is given by the formula $V = (\pi r^2)h$, where $r$ is the radius, $h$ is the height, and $\pi$ is 3.142.

## Questions to Ponder

63. Which is heavier, a pound of feathers or a pound of lead?

64. Is helium lighter than air? Explain your answer.

65. What are some advantages and problems associated with the United States switching to the metric system?

66. How would you explain why a helium balloon rises in air using your knowledge of density and gases?

# ATOMS

## 3-1 THE EVOLUTION OF AN IDEA

Matter is the concern of chemistry. The materials surrounding you, the earth beneath your feet, the air you breathe, and even the stars of the universe are forms of matter. One of the most important ideas in chemistry is that all matter consists of very tiny, characteristic particles called **atoms.** The atomic view of matter is not a new idea; it can be traced to ancient Greece. Around 500 B.C.E. the Greek philosopher, Democritus, extending the idea he learned from his teacher, suggested the particle nature of matter and called such particles atoms, from the Greek *atomos* 'that which cannot be divided or broken down'. The idea of the atom came from philosophical questions about matter. Democritus asked a hypothetical question: What would happen if a sample of matter were divided into smaller and smaller bits? Would such subdivision reach an ultimate particle, or could the subdivision occur indefinitely and still be characteristic of the sample? Democritus found the particle view to be more reasonable, and he developed the atomic view of matter as part of his philosophy. However, dominant Greek philosophers, such as Plato and Aristotle, considered matter to be continuous and not particulate.

The atomic view survived and was adopted by a later Greek philosopher, Epicurus (about 300 B.C.E.). Atoms as particles of matter became part of Epicurean philosophy, which influenced Roman thinkers. Lucretius in about 100 B.C.E. wrote *De Rerum Natura* (The Nature of Things), an epic poem espousing the atomic viewpoint. Apparently, a single hand-transcribed copy of this poem survived over the centuries. Gutenberg invented the printing press in the early 1400s. The epic Lucretian poem was one of the many works of literature that were set in print, and copies were widely distributed. This helped carry the hypothesis of atoms to modern times. The idea survived among many scientists, or natural philosophers as they were called in past times.

The philosophical idea of atoms was not based upon reproducible experimental evidence and measurements. Experimental science did not develop until the 1600s and 1700s. One result of this development was that experimental evidence for an atomic theory began to accumulate. In 1662 Robert Boyle, an English scientist, experimented with samples of air. To explain why air is easily compressed, he imagined an air sample to be composed of tiny, invisible particles separated by space. He used the hypothesis of atoms to explain the behavior of air. At the same time he planted seeds for the modern kinetic molecular theory that you learned about in Chapter 1.

More experimental evidence was collected by Antoine Lavoisier, a French chemist of the late 1700s. He is known as the father of modern chemistry since he developed careful methods of observation and measurement, and emphasized the use of precise balances to carefully weigh chemicals used in experiments. His emphasis on research and careful measurement established chemistry as a fundamental science. To fund his laboratory and scientific research, he invested in a tax collection agency in prerevolutionary France. During the French revolution he was arrested and executed in 1794. His death, at the peak of his scientific career, was a great loss for France and the world of science.

### 3-2  STATES OF MATTER

Matter typically occurs in one of three states or forms: the solid state, the liquid state, and the gas or vapor state. Solids include such materials as rocks, soils, minerals, ores, crystals, metals, glasses, plastics, and fibers. A **solid** has a definite volume and fixed shape. Solids are characteristically rigid and their shapes are changed by crunching them or reforming them. A **liquid** has a definite volume but flexible shape. Some familiar liquids are water, oils, ethyl alcohol, and gasoline. A liquid takes on the shape of the container in which it is placed, and can be poured and spread out on a flat surface. Some common gases are air, oxygen, helium, and carbon dioxide. A **gas** has neither a definite shape nor volume and occupies the entire container in which it is placed. If you fill a balloon with air the air sample fills the entire volume of the balloon.

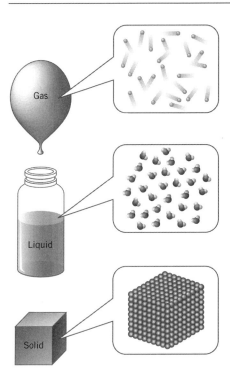

**FIGURE 3-1**
The kinetic molecular view
of gases, liquids, and solids.

The kinetic molecular theory provides a picture of a gas as a dynamic collection of high-speed particles with a relatively large amount of space between them. As shown in Figure 3-1, liquids and solids are pictured using a similar kinetic molecular view. Imagine a **liquid** as a teeming collection of particles in which there is relatively little space between them. Individual particles can move about in the collection, and the entire collection is quite flexible and takes the shape of the container. Picture the **solid** state as a collection of particles that have fixed positions in space and are arranged in some three-dimensional pattern. The particles can move or vibrate about their positions but cannot migrate like particles of liquids or gases. Upon heating, some materials burn or decompose. Many chemicals, however, change their physical state when heated or cooled. For instance, upon heating, ice melts to give liquid water. Further heating causes the liquid water to boil; this produces steam, which is water vapor. The various changes of state for materials are

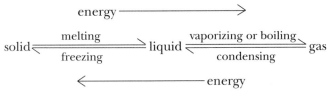

##  PHYSICAL AND CHEMICAL PROPERTIES

Chemists are interested in finding out what things are made of. To do this they probe and manipulate samples of matter and observe their properties. **Physical properties** are related to the physical nature of matter. Examples of physical properties are appearance, color, physical state, density, heat conductivity, electrical conductivity, and melting and boiling points. Water, for example, is typically a clear, colorless liquid with a density of about 1 g/mL, a melting point of 0°C, and a boiling point of 100°C at typical atmospheric pressure. Pure liquid water does not conduct electricity and is not a great conductor of heat. However, once heated, a sample of water retains its temperature and cools slowly.

Chemical properties refer to the chemical behavior of a form of matter. Chemical properties include how a chemical behaves when mixed with other chemicals. Does it burn in air? Does it decompose upon heating? Does it rust or corrode? Does it change chemically when mixed with water or other chemicals? These characteristics, or properties, serve to distinguish one type of matter from another. Table sugar, for example, readily dissolves in water and a sample decomposes when heated. In contrast, a chunk of iron does not burn when heated and does not dissolve or change when placed in water. The iron does, however, slowly corrode or rust when exposed to air or a moist environment. The ability to corrode or rust is a chemical property.

## MIXTURES AND SOLUTIONS

Many rocks appear to be a mixture of various materials, whereas a nickel coin appears uniform and, even when it is cut into bits, the pieces have the same uniform appearance. Some materials occur as heterogeneous mixtures. **Heterogeneous** means made of different parts. A mixture of materials can often be separated into homogeneous forms of matter. **Homogeneous** matter is uniform throughout, and any portion of a sample has the same properties and composition. Foods normally appear as mixtures of materials. A sample of milk, for example, appears to be uniform to the eye, but simple microscopic examination shows it to be a mixture. Milk, as a typical food, is a mixture containing water, proteins, carbohydrates, fats, and some vitamins and minerals. When a sample of fresh milk sits for a time, the less dense butterfat floats to the top and can be skimmed off to provide skimmed milk or nonfat milk. The fluid part of milk can be treated to coagulate the protein. The coagulated solid can be removed by filtration. This is done as a beginning step in making cheese. In a simple cheese, such as cottage cheese, you can see the coagulated curds of protein. The liquid remaining after removing the fat and most of the protein from milk is called whey. Whey is a dilute water solution of various chemicals including additional protein. It can be treated to isolate this protein as solids; the solids are used to make ricotta cheese. In addition, whey can be treated to isolate lactose, the common sugar in milk. Chemistry in Action 3-1 illustrates separating curds from whey.

A **solution** is called a homogeneous mixture since it is uniform throughout. This means that every portion of the solution is the same as any other portion; it will have the same density and the same appearance, even under microscopic examination. A solution consists of a **solvent** in which one or more substances are dissolved. The dissolved substances are called **solutes.** When salt is dissolved in water, the salt is the solute and the water is the solvent.

It is not possible to separate the components of a solution by simple filtration. They can, however, be separated in other ways. The salt in a salt–water solution is isolated by allowing the water to evaporate. Pure water is obtained by distilling it from water solutions. Distillation involves heating a water solution until it boils then condensing the vapors. This is how distilled water is prepared.

## 3-5 PURE SUBSTANCES OR CHEMICALS

With some effort, most heterogeneous mixtures and solutions can be separated into pure substances. **Pure substances** or **pure chemicals** are forms of matter that have the same properties throughout, and have definite and unchanging chemical composition. Pure substances are homogeneous. Water is an example of a pure chemical. No matter where a sample of pure water comes from, it has the same properties as any other sample of pure water. The key property of a pure substance is that it has fixed chemical composition. Water is always the same from a chemical perspective. In contrast, even though solutions are homogeneous forms of matter they do not have fixed unchanging chemical composition. Solutions of salt, for instance, are all homogeneous and contain salt mixed with water, but such solutions can range from slightly salty to very salty. They don't have fixed composition.

When the term pure is used with reference to a pure chemical it has a different meaning than you may realize. A **pure chemical** means one kind of chemical with little or no other chemicals mixed in. A few of the substances you can buy in the market are pure simple substances in the chemical sense. Sugar, salt, aluminum foil, and baking soda are examples of pure chemicals; most others, such as flour, butter, and milk, are complex mixtures. Commonly, pure is used to mean safe. Pure milk is milk that is free of harmful bacteria, but milk is a mixture of chemicals and is, therefore, not a pure chemical. Sometimes the term pure water means fresh water that is safe to drink. Referring to water as a pure chemical means water with little or no impurities.

It is important to realize that some mixtures are more complex than others. Pure water is easily obtained by distillation from water solutions. Pure table salt is obtained from the solid deposits left by evaporation of ocean water. However, it has to be separated from the mixture of chemicals making up the deposits. It requires greater effort and special techniques to separate the various mineral substances that make up rocks. Living systems are extremely complex mixtures containing thousands of chemicals. Analysis of even a relatively simple material such as milk is very complicated, time-consuming, and expensive.

## 3-6 ELEMENTS AND COMPOUNDS

Pure chemicals, like water and salt, are isolated by separating them from mixtures. See Figure 3-2. Using special techniques it is possible to decompose many pure substances into even simpler substances. Water, for instance, with some effort and energy, decomposes to form hydrogen gas and oxygen gas. Similarly, salt with some effort and energy, can be separated into sodium metal and chlorine gas.

Chemists have isolated and identified millions of pure chemicals. Close study and testing of substances have revealed there are just two kinds of pure chemicals: compounds and elements. **Compounds** are pure chemicals that can be separated into two or more even simpler chemicals. **Elements** are pure chemicals that cannot be further separated into simpler chemicals. Examples are hydrogen, oxygen, sodium, or chlorine. In general, any substance that cannot be broken down by chemical processes into simpler substances is an element, and any substance that can be broken down into simpler substances is a compound.

Liquid water cannot be decomposed by heating because it boils away. However, liquid water can be decomposed using electrical energy in a process called electrolysis. To observe this process you need: a 9-volt battery, some quarter inch by 1.5 inch strips of aluminum foil, tape, and a small glass of tap water.

1. Attach one end of each aluminum strip to the battery terminals using tape, then extend the strips upwards from the top of the battery. Do not let the strips touch one another or the metal case of the battery.

2. Fill a glass to the rim with tap water. Dip the ends of the aluminum strips in the water to a depth of about half an inch. Hold them in place for a few minutes. What do you observe?

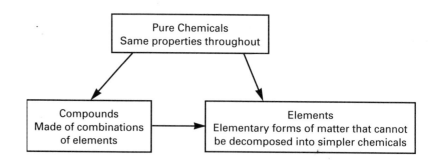

The word compound literally means made up of parts. **Chemical compounds** have definite properties and composition, but with some effort they can be separated into two or more chemical elements. The word ele-

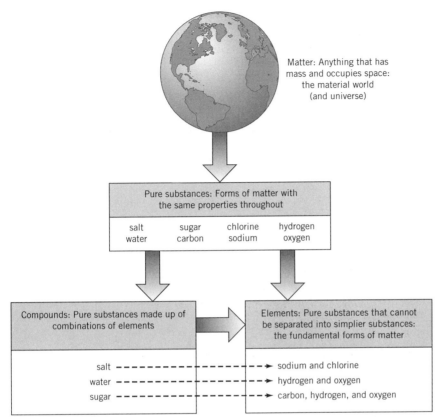

**FIGURE 3-2**
All matter is composed of compounds and elements.

ment comes from the Latin *elementum* 'fundamental' or 'basic principle.' Elements are the fundamental or simplest forms of matter. Water is a compound. It can be decomposed into the elements hydrogen and oxygen. The elements hydrogen and oxygen are fundamental forms of matter. They cannot be further separated into simpler chemicals. They cannot be decomposed into simpler forms.

A typical chemical compound is a combination of two or more elements and it always contains these elements in definite proportions by mass. Water always contains the elements hydrogen and oxygen, and the elements are always present in a proportion of 1.0 gram of hydrogen for every 8.0 grams of oxygen. Another way of describing the composition of water is to say that it is 11% hydrogen and 89% oxygen by mass. As we shall see, compounds are not just simple mixtures of elements. They are special chemical combinations of elements. Liquid water is obviously not a simple mixture of hydrogen gas and oxygen gas. How elements combine is discussed in Chapter 7, Chemistry in Action 3-2 illustrates forming an ele-

**ACTIVITY 3-2**

How many grams of chemically combined hydrogen are contained in a 500-g sample of water? (Hint: Use the percent given below as a multiplying factor to relate grams of H to grams of water.)

ment from a compound, and Chemistry in Action 3-3 and 3-4 illustrate compound formation.

Over several centuries, chemical elements were isolated from natural compounds, then added to the list of identified elements. Chemists have probed and investigated samples of matter from all regions of the earth. So far, 109* elements have been identified. Each element has unique properties that distinguish it from the other elements. For example, at room temperature the element oxygen is a colorless gas, and the element gold is a distinctly colored solid. Oxygen readily combines with other elements. Oxygen forms one or more compounds with all the known elements. Gold is not as chemically active and forms a limited variety of compounds.

## Percent

The term cent represents 100 as in the word century. Percent corresponds to parts per hundred. Percent is a common way of expressing what part one quantity is of a whole. Percent means parts per hundred, and is represented by the symbol %. Thus, 20% means 20 parts per 100 parts or 20/100. A percent can be expressed as a decimal fraction by dividing by 100. So 20% as a decimal fraction is 0.20.

If 59 g of water are found to contain 6.6 g of the element hydrogen, the percent by mass of hydrogen in water is found by dividing the mass of hydrogen by the mass of water, then multiplying by 100.

$$\%H = \frac{6.6\,g}{59\,g} \times 100 = 11\%$$

Generally, to calculate what percent one quantity, a, is of the whole, b, a is divided by b, then multiplied by 100.

$$\% = \frac{a}{b} \times 100$$

Percents are expressed using various bases of comparisons. For example, if it is found that in a large group of people three out of every 100 are left-handed, we say that 3% are left-handed. The basis of comparison is numbers of people. We'll most often use percents expressed on a basis of mass, referred to as percent by mass. For a given part of a whole, its percent by mass can be expressed as a factor in terms of grams of the part per 100 g of the whole. For example, the percent by mass of hydrogen in water is 11%. This percent can be expressed as a factor that relates mass of H to mass of water.

$$\frac{11\,g\,H}{100\,g\,water}$$

*Element 110 was reported in November 1994.

## 3-7 THE ELEMENTS

Since the elements are fundamental materials, which make up all matter, it is important to understand their similarities and differences. Much time and effort have been devoted to the isolation, purification, and description of chemical elements. Most of the elements occur in nature in various compounds, while a few occur as uncombined elements. Oxygen and nitrogen are found, for example, as free elements in the atmosphere. The elements copper, gold, silver, mercury, and sulfur are occasionally found in nature as free elements, but they are usually found in compounds.

Each element has a unique **name** and **symbol.** Since the elements have been isolated and identified over a period of centuries, their names and symbols have various historical origins. Inside the back cover of this book you'll find an alphabetical list of the element names and corresponding symbols. As you look over the list, note that most of the symbols come directly from the names of the elements. For example, the symbol for platinum is Pt. The name for this element comes from the Spanish name for silver, *platina.* When first discovered, this silvery-colored metal was mistaken for silver. Sometimes the symbol for an element comes from its Latin or Greek name. For instance, take potassium, symbol K; its name comes from the English *potash,* its symbol from the Latin *kalium.* Compounds containing potassium were first isolated from material obtained by extracting wood ashes with water, then boiling off the water in a large pot.

The symbol for the element silver is Ag, from the Latin *argentum.* The symbol for the element mercury is Hg, from the Greek *hydrargyros,* 'liquid silver', an apt name for this shiny liquid metal. The symbol for the element tungsten is W, from the German *Wolfram.* The name of the element *iron* has Anglo-Saxon origins but its symbol, Fe, comes from the Latin *ferrum.* The name for iron is *fer* in French, *hierro* in Spanish, and *Eisen* in German, but its international chemical symbol is Fe. Element symbols are agreed upon internationally.

The names of the elements can differ in various languages. The **symbol for an element,** however, is a specific letter or set of two letters used internationally to represent the element. The symbols always have a capital for their first letter; if there is a second letter, it is always lowercase. Knowledge of the names of the elements and the symbols that represent them is fundamental to any discussion of chemistry. To aid your study and learning of chemistry, it is good to memorize some element names and symbols. Table 3-1 is a list of some common elements.

Typically, elements occur in the environment as components of compounds. Of the 109 known elements, only 89 are found on earth in sufficient amounts so that samples can be isolated. The remaining elements are rare; some exist for only a short time because they are so unstable. A few elements do not exist in nature and have been synthesized using spe-

*Table 3-1* Some Important Elements

| ELEMENT | SYMBOL |
|---|---|
| Aluminum | Al |
| Arsenic | As |
| Barium | Ba |
| Boron | B |
| Bromine | Br |
| Cadmium | Cd |
| Calcium | Ca |
| Carbon | C |
| Chlorine | Cl |
| Chromium | Cr |
| Cobalt | Co |
| Copper | Cu |
| Fluorine | F |
| Gold | Au |
| Helium | He |
| Hydrogen | H |
| Iodine | I |
| Iron | Fe |
| Lead | Pb |
| Lithium | Li |
| Magnesium | Mg |
| Manganese | Mn |
| Mercury | Hg |
| Nickel | Ni |
| Nitrogen | N |
| Oxygen | O |
| Phosphorus | P |
| Potassium | K |
| Silicon | Si |
| Silver | Ag |
| Sodium | Na |
| Sulfur | S |
| Tin | Sn |
| Uranium | U |
| Zinc | Zn |

***Table 3-2***   Percent by Mass of Elements of the Earth's Surface

| Element | Percent | Element | Percent |
|---|---|---|---|
| Oxygen | 49.20 | Chlorine | 0.19 |
| Silicon | 25.67 | Phosphorus | 0.11 |
| Aluminum | 7.50 | Manganese | 0.09 |
| Iron | 4.71 | Carbon | 0.09 |
| Calcium | 3.39 | Sulfur | 0.06 |
| Sodium | 2.63 | Barium | 0.04 |
| Potassium | 2.40 | Fluorine | 0.03 |
| Magnesium | 1.93 | Nitrogen | 0.03 |
| Hydrogen | 0.87 | | |
| Titanium | 0.58 | Others | 0.47 |

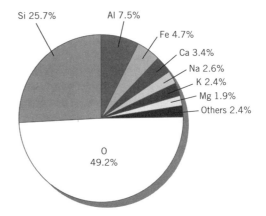

cial methods. About 40 of the elements have geological or biological importance. Table 3-2 gives the percents by mass of the 18 elements that make up 99.6% of the earth's environment. For comparison, Tables 3-3 and Table 3-4 list the percents by mass and atom number of the most common elements found in the human body. Some elements are vital to living systems. Carbon, hydrogen, oxygen, and nitrogen are the major elements included in compounds making up water, proteins, fats, and carbohydrates in the body. Other important elements in the body are described in Table 3-5. Trace elements are those elements needed in the diet in very small amounts. Table 3-6 describes important trace elements.

**Table 3-3**   Percent by Mass of Elements in the Human Body

| ELEMENT | PERCENT | ELEMENT | PERCENT |
|---------|---------|---------|---------|
| Oxygen | 64.6 | Sodium | 0.11 |
| Carbon | 18.0 | Magnesium | 0.03 |
| Hydrogen | 10.0 | Iron | 0.005 |
| Nitrogen | 3.1 | Zinc | 0.002 |
| Calcium | 1.9 | Copper | 0.0004 |
| Phosphorus | 1.1 | Tin | 0.0001 |
| Chlorine | 0.40 | Manganese | 0.0001 |
| Potassium | 0.36 | Iodine | 0.0001 |
| Sulfur | 0.25 | Others | 0.14 |

**Table 3-4**   Percent by Number of Atoms of Elements in the Human Body

| ELEMENT | PERCENT BY ATOM COUNT |
|---------|-----------------------|
| Hydrogen | 62.88 |
| Oxygen | 25.59 |
| Carbon | 9.50 |
| Nitrogen | 1.40 |
| Calcium | 0.30 |
| Phosphorus | 0.22 |
| Potassium | 0.06 |
| Sulfur | 0.05 |
| Sodium | 0.03 |
| Chlorine | 0.01 |
| Magnesium | 0.01 |
| Others | 0.01 |

**Table 3-5**   Some Important Elements in the Human Body

| ELEMENT | PERCENT OF BODY WEIGHT | ELEMENT | PERCENT OF BODY WEIGHT |
|---------|------------------------|---------|------------------------|
| Calcium | 1.5–2.2 | Chlorine | 0.15 |
| Phosphorus | 0.8–1.2 | Sodium | 0.15 |
| Potassium | 0.35 | Magnesium | 0.05 |
| Sulfur | 0.25 | Iron | 0.004 |

**Calcium:** Found in compounds in bones, teeth, and body fluids.

**Phosphorus:** About 85% found in combination with calcium in bones and teeth. The rest incorporated in compounds in body fluids and DNA and RNA in cells.

**Magnesium:** Found in compounds contained in bone and body fluids.

**Sodium:** Mainly found as dissolved salt contained in extracellular fluids. Also found in cellular fluids and involved in transmission of nerve impulses.

**Chlorine:** Mainly found as dissolved salt contained in extracellular fluids. Also found in gastric juices in the stomach.

**Potassium:** A major element found in compounds contained in cellular fluids. Also involved in transmission of nerve impulses.

**Sulfur:** Found in amino acids and proteins of the body.

**Iron:** An important component of blood hemoglobin and muscle myoglobin. Stored in compounds found in liver, spleen, and bone.

*Table 3-6*    Important Trace Elements

| ELEMENT | APPROXIMATE AMOUNT IN BODY (MILLIGRAMS PER KILOGRAM) | LOCATION OR FUNCTION IN THE BODY |
|---|---|---|
| Chromium | 0.08 | Related to function of insulin in glucose metabolism |
| Cobalt | 0.04 | Required for function of several enzymes, part of vitamin $B_{12}$ |
| Copper | 1.4 | Required for function of respiratory enzymes and some other enzymes |
| Iodine | 0.4 | Located in thyroid gland, needed for hormone thyroxine |
| Manganese | 0.3 | Required for function of several digestive enzymes |
| Molybdenum | 0.07 | Required for function of several enzymes |
| Zinc | 23 | Required for function of many enzymes |
| Fluorine | Trace | Found in bones and teeth; thought to be essential, but function unknown |
| Selenium | Trace | Essential for liver function |
| Silicon | Trace | May be essential in humans |
| Tin | Trace | May be essential in humans |

## 3-8 CHEMICAL EVIDENCE FOR ATOMS

Chemists over the ages have experimented with a variety of substances and identified many elements and compounds. They discovered ways to decompose compounds into their constituent elements and to form compounds from elements. Furthermore, with the development of precise weighing methods, they could weigh samples of elements and compounds before and after such transformations. This allowed precise measurement of the relative masses of elements combined in compounds and provided evidence for the development of an atomic theory.

Let us consider some of this evidence. Elements combine to form compounds, and compounds can be decomposed to form elements. When a sample of water is decomposed the liquid is transformed into gases. Extensive observations of chemical processes in which elements form compounds, or compounds decompose, reveal that no measurable mass is lost or gained. For example, the decomposition of 100 g of water yields 89 g of oxygen and 11 g of hydrogen, and 89 g of oxygen combine with 11 g of hydrogen to give 100 g of water. The fact that no mass change is observed in chemical processes is important. Experimental measurements show that in chemical processes or chemical reactions mass is conserved. To be conserved means that mass is not lost or gained. This idea became known as the law of conservation of matter as stated in the margin. The law of conservation of matter is another example of a physical law, an experimentally observed consistent pattern in nature.

Another significant observation about the nature of matter is the fact that a specific compound always contains the same elements. Let's consider this idea in more detail. As an example, common table salt (known chemically as sodium chloride), produced in the laboratory or obtained from a natural source, always is found to be a compound of the elements sodium and chlorine. Furthermore, pure samples of salt always have the same relative amounts of the two elements. Any salt sample contains 39.4 g of sodium and 60.6 g of chlorine for every 100 g of salt. In terms of percent, salt is always 39.4% sodium and 60.6% chlorine by mass. This is an example of a consistent fixed pattern in compounds. The pattern is generally true for all compounds and is expressed as the law of constant composition as given in the margin.

The laws of conservation of matter and constant composition provide evidence that elements are composed of some kind of characteristic building units or particles. These particles can join in definite patterns to form compounds. From another perspective, it is possible to explain these laws by assuming that elements are composed of characteristic particles: atoms. Experimental evidence related to the nature and behavior of elements and compounds suggests the existence of atoms. The experimental evidence provides a justification for the philosophical idea of atoms.

**Law of Conservation of Matter:**
*Matter is not created or destroyed in a chemical process.*

**Law of Constant Composition:**
*A compound always contains the same elements and they are present in definite, unchanging proportions by mass.*

## 3-9   ATOMIC THEORY

In 1803, John Dalton, drawing from the work of many early scientists, proposed a theory or model of the particle nature of matter. It is called the atomic theory and can be stated as follows:

1. Matter is composed of tiny, fundamental particles called atoms. (Dalton used the term atom in recognition of the ideas of Democritus 2200 years earlier.)

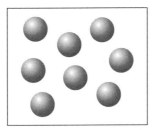

2. Atoms of an element are the same but differ from the atoms of all other elements. Each element has unique atoms.

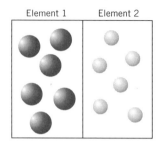

3. Atoms can combine to form compounds. Compounds contain atoms combined in definite whole number ratios.

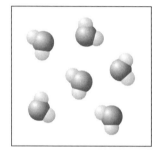

The atomic theory is very important; in addition to explaining many properties of matter, it provides a mental image of matter. An atom is pictured as a tiny spherical particle and matter is viewed as collections of atoms. An **atom** is the smallest representative particle of an element. Each kind of element has a unique kind of atom, and each atom has mass. A sample of an element contains only atoms of that element. The most obvious way in which the atoms of different elements differ is atomic mass. An aluminum atom has a different mass than an iron atom, an oxygen atom, or any other kind of atom. Look at a piece of aluminum. Imagine that it consists of a vast collection of tiny aluminum atoms. If the sample could be subdivided into smaller and smaller pieces, the smallest possible piece would be an aluminum atom. If the atom is subdivided, the pieces no longer represent the element. See Figure 3-3.

Dalton's atomic theory provided explanations of many chemical observations. Just as importantly, the theory produced a new way of thinking about chemical processes. Dalton used a set of wooden balls as atomic models and connected them in various ways to represent compounds. In addition, Dalton introduced the idea that symbols for the elements can be used to represent individual atoms so that compounds, as combinations of atoms, can be represented as combinations of symbols. For example, water

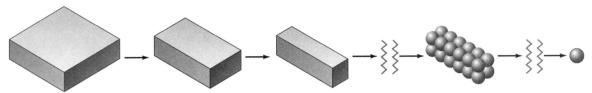

**FIGURE 3-3**
Idealized subdivision of a
piece of aluminum.

is known to contain two hydrogen atoms for each oxygen atom. Thus, it can be represented as $H_2O$. Such a representation is called a chemical formula.

## 3-10 USING THE ATOMIC THEORY

The atomic theory served as a foundation for the development of chemistry and is one of the most fundamental theories of chemistry. A good theory is a simple idea that provides many explanations of natural processes, introduces new ideas and concepts, and suggests some important questions.

### Explanations

Let's see how the atomic theory explains some observations mentioned in Section 3-8. We would expect elements to be distinct and unique since each element is composed of atoms that are unlike the atoms of any other element. Aluminum atoms differ from iron atoms, and iron atoms differ from gold atoms. The law of conservation of matter makes sense if we view compounds as collections of combined atoms. Each atom has mass. Thus, the mass of a sample of a compound is the sum of the masses of all of the atoms that compose the compound. Atoms are not destroyed when a compound is decomposed into its elements. Therefore, we would expect the total mass of the elements to be the same as the mass of the compound. In other words, the law of conservation of matter is a reflection of the fact that atoms are not created or destroyed. When elements combine to form compounds or compounds decompose to form elements, the atoms change their states of combination but are not created or destroyed. See Figure 3-4.

An explanation of the law of constant composition is that compounds involve definite and fixed combinations of atoms. That is, a given compound always includes the same kinds of atoms in the same relative numbers. The compound always contains the same proportions of the elements that make it up. See Figure 3-5. For instance, water always contains two combined hydrogen atoms for every combined atom of oxygen. This

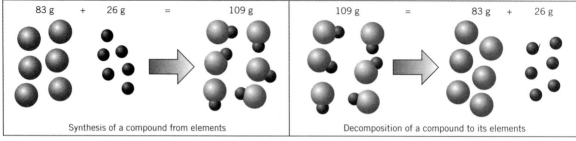

Synthesis of a compound from elements | Decomposition of a compound to its elements

**F I G U R E   3 - 4**    The synthesis and decomposition of a compound.

is why it is represented by the formula $H_2O$. Since the atoms of elements differ from one another, they form different kinds of compounds with other elements. Occasionally, the atoms of any two elements can combine in different ways to form two or more different compounds. For example, hydrogen and oxygen combine to form the common compound water and can also form the less common compound hydrogen peroxide, $H_2O_2$. See Figure 3-5.

**F I G U R E   3 - 5**
The hydrogen–oxygen compounds, water and hydrogen peroxide.

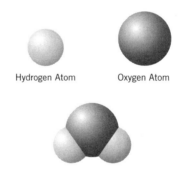

Hydrogen Atom        Oxygen Atom

Water is a very common solvent and is one of the most important compounds in the environment. It contains two combined hydrogen atoms for every one combined oxygen atom.

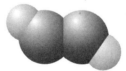

Hydrogen peroxide is a chemically active compound used as a paper pulp bleach, a hair bleach and a mild bactericide. It contains two combined atoms of combined hydrogen for every two combined atoms of oxygen.

## Ideas

Once the idea that elements were unique and had characteristic atoms was established, chemists carefully searched nature for more elements. Thirty-one elements were known when Dalton proposed the atomic theory. Since then 78 more have been added to the list.

The atomic theory states that atoms can combine to form compounds. Atoms are characteristic particles of elements, but compounds are composed of chemically combined atoms. In many compounds, atoms are combined in the form of molecules. Water is a compound composed of water molecules having two atoms of hydrogen combined with one atom of oxygen. A sample of water is a collection of a vast number of water molecules. (As discussed in Chapter 7, not all compounds contain molecules.)

The decomposition of a compound into its elements and the combination of elements to make a compound are examples of chemical processes. With the atomic theory, such processes are envisioned as changes in the combinations of atoms. Such transformations are called **chemical reactions.** The term reaction means to produce a result. A **chemical reaction** is a process in which one set of chemicals is mixed and transforms into a new set of chemicals. See Figure 3-6 for an illustration of a simple chemical reaction. Reactions are fundamental in chemistry and are discussed in more detail in Chapter 4.

As mentioned above, the law of conservation of matter is an expression of the fact that atoms are not created or destroyed in chemical processes. In other words, atoms are conserved in chemical reactions. A reaction is a special process in which certain chemical combinations of atoms transform into new combinations of atoms. In a reaction the total atom count does not change. Thus, it is not possible to destroy matter by chemical reactions. Burning a candle produces gases that are diluted and dispersed in the surrounding air, but the atoms are not destroyed. Solid chemicals in the candle are changed to gaseous forms of matter.

The atomic theory stimulated the imaginations of chemists and encouraged experimentation. The fact that compounds are combinations of elements suggested that many different compounds could exist. Chemists isolated and purified numerous naturally occurring compounds. Today,

**FIGURE 3-6**
A simple chemical reaction.

Carbon          Oxygen                    Carbon dioxide

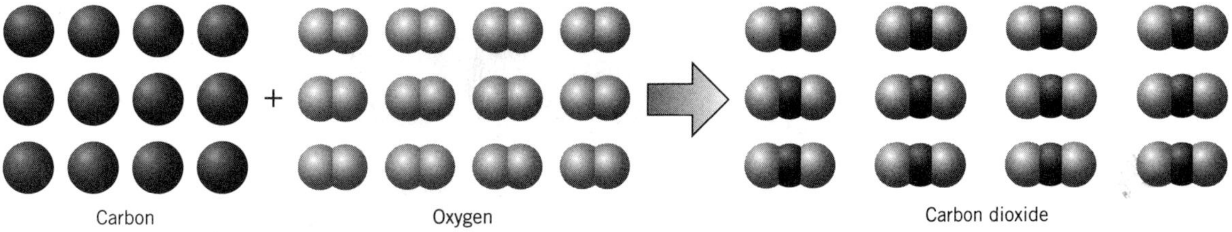

many naturally occurring compounds are isolated from plant and mineral sources and used in a variety of ways ranging from raw materials for industry to food and medicinal uses. As knowledge of chemistry developed, chemists began to explore possible combinations of elements that do not occur in nature. These combinations are made by a process called chemical synthesis. **Synthesis** means to make or put together. Today many unique methods of synthesis are used to produce a large variety of "synthetic" chemicals. Hundreds of different kinds of synthetic chemicals are used as medicines, pesticides, and plastics. Synthesis reactions are also used to make chemicals that are identical to naturally occurring compounds. They are identical because they contain the same elements combined in the same way. There is no difference between naturally occurring ammonia and ammonia that is synthesized by industrial processes, and there is no difference between pure synthetic vitamin C and pure natural vitamin C.

Some friends tell you that they only take natural vitamin C supplements. Explain to them why there is no difference between pure synthetic vitamin C and pure natural vitamin C.

An important consequence of the atomic theory is that it makes chemistry a conceptual science. Atoms are involved in chemical thinking. We can observe chemical processes with our senses or special instruments. We can witness chemical changes but we can't see atoms. Chemists often explain chemical processes on the atomic level and use the combination behavior of atoms in explanations. For example, for years chemists sought an explanation of burning or combustion in air. Burning, in the modern chemical view, is the rapid and vigorous chemical reaction of oxygen in the air with combustible materials. When charcoal, which is mostly elemental carbon, burns in air, the process is envisioned as carbon atoms combining with oxygen to produce molecules of carbon dioxide gas. See Fig. 3-6.

You probably have observed experimentally that aspirin can relieve a headache. A chemical explanation is that aspirin, as a chemical compound, somehow interacts with the chemicals involved in the transmission of pain in the body. In other words, the effect of aspirin is not magic but involves chemical reactions and the rearrangement of atoms. The precise chemical behavior of aspirin is still being investigated. Chemical thinking and atomic visions provide satisfactory explanations for chemical processes that take place in the laboratory, in our bodies and in the environment.

### Questions

The atomic theory suggests some important questions. A few are given here. Can you think of additional questions?

How are the atoms of a given element the same and how do atoms of the various elements differ from one another?

If atoms are characteristic chemical particles, are they composed of even more fundamental particles? If so, what kinds of particles structure atoms?

How are the formulas of compounds known?

How do atoms combine with one another?

Are there any patterns in the way in which elements combine to form compounds?

Answers to these questions are part of your continuing study of chemistry and are explored in future chapters.

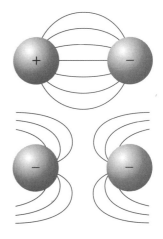

## 3-11  ELECTRICITY

In the early 1800s, atoms as proposed by Dalton were viewed as fundamental and indivisible particles. Experiments soon began to show that matter also had an electrical nature. A bolt of lightning is a dramatic example of the association of electricity and nature. You may have noticed the electrical effect of passing a comb through your hair, or you may have stuck a balloon to the wall after rubbing it on your hair. You even may have experienced static cling. Such behavior is related to the fact that objects can carry electrical charges. There are two types of electrical charge: positive and negative. The terms positive and negative refer to different types of charge and are different from positive and negative as used in arithmetic. However, positive charge is represented by the + symbol and negative charge by the − symbol. If an object has equal amounts of positive and negative charge, the charges balance one another and the object is electrically neutral. Recall that Coulomb's law, mentioned in Chapter 1, states that opposite electrical charges attract and like charges repel. Two pieces of clothing attract due to opposite electrical charges, while your hairs can repel one another because they carry like electrical charges.

What is electricity? Scientists, including the American Benjamin Franklin, developed ideas about the nature of electricity in the 1800s. It was theorized that electricity was related to tiny, discrete, negatively charged particles that carried electric current. The ancient Greeks had noted that when amber, a fossilized tree resin, was rubbed with cloth it would attract small bits of other materials. The Greek name for amber was *elektron*. When they were discovered in the late 1800s the characteristic particles of electricity were called electrons. Electrons are far too small to see but electricity can be envisioned as the movement or flow of numerous electrons. Electrical current flowing through a metal is viewed as the flow or movement of electrons.

Here is a simple exercise that illustrates that electrical charge is associated with matter. Cut 10 to 20 very thin slivers of paper having lengths of 1 cm. Place the slivers on a flat piece of paper. Blow up a balloon and tie it off. Rub the balloon on your dry hair or on a nylon garment. Lower the balloon until it is just above the paper slivers. Point a finger at some of the slivers that stick to the balloon. Use a toothpick to move the slivers around on the surface of the balloon. Record your observations. If you do not have a balloon use a clear plastic cup or a Styrofoam cup.

## 3-12 ELECTRICITY AND THE ATOM

In the 1830s an English chemist, Michael Faraday, experimented with the effect of electricity on matter. Using a strong battery as a source of electricity, he decomposed some compounds into elements. This was done by melting compounds, then passing electricity through the melt. Faraday coined the term electrolysis for the process of splitting compounds by electricity. As you know, water can be split into hydrogen and oxygen gas by electrolysis.

A solid compound that conducts electricity when melted or when dissolved in water is called an electrolyte. In electrolysis, a voltage source, such as a battery, maintains a voltage between two conductors called electrodes. One electrode, known as the anode, carries a positive charge; the other, known as the cathode, carries a negative charge. When the electrodes are dipped into the melt or solution, electricity flows. Faraday's chemical explanation was that the electrical current is carried through the molten compound or solution not by electrons, as is the case with metallic conductors, but by electrically charged atoms called ions. Normal atoms are electrically neutral, but ions are atoms that carry electrical charges.

Our word ion came from the Greek *ienai* 'to go'. The particles were moving or going toward the electrodes. Ions that move toward the cathode or negative electrode are called **cations.** Cations carry positive electrical charges. Ions that move toward the anode or positive electrode are called **anions.** Anions carry negative electrical charges. Electricity flowing in a metallic conductor is viewed as a stream of free electrons, but electricity flowing in a solution of an electrolyte is viewed as the movement of cations and anions. In any case, electricity represents the movement of charged particles.

We'll see how ions relate to the story of chemistry in Chapter 7, but first we'll consider the structure of atoms. Incidentally, the variety of ions found in body fluids are collectively called the body electrolytes. These electrolytes are responsible for the fact that electricity can flow through body tissue.

The particles that make up atoms are simply called subatomic particles. Experiments with electricity and matter suggest that electrons are one type of subatomic particle. However, electrons in atoms are bound to the atoms and are not free-moving as are the electrons involved in electrical current. It appears that under certain conditions, some of the electrons bound in atoms can leave the atoms and become free-moving electrons in electrical current. Normally atoms are electrically neutral. To be neutral, an atom must have a balance of positive and negative charges. If electrons with negative charges are found in atoms, some kind of positively charged particles must also be present. Such positive particles, called **protons,** were discovered in 1914. An electron carries a negative charge and a proton carries an equal amount of positive charge. Years later, in 1932, another major subatomic particle was discovered. It was electrically neutral and, so it was named the **neutron.**

## 3-13  SUBATOMIC PARTICLES

Atoms came to be viewed as collections of subatomic particles. How do subatomic particles structure atoms and how do atoms of the various elements differ from one another? Are atoms made of subatomic particles that arrange in different ways to form different types of atoms? Since the times of Dalton, scientists have carried out extensive experiments and observations related to questions about atomic structure and the nature of subatomic particles. Experiments suggest that atoms are composed of electrons, protons, and neutrons.

| PARTICLE | SYMBOL | CHARGE | MASS (g) | MASS RELATIVE TO PROTON |
|---|---|---|---|---|
| Electron | $e^-$ | 1 − | $9.110 \times 10^{-28}$ | 1/1835 |
| Proton | $p$ | 1 + | $1.672 \times 10^{-24}$ | 1 |
| Neutron | $n$ | 0 | $1.675 \times 10^{-24}$ | 1 |

**Electrons:**
*Subatomic particles that carry negative electrical charge*

**Protons:**
*Subatomic particles that carry positive electrical charge*

**Neutrons:**
*Subatomic particles that are electrically neutral*

Note that an electron has a relatively small mass and carries a fixed amount of negative charge. The electron carries the smallest possible unit of negative electrical charge, and we shall refer to its charge as 1− or simply −. It is interesting that each proton carries an amount of positive charge that is equal to the negative charge on an electron. That is, electrons and protons carry equal but opposite electrical charge. The charge of a proton is represented as a 1+ or simply +. Since an atom is electrically neutral it must contain an equal number of protons and electrons. Atoms of the various elements differ in the number of protons and electrons they contain. Even though electrons and protons are exact opposites with respect to electrical charge, they differ greatly in mass. An electron has a mass of $9.110 \times 10^{-28}$ g and a proton has a mass of $1.672 \times 10^{-24}$ g. This means that a proton has a mass that is 1835 times greater than the mass of an electron. Neutrons are also found in atoms, but since they are electrically neutral they do not contribute any electrical charge. A neutron has a mass of $1.675 \times 10^{-24}$ g, just slightly larger than a proton.

## 3-14  THE NUCLEAR ATOM

If atoms contain electrons, protons, and neutrons, the obvious question is how do these subatomic particles give structure to an atom? A partial an-

**FIGURE 3-7**
An atomic nucleus.

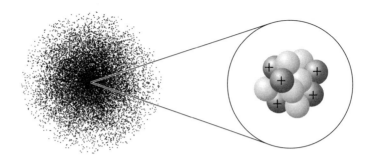

The ratio of the diameter of the atom to the diameter of the nucleus reveals how much larger the atom is compared to the nucleus. Calculate this ratio using the numbers given in Section 3-14.

swer to this question was given by the English scientist, Lord Rutherford, in the early 1900s. Experimental observations done in his research laboratory led him to theorize that an atom consists of a small, extremely dense, positively charged **nucleus** surrounded by a swarm of negatively charged electrons of relatively low mass. The term nucleus means center. This idea became known as the **nuclear model of the atom.** The vision of the atom became one in which very tiny electrons are in motion about a nucleus. The vision of the nucleus became a cluster of protons and neutrons (see Fig. 3-7).

Using unique experimental methods, Rutherford calculated the approximate sizes of atoms and nuclei. He calculated the diameter of an average atom to be about $2 \times 10^{-8}$ cm and the diameter of an average nucleus to be about $1 \times 10^{-13}$ cm. These numbers reveal that atoms are very tiny particles and that atomic nuclei are extremely small parts of atoms. This means that the nucleus of an atom makes up an extremely small portion of the entire atom and that the electrons populate a very large portion.

To get a feeling for, or picture of the size of an atom, consider that a speck of carbon barely visible to the eye contains $6 \times 10^{16}$ carbon atoms (60 million billion atoms). In comparison, the population of the earth is about 6 billion ($6 \times 10^9$) humans. It would take 10 million such populations to equal the number of atoms in this single speck of carbon.

Use a pencil or a pen to make a dot on a piece of paper. If a single atom could be expanded so that its nucleus had a diameter equal to this dot, the outer regions of the atom would reach about 100 meters from the dot. In contrast to the differences in atomic and nuclear sizes, almost all of the mass of an atom is concentrated in the nucleus. The nucleus makes up more than 99.9% of the mass of an atom. Nuclei are very dense concentrations of mass with positive electrical charge. They are dense clusters of protons and neutrons. The term **nucleon** is used to refer to protons and neutrons found in nuclei.

## 3-15 ATOMIC STRUCTURE

You may wonder why the atom is called the fundamental particle of matter. It is, after all, composed of subatomic particles. From a chemical point of view, atoms are fundamental. An **atom of an element** is the characteristic particle of the element. Since the behavior of atoms explains chemical processes, **chemistry** is the science of the nature and behavior of atoms. (Recall the definition of chemistry given in Chapter 1. Why is that definition essentially the same as the definition given here?)

An atom of a particular element differs from the atoms of all other elements in the numbers of subatomic particles it contains. Of course, since these particles have mass, the atoms of various elements also differ in mass. How do the structures of the atoms of the various elements differ? An atom of a specific element has a specific number of protons in the nucleus. For example, nuclei of oxygen atoms contain eight protons. This is characteristic of oxygen, so any atom that has a nucleus with eight protons is an oxygen atom. Thus, atoms with the same number of protons in the nuclei are atoms of the same element. Atoms with different numbers of protons in the nuclei are atoms of different elements. All hydrogen atoms have nuclei with one proton compared to oxygen atoms which have eight protons. A neutral atom has the same number of electrons as protons. Hydrogen atoms have one electron and one proton, and oxygen atoms have eight electrons and eight protons. For a neutral atom, it is always true that the number of protons equals the number of electrons.

The number of protons in a nucleus of an atom is called the **atomic number.** Each element has a unique atomic number. Look at the table inside the back cover of this book. The elements are listed alphabetically. Note that the atomic number is listed for each element. For instance, hydrogen has an atomic number of 1, oxygen 8, sulfur 16, and uranium 92. The atomic number of an element reveals the number of protons and the number of electrons in an atom of the element.

## 3-16 ISOTOPES

The fact that atoms of different elements have different numbers of protons and electrons partially accounts for the differences in masses of the atoms. The neutrons also contribute to the masses of atoms. The atoms of most elements have been found to have nuclei consisting of more than one combination of neutrons and protons. For example, study of chlorine shows that it is made up of two kinds of atoms. About 76% of chlorine atoms have nuclei with 18 neutrons and 17 protons, and 24% have nuclei with 20 neutrons and 17 protons. Of course, since the number of protons

| PRO-<br>TONS | NEU-<br>TRONS | ELEC-<br>TRONS |
|---|---|---|
| 8 | 8 | 8 |
| 8 | 9 | 8 |
| 8 | 10 | 8 |

**3-6**

The nucleus of an atom can be envisioned as a cluster of protons and neutrons. Hydrogen has the isotopes hydrogen-1, hydrogen-2, and hydrogen-3. Draw representations of the nuclei for each of these isotopes and show the number of protons and neutrons.

is 17 in each case, both types of atoms are chlorine atoms. Thus, there are two kinds of chlorine atoms found in nature, and they have slightly different masses since they contain different numbers of neutrons. Atoms of the same element with differing masses are called isotopes. **Isotopes** are atoms of the same element having the same number of protons but different numbers of neutrons.

The element oxygen has three natural isotopes. Let us see how they differ. The atomic number of oxygen is 8, so all oxygen atoms have eight protons and eight electrons. One isotope of oxygen has atoms with eight neutrons, the second isotope has nine neutrons, and the third has 10 neutrons. For any isotope, the sum of the number of protons and neutrons is called the **nucleon number** or **mass number.** The nucleon numbers of the oxygen isotopes are 16, 17, and 18, respectively. Nucleon numbers distinguish one isotope from another. Thus, the isotopes of oxygen are oxygen-16, oxygen-17, and oxygen-18. A summary of the composition of the oxygen isotopes is given in the margin. Incidentally, to find the number of neutrons in an isotope subtract the atomic number from the nucleon number. For example, since oxygen has an atomic number of 8 its isotopes have 8, 9, and 10 neutrons respectively.

Some elements found in nature exist in the form of only one type of atom. These elements have no natural isotopes. For instance, only one kind of aluminum atom occurs in nature. Other elements have two or more naturally occurring isotopes. Examples, shown in Figure 3-8, are hydrogen with two natural isotopes, carbon with two, oxygen with three, chlorine with two, and tin with 10. Notice that the figure shows the composition of each isotope and the percent each isotope contributes to the naturally occurring element.

The nucleon number of an isotope is a number representing the sum of the number of protons and the number of neutrons in the nucleus of an atom. Nucleon numbers are used to refer to a specific isotope of an element in order to distinguish it from other isotopes of the element. Some common examples are plutonium-239 used in nuclear weapons, uranium-235 used in nuclear reactors, americium-241 used in smoke detectors, cobalt-60 used in cancer therapy, and carbon-14 used in archaeological age dating. However, when we use the name or symbol of a specific element, we are referring to the naturally occurring element that is a collection of natural isotopes. That is, when we refer to oxygen, O, we mean that collection of atoms (oxygen-16, oxygen-17, and oxygen-18) which includes all the natural isotopes. When you breath oxygen, you are breathing all of the natural isotopes of oxygen. Oxygen-16 is the most abundant of these isotopes but the others are always present in natural samples of oxygen. It is important to realize that as far as chemical behavior and chemical properties are concerned all of these isotopes are the same. Typically all isotopes of an element have the same chemical behavior.

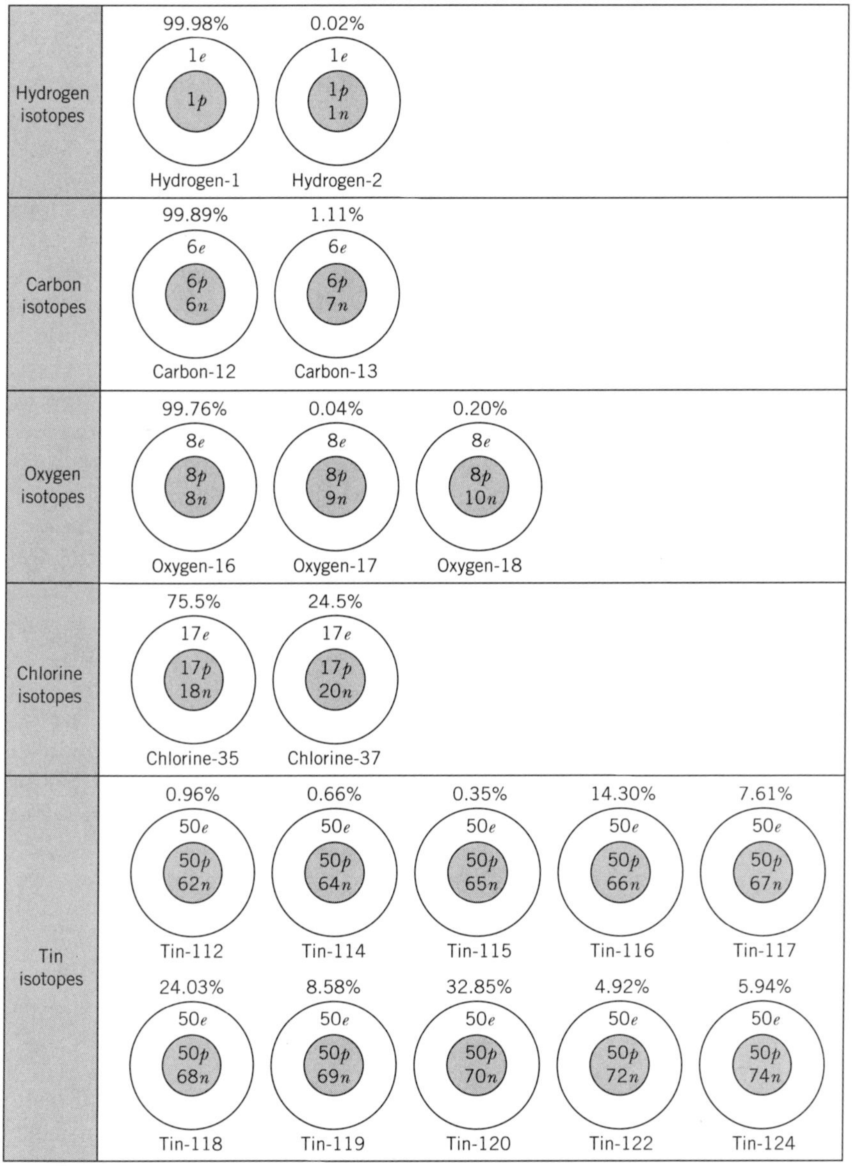

FIGURE 3-8
The naturally occurring isotopes of some elements.

## 3-17 ATOMIC MASSES

Atoms are very tiny and we cannot make the same kind of measurement on atoms that we can make on large samples that are collections of atoms. For example, we cannot weigh individual atoms. However, it is possible to measure the relative masses of the atoms of elements. By relative mass we

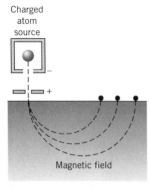

Charged
atom
source

Magnetic field

**FIGURE 3-9**
A type of mass
spectrometer: charged
atoms take on specific
circular paths depending
on their masses.

mean the mass of an atom of one element compared to the mass of an atom of another element. For example, a typical atom of carbon is about three times the mass of a typical helium atom. That is, the ratio of the mass of a helium atom to a carbon atom is 1 to 3.

The most accurate method of measuring the relative masses of atoms is mass spectroscopy. This very sensitive method uses an instrument called a mass spectrometer in which electrically charged atoms are forced to move through a magnetic field. As shown in Figure 3-9, the moving atoms take on curved paths directly related to their masses. As an analogy (see Fig. 3-10), imagine standing on a high bridge with a brisk wind blowing below. If you drop balls of different masses, say a tennis ball, a handball, and a baseball off the bridge, they will be deflected by the wind differently and take on different curved paths. By carefully measuring the curvature of the paths of atoms in a mass spectrometer, it is possible to measure the relative masses of the atoms. When a sample of a specific element is placed in a mass spectrometer, it is possible to measure the relative masses of its isotopes. Furthermore, mass spectrographic analysis reveals the percent of each of these isotopes in the sample. For instance, when chlorine is analyzed it is found to contain two isotopes having slightly different relative masses. The isotopes have nucleon numbers of 35 and 37 so their relative masses are close to one another but not equal. The analysis also reveals that chlorine is about 75.5% chlorine-35 and 24.5% chlorine-37.

Even though an element has isotopes, all of the atoms of that element have the same chemical behavior. Thus, from a purely chemical point of view, it is not important to know how many isotopes an element has or even whether it has isotopes at all. It is true, however, that the existence of

**FIGURE 3-10**
A mass spectrometer
analogy: balls of different
mass take on different
paths as they are dropped
into the wind.

Wind

isotopes is reflected in the masses of the atoms of elements. Consequently, when the relative masses of the atoms of various elements are determined, the existence of isotopes is important.

## 3-18  ATOMIC WEIGHTS

Atoms are very tiny particles. It is not convenient or useful to express masses of individual atoms in grams. A special unit of mass, called the atomic mass unit (u), is used as a unit to express the masses of atoms. To establish the atomic mass unit, an atom of the common isotope carbon-12 is defined as having a mass of exactly 12 u. Thus, one atomic mass unit is one-twelfth the mass of a carbon-12 atom. The atomic mass unit is defined using carbon-12 so that no atom has a mass less than 1 atomic mass unit. An atom of the lightest isotope of hydrogen, which is the least massive of all isotopes, has a mass around 1 u relative to carbon-12. The masses of isotopes of any element can be measured relative to carbon-12 and expressed in atomic mass units. Recall that the relative mass of a typical helium atom to a carbon atom is 1 to 3. Thus, a helium atom would have a mass in atomic mass units that is one-third the mass of a carbon-12 atom, or 4 u. Of course, the masses of atoms are typically measured to many more significant digits than given in this example. An aluminum atom has a mass that is 2.248462 times the mass of a carbon-12 atom. Thus, since carbon-12 has a mass of exactly 12 u, the atomic weight of aluminum is 26.98154 u (12 × 2.248462).

In chemistry it is very useful to have a numerical value for the average mass of the atoms of an element. This may seem difficult in view of the numerous isotopes of the elements. However, these average masses are easily calculated from the mass and the percent abundance of each natural isotope of an element. These data are available from mass spectroscopy experiments. For instance, using the masses of chlorine-35 and chlorine-37 and the percent distribution of each, the mass of an average atom of chlorine can be calculated. Each isotope contributes a share to the average mass of a chlorine atom. As an analogy, if we want to find the average weight of the linemen in a football team we consider the contributions of each individual. The **atomic weight** of an element is found by multiplying the mass of each natural isotope by its fractional abundance and adding these products. The fractional abundance is the percent divided by 100.

The mass of a typical or average atom of an element is called its atomic weight. The **atomic weight** is defined as the mass of a typical atom of an element determined by considering the contribution of its natural isotopes. Each element has a unique and characteristic atomic weight.

The list of elements inside the back cover of this book also gives the known atomic weights. The elements are listed alphabetically. Precise atomic weights for the elements are the result of many years of tedious ex-

Imagine a sample containing one thousand typical chlorine atoms. How many of these atoms are chlorine-35 and how many are chlorine-37? (Actually one thousand atoms is a very small sample, much too small to see, smell, or weigh.)

Use the following information to calculate the atomic weight of chlorine.

Chlorine-35
34.96885 u
75.53%

Chlorine-37
36.96590 u
24.47%

**Atomic Weight:**
*The average mass of an atom of an element determined by considering the contribution of each natural isotope.*

perimental work. You'll find a list of atomic weights to be very useful for some interesting kinds of chemical computations discussed in future chapters. When you need an atomic weight just look it up. You often see tables of elements posted on the walls of chemistry laboratories and lecture rooms. Such a table is called the **periodic table of the elements.** Elements in the table are listed from left to right by increasing atomic number. Typically the table gives the symbol for each element, its atomic number, and its atomic weight. A periodic table is given inside the front cover of this book. You can use a periodic table as a convenient and quick source of atomic weights. You'll have to wait until Chapter 6 to find out why the periodic table has such a curious shape.

A close look at a table of atomic weights shows that atomic weights vary in the number of significant digits. Atomic weights have been determined using experimental measurements of isotopic masses and percent abundances of the isotopes. Variations in these measurements for a given element dictate the number of significant digits in the atomic weight. Typically atomic weights have numerous significant digits. When we use atomic weights in computations we often round them to a useful number of digits. Four or five significant digits are sufficient since most calculations involving atomic weights include experimental data that are in this range of significant digits. A few atomic weight values are listed below. Note that atomic weights are not whole numbers and elements of higher atomic number typically have higher atomic weights.

**ACTIVITY 3-9**

Express the atomic weights of the following elements to four significant digits: carbon, nitrogen, sodium, aluminum, copper, zinc, and mercury.

| ATOMIC NUMBER | ELEMENT | ATOMIC WEIGHT | ROUNDED TO FOUR DIGITS |
|---|---|---|---|
| 1 | Hydrogen | 1.00794 | 1.008 |
| 8 | Oxygen | 15.9994 | 16.00 |
| 9 | Fluorine | 18.99840 | 19.00 |
| 16 | Sulfur | 32.066 | 32.06 |
| 17 | Chlorine | 35.4527 | 35.45 |
| 26 | Iron | 55.847 | 55.85 |
| 92 | Uranium | 238.0289 | 238.0 |

**FOCUS 3**

## *Some Facts About Atoms*

1. The atoms of a given element always have the same number of protons. A neutral atom contains equal numbers of protons and electrons. Example: Atoms of hydrogen have one proton and one electron.

2.  The atomic number of an element is the number of protons in the nuclei of the atoms of that element. If we know the atomic number of an element we know how many protons and electrons are in atoms of that element. Example: Iron has atomic number 26, so iron atoms must contain 26 protons and 26 electrons.

3.  The nucleus of an atom is a cluster of protons and neutrons. Isotopes are atoms of the same element that have different numbers of neutrons. The nucleon number of an isotope of an element is the number of protons plus the number of neutrons. To find the number of neutrons in an isotope subtract the atomic number from the nucleon number. Example: The isotope plutonium-239 has nucleon number 239. The atomic number of plutonium is 94, so there are 94 protons and 145 neutrons in an atom of the isotope ($239 - 94 = 145$).

4.  The atomic weight of an element is the mass of a typical or average atom of an element. The atomic weights of various elements are determined from experimental data. The table in the back cover lists values of the atomic weights of the elements and atomic weights are given in the periodic table shown in the front cover. Atomic weights can be rounded to a desired number of digits. Example: The listed atomic weight of silver, Ag, is 107.8682. Rounded to four digits this is 107.9.

## Seeing Atoms

Is it possible to see atoms? Atoms are submicroscopic in size and can't be seen, even by the most powerful conventional microscope. In one sense atoms are conceptual particles to be imagined in our minds. How do we know that they exist if we can't see them?

When a small piece of dust or pollen is suspended in a liquid and observed using a powerful microscope, it follows a random, erratic path as though it is being pushed about. This phenomenon is called Brownian motion after Robert Brown who first observed it in the early 1800s. In 1905, Albert Einstein explained this behavior by assuming the piece of dust to be continuously bombarded by atoms (or molecules) of the liquid. Consequently, the random motion of the small piece of dust provided a way to "see" atoms. Incidently, Einstein's explanation also serves as evidence for the kinetic molecular view of the incessant motion of particles of matter.

Another method used to see atoms is X-ray diffraction. X rays are a form of high energy light that move in a wave-like fashion. When materials are exposed to X rays, the X rays reflect or diffract as they encounter highly dense masses like nuclei. See Figure 3-11. As an analogy, note that water waves diffract as they pass through the pilings on a pier. If the pilings are equally spaced a diffraction pattern is observed. If the pilings were not present, the waves would not be reflected from them. Similarly, when X rays impinge upon material they diffract upon encountering nuclei. The resulting diffraction pattern reveals information about the spacing of the atoms containing the nuclei. If the atoms were not present, no diffraction would occur.

The most modern way to see atoms is by use of special microscopes called scanning tunneling microscopes. These devices have precise probes so fine that the tip of the probe is a single atom. The probe is moved over a surface. As it moves, a measure is made of the electrical cur-

**FIGURE 3-11**
The diffraction of
X-rays is evidence for
the location of atoms.

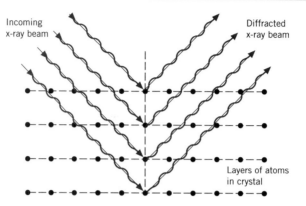

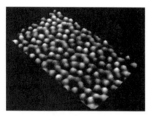

**FIGURE 3-12**
An image produced by a
scanning tunneling
microscope. (Courtesy
of IBM)

rent between the probe and atoms on the surface of a sample of material. Such microscopes allow us to see computerized images of atoms by visualizing the electrical force fields around them. The currents are analyzed by computer and atomic images are displayed on computer monitors. Specially designed scanning tunneling microscopes are used to pick up atoms and move them about. Figure 3-12 shows atomic images obtained from a scanning tunneling microscope.

How small are atoms? One nanometer is the name for one billionth of a meter ($10^{-9}$ m). Six carbon atoms could fit in a space about one nanometer wide. If six million carbon atoms could somehow be lined up, the line would be about one millimeter long. A sample of carbon, slightly larger than a sugar cube, with a mass of 12 grams, contains about 600 million-million-billion atoms ($6.0 \times 10^{23}$ atoms).

## 3-19 SOME COMMON ELEMENTS

Among the 109 different elements some are much more common and useful than others. Several of these elements are described below.

### Carbon, C

In the pure form, carbon occasionally occurs as diamond, but more commonly as graphite or in the amorphous forms of charcoal, coke (coal heated in the absence of air), and carbon black (a finely divided black soot-like powder). Carbon is also present in nature as the major component of compounds found in plants and animals. There are millions of chemical compounds that contain carbon in combination with hydrogen and oxygen, nitrogen, phosphorus, or sulfur. Such compounds are organic compounds and the study of the compounds of carbon is called **organic chemistry.** Carbon is an element essential to life, and there is an ample supply in the food that we eat since carbohydrates, proteins, and fats all contain carbon compounds.

Diamonds have many industrial uses that take advantage of the fact that this form of carbon is the hardest substance known. Graphite is used as a lubricant as well as the major component of pencil leads. The word graphite comes from the Greek *graphein* 'to write'. Charcoal and coke are used as fuels and carbon black is added to rubber to give it strength for use in tires. This is why tires are black.

### Chlorine, Cl

This element is a pale, greenish-yellow gas. It rarely occurs as the free element in nature, but it is obtained from the many compounds in which it is found. Combined chlorine is found in various minerals called chlorides, the most notable of which is common table salt or sodium chloride. Vast salt deposits are found in nature and dissolved salt is found in abundance in seawater. Elemental chlorine can kill bacteria, so it is used to treat drinking water and swimming pool water. Chlorine is used to make the plastic called PVC (polyvinyl chloride) and the plastic polyvinyldiene chloride, used in Saran wrap.

### Copper, Cu

Copper, a distinctive, reddish-colored metal, was known to ancient civilizations since it sometimes occurs free in nature and is rather easily extracted from minerals that contain it. The Stone Age man of Chapter 1 had an ax that was nearly pure copper. The ancients used copper for tools, weapons, and coins. They even discovered that when copper is mixed with some other metals, it becomes more durable and useful. Mixtures of metals that have properties that are different from the individual metals that make them up are called **alloys.** Bronze is an alloy of copper and tin (around 80% copper by mass), and brass is an alloy of copper and zinc (ranging from 60% to 85% copper by mass). Copper is a very good electrical conductor so it is often used in electrical wires.

### Gold, Au

The symbol for this enchanting metal comes from the Latin *aurum* 'shining dawn'. Relatively rare in nature, gold has been a precious and valuable element for millennia. It is used in jewelry and coins and hoarded as a symbol of wealth. This bright yellow metal can occur in nature as the free

Place a piece of white paper on a flat surface. Using a knife, scrape a pencil lead to make a small pile of graphite powder on the paper. Rub the graphite into the paper. Use a match to light another small piece of white paper and let it burn to form a small amount of carbon. Place the carbon on the first paper sheet and rub it into the paper. Record your observations.

A shiny penny is a good example of copper metal. Pennies made since 1982 are actually zinc metal with a thin layer of copper plate. If you scratch the surface of a penny with a knife or nail file or rub the edge on rough concrete you'll remove some copper and expose the zinc. See Activity 3-12.

**3-12**

Use a file, some sandpaper, or rough concrete to scrape the edge of a penny to remove the copper and expose zinc. Remove the copper from the entire edge of the penny. Place the penny in a glass containing some white vinegar. In a few minutes you'll see small bubbles of hydrogen gas. If possible leave the penny in the vinegar overnight, then inspect it. Put a warning label on the glass to identify it. Record your observations.

**3-13**

Typical paper clips are made of carbon steel. Observe a steel paper clip by straightening it. Rapidly bend the paper clip back and forth until it breaks. Record your observations.

metal and it is noted for its ability to keep its luster since it does not easily combine with oxygen in the air. It is also found in gold ore deposits. Gold is used in the manufacture of electronic devices and products, and dental crowns as well as its typical ornamental uses.

## Hydrogen, H

This element is a low-density, colorless, odorless gas. It is apparently the most abundant element in the universe. It is a component of stars and appears to be abundant in regions of interstellar space. Very little hydrogen is found in nature as the free element, but it is an important component of almost all of the chemicals found in plants and animals. It is also found, combined with carbon, in the fossil fuels, coal, petroleum, and natural gas. Petroleum and natural gas are the major industrial sources of elemental hydrogen. The name hydrogen literally means water former and, as you know, it is a component of water.

## Iron, Fe

This element is a silvery metal found compounded in a variety of minerals and, more important, in iron ores. Processing of the ores yields iron with its many uses ranging from tools to cars to massive structures. To be useful, iron is made into alloys that are strong, hard, and malleable. **Malleable** means able to be formed by bending or pounding into various shapes. A variety of iron alloys are made by mixing various metals and carbon with iron. The most common iron alloys are **carbon steel,** containing a small percent of carbon, and stainless steel. One of the most common types of **stainless steel** contains iron with around 18% chromium and 8% nickel by mass and some carbon.

## Nitrogen, N

This colorless, odorless gas is found as the free element in air. Nitrogen is the major component of air (78.08% by volume). Nitrogen compounds are found as minerals and are important constituents of plants and animals since they occur in amino acids that make up proteins. Nitrogen, along with phosphorus, P, and potassium, K, is an important plant nutrient. Some nitrogen-containing compounds, such as ammonia, ammonium sulfate, and urea, are used as plant fertilizers.

## Oxygen, O

Oxygen is the most abundant element on the surface of the earth. As a colorless, odorless gas, it is an important component of air (20.95% by volume). This gas is essential and fundamental to oxygen-breathing animal life. Many substances can react chemically with the oxygen of the air.

When the reaction is rapid and vigorous it is called burning or combustion. Oxygen is also found as a component of water and in compounds that compose rocks, minerals, plants, and animals. Recall that you made a sample of oxygen gas in Chemistry in Action 1-1.

### Silicon, Si

The element silicon is a shiny, bluish-gray, brittle solid. Silicon has unique electrical properties. Certain elements mixed with silicon, at the parts per million level, make it a semiconductor of electricity. Semiconducting silicon is used in transistors and other electronic components. Large numbers of these electronic components can be formed, at the microscopic level, on the surface of tiny chips of silicon. These chips contain integrated circuits having thousands of electronic components in an area measuring only a few millimeters on an edge. As you know, these chips are used in computers, calculators, and many other consumer products. Some experts claim that the development of the integrated circuit on a chip will affect society on a level rivaling that of the industrial revolution. Silicon chips are jokingly called educated grains of sand, since elemental silicon is obtained from common silica sand. Clays, sand, and other compounds containing silicon are widely used in making ceramics, ceramic glazes, and various types of glass.

Silicones are a group of compounds containing silicon, oxygen, carbon, and hydrogen, that range from oily liquids to rubbery plastics. Silicones are used for medical implants, as the centers of golf balls, and in various industrial products such as silicone caulking compounds. Some swimmers use soft pliable silicone ear plugs. The region of California in which many semiconductor and electronic companies are located is known as Silicon Valley, not Silicone Valley.

### Sodium, Na

Sodium is a soft, silvery metal that does not occur naturally as the free element. Its compounds are found in a variety of minerals, especially in vast deposits of salt formed by the evaporation of ancient seas. Salt is an essential compound in our diets. Normally we get sufficient salt from the foods that we eat. It is often added to foods to enhance flavors. The salt content of prepared foods is listed on the nutritional label in terms of milligrams of sodium. The recommended daily intake (called the daily value) of sodium is less than 2400 mg.

### Sulfur, S

Sulfur is a brittle, yellow solid sometimes called brimstone. It has been known for centuries because it sometimes occurs as the free element. Occasionally, it accumulates on the brim of water pools in volcanic re-

Elemental sulfur is sometimes used in agriculture to treat soils. Find a source or sample of sulfur and record a description of it. Sulfur is sold in garden supply stores but your instructor may supply a sample for you. Chemistry in Action 3-3 is about reacting copper with sulfur.

gions. In such regions, the "sulfur smell" is actually the odor of sulfur compounds, such as hydrogen sulfide (rotten egg odor) or sulfur dioxide (sharp choking odor sometimes produced upon striking a match). Sulfur is also found combined in minerals such as gypsum, used in wallboard, and iron pyrites, sometimes called fool's gold. Sulfur compounds are essential to plant and animal life since it is a component of some amino acids found in proteins. Sulfur has many industrial uses including the manufacture of matches. That's why you sometimes smell sulfur dioxide when you strike a match. The major use of sulfur is in the manufacture of sulfuric acid, one of the most important industrial chemicals.

In this exercise you are going to make curds and whey from milk. It is possible to treat milk chemically so that the majority of the suspended proteins coagulate into insoluble curds. The liquid that remains contains the rest of the suspended protein and other chemicals in water solution. This water solution with suspended proteins is called whey. You need: two small glasses, half a cup of warm nonfat milk (or any milk), white vinegar, paper towels or napkins, and a coffee filter. If you don't have a filter, shape a double layer of paper towel or napkin into a filter cone and secure its shape with tape.

1. Place half a cup of nonfat milk in a glass. Nonfat milk is used because the fat has already been removed and will not interfere with your observations. However, regular milk will work.

2. Add two tablespoons of white vinegar to the glass of milk, stir and let it sit for a few minutes. Record your observations.

3. Pour the coagulated milk into the filter and let the liquid flow into the second glass.

4. Reach into the filter and remove as much of the coagulated protein as you can. Place the curds on a paper towel or napkin and blot them.

5. Observe the curds by rubbing them with your fingers. Record your observations. Curds like these are similar to those found in cottage cheese and the curds used to make various kinds of cheeses. Roll the curds into a ball and set it aside to dry for a few days.

Carbohydrates are a group of compounds that contain carbon, hydrogen, and oxygen. Table sugar is a carbohydrate that can be decomposed by heat. You need: a 5 cm by 10 cm piece of aluminum foil, some granulated sugar, and matches (a small candle can be used in place of matches).

1. Obtain a sample of sugar about the size of one quarter of an aspirin tablet.

2. Hold the aluminum foil so that it is flat and gently press your finger on one end to make a slight indentation.

3. Place the sugar sample on the foil as you hold it flat.

4. Light a match and hold it under the foil just below the sugar. Allow the sugar to burn. Use more matches as needed to completely burn the sugar. Heat the sugar until it stops smoking and no more appears to decompose. Record your observations.

5. After the carbon-like material cools you can touch it and rub it with your fingers. Record its appearance and texture.

To illustrate compound formation, you are going to combine the elements copper and sulfur. You need: a shiny penny, a zip-lock bag (a sandwich bag will do), a cup or glass, some boiling hot water, and a small sample of powdered sulfur about the size of two aspirin tablets. Powdered sulfur is available in garden supply stores, but you may be able to get some from your teacher.

1. Put the sample of sulfur into the zip-lock bag so that it occupies one corner of the bag.

2. Add the penny to the bag and make sure that most of it is covered by the sulfur. Press the air from the bag and seal it.

3. Fill a cup or glass with boiling hot water and immerse the corner of the bag containing the penny and sulfur into the hot water. Leave the bag in the water for 5 to 10 minutes.

4. Remove the bag from the water and dry it. Open the bag and remove the penny. Record your observations.

Since sulfur is a plant nutrient, mix any extra sulfur into some soil. How do the appearances of the copper metal and the elemental sulfur differ from the copper–sulfur compound that formed on the surface of the penny?

To illustrate compound formation, you are going to combine the elements copper and chlorine. You need: two shiny pennies, a 2 inch by 2 inch piece of paper, a few drops of liquid bleach, an aspirin tablet, and an eyedropper. You can use a Q-tip in place of the eyedropper but be sure to rinse it with water and throw it away when you are done.

CAUTION: Although liquid bleach is a household chemical it is quite dangerous. Read the entire label on the bleach bottle. Do not spill the bleach or let it splash on your skin or clothing. If you spill or splash it on yourself quickly rinse the area with large amounts of water. Wear safety goggles.

1. Set a penny on a paper towel or napkin and place two or three small drops of liquid bleach on the surface of the penny. Set the other penny next to the first for comparison and put two or three drops of bleach on it.

2. Carefully place the aspirin tablet on top of the drops of bleach on one of the pennies and let it sit a minute or so until the liquid is absorbed by the tablet.

3. Slowly add a few drops of bleach on top of the tablet until you see a small amount of liquid forming on the penny beneath the tablet. Stop adding bleach and let the penny sit for five or more minutes.

4. Carefully remove as much of the aspirin tablet as you can from the penny and place it on the paper. After most of the solid white aspirin is removed record the appearance of the copper–chlorine compound that adheres to the surface of the penny.

The element chlorine is a greenish-yellow gas. In this exercise you could not see it's color since only a small amount was formed. Describe how the copper–chlorine compound differs from the copper metal and the chlorine gas.

## QUESTIONS

### Section 3-1

1. Make a time line from 500 B.C.E. to the present and show the dates given in Section 3-1 that represent the passage of the idea of an atom through time.

### Section 3-2

2. What are the different states of matter? Give an example of each type.

3. Draw sketches to illustrate the differences between solids, liquids, and gases according to the kinetic molecular view.

4. How are melting and freezing different and how are they similar? How are evaporation and condensation different and how are they similar?

### Sections 3-4 to 3-6

5. What does the term homogeneous mean? What does the term heterogeneous mean?

6. What is a solution?

7. What is a solvent? What is a solute?

8. What is a pure substance or chemical?

9. What is an element? What distinguishing characteristic applies to an element.

10. What is a compound? What distinguishing characteristic applies to a compound?

11. What is the difference between an element and a compound?

12. What is a percent?

13. *Express the following as percents by mass.
    (a) 56.0 g of baking soda contains 8.00 g of carbon
    (b) 237 g of water contains 26.3 g of hydrogen
    (c) 28.3 g of table sugar contains 11.9 g of carbon

14. Express the following as percents by mass.
    (a) 43.6 g of salt contains 26.41 g of Cl
    (b) 79.81 g of ammonia contains 65.66 g of N
    (c) 17.75 g of fructose contains 7.09 g of C

15. *Express the following percents as factors. For example, if water is 11.2% H, then there are (11.2 g H/100 g water).
    (a) acetic acid (vinegar) contains 40.0% C
    (b) table sugar contains 42.1% C
    (c) baking soda contains 33.3% C

16. Express the following percents as factors. For example, if water is 11.2% H, then there are (11.2 g H/100 g water).
    (a) glucose contains 53.3% O

(b) ammonia contains 82.3% N

(c) table sugar contains 6.48% O

17. How many (see Questions 15 and 16)

(a) grams of C are in 50.0 g of acetic acid?

(b) grams of C are in 50.0 g of table sugar?

(c) grams of O are in 25.0 g of glucose?

(d) grams of N are in 200 g of ammonia?

## Section 3-7

18. The most common elements found in the earth's crust are oxygen, silicon, aluminum, iron, calcium, sodium, potassium, and magnesium. Give the symbol for each of these elements.

19. The elements that make up your body include C, H, O, N, P, S, Ca, K, Cl, Na, Mg, and Fe. Name these elements.

20. Trace elements found in the body include iodine, copper, zinc, manganese, cobalt, chromium, selenium, molybdenum, fluorine, tin, silicon, and vanadium. Write the symbols for these elements.

21. As, Pb, Hg, Cd, Se, and Pu are some highly toxic elements. Give the names of these elements.

22. The following metals are used in jewelry and coins: Au, Ag, Cu, Zn, Al, Sn, Ni, and Pt. Give the names of these elements.

23. The elements nitrogen, phosphorus, potassium, and sulfur are important plant nutrients. Write the symbols for these elements.

24. Silicon, germanium, tellurium, gallium, indium, arsenic, antimony, and terbium are used in the semiconductor industry. Write the symbols for these elements.

25. List the elements that have symbols consisting of one letter only.

26. List the elements that have names ending in -*ine*.

27. List the elements that have symbols that are not derived from the English name of the element.

## Section 3-8

28. Give statements of the following:

(a) law of conservation of matter

(b) law of constant composition

29. *A widely traveled scientist collected water samples from many parts of the world. After returning to the lab, the scientist purified the samples, subjected them to elemental analysis, and recorded the data. What laws do these data support? Why?

| SOURCE | %H BY MASS | %O BY MASS |
|---|---|---|
| Water from the Antarctic ice cap | 11.2 | 88.8 |
| Water from the Stensvad water well in Montana | 11.2 | 88.8 |
| Water from the South China Sea | 11.2 | 88.8 |
| Water freshly synthesized from hydrogen and oxygen | 11.2 | 88.8 |

30. While the scientist of Question 29 was away, an assistant stayed in the laboratory and combined lithium metal and fluorine gas to make a compound. What laws do the assistant's data support? Why?

| MASS OF LITHIUM (g) | MASS OF FLUORINE (g) | MASS OF LITHIUM FLUORIDE COMPOUND (g) |
|---|---|---|
| 0.234 | 0.642 | 0.876 |
| 0.489 | 1.339 | 1.828 |
| 0.360 | 0.986 | 1.346 |

31. It is sometimes said that you can never get rid of wastes. What scientific law is the basis of this statement?

## Section 3-9

32. What is a scientific theory or model?

33. Describe or state the atomic theory. Use pictures to illustrate.

34. Why is the atomic theory fundamental to chemistry?

35. Describe an atom.

36. Who is noted for first suggesting the existence of atoms?

37. Look up John Dalton's atomic theory in an encyclopedia or a history of science text and describe the theory as Dalton stated it. Describe the historical background of the theory.

**Section 3-10**

38. Explain the law of conservation of matter in terms of the atomic theory.

39. Explain the law of constant composition in terms of the atomic theory.

40. What is a chemical reaction?

41. What makes one element different from other elements?

**Sections 3-11 and 3-12**

42. What is electricity?

43. What is electrolysis?

44. What are ions? What is the difference between a cation and an anion?

**Section 3-13**

45. Make a list of the three basic kinds of subatomic particles. How do they differ?

46. An electron was once described as "a negative twist of nothingness." Describe some properties of an electron.

**Section 3-14**

47. Describe the nuclear model of the atom.

48. What particles are present in atomic nuclei?

49. Imagine a typical atom enlarged to be the size of a beach ball having a radius of 8 inches. What radius in inches would the nucleus have in such an atom?

**Sections 3-15 and 3-16**

50. What is an atom in chemical terms?

51. Define or describe the following:
   (a) atomic number
   (b) nucleon number
   (c) isotope

52. *Describe the following isotopes in terms of the numbers of protons, neutrons, and electrons they contain (atomic numbers are found on the table inside the back cover of this book).
   (a) carbon-14      (d) O-18
   (b) chlorine-37    (e) phosphorus-32
   (c) U-234          (f) calcium-45

   (g) Sr-90          (i) I-125
   (h) iron-56        (j) Ba-137

53. Describe the following isotopes in terms of the numbers of protons, neutrons, and electrons they contain (atomic numbers are found on the table inside the back cover of this book).
   (a) hydrogen-3     (f) cobalt-60
   (b) potassium-40   (g) I-133
   (c) U-235          (h) helium-2
   (d) O-16           (i) Pu-239
   (e) cesium-137     (j) Ra-226

54. Hydrogen is the only element for which the isotopes have their own names. Hydrogen-1 is protium, hydrogen-2 is deuterium, and hydrogen-3 is tritium. Tritium is a radioactive isotope of hydrogen that does not occur naturally. Describe the difference between the three isotopes of hydrogen.

55. Three neutral atoms differ only in that one contains 23 electrons, one contains 24 electrons, and one contains 26 electrons. Are these atoms of the same or different elements? Why?

**Sections 3-17 and 3-18**

56. What is a mass spectrometer?

57. What is an atomic weight?

58. What information is needed to calculate the atomic weight of an element?

59. What is an atomic mass unit?

60. The ratio of the mass of a nitrogen atom to a carbon-12 atom is 1.167228 to 1. Calculate the atomic weight of nitrogen.

61. The ratio of the mass of a fluorine atom to a carbon-12 atom is 1.58320. Calculate the atomic weight of fluorine.

62. Use the following data to calculate the atomic weight of Li (see Activity 3-8).

| ISOTOPE | ISOTOPE MASS | % ABUNDANCE |
|---|---|---|
| Lithium-3 | 6.01513 amu | 7.42 |
| Lithium-7 | 7.01601 amu | 92.58 |

63. *Use the following data to calculate the atomic weight of silicon (see Activity 3-8).

| ISOTOPE | ISOTOPE MASS | % ABUNDANCE |
|---------|--------------|-------------|
| Silicon-28 | 27.98 | 92.21 |
| Silicon-29 | 28.98 | 4.70 |
| Silicon-30 | 29.97 | 3.09 |

## Section 3-19

64. Give brief descriptions of each of the following elements.

- (a) carbon
- (b) chlorine
- (c) copper
- (d) gold
- (e) hydrogen
- (f) iron
- (g) nitrogen
- (h) oxygen
- (i) silicon
- (j) sodium
- (k) sulfur

65. What are alloys? Give an example.

66. What is carbon steel? What is stainless steel?

67. What is burning or combustion?

68. Describe a major use of the element silicon.

## Questions to Ponder

69. Are atoms real? Explain your answer.

70. How would atoms and matter be different if electrons had positive electrical charge and protons had negative electrical charge?

71. Use any resources you can find to learn about theories that explain how elements in the universe were formed. Give brief statements of a theory.

72. Using your knowledge of the atomic nature of matter and the kinetic molecular theory, explain why materials differ in density.

# FORMULAS AND

# EQUATIONS

## 4-1 CHEMICAL FORMULAS

You know table salt as a white crystalline solid. As elements sodium is a silvery metal and chlorine is a greenish-yellow gas. Both elements are quite dangerous, yet they combine chemically to form the salt that we need in our diets to sustain life. The compound salt contains combined sodium and chlorine atoms and is called sodium chloride. When a sodium atom changes from an atom of sodium metal to a combined atom, it changes its properties dramatically. Chlorine atoms, similarly, change their properties as they go from the elemental form to the combined form. Atoms are the building blocks of nature and all compounds are composed of chemically combined atoms. The properties of a compound are quite different from the properties of the elements that make it up. In the elemental form, hydrogen and oxygen are both gases. We know that water, the compound formed by the chemical combination of hydrogen and oxygen, is typically a liquid. It has a specific density, freezing point, boiling point, and a characteristic appearance.

The fact that a given compound always contains the same elements present in the same percents by mass means that compounds contain atoms combined in definite proportions. Water contains two combined atoms of hydrogen for every one combined atom of oxygen and has the formula $H_2O$. This formula is a concise statement of the fact that the

$H_2O$
water

**105**

NaCl
sodium chloride
table salt

$C_{12}H_{22}O_{11}$
sucrose
table sugar

$NaHCO_3$
sodium
bicarbonate

$NH_3$
ammonia

$CO_2$
carbon dioxide

atomic proportion in the compound is two combined atoms of hydrogen for every one combined atom of oxygen. Any compound can be represented by a **formula** that shows which elements are present using their symbols and the relative number of combined atoms of each element using numerical subscripts. If the subscript is one, as in the case of oxygen in water, omit the number.

The term formula may conjure up images of a secret formula in the possession of a mad scientist. Mad scientists are quite rare, and a formula is not some obscure recipe but simply serves as a concise symbolic description of a compound. Consider the formulas of some common compounds. (See Fig. 4-1.) Table salt or sodium chloride has the formula NaCl. You can read a formula by just pronouncing the letters and any subscript numbers. NaCl reads N–A–C–L. Common table sugar is the chemical compound sucrose, $C_{12}H_{22}O_{11}$ (read C–12–H–22–O–11). Baking soda is the compound sodium hydrogen carbonate, $NaHCO_3$ read N–A–H–C–O–3). It is sometimes called bicarbonate of soda or sodium bicarbonate. Clear household ammonia is a water solution of the gaseous compound ammonia, $NH_3$ (read N–H–3). Soda water is a water solution of carbon dioxide gas, $CO_2$ (read C–O–2).

How do we go about finding the formula of a compound? How is it known that the formula of water is $H_2O$? In this chapter you'll learn about experimental methods that chemists have developed which allow the determinations of the formulas of compounds. These methods are based upon the quantitative analysis of a compound in terms of the percent by mass of its elements. For example, we can use the experimentally determined fact that water is 11.2% hydrogen and 88.8% oxygen by mass to find the formula.

## 4-2 THE MOLE

A one-carat diamond weighs 0.200 grams. A diamond is composed of a three-dimensional array of chemically bonded carbon atoms. How many carbon atoms are in a one-carat diamond? An answer to this question may seem impossible. You cannot count atoms one at a time. However, chemists have developed a way to find the number of atoms in a measured mass of an element. Balances are used to measure precise masses of samples. Atoms are very tiny particles. Even a sample of an element weighing less than a gram contains a vast number of atoms. How many? Each atom has mass and, as we know, the atomic weight serves as a measure of this mass. If we have a sample of an element, its mass must be the sum of the masses of the atoms that make it up. This is the key to counting the number of atoms in a sample of known mass.

To see how this works let's use an analogy. Suppose you are working in a nuts and bolts factory that makes one kind of nut and one kind of bolt.

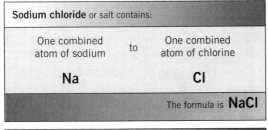

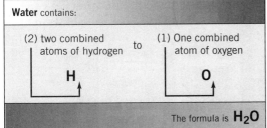

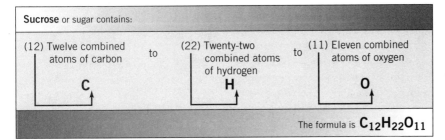

**FIGURE 4-1**
Formulas of some compounds.

These are specific nuts and bolts and by experiment you find that the weight of a nut is 12.0 grams and the weight of a bolt is 16.0 grams. These "hardware weights" can be expressed as the factors

$$\frac{12.0 \text{ g}}{1 \text{ nut}} \qquad \frac{16.0 \text{ g}}{1 \text{ bolt}}$$

These factors are used to count nuts and bolts by weighing samples. It is important to realize that samples of nuts and bolts, having the same numerical values as the relative weights given in the above factors, have equal numbers of nuts and bolts. For instance, if we weigh 12.0 pounds of nuts and 16.0 pounds of bolts the same number of nuts and bolts are in each sample.

$$12.0 \text{ lb} \times \frac{454 \text{ g}}{1 \text{ lb}} \times \frac{1 \text{ nut}}{12.0 \text{ g}} = 454 \text{ nuts}$$

$$16.0 \text{ lb} \times \frac{454 \text{ g}}{1 \text{ lb}} \times \frac{1 \text{ bolt}}{16.0 \text{ g}} = 454 \text{ bolts}$$

If we wanted a washer for each set of nuts and bolts we just need to know the weight per washer. For example, if there is 1.00 gram per washer (1.00 g/washer) we know that we need 1.00 pound of washers to give 454 washers. Perform the calculation that proves this conclusion.

Even though atoms are very tiny, we can count them using atomic weights in a way that is similar to the nut–bolt analogy. Of course, we typically express masses of chemicals in units of grams. Imagine that we had a sample of 12.0 grams of carbon and a sample of 16.0 grams of oxygen. Remember that the atomic weights of elements differ but they do reflect the relative masses of atoms. If we have 12.0 g of carbon and 16.0 g of oxygen the number of atoms of each element must be the same since the sample masses are numerically the same as their atomic weights. The atomic weights of these elements are

$$\frac{12.0 \text{ u}}{1 \text{ C atom}} \qquad \frac{16.0 \text{ u}}{1 \text{ O atom}}$$

Recall that u is the atomic mass unit. The number of atoms in each sample can be calculated using the atomic weights as factors. This is done in a similar way to the nut–bolt calculations shown above.

$$12.0 \text{ g} \times \frac{X \text{ u}}{1 \text{ g}} \times \frac{1 \text{ C atom}}{12.0 \text{ u}} = X \text{ C atoms}$$

$$16.0 \text{ g} \times \frac{X \text{ u}}{1 \text{ g}} \times \frac{1 \text{ O atom}}{16.0 \text{ u}} = X \text{ O atoms}$$

The factor $(X \text{ u}/1 \text{ g})$ is used here because we do not know the actual numerical value of this factor. So we use $X$ to represent the number of atomic mass units per gram. The point is that the same value of $X$ is the result in both calculations. This means that there are the same number of atoms in 16.0 g of oxygen as there are atoms in 12.0 g of carbon. Each element has a unique atomic weight. Thus, as a general principle we can say that a sample of any element having a mass in grams the same as its atomic weight would contain the same number of atoms as the above samples—$X$ atoms.

We commonly use special names for counting units. For instance, dozen is the name for 12 and gross is the name for 144. In chemistry the atomic counting unit is the mole. The term mole comes from the Latin word for heap or pile. The mole is a specific number of atoms. Counting atoms by weighing samples of elements is based upon the definition of the mole as given in the margin. The isotope carbon-12 is in the definition because it is also the reference used in the atomic weight scale. Twelve grams of carbon-12 is a mole of carbon atoms. The atomic weight of oxygen, to three digits, is 16.0. So we know that 16.0 g of oxygen contains as many atoms as 12 g of carbon-12. This number of atoms is a mole of oxygen. From the definition of the mole, the mass in grams of an element equal to

**Mole of an Element:** *The amount of an element that contains as many atoms as there are carbon atoms in exactly 12 grams of the isotope carbon-12.*

the numerical value of its atomic weight contains a mole of atoms. Furthermore, a mole of any element contains the same specific number of atoms called *X* in the above discussion. Soon we'll learn the actual numerical value of *X*.

If you look up the word mole in the dictionary you'll find that it has four meanings. It is interesting that these words are all related. A mole can be a stone structure used as breakwater, a pile of stones. A mole can be a slightly elevated colored spot on the skin, a pile of skin. A mole can be a small, burrowing animal that makes piles of dirt. And finally, of course, in chemistry a mole is a special kind of pile of atoms.

## 4-3 AVOGADRO'S NUMBER

How many atoms are in a mole of an element? In other words, what is the value of *X* referred to in the above discussion. The definition of the mole does not reveal this number. However, it can be determined using experimental data. A precise value comes from X-ray diffraction studies of samples of ultrapure silicon. Ultrapure silicon is material that has very low levels of impurities so that samples are assumed to be only silicon atoms. Using X-ray diffraction, it is possible to count atoms in a sample of carefully measured mass. This value is used, in a proportion, to calculate the number of atoms that are in a sample of silicon having a mass equal to its atomic weight. This experimentally determined number, as we might expect, turns out to be very large. The number of atoms in a mole of an element is found to be $6.0221367 \times 10^{23}$. This number is called **Avogadro's number** in honor of the Italian scientist, Amedeo Avogadro, who first suggested that it might be possible to count atoms in 1811. In normal form Avogadro's number is: 602,213,670,000,000,000,000,000. This is indeed a very large number that reflects the fact that atoms are very small. For convenience in calculations Avogadro's number can be rounded to four digits and expressed as

$$\frac{6.022 \times 10^{23} \text{ atoms}}{1 \text{ mol}}$$

$$\left( \frac{6.022 \times 10^{23} \text{ atoms}}{1 \text{ mol atoms}} \right)$$

To get an idea of how very large Avogadro's number is, imagine measuring the oceans of the world with a teaspoon. After measuring $6.022 \times 10^{23}$ teaspoons you would be half done. In other words, there are about two times Avogardro's number of teaspoons of water in the world's oceans. Don't try this experiment at home because, even if you measured one teaspoon each second, it would take you about $10^{16}$ years. Because atoms are very small particles, there are tremendous numbers of them in typical samples of matter. There are more atoms in a grain of sand than the estimated number of stars in the universe.

Referring to the discussion in Section 4-2, Avogadro's number corresponds to $X$, the number of atomic mass units per gram.

$$12\,g \times \frac{X\,u}{1\,g} \times \frac{1\ carbon\text{-}12\ atom}{12\,u} = X\ carbon\text{-}12\ atoms$$
$$= 6.022 \times 10^{23}\ carbon\text{-}12\ atoms$$

Twelve grams of carbon-12 contains Avogadro's number of atoms. A mole of any element is the mass that contains Avogadro's number of atoms. The mass, of course, is the atomic weight of the element in grams.

## 4-4  MOLAR MASS

The amount of an element in grams that contains as many atoms as there are in 12.0 g of carbon-12 is numerically equal to the atomic weight of the element. For any element this mass in grams is the mass of a mole of atoms. Thus, for any element we can express the **number of grams per mole** or the **molar mass** using the value of its atomic weight. Molar amounts of elements are in the range of a few grams to hundreds of grams. The molar mass of an element is useful as a unit factor. Four, or sometimes three, significant digits are normally sufficient for molar masses used in calculations. Some examples of molar masses as factors are given below. Note that the abbreviation for mole is mol and that these factors are expressed to four significant digits.

$$\frac{16.00\ g}{1\ mol\ O} \qquad \frac{1.008\ g}{1\ mol\ H} \qquad \frac{55.85\ g}{1\ mol\ Fe} \qquad \frac{238.0\ g}{1\ mol\ U}$$

To write the molar mass as a factor, just look up the appropriate atomic weight and round it off to the desired number of digits. A molar mass, as a factor, has a numerator that is the atomic weight expressed in grams and a denominator of one mole. Molar masses are very similar to densities as factors. They both relate two properties and are property factors. The density of a substance expresses the number of grams per one milliliter and relates mass and volume. The molar mass expresses the number of grams per one mole and relates mass to moles. The term molar mass is used because it refers specifically to the mass of a mole of an element.

## 4-5  THE MOLAR MASS AS A UNIT FACTOR

The amazing thing about the mole concept is that it allows easy counting of atoms. The molar mass of an element is used as a unit factor to accomplish this feat. For instance, if we have 24.0 g of carbon, we know that there are 2 moles of carbon atoms; or if we have 6.00 g, we know that there is 1/2 mole of carbon atoms. The number of atoms of an element is usually expressed in terms of moles instead of the absolute number of atoms. This is similar to the use of dozen for the number 12 or gross for the number 144. We commonly use the term dozen when referring to tortillas or eggs rather than saying 12. In a sense, the mole is the chemist's "dozen" since it is a name for the specific numerical value of $6.022 \times 10^{23}$. The molar mass as a property factor is used in the same way density is used as a property factor. A molar mass is used to find the number of moles of an element in any sample of known mass or the mass of any given number of moles. To find the number of milliliters of a given mass of a substance multiply by the inverse of the density:

$$X \, \text{mL} = Y \cancel{g} \times \frac{1 \, \text{mL}}{\# \cancel{g}}$$

The find the number of moles in a given mass of a substance multiply by the inverse of the molar mass:

$$X \, \text{mol} = Y \cancel{g} \times \frac{1 \, \text{mol}}{\# \cancel{g}}$$

To find the mass of a given number of milliliters of a substance multiply by the density:

$$Y \, \text{g} = X \cancel{\text{mL}} \times \frac{\# \, \text{g}}{1 \cancel{\text{mL}}}$$

To find the mass of a given number of moles of a substance multiply by the molar mass:

$$Y \, \text{g} = X \cancel{\text{mol}} \times \frac{\# \, \text{g}}{1 \cancel{\text{mol}}}$$

Suppose we had a chunk of iron and wanted to know how many iron atoms were in the sample. First we weigh the sample to find its mass in grams. Then, we look up the atomic weight of iron so that we have a value for the molar mass. The molar mass is used as a factor to find the number of moles of iron.

How many moles of iron atoms are in an 85.3-g sample of iron? According to its atomic weight the molar mass of iron is

$$\frac{55.85 \, \text{g}}{1 \, \text{mol Fe}}$$

Usually we use at least one more digit in the molar mass than is needed in the answer; this is to avoid potential calculation error due to premature roundoff. The molar mass of iron is inverted or "turned over" to give a factor that can be used to find the number of moles from the grams.

$$X \, \text{mol Fe} = 85.3 \, \text{g} \times \frac{1 \, \text{mol Fe}}{55.85 \, \text{g}} = 1.53 \, \text{mol Fe}$$

Note how the gram units cancel and the moles remain. Keep in mind that if we have a mass in grams and we want the number of moles, we have to multiply by the factor having moles in the numerator and grams in the denominator. In terms of numbers, this amount of iron is 1.53 times Avogadro's number of iron atoms. We typically use moles instead of the actual number of atoms.

---

**EXAMPLE 4-1**

How many moles of copper atoms are in an old copper penny (dated before 1982), which contains 3.2 g of copper?

We look up the atomic weight of copper and express the molar mass to three digits as (63.5 g/mol Cu). Using this as a unit factor gives

$$X \, \text{mol Cu} = 3.2 \, \text{g} \times \frac{1 \, \text{mol Cu}}{63.5 \, \text{g}} = 0.050 \, \text{mol Cu} \quad \text{or} \quad 5.0 \times 10^{-2} \, \text{mol Cu}$$

Note that it is good practice to include the units in the factor to ensure that the gram units cancel and we get moles. Remember, always include a unit in an answer.

---

The molar mass is also used to find the mass of an element corresponding to a specific number of moles. That is, if we know the number of moles in a sample of an element and want to know the corresponding mass, we multiply by the molar mass to convert moles to grams. Remember to use at least one more digit in the molar mass than needed in the final answer.

ACTIVITY
4-1

A one-carat diamond has a mass of 0.200 g. How many moles of carbon are in such a diamond? What is the mass of $1.00 \times 10^3$ moles of carbon?

**EXAMPLE 4-2**

What is the mass of a 0.525-mole sample of sulfur?

First look up the molar mass of sulfur. Multiplying the number of moles by the molar mass gives the number of grams.

$$X \, \text{g} = 0.525 \, \text{mol S} \times \frac{32.06 \, \text{g}}{1 \, \text{mol S}} = 16.8 \, \text{g}$$

## 4-6  NUMBER OF ATOMS IN A SAMPLE

The molar mass of an element serves as a factor that is used to calculate the number of moles of atoms in any sample of an element. We know that the number of atoms in a mole is Avogadro's number as represented by the factor

$$\frac{6.022 \times 10^{23} \text{ atoms}}{1 \text{ mol}}$$

This factor is used to find the actual number of atoms in a sample of an element. It can also be used to find the number of moles corresponding to a given number of atoms. As a factor, it is directly used to find the actual number of atoms in a sample if the number of moles is known.

---

**EXAMPLE 4-3**

Graphite, one form of carbon, is a powdery black solid. A speck of graphite powder just visible to the eye has a mass of about $1 \times 10^{-6}$ grams. How many carbon atoms are there in such a speck of graphite?

First we find the number of moles of carbon in the sample using the molar mass of carbon. The mass is multiplied by the inverted molar mass so that the grams cancel to give moles.

$$1 \times 10^{-6} \text{ g} \times \frac{1 \text{ mol C}}{12.01 \text{ g}}$$

Next, we multiply this product by Avogadro's number to give the actual number of carbon atoms.

$$X \text{ atoms C} = 1 \times 10^{-6} \text{ g} \times \frac{1 \text{ mol C}}{12.01 \text{ g}} \times \frac{6.022 \times 10^{23} \text{ atoms C}}{1 \text{ mol C}}$$

$$= 5 \times 10^{16} \text{ atoms C}$$

It is interesting that even a minute speck of an element contains millions upon billions of atoms. $5 \times 10^{16}$ is 50 million-billion.

ACTIVITY 4-2

A troy ounce of gold has a mass of 31.1 g. How many gold atoms are there in such a sample?

---

The absolute number of atoms in a sample can be found using the method given in the example. However, as we shall see, it is usually quite sufficient and convenient to express the number of atoms in terms of the number of moles rather than the number of atoms. We seldom deal with actual numbers of atoms but use the mole as a counting unit for atoms.

### 4-7 EMPIRICAL FORMULAS

Formulas are important in chemistry because they provide a concise way to describe the chemical composition of a compound. Formulas represent the symbolic language of chemistry. We use a formula to refer to a compound. How do we know the formula of a compound? The formula is not apparent from the appearance and properties of the compound. We can't look at a compound under a microscope and see little formulas. To find the formula, a compound is first analyzed chemically to discover which elements make it up. Following this analysis, the mass or percent of each element is determined. This is done by decomposing a carefully weighed sample of a compound into its constituent elements then weighing them, or by synthesizing the compound from weighed amounts of each element. As we know, the mass of an element can be used to find the number of moles of that element. The mole is the key to formula determination.

To illustrate how to find a formula, imagine that we do not know the "formula" of a bicycle. A shipment of 1.00 ton of parts contains 0.70 tons of frames and 0.30 tons of wheels. Thus, the mass composition is 30% wheels and 70% frames. We find that one wheel weighs 3.0 pounds and one frame weighs 14 pounds. The formula of a bicycle is found by changing the percents to pounds then determining the number of wheels and frames in these masses.

$$30\,\cancel{lb} \times \frac{1 \text{ wheel}}{3.0\,\cancel{lb}} = 10 \text{ wheels}$$

$$70\,\cancel{lb} \times \frac{1 \text{ frame}}{14\,\cancel{lb}} = 5 \text{ frames}$$

The formula comes from the ratio of wheels to frames.

$$\frac{10 \text{ wheels}}{5 \text{ frames}} = \frac{2 \text{ wheels}}{1 \text{ frame}} \quad \text{or} \quad \text{wheel}_2 \text{ frame}$$

The formula of a chemical compound gives the relative number of combined atoms of each element. Since moles of elements correspond to exact numbers of atoms, we can say that the same relative ratios exist for moles as exist for atoms. From an atomic view, water, for instance, has 2 parts hydrogen to 1 part oxygen. The subscripts in a formula reveal the ratio of combined atoms in the compound or the ratio of combined moles of atoms in the compound. Thus, a formula is interpreted as an expression of the molar relation between the constituent elements of the compound. The formula of water, $H_2O$, shows that there are 2 hydrogen atoms for one oxygen atom. Another interpretation is that water has 2 moles of combined hydrogen for every 1 mole of combined oxygen. In other words, if we know the formula then we know the **molar ratio** is

$$\frac{2 \text{ mol H}}{1 \text{ mol O}}$$

If the percent-by-mass composition or the mass of each element in a sample of a compound is known, the number of moles of each element in a given mass of the compound can be calculated. Dividing the number of moles of one element by the number of moles of another element gives the molar ratio. Molar ratios show the subscripts to be used in the formula of the compound. A formula found in this way is called an **empirical formula.** The term empirical means derived only from observations and experimental data.

## 4-8 EMPIRICAL FORMULA CALCULATIONS

The elemental composition of a compound is needed to find the empirical formula of the compound. The elemental composition is measured by experiment. Composition is expressed in terms of the percent of each element or the mass of each element in a sample of the compound. An example calculation illustrates the reasoning involved in the determination of a formula. Consider a compound of lead and sulfur. By experiment, it is found that 117 g of lead combine with 18.1 g of sulfur to produce 135 g of the compound. To perform the calculation, first find the number of moles of each element using the appropriate molar masses as unit factors. You need to consult a table of atomic weights for the molar masses. To change from grams to moles, use the inverted molar mass as a factor. In this way the grams cancel and give units of moles.

$$\text{lead } 117 \text{ g} \times \frac{1 \text{ mol Pb}}{207.2 \text{ g}} = 0.5647 \text{ mol Pb atoms}$$

$$\text{sulfur } 18.1 \text{ g} \times \frac{1 \text{ mol S}}{32.06 \text{ g}} = 0.5646 \text{ mol S atoms}$$

Dividing the number of moles of lead by the number of moles of sulfur gives the molar ratio.

$$\frac{0.5647 \text{ mol Pb}}{0.5646 \text{ mol S}} = \frac{1.00 \text{ mol Pb}}{1 \text{ mol S}}$$

The molar ratio is simply 1:1. Thus, the compound contains one combined lead atom for one combined sulfur atom. What is true for moles of atoms is true for individual atoms. Consequently, the empirical formula of the compound is PbS.

**EXAMPLE 4-4**

What is the empirical formula of a compound that contains 82.2% by mass nitrogen and 17.8% by mass hydrogen?

Since we want to deal with masses, we can assume that if we had 100 g of the compound, it would contain 82.2 g of combined nitrogen and 17.8 g of combined hydrogen. Any mass could be used, but 100 g is the most convenient. Find the number of moles of each element from its mass using the molar mass from a table of atomic weights. The number of moles of the elements are then used to find the molar ratio. The calculation sequence is summarized as:

|  | NITROGEN | HYDROGEN |
|---|---|---|
| Percent | 82.2% | 17.8% |
| Mass | 82.2 g | 17.8 g |
| Molar mass | 14.01 g/mol N | 1.008 g/mol H |
|  | Convert each mass to number of moles. | |
| Moles | 5.867 mol N | 17.66 mol H |
|  | Divide by smallest moles to give the molar ratio. | |
| Molar ratio | 1 | 3.01 |
| Formula |  | $NH_3$ (Use molar ratio for the formula.) |

The molar ratio reveals that the formula for the compound is $NH_3$. The molar ratio means that there are 3 moles of combined hydrogen per 1 mole of combined nitrogen or three combined hydrogen atoms for each nitrogen atom. When finding molar ratios, always divide by the smallest number of moles so that ratios are never less than 1. Notice that the molar ratio calculated above was 3.01 and we assumed it to be 3. It is reasonable to ignore such extra digits since it probably came from experimental errors in the data or roundoff errors in the calculations. Normally, the subscripts in formulas are small whole numbers, so we neglect these extra digits when deducing the formula from the molar ratios.

*"Percent to mass*
*Mass to mole*
*Divide by small*
*Multiply 'til whole."*

Anonymous

Commonly the molar ratios in empirical formula calculations are simple whole numbers. Occasionally some molar ratios are fractions rather than whole numbers and must be converted to the appropriate whole number. Let's consider an example of this situation. What is the formula for a compound that has the following percent-by-mass composition: 26.5% combined potassium, K; 35.4% combined chromium, Cr; and 38.1% combined oxygen, O?

We can express the number of grams of each element present in 100 g of the compound by simply using the percents. Next, the number of moles of each element is determined using their molar masses as factors. Then, the molar ratios are found from the numbers of moles.

|  | POTASSIUM, K | CHROMIUM, Cr | OXYGEN, O |
|---|---|---|---|
| Percent | 26.5% | 35.4% | 38.1% |
| Mass | 26.5 g | 35.4 g | 38.1 g |
| Molar mass | 39.10 g/mol K | 52.00 g/mol Cr | 16.00 g/mol O |
| Moles | 0.678 mol K | 0.681 mol Cr | 2.38 mol O |
| Divide by 0.678 for molar ratios | 1 | 1 | 3.51 |

We write the formula from the molar ratios, so the formula is $KCrO_{3.5}$. Note that 3.5 oxygen does not make sense from an atomic point of view. It is not possible to have 3.5 oxygen atoms. The value 3.5 comes from the fact that the calculations give the simplest ratio of the elements. By convention, only whole numbers are used in formulas, so any fractional ratios are changed to whole numbers. In this case, to obtain whole-number subscripts, multiply each of the subscripts by 2. Thus, the best empirical formula is $K_2Cr_2O_7$.

The reason fractions sometimes occur is that the method of calculation gives the simplest or most reduced molar ratios. These reduced ratios might be decimal fractions, such as 1.33, 1.25, 1.5, or 1.67. These decimal fractions can be expressed as fractions, such as 4/3, 5/4, 3/2, or 5/3. To obtain whole number subscripts all subscripts are multiplied by the appropriate numerical value (e.g., 3/2 is multiplied by 2).

**EXAMPLE 4-5**

A sample of a compound of iron and oxygen is found to contain 145 g of iron, Fe, and 55.3 g of oxygen, O. What is the empirical formula of the compound?

In this case, use the experimental data as mass composition and convert directly to number of moles using the molar masses as factors.

|  | IRON, Fe | OXYGEN, O |
|---|---|---|
| Mass | 145 g | 55.3 g |
| Molar mass | 55.85 g/mol Fe | 16.00 g/mol O |
| Moles | 2.60 mol Fe | 3.46 mol O |
| Divide by 2.60 to give molar ratios | 1 | 1.33 |

Thus, the inital formula is $FeO_{1.33}$. To make whole-number subscripts, it is best to deal in fractions. The subscript 1.33 is $1\frac{1}{3}$ or 4/3. Thus, the formula can be written as $FeO_{4/3}$. Multiplying each subscript by 3 gives the whole-number subscripts and a formula of $Fe_3O_4$.

(a) What is the empirical formula of a compound that contains 69.94% Fe and 30.06% O?

(b) What is the empirical formula of cholesterol? It contains 83.87% C, 11.99% H, and 4.14% O.

The empirical formula determined by the methods given here is always the simplest formula and may not be the most useful or representative formula of a compound. The empirical formula always reflects the simplest whole number ratio of elements. As we shall see, the formula of a compound depends on the way in which the atoms combine to form the compound. Consequently, the real formula may be some multiple of the simplest formula. However, the empirical formula is very useful and in a great many cases is the appropriate formula for a compound. See Section 4-10 for further discussion of compound formulas.

### 4-9 FORMULA UNITS AND MOLES OF COMPOUNDS

What does the formula of a compound represent other than the elemental composition? We can interpret the formula in terms of the number of atoms or the number of moles of each element. That is, the formula for sodium chloride, NaCl, shows that this compound contains one atom of combined sodium to one atom of combined chlorine. The formula for sucrose, $C_{12}H_{22}O_{11}$, shows 12 atoms of combined carbon to 22 atoms of combined hydrogen to 11 atoms of combined oxygen. Combined atoms make up compounds, but what about particles or units of compounds? How atoms combine in compounds is the subject of Chapter 7. For now we can view a compound as a collection of units corresponding to the formula.

A **formula unit** is the amount of a compound that contains the number of atoms of each element given in the formula. As an analogy, imagine you had a pile of nut–bolt combinations having two nuts on each bolt. A "formula unit" would be $nut_2bolt$. A formula unit of a compound is the smallest representative unit or portion of the compound. For compounds that are made of molecules the formula unit is the formula of the molecule. Just as it is not possible to deal with individual atoms in the laboratory, it is not possible to deal with or isolate a single formula unit. Samples of a compound measured in grams contain a vast number of formula units. The mole relates the mass of a sample of a compound to the number of formula units. It also relates the mass of a sample of a compound to the formula. See alternate definitions given in the margin.

A compound contains a specific number of combined moles of each element. Thus, a mole of a compound is viewed as the amount that contains the number of moles of each element given by the subscripts in the formula. Each compound has a unique number of grams per mole. A mole of sodium chloride, NaCl, is the amount that contains 1 mole of com-

**Mole of a Compound:**
*The amount of a compound that contains Avogadro's number of formula units.*

**Mole of a Compound:**
*The mass in grams of a compound that contains the number of moles of each combined element as given by the formula.*

bined sodium and 1 mole of combined chlorine. A mole of sucrose, $C_{12}H_{22}O_{11}$, is the amount that contains 12 moles of combined carbon, 22 moles of combined hydrogen, and 11 moles of combined oxygen.

Let us return to the nut–bolt analogy mentioned in Section 4-2. If we were using a compound made of two nuts and one bolt we could express the mass of the compound as a summation of the masses of the components that make it up. Since one nut is 12.0 grams and one bolt is 16.0 grams, then one nut₂bolt unit is 40.0 grams. Furthermore, we know that if we had a sample of nut₂bolt combinations weighing 40.0 pounds the sample would contain 454 nut₂bolt units. See Section 4-2. In a similar fashion we can refer to the molar mass of a compound. The **molar mass of a compound** expresses the number of grams per mole of the compound. Sometimes, chemists call molar masses of compounds molecular weights or formula weights. The term molar mass is used in this book. The molar mass of a compound is found from the formula of the compound. Each element in the compound contributes to the molar mass of the compound according to its subscript in the formula. To calculate the molar mass of a compound:

1. Multiply the molar mass of each element by its subscript in the formula.
2. Add the contributions of each element and express the result as the number of grams per mole of the compound.

---

**EXAMPLE 4-6**

Determine the number of grams per mole, or molar mass, of water, $H_2O$ (to four digits).

To find the molar mass of the compound, multiply the molar mass of each element by its subscript and add the products. Note, since we want a four-digit answer we need to use four or five digits (or more) in the element molar masses and then round off after the calculation.

$$\text{hydrogen} \quad 2(1.008) = 2.016$$

$$\text{oxygen} \quad 1(16.00) = \underline{16.00}$$

$$18.016 \quad \text{or} \quad 18.02$$

The molar mass is expressed as the number of grams per mole of the compound. When writing the molar mass as a factor, always show the formula of the compound following the word mole.

$$\text{Molar mass of water} = \frac{18.02 \text{ g}}{1 \text{ mol } H_2O}$$

Note that it is not necessary to carry the units through the calculation. We can include the units after we add up the numbers and round off. The units are grams in the numerator and 1 mole in the denominator. Remember to include the formula of the compound in the denominator.

---

**EXAMPLE 4-7**

Determine the molar mass of sucrose, $C_{12}H_{22}O_{11}$ (to four digits).

C    12(12.01) = 144.12

H    22(1.008) = 22.176

O    11(16.00) = 176.0

342.296    or    $\dfrac{342.3 \text{ g}}{1 \text{ mol } C_{12}H_{22}O_{11}}$

---

**ACTIVITY 4-4**

Determine the molar mass of sodium hydrogen carbonate (baking soda), $NaHCO_3$ (to four digits).

## 4-10  FORMULAS OF COMPOUNDS

The empirical formula of a compound expresses the simplest whole-number ratio of the atoms of the elements in the compound. As mentioned in Section 4-4, the empirical formula of a compound may not be the best representative formula. The real or representative formula of a compound depends on the way in which the atoms are combined in the compound. It may be the empirical formula or some simple whole-number multiple of the empirical formula. For instance, chemical analysis of the compound benzene reveals that it contains the elements carbon and hydrogen and has the empirical formula of CH. Based upon experimental measurements, benzene is known to have molecules containing chemical combinations of six atoms of carbon and six atoms of hydrogen. The molecular formula, $C_6H_6$, reflects this composition. The simplest ratio of carbon to hydrogen is 1 to 1 as the empirical formula shows, but the molecular formula is six times the empirical formula.

We have learned how to calculate the empirical formula from experimental data. Now let us see how the real formula can be found from the empirical formula. To do this, it is necessary to have an experimental determination of the molar mass of the compound. That is, we must have a measure of the molar mass of the compound that is found by experiment rather than by knowing the formula beforehand. The idea is that if we have a molar mass, we can use it to find the formula. There are several experimental methods available by which the molar mass of a compound can be found without first knowing its formula. The most accurate method is mass spectroscopy, discussed in Chapter 3. One identifying characteristic revealed by the mass spectroscopic analysis of a compound is the molar mass.

Suppose we do not know the molecular formula of benzene but only the empirical formula of CH. Experimental determination reveals the molar mass of benzene to be 78 g/mol benzene. Notice that the empirical formula is CH, with one mole of C and one of H, gives a molar mass for the empirical formula of 13 g/mol CH. To find the molecular formula of the compound, divide the molar mass of the compound by the molar mass of the empirical formula. This division shows how many empirical formulas there are in the molecular formula of the compound.

$$\frac{78 \text{ g/mol benzene}}{13 \text{ g/mol CH}} = \frac{6 \text{ mol CH}}{1 \text{ mol benzene}}$$

The result indicates that the formula of the compound contains six CH. This means that the molecular formula is six times the empirical formula. So the representative formula of benzene is $C_6H_6$ not 6CH. Notice that, as expected, the molar mass of benzene calculated from its formula is 78 g/mol ($6 \times 12 + 6 \times 1 = 78$).

The formula of a compound is either the same as the empirical formular or some simple whole number multiple of the empirical formula. Whenever you see a chemical formula you should appreciate that someone had to determine the formula using experimental data to find both the empirical formula and the representative or molecular formula.

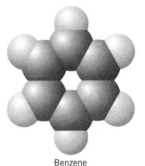

Benzene

---

**EXAMPLE 4-8**

Octane, a compound found in gasoline, has the empirical formula of $C_4H_9$. Experimental measurements give a molar mass value of about 120 g/mol octane. What is the molecular formula? Dividing the molar mass of the compound by the molar mass corresponding to the empirical formula, $C_4H_9$, we can deduce the formula of the compound. The first step is to find the molar mass corresponding to the formula $C_4H_9$.

$$C \quad 4(12.01) = 48.04$$

$$H \quad 9(1.008) = 9.072$$

$$57.112 \quad \text{or} \quad 57.11$$

Then, this molar mass is divided into the molar mass of the compound.

$$\frac{120 \text{ g/mol octane}}{57.11 \text{ g/mol } C_4H_9} = \frac{2.10 \text{ mol } C_4H_9}{1 \text{ mol octane}}$$

We know that a formula is a simple, whole-number multiple of the empirical formula so we round off the answer to the nearest whole number. The fraction is probably a result of experimental error. Therefore, the molecular formula is double the empirical formula or $C_8H_{18}$.

Glucose, a common sugar, has the empirical formula of $CH_2O$. What is the molecular formula if the experimental molar mass is 180 g/mole glucose?

Which element symbol is written first in a formula? By convention the elements in the formula of a compound are typically listed bottom to top in a vertical column of the periodic table (e.g. $SO_2$ rather than $O_2S$) and from left to right in a horizontal row of the table (e.g. $SCl_2$ rather than $Cl_2S$). One exception to this pattern is that hydrogen follows the vertical column starting with nitrogen and comes before the vertical column starting with oxygen (e.g. $NH_3$ and $H_2O$). Another exception is that oxygen follows all elements except fluorine (e.g. $Cl_2O$ and $OF_2$).

Sometimes formulas of compounds are written in alternative ways. For example, the formula of acetic acid is written as $HC_2H_3O_2$, $CH_3COOH$, or $CH_3CO_2H$. Note that each formula has 2 C, 4 H, and 2 O. Another example is ethyl alcohol written as $C_2H_6O$ or $C_2H_5OH$. Sometimes formulas are written using parentheses to reflect multiples of specific groups of atoms. Two examples are ammonium sulfate written as $(NH_4)_2SO_4$ and calcium acetate written as $Ca(C_2H_3O_2)_2$. To count the number of atoms of a given element within parentheses multiply its subscript by the subscript outside the parentheses. For instance, calcium acetate contains 1 Ca, 4 C, 6 H, and 4 O. The reason why parentheses are used in formulas is discussed in Chapters 7 and 8.

## 4-11 PERCENT-BY-MASS COMPOSITION

Sometimes we find it convenient to express the composition of a compound in terms of the percent of one or more elements in the compound. The compound $Fe_2O_3$ is the main component of certain iron ores. To four significant digits $Fe_2O_3$ contains 69.94% iron by mass. We have seen that it is possible to find the formula of a compound by using the experimentally determined percent-by-mass composition. Now let us consider how to calculate the percent of an element in a compound if we start with the formula of the compound.

The mass and the number of moles of a compound are related by the molar mass. Thus, the formula of the compound, which gives the number of moles of each constituent element, is used to determine the percent by mass of each element. Of course, it makes sense that if formulas are originally determined from the percent by mass of each element, then it should be possible to determine the percents from the formulas.

The compound $Fe_2O_3$ contains 2 moles of combined iron per mole of the compound. This can be expressed as a molar ratio.

$$\frac{2 \text{ mol Fe}}{1 \text{ mol Fe}_2O_3}$$

The percent of iron by mass is found by converting the moles of iron to mass and the mole of $Fe_2O_3$ to mass. Of course, the molar masses of each are used for this purpose. To begin, we need to calculate the molar mass of $Fe_2O_3$. Look up the molar mass of iron, multiply it by 2, then add this to three times the molar mass of oxygen. To calculate the percent of iron, multiply the above molar ratio by the molar mass of iron; this converts the 2 moles of iron to grams.

$$\frac{2 \text{ mol Fe}}{1 \text{ mol Fe}_2O_3} \times \frac{55.85 \text{ g}}{1 \text{ mol Fe}}$$

Next, convert the 1 mole of $Fe_2O_3$ to mass using its molar mass in the inverted form (note the units of the factor).

$$\frac{2 \text{ mol Fe}}{1 \text{ mol Fe}_2O_3} \times \frac{55.85 \text{ g}}{1 \text{ mol Fe}} \times \frac{1 \text{ mol Fe}_2O_3}{159.7 \text{ g}}$$

Canceling the units and multiplying by 100 gives the percent of iron in the compound. The first part of the calculation actually expresses the amount of the element as a fraction of a mole of the compound. To express as parts per hundred, it is necessary to multiply by 100. Canceling the units gives

$$\frac{2 \times 55.85}{159.7} \times 100 = 69.94\%$$

Note the general pattern revealed by this example. To find the percent by mass of an element, multiply the molar mass of the element by its subscript in the formula. Then, divide by the molar mass of the compound. Finally, multiply by 100 to express the percent. Units are not included in percents, so you don't have to include units in the calculation.

$$\text{percent element} = \frac{s \, (\text{molar mass element})}{\text{molar mass compound}} \times 100$$

where $s$ is the subscript of the element in the formula.

---

**EXAMPLE 4-9**

What is the percent-by-mass composition of water, $H_2O$ (to four digits)?

First, find the molar mass of water to five digits.

$$2(1.0079) + 15.9994 = 18.015$$

Use at least one more digit than the desired number of digits in the percent. Multiply the molar mass of hydrogen by its subscript in the formula and divide by the molar mass of the compound. Finally, multiply by 100 to give the percent of hydrogen.

$$\% \, H = 2 \times \frac{1.0079}{18.015} \times 100 = 11.19\%$$

**4-6**

The formula of urea, a compound used as fertilizer, is $N_2H_4CO$. What is the percent of nitrogen in urea to four digits?

Since there are only two elements in the compound, the percent by mass of oxygen is found by subtracting the percent by mass of hydrogen from 100.00.

$$\% \ O = 100.00\% - 11.19\% = 88.81\%$$

## 4-12 THE PHYSICAL STATES AND CHEMICAL FORMULAS OF THE ELEMENTS

Samples of most elements have been isolated from nature, so we know whether they occur in the solid, liquid, or gaseous physical state. Most elements are solids. Two elements, mercury and bromine, occur as liquids, and gallium and cesium become liquids at temperatures slightly above typical room temperatures. Eleven of the elements are gases. Elements that occur as gases include hydrogen, nitrogen, oxygen, fluorine, chlorine, and a group of elements called the noble gases (helium, neon, argon, krypton, xenon, and radon).

The metals copper and gold have notable colors. Other colored elements are fluorine, a pale yellow gas; chlorine, a greenish-yellow gas; bromine, a dark-red liquid; and iodine, a deep-purple solid. Nitrogen, hydrogen, oxygen, and the noble gases are colorless.

It is useful to know the appropriate chemical formula for an element when it is not combined with other elements. Most elements occur as collections of atoms that we can represent simply by using the elemental symbols. For instance, a sample of copper metal is represented as Cu. Among the elements there are seven that occur in unique chemical forms that we need to consider. The elements hydrogen, nitrogen, oxygen, fluorine, chlorine, bromine, and iodine occur as collections of molecules in which two atoms are bonded to one another. These are called diatomic molecules. Formulas for the **diatomic molecules** of these elements are $H_2$, $N_2$, $O_2$, $F_2$, $Cl_2$, $Br_2$, and $I_2$. Thus, for example, if we want to refer to oxygen in its natural elemental state (when it is not combined with other elements), we use the formula $O_2$. A sample of oxygen gas is pictured as a vast collection of diatomic molecules. This does not mean that oxygen is $O_2$ when it is combined with other elements in compounds. It does mean that when we refer to atmospheric oxygen we use the formula $O_2$. Incidentally, the two major components of air are elemental nitrogen, $N_2$, and elemental oxygen, $O_2$. A sample of air is pictured as a vast collection of nitrogen and oxygen molecules along with a variety of molecules of minor components.

All other elements in their elemental or uncombined state are normally represented only by the symbol, since they do not occur in diatomic forms. This is a very important point of symbolism in chemistry. That is, elements in the elemental uncombined state are represented by the element symbol unless the element is one of the seven diatomic elements.

**DIATOMIC ELEMENTS**

| | |
|---|---|
| Hydrogen | $H_2$ |
| Nitrogen | $N_2$ |
| Oxygen | $O_2$ |
| Fluorine | $F_2$ |
| Chlorine | $Cl_2$ |
| Bromine | $Br_2$ |
| Iodine | $I_2$ |

**4-7**

Draw a picture of a sample of air showing sketches of molecules of the four most common gases found in the atmosphere.

For example, we use the symbol Fe for a sample of iron metal and the symbol $Cl_2$ for a sample of chlorine. A sample of iron is envisioned as a collection of iron atoms and a sample of chlorine as a collection of diatomic $Cl_2$ molecules. When elements occur in compounds, we use the normal element symbols along with the appropriate subscript in the formula. For example, the formulas of some chlorine compounds are $KCl$, $MgCl_2$, $AlCl_3$, $SCl_4$, and $PCl_5$. The symbol $Cl_2$ by itself represents the uncombined element. It is useful to memorize the names and formulas of all of the seven diatomic elements. Notice the locations of these elements in the periodic table. Excluding hydrogen their locations form a shape similar to the figure seven.

## 4-13 CHEMICAL REACTIONS

When a silvery, solid sample of the element sodium is mixed with a sample of the greenish-yellow gas chlorine, a vigorous chemical transformation takes place producing common table salt, $NaCl$. Two chemicals having unique characteristics have interacted to form a new chemical having very different characteristics. When natural gas is mixed with air and ignited, vigorous combustion occurs with the evolution of heat and light. Methane, $CH_4$, the main component of natural gas, and oxygen in the air are transformed into carbon dioxide and water. One set of chemicals has changed into a new set of chemicals. The same elements are in each set of chemicals but new compounds are formed from the initial compounds and elements as the atoms recombine. New combinations of atoms have replaced the original combinations.

Transformations in which one set of chemicals is mixed and a new set of chemicals is produced are fundamental in chemistry. These transformations are called chemical reactions. A **chemical reaction** is defined as a process in which one set of chemicals is transformed into a new set of chemicals. A chemical reaction may involve elements combining to form a compound, or a compound decomposing to form elements, or chemical compounds interacting to form new chemical compounds.

Chemical reactions are truly amazing processes. They represent dynamic chemical actions. These actions range from the rusting of iron to the many complex reactions involved in life processes. In a chemical reaction the initial chemicals, called **reactants,** react to form the new chemicals called **products.** New substances are formed. Chemical reactions are occurring in and around us all of the time. The digestion and metabolism of food involve chemical reactions. The burning of fuels are chemical reactions. Experimental chemistry includes the observation and description of various chemical reactions. Reactions are also fundamental to industrial processes since they provide ways to extract elements from compounds and ways to synthesize new and useful compounds. Chemistry in Action 4-1 is about the reaction between vinegar and baking soda.

## 4-14 CHEMICAL REACTIONS AND CHEMICAL EQUATIONS

In a chemical reaction, the starting materials are those chemicals that react. The reactants have specific properties, such as color, texture, density, and physical state. A reaction produces chemical products. Reactants and products have different properties.

A chemical reaction involves the breaking of one set of atomic combinations in the reactants and the formation of a new set of combinations in the products. That is, a reaction involves the change of an initial arrangement of chemically bound atoms into a new arrangement of atoms. Many chemical reactions are observed by noting that the starting materials change into new materials. When sodium and chlorine react, a silvery solid and a greenish-yellow gas are transformed into the white crystals characteristic of salt. Reactions of colorless gases, such as the burning of methane in air, are not directly observed, but the release of heat and light in the reaction suggests that something is happening. It is possible to describe a chemical reaction by naming the reactants and products. When we burn charcoal briquettes in a barbecue, the main chemical reactants are carbon in the charcoal and oxygen gas in the air. The product of the reaction is carbon dioxide gas. The reaction is described by stating that carbon reacts with oxygen to form carbon dioxide. Note also that heat is released in the burning of charcoal. Many reactions release energy as heat, so heat can be evidence of a reaction.

A more succinct and useful way to describe a chemical reaction is to use a **chemical equation.** To write an equation we use precise formulas for the reactants and products. In the reaction of carbon and oxygen to give carbon dioxide, the formulas of the reactants are C for carbon and $O_2$ for oxygen. (Remember that oxygen is typically a diatomic element.) The formula of the product, carbon dioxide, is $CO_2$. The equation for the reaction is

$$C + O_2 \longrightarrow CO_2$$

By convention, the formulas of the reactants are separated by plus signs, which are read as "plus" or "and." Similarly, the formulas of the products, if there is more than one, are separated by plus signs. Typically, the reactants and products are separated by an arrow, which is read as "react to yield," or "react to give," or "react to produce," or "react to form." The equation above is read as "carbon and oxygen react to produce carbon dioxide."

A chemical equation is a symbolic description of a chemical reaction. The term equation is full of meaning. In mathematics the term equation means to make equal or have an equality on both sides of an equal sign. Of course, a chemical equation is not the same as a mathematical equation but there are some important equalities associated with chemical equations. The numbers of atoms of a specific element are equal on both sides of any equation. This makes sense if we remember that atoms are not created or destroyed in chemical processes. Of course a chemical reaction does not simply involve separate atoms so in an equation all chemicals are shown by the formulas of the substances involved in the reaction. It is important to realize that a chemical equation is a before-and-after description of a reaction. The chemical equation shows what the reactants are and what the products are. An equation does not tell how a reaction occurs.

## 4-15 COEFFICIENTS IN EQUATIONS

A chemical reaction is envisioned as one set of chemically combined atoms in the reactants changing to a new set of combinations in the products.

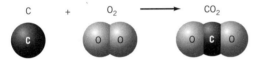

Atoms are conserved in a reaction, so in a chemical equation the number of atoms of each kind of element must be the same on the reactant and product side of the arrow. Atoms are not created or destroyed; they do, however, change their state of combination. The reaction that occurs when methane, $CH_4$, reacts with oxygen gas (burns in air), produces carbon dioxide, $CO_2$, and water, $H_2O$. The equation is written showing the formulas of the reactants and products separated by an arrow.

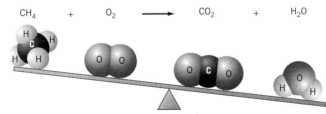

If you count the number of combined hydrogen atoms in the equation above you'll find four on the left side and two on the right of the arrow. Similarly the number of combined atoms of oxygen do not balance on both sides of the arrow. The equation is not correctly written. An equation

must be written to reflect the fact that atoms are conserved in chemical reactions. To show the conservation of atoms an equation needs to be balanced. A **balanced equation** has the same number of combined atoms of
each element on either side of the arrow. Let's consider how to balance
an equation.

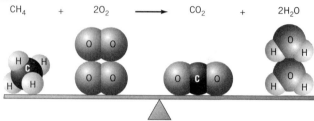

$$CH_4 \quad + \quad 2O_2 \quad \longrightarrow \quad CO_2 \quad + \quad 2H_2O$$

When **balancing an equation** never change the formulas of compounds or
any subscripts, since this would not represent the compounds correctly.
We can, however, change the number of formula units of any compound
that appears in the equation. We do this by placing a **numerical coefficient** in front of the formula of the compound. Refer to the reaction of
methane and oxygen. To get four hydrogen atoms on each side of the
equation, two $H_2O$ molecules must be produced; and to get four oxygen
atoms on each side, two $O_2$ molecules must react. This is represented by
placing the necessary numbers, called coefficients, in front of the formula
of the appropriate compounds or elements. Count the number of atoms
on each side of the arrow to confirm the equation is balanced.

## 4-16 BALANCING CHEMICAL EQUATIONS

In any chemical equation, the coefficients must be adjusted so that the
number of atoms of each element is the same on the left (reactant side) as
on the right (product side). An equation in which the number of atoms of
each element is the same on both sides of the arrow is a **balanced equation.** The process of adjusting the coefficients is called balancing the equation. Often, balancing is done by trial and error, that is, by trying various
coefficients and keeping a count of each kind of atom. You should never
change the formulas or subscripts to balance an equation. The compounds are treated as units, since they involve chemically combined
atoms. You can, however, use any numerical coefficient in front of a formula when balancing equations. To balance a chemical equation, follow
this pattern:

1. Write the correct formulas of the reactants and products. (For now
   you'll be given this information. You are not expected to predict the
   products of a reaction at this point in your study of chemistry.)
2. Count the number of atoms of each element on the reactant side and
   on the product side.

3. If the number of atoms of any element differs from side to side, add necessary coefficients in front of the appropriate formulas.
    (a) Work with one element at a time and leave diatomic elements until last.
    (b) Change coefficients to maintain the balance. Coefficients are changed as needed.
    (c) Never change a formula or a subscript in a formula.
    (d) Never add extra formulas to an equation.

Most equations are balanced using relatively small numbers for coefficients. Since balancing is a trial-and-error procedure, it is possible that your first attempt to balance an equation may not be successful. If you find that the coefficients are getting too large or that the equation does not balance, just start over or consult someone else.

Give the balanced equation for the reaction of iron metal and chlorine to give $FeCl_3$. Draw a picture to show the reaction of atoms of iron with molecules of chlorine to give formula units of the product. Chemistry in Action 4-2 is about reacting iron with chlorine.

**EXAMPLE 4-10**

When iron rusts, one of the reactions that takes place is that iron, Fe, reacts with oxygen, $O_2$, to form solid $Fe_2O_3$. Write a balanced equation.

The unbalanced equation is

$$Fe + O_2 \longrightarrow Fe_2O_3$$

There are two O atoms and one Fe atom on the reactant side and three O atoms and two Fe atoms on the product side. Balance the iron atoms by placing the coefficient of 2 in front of Fe on the reactant side.

$$2Fe + O_2 \longrightarrow Fe_2O_3$$

Note that there is an odd number of oxygen atoms on the product side and an even number on the reactant side. One way to balance the oxygen atoms is to use a coefficient of 3 for $O_2$ and a coefficient of 2 for $Fe_2O_3$. This gives an even number of oxygen atoms on both sides.

$$2Fe + 3O_2 \longrightarrow 2Fe_2O_3$$

This balances the oxygen atoms giving six on each side. Now, however, there are four Fe atoms on the product side and only two on the reactant side. To balance completely, we just change the coefficient of Fe on the left to 4.

$$4Fe + 3O_2 \longrightarrow 2Fe_2O_3$$

When working with a diatomic element sometimes there is an even number of atoms on one side and an odd number on the other side. One way to balance the equation is to make the odd side even by doubling the coefficient; the other side is adjusted as needed. Another approach is to balance the atoms of the diatomic element using a fraction as a coefficient then to multiply all coefficients in the equation by 2.

### EXAMPLE 4-11

Octane, $C_8H_{18}$, a component of gasoline, reacts with oxygen, $O_2$, to give carbon dioxide, $CO_2$, and water, $H_2O$. Balance the following unbalanced equation.

$$C_8H_{18} + O_2 \longrightarrow CO_2 + H_2O$$

First, balance the carbons using a coefficient of 8 for $CO_2$, then balance the hydrogens using a coefficient of 9 for $H_2O$. Note that 9 times the subscript of 2 for hydrogen gives a total of 18 hydrogens.

$$C_8H_{18} + O_2 \longrightarrow 8CO_2 + 9H_2O$$

Counting the oxygens, we note there are 25 oxygens on the product side ($8 \times 2 = 16$ and $16 + 9 = 25$) and 2 on the reactant side. Since we have $O_2$ on the reactant side, a way to give 25 oxygens on the left is to use a coefficient of 25/2 or 12.5.

$$C_8H_{18} + 12.5O_2 \longrightarrow 8CO_2 + 9H_2O$$

This is a balanced equation in terms of atoms, but usually we try to use whole-number coefficients. In this case, the coefficients are changed to whole numbers by doubling each coefficient, that is, multiplying each by 2. This gives the balanced equation

$$2C_8H_{18} + 25O_2 \longrightarrow 16CO_2 + 18H_2O$$

**ACTIVITY 4-9**

Balance the following unbalanced equations using whole number coefficients.

(a) The reaction for the decomposition of water to give oxygen gas and hydrogen gas

(b) $Ca + NH_3 \longrightarrow$
    $CaH_2 + Ca_3N_2$

(c) $C_2H_6O + O_2 \longrightarrow$
    $CO_2 + H_2O$

The last part of Example 4-11 illustrates another even–odd case that sometimes arises when a diatomic element is involved in a reaction. What happens is that during the balancing process, an odd number of atoms of the diatomic element results on the product side with two atoms of this element on the reactant side. When this occurs, the equation is balanced by using a fractional coefficient for the diatomic element (the odd number divided by 2). To get whole number coefficients double all of the coefficients in the equation as the last step of balancing. Occasionally, however, you may see such equations written with fractional coefficients.

### 4-17 CHEMICAL EQUATIONS WITH ADDED INFORMATION

When chemical equations are written, it is sometimes useful to show the physical states of the various chemicals. This is accomplished by using letter codes enclosed in parentheses following the formulas; (g) denotes gas, ($\ell$) denotes liquid, and (s) denotes solid. Thus, the equation

4-17 Chemical Equations with Added Information *131*

$$2HgO(s) \xrightarrow{\text{heat}} 2Hg(\ell) + O_2(g)$$

reveals that when solid HgO is heated, liquid mercury and oxygen gas form. Note that a special condition for a reaction can be written above or below the arrow in the equation as is done for heat in the above equation. The symbol (aq) is used to denote that a chemical is present in aqueous solution, which means dissolved in water. Many important reactions take place in water. For instance, the equation

$$2HCl(aq) + Mg(OH)_2(s) \longrightarrow MgCl_2(aq) + 2H_2O(\ell)$$

represents the reaction that takes place in the stomach when you swallow some milk of magnesia, a common antacid.

Some reactions do not occur readily unless a catalyst is present. A **catalyst** is a substance that increases the speed or rate of a reaction without undergoing a permanent chemical change. Since a catalyst is neither a reactant nor a product, its formula is sometimes written above the arrow in the equation. For instance, the catalyst $V_2O_5$ is required to make the compound $SO_3$ efficiently when $SO_2$ and $O_2$ react. This situation is represented by the equation

$$2SO_2(g) + O_2(g) \xrightarrow{V_2O_5} 2SO_3(g)$$

See Section 11-4 for a more detailed discussion of catalysts.

You'll encounter many equations in your study of chemistry. As you gain experience with elements and compounds, you can interpret some equations by naming the reactants and products involved. Until then, the best way to read an equation is to read the formulas involved in a literal manner. For example, the equation

$$Fe + S \longrightarrow FeS$$

is read as          F–E and S react to give F–E–S

The equation          $CH_4 + 2O_2 \longrightarrow CO_2 + 2H_2O$

is read as C–H–4 plus two O–2 react to form C–O–2 plus two H–2–O

The equation          $2SO_2 + O_2 \xrightarrow{V_2O_5} 2SO_3$

is read as

two S–O–2 plus O–2 react, using the catalyst V–2–O–5, to produce 2 S–O–3

Sometimes equations are used in a general way to succinctly give chemical information. For example, the fact that carbohydrates react with oxygen to give carbon dioxide and water is conveyed by a general equation:

$$\text{carbohydrates} + O_2 \longrightarrow CO_2 + H_2O$$

Such symbolic equations are not meant to be viewed as balanced equations.

ACTIVITY
**4-10**

The main ingredient in vinegar is acetic acid, $HC_2H_3O_2(aq)$. Write a balanced equation for the reaction of zinc metal, Zn, and acetic acid to give hydrogen gas, $H_2$, and $Zn(C_2H_3O_2)_2(aq)$. This is the reaction that occurred in Activity 3-12.

# Moles, Formulas, and Equations

Mole of an element—the amount of an element that contains as many atoms as there are carbon atoms in exactly 12 g of C-12.

Avogadro's Number—The number atoms in a mole of an element or the number of formula units in a mole of a compound: $6.022 \times 10^{23}$

Mole of a compound—the amount of a compound that contains Avogadro's number of formula units.

Molar Mass of an element—the number of grams per mole of an element as given by the numerical value of its atomic weight.

Molar Mass of a compound—the number of grams per mole found by multiplying the molar masses of elements in the compound by their subscripts in the formula and summing the products.

**Molar mass is used to find the number of moles given the mass or or the mass given the number of moles.**

| GIVEN THE MASS | FIND THE NUMBER OF MOLES | MULTIPLY THE MASS BY THE INVERSE OF THE MOLAR MASS |
|---|---|---|
| 24 g C | ? mol C | $24\,g \times \dfrac{1 \text{ mol C}}{12.0\,g} = 2.0 \text{ mol C}$ |

### FINDING THE EMPIRICAL FORMULA OF A COMPOUND:

Percent to mass    Percent composition ⟶ Mass composition

Mass to mole    Use the mass and the molar mass of each element to find the number of moles of each element.

Divide by small    Divide by the smallest number of moles to find the molar ratios.

Multiply 'til whole    The molar ratios reveal the subscripts. If necessary multiply each by the appropriate number to clear any fractions and to get whole number subscripts.

### STEPS USED TO BALANCE EQUATIONS:

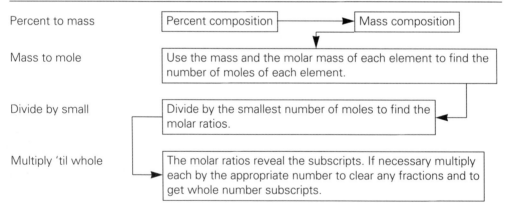

1. Start with the formulas of the reactants and products.

$$C_3H_8 + O_2 \longrightarrow CO_2 + H_2O$$

2. Count the number of atoms of each element on both sides.

$$C_3H_8 + O_2 \longrightarrow CO_2 + H_2O$$
3C 8H 2O    1C 2O 2H 1O
3O

3. Use coefficients in front of formulas to balance the atoms of each element.

$$C_3H_8 + O_2 \longrightarrow CO_2 + H_2O$$

(a) Change coefficients as needed.

(b) Work with one element at a time and leave diatomic elements until last.

$$C_3H_8 + O_2 \longrightarrow 3CO_2 + 4H_2O$$
$$C_3H_8 + 5O_2 \longrightarrow 3CO_2 + 4H_2O$$

(c) Never change formulas or add formulas.

Closely observe the chemicals you mix and the process that occurs. You need: a small zip-lock bag (a sandwich bag will do), a teaspoon, baking soda (sodium bicarbonate), and white vinegar (acetic acid solution).

1. Open the bag and place one heaping teaspoon of baking soda in one corner of the bag.

2. Hold the corner of the bag containing the baking soda so that it cannot move.

3. Tilt the bag so that the opposite corner is directed downward and pour one-fourth cup of vinegar into the corner of the bag.

4. Before releasing the baking soda seal the zip lock of the bag. Be sure that it is completely sealed.

5. Release your grip and tilt the bag so that the liquid and solid mix.

6. Record your observations. As the reaction occurs feel the outside of the bag in the region of the reaction.

What evidence did you note that suggest a reaction took place?

You need: a white napkin or some white tissue, two steel paper clips, some white vinegar, a small amount of liquid bleach (about 1 mL of household bleach), a flat plate or the bottoms cut off two plastic cups to make two shallow containers.

**CAUTION: Bleach is dangerous. Wear safety goggles and be especially careful. Do not spill it or splash it on your skin or face and use it in a well-ventilated space.**

1. Cut the napkin or tissue to make two pieces measuring 3 cm by 6 cm. Fold them twice to make two 3 cm by 2 cm pieces. Place the paper pieces on a flat surface or in shallow containers.

2. Pour a small amount of vinegar on one piece of napkin so that it is soaked with vinegar and pour a small amount of liquid bleach on the other piece so that it is soaked with bleach.

3. Place one iron paper clip on top of each piece of napkin. Allow the clips to sit a few minutes and record your observations.

4. Place several drops of vinegar on the napkin containing the bleach and paper clip. Allow it to sit for a few minutes and record your observations.

What is the color of the compound? Assuming that the iron–chlorine compound has the formula $FeCl_3$, write a balance equation for the reaction of iron metal and molecular chlorine to give this compound.

You need: a small glass, white vinegar, and a Tums tablet (several small bits of eggshell or an oyster-shell calcium tablet can be used in place of the Tums tablet).

1. Pour about one-fourth cup of vinegar into a glass.

2. Add a Tums tablet or some eggshell or a calcium tablet to the vinegar. Record your observations and let the mixture sit for about five minutes.

Write a balanced equation for the reaction. The reactants are solid $CaCO_3$ and $HC_2H_3O_2$(aq). The products are $Ca(C_2H_3O_2)_2$(aq), $CO_2$ gas and liquid $H_2O$.

## QUESTIONS

You will need to use a periodic table or a table of atomic weights for some of the following questions. In some of the questions the names of compounds are given for reference. You are not expected to name compounds at this time. Chemical nomenclature is discussed in Chapter 8.

### Section 4-1

1. What is a chemical formula? Give an example.

2. Describe in words the information that each of the following formulas contains in terms of atomic proportions of elements.

   (a) $ZnO$ (zinc oxide)

   (b) $HC_2H_3O_2$ (acetic acid)

   (c) $C_9H_8O_4$ (aspirin)

   (d) $SO_2$ (sulfur dioxide)

   (e) $MgCO_3$ (magnesium carbonate)

   (f) $CCl_2F_2$ (Freon a chlorofluorocarbon or CFC)

### Sections 4-2 to 4-6

3. Define or explain these terms.

   (a) mole of an element

   (b) molar mass

4. What is Avogadro's number and what units are used with it?

5. Discuss the formula of a compound in terms of the law of constant composition.

6. Explain why the molar mass of an element is numerically equal to the atomic weight of that element.

7. *Calculate the number of moles of atoms contained in each sample given below.

    (a) a 24.8-g silver bracelet

    (b) a 53.5-kg iron magnet

    (c) 6.71 g of powdered sulfur

    (d) a 0.0699-g tungsten filament

    (e) 235 metric tons of recycled aluminum (1 metric ton = 1000 kg)

    (f) a 4.81-g sample of helium gas

    (g) a 131-mg silicon chip

    (h) a 454-g spool of magnesium ribbon

    (i) a 2.76-kg copper kettle

    (j) an 875-g lead pipe

    (k) a 0.30-g sample of beryllium

8. Calculate the number of moles of atoms contained in each sample given below.

    (a) a 0.976-g sample of neon gas

    (b) a 14.7-g sample of mercury

    (c) a 0.354-g zinc wire

    (d) a 2.86-g chunk of carbon

    (e) a 32.01-g piece of gold

    (f) a 50.0-kg lead ball

    (g) an 8.92-g sample of calcium

    (h) a 237-g sample of liquid bromine

    (i) a 10.0-g sample of boron

    (j) a 69.3-g sample of nickel

    (k) a 1.00 mL sample of mercury

9. *Calculate the mass of the following samples of elements.

    (a) 124 mol Pt

    (b) $3.18 \times 10^{-6}$ mol Mo

    (c) $6.04 \times 10^{-7}$ mol Si

    (d) $4.24 \times 10^{5}$ mol Na

    (e) 11.9 mol Ar

    (f) $7.58 \times 10^{-10}$ mol Ga

10. Calculate the mass of the following samples of elements.

    (a) 84 mol K

    (b) $7.60 \times 10^{-3}$ mol U

    (c) 0.0395 mol Li

    (d) $2.25 \times 10^{6}$ mol Ba

    (e) 4.69 mol B

    (f) $5.78 \times 10^{-6}$ mol Y

11. *Calculate the number of atoms in each sample given below.

    (a) an 84.8-g sample of Hg in a barometer

    (b) a 0.249-g Ti metal heart valve

    (c) a 5.25-g sample of Kr gas

    (d) a 3709-g Ni meteor

    (e) an 8.23-mg sample of Ag

    (f) 1.00 troy ounce of Au (1 troy ounce = 31.1 g)

12. Calculate the number of atoms in each sample given below.

    (a) a 3.18-g Cu in a pure copper penny

    (b) a 0.171-g W filament

    (c) a 12.5-g sample of He

    (d) a 453.6-g chunk of iron

    (e) a 0.200-g or one-carat diamond (diamond is carbon)

    (f) a 1.31-g aluminum nail

13. *Use the molar mass of hydrogen and Avogadro's number to calculate the mass in grams of a typical H atom in units of grams per H atom. Use this result to calculate the mass in grams of a typical O atom. (Remember that atomic weights express the relative masses of atoms.)

## Sections 4-7 and 4-8

14. What is an empirical formula and what experimental data are needed to calculate the empirical formula of a compound?

15. *Calculate the empirical formulas for the following compounds for which the percent by mass or mass composition is given.

    (a) lysine: 19.1% N, 9.6% H, 22.0% O, 49.3% C

    (b) potassium tartrate found in baking powder: 20.8% K, 2.7% H, 25.5% C, 51.0% O

    (c) calcium fluoride found in the mineral fluorspar: 119.4 g Ca, 114.7 g F

    (d) hydroquinone used in photography: 72.06 g C, 6.05 g H, 32.0 g O

    (e) borax: 12.1% Na, 11.3% B, 71.4% O, 5.27% H

    (f) saccharin an artificial sweetner: 45.9% C, 2.76% H, 26.2% O, 7.65% N, 17.5% S

    (g) barium phosphate: 68.4% Ba, 10.3% P, 21.3% O

(h) vitamin C: 54.5% O, 40.9% C, 4.6% H.

16. Calculate the empirical formulas for the following compounds for which the percent by mass or mass composition is given.

(a) butane: 82.7% C, 17.3% H

(b) sodium nitrate: 27.0% Na, 16.5% N, 56.5% O

(c) adenosine triphosphate (ATP): 71.04 g C, 9.54 g H, 41.43 g N, 54.96 g P, 123 g O

(d) citric acid: 56.3 g C, 6.3 g H, 87.4 g O

(e) urea: 20.0% C, 46.8% N, 6.7% H, 26.5% O

(f) potassium chromate: 110.8 g K, 73.7 g Cr, 90.75 g O

(g) ethyl alcohol or ethanol: 52.2% C, 13.1% H, 34.7% O

(h) propane: 4.08 g C, 0.915 g H

17. *Determine the empirical formula of aspartame (Nutrasweet), an artificial sweetener made from amino acids, if a 250.0-g sample of the compound contains 142.85 g C, 15.40 g H, 23.80 g N, and the rest is O.

18. Cisplatin is a drug used in chemotherapy for some cancer patients. A 112.5-g sample contains 73.2 g Pt, 26.6 g Cl, 2.27 g H, and the rest is N. What is the empirical formula of cisplatin? If the molar mass cisplatin is 300 g/mol, what is the molecular formula of cisplatin?

19. *A 32.38-g sample of aluminum reacts with oxygen to form 61.18 g of an aluminum–oxygen compound. Calculate the empirical formula of the compound.

20. Nitroglycerine is used as an explosive and as a heart medicine. Determine the empirical formula if a sample is found to contain 19.88 g C, 2.75 g H, 23.13 g N, and 79.25 g O.

## Section 4-9

21. How is a mole of a compound defined?

22. What is a formula unit?

23. If you have one mole of water what does this mean in terms of Avogadro's number?

24. If you have a sample of 2.00 moles of NaCl, what does this mean in terms of Avogadro's number?

25. *Determine the molar mass of the following compounds to four significant digits.

(a) potassium oxalate, $K_2C_2O_4$

(b) pentane, $C_5H_{12}$

(c) methyl bromide, $CH_3Br$ (used as a soil fumigant but may be banned by the EPA)

(d) butane, $C_4H_{10}$ (used as a fuel)

(e) calcium fluoride, $CaF_2$

(f) acetylcholine (a neurotransmitter), $C_7H_{16}NO_2$

(g) trinitrotoluene, $C_7H_5N_3O_6$ (TNT an explosive)

(h) magnesium carbonate, $MgCO_3$

26. Determine the molar mass of the following compounds to four significant digits.

(a) isopropyl alcohol, $C_3H_8O$ (rubbing alcohol)

(b) barium sulfate, $BaSO_4$ (used as X-ray absorber during intestinal tract X rays)

(c) sodium nitrate, $NaNO_3$ (found in the mineral Chile saltpeter)

(d) potassium dichromate, $K_2Cr_2O_7$

(e) titanium tetrachloride (used in smokescreens), $TiCl_4$

(f) sodium cyanide, $NaCN$ (a deadly poison)

(g) sodium bisulfate, $NaHSO_4$ (used in some toilet bowl cleaners)

(h) ethane, $C_2H_6$

(i) sodium hydrogen carbonate or sodium bicarbonate, $NaHCO_3$

27. *Determine the number of moles of formula units in the following samples of compounds.

(a) 819 g $NH_3$

(b) 12.3 g $K_2Cr_2O_7$

(c) 0.1086 g NaCN

(d) 28.3 g $C_{12}H_{22}O_{11}$

(e) $2.19 \times 10^6$ g NaCl

(f) 6.37 g acetylene, $C_2H_2$

(g) 56.8 g phenylalanine, $C_9H_{11}NO_2$, an amino acid

28. Determine the number of moles of formula units in the following samples of compounds.

(a) 311 g $NaNO_2$ (sodium nitrite)

(b) 75.0 g of muskone, $C_{16}H_{30}O$ (a pheromone that acts as a sex attractant)

(c) $2.88 \times 10^5$ g of $SO_3$

(d) $2.24 \times 10^{-7}$ g $Au_2S_3$

(e) 9.10 g $H_2O_2$ (hydrogen peroxide)

(f) 250 g $C_6H_{12}O_6$ (glucose)

(g) 675 g $NaHCO_3$

29. *Determine the mass of each of the following samples of compounds. (When you have completed Question 25, you will have the molar masses needed for this

question.)

(a)  3.7 mol $K_2C_2O_4$

(b)  262 mol $C_5H_{12}$

(c)  5.87 mol $CH_3Br$

(d)  0.946 mol $C_4H_{10}$

(e)  $7.46 \times 10^{-3}$ mol $CaF_2$

(f)  0.215 mol acetylcholine, $C_7H_{16}NO_2$

(g)  $4.35 \times 10^5$ mol trinitrotoluene, $C_7H_5N_3O_6$

(h)  6.52 mol $MgCO_3$

30. Determine the mass of each of the following samples of compounds. (When you have completed Question 26, you will have the molar masses needed for this question.)

(a)  $2.13 \times 10^{-7}$ mol $C_3H_8O$

(b)  4.72 mol $BaSO_4$

(c)  $3.41 \times 10^{-2}$ mol $NaNO_3$

(d)  58 mol $K_2Cr_2O_7$

(e)  $6.9 \times 10^5$ mol $TiCl_4$

(f)  1.22 mol NaCN

(g)  $8.47 \times 10^4$ mol $NaHSO_4$

(h)  $7.29 \times 10^{-6}$ mol ethane, $C_2H_6$

(i)  9.05 mol $NaHCO_3$

31. *Epsom salt, $MgSO_4$, can be purchased in 5.00-lb boxes. How many moles of $MgSO_4$ are in such a box of Epsom salt? (453.6 g = 1 lb)

32. How many moles of iron are there in 2.00 pounds of iron nails? How many iron atoms are there in 2.00 pounds of nails?

33. *A box of baking soda contains 16 oz of sodium hydrogencarbonate, $NaHCO_3$. How many moles of $NaHCO_3$ are there in the box?

34. A nonaspirin tablet contains 5.0 grains of phenacetin, $C_{10}H_{13}N_2O$. How many moles of phenacetin are there in a tablet? (0.0648 g = 1 grain)

35. *A tablespoon of sugar has a volume of 15.0 mL. If the density of sugar is 1.59 $g/cm^3$, how many moles of sucrose, $C_{12}H_{22}O_{11}$, are in the sugar sample?

36. Ethanol, $C_2H_6O$, has a density of 0.79 g/mL. How many moles of ethanol are in a 4.00 oz-sample. (29.57 mL = 1 fluid oz)

### Section 4-10

37. *Calculate the empirical formulas of the compounds given below using the mass composition. Then use the molar mass given to calculate the representa-

tive or molecular formula of each compound.

(a)  tryptophan, an amino acid: 32.35 g C, 2.96 g H, 6.86 g N, 7.84 g O; molar mass 204 g/mol

(b)  Freon 12, a chlorofluorocarbon (CFC): 5.41 g C, 31.90 g Cl, 17.10 g F; molar mass 121 g/mol

(c)  vanillin, a flavoring agent: 50.56 g C, 4.21 g H, 25.28 g O; molar mass 152 g/mol

(d)  calomel: 12.75 g Hg, 2.25 g Cl; molar mass 472 g/mol

(e)  ribose sugar; 14.8 g C, 2.49 g H, 19.8 g O; molar mass 150 g/mol

(f)  octane, found in gasoline: 84.1% C, 15.9% H; molar mass 114.2 g/mol

38. Calculate the empirical formulas of the compounds given below using the mass composition. Then use the molar mass given to calculate the representative or molecular formula of each compound.

(a)  cholesterol: 36.03 g C, 5.05 g H, 1.78 g O; molar mass 386 g/mol

(b)  hydrogen peroxide: 5.93% H, 94.1% O; molar mass 34 g/mol.

(c)  sodium phosphate, TSP used to clean walls: 28.14 g Na, 12.66 g P, 26.13 g O; molar mass 164 g/mol

(d)  aluminum carbonate (antacid): 13.49 g Al, 9.01 g C, 36.00 g O; molar mass 234 g/mol

(e)  phenolphthalein, an acid–base indicator: 75.40% C, 4.43% H, 20.10% O; molar mass 318 g/mol

(f)  gold (III) sulfide: 40.20 g Au, 9.81 g S; molar mass 490 g/mol

### Section 4-11

39. *You are offered a choice between equal masses of two compounds of gold, $Au_2S_3$ and $Au_2S$. What is the percent by mass of gold in each compound? Which compound would you choose.

40. Two nitrogen-containing fertilizers are urea, $N_2H_4CO$, and ammonium nitrate, $N_2H_4O_3$. Which of these compounds has the greater percent by mass of nitrogen?

41. *Calculate the percent by mass of the following.

(a)  nitrogen in saccharin, $C_7H_5SNO_3$

(b)  iodine in thyroxine a thyroid hormone, $C_{15}H_{11}I_4NO_4$

(c)  sulfur in sulfur trioxide, $SO_3$

(d) carbon in the amino acid arginine, $C_6H_{16}N_4$

(e) sodium in baking soda, $NaHCO_3$

42. Use your answers to Question 41 to calculate the number of grams of

   (a) nitrogen in 0.186 g of saccharin

   (b) iodine in 425 g of thyroxine

   (c) sulfur in 250 g of sulfur trioxide

   (d) carbon in 75 mg of the amino acid arginine

   (e) sodium in 500 mg of baking soda

43. Calculate the percent by mass of the following.

   (a) oxygen in peroxyacetyl nitrate, found in smog, $C_2H_3NO_5$

   (b) silver in argentite a silver ore, $Ag_2S$

   (c) magnesium in chlorophyll, $C_{55}H_{72}MgN_4O_5$

   (d) nitrogen in nicotine, $C_{10}H_{14}N_2$

   (e) chromium in chromium (III) oxide, $Cr_2O_3$

44. *Determine the percent-by-mass composition of the following compounds (to four significant digits).

   (a) nitrogen oxide, NO

   (b) urea, $N_2H_4CO$

   (c) potassium chlorate, $KClO_3$

   (d) propane, $C_3H_8$

   (e) magnesium carbonate, $MgCO_3$

   (f) amphetamine, $C_9H_{13}N$

45. Determine the percent-by-mass composition of the following compounds (to four significant digits).

   (a) sulfur dioxide, $SO_2$

   (b) methyl chloride, $CH_3Cl$

   (c) sodium acetate, $NaC_2H_3O_2$

   (d) lead (II) oxide, $PbO$

   (e) copper (I) nitrate, $CuNO_3$

   (f) benzene, $C_6H_6$

46. *Which of the following compounds contains the highest percent by mass of fluorine?

   (a) LiF (b) $BaF_2$ (c) $CaF_2$ (d) NaF

47. Which of the following compounds contains the lowest percent by mass of cobalt?

   (a) CoO (b) $Co_2O_3$ (c) $Co_3O_4$

## Section 4-12

48. Give a list of the elements that are typically found as gases.

49. Which elements are normally found as liquids?

50. Give the names and molecular formulas for the elements that occur in the form of diatomic molecules.

## Sections 4-13 to 4-17

51. Define the following terms

   (a) chemical reaction

   (b) reactants

   (c) products

52. What is a chemical equation and how are the symbols + and $\longrightarrow$ used in equations?

53. What is a balanced chemical equation?

54. How is the law of conservation of matter related to the balancing of an equation?

55. Why is it not proper to balance an equation by changing the subscripts in the formulas of reactants or products?

56. *Balance the following equations.

   (a) $Zn + O_2 \longrightarrow ZnO$

   (b) $Cr + O_2 \longrightarrow Cr_2O_3$

   (c) $KClO_3 \longrightarrow KCl + O_2$

   (d) $Li + O_2 \longrightarrow Li_2O$

   (e) $B + F_2 \longrightarrow BF_3$

   (f) $C_3H_6 + O_2 \longrightarrow CO_2 + H_2O$

   (g) $Hg + N_2 \longrightarrow Hg_3N_2$

   (h) $N_2 + H_2 \longrightarrow NH_3$

   (i) $WO_3 + H_2 \longrightarrow W + H_2O$

   (j) $C_4H_{10} + O_2 \longrightarrow CO_2 + H_2O$

   (k) $Ca + NH_3 \longrightarrow CaH_2 + Ca_3N_2$

   (l) $P + S \longrightarrow P_2S_5$

   (m) $KNO_3 + C \longrightarrow K_2CO_3 + CO + N_2$

57. Balance the following equations.

   (a) $Ca + N_2 \longrightarrow Ca_3N_2$

   (b) $HgO \longrightarrow Hg + O_2$

   (c) $Fe + H_2O \longrightarrow Fe_3O_4 + H_2$

   (d) $N_2 + O_2 \longrightarrow N_2O_4$

   (e) $Pb_3O_4 \longrightarrow Pb + O_2$

   (f) $P + H_2 \longrightarrow PH_3$

   (g) $NH_3 + O_2 \longrightarrow NO + H_2O$

   (h) $Ba + O_2 \longrightarrow BaO$

   (i) $CuS + O_2 \longrightarrow CuO + SO_2$

(j)  $CaCO_3 + C \longrightarrow CaC_2 + CO_2$

(k)  $CaCN_2 + H_2O \longrightarrow CaCO_3 + NH_3$

(l)  $FeCr_2O_4 + C \longrightarrow Cr + Fe + CO$

(m) $MgO + Al \longrightarrow Mg + Al_2O_3$

58. Explain the meaning of the following when used in a chemical equation.

(a) (s)   (b) ($\ell$)   (c) (g)   (d) (aq)

## Questions to Ponder

59. If you had Avogadro's number of pennies and divided them among all of the 6 billion people on earth, how many dollars would each person have?

60. Use the following information to estimate the value of Avogadro's number. Explain any assumptions that you make in the estimation. The typical density of aluminum metal is 2.7 g/cm$^3$, the molar mass of aluminum is 27 g/mol Al and the volume of one atom of aluminum is $1.2 \times 10^{-23}$ cm$^3$.

61. Suppose you had a square piece of aluminum foil having a mass of 2.7 g. How many times would the foil need to be cut in half so that one atom of aluminum remains?

# CHEMICAL
# STOICHIOMETRY

**MOLES OF COMPOUNDS REVISITED**

A cup of water has a mass of about 237 g. How many molecules are in a cup of water? Since molecules are so small, we can assume that there is a huge number of molecules in a cup of water. We can't count the molecules one by one but we know that each molecule has mass and contributes to the total mass of the sample. In chemistry it is easy to weigh samples of chemicals but when we want to refer to the number of atoms or number of molecules in a sample of a chemical, **the mole is most often used.** As was the case with molar masses of elements, the molar mass of a compound is used as a unit factor to relate the mass of a compound to the number of moles of formula units it contains. The calculation approach is the same. That is, if we know the mass of a sample, the molar mass serves as a property factor to find the number of moles. The molar mass of a compound is deduced from its formula as discussed in Section 4-9. A molar mass is used as a factor in the following ways:

**MOLE**

*The amount of a chemical that contains Avogadro's number of chemical species. The species may be atoms, molecules, ions, or formula units.*

1. To find the number of moles in a measured mass of a chemical.
2. To find the mass of a given number of moles of a chemical.

---

**EXAMPLE 5-1**

How many moles of water, $H_2O$, are in a 237-g sample?

To relate mass to moles we need the molar mass of water. Use the periodic table to find the necessary molar masses of the elements. They are rounded off to convenient numbers of digits.

$$16.00 + 2(1.008) = 18.02 \quad \text{or} \quad 18.02 \text{ g/mol } H_2O$$

To find the number of moles in a given mass, we multiply the sample mass by the inverted molar mass.

$$237\,\cancel{g} \times \frac{1 \text{ mol } H_2O}{18.02\,\cancel{g}} = 13.2 \text{ mol } H_2O$$

Remember that a mole is Avogadro's number of particles. The sample has 13.2 times Avogadro's number of water molecules. A cup of water has 7,920,000,000,000,000,000,000,000 water molecules (7.92 million-billion-billion molecules). It is obviously more convenient to refer to the amount of water as 13.2 moles of $H_2O$.

### EXAMPLE 5-2

A sugar cube contains 5.15 g of sucrose, $C_{12}H_{22}O_{11}$. How many moles of sucrose are in this sample?

To relate mass to moles we need the molar mass of sucrose. From the formula the molar mass is

$$12(12.01) + 22(1.008) + 11(16.00) = 342.30 \quad \text{or} \quad 342.3 \text{ g/mol } C_{12}H_{22}O_{11}$$

To find the moles multiply the mass by the inverted molar mass

$$X \text{ mol } C_{12}H_{22}O_{11} = 5.15\,\cancel{g} \times \frac{1 \text{ mol } C_{12}H_{22}O_{11}}{342.3\,\cancel{g}} = 1.50 \times 10^{-2} \text{ mol } C_{12}H_{22}O_{11}$$

**ACTIVITY 5-1**

How many moles of ammonia, $NH_3$, are in a 525-g sample? How many moles of $NaHCO_3$ are in a 26.6-g sample?

Since the molar mass of a compound functions as a property factor, it is also used to find the mass of a given number of moles. Like any conversion factor the molar mass is used so that the unwanted units cancel and the desired unit remains.

### EXAMPLE 5-3

What is the mass in grams of 0.555 mole of $H_2O$? To find the mass in a given number of moles of a compound, multiply by the molar mass. See Example 5-1 for the molar mass of water.

$$X \text{ g } H_2O = 0.555 \cancel{\text{ mol } H_2O} \times \frac{18.02 \text{ g}}{1 \cancel{\text{ mol } H_2O}} = 10.0 \text{ g}$$

**EXAMPLE 5-4**

What is the mass in grams of 0.10 mole of NaCl?

To relate moles to mass we need the molar mass of NaCl

$$22.99 + 35.45 = 58.44 \quad \text{or} \quad 58.44 \text{ g/mol NaCl}$$

To find the mass in grams multiply the given number of moles by the molar mass.

$$X \text{ g NaCl} = 0.10 \text{ mol NaCl} \times \frac{58.44 \text{ g}}{1 \text{ mol NaCl}} = 5.8 \text{ g}$$

ACTIVITY 5-2

What is the mass of 202 moles of ammonia, $NH_3$? What is the mass of 0.049 mole of $NaHCO_3$?

## 5-2 CONSERVATION OF MASS IN CHEMICAL REACTIONS

Imagine what happens when a candle burns. The candle contains a variety of hydrocarbon compounds. Hydrocarbons are compounds containing only hydrogen and carbon. Combustion of these compounds involves their vigorous reaction with oxygen in the air to form gaseous carbon dioxide and water. Even though release of energy as heat and light accompanies this reaction, the mass is the same before and after the reaction. Suppose we place a candle in a container of air and weigh the container. Then we light the candle, seal the container, and let the candle burn until it goes out. When we reweigh the container we'll find that no detectable mass is lost or gained in the reaction (see Fig. 5-1.) Experimental measurements done on a variety of reactions reveal that mass is conserved in chemical reactions. Conservation of mass means that no detectable mass is lost or gained in a reaction.

The fact that mass is conserved in a chemical reaction is an expression of the law of conservation of matter. What significance does the law have when we work with equations? Mass is conserved in a reaction because the atoms are conserved. That is, as you know, the number of combined

**FIGURE 5-1**  An experimental confirmation of the conservation of mass.

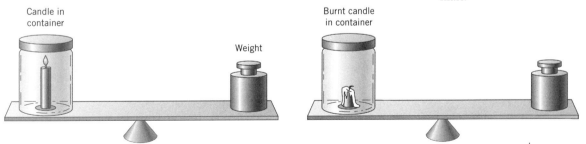

Candle in container

Weight

Burnt candle in container

**5-3**

Write the balanced equation for the reaction between hydrogen gas and oxygen gas to give water. To confirm the mass balance in the reaction, add the molar masses of the reactants multiplied by their coefficients in the balanced equation and compare to the molar mass of the product multiplied by its coefficient from the equation.

atoms in the products must be the same as the number of combined atoms in the reactants. Of course, the balanced equation for a chemical reaction reflects the conservation of atoms. Since atoms are not destroyed or created in a chemical reaction, it is reasonable that the total mass of the chemicals that react must equal the total mass of the products formed. This fact suggests another meaning for a balanced equation. The mass of the products formed in a reaction balances the mass of reactants used.

## 5-3 CHEMICAL STOICHIOMETRY

A chemical equation contains a wealth of information. An equation gives the formulas of the reactants and products, and the coefficients reveal the relative proportions of the reactants and products. Using an equation as a guide we can deduce the mass relations between individual reactants and products. For instance we can use an equation to predict amounts of products formed from specified amounts of reactants. Calculations involving mass relations are **stoichiometric** (pronounced stoy-key-oh-met'-rik) calculations. Stoichiometric is an interesting word. The -metric part refers to the measuring of masses. The term stoichio- refers to elements or parts of compounds; thus stoichiometric, taking measure of the masses of elements or compounds in chemical reactions. In other words, stoichiometric refers to how masses of elements and compounds involved in a reaction relate to one another. The methods of stoichiometric calculations are called **stoichiometry.**

Ammonia, $NH_3$, is an important industrial and agricultural chemical. Its main uses are in the manufacture of fertilizers, plastics, and explosives. The industrial synthesis of ammonia is accomplished by a process in which diatomic molecules of hydrogen and nitrogen combine.

$$3H_2 + N_2 \longrightarrow 2NH_3$$

This process is named the Haber process for Fritz Haber, a German chemist who developed it in the early 1900s. Air is the source of nitrogen used in the Haber process. The hydrogen typically comes from the reaction of methane and water when they are heated.

$$CH_4 + 2H_2O \longrightarrow 4H_2 + CO_2$$

The raw materials used in the industrial synthesis of ammonia are air, natural gas, and purified water.

Typically, for a reaction to take place it is necessary to mix the reactants, and adjust the reaction conditions in terms of temperature and sometimes pressure. Some reactions require the addition of a catalyst. See Section 11-4. For instance, finely divided iron is used as a catalyst in the Haber process. Some possible stoichiometric questions relating to the equation for formation of ammonia are shown on the next page.

1. How many grams of ammonia can be produced from specific masses of the reactants?
2. What mass of hydrogen or nitrogen are needed to produce a specific mass of ammonia?
3. What mass of hydrogen is needed to react with a specific mass of nitrogen?

Similar questions apply to any reaction. The names of chemicals given in the example problems in this chapter are for your reference. You are not expected to memorize chemical names at this time. The naming of chemicals is a topic covered in Chapter 8. However, you should become familiar with the names and formulas of a few very common chemicals, for example, ammonia, $NH_3$, methane, $CH_4$, and carbon dioxide $CO_2$.

## 5-4  MOLAR INTERPRETATION OF AN EQUATION

To represent a chemical reaction, we write an equation. When writing an equation be sure to use the proper formulas for all chemicals involved in the reaction. Once a balanced equation is written, the equation is useful for stoichiometric calculations. For example, the equation for the combustion of methane is

$$1CH_4 + 2O_2 \longrightarrow 1CO_2 + 2H_2O$$

Coefficients of 1 are shown for emphasis. They are normally not written in an equation. How can we interpret such an equation? We could say that the equation shows that one molecule of methane reacts with two molecules of oxygen to produce one molecule of carbon dioxide and two molecules of water. Of course, if this is true for a few molecules, it is also true for any number, even Avogadro's number.

1 molecule $CH_4$ + 2 molecules $O_2 \longrightarrow$ 1 molecule $CO_2$ + 2 molecules $H_2O$

1 dozen $CH_4$ + 2 dozen $O_2 \longrightarrow$ 1 dozen $CO_2$ + 2 dozen $H_2O$

1 gross $CH_4$ + 2 gross $O_2 \longrightarrow$ 1 gross $CO_2$ + 2 gross $H_2O$

1 million $CH_4$ + 2 million $O_2 \longrightarrow$ 1 million $CO_2$ + 2 million $H_2O$

1 billion $CH_4$ + 2 billion $O_2 \longrightarrow$ 1 billion $CO_2$ + 2 billion $H_2O$

$1(6.022 \times 10^{23})$ $CH_4$ + $2(6.022 \times 10^{23})$ $O_2 \longrightarrow 1(6.022 \times 10^{23})$ $CO_2$ + $2(6.022 \times 10^{23})$ $H_2O$

1 mole $CH_4$ + 2 moles $O_2 \longrightarrow$ 1 mole $CO_2$ + 2 moles $H_2O$

For stoichiometric purposes, it is best to interpret an equation on the basis of Avogadro's number of particles or moles of chemicals. Such an interpretation is called the **molar interpretation** (see Fig. 5-2).

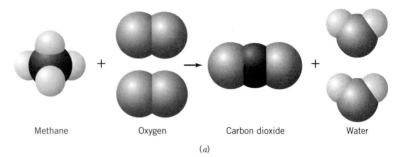

Methane          Oxygen          Carbon dioxide          Water

(a)

In this reaction one methane molecule reacts with two oxygen molecules to give one carbon dioxide molecule and two water molecules.

To interpret this reaction from a molar point of view it is assumed that the reaction that occurs with individual molecules will be the same if Avogadro's number of molecules are involved. The same reaction from a molar point of view is

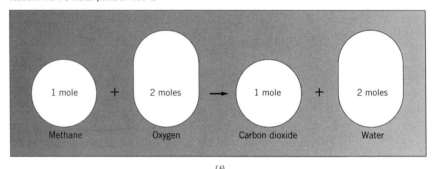

1 mole   +   2 moles   →   1 mole   +   2 moles

Methane          Oxygen          Carbon dioxide          Water

(b)

**FIGURE 5-2** The molar interpretation of an equation.

One mole of methane reacts with two moles of oxygen to give one mole of carbon dioxide and two moles of water.

From a molar point of view, this equation shows that 1 mole of methane reacts with 2 moles of oxygen to produce 1 mole of carbon dioxide and 2 moles of water. This means that the relative molar amounts of reactants and products must correspond to the coefficients. Thus, it follows that 0.5 mole of methane reacts with 1 mole of oxygen, or 2 moles of methane react with 4 moles of oxygen, or any other amounts that have a 1:2 molar ratio. The coefficients reveal the molar ratios relating reactants and products. It is very important to understand this molar interpretation because such a view is essential to stoichiometric calculations.

### 5-5 MOLAR RATIOS

In an equation, the coefficient in front of each formula shows the relative number of moles of each chemical involved in the reaction. In the absence of a coefficient, of course, a 1 is understood. It is possible to express the relation between any two chemicals involved in a reaction as a molar ratio. A **molar ratio** expresses the relation between chemicals as a proportion. For the reaction of methane and oxygen to give carbon dioxide and water, molar ratios relating the various reactants and products are

$$\frac{1 \text{ mol } CH_4}{2 \text{ mol } O_2} \qquad \frac{1 \text{ mol } CH_4}{1 \text{ mol } CO_2} \qquad \frac{1 \text{ mol } CH_4}{2 \text{ mol } H_2O}$$

$$\frac{2 \text{ mol } O_2}{1 \text{ mol } CO_2} \qquad \frac{2 \text{ mol } O_2}{2 \text{ mol } H_2O} \qquad \frac{1 \text{ mol } CO_2}{2 \text{ mol } H_2O}$$

Any of these ratios can be expressed in the inverse or reciprocal form. For instance, the molar ratio relating methane and oxygen can be written as

$$\frac{1 \text{ mol } CH_4}{2 \text{ mol } O_2} \qquad or \qquad \frac{2 \text{ mol } O_2}{1 \text{ mol } CH_4}$$

---

**EXAMPLE 5-5**

Give the molar ratios for the following reaction:

$$3H_2 + N_2 \longrightarrow 2NH_3$$

There are six possible ratios. Why? We can start with the ratios relating the reactants.

$$\frac{3 \text{ mol } H_2}{1 \text{ mol } N_2} \qquad and \qquad \frac{1 \text{ mol } N_2}{3 \text{ mol } H_2}$$

Next, we can write the ratios relating the reactants and product.

$$\frac{3 \text{ mol } H_2}{2 \text{ mol } NH_3} \qquad and \qquad \frac{2 \text{ mol } NH_3}{3 \text{ mol } H_2}$$

$$\frac{1 \text{ mol } N_2}{2 \text{ mol } NH_3} \qquad and \qquad \frac{2 \text{ mol } NH_3}{1 \text{ mol } N_2}$$

---

## 5-6 MOLAR RATIOS AS UNIT FACTORS

Molar ratios come from the coefficients in the balanced equation. They apply only to the specific reaction and are used as conversion factors to relate the number of moles of one chemical to the number of moles of another chemical. As an example, suppose we want to know how many moles of oxygen gas react with 426 moles of methane gas if the reaction is

$$CH_4 + 2O_2 \longrightarrow CO_2 + 2H_2O$$

We know from the coefficients of the equation that these chemicals react in a 2:1 molar ratio. We are given the number of moles of methane and want to find the corresponding number of moles of oxygen gas. Thus, to change to moles of oxygen, we need to multiply the given number of moles of methane by a molar ratio having moles of $O_2$ in the numerator and moles of $CH_4$ in the denominator.

$$426 \ \text{mol CH}_4 \ \times \ \frac{2 \ \text{mol O}_2}{1 \ \text{mol CH}_4} \ = \ 852 \ \text{mol O}_2$$

Molar ratios are used as **mole-to-mole factors** in stoichiometric calculations. Molar ratios serve as a bridge between the number of moles of any two chemicals involved in a reaction.

**EXAMPLE 5-6**

How many moles of oxygen gas can react with 3.27 moles of methane gas? Use the equation for the combustion of methane.

The question can be summarized as

$$CH_4 \qquad + \qquad 2O_2 \longrightarrow CO_2 + 2H_2O$$
$$3.27 \ \text{mol CH}_4 \qquad ? \ \text{mol O}_2$$

In this summary we write the given number of moles below $CH_4$ and indicate that we want to find the moles of $O_2$ using a ? mol under its formula. The appropriate molar ratio relating methane and oxygen comes from the balanced equation.

$$X \ \text{mol O}_2 \ = \ 3.27 \ \text{mol CH}_4 \ \times \ \frac{2 \ \text{mol O}_2}{1 \ \text{mol CH}_4} \ = \ 6.54 \ \text{mol O}_2$$

As a double check we note that with the appropriate molar ratio, the moles of methane cancel to give moles of oxygen.

**EXAMPLE 5-7**

How many moles of ammonia can form when 153 moles of hydrogen gas react with sufficient nitrogen gas? The reaction is

$$N_2 + 3H_2 \qquad \longrightarrow \qquad 2NH_3$$
$$153 \ \text{mol H}_2 \qquad ? \ \text{mol NH}_3$$

To find the moles of one chemical corresponding to the moles of another chemical, we use the appropriate molar ratio from the balanced equation.

$$X \ \text{mol NH}_3 \ = \ 153 \ \text{mol H}_2 \ \times \ \frac{2 \ \text{mol NH}_3}{3 \ \text{mol H}_2} \ = \ 102 \ \text{mol NH}_3$$

**ACTIVITY 5-4**

Find how many moles of hydrogen gas are formed from 895 moles of methane, $CH_4$, if the equation for the reaction is

$$CH_4 + 2H_2O \longrightarrow 4H_2 + CO_2$$

As we shall see, the molar ratios from a chemical equation serve as the basis for many types of stoichiometric calculations. Consequently, you need to have a correctly balanced equation to do calculations.

## 5-7 STOICHIOMETRIC MASS-TO-MOLE CALCULATIONS

If we know the formula of a chemical involved in a reaction, we can easily determine the molar mass of that chemical. The molar mass is useful for calculating the number of moles contained in a known mass of the chemical or vice versa. This technique is important in **mass-to-mole stoichiometric calculations.** See Figure 5-3 for an illustration.

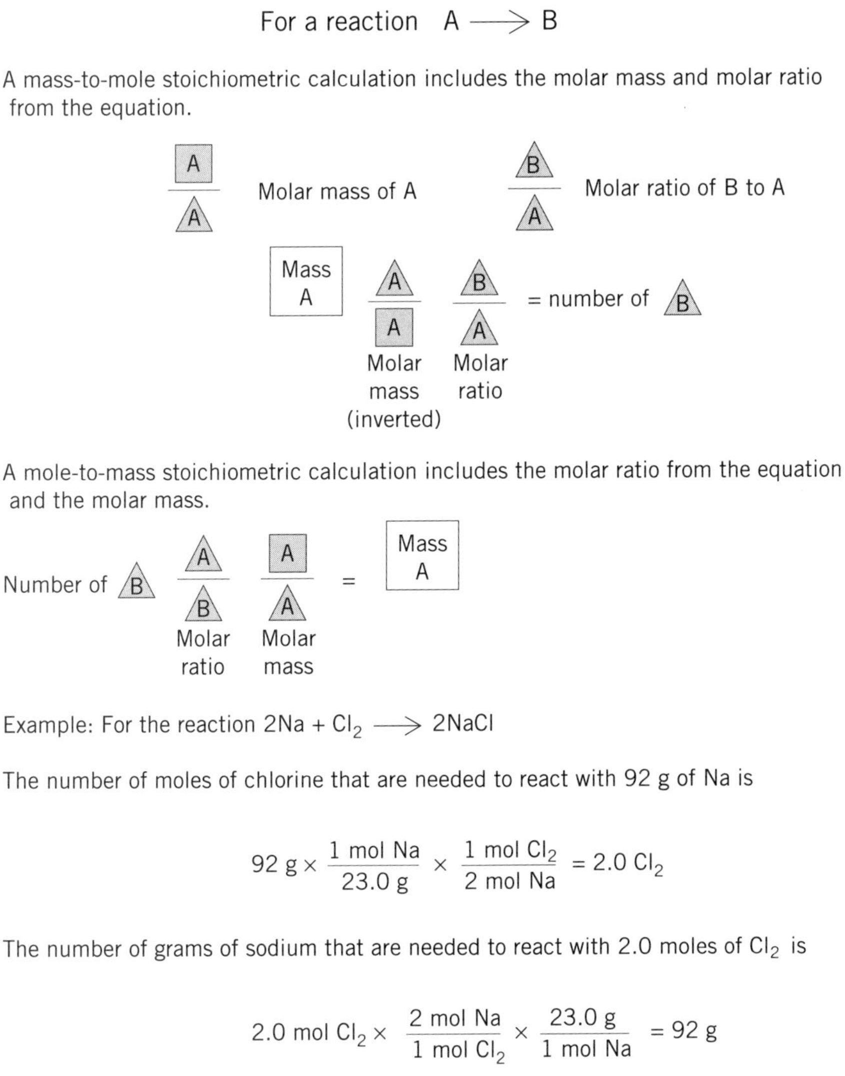

For a reaction   A $\longrightarrow$ B

A mass-to-mole stoichiometric calculation includes the molar mass and molar ratio from the equation.

$\dfrac{A}{A}$   Molar mass of A          $\dfrac{B}{A}$   Molar ratio of B to A

$\boxed{\text{Mass A}}$ $\dfrac{A}{A}$ $\dfrac{B}{A}$ = number of B

Molar    Molar
mass    ratio
(inverted)

A mole-to-mass stoichiometric calculation includes the molar ratio from the equation and the molar mass.

Number of B   $\dfrac{A}{B}$ $\dfrac{A}{A}$ = $\boxed{\text{Mass A}}$

Molar    Molar
ratio    mass

Example: For the reaction $2Na + Cl_2 \longrightarrow 2NaCl$

The number of moles of chlorine that are needed to react with 92 g of Na is

$$92 \text{ g} \times \frac{1 \text{ mol Na}}{23.0 \text{ g}} \times \frac{1 \text{ mol Cl}_2}{2 \text{ mol Na}} = 2.0 \text{ Cl}_2$$

The number of grams of sodium that are needed to react with 2.0 moles of $Cl_2$ is

$$2.0 \text{ mol Cl}_2 \times \frac{2 \text{ mol Na}}{1 \text{ mol Cl}_2} \times \frac{23.0 \text{ g}}{1 \text{ mol Na}} = 92 \text{ g}$$

**FIGURE 5-3** The conversion of mass to mole and mole to mass.

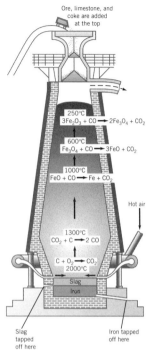

Ore, limestone, and
coke are added
at the top

250°C
$3Fe_2O_3 + CO \longrightarrow 2Fe_3O_4 + CO_2$

600°C
$Fe_3O_4 + CO \longrightarrow 3FeO + CO_2$

1000°C
$FeO + CO \longrightarrow Fe + CO_2$

Hot air

1300°C
$CO_2 + C \longrightarrow 2 CO$

$C + O_2 \longrightarrow CO_2$
2000°C

Slag

Iron

Slag
tapped
off here

Iron tapped
off here

A blast furnace used in refining iron

Consider an example of a mass-to-mole type of stoichiometric calculation. Iron ore is the industrial source of iron metal. The overall reaction in the refining of iron includes iron(III) oxide from iron ore reacting with carbon monoxide to give iron metal and carbon dioxide.

$$Fe_2O_3 + 3CO \longrightarrow 2Fe + 3CO_2$$

How many moles of iron form when 385 g of carbon monoxide react with sufficient iron(III) oxide? First, we decide what is wanted (the unknown). This question asks for the amount of iron in moles. Next, we note what is given (the knowns). In this question it is given that 385 g of carbon monoxide react. This information can be listed under the equation along with any molar masses that may be needed.

$$Fe_2O_3 + 3CO \longrightarrow 2Fe + 3CO_2$$

| | |
|---|---|
| mass | 385g |
| molar mass | 28.01 g/mol |
| moles | mol CO $\Longrightarrow$ ? mol Fe |

The top row is for mass; the bottom row is for moles. Since molar masses relate masses and moles, the middle row is for any molar masses. We enter the unknown as a "?" including its units. We fill the table with any given data, placing each item below the formula of the appropriate reactant or product in the equation. Molar masses can be found from the formulas of the reactants and products. Chemicals in reactions are connected only through moles, so in the moles row we draw an arrow connecting carbon monoxide to iron; this will emphasize the connection.

It is not possible to determine the number of moles of iron directly from the mass of carbon monoxide. Moles of chemicals relate through molar ratios, so we must work in moles. In other words, the equation gives the molar relations and the molar ratios serve as the bridge between the amounts of the two chemicals. The calculation sequence is

$$\text{g CO} \underset{\text{mass}}{\overset{\text{molar}}{\Longrightarrow}} \text{mol CO} \underset{\text{ratio}}{\overset{\text{molar}}{\Longrightarrow}} \text{mol Fe}$$

We find the moles of carbon monoxide from the grams of carbon monoxide then, using the appropriate molar ratios, we use the moles of carbon monoxide to find the moles of iron.

First, calculate the molar mass of carbon monoxide.

$$12.011 + 15.999 = 28.010 \text{ or } 28.01$$

Then, use it as a factor to determine the number of moles of CO. As we are given grams of CO, not moles, we use the molar mass in the form in which grams cancel and moles remain. The number of moles of CO is

$$385\text{g} \times \frac{1 \text{ mol CO}}{28.01 \text{g}}$$

To calculate the moles of iron, multiply the moles of CO by the molar ratio from the balanced equation. Choose the ratio that cancels moles of CO and gives moles of Fe.

$$385 \, g \times \frac{1 \, mol \, CO}{28.01 \, g} \times \frac{2 \, mol \, Fe}{3 \, mol \, CO} = 9.16 \, mol \, Fe$$

When working a problem of this type, first decide on a calculation sequence then carry out the calculation step by step. We do not have to do any numerical calculations until the final setup of the calculation sequence is written. Of course, as side calculations, we need to determine any necessary molar masses. In the preceding example, we did not calculate the numerical number of moles of carbon monoxide in the first step; instead, we just multiplied this product by the molar ratio. Once we have set up the calculations, including all factors, a numerical value can be calculated and rounded off to the correct number of digits. As a double check, the units should cancel properly; this is why units are important.

If you want, you can calculate an answer for each step in the calculation sequence. If you do, it is important to carry more digits than needed in each part and round off to the correct number of digits in the final answer. Carrying more digits prevents incorrect answers, which can result when intermediate answers are rounded off. As an example of step-by-step calculation, let us use the same question as given above. How many moles of iron form when 385 g of carbon monoxide react with sufficient iron(III) oxide? First, the molar mass of carbon monoxide is used to find the number of moles of CO.

$$385 \, g \times \frac{1 \, mol \, CO}{28.01 \, g} = 13.745 \, mol \, CO$$

To calculate the moles of iron, the above result is multiplied by the molar ratio from the equation and then rounded to the correct number of significant digits.

$$13.745 \, mol \, CO \times \frac{2 \, mol \, Fe}{3 \, mol \, CO} = 9.16 \, mol \, Fe$$

**EXAMPLE 5-8**

How many moles of ammonia can form when 56.0 g of nitrogen react with excess hydrogen? The equation for the reaction is

$$N_2 + 3H_2 \longrightarrow 2NH_3$$

Any molar masses that are needed are determined from the formulas and the atomic weights. The question is summarized as given on the next page.

$$N_2 + 3H_2 \longrightarrow 2NH_3$$

mass     56.0 g

molar mass    28.01 g/mol

moles    mol $N_2$ $\Longrightarrow$ ? mol $NH_3$

To convert from mass to moles we need to work on a molar basis. So use the molar mass to find the number of moles of $N_2$.

$$56.0\,\cancel{g} \times \frac{1 \text{ mol } N_2}{28.01\,\cancel{g}}$$

To find the moles of ammonia multiply by the molar ratio from the equation. The ratio is chosen so that the moles of $N_2$ cancel to give the answer in moles of $NH_3$.

$$X \text{ mol } NH_3 = 56.0\,\cancel{g} \times \frac{1 \text{ } \cancel{\text{mol } N_2}}{28.01\,\cancel{g}} \times \frac{2 \text{ mol } NH_3}{1 \text{ } \cancel{\text{mol } N_2}} = 4.00 \text{ mol } NH_3$$

The reverse problem of calculating the number of grams of a reactant or product corresponding to a given number of moles of another reactant or product (mole-to-mass calculation) is accomplished in a manner similar to the above examples. The difference is that the calculation sequence is reversed.

**ACTIVITY 5-5**

Find how many moles of hydrogen gas are formed from 825 grams of methane, $CH_4$, if the equation for the reaction is

$$CH_4 + 2H_2O \longrightarrow 4H_2 + CO_2$$

As a second part of this question calculate the number of grams of $CO_2$ produced when 51.4 moles of $CH_4$ react.

**EXAMPLE 5-9**

Ethane gas, $C_2H_6$, reacts with oxygen to give carbon dioxide and water. How many grams of ethane react with 14 moles of oxygen gas? Placing the information beneath the balanced equation for the reaction gives:

$$2C_2H_6 \quad + \quad 7O_2 \longrightarrow 4CO_2 + 6H_2O$$

mass    ? g

molar mass    30.1 g/mol

moles    ? mol $C_2H_6$ $\Longleftarrow$ 14 mol $O_2$

Since we are given moles and want grams, use the appropriate molar ratio and the molar mass of ethane ($C_2H_6$: $2(12.01) + 6(1.008) = 30.1$)

The molar ratio is used to change moles of $O_2$ to moles of $C_2H_6$, and, finally, the molar mass of $C_2H_6$ is used to cancel moles and gives the answer in grams.

$$X \text{ g } C_2H_6 = 14 \text{ } \cancel{\text{mol } O_2} \times \frac{2 \text{ } \cancel{\text{mol } C_2H_6}}{7 \text{ } \cancel{\text{mol } O_2}} \times \frac{30.1 \text{ g}}{1 \text{ } \cancel{\text{mol } C_2H_6}} = 120 \text{ g or } 1.2 \times 10^2 \text{ g}$$

## 5-8  STOICHIOMETRIC MASS-TO-MASS CALCULATIONS

Since we often work with masses of reactants and products in the laboratory or even on an industrial scale, the three most common stoichiometric questions are

1. How many grams of a product can be formed from a given number of grams of a reactant?
2. How many grams of a reactant are needed to form a given number of grams of a product?
3. How many grams of a reactant are needed to react with a given number of grams of another reactant?

**Mass-to-mass calculations** are stoichiometric calculations that provide solutions to these questions. Such calculations are simply combinations of mass-to-mole and mole-to-mole calculations. Mass-to-mass calculations are very important in chemistry because we usually work with masses of materials. Since an equation relates chemicals on a molar basis, we need to change mass to the number of moles.

Bauxite ore is the major source of natural aluminum oxide. Aluminum metal is made by dissolving aluminum oxide in a vat of molten cryolite. Then carbon rods are immersed into the vat and electrical current is passed through the melt. The electrical energy encourages a reaction in which the aluminum oxide combines with carbon to form aluminum metal and carbon dioxide gas. (See Section 13-11 for a further discussion of the process.) The equation for the overall reaction is

$$2Al_2O_3 + 3C \xrightarrow{\text{electrical energy}} 4Al + 3CO_2$$

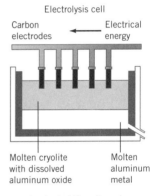

Electrolysis cell

Carbon electrodes ← Electrical energy

Molten cryolite with dissolved aluminum oxide

Molten aluminum metal

An electrolytic cell used to obtain aluminum from alumium ore.

As an example of a mass-to-mass question suppose we have a 204-g sample of aluminum oxide. How many grams of aluminum metal can be made from this mass? The information is summarized as

|  | $2Al_2O_3$ | $+ 3C \longrightarrow$ | $4Al$ | $+ 3CO_2$ |
|---|---|---|---|---|
| mass | 204 g |  | ? g |  |
| molar mass | 102.0 g/mol |  | 26.98 g/mol |  |
| moles | mol $Al_2O_3$ | $\Longrightarrow$ | ? mol Al |  |

It is not possible to convert directly from the mass of one chemical to the mass of another without working on a molar basis. Thus, the necessary calculation sequence is

$$\text{g } Al_2O_3 \xstackrel{\substack{\text{molar} \\ \text{mass}}}{\Longrightarrow} \text{mol } Al_2O_3 \xstackrel{\substack{\text{molar} \\ \text{ratio}}}{\Longrightarrow} \text{mol } Al \xstackrel{\substack{\text{molar} \\ \text{mass}}}{\Longrightarrow} \text{g } Al$$

According to this sequence, we need two molar masses and a molar ratio from the equation. The calculation involves multiplying the grams of aluminum oxide by the three appropriate factors. First, convert the grams of aluminum oxide to moles using its molar mass $(2(26.98) + 3(16.00) = 101.96$ or $102.0)$.

$$204 \text{ g} \times \frac{1 \text{ mol Al}_2\text{O}_3}{102.0 \text{ g}}$$

Multiplying by the molar ratio, from the equation, gives the corresponding number of moles of aluminum.

$$204 \text{ g} \times \frac{1 \text{ mol Al}_2\text{O}_3}{102.0 \text{ g}} \times \frac{4 \text{ mol Al}}{2 \text{ mol Al}_2\text{O}_3}$$

Finally, the molar mass of aluminum is used to find the mass of aluminum.

$$204 \text{ g} \times \frac{1 \text{ mol Al}_2\text{O}_3}{102.0 \text{ g}} \times \frac{4 \text{ mol Al}}{2 \text{ mol Al}_2\text{O}_3} \times \frac{26.98 \text{ g}}{1 \text{ mol Al}} = 108 \text{ g}$$

No numerical calculations need be done until the final setup is written. Check the setup by making sure the various units cancel and the answer has the desired unit. Note that in the final setup, the grams of aluminum oxide cancel, the moles of aluminum oxide cancel, and the moles of aluminum cancel, leaving grams of aluminum.

It is possible to calculate an intermediate answer for each step and use the answer for the next step. But remember to include enough digits and don't round off until the final answer. The number of moles of aluminum oxide is found in the first step.

$$204 \text{ g} \times \frac{1 \text{ mol Al}_2\text{O}_3}{102.0 \text{ g}} = 2.000 \text{ mol Al}_2\text{O}_3$$

The number of moles of aluminum is found in the second step.

$$2.000 \text{ mol Al}_2\text{O}_3 \times \frac{4 \text{ mol Al}}{2 \text{ mol Al}_2\text{O}_3} = 4.000 \text{ mol Al}$$

The mass of aluminum is found in the last step.

$$4.000 \text{ mol Al} \times \frac{26.98 \text{ g}}{1 \text{ mol Al}} = 108 \text{ g}$$

This example illustrates the general approach to a mass-to-mass stoichiometric calculation. We can recognize these problems by the context. If we are given the mass of one chemical and are asked to find the corresponding mass of a second chemical, this is a mass-to-mass question. Problems like this always involve multiplying the given mass by three factors: the molar mass of the given chemical, the molar ratio from the equation, and the molar mass of the second chemical. See Figure 5-4. A summary allows us

For a reaction   A ——→ B

A mass-to-mass stoichiometric calculation includes the molar masses and molar ratio from the equation.

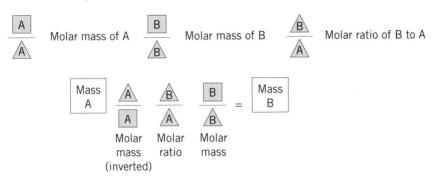

Example: For the reaction $2Na + Cl_2 \longrightarrow 2NaCl$

The number of grams of chlorine that are needed to react with 42 g of Na is

$$42 \text{ g} \times \frac{1 \text{ mol Na}}{23.0 \text{ g}} \times \frac{1 \text{ mol Cl}_2}{2 \text{ mol Na}} \times \frac{70.9 \text{ g}}{1 \text{ mol Cl}_2} = 65 \text{ g}$$

**FIGURE 5-4**   The conversion of mass to mass.

to visualize what is given and what is wanted. The summary also suggests a calculation path for the solution to the problem.

---

**EXAMPLE 5-10**

The element silicon can be made by the following reaction:

$$SiCl_4 + 2Mg \longrightarrow 2MgCl_2 + Si$$

How many grams of magnesium are needed to react completely with $6.25 \times 10^3$ grams of $SiCl_4$? The information is summarized as

|  | $SiCl_4$ | + | $2Mg \longrightarrow 2MgCl_2 + Si$ |
|---|---|---|---|
| mass | $6.25 \times 10^3$ g | | ? g |
| molar mass | 169.9 g/mol | | 24.31 g/mol |
| moles | mol $SiCl_4$ | $\Longrightarrow$ | ? mol Mg |

This is a mass-to-mass calculation. First, find the molar mass of $SiCl_4$ (28.09 + 4(35.45) = 169.89 or 169.9). Then use it to find the moles of $SiCl_4$.

$$6.25 \times 10^3 \text{ g} \times \frac{1 \text{ mol SiCl}_4}{169.9 \text{ g}}$$

**ACTIVITY 5-6**

Find how many grams of hydrogen gas can be produced from 825 grams of methane if the equation for the reaction is

$$CH_4 + 2H_2O \longrightarrow 4H_2 + CO_2$$

Next, use the proper molar ratio to find the corresponding number of moles of Mg.

$$6.25 \times 10^3 \, \cancel{g} \times \frac{1 \, \cancel{mol \, SiCl_4}}{169.9 \, \cancel{g}} \times \frac{2 \, mol \, Mg}{1 \, \cancel{mol \, SiCl_4}}$$

Finally, the mass of magnesium is found using its molar mass.

$$X \, g \, Mg = 6.25 \times 10^3 \, \cancel{g} \times \frac{1 \, \cancel{mol \, SiCl_4}}{169.9 \, \cancel{g}} \times \frac{2 \, \cancel{mol \, Mg}}{1 \, \cancel{mol \, SiCl_4}} \times \frac{24.31 \, g}{1 \, \cancel{mol \, Mg}}$$

$$= 1.79 \times 10^3 \, g$$

We have considered three common stoichiometric methods. A stoichiometry map, as described in Focus 5, summarizes these methods. Notice, that the solution to any stoichiometry problem is given by the map. The most convenient and useful way to go from the mass of one chemical to the mass of another is to use moles and molar ratios. The molar ratio is the bridge or connecting point of any stoichiometric calculation.

**FOCUS 5**

## *Stoichiometry Map*

The stoichiometry map applies to any balanced equation. The letters A and B refer to any chemicals. The factors shown are needed to move from one part of the map to another part. To find the mass of B related to the mass of A requires the use of the molar mass of A, the molar ratio from the equation and the molar mass of B.

Example: Hydrogen and oxygen gas react to give water, $2H_2 + O_2 \rightarrow 2H_2O$ The number of grams of hydrogen needed to react with 64 g of oxygen is

$$64 \, g \times \underbrace{\frac{1 \, mol \, O_2}{32.0 g}}_{\substack{molar \\ mass}} \times \underbrace{\frac{2 \, mol \, H_2}{1 \, mol \, O_2}}_{\substack{molar \\ ratio}} \times \underbrace{\frac{2.02 \, g}{1 \, mol \, H_2}}_{\substack{molar \\ mass}} = 8.1 \, g$$

The enthalpy change of a reaction is used to express appropriate heat ratios which are used to relate amounts of heat to the number of moles or mass of any species involved in a reaction.

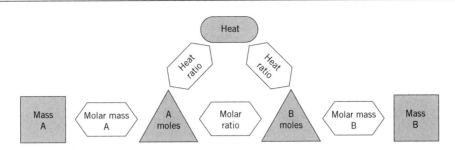

Example: For the reaction    $2H_2 + O_2 \rightarrow 2H_2O$

the enthalpy change is    $\Delta H = -121$ kJ/mol $H_2O$

The number of grams of oxygen that is needed to react with sufficient hydrogen to give 484 kJ of heat is found using the appropriate heat ratio obtained from the enthalpy change

$$\frac{121\,kJ}{1\,mol\,H_2O} \times \frac{2\,mol\,H_2O}{1\,mol\,O_2} = \frac{242\,kJ}{1\,mol\,O_2}$$

Following the stoichiometry map from heat to the mass of oxygen gives

$$484\,kJ \times \underbrace{\frac{1\,mol\,O_2}{242\,kJ}}_{\substack{heat \\ ratio}} \times \underbrace{\frac{32.00\,g}{1\,mol\,O_2}}_{\substack{molar \\ mass}} = 64.0\,g$$

## 5-9 LIMITING REACTANT

Gasoline is a complex mixture of organic chemicals, most of which are hydrocarbons. The chemical reactions that occur in an automobile engine involve these chemicals reacting with oxygen from the air. As long as there is an adequate supply of air, an engine runs until all the gasoline is used. The amount of gasoline is the limiting factor. On the other hand, if the air supply is cut off, the engine stops as soon as the supply of oxygen is depleted. In this case, the amount of oxygen limits the reaction. Chemistry in Action 5-1 is about the reaction between vinegar and baking soda; it investigates what happens with different amounts of reactants.

When reactants are mixed in reactions used in the laboratory or an industrial process the more expensive or least available reactant is normally the limiting factor. It is usually not convenient to measure out exact amounts of reactants so that they are present in the correct molar proportions. More commonly, reactants are mixed and the reaction occurs until one of the reactants is depleted. One of the reactants limits the reaction leaving an unreacted portion of the other reactant. The amount of the reactant that limits the reaction dictates the amount of product formed. We call this reactant the **limiting reactant.** The other reactant is in excess. Some of it is used but an unused portion remains after the reaction has taken place. As an analogy, suppose we are making cheese sandwiches and we have 10 pieces of cheese and 24 slices of bread. Since the "formula" of a sandwich is "cheese(bread)$_2$," we could make 10 sandwiches and have 4 slices of bread left over. The cheese is the limiting component. See Figure 5-5.

As a chemical example, suppose we mix 56.0 g of nitrogen gas and 12.0 g of hydrogen gas to make ammonia

$$N_2 + 3H_2 \longrightarrow 2NH_3$$

Which of the reactants is the limiting reactant? To find out, we calculate the number of grams of product that is potentially formed by the given amount of each reactant. The reactant that forms the lesser number of grams of product is the limiting reactant. Using the sandwich analogy from above we potentially could make 10 sandwiches from the cheese and 12 sandwiches from the bread, so the cheese must be the limiting component. Once all the cheese is used we can't make any more sandwiches and 4 bread slices remain unused. We use the normal mass-to-mass stoichiometric methods to find the potential number of grams of product from each reactant. In other words, we carry out two separate mass-to-mass calculations. In the example, we need the molar masses of N$_2$ (28.01), H$_2$ (2.016) and NH$_3$ (14.01 + 3(1.0079) = 17.03). First determine the potential mass of ammonia using the given mass of nitrogen.

**FIGURE 5-5**
Limiting reactant using bread and cheese.

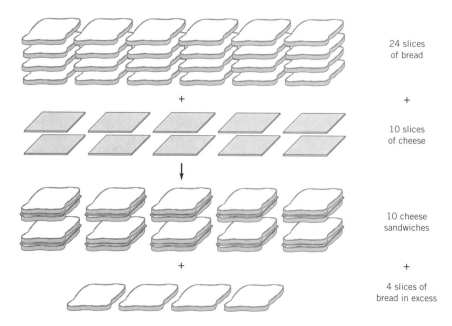

24 slices
of bread

+

10 slices
of cheese

10 cheese
sandwiches

+

4 slices of
bread in excess

$$X \text{ g NH}_3 = 56.0 \text{ g} \times \frac{1 \text{ mol N}_2}{28.01 \text{ g}} \times \frac{2 \text{ mol NH}_3}{1 \text{ mol N}_2} \times \frac{17.03 \text{ g}}{1 \text{ mol NH}_3}$$

$$= 68.1 \text{ g NH}_3$$

Next determine the potential grams of ammonia using the given mass of hydrogen.

$$X \text{ g NH}_3 = 12.0 \text{ g} \times \frac{1 \text{ mol H}_2}{2.016 \text{ g}} \times \frac{2 \text{ mol NH}_3}{3 \text{ mol H}_2} \times \frac{17.03 \text{ g}}{1 \text{ mol NH}_3}$$

$$= 67.6 \text{ g NH}_3$$

From the two answers, we conclude that hydrogen is the limiting reactant since it would give the least amount of product. Notice that it is very important to carry enough digits in these calculations so that roundoff error does not obscure the answer.

Since hydrogen is the limiting reactant the nitrogen is the reactant in excess. How many grams of nitrogen are used and how many grams remain unused? Since we know that the mass of hydrogen limits the reaction, we use it to find the mass of nitrogen that reacted. To find the mass of nitrogen corresponding to the mass of hydrogen requires another mass-to-mass calculation. The amount of nitrogen used is

$$X \text{ g N}_2 = 12.0 \text{ g} \times \frac{1 \text{ mol H}_2}{2.016 \text{ g}} \times \frac{1 \text{ mol N}_2}{3 \text{ mol H}_2} \times \frac{28.01 \text{ g}}{1 \text{ mol N}_2} = 55.6 \text{ g N}_2$$

To find the amount of nitrogen unused after the reaction has taken place we subtract the grams used from the grams present at the outset.

$$56.0 \text{ g} - 55.6 \text{ g} = 0.4 \text{ g N}_2$$

How do you recognize a limiting reactant question? Typically such a question gives the masses of two reactants and asks for the mass of product formed. When working limiting reactant problems, use the following approach:

1. Which reactant is the limiting reactant? Find the number of grams of product that are potentially formed by each reactant. The one that produces the least number of grams is the limiting reactant.
2. What mass of the reactant in excess was actually used and what mass remains? Use the number of grams of the limiting reactant to find the mass of the excess reactant used. Find the mass remaining by subtracting the mass used from the initial starting mass.

**EXAMPLE 5-11**

One step in the manufacture of methanol, $CH_3OH$, involves the reaction:

$$CO + 2H_2 \longrightarrow CH_3OH$$

How many grams of methanol can form when 72.8 g of CO and 10.6 g of $H_2$ react? How many grams of the reactant in excess are used and how many grams remain?

The first part of the problem is

$$
\begin{array}{lccl}
 & CO & + \quad 2H_2 \longrightarrow & CH_3OH \\
\text{mass} & 72.8\ g & 10.6\ g & ?\ g \\
\text{molar mass} & 28.01\ g/mol & 2.016\ g/mol & 32.04\ g/mol \\
\text{moles} & \text{mol CO} & \text{mol } H_2 \Longrightarrow & ?\ \text{mol } CH_3OH
\end{array}
$$

This summary suggests two possibilities. Potentially how much product forms from CO and how much from $H_2$? To find the limiting reactant, we calculate the potential number of grams of product obtained from each reactant using mass-to-mass calculations. We need the molar masses of CO (28.01), $H_2$ (2.016), and $CH_3OH$ (12.011 + 4(1.0079) + 15.999 = 32.04). The two mass-to-mass calculations are

$$X\ g\ CH_3OH = 72.8\ g \times \frac{1\ \text{mol CO}}{28.01\ g} \times \frac{1\ \text{mol } CH_3OH}{1\ \text{mol CO}} \times \frac{32.04\ g}{1\ \text{mol } CH_3OH}$$

$$= 83.3\ g$$

$$X\ g\ CH_3OH = 10.6\ g \times \frac{1\ \text{mol } H_2}{2.016\ g} \times \frac{1\ \text{mol } CH_3OH}{2\ \text{mol } H_2} \times \frac{32.04\ g}{1\ \text{mol } CH_3OH}$$

$$= 84.2\ g$$

Carbon monoxide is the limiting reactant since it gives the least number of grams of product.

In this reaction, hydrogen is in excess. For the second part of the question we use the number of grams of carbon monoxide to find the amount of hydrogen that reacted.

$$
\begin{array}{lccl}
 & CO & + \quad 2H_2 \longrightarrow CH_3OH \\
\text{mass} & 72.8\ g & ?\ g \\
\text{molar mass} & 28.01\ g/mol & 2.016\ g/mol \\
\text{moles} & \text{mol CO} & \Longrightarrow ?\ \text{mol } H_2
\end{array}
$$

$$X\ g\ H_2 = 72.8\ g \times \frac{1\ \text{mol CO}}{28.01\ g} \times \frac{2\ \text{mol } H_2}{1\ \text{mol CO}} \times \frac{2.016\ g}{1\ \text{mol } H_2} = 10.5\ g\ H_2$$

Then we subtract the mass of hydrogen actually used from the mass at the beginning to give the mass remaining.

$$10.6\ g - 10.5\ g = 0.1\ g\ H_2$$

ACTIVITY 5-7

The reaction that occurs when baking soda and vinegar are mixed is

$$NaHCO_3 + HC_2H_3O_2 \longrightarrow$$
$$H_2O + CO_2 + NaC_2H_3O_3$$

How many grams of carbon dioxide are formed when 4.28 g of $NaHCO_3$ and 3.00 g of $HC_2H_3O_2$ react? Which reactant is in excess and how many grams of it remain after the reaction?

## 5-10 THEORETICAL YIELD, ACTUAL YIELD, AND PERCENT YIELD

Often we carry out a stoichiometric calculation to find the amount of product formed by a specified amount of a reactant. Such a calculation gives the potential or theoretical amount of product that could be formed. This amount is called the thoretical yield of the reaction. Typically, in an actual synthesis reaction the product is not highly pure. It may be contaminated by some reactants or solvents used in the process or it may contain some impurities formed by side reactions that accompany the synthesis. Some of the product is lost in any purification process and some of the original reactant may be used in side reactions. The result is that the actual yield of product may be less than the theoretical (stoichiometric) yield. The percent yield for the product of any reaction is simply

$$\text{percent yield} = \frac{\text{actual yield}}{\text{theoretical yield}} \times 100$$

If we know the amount of reactant used in a reaction we can find the theoretical yield. The actual yield is then divided by the theoretical yield and the ratio is multiplied by 100.

---

**EXAMPLE 5-12**

The fertilizer urea is made by the following reaction:

$$2NH_3 + CO_2 \longrightarrow (NH_2)_2CO + H_2O$$

When $2.5 \times 10^3$ g of $NH_3$ reacted with sufficient $CO_2$ the actual yield of urea was $3.2 \times 10^3$ g. Calcualte the percent yield for the reaction. First we calculate the theoretical yield based on the starting amount of the reactant using mass-to-mass stoichiometry.

$$X \text{ g } (NH_2)_2CO = 2.5 \times 10^3 \text{ g} \times \frac{1 \text{ mol } NH_3}{17.04 \text{ g}} \times \frac{1 \text{ mol } (NH_2)_2CO}{2 \text{ mol } NH_3}$$

$$\times \frac{46.05 \text{ g}}{1 \text{ mol } (NH_2)_2CO}$$

$$= 3.4 \times 10^3 \text{ g}$$

The percent yield is then found from the actual yield and the theoretical yield.

$$\text{percent yield} = \frac{3.2 \times 10^3 \text{ g}}{3.4 \times 10^3 \text{ g}} \times 100 = 94\%$$

---

**5-11** **ENERGY**

When we do work we exert energy. However, energy is more than just physical work. The word energy developed as a composite of the Greek prefex *en* 'in' and the Greek word *ergon* 'work'. An object that possesses energy has within it the capacity to do work. **Energy** can be defined simply as the capacity to do work.

When an object has energy, there is the possibility of work in the form of some dynamic change. When you pick up an object, you must exert a force to overcome gravity. Your muscles have done work or expended energy. The object you picked up has stored energy or the potential for doing work. If you release the object, it expends the stored energy as it falls. The term work refers to a dynamic change; it may be useful, for example, picking something up or running a motor, but it does not have to be useful.

We experience various kinds of energy in energy flows, motion, and dynamic changes. Observations of energy changes show that energy can be stored and transformed from one form to another. The changes from one form of energy to another result in action, motion, chemical changes, and heat. Work is done when energy changes from one form to another. Some common types of energy are described below.

### Radiant Energy

Radiant energy is associated with light. Sometimes called electromagnetic radiation, it occurs in such forms as radiowaves, microwaves, infrared light, visible light, ultraviolet light, and X rays. Solar energy is radiant energy produced by processes on the sun. We experience solar energy as sunlight, which is mainly visible light along with some infrared and ultraviolet light.

### Heat

Heat is a mode of energy transfer between objects; it is sometimes called thermal energy. An object becomes hotter when it is heated by an energy source. We notice heat when a heat exchange causes objects to change temperature. When a colder object is in contact with a warmer object the colder object is heated and the warmer object is cooled (see Section 1-5).

### Potential Energy

Potential energy is stored energy or energy of position. Energy of this type has the potential to do work by undergoing a change or flow. Water behind a dam has potential energy.

## Kinetic Energy

Kinetic energy is energy of motion or energy associated with a moving object. A moving car, a moving molecule, in fact any moving object has kinetic energy.

## Mechanical Energy

Mechanical energy is energy associated with a motor, turbine, or engine. Mechanical energy is essentially a kind of kinetic energy associated with the motion of a flywheel or drive shaft of a motor, turbine, or engine.

## Electrical Energy

Electrical energy is energy that comes from the flow of electrical current. Electrical current passing through the filament of a light bulb changes the electrical energy to heat and light.

## Chemical Energy

Chemical energy is energy stored in chemicals. Chemical changes allow this energy to be released and converted to other forms of energy. When natural gas burns, energy is released as heat and light. A battery converts chemical energy to electrical energy and heat.

## 5-12   EXOTHERMIC AND ENDOTHERMIC REACTIONS

You might relish the warmth from the combustion of wood in a fire or use the heat from the combustion of methane to cook. Societies use tremendous amounts of chemical energy for industrial purposes and everyday living. **Exothermic** reactions are chemical reactions that release energy as heat. The term exothermic means heat out. An exothermic reaction can heat its surroundings as it occurs. In contrast, some reactions require energy to occur. **Endothermic** reactions absorb heat from their surroundings. The term endothermic means heat in.

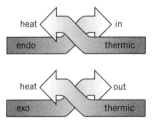

Potential energy is stored energy. Suppose we roll a boulder up a hill. At the top of the hill the boulder has potential energy with respect to ground level. With a slight nudge, the boulder tumbles down the hill, converting its stored energy into energy of motion or kinetic energy. Energy is associated with the bonding of atoms in compounds; each chemical has a kind of potential energy called chemical energy. Simply stated, chemical energy is energy stored within chemical compounds. Photosynthesis is the energy-capturing chemical process that occurs in all green plants. As

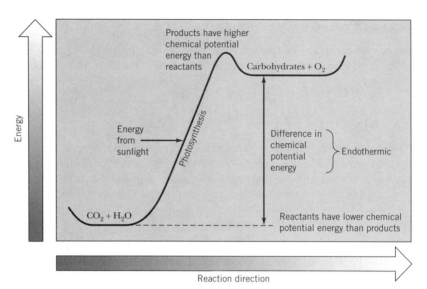

FIGURE 5-6
Change in chemical energy
during photosynthesis, an
endothermic reaction.

shown in Figure 5-6, plants synthesize carbohydrates and oxygen from carbon dioxide and water. Energy for this endothermic process comes from
sunlight. The chemical energies of the carbohydrates and oxygen are
higher than the chemical energies of the carbon dioxide and water.
However (somewhat like the boulder rolling down the hill), under the
correct conditions, carbohydrates can react with oxygen to give carbon
dioxide and water, and release the stored energy in the process. Figure 5-7

FIGURE 5-7
Change in chemical energy
during combustion of a
carbohydrate, an
exothermic reaction.

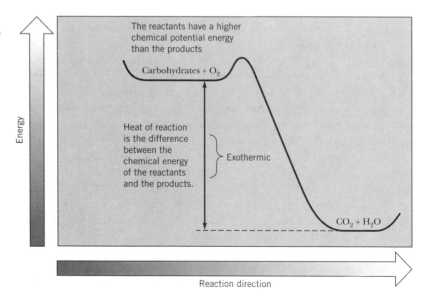

illustrates the energy change in such an exothermic reaction. Chemistry in Action 5-2 is about the energy changes that occur in a variety of reactions.

### 5-13 THE CALORIE AND THE JOULE

It is possible to measure amounts of heat exchanged in chemical processes. An exothermic reaction warms the surroundings and an endothermic reaction absorbs heat from the surroundings. Heat is thermal energy so its units are units of energy. A common unit of energy is the calorie (cal), often used for the energy content of food. One calorie is the amount of heat needed to change 1 g of liquid water by 1°C. Water is simply a convenient reference compound; the calorie can be used as a unit for heating any kind of material or as a unit to express amounts of any kind of energy. For example, it takes 0.2 calories to heat 1 g of aluminum by 1°C. Experimentally it is found that one gram of a carbohydrate typically releases 4000 calories when it reacts with oxygen. The calorie content of a food refers to the energy released when the food chemicals react with oxygen; it is expressed in terms of the dietetic or food Calorie, usually written with a capital C. One Calorie is equal to 1000 cal or 1 kilocalorie (kcal). The calorie content of carbohydrates, for instance, is 4 Calories per gram or 4 kcal per gram. The calorie is used to express the amounts of energy exchanged in chemical reactions. See Section 5-14 for more on food and calories.

A unit of energy used in the metric or SI system is the joule (abbreviated J and pronounced jewel). To understand this unit, imagine the physical work or energy expended in picking up and moving an object. To pick it up you have to apply a force, and to move it the force needs to be exerted over a distance. For instance, you have to use force to pick up a book and continue applying the force if you want to raise it above your head. Exerting a force through a distance requires work or the expenditure of energy. Forces are measured in the SI system in newtons, N. Newtons are kg m/s$^2$. One newton of force exerted over a distance of one meter, requires one joule of energy. That is, $1 \text{ N m} = 1 \text{ J} = 1 \text{ kg m}^2/\text{s}^2$. The joule is used as a unit for measuring any kind of energy. In this text we'll never have to use the detailed units for the joule so don't worry about them. Both the calorie and joule are used to express amounts of heat exchanged in chemical processes. The calorie and the joule are different units and the relation between them is: **1 cal = 4.184 J.** You may be familiar with some other units of energy like the kilowatt-hour (kwhr) or the British thermal unit (Btu). These units are seldom used in chemistry.

### 5-14 HEAT OF REACTION

Commonly, chemical reactions involve energy exchanges that occur as heat flows. Energy is exchanged quickly in fast chemical reactions. An **ex-**

**plosion** is a chemical reaction that releases relatively large amounts of energy in a short time. The particles of products move outward from the reaction quickly heating the surroundings and sometimes produce a shockwave that you can hear. For example, when a mixture of hydrogen and oxygen gas is ignited, the gases quickly react with a release of energy and a noticeable sound. Slow reactions exchange energy over a longer period of time. Skiers sometimes use instant hand warmers. These are plastic containers of chemicals that react slowly and release heat for a period of time. Let us look more closely at energy exchanges that accompany chemical reactions.

Endothermic reactions need energy to occur. As they occur, endothermic reactions absorb energy from the surroundings or from some specific energy source. Absorption of sunlight provides the energy for photosynthesis. The photosynthesis of the common carbohydrate glucose is represented by the equation:

$$energy + 6CO_2 + 6H_2O \longrightarrow C_6H_{12}O_6 + 6O_2$$

Since the synthesis is endothermic, energy is shown as a reactant. Energy is required for the reaction to produce the products, so "energy" is included on the reactant side of the equation.

Exothermic reactions release energy as they occur. When methane reacts with oxygen the reaction is exothermic, so energy is a product.

$$CH_4 + 2O_2 \longrightarrow CO_2 + 2H_2O + energy$$

Whether a reaction is exothermic or endothermic is determined by experiment. For instance, when carbon and oxygen react to give carbon dioxide, we notice energy is released in the form of heat and light. This is the main reaction that occurs when charcoal burns.

$$C + O_2 \longrightarrow CO_2 + energy$$

How much energy is exchanged in a reaction? This depends on the reaction, the reaction conditions, and the amounts of reactants. The amount of energy exchanged in a reaction is measured by allowing the reaction to occur under controlled conditions. The energy exchanged in a chemical reaction is called the **heat of reaction** or the **enthalpy change for the reaction.** The word enthalpy comes from the Greek *thalpein* 'to warm or heat'. The conventional symbol used to represent heat of reaction or enthalpy change is $\Delta H$. $\Delta$ is the Greek letter, delta, so the enthalpy change is pronounced delta H.

Imagine that we place 1 mole of carbon and 1 mole of oxygen gas in a reaction vessel at specific conditions. We initiate the reaction and 1 mole of carbon dioxide is produced. By appropriate design of the experiment, the total amount of heat released can be measured in terms of kilojoules or kilocalories. Heats of reaction usually are thousands of joules or calo-

ACTIVITY 5-8

You need two paper matches. Light one match and use it to ignite the second match. Record all that you see, hear, and feel when you ignite the match. Propose a hypothesis to explain why you hear the match igniting and burning.

ries, so they are normally measured in kilojoules or kilocalories. The measured heat of reaction or enthalpy change for the reaction of solid carbon and oxygen gas is 393 kJ/mol C. This can be shown as

$$C(s) + O_2(g) \longrightarrow CO_2(g) \quad \Delta H = -393 \text{ kJ/mol C}$$

By convention the enthalpy change of an exothermic reaction is given a negative sign corresponding to the fact that heat is released from the reaction system as the reaction occurs. The enthalpy change of an endothermic reaction is given a positive sign corresponding to the fact that heat is gained by the reaction system as the reaction occurs.

When 1 mole of carbon and 1 mole of oxygen gas react to form 1 mole of carbon dioxide under specified conditions, the reaction releases 393 kJ of heat. The negative enthalpy change reveals the heat released per mole of C that reacts. This fact can be shown by including the energy on the product side of the equation.

$$C(s) + O_2(g) \longrightarrow CO_2(g) + 393 \text{ kJ}$$

Each reaction has a unique enthalpy change that can be measured by experiment. Some chemical reference books list heats of reaction or enthalpy changes for many different reactions.

An example of an endothermic reaction is the reaction of gaseous hydrogen and iodine to give hydrogen iodide.

$$H_2(g) + I_2(g) \longrightarrow 2HI(g)$$

Heat energy is absorbed from the surroundings when this reaction occurs. That is, the surroundings lose an amount of energy equal to the absorbed energy. Experimentally, the enthalpy change is found to be $\Delta H = +25.9$ kJ/mol HI. The plus sign reveals that the reaction is endothermic; note that the value given is per mole of HI. For an endothermic reaction the energy can be included on the reactant side but in this case the value of the enthalpy change needs to be multiplied by 2 since the equation includes 2 moles of HI.

$$51.8 \text{ kJ} + H_2(g) + I_2(g) \longrightarrow 2HI(g)$$

ACTIVITY 5-9

Use your observations from Chemistry in Action 5-2. For each of the following reactions write the word energy on the reactant side or product side to indicate that the reaction is endothermic or exothermic. Some of the equations are not balanced. Balance them.

1. Reaction of baking soda and vinegar

$$NaHCO_3(s) + HC_2H_3O_2(aq) \longrightarrow H_2O(\ell) + CO_2(g) + NaC_2H_3O_3(aq)$$

2. The complete combustion of sugar

$$C_{12}H_{22}O_{11}(s) + O_2(g) \longrightarrow CO_2(g) + H_2O(g)$$

3. The complete combustion of aspirin

$$C_9H_8O_4(s) + O_2(g) \longrightarrow CO_2(g) + H_2O(g)$$

4. Iron reacting with oxygen

$$4Fe(s) + 3O_2(g) \longrightarrow 2Fe_2O_3(s)$$

## Food and Calories

Use a match to light a piece of popcorn, a potato chip, a corn chip, any kind of nut, or a cracker then watch it burn. The three major kinds of chemicals found in foods are fats, carbohydrates, and proteins. They occur in food as a variety of compounds. Carbohydrates, for instance, range from simple sugars, such as glucose, to complex carbohydrates, such as starch. It is possible to analyze a specific type of food to find the percent by mass of each of the three

*Table 5-1*   Composition of Some Foods

| FOOD | PERCENT BY MASS | | |
| --- | --- | --- | --- |
| | PROTEIN | FAT | CARBOHYDRATE |
| Meat (round steak) | 20.5 | 16.6 | – |
| Bacon | 9.9 | 65 | – |
| Milk (whole) | 3.5 | 3.7 | 4.8 |
| Eggs | 13.4 | 10.5 | – |
| Cheese (cheddar) | 27.7 | 36.8 | 4.1 |
| Butter | 1.0 | 85 | – |
| Peanuts | 25.8 | 38.6 | 21.9 |
| Beans (navy) | 22.5 | 1.8 | 55.2 |

kinds of food chemicals. Table 5-1 shows the composition of a few foods. Foods also contain varying amounts of water and small amounts of vitamins and minerals.

The three major food components are metabolized and serve as energy sources in our bodies. The metabolic use of foods for energy is represented by the following general equations (these are symbolic, not balanced, equations).

$$\text{fats} + O_2 \longrightarrow CO_2 + H_2O + \text{energy}$$

$$\text{carbohydrates} + O_2 \longrightarrow CO_2 + H_2O + \text{energy}$$

$$\text{proteins} + O_2 \longrightarrow CO_2 + H_2O + \text{urea} + \text{energy}$$

The amounts of energy released by reactions like these are measured by experiments using samples of fats, carbohydrates, and proteins. Typically fats provide 37 kJ/g or 9 kcal/g. In terms of the dietetic calorie, fats provide 9 Cal/g. Carbohydrates provide about 17 kJ/g or 4 Cal/g. Coincidentally, proteins also provide about 17 kJ/g or 4 Cal/g.

All our body energy comes from the metabolism of fats, carbohydrates, and proteins. However, much of the protein that we eat is used as a source for our body to make its own protein, and some of the fats that we eat are used to make body fat, which becomes fatty tissue. Nevertheless, in the determination of the Calorie content of a food it is assumed that all of the components of the food are used entirely for energy.

The body uses the energy from food for any endothermic metabolic reactions, physical work, and to maintain body temperature. The energy needs of a person depend upon many factors including basic metabolic processes to maintain life and energy for activity or physical work. As is expected, the typical number of Calories needed by a person depends upon their age, weight, gender, and kinds of work that they do. Calorie requirements vary depending upon the kinds of physical activity. A typical person requires about 2000 Calories per day.

The percent of a food that is carbohydrate, protein, and fat is used to determine the Calorie content of a food sample. You can use the following steps to find a sample's Calorie content.

1. Express the number of grams of the components carbohydrate, protein, and fat in 100 grams of the food sample. The label may give you grams or it may give you percents. You can convert percents to g/100 g. For instance, if a food is 27.7% protein the factor is (27.7 g protein/100 g food).
2. Multiply the total mass of the food sample by each of these g/100 g factors to find the mass of each component.
3. Multiply the mass of each food component by the factor expressing the number of Calories per gram.
4. Add the Calories from each component to give the total number of Calories.

As an example, let's calculate the Calorie content of 1.00 oz of cheddar cheese (1.00 oz is 28.3 g). See Table 5-1 for the composition of cheddar cheese. Typical cheddar cheese is 27.7% protein, 4.1% carbohydrate, and 36.8% fat. Using the method described above for a 28.3 g sample of cheese gives

$$\text{protein} \qquad 28.3\ \text{g} \times \frac{27.7\ \text{g protein}}{100\ \text{g}} \times \frac{4\ \text{Cal}}{1\ \text{g protein}} = 31.4\ \text{Cal}$$

$$\text{carbohydrate} \quad 28.3\ \text{g} \times \frac{4.1\ \text{g carbohydrate}}{100\ \text{g}} \times \frac{4\ \text{Cal}}{1\ \text{g carbohydrate}} = 4.6\ \text{Cal}$$

$$\text{fat} \qquad 28.3\ \text{g} \times \frac{36.8\ \text{g fat}}{100\ \text{g}} \times \frac{9\ \text{Cal}}{1\ \text{g fat}} = 93.7\ \text{Cal}$$

$$\text{Total Calories} = 31.4\ \text{Cal} + 4.6\ \text{Cal} + 93.7\ \text{Cal} = 129.7\ \text{or}\ 130\ \text{Cal}$$

The total Calorie content of 1.00 oz of cheddar cheese is 130 Cal. Notice that we can also calculate the percent of Calories provided by fat. Do this by dividing the Calories from fat by the total Calories and multiplying by 100. In cheddar cheese, fat contributes 72.2% of the Calories. The same methods can be used to estimate the Calorie content and percent of Calories from fat for any food given its composition. Prepared foods have the composition listed on the label expressed in terms of the stated serving size. The label lists the mass of each food component per serving size instead of its percent. Thus, to find the Calorie content just multiply the mass of each component by the appropriate Calorie per grams factor and add the results. Figure 5-8 shows an example food label. Check the Calorie content using the information on the label.

A nutrition fact label is required by the U.S. Food and Drug Administration to contain specific information. It must give the reasonable serving size and the number of servings in the container. The total calories per serving and the calories from fat must be listed. To find the percent of calories from fat, divide the fat calories by the total calories and multiply by 100. A label

**Nutrition Facts**

Serving Size 5 Crackers (14g)
Servings Per Container About 16

**Amount Per Serving**

**Calories** 60      Calories from Fat 15

| | % Daily Value* |
|---|---|
| **Total Fat** 1.5g | **3%** |
| Saturated Fat 0g | **0%** |
| Polyunsaturated Fat 0g | |
| Monounsaturated Fat 0.5g | |
| **Cholesterol** 0 mg | **0%** |
| **Sodium** 180mg | **7%** |
| **Total Carbohydrate** 10g | **3%** |
| Dietary Fiber Less than 1g | **2%** |
| Sugars 0g | |
| **Protein** 1g | |

| Vitamin A 0% | • | Vitamin C 0% |
|---|---|---|
| Calcium 2% | • | Iron 4% |

*Percent Daily Values are based on a 2,000 calorie diet. Your daily values may be higher or lower depending on your calorie needs:

| | Calories: | 2,000 | 2,500 |
|---|---|---|---|
| Total Fat | Less than | 65g | 80g |
| Sat Fat | Less than | 20g | 25g |
| Cholesterol | Less than | 300mg | 300mg |
| Sodium | Less than | 2400mg | 2400mg |
| Total Carbohydrate | | 300g | 375g |
| Dietary Fiber | | 25g | 30g |

must show the percent of the daily value of various food components. The daily values refer to number of grams or milligrams of these components recommended for a typical 2,000 Calorie diet. The grams of saturated fat are given as a subdivision of the total fat. Simple sugars and dietary fiber are given as subdivisions of total carbohydrates. These subtracted from the total carbohydrates give the grams of complex carbohydrates or starch. There is no daily value for protein but it should be consumed in moderate amounts. The content of combined sodium is given in milligrams along with the percent of the daily value of sodium. This is essentially the salt content of the food expressed in terms of sodium. Likewise, the cholesterol content is given in milligrams along with the percent of daily value. Only foods containing meat, fish, eggs and poultry can have cholesterol. Cholesterol never occurs in foods made only from vegetables or vegetable oils. Finally the content of Vitamin A, Vitamin C and the minerals calcium and iron supplied by the food are given in terms of the percent of the daily value.

**FIGURE 5-8**      A typical food label.

## 5-15  STOICHIOMETRY AND ENERGY EXCHANGES

The numerical value of the heat of reaction can be shown as a reactant or product in an equation. It is possible to relate the heat of reaction to the molar amounts of each chemical in the reaction. The equation for the combustion of methane can be expressed as

$$CH_4 + 2O_2 \longrightarrow 2H_2O + CO_2 \quad \Delta H = -803 \text{ kJ/mol } CH_4$$

The heat of reaction can be related to each chemical by expressing the following stoichiometric ratios or heat ratios. When writing heat ratios note the number of moles of the specific species as it relates to the enthalpy change.

$$\frac{803 \text{ kJ}}{1 \text{ mol } CH_4} \quad \text{or} \quad \frac{803 \text{ kJ}}{2 \text{ mol } O_2} \quad \text{or} \quad \frac{803 \text{ kJ}}{2 \text{ mol } H_2O} \quad \text{or} \quad \frac{803 \text{ kJ}}{1 \text{ mol } CO_2}$$

These ratios are used as stoichiometric factors to relate heat to the amount of any chemical in the reaction. Focus 5 shows a stoichiometry map that includes heat.

**EXAMPLE 5-13**

Find how many kilojoules of heat are produced when 15.3 g of oxygen gas react with sufficient methane by the reaction:

$$CH_4 + 2O_2 \longrightarrow 2H_2O + CO_2 \quad \Delta H = -803 \text{ kJ/mol } CH_4$$

A summary of the question is

$$CH_4 + 2O_2 \longrightarrow 2H_2O + CO_2$$

|  |  |
|---|---|
| mass | 15.3 g |
| molar mass | 32.00 g/mol |
| moles | mol $O_2$ |
| heat | ? kJ |

This type of question is similar to a mass-to-mass calculation since it is not possible to go directly from the mass of a chemical to heat. The heat of reaction relates kilojoules and moles so we need to work on a molar basis. First, the number of moles of oxygen gas is found from the mass using its molar mass. Then, the ratio relating the heat of reaction and moles of oxygen

$$\frac{803 \text{ kJ}}{2 \text{ mol } O_2}$$

is used as a factor to find the heat released.

$$15.3 \text{ g} \times \frac{1 \text{ mol } O_2}{32.00 \text{ g}} \times \frac{803 \text{ kJ}}{2 \text{ mol } O_2} = 192 \text{ kJ released}$$

**EXAMPLE 5-14**

A Btu or British thermal unit is a unit of heat equivalent to 1.054 kilojoules. Find how many grams of methane are needed to give 1.054 kilojoules of heat by the reaction:

$$CH_4 + 2O_2 \longrightarrow 2H_2O + CO_2 \quad \Delta H = -803 \text{ kJ/mol } CH_4$$

A summary is

$$CH_4 + 2O_2 \longrightarrow 2H_2O + CO_2$$

|  |  |
|---|---|
| mass | ? g |
| molar mass | 16.01 g/mol |
| moles | mol $CH_4$ |
| heat | 1.054 kJ |

Since we are given the number of kilojoules of heat, the appropriate heat ratio is used to find the number of moles of methane that produce this amount of heat. The molar mass of methane is then used to change the moles of methane to mass. Using the factors so that the kilojoules and moles cancel gives the result shown on the next page.

ACTIVITY
5-10

The overall reaction for the metabolism of sucrose in the body is

$$C_{12}H_{22}O_{11} + 12O_2 \longrightarrow$$
$$12CO_2 + 11H_2O$$
$$\Delta H = -1350 \text{ kcal/1 mol}$$
$$C_{12}H_{22}O_{11}$$

Calculate the number of kilocalories released by the metabolism of 1 gram of sucrose. Since sucrose is a carbohydrate, what do you notice about your answer? See the previous section on Food and Calories.

$$X\,\text{g CH}_4 = 1.054\,\cancel{kJ} \times \frac{1\,\cancel{\text{mol CH}_4}}{803\,\cancel{kJ}} \times \frac{16.01\,\text{g}}{1\,\cancel{\text{mol CH}_4}} = 2.10 \times 10^{-2}\,\text{g}$$

## Energy and Society

Fossil fuels (petroleum, coal, and natural gas) are the main source of energy in industrial societies. Figure 5-9 shows the sources and uses of energy in the United States. The percents are approximations since the energy mix changes yearly. Notice in Figure 5-9 that the fossil fuels provide about 90% of our energy. Coal provides 23%. More than one-half of this is used to produce electricity. Natural gas accounts for about 23% of our energy. Most of it is used directly as heating fuel. Petroleum and natural gas liquids (butane and propane) contribute about 44%. Of all petroleum used in the United States about 45% is imported. Another 6% of our energy comes from nuclear reactors and 4% from hydroelectric sources and other minor sources, such as wind farms.

Overall, around one-fourth of energy resources are used to produce electricity, and three-fourths is consumed directly as fuels. Some is used to make petrochemical products, such as plastics and chemicals, but most of it is used to produce energy. Industry uses 36% of the energy in producing consumer goods. The largest fraction of this energy is used in blast furnaces and metalworks to produce metals, such as iron and aluminum. Other large industrial energy consumers include oil refineries, mines, chemical plants, glass factories, food processing plants, paper plants, and the construction industry. Commercial establishments (stores, offices, hotels, etc.) account for about 15% of our energy use. One-half the energy used commercially is for space heating (heaters) and air conditioning. Transportation uses 28% of our energy. Most of this comes from the burning of gasoline and diesel oil in cars, buses, and trucks. Trains, ships, and airplanes are additional energy-consuming transportation modes. Household uses consume around 21% of our energy. Around one-half of this energy is for space heating and cooling using natural gas and heating oil. Other home uses include electricity for lighting, cooking, and various appliances.

Fossil fuels are useful since they release energy upon combustion. Fuels, however, vary in the amounts of energy that they release. One way to compare fuels is to calculate the fuel value. Fuel value is the number of kilocalories produced from the combustion of 1 gram of fuel. For example, the equation for the combustion of methane is

$$CH_4 + 2O_2 \longrightarrow 2H_2O + CO_2 \quad \Delta H = -192\,\text{kcal}/1\,\text{mol CH}_4$$

We can use stoichiometry to find the fuel value for methane by converting the number of kilocalories per mole to the number of kilocalories per gram.

$$\frac{192\,\text{kcal}}{1\,\cancel{\text{mole CH}_4}} \times \frac{1\,\cancel{\text{mole CH}_4}}{16.04\,\text{g}} = \frac{12.0\,\text{kcal}}{1\,\text{g}}$$

The table below shows the fuel values for some common fuels. An energy tax (sometimes called a Btu tax (1 Btu = 0.251 kcal) is a tax that could be levied on a fuel depending on its fuel value.

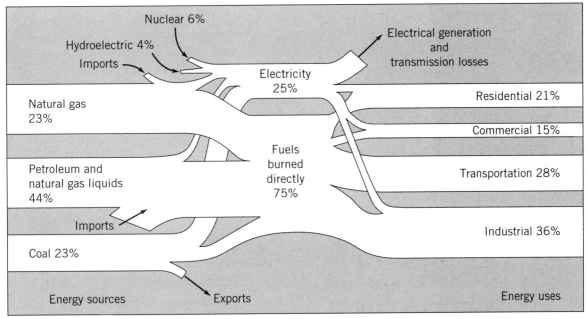

**FIGURE 5-9**   Approximate sources of energy uses and energy sources in the United States.

Higher fuel values represent more efficient fuels. More efficient fuels could be taxed at lower rates.

Fuel Values for Some Common Fuels

| | FUEL VALUE | |
|---|---|---|
| FUEL | kcal/g | btu/g |
| Methane | 12 | 48 |
| LPG (propane and butane) | 11 | 44 |
| Heating oils, kerosene, diesel oils, gasoline | 9–11 | 36–44 |
| Charcoal and coal | < 7.9 kcal/g | < 32 |
| Ethyl alcohol (ethanol) | 7 | 28 |

Use the following equation to calculate the fuel value for propane.

$$C_3H_8 + 5O_2 \longrightarrow 4H_2O + 3CO_2 \quad \Delta H = -489 \text{ kcal/1 mol } C_3H_8$$

You need: four zip-lock sandwich bags, a measuring cup or tablespoon, a teaspoon, baking soda, and white vinegar.

1.  Place one level teaspoon of baking soda in a zip-lock bag and leave it open. Place two level teaspoons of baking soda in the second bag, three level teaspoons in the third bag, and four level teaspoons in the fourth bag.
2.  Carefully pour one-fourth of a cup of vinegar or four tablespoons into the first bag, seal it, and mix the contents. Agitate the bag to be sure that the contents are well mixed.
3.  Repeat Step 2 for each of the other three bags.
4.  Compare the contents of the bags and record your observations.
5.  Feel the bags and note the temperatures. This observation is used in Activity 5-9.

Give the balanced equation for the reaction of acetic acid with sodium hydrogen carbonate (see Activity 5-9). If one-fourth of a cup of vinegar contains 0.049 mole of $HC_2H_3O_2$ how many grams of sodium hydrogen carbonate will react with this amount of acid? A level teaspoon of baking soda contains 4 g of sodium hydrogen carbonate. Use this fact and your calculated answer to explain your observations in this exercise.

You need: sugar, an aspirin tablet, matches, and fine steel wool. In place of fine steel wool you can use a small amount of steel wool obtained from a commercial scouring pad. Tear a small amount of steel wool from a pad, remove as much soap as possible and spread out the steel to make a fluffy sample. Remember that energy exchanges often involve heat and possibly flames and light.

1.  See Chemistry in Action 5-1. Record any evidence of energy exchange that occurred when baking soda and vinegar are mixed.
2.  Pour a pile of sugar about the size of an aspirin tablet on a flat surface.
3.  Light a match and dip the burning end into the sugar pile. When some sugar adheres to the match, remove the match and observe. If the match goes out in the pile, remove it and ignite it with another burning match. Record any evidence of energy exchange.
4.  Fold some paper around an aspirin tablet and crush the tablet to powder it. Light a match and dip the burning end into the aspirin pile. When some aspirin adheres to the match, remove the match and observe. If the match goes out in the pile, remove it and ignite it with another burning match. Record any evidence of energy exchange.
5.  Light a match. Pick up a fluffy wad of steel wool and hold the match near the other end of the sample. If the match goes out, use another. Record any evidence of energy exchange.

# QUESTIONS

See Chapter 3 for a review of the use of the mole.

## Section 5-1

1. *Liquid mercury has a density of 13.6 g/mL. How many moles of Hg atoms are in a thermometer that contains 0.055 mL of Hg?

2. Liquid bromine, $Br_2$, has a density of 2.90 g/mL. How many moles of $Br_2$ are in a 25-mL sample?

3. *A quartz crystal has a density of 2.65 $g/cm^3$. How many moles of $SiO_2$ are in a 37.5-$cm^3$ crystal?

4. A typical aspirin tablet contains 5.0 grains of acetylsalicylic acid, $C_9H_8O_4$ (180 g/mole). How many moles of acetylsalicyclic acid are in two tablets? (0.0648 g = 1 grain)

5. *Thyroxine is a hormone produced in the thyroid gland. It has the formula $C_{15}H_{11}O_4I_4N$. How many moles of combined iodine atoms are contained in 3.15 micrograms of thyroxine? (1 microgram = $10^{-6}$ g)

6. Lithium carbonate, $Li_2CO_3$, is used in the treatment of bipolar mental disorders. How many moles of lithium ion, $Li^+$, are contained in a 500-mg dose of lithium carbonate?

7. *Green vegetables are an important source of dietary magnesium because they contain chlorophyll, $C_{55}H_{72}MgN_4O_5$. If the recommended dietary allowance (RDA) for magnesium for adults is 280 mg/day, how many grams of chlorophyll meet this need?

8. Sodium fluoride is used as an anticavity agent in some toothpaste. How many moles of fluoride ion are in a 4.5-mg sample of sodium fluoride?

9. *Baking soda, $NaHCO_3$, is often used as a leavening agent in baked goods. If a recipe calls for 2 tsp of baking soda, how many formula units of $NaHCO_3$ would this be? (1 tsp = $\frac{1}{8}$ oz; give your answer to two significant digits)

10. Cotton is composed of cellulose that exists in the form of very large molecules having an average molar mass of $5.0 \times 10^5$ g. Assuming that a cotton ball is pure cellulose, determine the approximate number of cellulose molecules in a 0.32-g cotton ball.

11. *A sample of moon rock is found to contain 12.5% $Al_2O_3$. How many moles of $Al_2O_3$ are contained in 2.25 kg of the rock? (Hint: The rock contains 12.5 g $Al_2O_3$/100 g rock.) How many moles of oxide ion does the 2.25-kg sample contain?

12. Determine the mass or the number of moles for each of the following.
   (a) the number of moles of $O_2$ in a 25.0-g sample
   (b) the mass of 3.94 moles of fructose, $C_6H_{12}O_6$
   (c) the number of moles of calcium carbonate, $CaCO_3$, in a 125-mg sample
   (d) the mass of 0.445 mole of acetic acid, $HC_2H_3O_2$

13. Give a general definition of the mole.

14. How many grams of NaCl contains 2400 mg of sodium ion?

## Section 5-2

15. Why is mass conserved in a chemical reaction?

## Sections 5-3 to 5-6

16. *Acetylene, $C_2H_2$, burns in oxygen according to the following unbalanced equation:

$$C_2H_2(g) + O_2(g) \longrightarrow CO_2(g) + H_2O(g)$$

Balance the equation. Give the molar interpretation of the equation and give the molar ratios that relate the following pairs of reactants and products.
   (a) $C_2H_2$ and $O_2$
   (b) $C_2H_2$ and $CO_2$
   (c) $C_2H_2$ and $H_2O$
   (d) $O_2$ and $CO_2$

17. *Refer to the balanced equation in Question 16. Find the number of moles of the first chemical.
   (a) $O_2$, to react with 0.570 mole of $C_2H_2$
   (b) $C_2H_2$, to produce 0.400 mole of $CO_2$
   (c) $H_2O$, formed when 17.6 moles of $C_2H_2$ react
   (d) $CO_2$, produced when 3.89 moles of $O_2$ react with acetylene

18. Butane gas burns in air to form carbon dioxide and water as shown by the unbalanced equation:

$$C_4H_{10}(g) + O_2(g) \longrightarrow CO_2(g) + H_2O(g)$$

Balance the equation. Give the molar interpretation of this equation and give the molar ratios that relate the following pairs of reactants and products.
   (a) $C_4H_{10}$ and $O_2$
   (b) $C_4H_{10}$ and $CO_2$
   (c) $C_4H_{10}$ and $H_2O$
   (d) $O_2$ and $CO_2$

19. Refer to the balanced equation in Question 18. Find the number of moles of the first chemical.

(a) $O_2$, to react with 8.75 moles of $C_4H_{10}$

(b) $C_4H_{10}$, to produce 14.9 moles of $CO_2$

(c) $H_2O$, formed when $6.05 \times 10^3$ moles of $C_4H_{10}$ react

(d) $CO_2$, produced when 0.075 mole of $O_2$ reacts with butane

## Sections 5-7 to 5-9

20. *The ethanol or ethyl alcohol component of gasohol burns according to the equation:

$$C_2H_5OH + O_2 \longrightarrow CO_2 + H_2O$$

Balance the equation. Find the amount of the first chemical.

(a) Oxygen, to react with 52.6 g of $C_2H_5OH$; answer in moles

(b) Oxygen, to react with 52.6 g of $C_2H_5OH$; answer in grams

(c) $CO_2$, formed when 52.6 g of $C_2H_5OH$ react; answer in grams

(d) $CO_2$, produced when 52.6 g of $C_2H_5OH$ and 75.0 g of $O_2$ are mixed; answer in grams (limiting reactant problem)

21. Nitrogen dioxide can form nitric acid by reaction with water as shown by the equation:

$$NO_2 + H_2O \longrightarrow HNO_3 + NO$$

Balance the equation. Find the number of grams of the first chemical.

(a) $HNO_3$, formed from 50 moles of $NO_2$

(b) $H_2O$, to form 500 g of $HNO_3$

(c) $NO_2$, to form 250 g of $HNO_3$

(d) $HNO_3$, formed when 125.0 g of $NO_2$ and 95.0 g of water are mixed (limiting reactant problem)

22. *Antacids containing $CaCO_3$ react with "stomach acid" according to the equation:

$$CaCO_3(s) + HCl(aq) \longrightarrow CaCl_2(aq) + CO_2(g) + H_2O$$

Balance the equation. Find the amount of the first chemical.

(a) $CO_2$, formed from 500 mg of $CaCO_3$; answer in grams

(b) HCl, to react with 1.00 g of $CaCO_3$; answer in moles

(c) $CaCO_3$, to produce 1200 mg of calcium ion; answer in grams (some people use $CaCO_3$ as a supplemental Ca source)

(d) $CO_2$, formed when 9.45 g of HCl are mixed with 28.4 g of $CaCO_3$; answer in grams (limiting reactant problem)

23. The fermentation of glucose to form ethyl alcohol occurs according to the equation:

$$C_6H_{12}O_6 \longrightarrow C_2H_5OH + CO_2$$

Balance the equation. Find the amount of the first chemical.

(a) $C_6H_{12}O_6$, to form 500 g of ethyl alcohol; answer in grams

(b) $CO_2$, produced when 10.0 g of $C_6H_{12}O_6$ react; answer in moles

(c) $C_2H_5OH$, formed from 1.00 kg of $C_6H_{12}O_6$; answer in grams

24. *Calcium cyanamide, $CaCN_2$, is used as a fertilizer. It reacts with water to form $CaCO_3$, which counteracts excess acidity in the soil, and ammonia, $NH_3$, which fertilizes the soil, according to the equation:

$$CaCN_2 + 3H_2O \longrightarrow CaCO_3 + 2NH_3$$

(a) How many grams of $NH_3$ can be formed from 1.00 moles of $CaCN_2$?

(b) How many grams of water are needed to produce 625 g of $NH_3$?

(c) How many grams of $CaCN_2$ are needed to produce 500 kg of $NH_3$?

(d) If the $CaCN_2$ used in the reaction is only 70% pure, the rest being inert impurities, how many kilograms of $NH_3$ can be formed from 500 kg of impure $CaCN_2$?

25. Nitroglycerine, $C_3H_5(NO_3)_3$, is a powerful chemical explosive. Since this oily liquid can explode (even in the absence of atmospheric oxygen) by the slightest shock, it is very dangerous to handle. The equation for the reaction is

$$4\,C_3H_5(NO_3)_3 \longrightarrow$$
$$6\,N_2(g) + O_2(g) + 12\,CO_2(g) + 10\,H_2O(g) + energy$$

(a) How many moles of nitrogen can be produced from 15.5 moles of $C_3H_5(NO_3)_3$?

(b) How many grams of oxygen can be produced from 475 g of $C_3H_5(NO_3)_3$?

(c) If a vial of nitroglycerine contains 2.86 g of $C_3H_5(NO_3)_3$, how many grams of $CO_2$ can be formed?

(d) Dynamite is made by mixing nitroglycerine with inert diatomaceous earth. How many grams of water can form when 1.00 kg of dynamite, containing 50% nitroglycerine, explodes?

26. *Titanium metal is used to make relatively lightweight but high-strength alloys used in aircraft. It is obtained as the metal by the reaction

$$TiCl_4 + 4Na \longrightarrow 4NaCl + Ti$$

(a) How many moles of Ti can be formed when 11.6 g of Na react?

(b) How many grams of Ti can be formed from 385 g of $TiCl_4$?

(c) If you want to produce 5.0 kg of Ti, how many grams of Na are needed?

(d) How many grams of NaCl can be formed when 0.853 g of $TiCl_4$ reacts?

(e) How many grams of titanium can be formed when 965 g of $TiCl_4$ are mixed with 480 g of Na? (limiting reactant problem)

27. An important reaction in a blast furnace used to make iron is

$$Fe_2O_3 + 3CO \longrightarrow 2Fe + 3CO_2$$

(a) How many grams of $Fe_2O_3$ are needed to produce 887 g of Fe?

(b) How many kilograms of CO are needed to produce 750 kg of Fe?

(c) If the iron ore is 76% $Fe_2O_3$, how many grams of Fe can be formed from 945 g of ore?

(d) How many kilograms of Fe can be formed when $1.63 \times 10^4$ kg of $Fe_2O_3$ are mixed with 779 kg of CO? (limiting reactant problem)

28. *An important enzyme in your body, catalase, converts hydrogen peroxide, $H_2O_2$, to oxygen and water.

$$2H_2O_2 \xrightarrow{\text{catalase}} O_2 + 2H_2O$$

(a) How many grams of $H_2O_2$ are converted if $1.36 \times 10^{-6}$ mole of $O_2$ is formed?

(b) A single cell in your body produces $4.61 \times 10^{-12}$ g of hydrogen peroxide. How many

grams of water will be produced if all the $H_2O_2$ reacts?

(c) How many moles of $H_2O_2$ are required to produce 1.00 mg of water?

(d) How many grams of $O_2$ can be formed when $3.32 \times 10^{-8}$ g of $H_2O_2$ reacts?

29. The enzyme urease, isolated from jack beans, catalyzes the decomposition of urea according to the following reaction:

$$CN_2H_4O + H_2O \xrightarrow{\text{urease}} 2NH_3 + CO_2$$

(a) How many grams of urea are needed to produce 56.7 g of $NH_3$?

(b) How many moles of water are required to react with 118 mg of urea?

(c) How may grams of $NH_3$ can be produced from 7.62 g of urea?

(d) How many grams of urea are needed to react with 1.006 g of water?

30. *Wine can be made by fermenting sugar, such as glucose, by the action of enzymes found in yeast. The fermentation forms ethanol and carbon dioxide.

$$C_6H_{12}O_6 \xrightarrow{\text{yeast}} 2C_2H_5OH + 2CO_2$$

(a) How many grams of ethanol can be produced from 525 g of glucose?

(b) How many moles of $CO_2$ can be produced from 81.5 moles of glucose?

(c) If 255 kg of ethanol is produced, how many kilograms of carbon dioxide are also produced?

31. An astronaut excretes about $2.61 \times 10^3$ g of water a day. If lithium oxide is used in the spacecraft to absorb the water, how many kilograms of $Li_2O$ must be included for a 28-day space trip? The reaction involved is given by the equation:

$$Li_2O + H_2O \longrightarrow 2LiOH$$

32. A 8.15 g sample of salt water is analyzed by carrying out the following reaction:

$$Ag^+(aq) + Cl^-(aq) \longrightarrow AgCl(s)$$

If 0.910 g of AgCl is formed from the sample, what is the percent of chloride ion in the salt water?

33. *A quick bread recipe uses a mixture of baking

soda, $NaHCO_3$, and vinegar, which contains acetic acid, $HC_2H_3O_2$. This mixture serves as a leavening agent according to the equation:

$$NaHCO_3 + HC_2H_3O_2 \longrightarrow NaC_2H_3O_2 + CO_2 + H_2O$$

If 49.5 g of $NaHCO_3$ and 42.6 g of $HC_2H_3O_2$ are mixed, find

(a) how many grams of $CO_2$ can be formed

(b) which reactant is in excess and how many grams of this reactant remain after the reaction is complete

34. Sodium nitrite, $NaNO_2$, is added to packaged meats to prevent the growth of *Clostridium botulinum,* the organism that causes botulism poisoning. However, nitrites are potentially dangerous since they are converted in the stomach to nitrous acid, which can lead to the formation of carcinogenic nitrosamines.

$$NaNO_2 + HCl \longrightarrow HNO_2 + NaCl$$

If 3.75 mg of $NaNO_2$ is added to 15.0 mg HCl, find

(a) how many milligrams of $HNO_2$ can be formed

(b) which reactant is in excess and how many milligrams remain after the reaction is complete

35. *Aspirin, $C_9H_8O_4$, is synthesized from salicylic acid, $C_7H_6O_3$, and acetic anhydride, $C_4H_6O_3$, in a reaction represented by the following equation:

$$C_7H_6O_3 + C_4H_6O_3 \longrightarrow C_9H_8O_4 + HC_2H_3O_2$$

If 219 g of salicylic acid and 210 g of acetic anhydride are mixed, find

(a) how many grams of aspirin can be formed

(b) which reactant is in excess and how many grams remain after the reaction is complete

36. A method used to make uranium metal involves the reaction:

$$U_3O_8 + 4C \longrightarrow 3U + 4CO_2$$

If 11.16 kg of $U_3O_8$ and 0.0900 kg of C are mixed, find

(a) how many grams of uranium can be formed

(b) which reactant is in excess and how many grams remain after the reaction is complete

37. *Lime or calcium oxide, CaO, reacts with water to form calcium hydroxide, $Ca(OH)_2$. Write the balanced equation for the reaction. If 25.0 g of CaO is added to 21.0 g of water, find

(a) how many grams of $Ca(OH)_2$ can be formed

(b) which reactant is in excess and how many grams remain after the reaction is complete

38. Tungsten metal is used in light bulb filaments. One method used to make tungsten metal involves the reaction:

$$WO_3 + 3H_2 \longrightarrow W + 3H_2O$$

If 69.8 g of $WO_3$ and 3.90 g of $H_2$ are mixed, find

(a) how many grams of tungsten can be formed

(b) which reactant is in excess and how many grams remain after the reaction is complete

## Section 5-10

39. Define the terms theoretical yield and percent yield.

40. *Aspirin, $C_9H_8O_4$, is made from the reaction of salicylic acid, $C_7H_6O_3$, and acetic anhydride, $C_4H_6O_3$

$$C_7H_6O_3 + C_4H_6O_3 \longrightarrow C_9H_8O_4 + HC_2H_3O_2$$

When 69.2 g of salicylic acid reacts with acetic anhydride, 82.5 g of aspirin are obtained. What is the percent yield of aspirin?

41. The fermentation of glucose forms ethyl alcohol

$$C_6H_{12}O_6 \longrightarrow 2C_2H_5OH + 2CO_2$$

When 90 g of glucose reacts, 42 g of ethyl alcohol are obtained. What is the percent yield of ethyl alcohol?

## Sections 5-11 to 5-14

42. Define the following terms.

(a) energy

(b) kinetic energy

(c) chemical energy

(d) exothermic reaction

(e) endothermic reaction

(f) heat of reaction

(g) joule

(h) calorie

43. The label on an 8-oz container of yogurt lists 10 g protein, 46 g carbohydrate, and 4 g fat. What is the total Calorie content of the container? What percent of the Calories comes from fat?

44. Whole milk contains 3.5% protein, 3.7% fat, and 4.8% carbohydrates. If the density of milk is 1.1 g/mL what is the Calorie content of a cup of milk having a

volume of 237 mL? What percent of the Calories comes from fat?

45. Peanuts contain 25.8% protein, 36.8% fat, and 21.9% carbohydrates. What is the Calorie content of 10 grams of peanuts? What percent of the Calories comes from fat?

## Section 5-15

46. *Sucrose (table sugar) burns in oxygen to form carbon dioxide and water.

$$C_{12}H_{22}O_{11}(s) + 12O_2(g) \longrightarrow 11H_2O(\ell) + 12CO_2(g)$$
$$\Delta H = -5649 \text{ kJ}/1 \text{ mol } C_{12}H_{22}O_{11}$$

(a) How many grams of sucrose need to react to give 1.00 kcal of heat? (1 kcal = 4.184 kJ)

(b) When sucrose is metabolized in your body, the overall process can be represented by the above equation. How many kilojoules of heat could be produced by the metabolism of 1 cup, 220 g, of sugar? How many kilocalories or dietetic calories is this (see Section 5-14)?

(c) Sugar is a carbohydrate. Calculate the kcal produced per gram of sugar metabolized. (1 kcal = 4.184 kJ)

47. Glucose sugar burns in oxygen to form carbon dioxide and water.

$$C_6H_{12}O_6(s) + 6O_2(g) \longrightarrow 6H_2O(\ell) + 6CO_2(g)$$
$$\Delta H = -2816 \text{ kJ}/1 \text{ mol } C_6H_{12}O_6$$

(a) How many grams of glucose need to react to give 1.00 kcal of heat? (1 kcal = 4.184 kJ)

(b) When glucose is metabolized in your body, the overall process can be represented by the above equation. How many kilojoules of heat could be produced by the metabolism of 1 oz of glucose? (1 oz = 28.3 g) How many kcal or dietetic calories is this (see Section 5-14)?

(c) Glucose is a carbohydrate. Calculate the kcal produced per gram of glucose metabolized. (1 kcal = 4.184 kJ)

48. *When 1 mole of nitrogen gas and 1 mole of oxygen gas react to give 2 moles of nitrogen oxide gas, NO, 181 kJ of heat are absorbed. Write a balanced equation for the reaction which includes the heat of reaction. If 175.2 g of nitrogen react with sufficient oxygen, how many kilojoules of heat are required?

49. When magnesium metal, Mg, reacts with oxygen, magnesium oxide, MgO, is formed and intense heat and a bright flash of light accompany the reaction. When 1 mole of oxygen reacts with magnesium 2420 kJ of heat are released. Write a balanced equation for the reaction; include the heat of reaction. How many kJ of heat are released when 1.00 g of magnesium reacts with oxygen?

50. *The combustion reactions and the enthalpy changes for methane, propane, and butane are

$$CH_4(g) + 2O_2(g) \longrightarrow CO_2(g) + 2H_2O(g)$$
$$\Delta H = -803 \text{ kJ}/1 \text{ mol } CH_4$$

$$C_3H_8(g) + 5O_2(g) \longrightarrow 3CO_2(g) + 4H_2O(g)$$
$$\Delta H = -2046 \text{ kJ}/1 \text{ mol } C_3H_8$$

$$2C_4H_{10}(g) + 13O_2(g) \longrightarrow 8CO_2(g) + 10H_2O(g)$$
$$\Delta H = -2261 \text{ kJ}/1 \text{ mol } C_4H_{10}$$

Determine which is the best fuel in terms of the number of kilojoules produced per gram of fuel (this is the fuel value).

51. *Refer to the equations and enthalpy changes in Question 50 to answer the following.

(a) How many grams of methane, $CH_4$, are needed to produce $1.00 \times 10^3$ kJ of heat?

(b) A British thermal unit (Btu) is 1.05 kJ. How many Btu are produced from the combustion of 65.0 g of propane, $C_3H_8$?

(c) How many grams of butane, $C_4H_{10}$, are needed to produce 35.0 Btu of heat?

(d) If a methane heater produces 30,000 Btu of heat per hour ($30 \times 10^3$ Btu/hr) and 1 Btu = 1.05 kJ, how many grams of methane are burned per hour?

52. Gasohol is a mixture of ethanol and gasoline that can be found in some gas stations. The heats of reaction for the combustion of ethanol and octane (one component of gasoline) are

$$C_2H_6O(\ell) + 3O_2(g) \longrightarrow 2CO_2(g)$$
$$\Delta H = -1235 \text{ kJ}/1 \text{ mol } C_2H_6O$$

$$2C_8H_{18}(\ell) + 25O_2(g) \longrightarrow 16CO_2(g) + 18H_2O(g)$$
$$\Delta H = -5080 \text{ kJ}/1 \text{ mol } C_8H_{18}$$

(a) Calculate how much heat is produced by burning 1.00 gallons of ethanol (density = 0.789 g/mL).

(b) Calculate how much heat is produced by burning 1.00 gallons of octane (density = 0.702 g/mL).

(c) Compare octane and ethanol in terms of the number of kilojoules of heat produced per gram of each.

53. Refer to the equations and enthalpy changes in Question 52 to answer the following.

(a) How many grams of ethanol are needed to produce 7500 kcal of heat?

(b) How many grams of octane are needed to produce 1.00 kcal of heat?

(c) When 1.00 liters of ethanol are burned how many kilojoules of heat are released? (see Question 52(a))

(d) How many kilocalories of heat are released when 47.1 g of ethanol and 39.8 g of oxygen react?

54. Hydrogen is a potential fuel of the future. The combustion reaction of hydrogen and the heat of reaction are

$$2H_2(g) + O_2(g) \longrightarrow 2H_2O(g)$$
$$\Delta H = -121 \text{ kJ}/1 \text{ mol } H_2O$$

(a) How many grams of hydrogen are needed to produce $1.20 \times 10^3$ kJ of heat?

(b) How many kilojoules of heat are produced when 675 g of oxygen react with sufficient hydrogen?

(c) How many kilojoules of heat are produced per gram of hydrogen? How does hydrogen compare with methane in terms of kilojoules per gram of fuel? See Question 50.

## Energy and Society

55. What are the main sources and uses of energy in our society?

56. What fuels are commonly used as sources of energy?

57. What is fuel value?

## Questions to Ponder

58. Give some pros and cons for the use of a Btu tax.

59. Describe some energy sources that can substitute for fossil fuels. Give pros and cons for these alternate sources.

60. Considering the energy expended for various activities given below do you think that it is possible to lose weight by exercising? Give reasons why or why not.

| | | |
|---|---|---|
| Bicycling | slow: 210 Cal/hr | fast: 660 Cal/hr |
| Running | 900 Cal/hr | |
| Swimming | 300 Cal/hr | |
| Walking | slow: 210 Cal/hr | fast: 300 Cal/hr |
| Tennis | 420 Cal/hr | |

61. Make a list of all foods that you ate in one day. Estimate the number of Calories in the food that you ate and the percent of the Calories that came from fat.

# ATOMIC

# STRUCTURE

# AND THE

# PERIODIC

# TABLE

## 6-1 THE PERIODIC TABLE

In the 1800s, the atomic theory captured the imaginations of chemists. Many new elements were discovered and added to the list of previously known elements. During this time, chemists were successful in figuring out the relative atomic weights of the elements. Each new element was investigated and its properties were noted. Information about elements began to accumulate. Eventually it became apparent that the properties of some elements were quite similar to one another. Scientists noticed that some periodic or repeating pattern of properties existed among the elements. The term periodic means repeating again and again at regular intervals. A periodical is a magazine published weekly or monthly. A menstrual period typically follows a 28-day cycle.

   **Periodic behavior** of the elements means that there is a pattern in which elements of higher atomic number have chemical properties like those of elements of lower atomic number. Elements with similar properties undergo similar chemical combinations and form closely related compounds with other elements. Elements that have similar properties and form similar compounds are classified as **families or groups** of elements. Examples of similar compounds formed by some families of elements are given on the next page. (Atomic numbers are shown for reference and some of the formulas are empirical.)

181

| 3 | 4 | 5 | 6 | 7 | 8 | 9 |
|---|---|---|---|---|---|---|
| $Li_2O$ | $BeO$ | $B_2O_3$ | $CO_2$ | $N_2O_3$ | $O_3$ | $OF_2$ |

| 11 | 12 | 13 | 14 | 15 | 16 | 17 |
|---|---|---|---|---|---|---|
| $Na_2O$ | $MgO$ | $Al_2O_3$ | $SiO_2$ | $P_2O_3$ | $SO_2$ | $Cl_2O$ |

| 19 | 20 | 31 | 32 | 33 | 34 | 35 |
|---|---|---|---|---|---|---|
| $K_2O$ | $CaO$ | $Ga_2O_3$ | $GeO_2$ | $As_2O_3$ | $SeO_2$ | $Br_2O$ |

| 37 | 38 | 49 | 50 | 51 | 52 | 53 |
|---|---|---|---|---|---|---|
| $Rb_2O$ | $SrO$ | $In_2O_3$ | $SnO_2$ | $Sb_2O_3$ | $TeO_2$ | $I_2O$ |

| 55 | 56 | 81 | 82 | 83 | 84 | |
|---|---|---|---|---|---|---|
| $Cs_2O$ | $BaO$ | $Tl_2O_3$ | $PbO_2$ | $Bi_2O_3$ | $PoO_2$ | |

Observations of the similarities and differences in the behavior of elements stimulated the curiosity of many chemists. Was there a grand pattern to such similarities? If such a pattern existed, what message did it convey about the nature of matter?

When the elements are listed according to increasing atomic number, the eleventh element, sodium, is similar to the third element, lithium; the twelfth element, magnesium, is similar to the fourth element, beryllium; the thirteenth element, aluminum, is similar to the fifth element, boron; the fourteenth element, silicon, is similar to the sixth element, carbon; and so on. The pattern is repeated with the nineteenth element, potassium, having properties similar to the eleventh element, sodium, and so on. The elements can be arranged in columns according to this periodic repetition in properties. If an element has no similarities with any of the preceding elements, it is given a new column. The result is an arrangement called the **periodic table of elements** (see the inside front cover). Elements in the same vertical **column** in the table have similar properties and belong to the same group or family of elements. We'll use the word groups to refer to columns.

The periodic table, much as we know it today, was first proposed in 1869 by the Russian chemist, Dmitri Ivanovich Mendeleev. His table was based on studies of the similarities in the properties of the elements known at that time and represented a revolutionary step in the development of chemistry. Since then, many new elements have been discovered. In creating the periodic table Mendeleev left open positions in families that seemed to have missing members. In time, elements were discovered that fit into the open positions. In addition, elements having higher atomic numbers than any of those known to Mendeleev were discovered. These new elements fit nicely into the evolving periodic table. Today the periodic table is the central paradigm of chemistry.

Elements are listed in the periodic table from left to right by increasing atomic numbers. Each element has its own box in the table. A box usually contains the symbol of the element with the atomic number above the symbol and atomic weight below the symbol. You can use the table as a source of useful information. For instance, as you know, it is a convenient source of atomic weights.

## 6-2 METALS, NONMETALS, AND METALLOIDS

Elements are members of specific groups or families in the periodic table. Elements are also grouped into three important general classes or categories. Some elements have metallic properties. As **metals** they are

good conductors of electricity

good conductors of heat

malleable—able to be formed into a variety of shapes

ductile—able to be drawn into wires

shiny—have a metallic luster

Any element that has these properties is a metal. Figure 6-1 shows that more than three-fourths of the elements are metals.

Most metals have names that end with -ium. Some obvious exceptions are iron, cobalt, nickel, copper, silver, gold, zinc, mercury, tin, and lead. Aluminum is a good example of a metal. It conducts electricity and is used in a variety of electrical applications, such as long-distance power transmission lines. The use of aluminum in cookware shows that this metal is a good conductor of heat. Aluminum can be cast or formed into a variety of shapes and drawn into wires. It is formed into beams and tubes used in construction, and it is easily pressed into sheets and foil. The shiny, silvery appearance of aluminum is a metallic characteristic.

Other elements do not have metallic properties. They are poor conductors or nonconductors and occur as gases or as brittle, nonlustrous solids. These elements are the **nonmetals.** As shown in Figure 6-1, the nonmetals occupy the upper right-hand corner of the periodic table. Examples are the gases nitrogen, oxygen, fluorine, chlorine, and the solids carbon, sulfur, iodine.

Some elements display properties that are both metallic and nonmetallic. A good example is silicon. It is a shiny, bluish-gray metallic-appearing element, but it is quite brittle and cannot be formed into a variety of shapes like a metal. Silicon does not conduct electricity in the same way that metals do. It conducts in a special way; it is a semiconductor. Silicon mixed with small amounts of selected impurities acts as a semiconductor

| 1 | | | | | | | | | | | | | | | | | 18 NOBLE GASES |
|---|---|---|---|---|---|---|---|---|---|---|---|---|---|---|---|---|---|
| H H$_2$ | 2 | | | Nonmetals | | | | | | | | 13 | 14 | 15 | 16 | 17 | He |
| Li metal | Be metal | | | Metalloids | | | | | | | | B | C | N N$_2$ | O O$_2$ | F F$_2$ | Ne |
| Na metal | Mg metal | 3 | 4 | Metals | 6 | 7 | 8 | 9 | 10 | 11 | 12 | Al metal | Si | P | S | Cl Cl$_2$ | Ar |
| K metal | Ca metal | Sc metal | Ti metal | V metal | Cr metal | Mn metal | Fe metal | Co metal | Ni metal | Cu metal | Zn metal | Ga metal | Ge | As | Se | Br Br$_2$ | Kr |
| Rb metal | Sr metal | Y metal | Zr metal | Nb metal | Mo metal | Tc metal | Ru metal | Rh metal | Pd metal | Ag metal | Cd metal | In metal | Sn metal | Sb | Te | I I$_2$ | Xe |
| Cs metal | Ba metal | La metal | Hf metal | Ta metal | W metal | Re metal | Os metal | Ir metal | Pt metal | Au metal | Hg metal | Tl metal | Pb metal | Bi metal | Po | At | Rn |
| Fr metal | Ra metal | Ac metal | | | | | | | | | | | | | | | |

| Ce metal | Pr metal | Nd metal | Pm metal | Sm metal | Eu metal | Gd metal | Tb metal | Dy metal | Ho metal | Er metal | Tm metal | Yb metal | Lu metal |
|---|---|---|---|---|---|---|---|---|---|---|---|---|---|
| Th metal | Pa metal | U metal | Np metal | Pu metal | Am metal | Cm metal | Bk metal | Cf metal | Es metal | Fm metal | Md metal | No metal | Lr metal |

**FIGURE 6-1** A periodic table showing metals, nonmetals, and metalloids.

and is used to fabricate transistors and other electronic components. These electronic components can be made in miniature directly on the surfaces of small chips (about 5 mm squares of silicon) to form integrated circuits used in solid-state electronics. Elements, like silicon, which have some metallic and nonmetallic properties are called **metalloids** or **semimetals.** Figure 6-1 shows that the metalloids form a step-like demarcation between the metals and nonmetals.

## 6-3 THE EVOLUTION OF THE ATOM

Ideas about the structure of atoms have evolved since Dalton first proposed the atomic theory. Dalton's atoms were characteristic, indestructible particles that somehow combined to form compounds.

The discovery of electrons produced a view of the atom as some sort of collection of electrons and positive particles.

The atom, as suggested by Lord Rutherford in 1911, was visualized as electrons in motion about a positively charged nucleus. This view was called the nuclear model or the solar system atom.

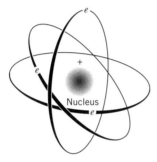

In 1913, Niels Bohr, a Danish physicist, theorized that an atom has electrons in motion about the nucleus in definite, fixed orbits.

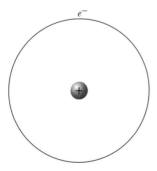

The modern model of the atom was developed in the 1920s by the efforts of many scientists; among the most notable are the Europeans Erwin Schrödinger, P. A. M. Dirac, and Werner Heisenberg. Their quantum mechanical atom views electrons in atoms as three-dimensional clouds surrounding the nucleus.

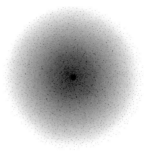

Atomic structure refers to the ways that electrons are distributed in atoms. Experimental evidence suggests that the chemical behavior of atoms is directly related to their atomic structures. So to understand the similarities and differences in the chemistry of various elements it is useful to know something about atomic structure.

Recall that each element has a unique atomic number, which reveals the number of protons and electrons in atoms of that element. Hydrogen, atomic number 1, has atoms containing one proton and one electron. Helium, atomic number 2, has atoms with two protons and two electrons. This pattern continues through the elements, with each subsequent element having atoms with one more proton and one more electron than the previous element of lower atomic number. Elements are arranged from left to right in the periodic table by atomic number. An element's atomic number tells us how many electrons there are in its atoms.

### 6-4  CATHODE-RAY TUBES

Much experimental evidence of atomic structure came from interesting scientific devices first constructed in the mid-1800s. These devices are sealed glass tubes that have the air pumped out and a metal electrode attached at each end (see Fig. 6-2). A high-voltage source is connected to the two electrodes. When the voltage is high enough, electricity passes through the tube, similar to a spark of electricity or a bolt of lightning. Within the tube electricity flows from one electrode to the other; it flows from the cathode to the anode. The flow of electricity in the tube is called a cathode ray, so the tube is known as a **cathode-ray tube (CRT).**

When a tube is placed between electrically charged plates, as illustrated in Figure 6-3, the cathode rays are attracted toward the positive plate. Furthermore, when the tube is placed between the poles of a strong magnet, the cathode-ray beam deflects from a straight-line path. Cathode rays were imagined to be composed of negatively charged particles that were the characteristic particles of electricity. These particles were named electrons. As we know, electrons are subatomic particles found in atoms. A **cathode-ray** beam is a free-moving beam of electrons. That is, electrons are not bound tightly to atoms but are free to move under the influence of voltage.

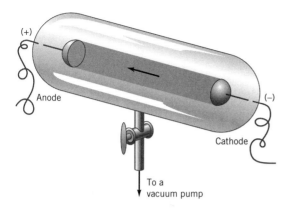

Cathode-ray tubes provide evidence for the existence of electrons and provide ways in which the properties of electrons are observed. Incidentally, tubes in television sets and computer terminals are cathode-ray tubes in which the cathode-ray beam is moved about by electromagnetic fields. The electron beam is focused on and scanned across a screen coated with phosphors. A phosphor is a chemical that glows temporarily when it is energized.

## 6-5 GAS DISCHARGE TUBES

Construction of a cathode-ray tube involves pumping the air out of the glass tube to which the electrodes are attached. Cathode-ray tubes are vacuum tubes; that is why they implode when they are broken. A device, known as a **gas discharge tube,** can be made by placing a small amount of a gas in the cathode-ray tube after the air is removed. A normal cathode-

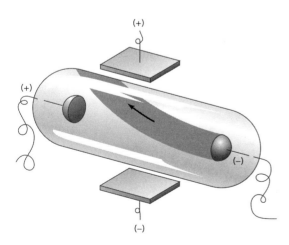

**FIGURE 6-3** Cathode-ray beam is deflected towards a positively charged plate.

ray tube has a partial vacuum in it, whereas a gas discharge tube has some kind of gas in it. When electricity flows through a gas discharge tube, a distinct light is produced. The light produced is of a unique color that depends upon the gas used in the tube. Discharge tubes are used in neon signs and mercury vapor and sodium vapor street lights. You probably have seen a hydrogen discharge tube as a "black" light source or as a sunlamp. When electricity is passed through a specific gas in a discharge tube, it gives off light of a characteristic color. Hydrogen gives off a distinct reddish-purple light; neon, a reddish-orange light.

Sunlight or white light from an incandescent light bulb passing through a prism or diffraction grating is dispersed into light having all the colors of the rainbow; red, orange, yellow, green, blue, indigo, and violet **(ROY G. BIV).** However, when the light from a gas discharge tube is passed through a prism or diffraction grafting the dispersion contains not all colors of the rainbow, but a unique mixture of certain colors (see Fig. 6-4). For instance, the light from a hydrogen discharge tube is revealed to be a distinct mixture of red, purple, and bluish-green. No other element gives the same colors. To the human eye the mixture of colors appears as reddish-purple light; we only see the separate colors when the light is diffracted. An element's fingerprint is a pattern of light from a discharge tube and is called its emission spectrum. It indicates an element's unique atomic structure and the arrangement of its electrons, a message from the inner spaces of atoms. Experimental evidence obtained from gas discharge tubes has been very important to the development of models or theories of atomic structure.

It is not possible to see inside an atom but it is possible to develop models of atomic structure using experimental evidence about the behavior of collections of atoms. A good model will answer many questions and contribute to chemical vision. How are the electrons distributed about the nucleus? How does atomic structure explain the properties of the various elements? How do atoms enter into chemical combination?

**FIGURE 6-4**   A light beam from a discharge tube passes through a diffraction grating and is separated into discrete bands of colored light.

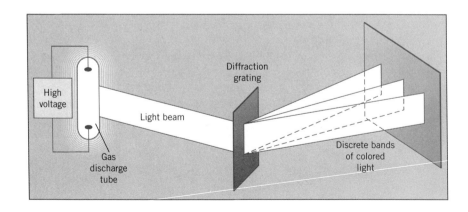

## 6-6 THE BOHR MODEL

The modern model of the atom has evolved from the ideas of the nuclear atom. Niels Bohr developed a model to explain how electrons orbit the nucleus. He developed his model in an attempt to try to explain why a hydrogen discharge tube produces light that is characteristic of hydrogen. According to the **Bohr model,** a hydrogen atom has a nucleus with a 1+ charge, around which an electron is moving in an orbit.

The electron has energy that results from its motion and position. This energy is somewhat analogous to the energy of a ball tied to a string and twirled in a circular motion. The ball has energy of motion (kinetic energy) and energy of position (potential energy) as it is held by the string. The electron has kinetic energy of motion and potential energy arising from attraction between the positively charged nucleus and the negative charges of the electron.

Bohr made the revolutionary assumption that an electron in an atom has quantized energy. This means that an electron can only have a fixed amount of energy or some multiple of that amount but cannot have any intermediate energies. In terms of orbits this means that an electron in a hydrogen atom is found only at specific distances from the nucleus in specific orbits corresponding to the allowable quantized energies.

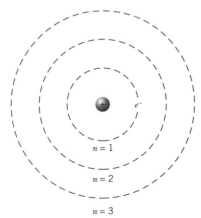

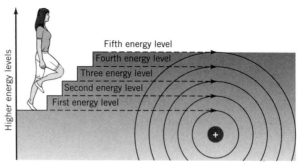

Higher energy levels

Fifth energy level
Fourth energy level
Three energy level
Second energy level
First energy level

**FIGURE 6-5**    While hopping up a staircase a person may take on only specific "quantized" energy positions. The steps represent energy levels. The person cannot stop between steps or energy levels. Likewise, electrons occupy specific quantized energy levels; they cannot exist between levels. An electron occupies a specific level corresponding to its energy.

The idea of quantized energy needs further explanation. As an analogy to the quantized energy of an electron, consider climbing a staircase by hopping on one leg. When hopping you can take on only "quantized" positions as you climb, as shown in Figure 6-5. That is, you cannot remain at any position between steps. As you climb the staircase, your energy (potential energy with respect to the bottom of the staircase) is of a certain value at the first step. Your energy increases by some fixed amount as you climb to higher steps.

The Bohr model gives a view of a hydrogen atom as a central nucleus orbited by an electron in an orbit that depends upon its energy. The electron can only have a specific energy out of a set of possible energy values. Each allowed energy corresponds to a specific orbit. Orbits corresponding to different energies are located at different distances from the nucleus. The possible quantized orbits of the electron are the **energy states** or **energy levels.** An electron is located in one of these possible orbits depending on its energy. It cannot be found anywhere between these orbits since the orbits represent the quantized energy states.

### 6-7  QUANTUM JUMPS

In the Bohr view, the single electron in a hydrogen atom normally occupies the lowest energy level. The lowest level is called the **ground state.** This level is the one closest to the nucleus. If the electron gains energy from some outside source, such as a flame or electricity, it can jump from a lower energy level to a higher level. Such a change in energy levels is a **quantum jump** or **quantum leap.** Similar to the stair-hopping analogy, the electron can hop from level to level but cannot stop anywhere between levels.

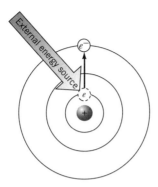

Since the development of modern models of atomic structure, quantum jump and quantum leap have become literary terms meaning a dramatic or profound change.

When an electron jumps to a higher energy level, it is said to be excited or in an **excited state.** If an electron gains enough energy it breaks away from the atom leaving a positive ion. Positive ions or cations are formed by the loss of electrons from atoms. Those excited electrons that do not break away from atoms occupy various higher energy levels. These excited electrons spontaneously fall back to lower energy levels. They fall to various lower levels ultimately reaching the ground state (see Fig. 6-6). As an excited electron falls from a higher energy level to a lower energy level, it releases its excess energy. This is analogous to what would happen if you were to leap down a staircase. You give off the potential energy as kinetic energy or energy of motion. As an electron jumps from a higher to a lower energy level it releases an amount of energy that corresponds to the difference in energy between the two levels.

The energy released by electrons falling to lower energy levels in atoms very often takes the form of radiant energy or light. Some of the light you see in a flame or the colored light given off by fireworks comes from the loss of energy of excited electrons. Since atoms of various elements differ in the number of electrons and the energies of the electron levels, each element emits different characteristic colors of light when energetically excited. In fireworks, compounds of specific elements are added to produce desired colors; sodium compounds for yellow, strontium compounds for red, copper compounds for blue, and barium compounds for green. Chemistry in Action 6-1 is about looking at the colors in a flame.

Bohr's model is used to explain why a hydrogen discharge tube gives off specific colors of light. Ground state hydrogen atoms have their electrons located in the lowest possible energy level. In the discharge tube electrons within the hydrogen atoms can gain energy from the electron

**FIGURE 6-6** Some possible quantum jumps in a hydrogen atom according to the Bohr model.

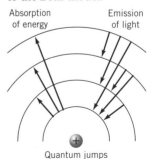

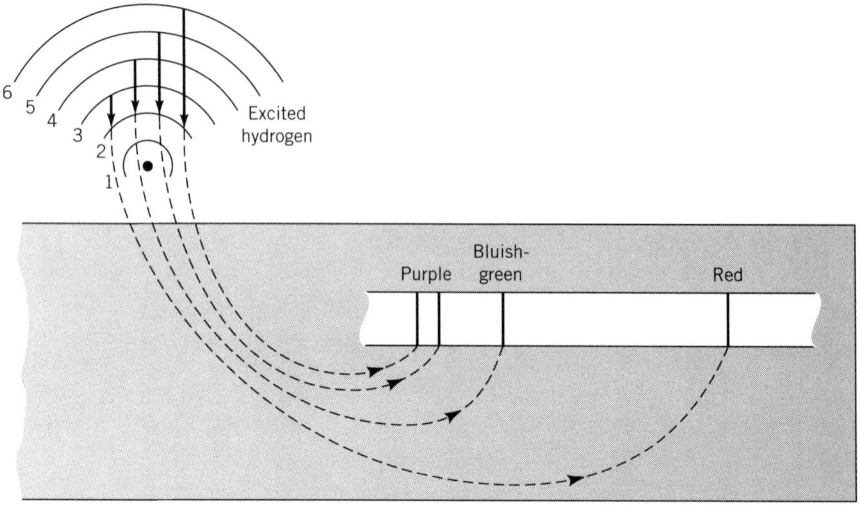

**FIGURE 6-7** An interpretation of the quantum jumps in excited hydrogen atoms. Jumps from higher levels to level 2 produce red, blue-green, and purple light. Jumps from and to other levels produce ultraviolet and infrared light, which is invisible.

beam in the tube and become excited to higher energy levels. The excited electrons spontaneously undergo quantum jumps in which they move from the higher energy levels to the lower levels. As the electrons drop to lower energy levels their excess energy is released as light energy. The hydrogen atom has a specific set of allowable energy levels for its electron. Thus, only a fixed set of quantum jumps are possible for electrons in hydrogen atoms. Consequently, specific colors of light corresponding to the possible quantum jumps are expected. Bohr used his model to show that hydrogen is expected to emit the specific colors of light as seen in its emission spectrum (see Fig. 6-7).

In neon discharge tubes the neon gas is energetically excited by passing electricity through the tube. Some of the electrical energy is absorbed by the neon atoms as the electrons are excited. As the excited electrons fall back to lower energy levels, the characteristic reddish-orange light of a neon tube is emitted. As long as the electricity flows, the atoms become excited and, as they relax and lose their energy, the tube emits light. A different gas gives a different color. Helium gives a yellowish-pink light, argon a lavender light, krypton a silvery-white light, and xenon a blue light. Hydrogen gives the distinctive reddish-purple light. Each color is characteristic of a particular gas. However, a hydrogen discharge tube also emits invisible ultraviolet light similar to the ultraviolet light in sunlight. This is why hydrogen discharge tubes can be used as sunlamps.

## 6-8 THE QUANTUM MECHANICAL MODEL

The Bohr model gives a description of a simple hydrogen atom. Alas, it was not a good model for complex atoms with many electrons. To encourage further development of theories of atomic structure Bohr established a scientific institute in Copenhagen, Denmark. In the 1920s and 1930s many scientists contributed to the work of this institute. As a result, one of the most important scientific models of the 20th century developed: the **quantum mechanical model** of atomic structure.

In the Bohr model an electron follows a determined circular orbit. The quantum mechanical model views electrons in a significantly different way. An electron in an atom has energy because of its motion and position. In the quantum mechanical model an electron of a specific energy takes the form of a wave called an electron orbital. Energy is carried through water as a wave; when a pebble is dropped into a pool of still water, some of the energy forms waves that move out from the point of impact. Electron orbitals are wave patterns that surround the nucleus of an atom. The electron energy waves remain in place around the nucleus; they do not move and dissipate like energy waves in water. Different electron energies correspond to specific wave patterns.

The mathematical methods used in the model are called quantum mechanics or wave mechanics. Using quantum mechanics, a complex set of equations predicts the number and nature of the quantized energy levels of electrons in atoms. The equations are sometimes called wave equations because they relate to electron waves. To illustrate how an electron can be visualized by the quantum mechanical model consider the bicycle wheel shown in Figure 6-8. At rest you can clearly see all of the spokes and you can easily stick your finger between the spokes. When the wheel is moving rapidly the images of the spokes appear to blur and you would hurt yourself if you tried to put your finger between them. Thus, the moving spokes occupy or fill the entire volume of the wheel. Just as the moving spokes define a volume in which there is a high probability of finding a spoke, an electron defines a certain wave pattern in which it occupies or fills the en-

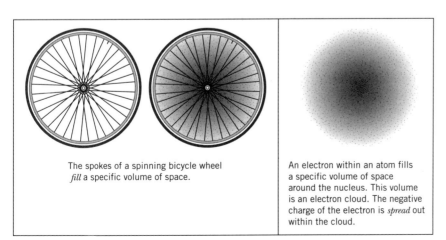

The spokes of a spinning bicycle wheel *fill* a specific volume of space.

An electron within an atom fills a specific volume of space around the nucleus. This volume is an electron cloud. The negative charge of the electron is *spread* out within the cloud.

**FIGURE 6-8**    A bicycle wheel cloud and an electron cloud.

tire volume of the pattern. The electron orbitals represent three-dimensional probability volumes of electrons.

The electron is not considered as a particle in a specific orbit following a determined path, but rather as a cloud filling a volume about the nucleus just as the spokes fill a "spoke cloud" in the moving wheel. The term **electron cloud** comes from the idea that the negatively charged electron is filling the volume as a wave. It spreads over the volume of space. The shapes of these orbitals correspond to the regions of space "filled" by the rapid motion of the electron. The term **electron orbital** emphasizes that these are not orbits like those of the Bohr model but rather electron clouds. The wave equations of the model define a set of possible wave patterns that an electron may occupy. Thus, within an atom, there is a set of electron orbitals that represent the quantized energies of the various electrons in the atom. Electrons filling the orbitals give shape and substance to atoms.

The quantum mechanical model is more abstract than the Bohr model. However, it does allow the visualization of atoms in three-dimensional space. Atoms are visualized as diffuse fuzzy balls of electron clouds surrounding tiny nuclei. Furthermore, the model also defines a pattern for the electron distribution in the various energy levels of a hydrogen atom. It is assumed that this pattern is the same for all atoms and it is used to describe how electrons distribute in atoms with many electrons.

## 6-9 ENERGY STATES

According to the quantum mechanical model, there are several possible **main or principal energy states** in which the electrons can be found. These main energy states are the energy levels or energy shells. Each main energy level has one or more energy states called **energy sublevels** or **energy subshells.** Furthermore, each energy sublevel consists of one or more specific electron energy states. These specific energy states within the energy sublevels are the **electron orbitals.** Electrons can occupy these orbitals. We can think of the main energy levels as a set of large boxes, one box for each level. Opening the boxes, we find that each contains a certain number of smaller boxes, the energy sublevels. Inside the smaller boxes we find a certain number of compartments that serve as positions for the electrons. These compartments are the electron orbitals. Of course energy levels are not boxes but they are shells of electron clouds of varying complexity. See Figure 6-9 for a summary of the quantum mechanical view of the energy states in an atom. The description of an atom in terms of energy states does not provide us with a picture or mental image of the atom. However, as shown in Figure 6-10, the theory does provide good images of atoms using three-dimensional shapes and sizes of the electron clouds or orbitals.

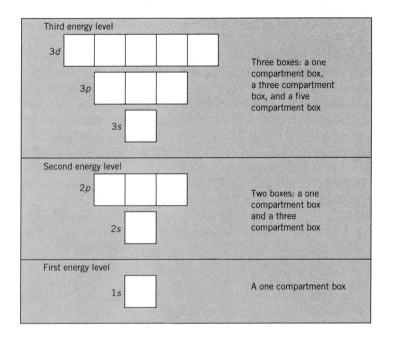

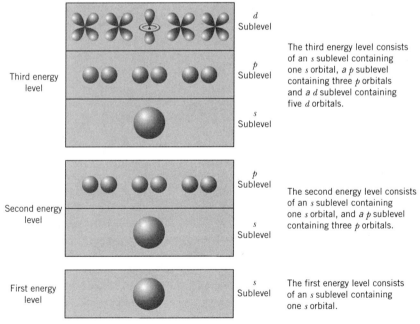

**FIGURE  6-9**    Representation of the quantum mechanical model of the atom.

**FIGURE 6-10**
Representations of some atomic orbitals. The x, y, and z axes are given to show the three-dimensional spatial distribution of the orbitals. The first energy level contains only the 1s orbital; the second energy level contains the 2s orbital and the three 2p orbitals. The p orbitals are sometimes distinguished as $p_x$, $p_y$, and $p_z$ to emphasize their different spatial orientations.

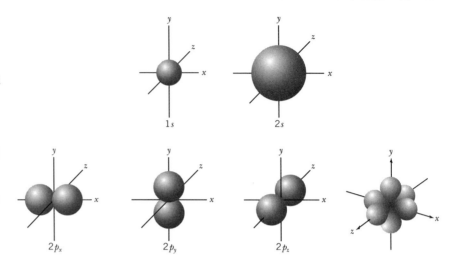

## 6-10 ENERGY LEVEL STRUCTURE

An atom of a given element consists of a nucleus, with a charge corresponding to the atomic number, surrounded by electron clouds that overlap and extend outward from the nucleus. The atomic number of an element tells us how many electrons are in its atoms. The electrons occupy the lowest available energy states. However, several questions arise at this point. How many electrons can fit in each main energy level and in each sublevel? How many electrons can fit in each electron orbital? How do the energies of the various orbitals and sublevels differ with respect to one another?

With answers to these questions it is possible to deduce the arrangement of electrons in the atom of an element. Since we know the number of electrons in the atom, we can hypothetically build an atom by placing electrons in the orbitals starting at the lowest energy position and working up to higher positions. The electronic arrangement of an atom of an element is its electronic structure or **electronic configuration.** Each element has a unique configuration.

Since there are a variety of electron energy levels, it is useful to distinguish one from another by numbers. The numbers used to designate the main energy levels are

1    First energy level
2    Second energy level
3    Third energy level
4    Fourth energy level, and so on.

The energy-level number is called the **principal quantum number.**

energy level    1   2   3   4   5   6   7 . . .

distance from nucleus increases
$\longrightarrow$
energy increases

Each energy level consists of a specific number of sublevels. It is interesting that the number of sublevels in an energy level is the same as the energy level number. That is, the first energy level has one sublevel, the second energy level has two sublevels and so on. The letters *s, p, d,* and *f* are used to reference sublevels. Sublevel structure is more complex for higher levels.

1      The first energy level has one sublevel, an *s* sublevel

2      The second energy level has two sublevels, an *s* sublevel and a *p* sublevel

3      The third energy level has three sublevels, an *s* sublevel, a *p* sublevel, and a *d* sublevel

4      The fourth energy level has four sublevels, an *s* sublevel, a *p* sublevel, a *d* sublevel, and an *f* sublevel

Within a given energy level the sublevel energies differ. The order of sublevel energies is

sublevel    *s*    *p*    *d*    *f* . . .

increasing energy
$\longrightarrow$

This means that within an energy level, an *f* sublevel is of higher energy than a *d* sublevel; a *d* sublevel is of higher energy than a *p* sublevel; and a *p* sublevel is of higher energy than an *s* sublevel.

A specific sublevel is identified by giving the main energy-level number followed by the letter corresponding to the sublevel. So we refer to the 1*s* (pronounced one-ess) sublevel, the 2*p* (pronounced two-pee) sublevel, or the 3*d* (pronounced three-dee) sublevel. The sublevel structure of the first four energy levels is

| ENERGY LEVEL | SUBLEVELS |
|---|---|
| First | 1*s* |
| Second | 2*s*, 2*p* |
| Third | 3*s*, 3*p*, 3*d* |
| Fourth | 4*s*, 4*p*, 4*d*, 4*f* |

Notice, as shown in Table 6-1, that as we go to higher energy levels they become larger and more complex. Thus, they can hold more electrons. Each sublevel consists of a certain number of electron orbitals. See Figure 6-10 for illustrations of the shapes of orbitals.

An *s* sublevel has one orbital—the *s* orbital.

A *p* sublevel has three orbitals—the *p* orbitals.

A *d* sublevel has five orbitals—the *d* orbitals.

An *f* sublevel has seven orbitals—the *f* orbitals.

This information is summarized in Table 6-1. Notice that the pattern is that the number of orbitals in a sublevel increases as a sequence of odd numbers (1, 3, 5, 7 . . .).

Orbitals within a given sublevel have the same electron energies. For example, the three *p* orbitals of a *p* sublevel are equivalent in electron energy but the orbitals are arranged in different spatial positions as shown in Figure 6-10.

**Table 6-1**   *Sublevel Structure of the First Four Energy Levels*

| ENERGY LEVEL | TYPE OF SUBLEVEL | NUMBER OF ORBITALS OF GIVEN TYPE | ELECTRON STATE NOTATION |
|---|---|---|---|
| First | *s* | One *s* | 1*s* |
| Second | *s* | One *s* | 2*s* |
|  | *p* | Three *p* | 2*p* |
| Third | *s* | One *s* | 3*s* |
|  | *p* | Three *p* | 3*p* |
|  | *d* | Five *d* | 3*d* |
| Fourth | *s* | One *s* | 4*s* |
|  | *p* | Three *p* | 4*p* |
|  | *d* | Five *d* | 4*d* |
|  | *f* | Seven *f* | 4*f* |

## 6-11  ELECTRON SPIN

The important question at this point is: How many electrons can a single orbital hold? Electrons have negative charges and repel one another. Thus, we might expect that an orbital can only accommodate one electron. However, electrons have another property that makes it possible for two electrons to occupy a given orbital. Electrons have a property called spin that is analogous to the spinning of a top. A top can spin in one direction or in the opposite direction. Similarly, a given electron can have only one of two possible spin states that we can represent symbolically as spin up or spin down arrows ( ↑ or ↓ ). The spin of electrons gives them properties somewhat like tiny bar magnets. If two bar magnets are forced together so that their like poles align, they repel one another. When they arrange with their opposite poles aligned, they do not repel one another. In a similar fashion two electrons cannot occupy the same orbital if they have the same spin but can occupy the same orbital if they have opposite spins. Wolfgang Pauli, a German scientist involved in the development of the quantum mechanical model, first expressed this idea, which is now called the Pauli Exclusion Principle. See margin

An orbital that contains two electrons is a **filled orbital** ⬆⬇, while an orbital with one electron is a **half-filled orbital** ⬆. Electrons repel one another and tend to occupy single orbitals whenever possible. This becomes important when there are several equivalent orbitals within a sublevel. Friedrich Hund, another German scientist, described the pattern electrons follow in such cases. See margin

⬆|⬆|   rather than   ⬆⬇|   |   and   ⬆|⬆|⬆   rather than   ⬆⬇|⬆|

This pattern is called Hund's rule and it has significance when deciding where the electrons are located within the energy sublevels of atoms. Incidently, these somewhat abstract ideas of electron spin and the Pauli exclusion principle do have practical significance. The magnetic properties of spinning electrons are responsible for the magnetic properties of matter. Chemistry in Action 6-2 considers these properties by working with an electromagnet. This property is most notable with iron. Pieces of iron can be made into permanent magnets that take advantage of electron spins in iron atoms. Furthermore, moving electrons in a wire can produce a magnetic field used to make electric motors work. An obvious, but notable, illustration of the Pauli exclusion principle is that two solids cannot occupy the same space at the same time. When the solids are forced together, the electrons repel one another with great force.

**Pauli Exclusion Principle:**

*An orbital can contain a maximum of two electrons and those two electrons must have opposite spin states.*

**Hund's Rule:**

*Electrons in a set of orbitals within a sublevel tend to occupy empty orbitals and pair only when all orbitals have one electron.*

ACTIVITY
6-1

Why is it not possible to walk through a brick wall?

## 6-12  RELATIVE ENERGIES OF THE SUBLEVELS

The maximum capacity of an orbital is two electrons. Since we know how many orbitals are in the various sublevels we can easily predict the maximum electron capacity of the sublevels.

| | | | |
|---|---|---|---|
| *s* sublevel | one *s* orbital | 2 electrons | an *s* sublevel can hold 1 or 2 electrons |
| *p* sublevel | three *p* orbitals | 6 electrons | a *p* sublevel can hold 1 to 6 electrons |
| *d* sublevel | five *d* orbitals | 10 electrons | a *d* sublevel can hold 1 to 10 electrons |
| *f* sublevel | seven *f* orbitals | 14 electrons | an *f* sublevel can hold 1 to 14 electrons |

To figure out how the electrons are distributed in an atom of an element, we hypothetically place the electrons of the atom in the various energy sublevel positions. To do this we start at the lowest position and move to a higher position when the lower level is full. Recall from the box analogy, referred to in Section 6-9, that the orbitals are like small boxes within the sublevel boxes. Each main energy-level box contains one or more sublevel boxes. The electrons are distributed in the boxes with the lowest sublevel boxes filled first. Since the boxes have limited capacities, they fill with electrons and any additional electrons have to be placed in the next-higher-sublevel boxes. The process continues from level to level until all of the electrons have been placed.

To figure out the electronic configurations of the atoms of elements it is important to know the order of **increasing energy of the sublevels.** That is, we need to know which sublevel is the lowest energy, which is the next lowest, and so on. This order is generally

$$1s\ 2s\ 2p\ 3s\ 3p\ 4s\ 3d\ 4p\ 5s\ 4d\ 5p\ 6s\ 4f\ 5d\ 6p\ 7s\ 5f\ 6d$$

The order follows a pattern, and Figure 6-11 shows a way to remember it.

**FIGURE 6-11**
Relative energies of the sublevels.

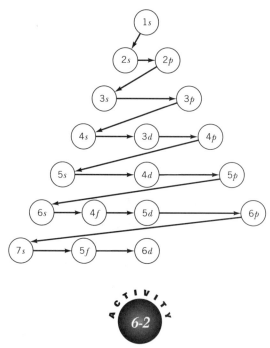

ACTIVITY
6-2

The maximum electron capacity of an energy level is given by the expression: $2n^2$, where $n$ is the energy level number. Confirm this expression by filling the number of electrons in the following table and counting the total electrons in each level.

| | | | | |
|---|---|---|---|---|
| Level 1 | 1s sublevel ____ electrons | | | total ____ |
| Level 2 | 2s sublevel ____ electrons | 2p sublevel ____ electrons | | total ____ |
| Level 3 | 3s sublevel ____ electrons | 3p sublevel ____ electrons | 3d sublevel ____ electrons | total ____ |
| Level 4 | 4s sublevel ____ electrons | 4p sublevel ____ electrons | 4d sublevel ____ electrons | 4f sublevel ____ | total ____ |

## 6-13  ENERGY OVERLAP

Note that, according to the sublevel energy pattern, the 4*s* sublevel is of lower energy than the 3*d* sublevel. This is an example in which a sublevel of a higher main energy level has lower energy than a sublevel of a lower main energy level. **Energy overlap** is the term often used for this situation. It is not possible to draw a picture of energy to show the overlap. However, as shown in Figure 6-12, you could imagine the energy levels as a series of ladders extending upwards. Where one ladder ends the next begins. The ladders, representing various energy levels, have a number of rungs corresponding to the sublevels. Energy overlap occurs when the higher rungs of a ladder extend beyond the lower rungs of the next ladder. The higher rungs of the first ladder are of higher energy than the lower rungs of the second ladder and correspond to an energy overlap.

As a simple explanation of energy overlap consider that the sublevels within a main level increase in energy in the order

$$\text{-----------------} \rightarrow \text{energy}$$
$$s \quad p \quad d \quad f$$

and the levels increase in energy in the order

$$\text{-----------------} \rightarrow \text{energy}$$
$$1 \quad 2 \quad 3 \quad 4 \quad 5 \quad 6 \quad 7$$

In some cases the increase in energies of sublevels causes the higher-energy sublevels to exceed the lower-energy sublevels of the next main energy level. For example, the sublevel energy pattern for the third and fourth main energy level are shown on the next page.

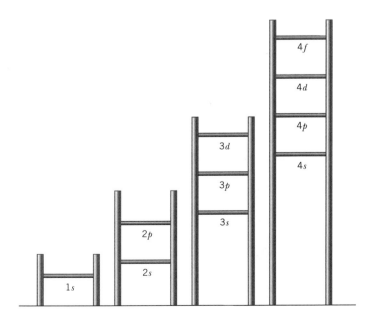

**FIGURE 6-12**
Ladder analogy of the sublevels showing energy overlap.

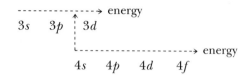

The 3*d* sublevel overlaps the 4*s* sublevel. List three other energy sublevel overlaps by indicating which sublevels are involved.

Thus, the energy of the 3*d* sublevel is higher than the energy of the 4*s* sublevel and we say that the 3*d* sublevel overlaps the 4*s* sublevel. Figure 6-11 shows that there are several energy sublevel overlaps between various main levels. It is important to realize that energy overlap does not mean that the higher energy sublevel is physically further from the nucleus than the lower energy sublevel. That is, the 4*s* orbital is not necessarily located closer to the nucleus than the orbitals of the 3*d* sublevel although the 4*s* sublevel is lower in energy than the 3*d* sublevel. The point is that the 4*s* sublevel is actually further from the nucleus than the 3*d* sublevel but the 4*s* fills with electrons before the 3*d*.

## 6-14 THE ELECTRONIC CONFIGURATIONS OF THE ELEMENTS

We know that the atomic number of an element is equal to the number of electrons in neutral atoms of that element. Elements are listed in the periodic table by increasing atomic numbers. Atoms of an element in the table have one more electron than the atoms of the preceding element. Hydrogen, atomic number 1, has one electron. Helium, atomic number 2, has two electrons, and so on.

If we keep in mind the number of electrons that various sublevels hold, we can use the order of sublevel energies to write the electronic structure of an element. **An *s* sublevel holds 1 or 2 electrons. The *p* sublevel holds from 1 to 6 electrons; the *d* sublevel holds from 1 to 10 electrons; and the *f* sublevel holds from 1 to 14 electrons.** The electronic structure of an element is represented by the sublevel distribution of the electrons. To do this, we distribute electrons in the lowest-energy sublevel first, and when it is full, we use the next-higher-energy sublevel. We continue until all of the electrons are placed in sublevels. To illustrate the process, we can use the ladder analogy shown in Figure 6-13. Imagine birds landing on the rungs of the ladders. Each rung can hold a specific number of birds. Suppose the birds prefer the lower rungs (lower-energy sublevels). They occupy these rungs first. As the lower rungs are filled, the birds locate in the next higher rungs. Also when sublevel energy overlap occurs, the birds would prefer the lower-energy rung even when it is in the next higher ladder (next higher main energy level).

The one electron in a hydrogen atom locates in the *s* sublevel of the first main energy level. The standard method of showing the **electronic configuration** is shown on the next page.

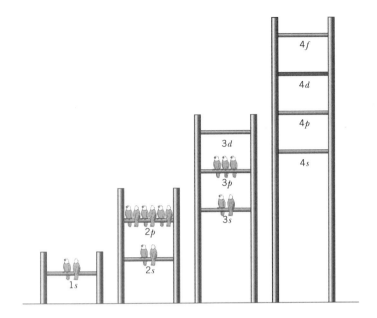

**FIGURE 6-13**   Bird roost analogy for electrons in sublevels.

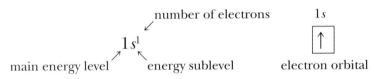

This sublevel or orbital notation shows one electron in the $s$ sublevel of the first main energy level. The number on the left is the main energy level and the letter is the energy sublevel. The superscript digit shows the number of electrons in that sublevel. Note, on the right above, the electron in the $1s$ orbital is represented as an arrow in a box. This box is given for comparison and is not typically shown as part of an electronic configuration. The two electrons in a helium atom are located in the lowest-energy position; that is, they are located in the $s$ sublevel of the first main energy level. The electronic configuration of helium is

$$1s^2 \qquad \begin{array}{c} 1s \\ \boxed{\uparrow\downarrow} \end{array}$$

The next element, lithium, has atoms with three electrons. Two of these electrons occupy the lowest energy position, the $1s$ sublevel, and the third electron must occupy the next-lowest-energy position, the $2s$ sublevel. Thus, the electronic structure of lithium is

$$1s^2 2s^1 \qquad \begin{array}{cc} 1s & 2s \\ \boxed{\uparrow\downarrow} & \boxed{\uparrow} \end{array}$$

To figure out the electronic configuration of an element, all we have to do is decide how many electrons are involved then distribute these electrons in the sublevels; we start in the lowest-energy sublevel and continue until all of the electrons are distributed.

| ELEMENT | ATOMIC NUMBER | ELECTRONIC CONFIGURATION | ELECTRON ORBITAL BOX NOTATION | | |
| --- | --- | --- | --- | --- | --- |
| | | | 1s | 2s | 2p |
| Beryllium, Be | 4 | $1s^2 2s^2$ | ↑↓ | ↑↓ | ☐ ☐ ☐ |
| Boron, B | 5 | $1s^2 2s^2 2p^1$ | ↑↓ | ↑↓ | ↑ ☐ ☐ |
| Carbon, C | 6 | $1s^2 2s^2 2p^2$ | ↑↓ | ↑↓ | ↑ ↑ ☐ |
| Nitrogen, N | 7 | $1s^2 2s^2 2p^3$ | ↑↓ | ↑↓ | ↑ ↑ ↑ |
| Oxygen, O | 8 | $1s^2 2s^2 2p^4$ | ↑↓ | ↑↓ | ↑↓ ↑ ↑ |
| Fluorine, F | 9 | $1s^2 2s^2 2p^5$ | ↑↓ | ↑↓ | ↑↓ ↑↓ ↑ |
| Neon, Ne | 10 | $1s^2 2s^2 2p^6$ | ↑↓ | ↑↓ | ↑↓ ↑↓ ↑↓ |

The box notation shows that, as the $2p$ sublevel fills with electrons, one electron is placed in each of the three $p$ orbitals in accordance with Hund's rule. Once again, we can use the ladder analogy of Figure 6-13. The ladder rungs have specific numbers of perching spots. An $s$ rung has one, a $p$ rung three, a $d$ rung five, and an $f$ rung seven. There is room for two birds on each spot. As the birds occupy a rung, they prefer their own spot if possible, so they occupy empty spots on a rung. If all spots are used, they will pair up with another bird. This happens until the rung is filled, then the next higher rung is used.

The boxes shown along with the electronic configurations represent the electrons in the orbitals. Notice where the electrons are located in the $p$ orbitals of a $p$ sublevel. Nitrogen has three unpaired electrons in the $2p$ sublevel, one in each orbital. As more electrons are added, they must be placed in half-filled orbitals. This results in neon having a completely filled $2p$ sublevel with three filled $p$ orbitals. To write the electronic configuration of an element, distribute its electrons within the sublevels using the order of their energies and their electron capacities ($s$ 2-electron capacity, $p$ 6-electron capacity, $d$ 10-electron capacity, $f$ 14-electron capacity).

### EXAMPLE 6.1

What is the electronic configuration of phosphorus, atomic number 15? Phosphorus atoms have 15 electrons. The order of energy of the sublevels is

$$1s\ 2s\ 2p\ 3s\ 3p \ldots$$

Distributing the 15 electrons in these sublevels so that the lower-energy sublevels are filled gives the electronic configuration

$$1s^2\, 2s^2\, 2p^6\, 3s^2\, 3p^3$$

ACTIVITY 6-4

What is the electronic configuration of zinc, atomic number 30?

## 6-15 OUTER-LEVEL ELECTRONS

Vanadium has an atomic number of 23. Note that in the electronic configuration of vanadium $(1s^2 2s^2 2p^6 3s^2 3p^6 4s^2\, 3d^3)$ the $4s$ sublevel comes before the $3d$ sublevel. However, when writing electronic configurations, the distribution is often written by grouping all the sublevels of each energy level. The configuration of vanadium is rewritten as

$$1s^2 2s^2 2p^6 3s^2 3p^6 3d^3 4s^2$$

Such notation emphasizes the number of outer-energy-level electrons. **Outer-energy-level electrons** are those in the highest-numbered energy levels (e.g., vanadium has two outer-energy-level electrons, the $4s$ electrons). We shall soon see why the notation that shows the number of outer-level electrons is useful. Just remember that outer-level electrons are those electrons with the highest-numbered energy level.

## 6-16 THE PERIODIC TABLE AND ELECTRONIC CONFIGURATIONS

The periodic table lists the elements according to increasing atomic number and its groups contain families of elements having similar properties. It was not until the properties of the elements were related to electronic configurations that an explanation of the unique shape of the periodic table became apparent. It was found that the elements having similar electronic configurations in their outermost energy levels are those with similar properties. For example, the electronic configurations of lithium, sodium, potassium, rubidium, cesium, and francium are

| ATOMIC NUMBER | | |
|---|---|---|
| 3 | Li | $1s^2 2s^1$ |
| 11 | Na | $1s^2 2s^2 2p^6 3s^1$ |
| 19 | K | $1s^2 2s^2 2p^6 3s^2 3p^6 4s^1$ |
| 37 | Rb | $1s^2 2s^2 2p^6 3s^2 3p^6 3d^{10} 4s^2 4p^6 5s^1$ |
| 55 | Cs | $1s^2 2s^2 2p^6 3s^2 3p^6 3d^{10} 4s^2 4p^6 4d^{10} 5s^2 5p^6 6s^1$ |
| 87 | Fr | $1s^2 2s^2 2p^6 3s^2 3p^6 3d^{10} 4s^2 4p^6 4d^{10} 4f^{14} 5s^2 5p^6 5d^{10} 6s^2 6p^6 7s^1$ |

| Period | 1 | 2 | 3 | 4 | 5 | 6 | 7 | 8 | 9 | 10 | 11 | 12 | 13 | 14 | 15 | 16 | 17 | 18 NOBLE GASES |
|---|---|---|---|---|---|---|---|---|---|---|---|---|---|---|---|---|---|---|
| 1 | H $1s^1$ | | | | | | | | | | | | | | | | | He $1s^2$ |
| 2 | Li $2s^1$ | Be $2s^2$ | | | | | | | | | | | B $2s^22p^1$ | C $2s^22p^2$ | N $2s^22p^3$ | O $2s^22p^4$ | F $2s^22p^5$ | Ne $2s^22p^6$ |
| 3 | Na $3s^1$ | Mg $3s^2$ | | | | | | | | | | | Al $3s^23p^1$ | Si $3s^23p^2$ | P $3s^23p^3$ | S $3s^23p^4$ | Cl $3s^23p^5$ | Ar $3s^23p^6$ |
| 4 | K $4s^1$ | Ca $4s^2$ | Sc $3d^14s^2$ | Ti $3d^24s^2$ | V $3d^34s^2$ | Cr $3d^54s^1$ | Mn $3d^54s^2$ | Fe $3d^64s^2$ | Co $3d^74s^2$ | Ni $3d^84s^2$ | Cu $3d^{10}4s^1$ | Zn $3d^{10}4s^2$ | Ga $4s^24p^1$ | Ge $4s^24p^2$ | As $4s^24p^3$ | Se $4s^24p^4$ | Br $4s^24p^5$ | Kr $4s^24p^6$ |
| 5 | Rb $5s^1$ | Sr $5s^2$ | Y $4d^15s^2$ | Zr $4d^25s^2$ | Nb $4d^45s^1$ | Mo $4d^55s^1$ | Tc $4d^55s^2$ | Ru $4d^75s^1$ | Rh $4d^85s^1$ | Pd $4d^{10}$ | Ag $4d^{10}5s^1$ | Cd $4d^{10}5s^2$ | In $5s^45p^1$ | Sn $5s^25p^2$ | Sb $5s^25p^3$ | Te $5s^25p^4$ | I $5s^25p^5$ | Xe $5s^25p^6$ |
| 6 | Cs $6s^1$ | Ba $6s^2$ | La $5d^16s^2$ | Hf $5d^26s^2$ | Ta $5d^36s^2$ | W $5d^46s^2$ | Re $5d^56s^2$ | Os $5d^66s^2$ | Ir $5d^76s^2$ | Pt $5d^96s^1$ | Au $5d^{10}6s^1$ | Hg $5d^{10}6s^2$ | Tl $6s^26p^1$ | Pb $6s^26p^2$ | Bi $6s^26p^3$ | Po $6s^26p^4$ | At $6s^26p^5$ | Rn $6s^26p^6$ |
| 7 | Fr $7s^1$ | Ra $7s^2$ | Ac $6d^17s^2$ | | | | | | | | | | | | | | | |

**FIGURE 6-14**
Similarities in outer-level configurations of elements. Members of the same group in the table have similar electronic configurations. This is not a complete table.

These six elements have very similar chemical properties, and note that each has one electron in the outermost or highest-numbered main energy level. This similarity in configuration is denoted generally as $ns^1$ where $n$ refers to the number of the outer main energy level. These elements share a similar $ns^1$ outer-level configuration with $n$ ranging from 2 to 7. As expected, these elements make up one of the groups of the periodic table.

Elements can be arranged by increasing atomic number so that the elements of similar electronic configuration are in groups. Such an arrangement gives the **periodic table.** Ordering elements by similarities in electronic configuration gives essentially the same table that Mendeleev created over one hundred years ago. It is impressive that he did it long before any understanding of atomic structure existed. His creative arrangement anticipated and now confirms the quantum mechanical model of the atom. And in turn the model predicts the shape of the table in terms of the sublevel structures of atoms. Figure 6-14 shows the electronic configurations of most of the elements and emphasizes that elements of similar configuration are located in the same group of the table. Since the table is based upon the electronic configuration of elements, its shape corresponds to the filling of the energy sublevels. Recall the sublevel filling pattern in Figure 6-11. An $s$ sublevel can hold 2 electrons, so the $s$-sublevel groups are 2 elements wide. A $p$ sublevel can hold 6 electrons, so the p-sublevel groups are 6 elements wide. A $d$ sublevel has can hold 10 electrons, so the d-sublevel part of the table is 10 elements wide. An $f$ sublevel can hold 14 electrons, so the f-sublevel part of the table is 14 elements wide. Once we realize that the elements are arranged in the table according to the filling pattern of the various sublevels, we have an explanation for the shape of the periodic table. In summary, the periodic table has its curious shape because the electron energy sublevels in atoms have specific arrangements and capacities. See Figure 6-15 on the next page.

## 6-17  PARTS OF THE PERIODIC TABLE

The elements are sometimes classified according to the part of the periodic table in which they are found. Figure 6-15 shows such a classification. The elements are classified into four blocks, the **s-block elements,** the **p-block elements,** the **d-block elements,** and the **f-block elements.** Note that the s block is 2 elements wide, the p block 6 elements wide, the d block 10 elements wide, and the f block 14 elements wide. Confirm this by counting them yourself. The f-block elements actually should be located within the d block, but this would make the table inconveniently wide. To understand what this means look closely at the atomic numbers of the elements in the f block and those in the lower parts of the d block. To shorten the space occupied by the periodic table, the f-block elements are normally placed below the d-block elements.

The right-hand group of the periodic table contains a unique collection of chemically inactive gases known as the **noble gases.** These gases are helium, neon, argon, krypton, xenon, and radon. Helium has the configuration $1s^2$; all the other noble gases have eight outer-level electrons corresponding to the configuration $ns^2np^6$. Confirm this for yourself by writing the electronic configurations for Ne (atomic number 10), Ar (18), and Kr (36). For years chemists thought these gases were totally chemically inert, so they called them inert gases. After a few compounds of xenon and krypton were synthesized, chemists realized they were not completely inert and renamed them noble gases, indicating a reluctance to form compounds. The lack of much chemical activity for the noble gases suggests that the electron configuration of eight outer-level electrons is quite stable. Note that $ns^2np^6$ is eight electrons, two s and six p electrons.

The elements that occupy the s and p blocks are called the **representative elements.** The s-block representative elements have the outer-energy-level configuration $ns^1$ or $ns^2$, and the p-block representative elements have an outer-energy-level configuration ranging from $ns^2np^1$ to $ns^2np^6$. The d-block elements have electronic configurations that correspond to the filling of the d sublevels. These elements are called the **transition elements.** They represent a transition between the s block and the p block elements in a period. The f-block elements have electronic configurations that correspond to the filling of the f sublevels; they are called the **inner transition elements** or the rare earths. The inner transition elements involving the 4f sublevel start after lanthanum and are called the lanthanide series; those that involve the 5f sublevel start after actinium and are the so-called actinide series.

ACTIVITY
6-5

Without looking at a periodic table use the order of energy sublevels and the electron capacities of sublevels to sketch a periodic table. You can peek at a table if you need to but do not copy it. See if you can sketch a copy by following the order and sublevel capacity. You are not expected to include the symbols of the elements.

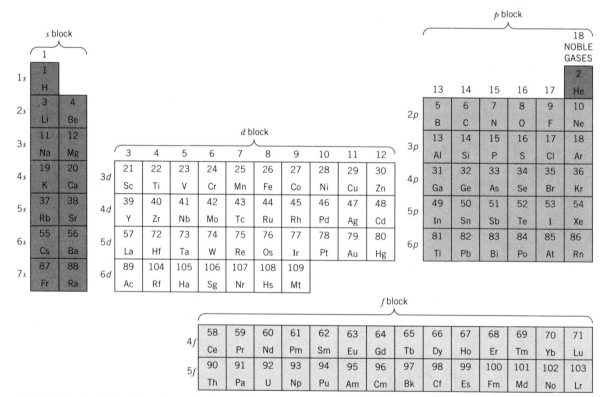

**FIGURE 6-15**   The shape of the periodic table corresponds to the filling of the sublevels.

## 6-18  GROUPS AND CLASSES OF ELEMENTS

Not counting the *f* block their are 18 columns or **groups** in the periodic table. Often these groups are simply numbered left to right as 1 to 18. The representative elements are groups 1, 2, 13, 14, 15, 16, 17, and 18; in a slightly different notation these correspond to IA, IIA, IIIA, IVA, VA, VIA, VIIA, and VIIIA; VIIIA is sometimes labeled "noble gases." These groups include many of the most common elements. Group 1 (IA) are the **alkali metals** and group 17 (VIIA) elements are the **halogens.** Alkali comes from the Ancient Arabic name for potash, and potassium is one of the members of this group. Halogen comes from the Greek *hals* 'salt' and *genes* 'born' since all members of the family form salt-like compounds with sodium. The transition metal groups are sometimes numbered in sequence as IIIB, IVB, VB, VIB, VIIB, VIII, VIII, VIII, IB, and IIB. The B designations indicate a slight similarity in properties to the corresponding A groups of representative elements. The three transition metal groups designated as VIII have no corresponding representative element groups.

The left-to-right horizontal sequences or rows of elements in the table are the **periods.** The six complete periods contain 2, 8, 8, 18, 18, and 32 elements; the seventh period is incomplete and contains 23 elements. The periods correspond to the filling of the electronic sublevels and the last period is incomplete because currently there are not enough known elements to fill the 6*d* and 7*p* sublevels.

In the periodic table the **nonmetals** include hydrogen and those elements in the upper right-hand portion of the periodic table. The **metalloids** separate the metals and nonmetals in a step-like fashion. The **metals** occupy the left-hand portion of the table including the *s* block, the *d* block (transition metals), and the *f* block. There are even some metals, such as

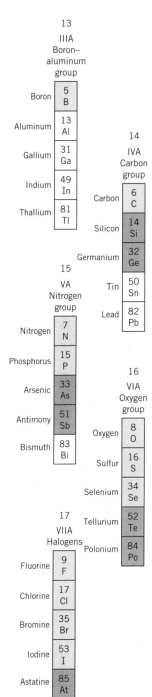

tin, lead, and bismuth, located in the $p$ block. Generally metals are those elements with few (one, two, or three) outer-level electrons. Tin, lead, and bismuth have atoms with four or five outer-level electrons but they are classified as metals because they have metallic properties. Nonmetals include those elements with greater numbers (4, 5, 6, 7, and 8) of outer-level electrons. As we shall see, the chemical behaviors and similarities of elements are directly related to the number of outer-level electrons.

## 6-19 SHORTHAND ELECTRONIC CONFIGURATIONS

For most elements, it is possible to read the electronic configuration directly from the periodic table (see Fig. 6-15). Since the table reflects electronic configurations, it makes sense that configurations can be read from the table. Do this by noting the pattern of sublevel filling up to the position of an element in the table. For the element boron, we note in the table that the five electrons in boron involve the filling of the $1s$ sublevel, the $2s$ sublevel, and one electron in the $2p$ sublevel. Thus, the configuration of boron is $1s^2 2s^2 2p^1$. For zinc (atomic number 30), we fill the $1s$, $2s$, $2p$, $3s$, $3p$, $4s$, and $3d$ sublevels to reach zinc. So the configuration is $1s^2 2s^2 2p^6 3s^2 3p^6 4s^2 3d^{10}$ or $1s^2 2s^2 2p^6 3s^2 3p^6 3d^{10} 4s^2$ as we normally will write it. Using this counting method, the periodic table is a quick source of the electronic configuration of an element.

It is often convenient to read the electronic configuration of an element directly from the table and write it in an abbreviated or shorthand fashion. Look at a periodic table. Each period ends with a noble gas. The next period that begins after a noble gas represents the next energy level. For instance, the third period ends with argon (18) and the next period begins with potassium (19). The outer-level electron in an atom of potassium begins the fourth energy level. This pattern repeats from period to period. To write the **shorthand electronic configuration** of an element, find the element in the table and trace back to the previous noble gas. The electronic configuration of the element is the same as that noble gas plus the additional electrons the element has beyond it. Write an abbreviated configuration by giving the symbol of the noble gas, enclosed in square brackets, and then the various extra sublevels with the proper numbers of electrons. To do this, pay attention to the parts of the table that correspond to the filling of various sublevels (see Fig. 6-15). For example, after locating sodium, Na, in the table, we note that neon, Ne, is the previous noble gas. Sodium has one more electron than neon, and this electron is in the $3s$ sublevel, so the electronic configuration of sodium is

$$[Ne]3s^1$$

In the same fashion, the electronic configuration of zinc would include the structure of argon plus electrons in the $4s$ and $3d$ levels. We count back 12 elements from zinc to argon; that represents 12 electrons. Starting with argon [Ar] we add the 12 electrons, 2 in the $4s$ sublevel and 10 in the $3d$ sublevel. Thus, the shorthand configuration of zinc is

$$[\text{Ar}]4s^23d^{10} \quad \text{or} \quad [\text{Ar}]3d^{10}4s^2$$

Now consider lead, Pb, atomic number 82. The preceding noble gas is xenon, atomic number 54. So, we need $82 - 54 = 28$ more electrons. We place these electrons in the sublevels corresponding to the period of the element, filling each sublevel then moving to new sublevels as needed. Thus, the shorthand configuration of lead is

$$[\text{Xe}]6s^24f^{14}5d^{10}6p^2 \quad \text{or} \quad [\text{Xe}]4f^{14}5d^{10}6s^26p^2$$

This example shows that we must remember to fill the $f$ sublevels when writing electronic configurations of elements in this part of the table. Working with elements in the $f$ block the pattern is that one electron goes in the $5d$ (or $6d$) then the $4f$ (or $5f$) gets electrons. After the $f$ sublevel fills, continue filling the $d$ sublevel. You can always remember this pattern by noting that La (element 57) is in the $d$ block followed by elements 58 to 71 in the $f$ block, and Ac (element 89) is in the $d$ block followed by elements 90 to 103 in the $f$ block. Consult a table to confirm this pattern. Following the pattern, the electronic configuration of curium, Cm, atomic number 96 is

$$[\text{Rn}]7s^26d^15f^7 \quad \text{or} \quad [\text{Rn}]5f^76d^17s^2$$

The counting method works quite well for most elements but a close look at Figure 6-14 shows that some $d$ block configurations do not fit this method exactly. But there are only a few and we need not bother with the exceptions.

Write the shorthand electronic configurations for

(a) mercury, Hg,

(b) barium, Ba, and

(c) iodine, I.

## 6-20  IONIZATION ENERGY

The term periodic refers to the fact that elements of higher atomic number have properties similar to those elements of lower atomic number. These similarities follow a pattern that repeats several times through the

**FIGURE 6-16**

Ionization of an atom by an electron beam. An electron in a cathode-ray beam interacts with the orbital electron causing it to gain energy and break away from the atom, leaving a positive ion.

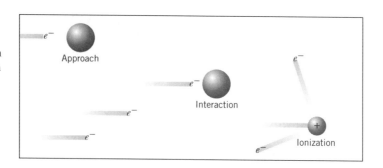

periods of elements. Two properties that follow this **periodic behavior** are discussed in the next few sections.

It requires energy to remove an electron from an atom to form an ion. The process of forming an ion from an atom is called ionization. It is possible to measure the amount of energy required to remove one electron from a neutral atom. This energy is called the first **ionization energy.** A specially designed instrument using a device similar to a cathode-ray tube is used to measure ionization energies. An ionization energy is measured by bombarding atoms of an element in the gaseous state with a beam of free electrons. If the beam of electrons is of sufficient energy, bombarding electrons will cause the loss of the most loosely bound outer-energy-level electron of the atom. This results in the formation of a singly charged ion (see Fig. 6-16). The general ionization process is shown as

$$\text{energy} + X(g) \longrightarrow X^+(g) + e^-$$

where X represents an atom of a specific element and $e^-$ represents the electron removed to form the $X^+$ ion. Since the neutral atom loses an electron the resulting ion has a positive charge. In the positive ion there is one less electron, which results in an excess of one proton and a net positive charge.

The ionization energies of the elements relate to their atomic structures. Figure 6-17 gives a plot of these ionization energies versus atomic number. The plot illustrates the periodic behavior of this property quite well. The sequence is a series of increasing ionization energies followed by sharp decreases in a repeating, periodic fashion. The ionization energies of the elements generally decrease from the top to the bottom of a given group and increase from left to right within a given period. This pattern is reasonable if we consider the factors that affect the ionization energies. Two important generalizations are that metals lose electrons more readily than nonmetals, and the noble gases do not readily lose electrons.

Two factors can account for the general decrease in the ionization energies of the elements within a group. One factor is the increased distance of the outer-energy-level electrons from the nucleus. In addition, the in-

Read the plot of ionizations energies and record the approximate values for Li, Na, K, Rb, and Cs.

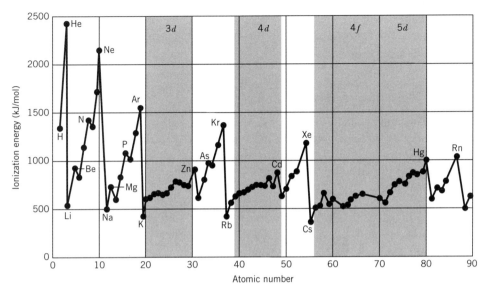

FIGURE 6-17 Plot of the ionization energies of elements versus their atomic numbers.

ner core of the lower-energy-level electrons shield the outer-energy-level electrons from the positive nucleus. The **shielding effect** tends to decrease the attraction of the nucleus for the outer-energy-level electrons, and they are more loosely held (see Fig. 6-18). As we move down a given group, the outer-energy-level electrons in the atoms of the elements are further from the nucleus, and the attraction by the nucleus is diminished by shielding. The result is that the electrons are less tightly bound to the atoms and thus more easily removed.

The increase in ionization energy from left to right within a given period of the table occurs for a different reason. This increase results from an increase in nuclear charge without an increase in shielding. No increase in shielding occurs because electrons are added to the same outer energy level and, thus, do not shield one another. The increase in nuclear charge, as we go from left to right in a given period, usually causes an in-

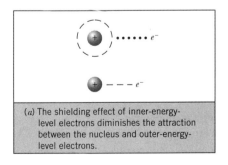

(a) The shielding effect of inner-energy-level electrons diminishes the attraction between the nucleus and outer-energy-level electrons.

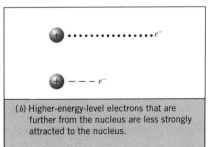

(b) Higher-energy-level electrons that are further from the nucleus are less strongly attracted to the nucleus.

FIGURE 6-18 Factors that influence ionization energies.

crease in the attraction between the nucleus and the outer-level electrons. The result is that these electrons are more tightly bound and to remove one requires more energy.

## 6-21 Atomic Sizes

In the light of the quantum mechanical model, how do we visualize an atom? Imagine slicing into one. Inside there is a tiny central nucleus with surrounding layers of electron clouds extending outward, starting with the lower-energy sublevels and ending with the outer-energy sublevels. Of course, we never deal with such atomic slices. How can we imagine an atom if we look at it from the outside? Generally, we view an atom as essentially a spherical distribution of electron clouds. That is, we imagine them as small spheres or possibly bumpy-surfaced spheres since the outer level may consist of several orbitals.

As part of our chemical vision we picture atoms as various-sized fuzzy balls with the outer-level electrons near the outer surfaces. As we shall see, it is these electrons that normally are responsible for the chemical behavior of an atom. Average atomic sizes are found indirectly from experimental information concerning molecules. One method used to find atomic sizes is X-ray diffraction (see Section 3-18). X rays are reflected from a sample of an element or compound. The pattern of X-ray reflection gives information about the distances between atoms and, therefore, atomic sizes. As an analogy, if we know the distance between the centers of two balls that are touching we know the radius of each ball.

If we assume atoms to be essentially spherical in shape, we can express sizes in terms of atomic radii. The radius of an atom would be the distance from the center of the nucleus to the outer electrons. Atomic radii are quite small and are in the range of about $10^{-10}$ meters. Since atomic radii are so small they are expressed in units of nanometers; 1 nm = $10^{-9}$ m. Figure 6-19 shows atomic radii of most of the representative elements. Hydrogen has the smallest atoms of all elements.

As shown in Figure 6-19, the sizes of atoms within a group generally increase down the group from top to bottom. The sizes of atoms within a period generally become smaller from left to right. The periodic relations between sizes are those we would expect, since the factors affecting the sizes are essentially the same as those discussed previously for ionization energies. Atomic sizes increase down a given group because higher-energy-level electrons are located further from the nucleus. The shielding effect decreases the attraction between outer-energy-level electrons and the nucleus. The result is that these loosely held electrons occupy greater volumes about

The relative sizes of some common atoms

| Period | 1 IA | 2 IIA | 13 IIIA | 14 IVA | 15 VA | 16 VIA | 17 VIIA |
|--------|------|-------|---------|--------|-------|--------|---------|
| 1 | H 0.037 | | | | | | |
| 2 | Li 0.152 | Be 0.111 | B 0.088 | C 0.077 | N 0.070 | O 0.066 | F 0.064 |
| 3 | Na 0.186 | Mg 0.160 | Al 0.143 | Si 0.118 | P 0.110 | S 0.104 | Cl 0.099 |
| 4 | Na 0.227 | Ca 0.197 | Ga 0.122 | Ge 0.122 | As 0.121 | Se 0.117 | Br 0.114 |
| 5 | Rb 0.248 | Sr 0.215 | In 0.162 | Sn 0.140 | Sb 0.141 | Te 0.137 | I 0.133 |
| 6 | Cs 0.265 | Ba 0.217 | Tl 0.171 | Pb 0.175 | Bi 0.155 | Po 0.140 | At 0.140 |

**FIGURE 6-19**
Variations in the sizes of atoms. The atomic radius of each element is given in nanometers. The noble gases, the transition elements, and the inner transition elements are not included in this table.

the nucleus. The decrease in atomic sizes from left to right within a period results from an increase in nuclear charge in a period. Each subsequent element in a period has one more proton and one more electron than the previous element. Since there is an increase in nuclear charge with no increase in shielding the result is a greater attraction for the outer-energy-level electrons and, thus, smaller atoms. Figure 6-20 illustrates these factors.

ACTIVITY 6-8

Using the radii values given in Figure 6-19 draw circles to show the relative sizes of atoms of hydrogen, oxygen, sodium, and chlorine.

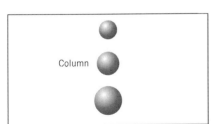

Column

The shielding effect of the inner-level electrons diminishes the attraction between the nucleus and outer-energy-level electrons. As the number of inner-energy-level electrons increases the outer-energy-level electrons occupy greater volumes. This accounts for the increase in atomic sizes within a group of elements.

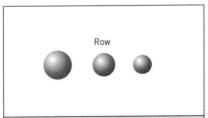

Row

The increase in nuclear charge with no corresponding increase in the energy levels of the electrons, within a row of representative elements, results in the outer-energy-level electrons being more strongly attracted to the nucleus. Thus, since the nuclear charge increases from left to right, the atomic sizes will decrease from left to right.

**FIGURE 6-20**
Effect of number of electrons, shielding, and nuclear charge on atomic size.

## Reading the Periodic Table

Much information about an element can be read from the periodic table. The box of a given element in the table usually contains the atomic number, the symbol, and the atomic weight. The position of an element in the table reveals its group or family, whether it belongs to the *s* block, *p* block, *d* block, or *f* block, and whether it is a representative, transition, or inner transition element. Its position also shows an element as a metal, nonmetal, or metalloid. For the representative elements the last digit of the group number reveals the number of outer-level electrons. For example, group 2 elements have two outer-level electrons, group 13 elements have three outer-level electrons, and group 17 elements have seven outer-level electrons. Roman numeral group numbers for the A groups also reveal the number of outer-level electrons within a group.

As an example of the use of the table, let us locate the element boron, B. Finding B in the table reveals the following:

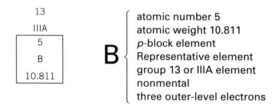

The position of an element in the table reveals its electronic configuration. The shorthand configuration is deduced by tracing back to the previous noble gas then counting to the element's location in the table. Thus, the shorthand configuration of boron is [He]$2s^2 2p^1$. The counting method does not give the correct configuration for some transition and inner transition elements because their atoms exhibit such small differences in the relative energies of their sublevels. These exceptions are rare and of little concern to us.

The position of a representative element in the table also reveals how it differs in atomic size from neighbors in its period or group. For example, we predict that boron has smaller atoms than beryllium and larger atoms than carbon. In addition, we predict boron has smaller atoms than aluminum or any of the other elements in its group. Furthermore, we predict that boron has a higher ionization energy than aluminum or any other members of its group.

 **FAMILY RELATIONSHIPS**

The periodic table is a valuable source of information about the elements. The idea that the location of an element in the table corresponds to its electronic configuration is insightful. The families of elements that occupy the various columns of the table have similar properties because they have **similar outer-level electronic configurations.** As discussed in Section 6-1, we expect the elements in families to form similar kinds of compounds with other elements. However, this does not mean that elements in a family form the same number and kinds of compounds. Each element has its unique and characteristic chemistry. The chemistries of the transition and inner transition elements are related to their outer-level electronic configurations and sometimes even to their next-to-outer-level electronic configurations. As a result there are similarities and distinct differences among the elements and the periodic table is a fundamental guide to making sense of the variety of chemical compounds they form.

In this simple exercise you are going to look closely at the colors in a flame. You need: two small glasses, commercial blue food color, commercial yellow food color, and a book of paper matches.

1. Fill each glass with water.
2. Add six drops of blue food color to one glass and stir.
3. Add six drops of yellow food color to the other glass and stir.
4. Light a match, look closely at the flame, and record your observations.
5. Light another match and hold it behind the glass with the blue solution. Record your observations.
6. Light a third match and hold it behind the glass with the yellow solution. Record your observations.

Explain why the colored solutions were useful for observing the flame.

The spins of electrons are related to magnetic behavior. In this exercise you are going to make an electromagnet. You need: a 9-volt alkaline battery, a twist tie, tape, two 3 cm lengths of steel paper clip, and a few 5 mm lengths of paper clip or a few iron staples. Make the steel lengths by straightening and breaking paper clips.

1. Peel 1 cm of paper from each end of a twist tie and coil the covered portion around one of the paper clip lengths in a spiral or helical fashion.
2. Twist the bare ends of the twist tie around the contacts of the battery and tape them in place.

3. Pick up the short paper clip pieces or staples with one end of your electromagnet.

4. Allow the magnet to remain connected for about one minute then disconnect it and remove the twist tie from the length of paper clip.

5. Test the length of paper clip to see if it behaves as a magnet. Compare the magnetized length of paper clip to the nonmagnetized length of paper clip by trying to pick up the short paper clip lengths or staples.

How many unpaired electrons are there in one atom of iron?

# QUESTIONS

## Sections 6-1 and 6-2

1. What is the periodic table of elements? What is a column in the table called?

2. What is a metal and where are the metals located in the periodic table?

3. What is a nonmetal and where are the nonmetals located in the periodic table?

4. What is a metalloid and where are the metalloids located in the periodic table?

## Section 6-4

5. Describe a cathode-ray tube.

## Section 6-5

6. Describe a gas discharge tube.

## Section 6-6

7. What factors contribute to the energy of an electron in an atom?

8. Give a brief description of the Bohr model of the atom.

9. Explain what is meant by the quantized energy of electrons.

10. What is an energy state or energy level of an electron in an atom?

11. What is a quantum jump or a quantum leap?

## Sections 6-7 to 6-13

12. Describe the quantum mechanical model of the atom and include the sublevel and orbital structure of energy levels.

13. What is an electron orbital? What is an electron cloud?

14. What is meant by the energy overlap of energy levels? Give an example of an energy overlap.

15. What is electron spin?

16. Using the pattern shown below as a beginning, construct a diagram of the electron energy sublevels in increasing order of energy. Be sure to indicate the positions of energy overlap (e.g., the $3d$ sublevel overlaps the $4s$ sublevel).

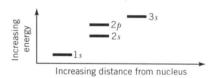

17. *Which of the following electron states do not exist?

$$2d, 3p, 3f, 2p, 7s, 4d, 1p$$

18. Which of the following electron states do not exist?

$$1p, 2s, 2f, 3f, 2p, 3s, 4f$$

19. What is the maximum number of electrons that can be contained in

   (a) any electron orbital   (b) a $p$ orbital

   (c) an $s$ orbital   (d) an $s$ sublevel   (e) a $p$ sublevel

   (f) a $d$ sublevel   (g) an $f$ sublevel   (h) an $f$ orbital

20. In what ways does the second main energy level of an atom differ from the fourth main energy level?

21. Consider the capacities of the sublevels. Given these capacities, determine the maximum number of electrons that can be contained in

(a)  the first main energy level

(b)  the second main energy level

(c)  the third main energy level

(d)  the fourth main energy level

## Sections 6-14 and 6-15

22. *Give the electronic configuration in sublevel notation for the following elements for which the atomic numbers are given in parentheses.

    (a)  H (1)      (b)  Al (13)     (c)  Ga (31)

    (d)  Se (34)    (e)  Kr (36)     (f)  Zr (40)

    (g)  Sn (50)    (h)  O (8)       (i)  Pt (78)

    (j)  P (15)     (k)  I (53)

23. Give the electronic configuration in sublevel notation for the following elements for which the atomic numbers are given in parentheses.

    (a)  C (6)      (b)  F (9)       (c)  Ca (20)

    (d)  As (33)    (e)  Ne (10)     (f)  Ba (56)

    (g)  Fr (87)    (h)  Ni (28)     (i)  Pb (82)

    (j)  Br (35)    (k)  Cl (17)

24. What are outer-energy-level electrons? How is the number of outer-level electrons determined for a specific element?

## Sections 6-16 to 6-18

25. What is the periodic table of elements and why is it useful?

26. What is meant by the term periodic in the periodic table?

27. Describe the shape of the periodic table and indicate how many elements compose the various parts of the table. Relate the shape of the table, and the number of elements in each part of the table, to the electronic configuration of the elements.

28. How many elements wide are each of the following?

    (a)  the s block    (b)  the p block    (c)  the d block

    (d)  the f block

29. Sketch an outline of the periodic table and indicate the following.

    (a)  s block, p block, d block, and f block

    (b)  Label the portions of the table corresponding to the filling of the 2s sublevel, the 2p sublevel,

the 4d sublevel, the 5p sublevel, and the 4f sublevel

30. What is a period in the periodic table? What are rows in the periodic table? What is a group in the periodic table? Relate periods and groups in the periodic table to the electronic configurations of the elements.

31. Explain why elements in the same group have similarities in chemical behavior.

32. Explain why the numbers of elements in the periods of the periodic table follow the pattern 2, 8, 8, 18, 18, 32, 23.

33. Write the electronic configurations of carbon (6), silicon (14), germanium (32), and tin (50). What do you notice about the number of electrons in the outer energy level of each of these atoms. The outer energy level has the highest energy-level number.

34. Write the electronic configurations for the following (the atomic number is given in parentheses).

    (a)  Na (11)    (b)  K (19)    (c)  Rb (37)    (d)  Cs (55)

What do the electrons of the outermost energy level of each of these elements have in common? What do these configurations reveal about the positions of these elements in the periodic table? Would you expect them to have chemical properties similar to one another? Why?

35. Write the electronic configurations for the following (the atomic number is given in parentheses).

    (a)  O (8)    (b)  S (16)    (c)  Se (34)    (d)  Te (52)

What do the electrons of the outermost energy level of each of these elements have in common? What do these configurations reveal about the positions of these elements in the periodic table? Would you expect them to have chemical properties similar to one another? Why?

36. Write the electronic configurations of the elements having atomic numbers 29, 47, and 79. What do these configurations tell you about the positions of these elements in the periodic table?

37. Suppose that a new element, gandhium, had recently been made by nuclear scientists. If the new element has an atomic number of 116, to what group in the periodic table would you assign the element?

38. List some properties associated with metals. Where are the metals located in the periodic table?

39. Describe nonmetals and metalloids. Where are the nonmetals located in the periodic table?

40. Sketch an outline of the periodic table and indicate the positions of the following.

 (a) the noble gases, the representative elements, the transition elements, and the inner transition elements

 (b) the alkali metals, the halogens, the lanthanides, and the actinides

41. *Refer to the periodic table inside the front cover of this book and classify the following elements as representative element, transition element, inner transition element, metal, nonmetal, or noble gas.

 (a) scandium, Sc   (b) argon, Ar

 (c) radium, Ra     (d) radon, Rn

 (e) cadmium, Cd    (f) antimony, Sb

 (g) thorium, Th    (h) chlorine, Cl

 (i) osmium, Os     (j) indium, In

42. Refer to the periodic table inside the front cover of this book and classify the following elements as representative element, transition element, inner transition element, metal, nonmetal, or noble gas.

 (a) molybdenum, Mo   (b) helium, He

 (c) rubidium, Rb     (d) iodine, I

 (e) zirconium, Zr    (f) gallium, Ga

 (g) neptunium, Np    (h) astatine, At

 (i) lead, Pb         (j) phosphorus, P

**Section 6-19**

43. *Answer Question 22, but give the shorthand electronic configurations.

44. Answer Question 23, but give the shorthand electronic configurations.

**Section 6-20**

45. Define ionization energy.

46. Sketch an outline of the main portion of the periodic table ($s$, $p$, and $d$ blocks) and indicate with arrows how the ionization energies of the elements vary from left to right in periods and from top to bottom in groups.

47. Describe the shielding effect and how it influences the attraction of a nucleus for outer-level electrons.

48. For each of the following sets of three elements, indicate which has the lowest and which has the highest ionization energy.

 (a) barium, Ba; calcium, Ca; magnesium, Mg

 (b) carbon, C; nitrogen, N; fluorine, F

 (c) chlorine, Cl; argon, Ar; potassium, K

 (d) neon, Ne; copper, Cu; cesium, Cs

**Section 6-21**

49. Explain how the sizes of the atoms vary within the periods and groups of the periodic table. In which part of the table are the elements with the smallest atoms found? In which part of the table are the elements with the largest atoms found?

50. What is a nanometer? Express the relation between centimeters and nanometers in terms of the number of centimeters per one nanometer.

51. The element with the largest atoms is francium. Explain why.

52. *A sulfur atom has a diameter of 0.208 nm. Calculate the diameter of the sulfur atom in inches. (1 in. = 2.54 cm and see Question 50.)

53. An old copper penny is 1.0 mm thick. Assuming that the copper atoms are stacked on top of one another, calculate how many copper atoms make up the thickness of a penny. The diameter of a copper atom is 0.256 nm.

54. *A carbon atom has a diameter of about 0.154 nm. If Avogadro's number of carbon atoms were stacked one on top of the other, how many meters high would the stack be? For comparison note that the sun is about $1 \times 10^{11}$ meters from earth. (It is not possible to stack atoms in this way.)

55. The volume of a sphere is given by the formula: $V = d^3/6$, where $d$ is the diameter of the sphere. Calculate the volume of a single calcium atom in cubic centimeters if its diameter is 0.394 nm and $\pi = 3.141$. (See Question 50.)

56. For each of the following sets of elements, indicate which element has the smallest atoms.

 (a) calcium, Ca; bromine, Br; germanium, Ge

 (b) carbon, C; silicon, Si; tin, Sn

 (c) rubidium, Rb; sodium, Na; francium, Fr

 (d) fluorine, F; nitrogen, N; boron, B

57. *Gold is the most malleable metal known. It can be pressed or beaten into very thin sheets called gold leaf. The diameter of a gold atom is 0.29 nm. Assuming the atoms are stacked one on top of the other, calculate the number of atoms of gold that make up the thick-

ness of a 0.000005-inch-thick gold leaf. (1 in. = 2.54 cm and see Question 50.)

## Section 6-22

58. *Using the periodic table, give the atomic number, atomic weight, block, type of element, shorthand electronic configuration, and number of electrons in the outer level for the following elements.

   (a)  nitrogen, N     (b)  sodium, Na

   (c)  iron, Fe       (d)  nickel, Ni

   (e)  silicon, Si      (f)  nobelium, No

   (g)  barium, Ba    (h)  neon, Ne

   (i)  phosphorus, P  (j)  oxygen, O

59. Using the periodic table, give the atomic number, atomic weight, block, type of element, shorthand electronic configuration, and number of electrons in the outer level for the following elements.

   (a)  titanium, Ti     (b)  tantalum, Ta

   (c)  magnesium, Mg  (d)  hydrogen, H

   (e)  boron, B       (f)  tin, Sn

   (g)  lawrencium, Lr  (h)  fluorine, F

   (i)  tellurium, Te   (j)  tungsten, W

## Section 6-23

60. Why are groups in the periodic table sometimes called families?

## Questions to Ponder

61. Consult some reference books and give a brief description of how a laser works at the atomic level.

62. Using the patterns of compounds formed by families and oxygen in Section 6-1, predict the formulas for the chlorine compounds of all of the elements shown in that section. (Some of them may not exist.)

63. Calculate the number of electrons in 1.0 tons of aluminum metal. What percent of the total mass of 1.0 tons of aluminum is represented by electrons?

64. How many electrons are in the head of a pin?

# CHEMICAL

# BONDS

## 7-1 THE NATURE OF THE CHEMICAL BOND

**Chemical bonds** are forces that hold atoms in combination. They are attractions between opposite charges, the same forces that hold electrons in
atoms. When atoms form compounds the outer-level electrons are involved. If atoms did not interact with one another to form compounds all
elements would be monoatomic gases. It is interesting that the noble gases
(group 18 elements) all occur as monoatomic gases at normal earth conditions. For example, a sample of helium gas is just a collection of separate helium atoms. In contrast, elemental gases, such as hydrogen, nitrogen, oxygen, fluorine, and chlorine, occur as diatomic molecules. In
diatomic molecules atoms are held in combination by chemical bonds.

Sugar and salt are examples of common chemical compounds; some of
their properties are investigated in Chemistry in Action 7-1. Both salt and
sugar occur as crystalline solids. Chemical investigations reveal a distinct
difference in the combination of sodium and chlorine in salt and the combination of carbon, hydrogen, and oxygen in sucrose. Figure 7-1 shows the
fundamental particles that make up sodium chloride are not ordinary
atoms. The particles are ions, which carry positive and negative charges.
Positively charged sodium ions and negatively charged chloride ions compose salt. In sugar, as illustrated in Figure 7-2, the fundamental particles
are not independent atoms but groups of atoms linked by the sharing of

223

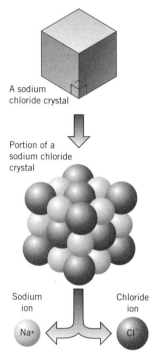

Sodium
ion

Chloride
ion

Na+    Cl⁻

**FIGURE 7-1** The chemical particles in sodium chloride are sodium ions and chloride ions.

electrons. These bonded groups of atoms are sucrose molecules. Sugar and salt are examples of the two kinds of chemically combined elements found in compounds. One type involves ion formation, the other involves molecule formation. Atoms interact through their outer-level electrons to form ions or molecules. The **number of outer-level electrons** an atom holds dictates the kinds of chemical compounds it forms.

### 7-2 ELECTRON DOT SYMBOLS

When atoms of elements form compounds, the outer-energy-level electrons are normally the only electrons involved. These outer-energy-level electrons are called **valence electrons.** Valence comes from the Latin *valentia* 'capacity'. For atoms, valence refers to the capacity to form bonds. Rather than write the electronic configurations for elements to show the number of valence electrons special symbols are used. The valence electrons of each element are shown by an electron dot symbol. To write an **electron dot symbol,** give the usual symbol of the element surrounded by dots representing the valence electrons. For example, the dot symbol for hydrogen, with its one electron, is

$$H\cdot$$

Electron dot symbols are normally used for the representative elements. For these elements the electron dot symbol is the symbol of the element surrounded by a number of dots corresponding to the number of valence electrons. The symbol represents the nucleus and the core electrons and the dots represent the outer-level electrons. When writing a dot symbol, imagine a square around the symbol of the element and put a dot on each side until all the valence electrons are used. Once there is one dot on each side electrons may be doubled up as necessary. For example, symbols for the elements sodium, magnesium, boron, carbon, nitrogen, oxygen, fluorine, and neon are

| Group Number | 1 IA | 2 IIA | 13 IIIA | 14 IVA | 15 VA | 16 VIA | 17 VIIA | 18 noble gas |
|---|---|---|---|---|---|---|---|---|
| | Na· | ·Mg· | ·Ḃ· | ·Ċ· | ·N̈· | :Ö· | :F̈· | :N̈e: |

The side used for pairing dots is not important, so the dot symbol for oxygen could be written

$$:\ddot{O}\cdot \quad \cdot\ddot{O}: \quad :\dot{O}\cdot \quad :\dot{O}\cdot \quad \cdot\ddot{O}\cdot$$

You can interpret a dot symbol for groups 15, 16, and 17 elements by imagining that the four sides around the element symbol represent the *s* orbital and the three *p* orbitals of the outer energy level.

A sugar (sucrose) crystal

$$_s \atop p\,\mathbf{X}\,p \atop p$$

The outer-level electrons are located within these orbitals; the $s$ orbital is filled first then the $p$ orbitals. According to Hund's rule whenever possible each $p$ orbital contains no more than one electron. Eventually all three $p$ orbitals do contain one electron, then each additional electron fills a $p$ orbital until the noble gas configuration is reached.

Table 7-1 gives the dot symbols for most of the representative elements. Note carefully that the dot symbols for all the elements with the same outer-level configurations are the same, since the number of valence electrons is the same. Although these dot symbols are simple representations of the distributions of valence electrons, they can be very useful in the discussion of chemical bonding. Remember that the number of outer-level electrons for a representative element is given by its group number.

Portion of sugar crystal

*Table 7-1*  Electron Dot Symbols of Some Representative Elements

|  | \multicolumn{7}{c}{Group} | | | | | | |
|---|---|---|---|---|---|---|---|
|  | IA 1 | IIA 2 | IIIA 13 | VIA 14 | VA 15 | VIA 16 | VIIA 17 |
|  | $ns^1$ | $ns^2$ | $ns^2np^1$ | $ns^2np^2$ | $ns^2np^3$ | $ns^2np^4$ | $ns^2np^5$ |
| H· | Li· | ·Be· | ·Ḃ· | ·Ċ· | ·Ṅ· | :Ö· | ·F̈· |
|  | Na· | ·Mg· | ·Äl· | ·S̈i· | ·P̈· | :S̈· | :Cl̈· |
|  | K· | ·Ca· |  | ·Ġe· | ·Äs· | :S̈e· | :B̈r· |
|  | Rb· | ·Sr· |  |  |  | :Te· | :Ï· |

## 7-3 METALLIC BONDING

Metals characteristically have atoms with few valence electrons; most of them have two, some have one, and a few have more than two. Metals are

Sucrose molecule
$C_{12}H_{22}O_{11}$

**FIGURE 7-2** The chemical particles in table sugar are $C_{12}H_{22}O_{11}$ molecules.

ACTIVITY 7-1

Consult a periodic table and list those metals that have one valence electron and those that have more than two valence electrons.

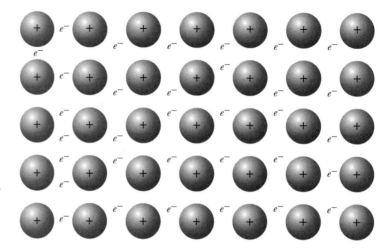

**FIGURE 7-3** In metallic bonding the inner cores of metal atoms are immersed in a "sea" of outer-level electrons.

typically solids that are good conductors of electricity and heat. Normally they are easily stretched into wires, pressed into sheets, and cast or molded into various shapes.

How are metal atoms held in combination in metallic solids? A simple picture is that positively charged atomic cores are surrounded by a "sea" of loosely held valence electrons. See Figure 7-3. Sodium metal can be envisioned as

Na    •    Na    •    Na    •    Na    •    Na    •    Na    •

•    Na    •    Na    •    Na    •    Na    •    Na

Na    •    Na    •    Na    •    Na    •    Na    •    Na

•    Na    •    Na    •    Na    •    Na    •    Na

Electrons can move through a sample of a solid metal. This is why metals are good conductors. The various atoms in a metallic solid are held in place by relatively strong electrostatic forces of attraction. The positive atomic cores and outer sea of electrons attract one another. These attractions are nondirectional and act in three dimensions. As a consequence, metals are typically crystalline solids in which the atoms are stacked in a three-dimensional pattern. The relative positions of atoms in such solids are readily changed so metals are flexible and malleable. This is why they can be bent, stretched, shaped, molded, and cast.

## 7-4  THE OCTET RULE

Ideas concerning chemical bonding were developed by observing the nature and behavior of many chemical compounds. The ability of an atom to form chemical bonds relates to the distribution of electrons in that atom.

Knowledge of the electronic configurations of elements is the basis of understanding chemical bonding between elements. In the 1920s the American chemist, Gilbert Lewis, proposed the theory that elements tend to enter into chemical combinations involving the loss, gain, or sharing of electrons. The study of the properties of many compounds of the representative elements has resulted in an important generalization called the **octet rule.** This rule states that atoms tend to lose, gain, or share electrons so that they attain a total of **eight (an octet) outer-energy-level electrons.** To show an octet of electrons, eight dots can be written around the symbol. For example, the octet of argon is

$$:\overset{..}{\underset{..}{Ar}}: \qquad [Ne]\ 3s^2\ 3p^6$$

It is important to note that an octet of electrons corresponds to filled $s$ and $p$ sublevels ($ns^2np^6$) in the outer level. An octet of electrons in the outer level of an atom is a particularly stable arrangement. Hydrogen has only one valence electron. It gains stability when it has two outer-level electrons. These two electrons correspond to a filled first energy level ($1s^2$). Hydrogen follows a duet rule rather than the octet rule since there is no room in the first energy level for eight electrons. The noble gases are chemically stable and they form very few compounds. Unlike other elements, all of the noble gases except one have an octet of outer-level electrons. The exception is helium, which has atoms with two outer-level electrons.

Noble gases only form compounds that do not fit the octet rule. Atoms of other elements have fewer than eight outer-level electrons and tend to combine in ways through which they attain eight outer-level electrons. The octet rule suggests that atoms combine not only in ways governed by electrostatic attractive forces, but also in ways that allow them to be surrounded by a stable arrangement of electrons. Apparently, a stable arrangement of electrons facilitates chemical bonding.

The octet pattern is called a rule rather than a law since it is not followed in every case of compound formation. Nevertheless, it does apply to a great number of compounds. It serves as a useful guide to the way in which atoms of elements enter into combination. There are two basic types of chemical bonds. An **ionic bond** results when electrons transfer from one atom to another creating ions. In this case the octet rule is followed by the loss and gain of electrons. A **covalent bond** results when two atoms share electrons between them. In this case the octet rule is followed by mutual sharing of electrons.

## 7-5   ION FORMATION

When sodium metal and chlorine gas mix, they readily react to form salt. Imagine a reaction between sodium atoms and chlorine atoms. The sodium atoms each have one outer-level electron. In the reaction they lose these electrons to form positive sodium ions.

$$\text{Na} \longrightarrow \text{Na}^+ + 1\ e^-$$
$$[\text{Ne}]\ 3s^1 \longrightarrow [\text{Ne}]\ \text{or}\ [\text{He}]2s^22p^6$$

The electron lost by a sodium atom is gained by a chlorine atom to form a negative chloride ion.

$$\text{Cl} + 1\ e^- \longrightarrow \text{Cl}^-$$
$$[\text{Ne}]\ 3s^23p^5 \longrightarrow [\text{Ar}]\ \text{or}\ [\text{Ne}]\ 3s^23p^6$$

The transfer of the electron allows both kinds of atoms to get an octet of outer-energy-level electrons. The sodium atom loses an electron and ends up with an octet in the next lowest energy level. The chlorine atom gains an electron to complete an octet in the outer-energy level.

When an atom loses electrons, negative charges are removed (each electron carries a single negative charge), leaving behind a positively charged particle consisting of a nucleus and the remaining electrons. Such a charged particle is a **positive ion or a cation** (pronounced cat-ion). See Figure 7-4. The sodium atom loses an electron to form a singly charged sodium cation.

**FIGURE 7-4**   Ion formation by the loss and gain of electrons.

11 protons          11 protons

$$\text{Na} \longrightarrow \text{Na}^+ \quad + 1e^-$$

11 electrons        10 electrons

An atom that gains electrons takes on the negative charge of the electrons. Such a negatively charged particle is also an ion, but in this case, a **negative ion or an anion** (pronounced an-ion). See Figure 7-4. The chlorine atom gains an electron to form a singly charged anion.

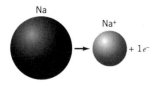
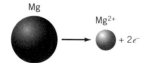

Positive ions are formed by loss of electrons.

17 protons          17 protons

$$1e^- + \text{Cl} \longrightarrow \text{Cl}^-$$

17 electrons        18 electrons

Ions are fundamental chemical species or particles found in compounds that are appropriately called ionic compounds. When ionic compounds form from elements the atoms lose and gain electrons to form ions. The number of protons in an atom does not change when it forms an ion so the charge of the ion is a result of excess or deficient electrons. It is important to realize that ions are very different chemical particles compared to the neutral atoms from which they form, and they have significantly different properties. We sprinkle salt on our food and eat it. In fact, we need it in small amounts to keep our bodies functioning correctly. No one would dare eat either sodium metal or chlorine gas. Both are highly toxic and dangerous. Chemists and professionals sometimes use terminology that may be misleading to a beginner. For instance, a doctor may refer to a low-sodium diet, or a nutritionist may claim that we need a certain

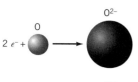
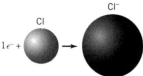

Negative ions are formed by gain of electrons.

amount of calcium in our diets. These professionals know that they are referring to sodium ions and calcium ions, not sodium or calcium metal. As
you learn more chemistry, you'll become more aware of the use of this terminology.

## 7-6  FORMULAS FOR IONS

An ion is represented by a formula showing the symbol of the element
from which the ion is formed with the charge of the ion as a superscript.
The formula for a sodium ion is $Na^+$ (read N–A–plus), and the formula
for the chloride ion is $Cl^-$ (read C–L–minus). Notice that a simple negative ion has a name that uses the name of the element with the ending
changed to -ide. A simple positive ion has a name that is the same as the
element name. The formula for the calcium ion, $Ca^{2+}$ (read
C–A–two–plus), shows that calcium ion carries a double positive charge.
The formula for the oxide ion, $O^{2-}$ (read O–two–minus), shows that the
oxide ion has a double negative charge.

Which elements form positive ions and which form negative ions? The
metals with few (one, two, or three) outer-level electrons tend to lose electrons. Metals generally have lower ionization energies and so they form
cations readily. The nonmetals, with many outer-level electrons (five, six,
or seven), tend to gain electrons. Nonmetals generally have higher ionization energies and do not lose electrons readily. Nonmetals can gain electrons to form anions.

Table 7-2 shows the ions formed by common metals and nonmetals. It

Consult a periodic table and
predict the cation formed by
the element barium and the
anion formed by the ele-
ment sulfur.

*Table 7-2*   Names and Formulas of Some Common Cations and Anions

| CATIONS | | ANIONS | |
|---|---|---|---|
| Group 1 (IA) | | Group 15 (VA) | |
| $Na^+$ | sodium ion | $N^{3-}$ | nitride ion |
| $K^+$ | potassium ion | $P^{3-}$ | phosphide ion |
| Group 2 (IIA) | | Group 16 (VIA) | |
| $Ca^{2+}$ | calcium ion | $O^{2-}$ | oxide ion |
| $Mg^{2+}$ | magnesium ion | $S^{2-}$ | sulfide ion |
| $Sr^{2+}$ | strontium ion | $Se^{2-}$ | selenide ion |
| $Ba^{2+}$ | barium ion | $Te^{2-}$ | telluride ion |
| Group 13 (IIIA) | | Group 17 (VIIA) | |
| $Al^{3+}$ | aluminum ion | $F^-$ | fluoride ion |
| | | $Cl^-$ | chloride ion |
| | | $Br^-$ | bromide ion |
| | | $I^-$ | iodide ion |

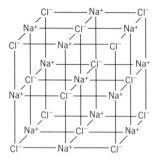

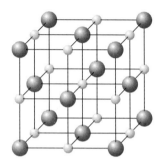

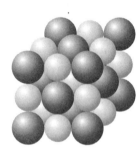

Sodium         Chloride
ion            ion

**FIGURE 7-5**

Sodium ions and chloride ions are held in chemical combination by ionic bonds, the electrostatic attraction between ions of opposite charge.

**Ionic Bond:**

*The electrostatic force of attraction between oppositely charged ions.*

is important to notice the pattern. The group number from the periodic table reveals the number of valence electrons. For representative metals, the charge on the positive ions they form is the number of valence electrons. Sodium (group 1 or IA) forms $Na^+$, magnesium (Group 2 or IIA) forms $Mg^{2+}$, and aluminum (group 13 or IIIA) forms $Al^{3+}$. For nonmetals, the number of electrons needed to make an octet is found by subtracting the number of valence electrons from 8. Thus, nitrogen (group 15 or VA) forms $N^{3-}$, nitride ion; oxygen (group 16 or VIA) forms $O^{2-}$, oxide ion; and fluorine (group 17 or VIIA) forms $F^-$, fluoride ion. Such negative ions have names that end in -ide.

## 7-7 THE IONIC BOND

We return to the reaction between sodium metal and chlorine gas, which forms positive sodium ions and negative chloride ions. Of course, the sodium atoms lose one electron each and the chlorine atoms gain one electron each. In this compound, the number of positive ions equals the number of negative ions. As these ions form, they are in close proximity and intermingle. Since the ions are of opposite charge, they are attracted to one another by electrostatic attractive forces. The attractive forces cause the ions to group into a three-dimensional pattern of alternating ions, as illustrated in Figure 7-5. This three-dimensional accumulation of ions forms a crystal. The resulting compound is crystalline. In the reaction of sodium and chlorine, the reactants have changed from the soft metal sodium and gaseous chlorine to a product that is a colorless crystalline solid. The force of attraction between the ions that allows a compound to form is the ionic bond as defined in the margin.

Even a tiny crystal of salt contains a vast number of sodium ions and chloride ions stacked in three-dimensional space. You may notice that the salt grains are tiny cubes, which is the pattern in which the ions happen to aggregate, as shown in Figure 7-5.

Compounds containing ions held together by ionic bonds are called **ionic compounds.** When a metal reacts with a nonmetal, a transfer of electrons occurs; this results in the formation of ions that combine through ionic bonds. Thus, we can say that binary or two-element compounds involving a metal and a nonmetal are typically ionic compounds; they contain metal cations and nonmetal anions. Figure 7-6 illustrates a few examples of ionic compound formation. Ionic compounds are **crystalline solids** and have ions stacked in three-dimensional space. The three-dimensional pattern of stacked ions is called a **crystal lattice.** Because of differences in sizes, shapes, and charges of ions, not all ionic compounds have cubic crystals like sodium chloride.

Ionic bonds are relatively strong, making ionic compounds hard solids. Ionic bonds are nondirectional; negative ion attracts positive ions in all directions and a positive ion attracts negative ions in all directions. This results in ions arranged in space as a specific three-dimensional crystalline solid. Typically ionic solids can be easily cracked or fractured. As shown in Figure 7-7, when a portion of an ionic solid is pushed out of position, like charges can align and their mutual repulsion pushes the crystal apart, thereby cracking it.

## 7-8 FORMULAS OF BINARY IONIC COMPOUNDS

An ionic compound contains oppositely charged ions. However, all ionic compounds are overall electrically neutral. This is because the compound always contains as many positive charges as negative charges. The positive ions balance the negative ions. The formulas of ionic compounds reflect this balance. The formula of sodium chloride is simply NaCl, since a sodium ion carries one positive charge and a chloride ion carries one negative charge. The formula could be written as $(Na^+)(Cl^-)$ or $Na^+Cl^-$. However, the charges are not normally included in the formulas of ionic compounds; they are implied by the number of combined atoms shown.

How do we know how to write the likely formula for a **binary ionic compound?** To make the amount of positive charge equal the amount of negative charge, the appropriate number of ions of each type is needed. Take, as an example, the compound involving calcium and chlorine. Calcium is a group 2 element, so it forms $Ca^{2+}$ ions; chlorine is a group 17 element, so it forms $Cl^-$ ions. Group numbers and, thus, the expected formulas of the ions can be found from a periodic table. Two chloride ions balance one calcium ion since the number of negative charges must equal the number of positive charges as shown on the next page.

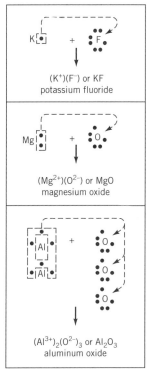

$(K^+)(F^-)$ or KF
potassium fluoride

$(Mg^{2+})(O^{2-})$ or MgO
magnesium oxide

$(Al^{3+})_2(O^{2-})_3$ or $Al_2O_3$
aluminum oxide

**FIGURE 7-6**  The idealized formation of some ionic compounds.

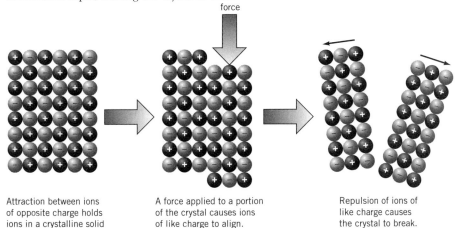

**FIGURE 7-7**  Ionic compounds are typically brittle; when a shearing force is exerted on a portion of an ionic crystal, ions of the same charge come into contact and repel, causing the crystal to break.

force

Attraction between ions of opposite charge holds ions in a crystalline solid

A force applied to a portion of the crystal causes ions of like charge to align.

Repulsion of ions of like charge causes the crystal to break.

$$(\text{Cl})^- (\text{Ca}^{2+}) (\text{Cl}^-) \quad \text{giving the formula } \text{CaCl}_2$$

By convention the metal always comes first in the formula of an ionic compound.

---

**EXAMPLE 7-1**

What is the likely formula for a binary compound of potassium and sulfur ions?

Potassium (group 1) forms the ion $K^+$ and sulfur (group 16) forms the sulfide ion, $S^{2-}$. Thus, the formula contains two $K^+$ for each $S^{2-}$.

$$\left.\begin{array}{c} K^+ \\ K^+ \end{array}\right\} S^{2-} \text{ gives the formula } K_2S$$

---

**EXAMPLE 7-2**

What is the likely formula for a binary compound of aluminum and oxygen?

Aluminum (group 13) forms the ion $Al^{3+}$ and oxygen (group 16) forms the oxide ion, $O^{2-}$. Thus, the formula contains two aluminum ions (to give charge 6+) and three oxide ions (to give charge 6–).

$$6+ \left.\begin{array}{c} Al^{3+} \\ \\ Al^{3+} \end{array}\right\} \left\{\begin{array}{l} O^{2-} \\ O^{2-} \\ O^{2-} \end{array}\right. 6- \text{gives the formula } Al_2O_3$$

---

**ACTIVITY 7-3**

What is the likely formula of the binary compound of magnesium and nitrogen?

## 7-9 THE COVALENT BOND

Many compounds contain molecules rather than ions. Molecules involve the sharing of electrons between bonded atoms. We consider how this sharing occurs by using, as a simple example, the formation of a compound of hydrogen and chlorine. Suppose an atom of hydrogen and an atom of chlorine approach one another as shown in Figure 7-8. The stable electron arrangement for a hydrogen atom is two electrons. A chlorine atom is stabilized with eight electrons in the outer level. As shown in the figure, the atoms attain stability by sharing electrons. When an electron pair is located between the nuclei, the electrostatic forces of the electron–nucleus attractions hold the atoms in combination.

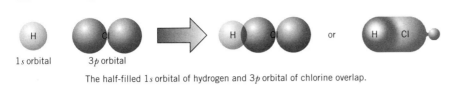

The half-filled 1s orbital of hydrogen and 3p orbital of chlorine overlap.

The unpaired electrons of hydrogen and chlorine are shared by both atoms.

The spherical atoms of hydrogen and chlorine interpenetrate by the mutual possession of an electron pair

**FIGURE 7-8** Three idealized views of covalent bond formation in hydrogen chloride, HCl.

The result of sharing is that both atoms attain a stable electronic configuration by mutual possession of electrons. Each nucleus attracts the shared pair of electrons and no transfer of electrons occurs. The sharing produces a bonded group of atoms called a molecule.

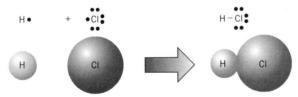

The sharing of an electron pair gives hydrogen two electrons and chlorine an octet of outer-level electrons. A duet of electrons for hydrogen and an octet for chlorine are attained when they share electrons rather than lose or gain them. Notice that the shared electrons "belong" to both the hydrogen and the chlorine atoms.

When nonmetals or metalloids form compounds with other nonmetals, the atoms attain stability by sharing outer-level electrons. The sharing of an electron pair between atoms acts as a bonding force to link the atoms. This type of chemical bond is the **covalent bond** (co- as in cooperation and -valent from valence or combining capacity).

Covalently bonded atoms form specific and discrete groups called molecules. A **molecule** is defined as a group of two or more covalently bonded atoms. A simple way to picture the formation of a covalent bond is to imagine the valence electron orbitals of the bonded atoms. Recall that an orbital is filled when it contains two electrons and an orbital containing

**Covalent Bond:**

*The force of attraction between two atoms resulting from the sharing of a pair of electrons.*

one electron is half-filled. As two atoms covalently bond, the half-filled orbitals overlap so that the electron pair is shared. In other words, a covalent bond can result from the **overlap of half-filled atomic orbitals** that allow the electrons to pair.

Figure 7-8 illustrates the formation of a compound of hydrogen and chlorine. In this case, the overlap of the half-filled $1s$ atomic orbital of hydrogen and the half-filled $3p$ orbital of chlorine is the covalent bond. The overlap of half-filled orbitals makes it possible for the atoms to share the electrons.

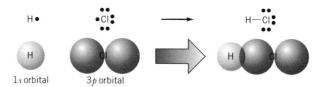

## 7-10 POLYATOMIC IONS

As we have seen, compounds are chemical combinations of atoms that take the form of ions or molecules. Atoms are the fundamental and basic type of chemical particles. They are the building blocks of compounds. Once combined, atoms become parts of the other kinds of chemical particles or chemical species. Thus, there are three kinds of **chemical particles or chemical species: atoms, ions, and molecules.**

The ions discussed in Section 7-5 are those that result when atoms lose or gain electrons. These types of ions are simply charged atoms. They are called **monoatomic ions** or **simple ions.** Some natural ionic compounds and some manufactured ionic compounds contain another type of ion. For example, a piece of limestone or marble is an ionic compound that contains positive calcium ions, $Ca^{2+}$, in combination with unique kinds of negative ions. These negative ions are not simple ions but rather polyatomic ions containing more than one atom.

A **polyatomic ion** is a group of two or more covalently bonded atoms that carries an electrical charge. The term poly means many, so a polyatomic ion contains many (two or more) atoms whereas a simple ion is a single charged atom or a monoatomic ion. Limestone and marble contain the compound calcium carbonate made up of calcium ions in combination with the polyatomic carbonate ion, $CO_3^{2-}$ (read C–O–three–two–minus). The carbonate ion is an ion composed of three oxygen atoms covalently bonded to one carbon atom. The four-atom particle carries a double negative charge.

$$CO_3^{2-}$$

*Table 7-3*   Names and Formulas of Some Common Ions

| CATIONS | | ANIONS | |
|---|---|---|---|
| Ammonium ion | $NH_4^+$ | Acetate ion | $C_2H_3O_2^-$ |
| Copper(I) ion (cuprous ion) | $Cu^+$ | Bromide ion | $Br^-$ |
| | | Chlorate ion | $ClO_3^-$ |
| Hydronium ion[a] | $H_3O^+$ | Chloride ion | $Cl^-$ |
| Potassium ion | $K^+$ | Cyanide ion | $CN^-$ |
| Silver ion | $Ag^+$ | Fluoride ion | $F^-$ |
| Sodium ion | $Na^+$ | Hydrogen carbonate ion (bicarbonate ion) | $HCO_3^-$ |
| Barium ion | $Ba^{2+}$ | | |
| Calcium ion | $Ca^{2+}$ | Hydrogen sulfate ion | $HSO_4^-$ |
| Cobalt(II) ion | $Co^{2+}$ | Hydroxide ion | $OH^-$ |
| Copper(II) ion (cupric ion) | $Cu^{2+}$ | Iodide ion | $I^-$ |
| | | Nitrate ion | $NO_3^-$ |
| Iron(II) ion (ferrous ion) | $Fe^{2+}$ | Nitrite ion | $NO_2^-$ |
| | | Permanganate ion | $MnO_4^-$ |
| Lead(II) ion | $Pb^{2+}$ | Carbonate ion | $CO_3^{2-}$ |
| Magnesium ion | $Mg^{2+}$ | Chromate ion | $CrO_4^{2-}$ |
| Manganese(II) ion | $Mn^{2+}$ | Dichromate ion | $Cr_2O_7^{2-}$ |
| Mercury(I) ion (mercurous ion) | $Hg_2^{2+}$ | Oxalate ion | $C_2O_4^{2-}$ |
| | | Oxide ion | $O^{2-}$ |
| Mercury(II) ion (mercuric ion) | $Hg^{2+}$ | Sulfate ion | $SO_4^{2-}$ |
| | | Sulfide ion | $S^{2-}$ |
| Nickel(II) ion | $Ni^{2+}$ | Sulfite ion | $SO_3^{2-}$ |
| Tin(II) ion | $Sn^{2+}$ | Phosphate ion | $PO_4^{3-}$ |
| Zinc ion | $Zn^{2+}$ | | |
| Aluminum ion | $Al^{3+}$ | | |
| Chromium(III) ion | $Cr^{3+}$ | | |
| Iron(III) ion (ferric ion) | $Fe^{3+}$ | | |

[a]The hydronium ion exists only in water solutions, not in compounds.

Table 7-3 includes the names and formulas of some common polyatomic ions (a similar table is inside the front cover); most polyatomic ions have negative charges but the ammonium ion, $NH_4^+$, is a common polyatomic ion that has a positive charge. It is found in combination with negative ions, for instance, with the chloride ion, $Cl^-$, it forms ammonium chloride, $NH_4Cl$, and with the sulfate ion, $SO_4^{2-}$, it forms ammonium sulfate, a common fertilizer. Note, as shown on the next page, that since the charge of the ammonium ion is 1+ and the charge of a sulfate ion is 2−, two ammonium ions, $NH_4^+$, are needed to balance one sulfate ion, $SO_4^{2-}$.

$$\left.\begin{array}{l} NH_4^{+} \\[1ex] NH_4^{+} \end{array}\right\} SO_4^{2-} \text{ resulting in the formula } (NH_4)_2SO_4$$

To represent two ammonium ions, the formula for the ammonium ion is enclosed in parentheses with a subscript 2, $(NH_4)_2SO_4$ (read N–H–four–taken–twice S–O–four). The parentheses provides a way for the subscript to indicate two ammonium ions.

Polyatomic ions are always found near oppositely charged ions; they never occur on their own. It is not possible, for example, to have a sample of pure $SO_4^{2-}$ ions; they occur only when combined with or proximal to positive ions. Polyatomic ions occur in ionic compounds found in nature as well as some manufactured chemicals. Some common polyatomic ions are found dissolved in natural waters and cellular fluids along with simple ions.

Although polyatomic ions contain several combined atoms they are components of ionic compounds just like simple ions. Negative polyatomic ions are found in combination with simple positive ions or with positive polyatomic ions, such as the ammonium ion. The formula of a compound containing polyatomic ions reflects a balance of positive and negative charges. When we give the name and formula of polyatomic ions, the charge is part of the formula. The sulfate ion is not $SO_4$ but $SO_4^{2-}$. These charges are important when we want to know the formula of a compound containing ions. However, the charges are not written as part of the compound formula.

---

**EXAMPLE 7-3**

What is the formula of the compound containing potassium ions and carbonate ions?

Potassium is a group 1 metal, so it forms the $K^+$ ion. The carbonate ion is $CO_3^{2-}$. It requires two $K^+$ to balance its charge.

$$\left.\begin{array}{l} K^+ \\[1ex] K^+ \end{array}\right\} CO_3^{2-} \text{ gives the formula } K_2CO_3$$

---

**EXAMPLE 7-4**

What is the formula of the compound of aluminum ions and sulfate ions?

Aluminum is a group 13 metal that forms the $Al^{3+}$ ion. The sulfate ion is $SO_4^{2-}$. To have the same amount of positive and negative charge the compound must contain two aluminum cations (charge 6+) balanced with three sulfate anions (charge 6–).

$$
\left.\begin{array}{l} Al^{3+} \\ \\ Al^{3+} \end{array}\right\} \left\{\begin{array}{l} SO_4^{2-} \\ SO_4^{2-} \\ SO_4^{2-} \end{array}\right. \text{ giving the formula } Al_2(SO_4)_3
$$

We need parentheses around the sulfate to show that the formula is A–L–two S–O–four–taken–three–times.

## 7-11 MOLECULES AND IONS COMPARED

An ionic compound consists of a three-dimensional stack of ions held in place by the electrostatic attractive forces between the ions. The ionic bond is the force holding the ions in place. Positive and negative ions are the characteristic particles that compose ionic compounds, which are typically crystalline solids. Many ionic solids are water soluble. Because ionic compounds have salt-like crystalline forms, they are sometimes collectively called salts. That is, the generic meaning of **salt** is an ionic compound. Compounds involving covalent bonds are quite different. Covalent compounds have molecules as fundamental particles. The covalent bond is the force holding atoms in molecules.

It is important to appreciate the differences between ions and molecules. Ions carry charges, and ionic compounds consist of positive ions balanced with negative ions. Positive and negative ions always exist in proximity. It is not possible to have a sample of a pure negative ion or a pure positive ion. Ions are found as components of ionic compounds. However, since many ionic compounds are soluble in water, ions are also found dissolved in natural waters, such as ocean water, drinking water, and body fluids. Ocean water is salty due to dissolved ionic compounds. The term body electrolytes refers to various ions dissolved in the fluids of our bodies. Ions can exist in solution as long as there are as many negative charges as positive charges.

The **formula unit of an ionic compound** is the simplest combination of ions that satisfies the charges. The formula unit is not a discrete particle since it represents only a part of an entire crystal lattice. For instance, the formula unit for sodium chloride is NaCl since this is the simplest ratio of ions in which there is a balance of charge. A sample of sodium chloride is a vast collection of interconnecting NaCl units. The formula unit for the ionic compound of aluminum and oxygen is $Al_2O_3$ since the charge is balanced by two aluminum ions and three oxide ions. **Ionic compounds** normally occur as pieces of crystalline solid, sometimes as beautifully formed crystals, but more often as irregular chunks or even finely divided or powdered solids. Some are found concentrated in nature as mineral deposits. Other ionic compounds are manufactured by industrial chemical processes.

ACTIVITY

7-4

Give the formulas of the compounds made up of the following sets of ions.

(a) ammonium ion and phosphate ion

(b) calcium ion and nitrate ion

(c) sodium ion and oxalate ion

(d) zinc ion and acetate ion

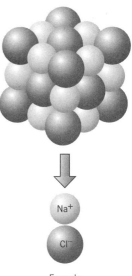

Formula unit

Solid iodine          $I_2$

Powered (ground up crystals)
calcium oxide

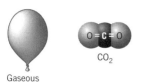

Liquid
carbon tetrachloride          $CCl_4$

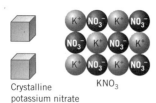

Gaseous
carbon dioxide          $CO_2$

Crystalline          $KNO_3$
potassium nitrate

**FIGURE 7-9** Some
examples of molecules and
ions.

**Ionic Compound:**

*A compound that has a formula starting with a metal.*

**Molecular Compound:**

*A compound containing only nonmetals or metalloids and nonmetals.*

A molecule is a unique chemical species or particle that is electrically neutral. It consists of two or more atoms in close contact held together by covalent bonds. A **molecular compound** is a collection of molecules each containing atoms of elements shown in the formula (e.g., $H_2O$, $CH_4$, and $C_{12}H_{22}O_{11}$). In a sample of a pure covalent compound all the molecules are alike, and the molecule is a representative particle of the compound. Molecules are stable particles and it requires energy to break the covalent bonds within molecules.

A molecule is a discrete particle, so the formula unit of a molecular compound is the formula of the molecule. The formula tells how many combined atoms of each type are in the molecule. The compound involving hydrogen and chlorine, called hydrogen chloride, is composed of HCl molecules. Molecules as formula units of compounds give us insight into why the empirical formulas of some compounds are different from the actual molecular formulas (see Section 4-10). The compound butane has molecules containing 4 carbon atoms and 10 hydrogen atoms. The molecular formula of butane is $C_4H_{10}$, which reflects the nature of the molecules. However, the simplest ratio of carbon to hydrogen in this compound is 2 to 5 ($4/10 = 2/5$). Thus, when the formula of butane is determined by experiment, it has an empirical formula of $C_2H_5$. Since butane contains molecules having 4 carbons and 10 hydrogens it is represented by its actual molecular formula, $C_4H_{10}$.

**Covalent compounds** occur as gases, liquids, and solids. Generally, those molecular compounds with relatively low molar mass and molecules with few atoms are gases at room temperature. For instance, carbon dioxide, $CO_2$, and chlorine, $Cl_2$, are gases. Covalent compounds with more complex molecules or higher molar masses are liquids. Bromine, $Br_2$, and carbon tetrachloride, $CCl_4$, are examples of molecular liquids. Those molecular compounds with relatively high molar masses or many atoms in their molecules are usually solids. Iodine, $I_2$, and sucrose, $C_{12}H_{22}O_{11}$, are solids.

Figure 7-9 shows that when we have a sample of a compound, it is imagined to contain ions if it is ionic. If it is molecular it is imagined to contain molecules. This is an important part of a chemical vision. How do we know whether a compound is ionic or molecular? Generally we can use the guidelines shown in the margin, though there may be some exceptions.

Classify the following compounds as ionic or molecular. Explain why in each case.

KBr, $CH_4$, FeO, $H_2S$, MgS, $SiF_4$, $C_3H_8$, $ZnCl_2$

## 7-12 BONDING ABILITIES OF NONMETALS

Molecular compounds normally involve nonmetals bonded to nonmetals. Recall the locations of nonmetals in the periodic table. How many covalent bonds do the atoms of the various nonmetals commonly form? A hydrogen atom, with only one electron to share, forms one covalent bond. The other nonmetals generally tend to form as many covalent bonds as there are single or unpaired electrons in the dot symbol (see Table 7-4). Some examples are

$$
\begin{array}{cccc}
\overset{\displaystyle|}{\underset{\displaystyle|}{-\text{C}-}} & \overset{\displaystyle|}{-\ddot{\text{N}}-} & :\ddot{\text{O}}- & :\ddot{\ddot{\text{F}}}- \\[6pt]
\text{4 bonds} & \text{3 bonds} & \text{2 bonds} & \text{1 bond}
\end{array}
$$

Since hydrogen forms one bond and oxygen two, the compound water involves the combination of two hydrogen atoms and one oxygen atom.

$$
\begin{array}{c}
:\overset{..}{\underset{.}{\text{O}}}\cdot\ \cdot\text{H} \\
\text{H}
\end{array}
\qquad
\begin{array}{c}
:\ddot{\text{O}}\ (:)\text{H} \\
\text{H}
\end{array}
$$

Notice that oxygen follows the octet rule by sharing two pairs of electrons, and each hydrogen has two electrons.

***Table 7.4***   Typical Numbers of Covalent Bonds Formed by Nonmetals[a]

| | | | *Group* | | |
|---|---|---|---|---|---|
| **IIIA**<br>**13** | **IVA**<br>**14** | **VA**<br>**15** | **VIA**<br>**16** | **VIIA**<br>**17** | **IA**<br>**1** |
| $-\text{B}-$ | $-\text{C}-$ | $-\ddot{\text{N}}-$ | $:\ddot{\text{O}}-$ | $:\ddot{\ddot{\text{F}}}-$ | $\text{H}-$ |
| 3 Bonds | 4 Bonds | 3 Bonds | 2 Bonds | 1 Bond | 1 Bond |
| | $-\text{Si}-$ | $-\ddot{\text{P}}-$ | $:\ddot{\text{S}}-$ | $:\ddot{\ddot{\text{Cl}}}-$ | |
| | 4 Bonds | 3 Bonds | 2 Bonds | 1 Bond | |
| | | | $:\ddot{\text{Se}}-$ | $:\ddot{\ddot{\text{Br}}}-$ | |
| | | | 2 Bonds | 1 Bond | |
| | | | | $:\ddot{\ddot{\text{I}}}-$ | |
| | | | | 1 Bond | |

[a]Some can form more or less bonds than the typical number in certain compounds.

A molecule of water is sometimes represented by a **structural formula** such as

$$:\overset{..}{O}-H$$
$$|$$
$$H$$

where the lines represent the electron pairs of covalent bonds. Structural formulas show the sequences of covalently bonded atoms in molecules. More commonly, the **molecular formula** $H_2O$ is used for water. Molecular formulas take up less room but they do not show the bonding sequence. The formula $H_2O$ suggests that water consists of molecules in which two atoms of hydrogen bond to one oxygen atom. To show how the atoms are bonded, we need a structural formula.

Molecular compounds may involve two, three, or more elements. **Binary molecular compounds** are those that contain only two elements. Many molecular compounds involve a few atoms bonded to form simple molecules. However, molecules can consist of several atoms, tens of atoms, hundreds of atoms, or thousands of covalently bonded atoms. Several million different molecular compounds are known. Of course, learning chemistry does not mean that we have to learn about millions of different compounds. The bonding patterns that apply to simple molecules also apply to complex molecules.

### 7-13 THE DIATOMIC MOLECULAR ELEMENTS

**DIATOMIC ELEMENTS**

| Hydrogen | $H_2$ |
| Nitrogen | $N_2$ |
| Oxygen | $O_2$ |
| Fluorine | $F_2$ |
| Chlorine | $Cl_2$ |
| Bromine | $Br_2$ |
| Iodine | $I_2$ |

A sample of an uncombined metallic element is viewed simply as a collection of metal atoms as described in Section 7-3. For example, a sample of iron consists of a collection of Fe atoms. Some nonmetals have such a great tendency to form covalent bonds that the atoms covalently bond with one another in the elemental state. This means that when the element is not combined with other elements, it bonds to itself. Recall that seven of the nonmetals occur in the elemental state in the form of **diatomic molecules** in which two atoms are covalently bonded. The formation of such molecules satisfies the octet rule for each atom. Two examples are

| | | | |
|---|---|---|---|
| chlorine | $Cl_2$ | $:\!\overset{..}{\underset{..}{Cl}}\!\!\bigcirc\!\!\overset{..}{\underset{..}{Cl}}\!:$ | $:\!\overset{..}{\underset{..}{Cl}}\!-\!\overset{..}{\underset{..}{Cl}}\!:$ |
| bromine | $Br_2$ | $:\!\overset{..}{\underset{..}{Br}}\!\!\bigcirc\!\!\overset{..}{\underset{..}{Br}}\!:$ | $:\!\overset{..}{\underset{..}{Br}}\!-\!\overset{..}{\underset{..}{Br}}\!:$ |

When we have a sample of a diatomic molecular element, the element is represented by the formula of the diatomic molecule. For example, a sample of chlorine is a collection of $Cl_2$ molecules. However, these elements do not remain as diatomic molecules when they combine with other elements. When this happens, the diatomic molecules break up and the atoms form new bonds with other elements.

## 7-14 ELECTRON DOT OR LEWIS STRUCTURES OF MOLECULES

The formula of a molecule shows the number of atoms of each element but does not show the bonding arrangement of the atoms. To represent the bonding sequences in a molecule, the electron dot symbols of the elements are arranged so that the shared pairs are shown and the octet rule (or duet for hydrogen) is satisfied. For instance, a molecule of fluorine, $F_2$, is shown as

$$:\!\overset{..}{\underset{..}{F}}\!:\!\overset{..}{\underset{..}{F}}\!:  \quad \text{or} \quad :\!\overset{..}{\underset{..}{F}}\!-\!\overset{..}{\underset{..}{F}}\!:$$

and a molecule of hydrogen fluoride, HF, is shown as

$$H\!:\!\overset{..}{\underset{..}{F}}\!:  \quad \text{or} \quad H\!-\!\overset{..}{\underset{..}{F}}\!:$$

Arrangements of dot symbols used to represent molecules are called **Lewis structures** in honor of Gilbert Lewis who first suggested their use. These Lewis structures don't convey any information concerning the shapes of the electron clouds nor do they suggest that electrons are dots. They merely serve as convenient representations of molecules. Usually the **shared pairs** of electrons are represented by lines between atoms. Any unshared electron pairs are shown as dot pairs. These **unshared pairs** are also called **lone pairs** or **nonbonding pairs** since they are not directly involved in covalent bonds.

Lewis structures are written by fitting the element dot symbols together to show shared electron pairs and to satisfy the octet rule. Some examples are

1. Water, $H_2O$, two H· and one $:\!\overset{..}{O}\!\cdot$ fit together as

$$\overset{..}{\underset{|}{:O}}\!-\!H$$
$$H$$

2. Ammonia, $NH_3$, three H· and one $\cdot\!\overset{..}{N}\!\cdot$ fit together, with the three hydrogens bonded to the nitrogen

$$H\!-\!\overset{..}{\underset{|}{N}}\!-\!H$$
$$H$$

3. Carbon tetrachloride, $CCl_4$, four $:\!\overset{..}{Cl}\!\cdot$ and one $\cdot\!\overset{.}{C}\!\cdot$ fit together, with the four chlorines bonded to the carbon as shown on the next page.

$$:\ddot{C}l: \\ | \\ :\ddot{C}l-C-\ddot{C}l: \\ | \\ :\ddot{C}l:$$

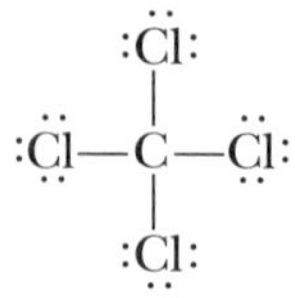

When writing dot structures for molecules, first try to fit the dot symbols together to try to satisfy the octet rule. Remember hydrogen forms one bond, oxygen forms two bonds, and carbon forms four bonds. If the octet rule cannot be satisfied, another approach is used as discussed in the next section. Keep in mind, however, that there are exceptions to the octet rule. For some molecules, the dot structures do not satisfy the rule. In other cases, it is not possible to write a simple dot structure for a molecule. These exceptions are not important to us. We'll consider molecules with elements that follow the octet rule.

**EXAMPLE 7-5**

What is the Lewis structure of hydrogen fluoride, HF? Fitting a hydrogen, H·, and fluorine, ·$\ddot{\underset{..}{F}}$:, gives the Lewis structure.

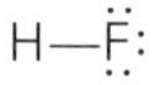

$$H-\ddot{\underset{..}{F}}:$$

**EXAMPLE 7-6**

What is the Lewis structure of sulfur dichloride, $SCl_2$? Using the dot symbol of sulfur, :$\ddot{S}$·, with two chlorines, ·$\ddot{\underset{..}{Cl}}$:, gives a possible structure of

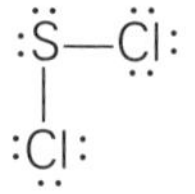

$$:\ddot{S}-\ddot{\underset{..}{Cl}}: \\ | \\ :\ddot{\underset{..}{Cl}}:$$

We can check the structure by noting that each atom has an octet of electrons when the shared pairs are counted for both atoms sharing.

**EXAMPLE 7-7**

What is the Lewis structure of iodoform, $CHI_3$?

This compound contains one carbon, ·$\dot{C}$·, one hydrogen, H·, and three iodines, ·$\ddot{\underset{..}{I}}$:. As noted in Table 7-3 carbon tends to form four bonds and hydrogen and iodine each tend to form one bond. We can fit the dot symbols together with the carbon in the center.

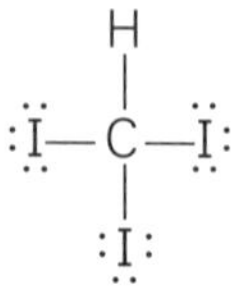

$$H \\ | \\ :\ddot{\underset{..}{I}}-C-\ddot{\underset{..}{I}}: \\ | \\ :\ddot{\underset{..}{I}}:$$

It does not matter which position we use for the hydrogen; we could just as easily draw the Lewis structure as

$$:\ddot{I}:$$
$$|$$
$$:\ddot{I}-\underset{|}{\overset{|}{C}}-H$$
$$:\ddot{I}:$$

Write Lewis structures for the following molecules.

$PBr_3$, $H_2S$, $CF_4$, $OF_2$

## 7-15 MULTIPLE BONDS

When we attempt to write the Lewis structures of some molecules and try to obey the octet rule, it is necessary to move electron pairs around so that more than one pair is shared between atoms. For instance, the electron dot symbol for nitrogen is $:\dot{\ddot{N}}\cdot$ Attempting to write the Lewis structure for molecular nitrogen, $N_2$, by just fitting the dot symbols together gives

$$:\dot{N}-\dot{N}:$$

This structure provides for only six electrons around each nitrogen. Remember, the shared pair counts for both atoms. If we move one electron from each atom to make another shared pair, this gives

$$:\dot{N}=\dot{N}:$$

This structure gives only seven electrons around each nitrogen. However, if we move one more electron from each atom to make a third shared pair, we have a Lewis structure in which each nitrogen has an octet.

$$:N{\equiv}N:$$

Note that each nitrogen has an octet when the three shared pairs are counted for each nitrogen. Since there is a total of 10 outer-energy-level or valence electrons (five from each nitrogen) involved in diatomic nitrogen, it is necessary to have the two combined nitrogen atoms share three pairs of electrons to satisfy the octet rule.

As another example, when we try to write the Lewis structure of carbon dioxide, $CO_2$, by connecting dot symbols, we obtain the incorrect structure of

$$:\dot{O}-\dot{C}-\dot{O}:$$

However, we achieve a satisfactory structure by having each oxygen share two pairs of electrons with the carbon.

$$:\ddot{O}=C=\ddot{O}:$$

Since there are 16 valence electrons (six from each oxygen and four from the carbon), it is necessary to place two pairs of electrons between the carbon and each oxygen. This satisfies the octet rule for both atoms.

The sharing of more than one pair of electrons by atoms is entirely possible. It is called **multiple bonding.** The sharing of two pairs of electrons between two atoms is a **double bond,** and the sharing of three pairs of electrons between two atoms is a **triple bond.**

### 7-16  WRITING LEWIS STRUCTURES

To predict a possible Lewis structure for a molecule, it is necessary first to decide which atoms are bonded. Which atoms bond to one another is sometimes obvious, but often, we need more information besides the formula to establish the bonding sequence. Next, we write the dot symbols of each atom and bond the atoms in the sequence by single covalent bonds. Finally, we move any extra electrons about to form multiple bonds needed to satisfy the octet rule and the bonding tendencies of the atoms. The steps involved in this pattern are:

1.  Select the appropriate dot symbols for each element. If you are not sure consult a periodic table to see in which groups the elements are found.
2.  Fit the dot symbols together according to the bonding sequence so that there is one single bond between each of the bonded atoms.
3.  If the octet rule is not satisfied for any atoms, move electrons about to make multiple bonds as needed. Hydrogen is satisfied with two electrons.
4.  For some molecules it is not possible to write simple Lewis structures that obey the octet rule.

---

**EXAMPLE 7-8**

What is the Lewis structure of formaldehyde, $H_2CO$, in which the carbon bonds to the hydrogens and the oxygen?

The bonding sequence gives this arrangement of dot symbols:

Forming single covalent bonds gives the structure shown on the next page.

$$\begin{array}{c} H \\ | \\ \cdot\overset{}{C}-\ddot{O}: \\ | \\ H \end{array}$$

The unpaired electrons of carbon and oxygen are moved so that carbon and oxygen are double bonded.

$$\begin{array}{ccc} H & & H \\ | & & | \\ \cdot\overset{}{C}\overset{\frown}{-}\overset{..}{O}: & & C=\ddot{O}: \\ | & & | \\ H & & H \end{array}$$

**EXAMPLE 7-9**

What is the Lewis structure of acetylene, $C_2H_2$, in which a hydrogen is bonded to each carbon and the carbons are bonded to one another?

The dot symbols are written in the bonding sequence so that each atom is bonded to the next by single covalent bonds which gives

$$H-\overset{.}{C}-\overset{.}{C}-H$$

To have an octet around each carbon, the unpaired electrons are moved to form a triple bond between the carbons

$$H-\overset{\frown}{C}\overset{}{\vdots}C-H \qquad H-C\equiv C-H$$

**ACTIVITY 7-7**

What is the Lewis structure of hydrogen cyanide, HCN, in which a hydrogen bonds to carbon that, in turn, is bonded to nitrogen? What is the Lewis structure of carbon disulfide, $CS_2$, in which two sulfur atoms are bonded to one carbon atom?

## 7-17 LEWIS STRUCTURES OF POLYATOMIC IONS

Since polyatomic ions are covalently bonded groups of atoms, it is often possible to write Lewis structures for them. The same rules for Lewis structures of molecules are used for writing Lewis structures of polyatomic ions except that the charge of the ion must be considered. The charge of a polyatomic ion changes the way we count the valence electrons. A positive polyatomic ion has fewer valence electrons than the total count of valence electrons of the atoms in the ion. A negatively charged polyatomic ion has excess valence electrons corresponding to the charge. (i.e., a 1− ion has one extra valence electron, and a 2− ion has two extra valence electrons). Consider, for example, the chlorate ion, $ClO_3^-$, in which the chlorine has three oxygen atoms bonded to it.

$$\begin{array}{ccc} & O & \\ O & Cl & O \end{array}$$

First, put in the valence electrons according to the dot symbols of the elements.

$$\cdot \ddot{\text{O}}:$$
$$:\ddot{\text{O}}::\ddot{\text{Cl}}:\cdot\ddot{\text{O}}:$$

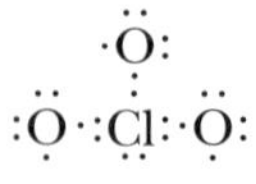

Since the charge is 1−, add one more valence electron, usually on the central atom, which is chlorine in this case.

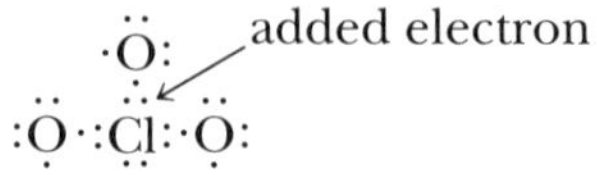

$$\cdot \ddot{\text{O}}: \quad\nearrow \text{added electron}$$
$$:\ddot{\text{O}}::\ddot{\text{Cl}}:\cdot\ddot{\text{O}}:$$

Next, move the electrons about to make at least one single covalent bond to each oxygen. When writing Lewis structures, any of the electrons of the dot symbols can be moved about as needed to satisfy the octet rule.

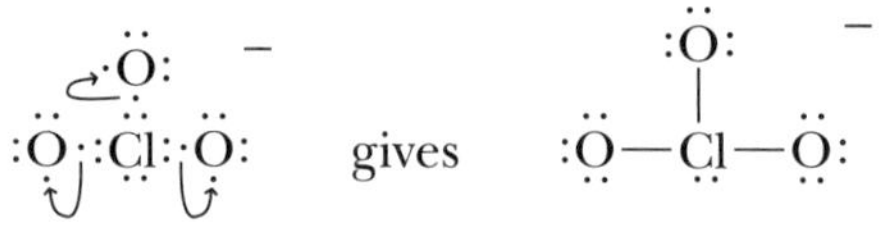

$$\begin{array}{ccc} \overset{\curvearrowleft\ddot{\text{O}}:}{} & & \ddot{\text{O}}: \quad^{-} \\ :\ddot{\text{O}}::\ddot{\text{Cl}}:\cdot\ddot{\text{O}}: & \text{gives} & :\ddot{\text{O}}-\text{Cl}-\ddot{\text{O}}: \end{array}$$

Note that since an extra electron was added, the group of atoms has a charge (one more electron than protons) and the octet rule is satisfied for each atom. Show the charge as part of the Lewis structure to emphasize that the species is an ion, not a molecule.

---

**EXAMPLE 7-10**

Give a Lewis structure for the ammonium ion, $NH_4^+$, in which the nitrogen is bonded to four hydrogens. First, write each atom in sequence as its dot symbol.

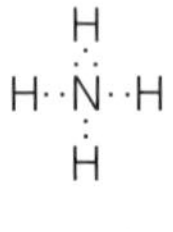

$$\begin{array}{c} \text{H} \\ \ddot{} \\ \text{H}\cdot\cdot\text{N}\cdot\cdot\text{H} \\ \ddot{} \\ \text{H} \end{array}$$

As each atom of hydrogen joins to the nitrogen with a single bond, we notice that there is one too many electrons. Since the charge of the ion is 1+, we remove one valence electron to give the Lewis structure.

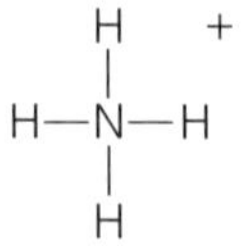

$$\begin{array}{c} \text{H} \quad + \\ | \\ \text{H}-\text{N}-\text{H} \\ | \\ \text{H} \end{array}$$

---

**EXAMPLE 7-11**

Give a Lewis structure for the cyanide ion, $CN^-$. First, write the dot symbols of the elements and add one extra electron for the 1− charge.

$$\cdot \overset{\cdot}{\underset{\cdot}{C}} \cdot \cdot \overset{\cdot}{\underset{\cdot}{N}} : \qquad : \overset{\cdot}{\underset{\cdot}{C}} \cdot \cdot \overset{\cdot}{\underset{\cdot}{N}} :$$

added electron

Next, try a single bond:

$$: \overset{\cdot}{\underset{\cdot}{C}} - \overset{\cdot}{\underset{\cdot}{N}} :^{-}$$

This does not satisfy the octet rule so we try a double bond.

$$: C = N :^{-}$$

In this case we need one more bond to satisfy the octet rule. This gives the Lewis structure

$$: C \equiv N :^{-}$$

## 7-18  SHAPES OF MOLECULES

Unlike ionic bonds, covalent bonds are directional. Covalent bonds occur between atoms, and when atoms have more than one single bond the bonds point in different directions in three-dimensional space. As a result molecules have three-dimensional shapes. A molecule is visualized as a group of atoms bonded through the merging of their electron orbitals. The orbitals have specific spatial orientations. Thus, a water molecule has two hydrogen atoms bonded to an oxygen atom in an arrangement that gives a specific shape to the molecule. We can describe the shape of a molecule by imagining straight lines connecting the centers of the bonded atoms. These lines define a skeletal structure for a molecule that allows a description of the molecular shape. In a water molecule the hydrogen atoms bond to the oxygen atom so that the skeletal structure is angular or bent. Thus, the shape of a water molecule is angular, bent, or V-shaped.

The Lewis structure of a molecule shows the number of electrons distributed about each atom. Within a given molecule in which the octet rule is satisfied, a specific atom occupies a somewhat central position. This is the atom to which other atoms are bonded; we call it the **central atom.** For example, in methane

$$\begin{array}{c} \text{H} \\ | \\ \text{H} - \text{C} - \text{H} \\ | \\ \text{H} \end{array}$$

the carbon is the central atom; around it are four pairs of electrons involved in bonding. If we figure out how the electron pairs arrange in

***Table 7-5***   Common Molecular Shapes

| Formula (A is central atom) | Electron dot structure | Pair Arrangement | Shape |
|---|---|---|---|
| $AX_4$ Four bonding pairs No lone pairs | | | Tetrahedral |
| $AX_3$ Three bonding pairs One lone pair | | | Triangular pyramid |
| $AX_2$ Two bonding pairs Two lone pairs | | | Angular |

ACTIVITY 7-8

What is the likely arrangement of four pairs of electrons about a central atom? You can answer this question yourself by using four small balloons of the same size. Inflate each balloon to the same volume and tie the stems. Tie or twist the necks of all the balloons together. The balloons will arrange as far apart as possible. Instead of balloons you could use four straight pins and a small eraser removed from the end of a pencil. Push the ends of the pins into the eraser and try to arrange them in a three-dimensional pattern so that each pin is at an equal distance from the others. In other words, try to find the arrangement in which the angle between any two pins is the same and the pins are as far apart as possible. Remember to work in three-dimensional space; you should not make a flat square shape.

space, we can visualize the shape of the molecule. What is the likely arrangement of four pairs of electrons about the central atom? The **valence shell electron pair repulsion theory (VSEPR theory)** gives an answer. This theory is based on the fact that electron pairs repel each other and are arranged about the central atom so that they are as far apart as possible.

The likely arrangement of four pairs of electrons around a central atom is based upon the idea that the pairs are located as far from each other as possible. This maximizes the space the orbitals can occupy and ensures they do not crowd each other. The arrangement of four electron pairs is **tetrahedral,** with the pairs directed toward the corners of a **tetrahedron,** as shown in Table 7-5. In a tetrahedral distribution, the angle between two pairs of electrons is about 109.5°. This is the internal tetrahedral angle. A **tetrahedral distribution** is represented in two dimensions by a sketch such as

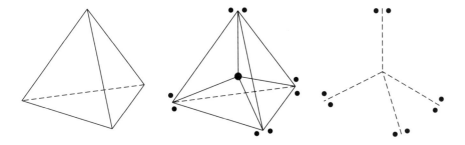

## Tetrahedral Molecules

Knowing that four pairs of electrons have a tetrahedral distribution, we can predict the likely shape of many molecules, as shown in Table 7-5. Accordingly we predict that a molecule of methane has a tetrahedral shape.

$$
\begin{array}{c}
H \\
| \\
C \\
H \diagup \ \diagdown H \\
H
\end{array}
$$

## Triangular Pyramid Molecules

The Lewis structure for ammonia, $NH_3$, is

$$
\begin{array}{c}
H—\ddot{N}—H \\
| \\
H
\end{array}
$$

The central nitrogen has four pairs of valence electrons, which would be distributed tetrahedrally. However, the shape of ammonia relates to the locations of the centers of the various atoms in the molecule. How do the four atoms arrange in space? The tetrahedral arrangement of the valence shell electron pairs in ammonia results in the molecule having a **triangular pyramid** shape (see Table 7-5). Triangular pyramid refers to the fact that the nitrogen atom is located at the peak of a pyramid and the hydrogen atoms have a triangular arrangement at the base of the pyramid.

$$
\begin{array}{c}
\ddot{N} \\
\diagup \ \diagdown \\
H \quad H \ H
\end{array}
$$

If one of the balloons in your balloon model is viewed as a nonbonding pair or if you remove one pin from your tetrahedral model, you'll have a triangular pyramid.

### Angular or Bent Molecules

The Lewis structure for water is

$$\ddot{\text{O}}\text{—H}$$
$$|$$
$$\text{H}$$

The central oxygen has four valence electron pairs, which would be distributed tetrahedrally. However, the shape of water depends on how the centers of the hydrogen atoms are located about the oxygen. As shown in Table 7-5, the water molecule is expected to have an **angular shape.** This is also called a V-shape or bent shape. Within the molecule there are two bonding pairs of electrons and two lone pairs of valence electrons. The shape is represented as

If two of the balloons in your balloon model are viewed as nonbonding pairs or if you remove two pins from your original tetrahedral model, you'll have an angular or bent shape.

### Linear Molecules

When a molecule consists of only two atoms with one bonding electron pair and three lone pairs around the "central" atom, the molecule is **linear** in shape. All diatomic molecules are linear. Linear means all of the atoms are arranged along a straight line. For example, a molecule of hydrogen fluoride, HF, is linear even though the four pairs of electrons are arranged in a tetrahedral fashion around the fluorine.

$$\text{H—}\ddot{\text{F}}\text{:}$$

### Some Other Shapes

The atoms involved in a double bond or a triple bond are arranged in a linear sequence. Thus, a molecule of carbon dioxide with two double bonds is linear.

$$\text{:}\ddot{\text{O}}\text{=C=}\ddot{\text{O}}\text{:}$$

And a molecule of acetylene is linear.

$$\text{H—C}\equiv\text{C—H}$$

Consider a central atom that has one double bond and two single bonds. In such a case, the central atom and the three other atoms to which it is bonded are located in the same plane and arranged in a **triangular shape** around the center. Thus, in a molecule of formaldehyde, $H_2CO$, all of the atoms are in the same plane and the molecule has a triangular planar shape. Triangular planar means that all of the atoms lie in a plane with the outer atoms on the corners of a triangle with angles of 120°.

More information and examples of VSEPR theory are given in Appendix 3.

FOCUS 7

## Ions, Molecules, and Bonding

Octet rule: Atoms tend to lose, gain, or share electrons so that they attain a total of eight outer-energy-level electrons.

Ion: A chemical particle that carries an electrical charge.

Cation: An ion having a positive charge. Anion: an ion having a negative charge.

Simple ion: An atom of an element that has an electrical charge resulting from the loss or gain of electrons.

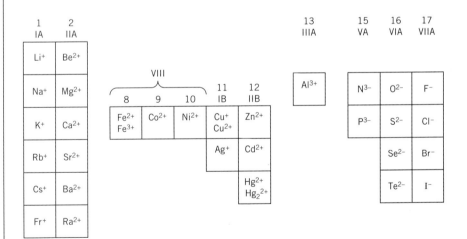

Ionic bond: The electrostatic force of attraction between ions that holds them in combination.

Ionic compounds: Compounds containing cations and anions. Typical ionic compounds contain metals in combination with nonmetals and have formulas that start with a metal.

Formulas of ionic compounds reflect the balance of positive and negative charges

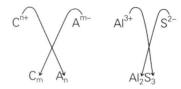

Covalent bond: The force of attraction between atoms resulting from the sharing of an electron pair.

Molecule: A chemical particle composed of two or more covalently bonded atoms.

Molecular compound: A compound that consists of molecules. Typical molecular compounds contain only nonmetals or metalloids and nonmetals.

Polyatomic ion: A charged particle composed of two or more covalently bonded atoms. Many ionic compounds contain polyatomic anions in combination with metallic cations.

Molecules and polyatomic ions are represented by Lewis dot structures that show which atoms are bonded. The pattern for writing a simple Lewis dot structure is as follows:

1. Select the appropriate dot symbols for each element.
2. Fit the dot symbols together according to the bonding sequence so that there is one single bond between each of the bonded atoms. Use lines to represent single covalent bonds and pairs of dots to represent electron pairs not involved in covalent bonds.
3. If the octet rule is not satisfied for an atom, move electrons about to make double bonds or triple bonds as needed. Hydrogen does not obey the octet rule. It forms one single bond.
4. For some molecules it is not possible to write simple Lewis structures that obey the octet rule.

The shapes of simple molecules and polyatomic ions can be predicted by using their Lewis dot structures. Identify the central atom or atoms in the structure and count the number of bonding pairs of electrons (BP) and nonbonding or lone pairs of electrons (LP) around the central atom. Typical shapes are:

4 BP = tetrahedral

3 BP and 1 LP = triangular pyramid

2 BP and 2 LP = bent, v-shape or angular

1 double bond and 2 single bonds = triangular planar

$$\begin{array}{c} H \\ \diagdown \\ \phantom{H}C{=}\ddot{O} \\ \diagup \\ H \end{array}$$

2 double bonds = linear

$$:\ddot{O}{=}C{=}\ddot{O}:$$

1 single bond and 1 triple bond = linear

$$H{-}C{\equiv}N:$$

## Chemical Vision of Solids, Liquids, and Gases

Now that we have useful mental images of atoms, molecules, and ions we can better understand the various physical states of matter. Solids, liquids, and gases are obviously different. How can we explain the differences? A sample of a **solid** occupies a fixed volume and is rigid. The shape of a solid can change only with some effort. A sample of a **liquid** has a fixed volume and is quite flexible. A liquid can be poured from one container to another or spread out on a flat surface. A sample of a **gas** is quite flexible and needs a container. A gas completely occupies any container in which it is placed and escapes into the atmosphere when released. Gases are typically colorless but some, such as chlorine, $Cl_2$, and nitrogen dioxide, $NO_2$, have distinct colors. In any case, gases are always transparent, which means it is possible to see through them. A gas sample can be easily compressed and it has a measurable pressure, and can leak through a hole in its container

An explanation of the differences between gases, liquids, and solids comes from considering not only the chemical particles that compose them but the behavior of these particles. Figure 7-10 illustrates the gaseous, liquid, and solid states as collections of particles. Using the kinetic

**FIGURE 7-10** Kinetic molecular view of a gas, a liquid, and a solid.

  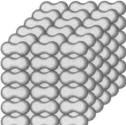

Solids as three dimensional spatial arrangements of ions, atoms or molecules

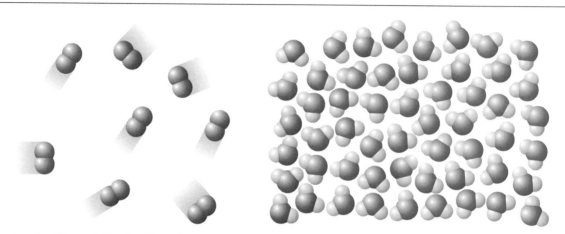

A gas is a diffuse collection of rapidly moving molecules    A liquid is a collection of closely packed molecules in random motion

**FIGURE 7-10**    (*Continued*)

molecular theory, discussed in Chapter 1, we view a gas as a diffuse collection of molecules that are in random high-speed motion. On the average, the molecules are far apart and occupy the volume by their incessant motion. In contrast, the liquid state consists of a collection of particles in a more condensed form. In a liquid, the particles are close together and are moving or wandering about within the collection. The solid state consists of particles arranged in fixed spatial positions. However, the particles in the solid state still vibrate about their fixed positions. The solid state is a highly condensed and ordered form of matter. The particles are packed very close to one another.

The motion of chemical particles gives them kinetic energy (KE). The kinetic energy of a moving object or particle, such as a molecule, is given by the expression: $KE = \frac{1}{2}mv^2$ where $m$ is its

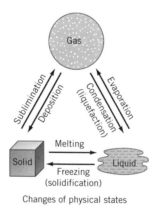

Changes of physical states

mass and $v$ is its speed. There is a direct relation between kinetic energy and the speed of gas particles; higher speeds correspond to higher energies, and higher energies correspond to higher temperatures. The temperature of a sample of a material is directly related to the average kinetic energy of its particles. Heating a substance causes the chemical particles to move more rapidly.

## The Liquid State

When a gas is cooled, the particles slow down and the average kinetic energy of the particles decreases. When a gas is compressed, the gaseous particles move closer together. Under the proper conditions of low temperature and high pressure a gas liquefies, that is, it converts from a gas to the liquid state. The kinetic motions of the molecules tend to make them move apart from one another. **Liquefaction** occurs when the molecules move closer together and condense to the liquid state. As they condense, the molecules aggregate into a group giving the characteristic appearance of a liquid. See Figure 7-11.

The liquid state consists of molecules or, sometimes, atoms or ions that are packed in a relatively close manner with little space between particles. However, the liquid state is dynamic and

**FIGURE 7-11**   Liquefaction of a gas.

(a)

(b)

(c)

In the gaseous state the molecules are rapidly moving and are far apart.

As the pressure is increased and the temperature decreased the molecules slow down and are forced closer together.

Finally at the correct combination of temperature and pressure most of the molecules aggregate into a collection of molecules characteristic of the liquid state; liqufaction has taken place.

the molecules move and wander about the entire collection. The molecules move about randomly with short distances between collisions with other molecules and the container. These collisions result in no net loss in kinetic energy, but the molecules can exchange kinetic energy by collision.

The average kinetic energy of the molecules is directly related to the temperature of the liquid. Heating a liquid increases the average speed and kinetic energy as the temperature increases. Cooling a liquid decreases the average kinetic energy as the temperature decreases. Molecules do not move around as rapidly at lower temperatures. A useful image of the liquid state is to view a liquid as a teeming collection of closely packed particles in which individual particles are in constant random motion within the collection. See Figure 7-10.

The liquid state typically consists of a definite collection of molecules in which the individual molecules have freedom of motion within the collection. Heating a liquid causes the average kinetic energy of the molecules to increase. The molecules move more rapidly. A liquid can be heated so that the molecules break away from the liquid state and enter the gaseous state. The change from a liquid to a gas can be accomplished by heating a liquid until it boils; the characteristic temperature at which a given liquid boils is its **boiling point.**

*The Solid State*

Consider what happens when a liquid is cooled rather than heated. As the liquid is cooled, the kinetic motion of the particles decreases. As cooling continues the particles begin to occupy relatively fixed positions in space. In fact, as the liquid cools, a point is reached when the particles are arranged in a three-dimensional pattern in which they occupy definite spatial positions. When the particles are arranged in fixed positions, the liquid has changed to a solid. The change from the liquid to the solid state is called **solidification, crystallization, or freezing.** See Figure 7-12. The temperature at which a given liquid changes to a solid is its **freezing point.**

A crystalline solid is characterized by a definite three-dimensional arrangement of particles occupying relatively fixed positions in space. Such a three-dimensional arrangement is a **crystal lattice** or simply a **crystal.** Metals, ionic compounds, and molecular compounds can form crystalline solids. Crystals occur in a variety of shapes and sizes. Some crystals are fine powders, some are rough chunks and bits, and some have beautiful geometrical shapes. The positions that the particles occupy in the crystal lattice are crystal lattice sites. Molecules occupy these sites in a molecular solid. In metals the metal atoms occupy lattice sites; in ionic compounds cations and anions occupy lattice sites. See Figure 7-10.

The particles occupying the lattice sites in a crystalline solid are held in a fixed arrangement in space. However, these particles still have some kinetic energy, which results from the particles vibrating about their positions in the crystal lattice (see Fig. 7-13). This vibratory motion allows the transfer of kinetic energy throughout the solid by collision of a particle with its neighboring particles. The vibrations allow particles in solids to exchange kinetic energy and conduct heat. When a solid is heated the particles gain kinetic energy. As they gain enough kinetic en-

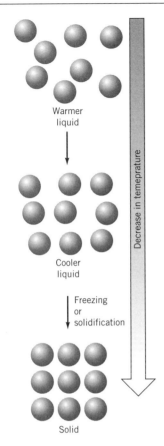

Warmer
liquid

Decrease in temeprature

Cooler
liquid

Freezing
or
solidification

Solid

**FIGURE 7-12** The formation of a solid from a liquid. As a liquid is cooled, the particles ultimately arrange in a definite three-dimensional pattern in which they have fixed positions in space.

**FIGURE 7-13** Particles occupying lattice sites in a crystalline solid vibrate within these positions. These vibrations and the resulting collisions between particles can transfer kinetic energy throughout the crystal.

ergy they break from the crystal lattice and enter the liquid state. In this way the solid melts into a liquid. The characteristic temperature at which a solid melts is its **melting point.**

Because each pure chemical has a unique melting point, the melting point can be used to help identify the chemical. For instance, pure aspirin, acetylsalicylic acid, has a melting point of 135°C. Suppose we measure the melting point of a substance that is thought to be pure aspirin and find that it does not have a melting point of 135°C. We can conclude that either the substance is impure aspirin mixed with other chemicals or that is not aspirin at all.

Make drawings or sketches of the following using your knowledge of the natures of the solid, liquid, and gaseous states and your knowledge of atoms, molecules, and ions. Show the chemical particles that make up the materials.

(a) a piece of aluminum foil

(b) a crystal of sodium bicarbonate

(c) a crystal of table sugar

(d) a sample of liquid ethyl alcohol

(e) a sample of carbon dioxide gas.

Make drawings or sketches of molecules showing the melting of ice to form liquid water and the boiling of liquid water to form steam.

The taste buds on our tongues allow us to distinguish certain tastes. You know that some foods taste salty and some taste sweet. Table salt is salty and table sugar is sweet. In this exercise you are going to observe how salt (sodium chloride) and sugar (sucrose) are similar and how they differ. For the first part you need: granulated sugar, salt, two sheets of paper, a plastic knife, and a dry cloth towel

1. Place a small amount (about the size of half an aspirin tablet) of dry salt on a flat piece of paper and spread it out to make a flat layer.
2. Place a similar amount of dry granulated sugar on another flat piece of paper and spread it out.
3. Look very closely at each sample and record your observations.
4. Rub the plastic knife with a dry towel about 20 times to give it an electrostatic charge. Lower the knife so that it touches the center of the sample of salt. Note the scattering pattern. Record your observations.
5. Once again put a charge on the knife and, this time, lower it so that it touches the center of the sugar. Note the scattering pattern. Record your observations.

For the second part of this exercise you need: a book of paper matches, granulated sugar, salt, and a 5 cm by 10 cm piece of aluminum foil.

1. Hold the aluminum foil so that it is flat and gently press your finger on one end to make a slight indentation.
2. Place a sample of salt about the size of half an aspirin tablet in the indentation.
3. Light a match and hold it under the aluminum just below the salt. Record your observations.
4. Remove the salt from the foil and add a sample of sugar about the size of half an aspirin tablet.
5. Light a match and hold it under the aluminum just below the sugar. Record your observations. If the sugar begins to burn and char remove the match.

For the third part you need: granulated sugar, salt, two sheets of white paper (the copier kind), distilled water (tap water will do), two wire twist ties, tape, and a 9-volt battery.

1. Strip the paper from the twist ties and make sure that the wire is clean. Twist the end of one of the wires around a terminal of the battery and tape it in place. Twist the second wire around the other terminal and tape it in place. Arrange the wires so that they extend straight up from the top of the battery. These wires are the electrodes.
2. You need two paper dishes. Make a paper dish by folding a sheet of paper in quarters and crimping the edges to form the dish.
3. Place a sample of salt about the size of an aspirin tablet in the middle of one paper dish. Add a small amount of water on top of the salt. Test the resulting solution with your conductivity apparatus by dipping the electrodes into the solution. Look closely at the electrodes and record your observations.
4. Place a sample of sugar about the size of an aspirin tablet in the middle of the second paper dish. Add a small amount of water on top of the sugar. Test the resulting solution with your conductivity apparatus. Record your observations.

Make a table to summarize your observations of salt and sugar and include the chemical names and formulas in the table.

# QUESTIONS

## Section 7-1

1. What is the force that holds atoms in combination?

## Section 7-2

2. What are valence electrons?

3. What is an electron dot symbol? Give an example.

4. Give the electron dot symbols for these elements (refer to a periodic table).

(a) Be  (b) Ne  (c) He  (d) S  (e) Al
(f) N  (g) Te  (h) Rb  (i) I  (j) Xe  (k) C

5. Select two elements from the list in Question 4 that you suspect would have similar chemical behavior based on their electron dot symbols.

## Section 7-3

6. Describe how metal atoms bond to one another in metallic solids.

## Section 7-4

7. Give a statement of the octet rule.

## Section 7-5

8. What is an anion? What is a cation?

9. What is the difference between an atom and an ion?

10. A negative ion is larger than the atom from which it is formed. Why? A positive ion is smaller than the atom from which it is formed. Why?

## Section 7-6

11. *Give the formulas of the simple cations formed by the following elements (refer to a periodic table for group numbers).

    (a) H   (b) K   (c) Be   (d) Sr   (e) Ga
    (f) Fr   (g) Ca   (h) Rb

12. Give the formulas of the simple cations formed by the following elements (refer to a periodic table for group numbers).

    (a) Li   (b) Mg   (c) Na   (d) Cs   (e) Ba
    (f) Al   (g) Zn

13. *Give the formulas of the simple anions formed by the following elements (refer to a periodic table for group numbers).

    (a) S   (b) I   (c) P   (d) Br   (e) Se   (f) As

14. Give the formulas of the simple anions formed by the following elements (refer to a periodic table for group numbers).

    (a) F   (b) Te   (c) O   (d) N   (e) Cl

15. Describe the kinds of simple ions formed by elements of the following groups.

    (a) group 1   (b) group 2   (c) group 13
    (d) group 15   (e) group 16   (f) group 17

## Section 7-7

16. Describe the ionic bond. How does it differ from a covalent bond?

17. What physical form do ionic compounds usually have? Why?

18. Give definitions or descriptions of the following.

    (a) ionic compound   (b) crystal lattice
    (c) electrostatic force

## Section 7-8

19. *Give the formulas of the ionic compounds containing the following ions.

    (a) $Na^+$ and $S^{2-}$   (b) $Li^+$ and $Br^-$
    (c) $Ca^{2+}$ and $O^{2-}$   (d) $Mg^{2+}$ and $F^-$
    (e) $Al^{3+}$ and $O^{2-}$   (f) $Al^{3+}$ and $Br^-$

20. *Give the formulas of the ionic compounds containing the following ions.

    (a) $K^+$ and $S^{2-}$   (b) $Na^+$ and $F^-$
    (c) $Ba^{2+}$ and $Cl^-$   (d) $Be^{2+}$ and $O^{2-}$
    (e) $Al^{3+}$ and $P^{3-}$   (f) $Al^{3+}$ and $S^{2-}$

21. Give the formulas of the ionic compounds formed between the following metals and nonmetals. Use the expected positive ions for the metals, and the expected negative ions for the nonmetals (refer to a periodic table for group numbers).

    (a) lithium and chlorine
    (b) magnesium and nitrogen
    (c) barium and chlorine
    (d) strontium and oxygen
    (e) sodium and sulfur
    (f) aluminum and sulfur
    (g) calcium and arsenic
    (h) sodium and phosphorus
    (i) potassium and oxygen

22. Give the formulas of the ionic compounds formed between the following metals and nonmetals. Use the expected positive ions for the metals, and the expected negative ions for the nonmetals (refer to a periodic table for group numbers).

    (a) aluminum and bromine
    (b) magnesium and phosphorus
    (c) calcium and sulfur
    (d) strontium and fluorine
    (e) sodium and nitrogen
    (f) gallium and oxygen
    (g) barium and arsenic
    (h) lithium and nitrogen
    (i) potassium and iodine

## Section 7-9

23. Describe the covalent bond. How does it differ from an ionic bond?

24. Which elements tend to form covalent molecules? Which elements tend to form binary ionic compounds?

25. Give definitions or descriptions of the following.

    (a) molecule

(b) shared electron pair

(c) molecular compound

26. Explain the sharing of electrons between two atoms of hydrogen in terms of electron orbitals.

## Section 7-10

27. What is a polyatomic ion?

28. Give the names of the following anions.

(a) $C_2H_3O_2^-$  (b) $NO_2^-$  (c) $Br^-$  (d) $CO_3^{2-}$

(e) $CrO_4^{2-}$  (f) $HSO_4^-$  (g) $I^-$  (h) $OH^-$

(i) $Cr_2O_7^{2-}$  (j) $SO_3^{2-}$

29. Give the formulas of the following polyatomic ions. Where they are commonly found is shown in parentheses.

(a) sulfate ion (ammonium sulfate fertilizer)

(b) hydroxide ion (sodium hydroxide)

(c) bicarbonate ion (baking soda)

(d) nitrate ion (saltpeter)

(e) oxalate ion (spinach and rhubarb)

(f) nitrite ion (sodium nitrite, used to preserve processed meats)

(g) chlorate ion (fireworks)

(h) phosphate ion (cola beverages and nondairy coffee creamers)

(i) dichromate ion (safety matches)

(j) cyanide ion (sodium cyanide, a violent poison)

(k) permanganate ion (some antifungal medicines)

30. The formula for the ionic compound calcium phosphate is $Ca_3(PO_4)_2$. Explain why the parentheses are needed in this formula.

31. *Give the formulas for the ionic compounds formed between the following pairs of ions.

(a) $K^+$ and $SO_3^{2-}$  (b) $Al^{3+}$ and $NO_2^-$

(c) $Mg^{2+}$ and $PO_4^{3-}$  (d) $K^+$ and $C_2H_3O_2^-$

(e) $Ca^{2+}$ and $HCO_3^-$  (f) $Ag^+$ and $I^-$

(g) $Hg^{2+}$ and $Br^-$  (h) $NH_4^+$ and $SO_4^{2-}$

32. Give the formulas for the ionic compounds formed between the following pairs of ions.

(a) $Na^+$ and $SO_4^{2-}$  (b) $Al^{3+}$ and $SO_4^{2-}$

(c) $Ba^{2+}$ and $NO_3^-$  (d) $K^+$ and $ClO_3^-$

(e) $Na^+$ and $C_2O_4^{2-}$  (f) $Cu^+$ and $O^{2-}$

(g) $Zn^{2+}$ and $Br^-$  (h) $NH_4^+$ and $PO_4^{3-}$

## Section 7-11

33. The following are among the most common chemicals manufactured in the United States. Indicate which of them are ionic and which are molecular.

(a) sulfuric acid, $H_2SO_4$

(b) ammonia, $NH_3$

(c) lime, CaO

(d) oxygen, $O_2$

(e) nitrogen, $N_2$

(f) ethylene, $C_2H_4$

(g) sodium hydroxide, NaOH

(h) chlorine, $Cl_2$

(i) phosphoric acid, $H_3PO_4$

(j) ammonium nitrate, $NH_4NO_3$

(k) sodium carbonate, $Na_2CO_3$

(l) urea, $CH_4N_2O$

(m) propylene, $C_3H_6$

(n) toluene, $C_7H_8$

(o) benzene, $C_6H_6$

(p) ethylene dichloride, $C_2H_4Cl_2$

(q) ethyl benzene, $C_8H_{10}$

(r) carbon dioxide, $CO_2$

(s) methanol, $CH_3OH$

(t) nitric acid, $HNO_3$

34. For all the ionic compounds listed in Question 33, write the formulas with charges of the individual ions that make up the compounds.

## Section 7-12

35. How does the bonding tendency of a nonmetal correspond to its group position in the periodic table?

## Section 7-13

36. What does it mean to say that an element is diatomic?

37. List the names and formulas of those elements that occur in the form of diatomic molecules when they are not combined with other elements.

## Sections 7-14 to 7-17

38. *Give Lewis structures for the following.

(a) chlorine, $Cl_2$

(b) hydrogen iodide, HI

(c) chloroform, $CHCl_3$ (1 H and 3 Cl bonded to a central C)

(d) water, $H_2O$

(e) hydrogen telluride, $H_2Te$

(f) dichlorine oxide, $Cl_2O$ (2 Cl bonded to a central O)

(g) perchloric acid, $HClO_4$ (4 O bonded to Cl; H bonded to one of the O)

(h) hydrazine, $N_2H_4$ (2 H bonded to each N, which are bonded to one another)

(i) phosphorus trichloride, $PCl_3$

(j) boron trifluoride, $BF_3$ (exception to octet rule)

(k) ethane, $C_2H_6$ (3 H bonded to each C, which are bonded to one another)

(l) cyclobutane, $C_4H_8$ (4 C bonded in a square; 2 H bonded to each C)

(m) carbon tetrachloride, $CCl_4$

(n) hydrogen peroxide, $H_2O_2$ (1 H bonded to each O, which are bonded to one another)

(o) hydroxide ion, $OH^-$

(p) ammonium ion, $NH_4^+$

(q) phosphate ion, $PO_4^{3-}$

(r) nitrate ion, $NO_3^-$

39. Give Lewis structures for the following.

(a) hydrogen selenide, $H_2Se$

(b) dibromine oxide, $Br_2O$ (2 Br bonded to a central O)

(c) iodine, $I_2$

(d) hydrogen chloride, HCl

(e) methyl bromide, $CH_3Br$

(f) silicon tetrachloride, $SiCl_4$

(g) nitrogen trifluoride, $NF_3$

(h) stibine, $SbH_3$

(i) chloric acid, $HClO_3$ (3 O bonded to Cl; H bonded to one of the O)

(j) hydroxyl amine, $NH_2OH$ (2 H and O bonded to N; 1 H bonded to O)

(k) chloroethane, $C_2H_5Cl$ (2 C bonded and 3 H bonded to one C; 2 H and 1 Cl bonded to the other C)

(l) cyclopropane, $C_3H_6$ (3 C bonded in a triangle; 2 H bonded to each C)

(m) carbon tetrafluoride, $CF_4$

(n) methyl alcohol, $CH_3OH$ (3 H and 1 O bonded to C; 1 H bonded to O)

(o) permanganate ion, $MnO_4^-$ ($:\dot{\underset{\cdot}{Mn}}\cdot$)

(p) chlorate ion, $ClO_3^-$

(q) peroxide ion, $O_2^{2-}$

(r) hydronium ion, $H_3O^+$

40. Define a multiple covalent bond. Give an example of a molecule that has such a bond.

41. What is the difference between a double bond and a triple bond?

42. *Give the Lewis structures for the following molecules; all contain multiple bonds.

(a) phosgene, a nerve gas, $COCl_2$ (2 Cl and 1 O bonded to a central C)

(b) acetylene, a fuel, $C_2H_2$ (1 H bonded to each C, which are bonded to one another)

(c) carbon dioxide, used in cold storage of produce, $CO_2$ (2 O bonded to C)

(d) hydrogen cyanide, a poison, HCN (H and N bonded to a central C)

(e) urea, a product of protein metabolism, $CN_2H_4O$ (1 O and 2 N bonded to a central C; 2 H bonded to each N)

(f) acetone, a solvent, $C_3H_6O$ (2 C and 1 O bonded to a central C; 3 H bonded to each of the outer C's)

(g) tetrabromoethene, $C_2Br_4$ (2 Br bonded to each C)

(h) vinyl chloride, $C_2H_3Cl$ (2 H bonded to one C; 1 H and 1 Cl bonded to the other C)

43. Give the Lewis structures for the following molecules; all contain multiple bonds.

(a) silicon dioxide, $SiO_2$ (2 O bonded to Si)

(b) cyanogen, $C_2N_2$ (1 N bonded to each C, which are bonded to one another)

(c) nitrous acid, $HNO_2$ (2 O bonded to N; H bonded to an O)

(d) trichloroethylene, $C_2HCl_3$ (2 C bonded and 2 Cl bonded to one C; 1 H and 1 Cl bonded to the other C)

(e) dichloroethyne, $C_2Cl_2$ (1 Cl bonded to each C, which are bonded to one another)

(f) allene, $C_3H_4$ (3 C bonded; 2 H bonded to each of the end C's)

(g) propyne, $C_3H_4$ (3 C bonded; 3 H bonded to one end C and 1 H bonded to the other end C)

(h) formaldehyde, $CH_2O$ (2 H and 1 O bonded to C)

44. Write Lewis structures for (b), (e), (f), (h), (r), and (s) in Question 33. In methanol 3 H and 1 O are bonded to C; 1 H is bonded to O.

## Section 7-18

45. *Describe the shapes and give Lewis structures that suggest the shapes of the following (the central atoms are underlined).

(a) $\underline{Si}Cl_4$, silicon tetrachloride

(b) $\underline{P}Cl_3$, phosphorus trichloride

(c) $H_2\underline{S}$, hydrogen sulfide

(d) $\underline{O}F_2$, oxygen difluoride

(e) $\underline{C}H_3Br$, methyl bromide

(f) $Br_2\underline{O}$, dibromine oxide

(g) $\underline{C}H_2F_2$, difluoromethane

(h) $\underline{N}H_3$, ammonia

(i) $H\underline{I}$, hydrogen iodide

(j) $H_2O_2$, hydrogen peroxide (both oxygens can be viewed as central atoms)

(k) $C_2H_6$, ethane (both carbons can be viewed as central atoms)

46. Describe the shapes and give Lewis structures that suggest the shapes of the following (the central atoms are underlined).

(a) $\underline{C}Cl_2F_2$, dichlorodifluoromethane

(b) $\underline{N}F_3$, nitrogen trifluoride

(c) $Cl_2\underline{O}$, dichlorine oxide

(d) $\underline{C}F_4$, carbon tetrafluoride

(e) $H_2\underline{Se}$, hydrogen selenide

(f) $\underline{Si}H_4$, silane

(g) $\underline{As}H_3$, arsine

(h) $H\underline{F}$, hydrogen fluoride

(i) $\underline{Cl}O_2^-$, chlorite ion

(j) $\underline{N}H_4^+$, ammonium ion

(k) $H_3\underline{O}^+$, hydronium ion

47. *Describe the shapes of the molecules or ions in parts (b), (c), (d), (e), (f), (i), (j), (m), (o), (p), and (r) of Question 38.

48. Describe the shapes of the molecules or ions in parts (a), (b), (e), (f), (g), (h), (o), (p), (q), and (r) of Question 39.

49. How do solids, liquids, and gases differ? Draw pictures to illustrate your answer.

50. How do the liquid and gaseous states of matter differ in terms of the arrangement and motion of the particles that make them up?

51. Explain the process of liquefaction of a gas in terms of the temperature and pressure and the motion of molecules.

52. Give a description of the liquid state at the molecular level.

53. What is a boiling point for a liquid?

54. Why do some liquids have odors that can be detected at some distance from an open container of the liquid?

55. Describe what happens, in terms of molecular motion and particle arrangement, when a liquid freezes to form a solid.

56. Describe a crystalline solid.

57. What is the melting point of a solid?

## Questions to Ponder

58. Draw a picture to show how the electron orbitals in an oxygen atom and the electron orbitals in two hydrogen atoms overlap to form a water molecule.

59. Use some creative materials to make a model of a tetrahedron.

# CHEMICAL

# NOMENCLATURE

## 8-1 CHEMICAL NAMES

What's in a chemical name? Hopefully, some chemical information and often a story. After Alexander the Great conquered parts of Egypt around 300 B.C.E., the Greeks built a temple dedicated to the god Ammon. Fires built in this temple used dried camel droppings as fuel. Over many years a white crystalline material built up on the walls of the temple along with soot from the fires. This white salt-like material was known as *sal ammoniac*, 'salt of Ammon'. Today, it is obtained from other sources and is usually named ammonium chloride, but sometimes the old name is used. In the early 1700s, sal ammoniac was used to produce a gaseous compound named ammonia, (the spirit of Ammon) after its source. The compound is $NH_3$ and centuries later ammonia is still its official name.

We read a formula by pronouncing the element symbols as letters and the subscripts as numbers.

| | |
|---|---|
| $Na_2SO_3$ | N–A–two–S–O–three |
| $Ca(NO_3)_2$ | C–A N–O–three–taken–twice |
| $K_2O$ | K–two–O |
| $PBr_3$ | P–B–R–three |
| $N_2O_5$ | N–two–O–five |

**265**

Another approach is to translate the formula to the chemical name. Admittedly, this requires some knowledge of **chemical nomenclature.** Official systems of nomenclature are used to name various kinds of compounds. Names that convey information about the composition of compounds are **systematic names.** Many chemical compounds were named before there were any nomenclature rules. These names generally are not systematic and do not convey any information about the structure of the compound. They are **common** or **trivial names.** Some are so familiar that chemists normally use them, too. For example, the name water for $H_2O$ is a common name that is invariably used. A few important compounds that have common names are

| | | | |
|---|---|---|---|
| $CH_4$ | methane | $C_2H_6$ | ethane |
| $C_3H_8$ | propane | $C_4H_{10}$ | butane |
| $NH_3$ | ammonia | $H_2O_2$ | hydrogen peroxide |
| $C_6H_6$ | benzene | $C_6H_{12}O_6$ | glucose |
| $C_{12}H_{22}O_{11}$ | sucrose | $CH_3OH$ | methyl alcohol |
| $CH_3CH_2OH$ | ethyl alcohol | $C_2H_2$ | acetylene |

Table 8-1 lists the trivial and chemical names for some other familiar compounds. Occasionally you may hear mineral names used for some chemicals. Examples are quartz ($SiO_2$), galena (a lead ore), bauxite (an aluminum ore), and hematite and magnetite (iron ores). Mineral names have developed historically and are not systematic chemical names. Most chemicals used as medicines and drugs are complex organic compounds. Normally they are not referred to by their complex chemical names. Most have common names, trade names, or cryptic names. For example, the pain reliever with the chemical name of acetylsalicylic acid is simply called aspirin. The pain killer obtained from the juice of opium poppies is called morphine and the active ingredient in marijuana is called THC, a cryptic form of the chemical name delta-9-tetrahydrocannabinol.

## Generic Drug Names

Prescription and over-the-counter medicines contain chemicals that are complex organic compounds. These chemicals are sold under a variety of trade names or brand names. Many common drugs are marketed under specific brand names, but all brands contain the same drug. The generic name is the common chemical name that identifies the drug. Generic names cannot be used as exclusive brand names. Very often a less advertised over-the-counter drug or a generic prescription drug has the same active ingredient but a lower price than a name brand. Ask a pharmacist for advice and you may save some money.

***Table 8-1***   Trivial and Systematic Names for Some
Common Compounds

| FORMULA | TRIVIAL NAME | SYSTEMATIC NAME |
| --- | --- | --- |
| $C_2H_2$ | Acetylene | Ethyne |
| $NaHCO_3$ | Baking soda | Sodium hydrogen carbonate |
| $CaCO_3$ | Calcite, marble, limestone, chalk | Calcium carbonate |
| $KHC_4H_4O_6$ | Cream of tartar | Potassium hydrogen tartrate |
| $C_2H_5OH$ | Grain alcohol, drinking alcohol | Ethyl alcohol or ethanol |
| $Na_2S_2O_3$ | Hypo | Sodium thiosulfate |
| $N_2O$ | Laughing gas, nitrous oxide | Dinitrogen oxide |
| $PbO$ | Litharge | Lead(II) oxide |
| $CaO$ | Quick lime | Calcium oxide |
| $NaOH$ | Lye, caustic soda | Sodium hydroxide |
| $K_2CO_3$ | Potash | Potassium carbonate |
| $NH_4Cl$ | Sal ammoniac | Ammonium chloride |
| $NaNO_3$ | Saltpeter (Chile) | Sodium nitrate |
| $Ca(OH)_2$ | Slaked lime | Calcium hydroxide |
| $C_{12}H_{22}O_{11}$ | Sugar | Sucrose or alpha-D-glucopyranosyl-beta-D-fructofuranoside |
| $NaCl$ | Salt | Sodium chloride |
| $(CH_3)_2CHOH$ | Rubbing alcohol | Isopropyl alcohol, 2-propanol |
| $CH_3OH$ | Wood alcohol | Methyl alcohol or methanol |

The most commonly used over-the-counter drugs are the pain relievers aspirin, ibuprofen, acetaminophen and naxproxen sodium. Aspirin relieves pain and reduces fever and inflammation. It is not useful for sharp, stabbing, intense pain caused by direct stimulation of nerves, but is effective for dull, throbbing pain of inflammation or headache. Although aspirin was originally a brand name belonging to the Bayer corporation, it has, over time, become a generic name. Any brand of aspirin marked as USP (United States Pharmacopeia, a standard of purity) meets the same standards and has the same effect as any of the most heavily advertised brands.

Aspirin use does have side effects. Typically some stomach bleeding occurs when you take an aspirin. People with certain allergies, asthma, diabetes, and peptic ulcers should not take aspirin. Aspirin should not be used during the last three months of pregnancy since it "thins" the blood and can interfere with clotting. Children with flu-like symptoms or with chicken pox should not be given aspirin since aspirin use is somehow related to Reye syndrome, a serious but rare blood illness. Aspirin is poisonous in large amounts and thousands of cases of serious overdose occur each year. Young children are particularly sensitive to overdose. Safely store aspirin and any over-the-counter or prescription drugs and keep them in childproof containers.

Acetaminophen is an aspirin substitute. It is in products like Tylenol, Anacin-3, and Midol. It can, however, be purchased under the generic name. Acetaminophen has fever- and pain-reducing effects but does not irritate the stomach as much as aspirin. Its minimal anti-inflammatory effect recommends against its use for treating arthritis and rheumatism. Chronic use of acetaminophen can cause liver and kidney damage, so prolonged use is inadvisable without consulting a physician.

Ibuprofen, another aspirin substitute, is a pain reliever and reduces fever and inflammation. It can be purchased under the generic name and under brand names such as Advil, Motrin IB, Midol 200, and Nuprin. Check the ingredients on the label when buying a pain reliever. Some people with arthritis use ibuprofen rather than aspirin. Chronic use, however, can cause serious kidney damage in persons having certain kidney diseases. Ibuprofen also "thins" the blood just like aspirin. Women should not use it during the last four months of pregnancy.

Naxproxen sodium is an aspirin substitute sold under the brand name of Aleve. It is useful for pain from headache, menstrual cramps, or arthritis. One tablet provides relief for 8 to 12 hours. It can cause stomach irritation and allergic reactions. Furthermore, the recommended dose is not effective as an anti-inflammatory. It should not be taken in doses above the recommended dose without consulting a pharmacist or doctor.

## 8-2 CLASSIFYING COMPOUNDS

This chapter concerns the naming of simple ionic and molecular compounds, not complex organic compounds. Different methods of nomenclature are used for ionic and molecular compounds. Consequently, to name a compound it is necessary to classify it as ionic or molecular. Fortunately there is a simple pattern.

1. Ionic Compounds
   A. Metal–nonmetal binary compounds (e.g., sodium chloride, NaCl)
   B. Metal ion–polyatomic ion compounds (e.g., potassium nitrate, $KNO_3$)
   C. Ammonium ion–polyatomic ion or nonmetal ion compounds [e.g., ammonium sulfate, $(NH_4)_2SO_4$, and ammonium chloride, $NH_4Cl$]
2. Molecular Compounds
   A. Nonmetal–nonmetal binary compounds (e.g., carbon dioxide, $CO_2$)
   B. Metalloid–nonmetal binary compounds (e.g., silicon tetrafluoride, $SiF_4$)
   C. Compounds including three or more nonmetals (e.g., nitric acid, $HNO_3$, and glucose, $C_6H_{12}O_6$

Generally, if the formula of a compound starts with a metal or ammonium ion it is likely to be ionic. If the formula of a compound includes only nonmetals or metalloids it is likely to be molecular. Consult a periodic table to see whether an element is a metal, nonmetal, or metalloid. Some examples illustrate the classification of compounds.

| | |
|---|---|
| CaS | This metal–nonmetal compound is ionic. |
| $Al_2(SO_4)_3$ | This metal–polyatomic ion compound is ionic. |
| $NH_4Br$ | This ammonium ion containing compound is ionic. |
| $P_4O_{10}$ | This nonmetal–nonmetal compound is molecular. |
| $H_2SO_4$ | This nonmetal compound is molecular. |
| $AsCl_3$ | This metalloid–nonmetal compound is molecular. |

Classify the following compounds as ionic or molecular.

$Ca_3N_2$, $NH_4Br$, $C_{12}H_{22}O_{11}$, $H_2S$, $K_3PO_4$, $NO$

## 8-3 NAMING IONS

The name of an ionic compound comes from the names of the ions that compose it. Metals form positive ions by loss of electrons. Normally, a **metal ion** is named by using the name of the metal followed by the word ion. Some examples are

| | |
|---|---|
| $Na^+$ | sodium ion |
| $Ca^{2+}$ | calcium ion |
| $Zn^{2+}$ | zinc ion |

Those metals that form only one kind of ion include group 1 (IA) metals, group 2 (IIA) metals, and zinc ($Zn^{2+}$), silver ($Ag^+$), cadmium ($Cd^{2+}$), and aluminum ($Al^{3+}$). Some transition metals and the group 14 (IVA) metals, tin and lead, can form more than one type of metal ion. For example, iron forms the $Fe^{2+}$ and $Fe^{3+}$ ions and lead forms the $Pb^{2+}$ and $Pb^{4+}$ ions. Name such ions by using the name of the metal followed by a Roman numeral in parentheses. The Roman numeral shows the charge.

| | |
|---|---|
| $Fe^{2+}$ | iron(II) ion (read as iron–two–ion) |
| $Fe^{3+}$ | iron(III) ion (read as iron–three–ion) |
| $Pb^{2+}$ | lead(II) ion (read as lead–two–ion) |
| $Pb^{4+}$ | lead(IV) ion (read as lead–four–ion) |

Table 8-2 gives the formulas and names of more metal ions. Some have alternate names, for example, $Fe^{2+}$ is the ferrous ion and $Fe^{3+}$ is the ferric ion. These names have been superseded by Roman numerals but they are still used particularly in medicine, agriculture, and other areas of applied chemistry. You may have noticed ferrous sulfate as an ingredient in some vitamin–mineral supplements or food products containing iron-fortified

**Table 8-2** Names and Formulas of Some Metal Cations

| FORMULA | NAME |
|---------|------|
| $Li^+$ | lithium ion |
| $Na^+$ | sodium ion |
| $K^+$ | potassium ion |
| $Mg^{2+}$ | magnesium ion |
| $Ca^{2+}$ | calcium ion |
| $Sr^{2+}$ | strontium ion |
| $Ba^{2+}$ | barium ion |
| $Cr^{3+}$ | chromium(III) ion (chromic ion) |
| $Mn^{2+}$ | manganese(II) ion (manganous ion) |
| $Fe^{3+}$ | iron(III) ion (ferric ion) |
| $Fe^{2+}$ | iron(II) ion (ferrous ion) |
| $Co^{2+}$ | cobalt(II) ion (cobaltous ion) |
| $Ni^{2+}$ | nickel(II) ion (nickelous ion) |
| $Cu^+$ | copper(I) ion (cuprous ion) |
| $Cu^{2+}$ | copper(II) ion (cupric ion) |
| $Ag^+$ | silver ion |
| $Zn^{2+}$ | zinc ion |
| $Cd^{2+}$ | cadmium ion |
| $Hg^{2+}$ | mercury(II) ion (mercuric ion) |
| $Hg_2^{2+}$ | mercury(I) ion (mercurous ion) |
| $Al^{3+}$ | aluminum ion |

flour. In the older nomenclature, the ion of lower charge is the -ous ion and the ion of higher charge is the -ic ion. The ferrous and ferric ions illustrate this point.

Nonmetals form simple negative ions that are found in combination with positive metal ions in ionic compounds. Name **simple negative ions** by using the root of the name of the nonmetal with an -ide ending.

| | | |
|---|---|---|
| $N^{3-}$ | nitride ion | nitrogen is a group 15 (VA) element. |
| $S^{2-}$ | sulfide ion | sulfur is a group 16 (VIA) element. |
| $F^-$ | fluoride ion | fluorine is a group 17 (VIIA) element. |

Table 8-3 lists the formulas and names of some nonmetal ions and Table 8-4 lists the roots of the names of some nonmetals. Notice how the charges of these simple ions relate to the group numbers of the elements in the periodic table. Group 15 nonmetals form 3− ions, group 16 nonmetals form 2− ions, and group 17 nonmetals form 1− ions.

Now consider the names of **polyatomic ions.** The table inside the front cover lists the formulas and names of the most common polyatomic ions. When dealing with polyatomic ions, you must learn the names, formulas, and charges of the ions. The names of polyatomic ions are common names that do not reveal the formulas. Some names, however, suggest the presence of certain elements. The sulfate ion, $SO_4^{2-}$, is a polyatomic ion containing sulfur. In fact, you may notice that many polyatomic ions have names that end in -ate. What do these ions have in common?

The one common positively charged polyatomic ion is the ammonium ion, $NH_4^+$. In a sense, the mercury(I) ion, $Hg_2^{2+}$, is a polyatomic ion. We might expect the mercury(I) ion to be $Hg^+$ but it actually occurs as a covalent combination of two mercury atoms that carries a double positive charge. The mercury(I) ion always occurs as $Hg_2^{2+}$ in ionic compounds.

Most of the polyatomic ions are negative ions and most contain oxygen and another nonmetal. Note that some nonmetals form two different ions with oxygen. The name of the ion with the lesser number of oxygen atoms ends in **-ite** whereas the name of the ion with the greater number of oxygen atoms ends in **-ate.** When nomenclature rules were first devised slight variations in endings were used much like various endings are used in the conjugation of verbs in many languages. The endings used in nomenclature come from Latin. It is interesting that similar endings occur in some common words, such as arsenic, sedate, gracious (Spanish), and tripartite. Examples of ions with different endings in their names are

| | | | |
|---|---|---|---|
| $SO_4^{2-}$ | sulfate ion and | $SO_3^{2-}$ | sulfite ion |
| $NO_3^-$ | nitrate ion and | $NO_2^-$ | nitrite ion |

##  FORMULAS OF IONIC COMPOUNDS FROM ION FORMULAS

All ionic compounds are electrically neutral so the total positive charge always equals the total negative charge. This fact is reflected in the formula of an ionic compound. Thus, it is easy to predict the formula of an ionic compound if we know the formulas of the ions that make it up. For example, the formula for the compound containing sodium ions, $Na^+$, and fluoride ions, $F^-$, is simply

$$Na^+ \quad F^-$$

NaF (one sodium ion for each fluoride ion)

The formula for the compound containing potassium ions, $K^+$, and sulfide ions, $S^{2-}$, is

$$K^+ \qquad S^{2-}$$

$K_2S$ (Two potassium ions for each sulfide ion gives 2 + charges for 2 − charges)

The arrow indicates that two $K^+$ ions are needed to balance one $S^{2-}$ ion. In other words, the charge of each ion indicates the number of oppositely charged ions that are needed in the formula.

The formula for the compound containing aluminum ions, $Al^{3+}$, and oxide ions, $O^{2-}$, is

$$Al^{3+} \qquad O^{2-}$$

$Al_2O_3$ (Two aluminum ions for three oxide ions gives 6 + charges for 6 − charges)

You can deduce the formulas for metal–polyatomic ion compounds in the same way. The formula for the compound containing sodium ions and nitrate ions, $NO_3^-$, is

$$Na^+ \qquad NO_3^-$$

$$NaNO_3$$

The formula for the compound of calcium ions, $Ca^{2+}$, and hydroxide ions, $OH^-$, is

$$Ca^{2+} \qquad OH^-$$

$Ca(OH)_2$ (Two hydroxide ions are needed for each calcium ion, shown by enclosing the hydroxide in parentheses and using a subscript)

**Table 8-3**   Names and Formulas of Some Nonmetal Anions

| FORMULA | NAME |
| --- | --- |
| $N^{3-}$ | Nitride ion |
| $P^{3-}$ | Phosphide ion |
| $As^{3-}$ | Arsenide ion |
| $O^{2-}$ | Oxide ion |
| $S^{2-}$ | Sulfide ion |
| $Se^{2-}$ | Selenide ion |
| $Te^{2-}$ | Telluride ion |
| $F^-$ | Fluoride ion |
| $Cl^-$ | Chloride ion |
| $Br^-$ | Bromide ion |
| $I^-$ | Iodide ion |

**Table 8-4**   Roots of the Names of Some Common Elements

| ELEMENT | ROOT |
| --- | --- |
| Hydrogen | Hydr- |
| Boron | Bor- |
| Carbon | Carb- |
| Silicon | Silic- |
| Nitrogen | Nitr- |
| Phosphorus | Phosph- |
| Arsenic | Arsen- |
| Antimony | Antimon- |
| Oxygen | Ox- |
| Sulfur | Sulf- or Sulfur- |
| Selenium | Selen- |
| Tellurium | Tellur- |
| Fluorine | Fluor- |
| Chlorine | Chlor- |
| Bromine | Brom- |
| Iodine | Iod- |

The formula for the compound of magnesium ions, $Mg^{2+}$, and phosphate ions, $PO_4^{3-}$, is

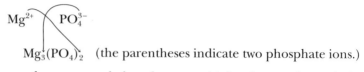

$Mg_3(PO_4)_2$    (the parentheses indicate two phosphate ions.)

Use parentheses as needed to show a multiple of any polyatomic ion in a compound. However, don't use parentheses if they are not needed.

## 8-5 NAMING IONIC COMPOUNDS

As mentioned in Section 8-2, classes of ionic compounds include metal–nonmetal binary, metal–polyatomic ion, and ammonium ion–nonmetal or polyatomic ion compounds. We can recognize from the formula that a compound is ionic. We name it simply by giving the name of the positive ion (cation) followed by the name of the negative ion (anion). The charge on a simple nonmetal anion is deduced by locating its group in the periodic table. Group 17 elements form singly charged negative ions. Group 16 nonmetals form doubly charged negative ions. Group 15 nonmetals form triply charged negative ions. Consider some examples of each type of ionic compound.

### Metal–Nonmetal Binary

We name these binary compounds by using the name of the metal followed by the root of the name of the nonmetal with an -ide ending. See Table 8-4 for the roots of nonmetal names.

KBr contains potassium and bromine; thus its name is potassium bromide.

$Ca_3N_2$ contains calcium and nitrogen; thus its name is calcium nitride.

If the metal in the compound is an element that forms more than one kind of ion, its charge is deduced from the charge of the nonmetal ion. The positive charge is shown in the ion name by using a Roman numeral. The charge of the nonmetal is related to its group.

FeS is iron(II) sulfide; sulfide is $S^{2-}$, so the iron must be $Fe^{2+}$ to balance the charge.

$PbO_2$ is lead(IV) oxide; oxide is $O^{2-}$, so the lead must be $Pb^{4+}$ to balance the charge of the two oxide ions in the compound.

### Metal–Polyatomic Ion

To name these compounds we must recognize what polyatomic ion is present. Furthermore, for metals that form more than one kind of ion, the charge of the metal must be deduced from the charge of the polyatomic ion.

$NaNO_3$      $NO_3^-$ is the nitrate ion so the name is sodium nitrate.

$Al(OH)_3$      $OH^-$ is the hydroxide ion so the name is aluminum hydroxide.

$K_2SO_4$      $SO_4^{2-}$ is the sulfate ion so the name is potassium sulfate.

$HgCO_3$      $CO_3^{2-}$ is the carbonate ion; thus the mercury ion must be $Hg^{2+}$, which is the mercury(II) ion, so the name is mercury(II) carbonate.

$Cu_3(PO_4)_2$      $PO_4^{3-}$ is the phosphate ion; thus the copper ion must be $Cu^{2+}$ to balance the charge. That is, since two $PO_4^{3-}$ are in the compound, there must be three copper(II) ions and the name is copper(II) phosphate.

### Ammonium Ion–Nonmetal or Polyatomic Ion

We name these compounds "ammonium" followed by the name of the anion.

    $NH_4Cl$ is ammonium chloride.

    $(NH_4)_2SO_4$, $SO_4^{2-}$ is the sulfate ion so the name is ammonium sulfate.

ACTIVITY 8-2

Name the following compounds.

$SnF_2$, $Ca_3N_2$, $K_3PO_4$, $NH_4Br$

### 8-6  NAMING NONMETAL–NONMETAL BINARY COMPOUNDS

Glance at a periodic table and recall that the nonmetals populate the upper right-hand region of the table. The metalloids separate the metals and nonmetals. Although there are fewer nonmetals than metals, there is a large number of known nonmetal–nonmetal compounds. Note that this class of compounds contains only two different kinds of elements (binary) and the elements must be nonmetals or metalloids. Some nonmetals can form more than one compound with one another, for instance, nitrogen and oxygen form $N_2O$, $NO$, $NO_2$, $N_2O_3$, $N_2O_4$, and $N_2O_5$.

An example of a binary nonmetal–nonmetal compound is $CO_2$, which has the name carbon dioxide. We name binary nonmetal–nonmetal and metalloid–nonmetal compounds according to the following general pattern:

| prefix | name of first element | prefix | root of name of second element | ide suffix |
|--------|----------------------|--------|-------------------------------|-----------|
| $CO_2$ | carbon | di | ox | ide |

To denote the number of combined atoms of each element in the compound we begin each part of the name with an appropriate prefix. For example, the compound $BF_3$ is named boron trifluoride to indicate one boron in combination with three fluorines. Table 8-4 lists the roots used for the nonmetals. Below is a list of common prefixes used to name binary compounds.

| | |
|---|---|
| **mono-** | one (this prefix is usually omitted except in carbon monoxide, CO) |
| **di-** | two |
| **tri-** | three |
| **tetra-** | four |
| **penta-** | five |
| **hexa-** | six |
| **hepta-** | seven |
| **octa-** | eight |
| **nona-** | nine |
| **deca-** | ten |

Which nonmetal is named first in a binary compound and which is given the -ide ending? The preferred order of elements in nonmetal–nonmetal or metalloid–nonmetal compounds is

B, Ge, Si, C, Sb, As, P, N, H, Te, Se, S, I, Br, Cl, O, and F

Notice that except for hydrogen and oxygen, this order is from bottom to top of each group and from left to right in the periodic table. By convention hydrogen is written after any element in group 15 or lower and before any element in group 16 or higher (e.g., $NH_3$ and $H_2O$). Naming compounds by the prefix method relates directly to the formula of the compound; if we know the formula the name is obvious; if we know the name the formula is obvious. Here are some examples.

| | |
|---|---|
| $N_2O$ | dinitrogen oxide |
| $NO_2$ | nitrogen dioxide |
| $N_2O_5$ | dinitrogen pentoxide (the "a" of penta is omitted for easier pronunciation) |
| $P_4O_{10}$ | tetraphosphorus decoxide (the "a" of deca is omitted for easier pronunciation) |
| $OF_2$ | oxygen difluoride |
| $P_4S_7$ | tetraphosphorus heptasulfide |
| $PCl_3$ | phosphorus trichloride |
| $PCl_5$ | phosphorus pentachloride |
| $CO$ | carbon monoxide |
| $CS_2$ | carbon disulfide |

**A C T I V I T Y**
**8-3**

Name the following compounds.

$SiF_4$, $P_4O_6$, $N_2O_3$, $KNO_3$, $HBr$

## 8-7   NAMING BINARY ACIDS

Acids are an important set of chemicals that have their own unique method of nomenclature. The binary compounds involving hydrogen and the halogens (group 17 or VIIA) are named according to the nonmetal–nonmetal nomenclature rules. For example, HCl is hydrogen chloride and HI is hydrogen iodide. When these compounds dissolve in water, the resulting solutions display specific properties called **acidic properties**. Some common properties of acidic solutions are that they taste sour, dissolve many metals, can burn the skin, and turn blue litmus paper red. Acids are described in detail in Chapter 12.

The water solutions of these compounds are called **acids** and these solutions are named accordingly. Normally we give names only to pure compounds. However, since these acid solutions are very important in chemistry, they are assigned special names. Their prefix is hydro- and their root is the name of the halogen with an -ic ending. This first part of the name is followed by the word acid.

hydro (root of halogen name) ic acid

The names of the four **hydrogen-halide acids** are given in the following table.

| FORMULA OF PURE COMPOUND | NAME OF PURE COMPOUND | NAME OF WATER SOLUTION |
|---|---|---|
| HF | Hydrogen fluoride | Hydrofluoric acid |
| HCl | Hydrogen chloride | Hydrochloric acid |
| HBr | Hydrogen bromide | Hydrobromic acid |
| HI | Hydrogen iodide | Hydroiodic acid |

The water solutions of acids can be denoted by using the formula of the compound followed by (aq) to suggest aqueous or water solution, for example, HCl(aq), HF(aq). Note that HF is named hydrogen fluoride and HF(aq) is named hydrofluoric acid. We use the same pattern for any of the four hydrogen-halide acids. Hydrochloric acid is the most common of these acids; it is typically found in chemistry laboratories and has many industrial uses. Its trivial name is muriatic acid.

## 8-8   NAMING OXOANIONS AND OXOACIDS

Many nonmetals form compounds that include hydrogen and oxygen. Most of these compounds and their corresponding water solutions have acidic properties. The compounds that contain hydrogen, oxygen, and another element are **oxoacids** or **oxyacids**. For a given oxoacid the same name is usually used to refer to both the pure oxoacid and its water solutions. For example, $H_2SO_4$ and $H_2SO_4(aq)$ are both called sulfuric acid. Very often you can recognize an oxoacid if it has a formula that begins with hydrogen and contains another element and oxygen. But sometimes formulas are given in different ways, for instance, acetic acid is written $HC_2H_3O_3$ or $CH_3CO_2H$.

**Oxoanions** or **oxyanions** are negative polyatomic ions related to the oxoacids. The formulas of these ions follow the same pattern as the formulas of the oxoacids except that hydrogen is not present and the ions carry a corresponding number of negative charges (i.e. $H_2SO_4$ and $SO_4^{2-}$) Some nonmetals form more than one oxoanion, so they are distinguished by use of different suffixes or prefixes in their names. The different oxoanions of a given element differ in the number of combined oxygen atoms. Nitrogen forms two oxoanions.

$NO_3^-$    nitrate ion         $NO_2^-$    nitrite ion

When there are two oxoanions for an element, the one with the greater number of oxygen atoms has the -ate suffix and the one with the lesser number of oxygen atoms has the -ite suffix. The name always includes the word ion. The oxoanions of group 16 (VIA) elements (sulfur, selenium, and tellurium) are exemplified by those of sulfur.

$SO_4^{2-}$    sulfate ion         $SO_3^{2-}$    sulfite ion

All of the halogens form four oxoanions except fluorine, which forms none. These ions are exemplified by those of chlorine. Since there are four, special prefixes are needed in their names.

$ClO_4^-$    perchlorate ion
$ClO_3^-$    chlorate ion
$ClO_2^-$    chlorite ion
$ClO^-$    hypochlorite ion

$ClO_3^-$ is the chlorate ion and $ClO_4^-$, which has one more oxygen, is the perchlorate ion. The **prefix per-** comes from hyper meaning above or elevated (e.g., hyperactive). $ClO_2^-$ is chlorite ion and $ClO^-$, which has one less oxygen, is hypochlorite ion. The **prefix hypo-** means below or under (e.g., hypodermic means under the skin).

Oxoanions and oxoacids are related in name and formula. We name an oxoacid by taking the name of the corresponding oxoanion, changing the suffix, and adding the word acid. The **-ate** ending of the ion name is changed to **-ic** for the acid name, and the **-ite** ending of the ion name is changed to **-ous** for the acid name. Some examples of polyatomic oxoacids related to oxoanions are

| | | | | | **COMMON OXOACIDS** | |
|---|---|---|---|---|---|---|
| $NO_3^-$ | nitrate ion | | $HNO_3$ | nitric acid | | |
| $NO_2^-$ | nitrite ion | | $HNO_2$ | nitrous acid | $H_2SO_4$ | sulfuric acid |
| $SO_4^{2-}$ | sulfate ion | | $H_2SO_4$ | sulfuric acid | $HNO_3$ | nitric acid |
| $SO_3^{2-}$ | sulfite ion | | $H_2SO_3$ | sulfurous acid | $H_3PO_4$ | phosphoric acid |
| $ClO_4^-$ | perchlorate ion | | $HClO_4$ | perchloric acid | | |
| $ClO_3^-$ | chlorate ion | | $HClO_3$ | chloric acid | $CH_3CO_2H$ | acetic acid |
| $ClO_2^-$ | chlorite ion | | $HClO_2$ | chlorous acid | or | |
| $ClO^-$ | hypochlorite ion | | $HClO$ | hypochlorous acid | $HC_2H_3O_2$ | |

Note that the syllable -ur- is added in the name of sulfur acids by convention.

The names of oxoacids do not directly reveal their formulas, so we need to memorize the names and formulas of the common acids. By convention, the formula of an acid begins with hydrogen (i.e., HCl, $H_2SO_4$). Thus, if we encounter such a formula we normally classify the compound as an acid; water, $H_2O$, is a notable exception.

Binary acids and oxoacids with more than one hydrogen can form polyatomic ions that result from the loss of one or more hydrogen ions ($H^+$) from the parent acid. For instance, when sulfuric acid, $H_2SO_4$, loses one hydrogen ion it forms the hydrogen sulfate ion, $HSO_4^-$. The charges of these ions correspond to the absence of the hydrogens. We name these ions by using the anion name preceded by the word hydrogen. If the ion has two hydrogens the prefix di- is used with the word hydrogen to tell that two hydrogens are associated with the ion.

Give the names and formulas for the oxoanions of selenium and bromine.

Give the names and formulas for the oxoacids of selenium and bromine.

| | | |
|---|---|---|
| $H_2PO_4^-$ | dihydrogen phosphate ion | ($H_3PO_4$ less one $H^+$) |
| $HPO_4^{2-}$ | hydrogen phosphate ion | ($H_3PO_4$ less two $H^+$) |
| $HSO_4^-$ | hydrogen sulfate ion | ($H_2SO_4$ less one $H^+$) |
| $HS^-$ | hydrogen sulfide ion | ($H_2S$ less one $H^+$) |
| $HCO_3^-$ | hydrogen carbonate ion | ($H_2CO_3$ less one $H^+$), often called bicarbonate ion; "bi" refers to hydrogen, not two of something. |

## 8-9 NAMES FROM FORMULAS

We have considered the nomenclature of common ionic compounds and some molecular compounds; these include most of the compounds we shall encounter in this book. But there are other compounds named according to other nomenclatures. Compounds of carbon, organic compounds, fall into many classes, which follow prescribed methods of nomenclature. However, we shall only be concerned with the nomenclature of the types of compounds described in this chapter.

To name a compound given its formula, we first classify the compound. If it is classed as ionic or binary molecular, we can name it. We use the prefix method only for binary molecular compounds. Never use the prefix method to name an ionic compound or a complex molecular compound. If a compound does not belong to one of the classes we have considered, we can name it only if we recognize it as a common compound or an acid. A flowchart for nomenclature is shown in Focus 8.

Once classified, we name a compound using the appropriate nomenclature. Using the flowchart in Focus 8, we can name some examples. The compound CaS starts with a metal so it classified as an ionic compound. It contains only two elements as the calcium ion and the sulfide ion, so its name is

calcium sulfide

The compound $Cu(OH)_2$ contains a metal and is classified as an ionic compound. It contains the hydroxide ion, $OH^-$, so it also has the $Cu^{2+}$ or copper(II) ion. Its name is

copper(II) hydroxide

The compound $Cl_2O_5$ is classified as a binary nonmetal–nonmetal compound, so by the prefix method its name is

dichlorine pentoxide

The compound $SiF_4$ is classified as a binary metalloid–nonmetal compound, so by the prefix method its name is

silicon tetrafluoride

The compound $NH_3$ is classified as a binary nonmetal–nonmetal compound but it should be recognized as having the common name

ammonia

The compound $NaHCO_3$ contains a metal and is classified as an ionic compound. It contains the sodium ion and the hydrogen carbonate ion, $HCO_3^-$. Thus, its name is

sodium hydrogen carbonate

The compound $H_3PO_4$ is classified as an oxoacid and has the name

phosphoric acid

ACTIVITY
8-6

Name the following compounds.

$Ag_2S$, $PbCl_4$, $SiC$, $I_2O_5$, $HNO_3$, $H_2O_2$

## 8-10 FORMULAS FROM NAMES

Another aspect of nomenclature is to deduce the formula of a compound from its name. This is quite straightforward for binary nonmetal compounds. For these compounds, the name gives the precise formula. Some examples are

dinitrogen trioxide $N_2O_3$
phosphorus pentafluoride $PF_5$

The name of an ionic compound suggests which ions make up the compound. To deduce the formula from the name it is necessary to consider the precise formula of each ion and the charges of the ions. You need to memorize the formulas of common polyatomic ions or look them up in a table of ions. Consulting a periodic table can help you recall the charges of many simple ions. Also remember that the Roman numeral in an ion name tells the charge of the ion. The formula is deduced by considering the number of positive and negative ions needed to make sure that the charges balance, that is, the number of ions needed to make the ionic formula unit neutral. Magnesium fluoride has magnesium ions and fluoride ions

$$Mg^{2+} \quad F^-$$

so the formula is $MgF_2$. Calcium phosphide is composed of calcium ions and phosphide ions

$$Ca^{2+} \quad P^{3-}$$

so the formula is $Ca_3P_2$. Iron(II) nitrate is composed of iron(II) ions and nitrate ions

$$Fe^{2+} \quad NO_3^-$$

so the formula is $Fe(NO_3)_2$. Ammonium phosphate is composed of ammonium ions and phosphate ions

$$NH_4^+ \quad PO_4^{3-}$$

so the formula is $(NH_4)_3PO_4$. Lead(II) acetate is composed of lead(II) ions and acetate ions

$$Pb^{2+} \quad C_2H_3O_2^-$$

so the formula is $Pb(C_2H_3O_2)_2$. Sulfuric acid is one of the common oxoacids. We associate it with the formula $H_2SO_4$.

Give the formulas for the following compounds.

ammonium iodide,
boron trifluoride,
potassium hydrogen sulfate,
acetic acid, calcium nitrate

FOCUS
8

## A Flowchart for Nomenclature

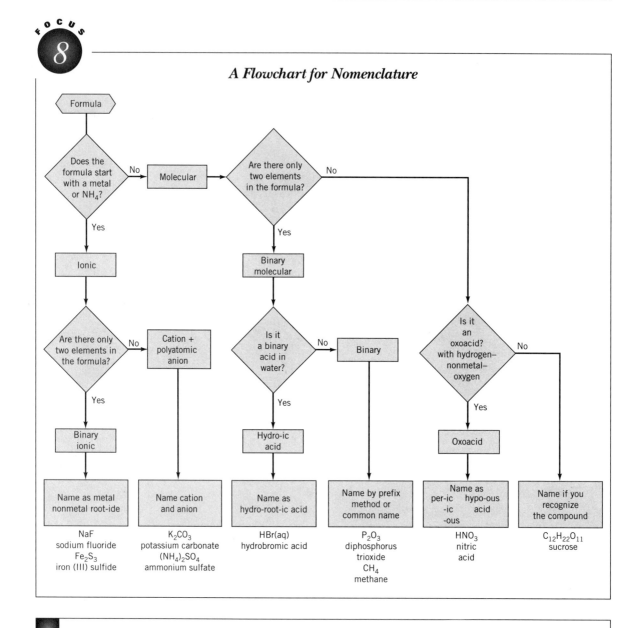

---

■ ## Reading Labels

Many over-the-counter medicines, cosmetics, household products, and prepared food products have the ingredients listed on the label. Some products list the ingredients as required by law and some list the ingredients as consumer information. Occasionally a chemical ingredient is listed by its systematic names. Most ingredients, however, are listed by common names or trade names.

INGREDIENTS SALT, SODIUM SILICO-
ALUMINATE, DEXTROSE, POTASSIUM IODIDE
0.01% AND SODIUM BICARBONATE

DISTRIBUTED BY UNITED GROCERS LTD
RICHMOND, CA 94804 • PRODUCT OF U S A

When you read a product label, you may recognize an occasional chemical name. Be content if you can interpret a few chemical names on labels. Keep in mind that even a trained chemist would have to use reference books to identify many ingredients listed on labels.

**FIGURE 8-1** The label from a vitamin-mineral supplement.

## EACH TABLET CONTAINS

For Adults-Percentage of U.S.
Recommended Daily Allowance (U.S. RDA)

| | |
|---|---|
| Vitamin A (as Acetate and Beta Carotene) | 6000 IU (120%) |
| Vitamin $B_1$ (as Thiamine Mononitrate) | 1.5 mg (100%) |
| Vitamin $B_2$ (as Riboflavin) | 1.7 mg (100%) |
| Vitamin $B_6$ (as Pyridoxine Hydrochloride) | 3 mg (150%) |
| Vitamin $B_{12}$ (as Cyanocobalamin) | 25 mcg (416%) |
| Biotin | 30 mcg (10%) |
| Folic Acid | 200 mcg (50%) |
| Niacinamide | 20 mg (100%) |
| Pantothenic Acid (as Calcium Pantothenate) | 10 mg (100%) |
| Vitamin C (as Ascorbic Acid) | 60 mg (100%) |
| Vitamin D | 400 IU (100%) |
| Vitamin E (as di-Alpha Tocopheryl Acetate) | 45 IU (150%) |
| Vitamin $K_1$ (as Phytonadione) | 10 mcg* |
| Calcium (as Dibasic Calcium Phosphate and Calcium Carbonate) | 200 mg (20%) |
| Copper (as Cupric Oxide) | 2 mg (100%) |
| Iodine (as Potassium Iodide) | 150 mcg (100%) |
| Iron (as Ferrous Fumarate) | 9 mg (50%) |
| Magnesium (as Magnesium Oxide) | 100 mg (25%) |
| Phosphorus (as Dibasic Calcium Phosphate) | 48 mg (5%) |
| Zinc (as Zinc Oxide) | 15 mg (100%) |
| Chloride (as Potassium Chloride) | 72 mg* |
| Chromium (as Chromium Chloride) | 100 mcg* |
| Manganese (as Manganese Sulfate) | 2.5 mg* |
| Molybdenum (as Sodium Molybdate) | 25 mcg* |
| Nickel (as Nickelous Sulfate) | 5 mcg* |
| Potassium (as Potassium Chloride) | 80 mg* |
| Selenium (as Sodium Selenate) | 25 mcg* |
| Silicon (as Sodium Metasilicate) | 10 mcg* |
| Vanadium (as Sodium Metavanadate) | 10 mcg* |

ACTIVITY 8-8

A label from an over-the-counter mineral supplement is shown in Figure 8-1. Using the methods discussed in this chapter write the formulas of any of the compounds on the label that you recognize as ionic compounds.

## QUESTIONS

### Section 8-1

1. What does the term generic mean in naming drugs and medicines?

### Section 8-2

2. Make a diagram that starts with the title "common compounds." Underneath the title list the three types of ionic compounds and two types of covalent compounds discussed in this chapter. Next to each type of compound indicate how a compound of this type can be recognized.

### Section 8-3

3. Explain how simple positive and simple negative ions are named.

4. Give the names of the following ions.

| | | | |
|---|---|---|---|
| (a) $Al^{3+}$ | (b) $Br^-$ | (c) $CO_3^{2-}$ | (d) $Ag^+$ |
| (e) $S^{2-}$ | (f) $Cl^-$ | (g) $CrO_4^{2-}$ | (h) $Cu^+$ |
| (i) $Ni^{2+}$ | (j) $PO_4^{3-}$ | (k) $Cr_2O_7^{2-}$ | (l) $Fe^{2+}$ |
| (m) $HSO_4^-$ | (n) $MnO_4^-$ | (o) $HCO_3^-$ | (p) $OH^-$ |
| (q) $Mg^{2+}$ | (r) $NH_4^+$ | (s) $Ca^{2+}$ | (t) $K^+$ |

5. Give the formulas for the following ions.

| | |
|---|---|
| (a) hydrogen sulfate ion | (b) copper(II) ion |
| (c) chlorate ion | (d) fluoride ion |
| (e) nitrate ion | (f) sulfite ion |
| (g) iron(III) ion | (h) acetate ion |
| (i) oxalate ion | (j) cyanide ion |
| (k) chlorite ion | (l) chromium(III) ion |

### Section 8-4

6. Explain how the formula of an ionic compound is written to convey the electrical neutrality of the compound.

7. The following compounds contain parentheses. In each case, explain why parentheses are needed.

(a) calcium hydroxide, $Ca(OH)_2$

(b) iron(III) sulfate, $Fe_2(SO_4)_3$

(c) mercury(I) phosphate, $(Hg_2)_3(PO_4)_2$

(d) strontium acetate, $Sr(C_2H_3O_2)_2$

8. Give the correct formulas for the compounds formed by combinations of the positive and negative ions given in the following table. (NaCl is given as an example).

| | $Cl^-$ | $CO_3^{2-}$ | $OH^-$ | $NO_3^-$ | $SO_4^{2-}$ | $PO_4^{3-}$ |
|---|---|---|---|---|---|---|
| $Na^+$ | NaCl | | | | | |
| $Ba^{2+}$ | | | | | | |
| $Al^{3+}$ | | | | | | |
| $Cu^{2+}$ | | | | | | |
| $Fe^{2+}$ | | | | | | |
| $Fe^{3+}$ | | | | | | |
| $K^+$ | | | | | | |

### Sections 8-5 to 8-10

9. *Classify each of the following compounds as metal–nonmetal, metal–polyatomic ion, nonmetal–nonmetal, or acid and name each by the appropriate nomenclature method. For example, $CO_2$, nonmetal–nonmetal, carbon dioxide.

| | | |
|---|---|---|
| (a) $PBr_5$ | (b) $ClO_2$ | (c) $Cu_2S$ |
| (d) $AlN$ | (e) $N_2O$ | (f) $Ba(OH)_2$ |
| (g) $Ni(C_2H_3O_2)_2$ | (h) $SnCl_4$ | (i) $BCl_3$ |
| (j) $Na_2CrO_4$ | (k) $H_3PO_4$ | (l) $HgCl_2$ |
| (m) $Hg_2Br_2$ | (n) $Cr_2O_3$ | (o) $Fe(CN)_3$ |
| (p) $Ca_3P_2$ | (q) $KClO_3$ | (r) $HF(aq)$ |
| (s) $Sn(NO_2)_2$ | (t) $H_2SeO_4$ | (u) $NaCN$ |
| (v) $N_2O_3$ | (w) $SF_6$ | (x) $HNO_3$ |
| (y) $CuO$ | (z) $NaOH$ | (aa) $CCl_4$ |
| (bb) $I_2O_5$ | (cc) $H_2S$ | (dd) $H_2Se$ |
| (ee) $Na_2CO_3$ | (ff) $SbBr_5$ | (gg) $LiNO_2$ |

10. Classify each of the following compounds as metal-nonmetal, metal-polyatomic ion, nonmetal-nonmetal, or acid, and name each by the appropriate nomenclature method. For example, $CO_2$, nonmetal–nonmetal, carbon dioxide.

| | | |
|---|---|---|
| (a) $H_3PO_4$ | (b) $Cu(NO_3)_2$ | (c) $ZnCl_2$ |
| (d) $Cr_2S_3$ | (e) $Cd(CN)_2$ | (f) $Mg_3P_2$ |
| (g) $NaIO_3$ | (h) $PCl_5$ | (i) $OF_2$ |
| (j) $Ag_2S$ | (k) $Ca(OH)_2$ | (l) $Co(C_2H_3O_2)_2$ |
| (m) $PbCl_4$ | (n) $K_2Cr_2O_7$ | (o) $HF(aq)$ |
| (p) $SnF_2$ | (q) $H_2SO_4$ | (r) $AlN$ |
| (s) $N_2O_3$ | (t) $MgCO_3$ | (u) $SF_4$ |
| (v) $HNO_2$ | (w) $ZnO$ | (x) $AuBr_3$ |

(y)  $I_2O_5$    (z)  SiC    (aa)  $H_2Te$

(bb)  $CF_4$    (cc) $PCl_3$    (dd)  $NaH_2PO_4$

(ee)  $GeCl_4$

11. *Lead compounds are used in some commercial products. The following are typical compounds of lead for which the trivial names are given. galena, PbS anglesite, $PbSO_4$ cerusite, $PbCO_3$ litharge, PbO chrome yellow, $PbCrO_4$. Name these compounds of lead using systematic names.

12. *Freon 12, a refrigerant with the formula $CCl_2F_2$, is made by reacting carbon tetrachloride with antimony trifluoride. Freons can damage the environment so their use is being curtailed. Give the formulas of the two compounds used to make Freon 12.

13. Sulfur compounds in coal produce $SO_2$ when the coal is burned. The $SO_2$ is an undesirable air pollutant. One possible method of removing $SO_2$ in the waste gases is the limestone process, which involves mixing $SO_2$ with $O_2$ in air and passing it over solid $CaCO_3$. A reaction occurs forming $CaSO_4$ and $CO_2$. Name all the compounds given as formulas.

14. Nitric acid is manufactured by first reacting ammonia with oxygen gas to produce nitrogen oxide and water. The nitrogen oxide is reacted with more oxygen gas to produce nitrogen dioxide. Finally, nitrogen dioxide is mixed with water to form nitric acid and nitrogen oxide. Give the formulas of all the compounds named.

15. *Give the formulas for the following compounds:

(a)  sodium nitrate    (b)  tin(II) acetate

(c)  manganese(II) sulfate

(d)  ammonium acetate

(e)  chlorine trifluoride

(f)  diarsenic pentasulfide

(g)  dinitrogen trioxide

(h)  calcium fluoride    (i)  iron(II) nitrate

(j)  magnesium carbonate    (k)  methane

(l)  copper(I) oxide    (m) silver phosphate

(n)  strontium cyanide    (o)  nitric acid

(p)  ammonium sulfide    (q)  phosphoric acid

(r)  magnesium oxalate    (s)  zinc chromate

(t)  mercury(II) iodide    (u)  potassium sulfite

(v)  sodium dichromate    (w) carbon disulfide

(x)  dinitrogen pentoxide

(y)  magnesium chloride    (z)  hydrogen fluoride

16. Give the formulas of the following compounds:

(a)  barium carbonate    (b)  mercury(I) chloride

(c)  chloric acid    (d)  copper(I) sulfide

(e)  acetic acid

(f)  potassium dihydrogen phosphate

(g)  carbon monoxide    (h)  nitrous acid

(i)  chromium(III) fluoride    (j)  sodium cyanide

(k)  hydrogen iodide    (l)  magnesium hydroxide

(m) sodium oxalate    (n)  sodium phosphate

(o)  lead(II) chromate    (p)  manganese(II) sulfide

(q)  ammonium carbonate    (r)  boron trifluoride

(s)  diantimony pentasulfide    (t)  iron(II) oxide

(u)  dinitrogen oxide

(v)  potassium hydrogen phosphate

(w) calcium hydrogen carbonate

(x)  dichlorine trioxide    (y)  magnesium chloride

(z)  potassium permanganate

## Sections 8-1 to 8-10

17. Give the chemical name for each of the following compounds, which have the trivial names given.

(a)  $NaHCO_3$, baking soda    (b)  $CaCO_3$, marble

(c)  $Hg_2Cl_2$, calomel    (d)  $N_2O$, laughing gas

(e)  CaO, lime    (f)  NaOH, lye

(g)  $K_2CO_3$, potash    (h)  $NaNO_3$, saltpeter

(i)  $Ca(OH)_2$, slaked lime    (j)  $Al_2O_3$, alumina

(k)  $(NH_4)_2CO_3$, smelling salts

18. List the names and formulas of three compounds with common names that do not describe their chemical compositions.

19. *Classify and name each of the following compounds

(a)  $CCl_4$    (b)  $FeSO_4$    (c)  MgO

(d)  $HgBr_2$    (e)  KI    (f)  $Li_3As$

(g)  $NaC_2H_3O_2$    (h)  $KMnO_4$    (i)  $PF_3$

(j)  HCl    (k)  $UF_6$    (l)  $CoCl_2$

(m) $BaSO_3$    (n)  $CaF_2$    (o)  $SnBr_2$

(p)  $Zn(OH)_2$    (q)  NaBr    (r)  $Al_2S_3$

(s)  AgCl    (t)  $Ni(OH)_2$    (u)  $CrO_3$

(v)  $HC_2H_3O_2$    (w)  $NH_4I$    (x)  $CaC_2O_4$

20. Classify and name each of the following compounds.

(a) $FePO_4$    (b) $ZnC_2O_4$    (c) $AgNO_3$

(d) $SeF_4$    (e) $NH_3$    (f) $H_2SO_3$

(g) $PbO_2$    (h) $HgS$    (i) $HClO_2$

(j) $Ba_3N_2$    (k) $Fe(OH)_3$    (l) $Hg_2I_2$

(m) $KBr$    (n) $Fe_2O_3$    (o) $CsC_2H_3O_2$

(p) $CdSO_4$    (q) $CrI_3$    (r) $K_2O$

(s) $Na_2SO_3$    (t) $KOH$    (u) $Cl_2O$

(v) $Mg(HSO_4)_2$    (w) $UF_6$    (x) $(NH_4)_3PO_4$

(y) $AlP$    (z) $KHSO_3$

## Section 8-8

21. What are oxoacids? What are oxoanions?

22. *Give the names of the missing oxoacids or oxoan-ions.

(a) $HNO_3$ _____, $NO_3^-$ nitrate ion

(b) $HNO_2$ nitrous acid, $NO_2^-$ _____

(c) $H_3PO_4$ _____, $PO_4^{3-}$ phosphate ion

(d) $HBrO_3$ bromic acid, $BrO_3^-$ _____

(e) $H_2SeO_4$ _____, $SeO_4^{2-}$ selenate ion

(f) $H_3AsO_4$ arsenic acid, $AsO_4^{3-}$ _____

(g) $H_2SO_3$ _____, $SO_3^{2-}$ sulfite ion

23. Give the names of the following ions.

(a) $HSO_4^-$    (b) $HSO_3^-$    (c) $HCO_3^-$

(d) $HS^-$    (e) $H_2PO_4^-$    (f) $HPO_4^{2-}$

## Reading Labels

24. Read the labels on two foods, over-the-counter medicines, or other products. List the names and write the formulas of any compounds that follow nomenclature discussed in this chapter.

25. Read the label on a bottle of commercial soda water and give the formulas of the chemicals it contains.

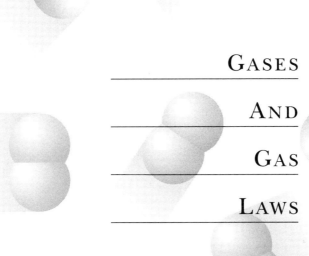

# GASES
# AND
# GAS
# LAWS

## 9-1 PROPERTIES OF GASES

When riding a bicycle you notice that the rear tire needs some air. Sitting on the bicycle increases the pressure on the tire and it flattens as its volume decreases. You add air with a pump and you notice that the volume increases until the tire becomes rigid, then the pressure increases as you add more air. Riding the bicycle on a hot day, you notice that as the tires heat up, they increase in volume and become more rigid, reflecting an increase in pressure. Your experiences with the bicycle are confirmations of the kinetic molecular theory of gases and the behavior of gases as predicted by the gas laws.

Gases have interested and stimulated the imaginations of scientists for centuries. Gases are fascinating because we can experience them without seeing them. Review the kinetic molecular theory discussed in Section 1-4. According to the theory, the molecules of a gas are darting about at high speeds, which contributes to their interesting properties.

The gaseous state is a very curious state of matter. A gas occupies the entire volume of any container in which it is placed. To prevent a gas from escaping into the atmosphere, we have to keep it in a sealed container. From experience with gas-filled balloons we know that a gas is very flexible and the volume of a balloon changes when it is squeezed or heated. Furthermore, we know that a gas readily leaks from a balloon even through very tiny holes.

Sometimes gases are called vapors. Actually the terms gas and vapor have different meanings. The term gas refers to any substance in the gaseous state. A **vapor** is a gas that is easily changed to a liquid by a slight change in its pressure or temperature. Air is a mixture of gases that includes some water vapor. The water vapor in the air can easily condense to form liquid water. This happens when dew forms. It is not easy, however, to liquify the nitrogen and oxygen gases in air. It takes a combination of very low temperature and high pressure to cause these components of air to become liquids. Nevertheless, liquefaction of air is done on an industrial scale and is the major source of industrial nitrogen and oxygen.

Some compounds occur as gases at typical earth temperatures and pressures. The noble gases are minor components of the atmosphere; they exist as monoatomic gases. All other gases occur as collections of molecules; hydrogen, nitrogen, oxygen, fluorine, and chlorine gases consist of collections of diatomic molecules. For example, a sample of hydrogen gas contains $H_2$ molecules and a sample of chlorine gas contains $Cl_2$ molecules. Some other common gases are carbon dioxide, $CO_2$, ammonia, $NH_3$, nitrogen dioxide, $NO_2$, and the hydrocarbons, methane, $CH_4$, ethane, $C_2H_6$, propane, $C_3H_8$ and butane, $C_4H_{10}$.

## 9-2 PRESSURE

If we capture a gas in a container and keep the container closed, it can be kept as a gas indefinitely. As long as the sample container is not opened or the gas does not change chemically the amount of gas in terms of the mass or number of moles will not change. We can describe a gas sample by stating the volume of the container and the temperature. Since a gas occupies the entire container, the volume of the container is the volume of the gas. A gas sample also exerts a pressure on the container. A pressure-measuring device attached to the container registers a specific pressure.

**Pressure** is defined as force per unit area. The molecules of a gas are in constant motion. The pressure exerted on objects by a gas results from the molecules of the gas constantly colliding with their surfaces. A continuous series of collisions exert a specific force per unit area on any object.

## 9-3 MEASUREMENT OF GAS PRESSURES

Sometimes when you dive deeply in water you can feel the pressure on your eardrums. In deep water you feel the pressure of the layer of water above. The atmosphere is the layer of air contained about the earth by gravitational forces. The molecules and atoms present in the atmosphere exert pressures on all objects in the atmosphere; this is the air pressure or atmospheric pressure. It arises from the continuous bombardment of ob-

ACTIVITY 9-1

To observe the effects of atmospheric pressure you need a small, flexible plastic bottle with a tight lid such as an empty juice bottle or water bottle. You'll also need some boiling hot water. Remove the lid from the bottle and pour in enough boiling hot water to fill the bottle to a level about one inch high. Swirl the bottle with your hand until you feel that it is quite hot and then, quickly put the cap on the bottle. Run cold tap water over the sealed bottle for a few minutes to cool the contents. Explain your observations and draw pictures to aid your explanation.

jects by molecules and atoms in the air. Since gravitational forces decrease with altitude, the atmosphere is more dense near the lower surfaces of the earth and becomes less dense at distances removed from the earth's surface. Atmospheric pressure decreases with altitude because the air is less dense at higher altitudes and fewer molecules are available to exert pressure.

Atmospheric pressure is measured with a device known as a barometer, invented by Evangelista Torricelli in the 1600s. A **barometer** is constructed by first filling a glass tube, one end of which is sealed, with liquid mercury. This tube is then inverted and supported in a container of mercury that is open to the atmosphere (see Figure 9-1). The pressure exerted on the surface of the mercury by the atmosphere supports a column of mercury in the tube. The tube must be long enough so that some space remains at the top of the tube. The atmospheric pressure supports the mercury column and the space at the top of the column is essentially a vacuum. Of course, the pressure results from the continuous bombardment of the surface of the mercury by air molecules. The height of the mercury column is directly proportional to the atmospheric pressure. As the atmospheric pressure changes due to changes in temperature and weather conditions, the height of the mercury column fluctuates accordingly. A falling barometer, that is, a decreasing barometer reading, usually foretells an approaching storm. Can you explain why?

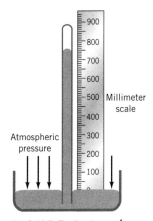

**FIGURE 9-1**   A mercury barometer.

### 9-4  UNITS FOR PRESSURE

The height of the mercury column supported by any gas is proportional to the pressure of the gas. Consequently, gas pressures are often expressed in terms of the height of a mercury column as measured with a barometer. This height is usually expressed in **millimeters of mercury** or **mm Hg.** An alternate unit for pressure is the **torr,** named in honor of Torricelli. One torr is defined as one millimeter of mercury. In U.S. weather reports atmospheric pressures are usually given in inches of mercury. Since 1 inch = 25.4 mm, it then follows that 1 inch Hg = 25.4 torr.

**1 torr = 1 mm Hg**

Another unit of pressure is the **atmosphere, atm,** defined as the pressure equal to 760 torr. The term atmosphere comes from the fact that the atmospheric pressure around sea level is typically around 760 torr. The atmosphere as a unit of pressure is often used to express pressures common to industrial and experimental work. For instance, the process used to make synthetic diamonds includes subjecting graphite to temperatures of about 2000°C and pressures around 70,000 atm.

**1 atm = 760 torr**

The pressure units torr and atm are associated with air pressures. They are also used to express the pressure of any sample of gas in a closed container. The relation between the torr and the atmosphere can be written as a conversion factor as shown in the margin.

$$\frac{\textbf{1 atm}}{\textbf{760 torr}}$$

ACTIVITY
9-2

One day in Death Valley, California, the atmospheric pressure was found to be 772 torr. What is this pressure in atmospheres? What is this pressure in inches of mercury?

**EXAMPLE 9-1**

The measured pressure in the eye of a specific hurricane was 669 torr. What is this pressure in atmospheres? Use the factor relating torr and atm as a conversion factor.

$$669 \text{ torr } \frac{1 \text{ atm}}{760 \text{ torr}} = 0.880 \text{ atm}$$

The SI unit of force is the newton, N. A pressure can be expressed in terms of newtons per square meter ($N/m^2$), which is a force per unit area. The SI unit for pressure is the pascal (Pa). One pascal is equivalent to one $N/m^2$. The pascal represents a small unit of pressure. One torr is 133 Pa. Consequently, pressures in SI units are often expressed in units of kilopascals. Another SI pressure unit is the bar. One bar is defined as $10^5$ Pa. Thus, the relation between the bar and the atmosphere is 1.01 bar = 1 atm.

Gas pressures in chemistry are typically measured and expressed in torr and atm and, occasionally, in units of kPa or bar. You may be familiar with barometric pressures expressed in inches of mercury (1 atm = 29.92 in. Hg) and pressures expressed in pounds per inch (1 atm = 14.696 psi) used by engineers, but these units are seldom used in chemistry. Incidently, about 34 feet of water is equivalent to 1 atm. So if you dive in water to 34 feet you are subjected to a pressure of about 1 atm over the prevailing atmospheric pressure.

### PRESSURE UNITS

| | |
|---|---|
| inches of mercury (in. Hg) | 1 atm = 29.92 in. Hg |
| millimeter of mercury (mm Hg) | 1 atm = 760 mm Hg |
| torr (torr) | 1 atm = 760 torr |
| pound per square inch (lb/in.²) psi | 1 atm = 14.696 lb/in.² |
| newton per square meter ($N/m^2$) | 1 atm = 101,325 $N/m^2$ |
| pascal (Pa) | 1 atm = 101,325 Pa |
| kilopascal (kPa) | 1 atm = 101.325 kPa |
| bar (bar) | 1 atm = 1.01325 bar |

### 9-5  CHARLES' LAW OR THE VOLUME–TEMPERATURE LAW

As you learned in Chapter 1, Jacques Charles, in 1887, carefully measured the relation between the volume and temperature of a sample of a gas at constant pressure. When the pressure is constant, the volume is affected only by the temperature change. Charles observed that when the temperature of such a gas sample is increased, the volume increases; and when the temperature decreases, the volume decreases. Specifically, he noted a direct linear relation between the volume and temperature. His observation has become known as Charles' law as stated in the margin. (Remember that a scientific law is a statement of some consistent behavior in nature.)

A **direct proportion** between two quantities means that when one is increased the other increases, and as one is decreased the other decreases. A way to remember Charles' law is to recall that a filled balloon increases in volume when warmed and decreases in volume when cooled. The proportionality between volume and temperature can be expressed as

$$V \propto T$$

where $\propto$ represents "proportional to." The proportionality represented by Charles' law can also be written in an algebraic form as

$$V = kT \quad \text{or} \quad V/T = k$$

where $k$ is a proportionality constant that relates the volume and temperature. The proportionality constant has a constant value for a sample of a gas at a fixed pressure. When a gas sample of volume $V_1$ at temperature $T_1$ has its temperature changed to $T_2$ the volume changes to $V_2$.

$$\frac{V_1}{T_1} = k \qquad \frac{V_2}{T_2} = k$$

Note that $k$ does not change. Since, by the rules of algebra, we know that terms equal to the same term are equal to one another it follows that

$$\frac{V_2}{T_2} = \frac{V_1}{T_1}$$

This alternate version of Charles' law is used to predict the new volume that results when the temperature of a given volume of gas changes.

Remember that a gas sample represents a fixed amount in terms of grams or moles of gas. A sample volume can be measured in any convenient unit but to use the equation we need to express the temperatures in kelvins. Why can't Celsius temperatures be used in the algebraic form of the law?

**Charles' Law:**

*The volume of a gas sample at constant pressure is directly proportional to the temperature.*

Cooling

Heating

## 9-6 BOYLE'S LAW OR THE VOLUME–PRESSURE LAW

In 1662 Robert Boyle, an English scientist, experimented with the compressibility of air. Boyle, using an apparatus similar to that shown in Figure 9-2, measured the relation between the volume and the pressure of a sample of gas. He observed that if a sample of gas is kept at a constant temper-

**FIGURE 9-2**
Boyle's law apparatus.

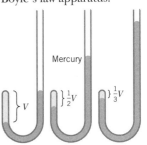

ature, then the volume of the gas decreases with increasing pressure. Boyle observed that if the pressure of the gas doubles, the gas compresses to occupy one-half its original volume; if the pressure triples, the gas compresses to occupy one-third its original volume. Boyle also noted that relieving the pressure increases the volume of the gas sample. For instance, if the pressure decreases by one-half, the gas expands to occupy twice its original volume. Today, after many confirmations of Boyle's work and additional experiments, a generalization concerning the relation between the volume and the pressure of a gas is known as Boyle's law as shown in the margin.

**Boyle's Law:**

*The volume of a gas sample at constant temperature is inversely proportional to the pressure.*

This law relates to simple observations of gas samples in flexible containers. Squeezing a balloon filled with a gas increases the pressure and decreases the volume. Releasing the balloon decreases the pressure and increases the volume. Figure 9-3 is a graph that shows that as the pressure of a gas sample increases, the volume decreases proportionally; and as the pressure decreases, the volume increases. Two quantities are **inversely proportional** when one increases as the other decreases or vice versa. The volume and pressure of a gas are inversely proportional.

A graph or plot is commonly used to show the relation between two interdependent properties. Such a graph shows the functional dependency of the properties or how the properties vary with respect to one another. Experimental data comes from measuring the volume of a gas sample at various pressures. The data are plotted as volume versus pressure. See Figure 9-3 for a graph or plot illustrating Boyle's law. Following the pressure axis of the graph, smaller volumes correspond to higher pressures, and larger volumes correspond to lower pressures. The inverse proportion between volume and pressure is

$$V \propto \frac{1}{P}$$

**FIGURE 9-3**
Boyle's law.

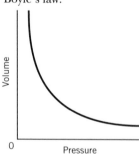

0         Pressure

$V = k/P$ where $k$ depends upon the mass and temperature of the gas

Boyle's law in an algebraic form using a proportionality constant is

$$V = k/P \quad \text{or} \quad PV = k$$

where $V$ is the volume, $P$ is the pressure, and $k$ is a proportionality constant that relates pressure and volume. The proportionality constant has a constant value for a sample of a gas at a fixed temperature. When a gas sample of volume $V_1$ at pressure $P_1$ changes pressure to $P_2$ the volume changes to $V_2$.

$$P_1 V_1 = k \qquad P_2 V_2 = k$$

Note that since $k$ does not change it follows that

$$P_2 V_2 = P_1 V_1$$

This alternate version of Boyle's law is used to predict the new volume when the pressure of a given volume of gas changes.

## 9-7 THE PRESSURE-TEMPERATURE LAW

When you use a pressure cooker in cooking, the inside pressure increases as the temperature increases. Experimental observations of gas samples in rigid, inflexible containers reveal that pressure varies directly with the temperature. When the temperature of a gas sample of fixed volume is increased, the pressure increases; when the temperature is decreased, the pressure decreases. This behavior is stated in the pressure–temperature law as defined in the margin. The pressure–temperature law is illustrated in Figure 9-4 and can be expressed as

$$P \propto T$$

or algebraically as

$$P = kT \quad \text{or} \quad P/T = k$$

**Pressure–Temperature Law:** *The pressure of a gas sample at constant volume is directly proportional to the temperature.*

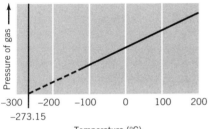

A plot of the pressure of a sample of a gas versus the temperature. (The temperature at which the pressure of the gas becomes zero is absolute zero or zero kelvin.)

**FIGURE 9-4** Relationship between the pressure and temperature of a gas at constant volume.

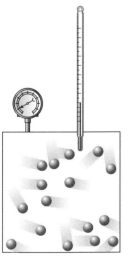

Lower temperature

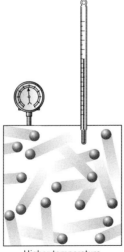

Higher temperature

As a practical example, consider what happens when a gas-filled aerosol can is heated in a fire. The volume of the can is fixed; thus, as the temperature increases, the pressure of the gas increases. When the temperature is high enough, the pressure of the gas is so great that the can ruptures and an explosion results. This is why you should never put an aerosol can into an open fire.

### 9-8  THE COMBINED GAS LAW

The gas laws are used to determine how the volume of a gas sample changes if the temperature or the pressure changes. By combining the gas laws it is possible to determine how the volume of a gas sample changes when both the temperature and the pressure change. The gas laws can be expressed in the form of a combined gas law equation.

$$\frac{V_2 P_2}{T_2} = \frac{V_1 P_1}{T_1}$$

This equation is used to calculate a new volume, pressure or temperature corresponding to specific changes in the other terms. In other words, if you know any five terms, it is possible to calculate the sixth term. Note also that for a sample of constant pressure the equation becomes Charles' law and for a sample of fixed temperature the equation becomes Boyle's law.

The new volume of a gas sample that results from a pressure change and a temperature change is found by multiplying the original volume by both a pressure ratio and a temperature ratio.

$$V_2 = V_1 \frac{P_1}{P_2} \frac{T_2}{T_1}$$

where subscript 1 indicates an initial value and subscript 2 indicates a final value. Suppose we had a balloon of gas that occupies 2.0 L of volume at 20°C, typical room temperature. What volume does the balloon occupy at a temperature of 50°C, if the pressure is constant. When using temperatures in gas laws they must always be expressed in kelvins rather than degrees Celsius. That's easily done since we just have to add 273 to each Celsius temperature:

$$20°C + 273 = 293 \text{ K} \qquad \text{and} \qquad 50°C + 273 = 323 \text{ K}$$

The information given in the question is summarized as:

| $V_1$ | $P_1$ | $T_1$ | $V_2$ | $P_2$ | $T_2$ |
|-------|-------|-------|-------|-------|-------|
| 2.0 L | constant | 293 K | ? | constant | 323 K |

The new volume is found by multiplying the initial volume by the appropriate ratio of temperatures:

$$V_2 = 2.0 \text{ L} \times \frac{323 \cancel{K}}{293 \cancel{K}} = 2.2 \text{ L}$$

Since volume is directly proportional to the temperature, the temperature ratio should be greater than one for an increase in temperature and less than one for a decrease in temperature. We know that the volume and temperature are directly proportional. If the temperature increases the volume must increase. The new volume is the initial volume multiplied by a ratio of the temperatures which is greater than one. We just use the ratios deduced from the proportions corresponding to the changes. Using this kind of reasoning we do not have to substitute into an algebraic equation. Of course we get the same result if we substitute the given values into the gas law equation.

As another example, suppose we have a 4.5-L sample of a gas at a pressure of 760 torr. What is the new volume when the pressure changes to 790 torr and the temperature remains constant? A summary of the data is

| $V_1$ | $P_1$ | $T_1$ | $V_2$ | $P_2$ | $T_2$ |
|-------|-------|-------|-------|-------|-------|
| 4.5 L | 760 torr | constant | ? | 790 torr | constant |

The new volume is found by multiplying the initial volume by the appropriate ratio of pressures:

$$V_2 = 4.5 \text{ L} \times \frac{760 \cancel{\text{ torr}}}{790 \cancel{\text{ torr}}} = 4.3 \text{ L}$$

If the pressure of a gas sample at a constant temperature increases, the original volume decreases. To find the new volume we multiply the original volume by a ratio of the pressures which is less than one. If the pressure decreases, the original volume increases. To find the new volume we multiply the original volume by a ratio of the pressures which is greater than one. We can use any units of pressure as long as we express both pressures in the same unit, then the units cancel.

---

**EXAMPLE 9-2**

A 325-mL sample of a gas in a balloon is maintained at 25°C and 750 torr pressure. The temperature changes to 20°C and the pressure changes to 760 torr. What is the new volume?

First, summarize the data:

| $V_1$ | $P_1$ | $T_1$ | $V_2$ | $P_2$ | $T_2$ |
|-------|-------|-------|-------|-------|-------|
| 325 mL | 750 torr | 298 K (25 + 273) | ? | 760 torr | 293 K (20 + 273) |

Both the temperature and pressure are changing. A pressure increase (750 torr to 760 torr) gives a smaller volume, so the pressure ratio should be less

A 50.5-mL sample of a gas is in a syringe with a pressure gauge attached. Initially, the gauge shows a pressure of 1.00 atm. The plunger of the syringe is pushed so that the pressure reads 1.45 atm. What is the new volume of the gas if the temperature is constant?

than one. A temperature decrease (298 K to 293 K) gives a smaller volume, so the temperature ratio should also be less than one:

$$325 \text{ mL} \times \frac{750 \text{ torr}}{760 \text{ torr}} \times \frac{293 \text{ K}}{298 \text{ K}} = 315 \text{ mL}$$

### EXAMPLE 9-3

A balloon occupies a volume of 625 mL at 25°C. It is placed in a freezer and the volume decreases to 552 mL. What is the temperature of the balloon in the freezer in degrees Celsius if the pressure remains constant?

Tabulate the data as

| $V_1$ | $P_1$ | $T_1$ | $V_2$ | $P_2$ | $T_2$ |
|---|---|---|---|---|---|
| 625 mL | constant | 298 K (25 + 273) | 552 mL | constant | ? |

Note that we want to find the final temperature in degrees Celsius. Nevertheless, we must use Kelvin temperatures in the calculations. The final Kelvin temperature is changed to give the final temperature in degrees Celsius.

Since the volume decreases the temperature must decrease, so to find the final temperature we multiply the initial temperature by a volume ratio of less than one.

$$298 \text{ K} \times \frac{552 \text{ mL}}{625 \text{ mL}} = 263 \text{ K}$$

To find the temperature in Celsius we need to subtract 273 K from this Kelvin temperature.

$$263 \text{ K} - 273 \text{ K} = -10°\text{C}$$

### EXAMPLE 9-4

A 4.25-L sample of a gas is contained in a piston chamber at 264 K and 0.989 atm pressure. If the piston moves to compress the gas sample to 1.32 L and the temperature changes to 395 K what is the new pressure of the gas sample?

First, tabulate the data.

| $V_1$ | $P_1$ | $T_1$ | $V_2$ | $P_2$ | $T_2$ |
|---|---|---|---|---|---|
| 4.25 L | 0.989 atm | 264 K | 1.32 L | ? | 395 K |

To find the new pressure multiply the initial pressure by a volume ratio and a temperature ratio. A decrease in volume increases the pressure so the volume ratio is greater than one. An increase in temperature increases the pressure so the temperature ratio is greater than one.

$$P_2 = 0.989 \text{ atm} \times \frac{4.25\,\cancel{L}}{1.32\,\cancel{L}} \times \frac{395\,\cancel{K}}{264\,\cancel{K}} = 4.76 \text{ atm}$$

## 9-9 RELATING *P*, *V*, *T*, AND *n* FOR GASES

We can picture a gas sample using the kinetic molecular theory. The theory describes an idealized gas, referred to as an ideal gas, which behaves according to the theory. The theory states that there are no significant attractive forces between the molecules of a gas. Actually weak attractive forces do exist between molecules. This makes real gases deviate from ideal behavior but most real gases behave as though they are ideal gases within certain ranges of temperature and pressure. For our purposes, we can assume that real gases are close to behaving as ideal gases. Thus, it is reasonable to treat them as ideal and expect them to behave according to the gas laws.

The properties of pressure, volume, and temperature of a sample of gas relate to the dynamic nature of the gas. These properties are not independent of one another and interrelate as expressed by the gas laws. Boyle's law states that the volume of a gas sample is inversely proportional to the pressure.

$$V \propto 1/P$$

According to Charles' law, the volume of a gas sample is directly proportional to its Kelvin temperature.

$$V \propto T$$

Another proportionality that is relevant is how the volume of a gas sample relates to the number of moles of gas. The number of moles, *n*, is an expression of the number of particles in a sample. Imagine a balloon filled with a gas at a constant temperature and pressure. If we blow air into the balloon we increase the number of moles of gas and the volume increases. Allowing some of the gas to escape decreases the number of moles. The result is a decrease in volume. Generally, the volume of a gas at constant temperature and pressure is directly proportional to the number of moles of gas.

$$V \propto n$$

It is possible to combine the three proportions into a general form that relates volume to any of the other three factors.

$$V \propto nT/P$$

This combined relation indicates that the volume is directly proportional to the number of moles and the temperature and inversely proportional

A potato chip bag occupies a volume of 1.4 L at 760 torr and 20°C. If the bag is transported to the mountains, what volume does it occupy at 725 torr and 18°C?

Under what conditions of temperature and pressure do you think gases deviate the most from ideal behavior? Why?

to the pressure. Another way to express this relation is to rearrange it so that pressure and volume are on one side and the number of moles and temperature are on the other side; to do this we multiply both sides by the pressure, $P$, giving the proportionality

$$PV \propto nT$$

This relation states that the product of the pressure and volume of a gas sample is directly proportional to the product of the number of moles and temperature of the sample. The next section shows how this proportion can be written as an algebraic equation using a proportionality constant.

## 9-10 THE IDEAL GAS LAW

The proportionality between the pressure, volume, number of moles, and temperature of a gas can be written in an algebraic form as

$$PV = knT$$

This relation can also be expressed with the constant on one side and $P$, $V$, $n$, and $T$ on the other side.

$$k = PV/nT$$

**$PV = nRT$**

In this form the relation reveals that for any ideal gas the product of the pressure and volume divided by the product of the number of moles and temperature is a constant. This is true for any ideal gas and any set of conditions. A value for the proportionality constant is found by using a sample containing a known number of moles of gas and measuring its pressure, volume, and temperature. The proportionality constant is the universal gas constant or ideal gas constant and is represented by the symbol $R$. Using $R$ in the proportion gives the expression shown in the margin. This expression relates the pressure, volume, temperature, and number of moles of a gas and is called the **ideal gas law.** $R$ is a universal constant and the same numerical value of $R$ applies to any ideal gas. We can use the ideal gas law with any gas as long as we assume it is ideal.

The ideal gas law can be very useful. Since $R$ is a known constant, if we know any three of the properties associated with a gas sample, the fourth can easily be calculated. For example, if we measure the pressure, volume, and temperature of a sample of a gas, the ideal gas law can be used to find the number of moles of gas in the sample. In other words, we can count the molecules in a sample of a gas by just measuring the volume, pressure, and temperature.

## 9-11 THE GAS CONSTANT, *R*

The numerical value and units of the gas constant, $R$, depend on the units used to measure the pressure, volume, and temperature of the gas sample.

The temperature used in the ideal gas law is always the Kelvin temperature. The volume of a sample of a gas is usually expressed in liters and sometimes in milliliters. Its pressure is usually expressed in atmospheres and sometimes in torr.

The experimental determination of the value of $R$ is accomplished by measuring the pressure, volume, temperature, and number of moles for gas samples. Suppose we find that 1.54 mol of gas occupy 37.9 L at 298 K and 0.994 atm. To calculate $R$, we first solve $PV = nRT$, for $R$ by dividing both sides by $n$ and $T$.

$$R = VP/Tn$$

The values $V = 37.9$ L, $P = 0.994$ atm, $T = 298$ K, and $n = 1.54$ mol are substituted into the expression to find $R$.

$$R = \frac{(37.9 \text{ L})(0.994 \text{ atm})}{(298 \text{ K})(1.54 \text{ mol})} = \frac{0.0821 \text{ L atm}}{\text{K mol}} \qquad R = \frac{0.0821 \text{ L atm}}{\text{K mol}}$$

The same value of $R$ is found using any set of related $P$, $V$, $n$, and $T$ values for an ideal gas. You can calculate a value for $R$ in Chemistry in Action 9-1. $R$ serves as a known constant in all ideal gas law computations. Different units for the volume or pressure give different values for the gas constant $R$. The value of $R$ given above is convenient for most calculations. When using $R$ in calculations be sure to include the units.

**VARIOUS VALUES OF R**

0.0821 L atm/K mol

82.1 mL atm/K mol

62.4 L torr/K mol

62,400 mL torr/K mol

$8.31 \times 10^3$ L Pa/K mol

## 9-12 CALCULATIONS WITH THE IDEAL GAS LAW

Calculations with the ideal gas law usually involve finding the pressure, volume, temperature, or number of moles of a gas. To calculate any one of these, we need values for the other three. The gas law is rearranged algebraically to solve for any one of the four terms $P$, $V$, $n$, or $T$.

---

**EXAMPLE 9-5**

A weather balloon contains 4.73 moles of helium gas. What volume does the gas occupy at an altitude of 4300 m if the temperature is 0°C and the pressure is 0.595 atm? First, tabulate the data.

| $P$ | $V$ | $n$ | $T$ |
|---|---|---|---|
| 0.595 atm | ? | 4.73 mol | 273 K |

The ideal gas law is solved for volume by dividing both sides by $P$.

$$\frac{PV}{P} = \frac{nRT}{P}$$

which gives

$$V = \frac{nRT}{P}$$

Find the volume using the data in the expression.

$$V = \frac{nRT}{P} = \frac{4.73 \text{ mol}}{0.595 \text{ atm}} \times \frac{0.0821 \text{ L atm}}{\text{K mol}} \times (273 \text{ K}) = 178 \text{ L}$$

Notice how the units of $R$ cancel the other units giving liters as the final unit.

---

### EXAMPLE 9-6

A used aerosol can contains 0.0104 mole of gas and has a volume of 255 mL. Calculate the pressure of the gas in the can if the can is accidentally heated to 400°C.

**DANGER: NEVER HEAT AN AEROSOL CAN! IT MAY EXPLODE.**

First, tabulate the data.

| $P$ | $V$ | $n$ | $T$ |
|---|---|---|---|
| ? | 255 mL | 0.0104 mol | 673 K(400 + 273) |

To use the volume in the ideal gas law, we must change the original volume to units of liters to match the units used in $R$.

$$255 \text{ mL} \times \frac{1 \text{ L}}{1000 \text{ mL}} = 0.255 \text{ L}$$

Next, we solve the ideal gas law for pressure by dividing both sides by $V$.

$$\frac{PV}{V} = \frac{nRT}{V}$$

which gives

$$P = \frac{nRT}{V}$$

We use the data in this relation to calculate the pressure.

$$P = \frac{nRT}{V} = \frac{0.0104 \text{ mol}}{0.255 \text{ L}} \times \frac{0.0821 \text{ L atm}}{\text{K mol}} \times (673 \text{ K}) = 2.25 \text{ atm}$$

---

### EXAMPLE 9-7

A tire with an interior volume of 3.60 L contains 0.367 mole of air at a pressure of 2.49 atm. What is the temperature in kelvins of the air in the tire?

First, tabulate the data.

|   P   |   V   |   n   |   T   |
|-------|-------|-------|-------|
| 2.49 atm | 3.60 L | 0.367 mol | ? |

Next, solve the ideal gas law for temperature by dividing both sides by $nR$.

$$\frac{PV}{nR} = \frac{nRT}{nR}$$

which gives

$$T = \frac{PV}{nR}$$

Use the relation to calculate the temperature.

$$T = \frac{PV}{nR} = \frac{2.49 \text{ atm}}{0.367 \text{ mol}} \times \frac{3.60 \text{ L}}{0.0821 \text{ L atm}/\text{K mol}} = 298 \text{ K}$$

Note that when the units of $R$ cancel the other units the result is

$$\frac{1}{1/K} \quad \text{or} \quad \frac{1}{\frac{1}{K}} = K$$

To divide 1 by 1/K we invert the fraction and multiply, which gives the unit of K.

ACTIVITY 9-6

How many moles of gas are in 1.00 cubic feet of natural gas at 273 K and 1.00 atm ($1 \text{ ft}^3 = 28.3$ L)?

## 9-13  STP AND MOLAR VOLUME

Suppose you read that the annual commercial production of nitrogen gas in the United States is 680 billion cubic feet or 19 trillion liters. What is wrong with this statement? We have learned that the volume of a gas sample depends on the pressure and temperature. So the statement has no meaning unless it is accompanied by some specific temperature and pressure conditions.

Specific values of temperature and pressure are defined as standard reference conditions for gases. These defined conditions are

**0°C**   or   **273 K**

**760 torr**   or   **1 atm**

These values are arbitrarily chosen as reference conditions of gases and they are called **standard temperature and pressure, abbreviated STP.**

What volume does a gas occupy at STP? This is not a meaningful question because, according to the ideal gas law, it is not possible to calculate

**STP**

0°C or 273 K and
760 torr or 1.00 atm

the volume of a gas sample at a given temperature and pressure unless we know the number of moles of gas. A more meaningful question is: What volume does one mole of a gas occupy at STP? This is easily calculated by substituting a $P$ of 1.00 atm, a $T$ of 273 K, an $n$ of 1.00 mol, and $R$ into the relation $V = nRT/P$.

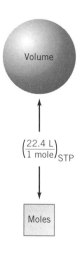

Volume

$$V = \frac{1.00 \cancel{mol}}{1.00 \cancel{atm}} \times \frac{0.0821 \text{ L} \cdot \cancel{atm}}{\cancel{K} \cdot \cancel{mol}} \times (273 \cancel{K}) = 22.4 \text{ L}$$

Thus, one mole of an ideal gas occupies a volume of 22.4 L at STP. This fact can be expressed as the factor

$$\left(\frac{22.4 \text{ L}}{1 \text{ mol}}\right)_{STP}$$

$\left(\frac{22.4 \text{ L}}{1 \text{ mole}}\right)_{STP}$

Moles

The subscript denotes the conditions (1 atm and 273 K) at which the relation is valid. This factor is called the molar volume of a gas and is valid only at STP. A volume of 22.4 L is equal to the volume of a cube measuring about 28 cm or 11 in. on an edge. The value of 22.4 L/mol applies only to ideal gases, not to liquids or solids. Use the molar volume as a simple conversion factor to calculate the number of moles of a gas in a given volume at STP or vice versa, but remember that it only applies to gases at STP.

ACTIVITY 9-7

The molar mass of a gas has units of grams per mole. The molar volume of a gas has units of moles per liter. How can the molar mass and molar volume be used to calculate the density of a gas at STP? To figure this out pay attention to units. Use your answer to calculate the density of methane gas, $CH_4$, at STP and the density of carbon dioxide gas, $CO_2$, at STP.

**EXAMPLE 9-8**

One cubic foot of volume equals 28.3 L. How many moles of nitrogen are in 1.00 cubic feet of nitrogen gas at STP?

Use the molar volume in the inverted form to convert the volume of a gas at STP to the number of moles.

$$28.3 \cancel{L} \times \frac{1 \text{ mol } N_2}{22.4 \cancel{L}} = 1.26 \text{ mol } N_2$$

## 9-14 DALTON'S LAW OF PARTIAL PRESSURES

So far, we have been dealing with samples of gases without being concerned with their chemical composition. Now we consider the relation that exists between the components of a mixture of gases. If a mixture of three gases is placed in a container of fixed volume, $V$, each of the gases is viewed as occupying the entire volume (see Fig. 9-5). Likewise all three gases must be at the same temperature, $T$.

The pressure exerted by a component of the mixture, if it alone were to occupy the container, is called the **partial pressure.** Since the particles

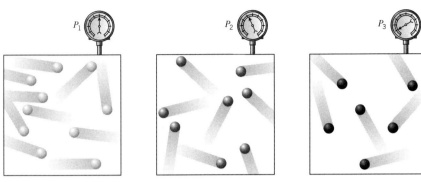

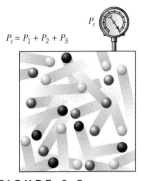

**FIGURE 9-5**
Dalton's law of partial pressures. Each gas exerts a pressure according to its amount. When the gases are in the same container, each contributes to the total pressure.

of each component behave independently, the total pressure exerted by the mixture is a result of all the particles in the mixture. Consequently, the total pressure of the mixture, $P_t$, is simply the sum of the partial pressures of the components of the mixture. Expressed algebraically, the total pressure is

$$P_t = P_a + P_b + P_c + \dots$$

where $P_a$, $P_b$, $P_c$, and so on refer to the partial pressures of the components of a mixture. This relation is **Dalton's law of partial pressures,** named after John Dalton of atomic theory fame. Dalton's law is quite convenient when we want to relate the partial pressures of the components of a mixture to the total pressure.

In the laboratory, samples of gases are often collected by displacement of water. As the gas forms, it is bubbled into an inverted container of water. The accumulating gas displaces the liquid water. However, as the gas bubbles through the water it becomes saturated with gaseous water vapor. This means that a gas collected by water displacement is actually a mixture of the gas and water vapor. The total pressure of the sample is the sum of the partial pressure of the gas and the partial pressure of the water vapor. According to Dalton's law, $P_t = P_g + P_{water}$, so to find the actual pressure of the gas we subtract the partial pressure of water from the total pressure of the mixture: $P_g = P_t - P_{water}$

Reference tables of the vapor pressures of water at various temperatures are readily available. Thus, to find the actual pressure of a gas collected by water displacement, subtract the vapor pressure of water at the temperature of the sample from the total pressure of the sample.

**EXAMPLE 9-9**

A sample of hydrogen gas is collected by water displacement. The container of gas collected has a volume of 645 mL at 25°C and a total pressure of 758 torr. How many moles of hydrogen are in the sample?

We know that we can use the ideal gas law to find the number of moles of gas from the measured volume, temperature, and pressure. In this sample, however, water vapor contributes to the total pressure of the "wet" hydrogen. The vapor pressure of water at the measured temperature is the partial pressure of water in the sample. Since the gas sample is a mixture, the partial pressure of hydrogen can be calculated using Dalton's law of partial pressures. The total pressure is the result of two components.

$$P_t = P_{H_2} + P_{H_2O}$$

A table of vapor pressures of water shows that the vapor pressure of water at 25°C is 24 torr.

$$P_t = 758 \text{ torr} \quad \text{and} \quad P_{H_2O} = 24 \text{ torr}$$

Thus

$$P_{H_2} = P_t - P_{H_2O} = 758 \text{ torr} - 24 \text{ torr} = 734 \text{ torr}$$

The actual pressure of the hydrogen gas is used with the volume and temperature of the sample to find the number of moles of hydrogen gas in the sample. The units of the data must be adjusted to be compatible with the units of the gas constant $R$. We need pressure in atmospheres, volume in liters, and temperature in kelvins.

$$P = 734 \text{ torr} \times \frac{1 \text{ atm}}{760 \text{ torr}} = 0.9658 \text{ atm}, \quad V = 645 \text{ mL or } 0.645 \text{ L}$$

$$T = 25°C \text{ or } 298 \text{ K}$$

$$n = \frac{PV}{RT} = \frac{0.9658 \text{ atm}}{0.0821 \text{ L atm}/K \text{ mol } H_2} \times \frac{0.645 \text{ L}}{298 \text{ K}} = 0.0255 \text{ mol } H_2$$

## 9-15 THE MOLAR MASS OF A GAS

Suppose we had a sample of a pure gas and we wanted to know the **number of grams per mole** or the **molar mass** of the gas. Of course, if we knew the formula of the gas we could easily find the molar mass. However, it is possible to find the molar mass of a pure gas by a simple experiment without knowing its formula. This is done by measuring the pressure, volume, temperature, and mass of a sample of the gas. Furthermore, the molar mass of a volatile liquid is found by boiling sample of the liquid and determining the volume, pressure, and temperature of a known mass of vapor (see Fig. 9-6). The key to this process is that it is possible to find the number of moles of a gas by simply measuring the volume, pressure, and temperature of the sample.

To calculate the number of grams per mole or molar mass of a gas from its properties, we use the ideal gas law, $PV = nRT$. The number of

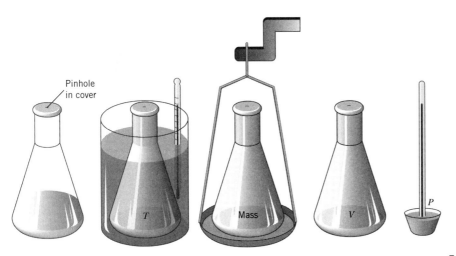

Pinhole
in cover

*T*

Mass

*V*

*P*

**FIGURE 9-6**
Experimental
determination of the
number of grams per mole
of a liquid.

moles of gas is found from the pressure, volume, and temperature using $n = PV/RT$. Then, the mass of the sample is divided by the number of moles to find the number of grams per one mole. The resulting ratio is the molar mass.

---

**EXAMPLE 9-10**

What is the molar mass of a 0.935-g sample of gas that occupies 0.515 L at 20°C and 776 torr? First, tabulate the data with mass in grams, pressure in atmospheres, volume in liters, and temperature in kelvins.

| Mass | P | V | T |
|------|---|---|---|

$$0.935 \text{ g} \quad 766 \text{ torr} \left( \frac{7 \text{ atm}}{760 \text{ torr}} \right) = 1.021 \text{ atm} \quad 0.515 \text{ L} \quad 293 \text{ K}(20 + 273)$$

The number of moles in the sample is found from the ideal gas law.

$$n = \frac{PV}{RT} = \frac{1.021 \cancel{\text{ atm}}}{0.0821 \cancel{L} \cancel{\text{ atm}}/K \text{ mol}} \times \frac{0.515 \cancel{L}}{293 \cancel{K}} = 0.02186 \text{ mol}$$

Carry an extra digit in the pressure and number of moles. Do not round off until the final molar mass calculation.

Finally, divide the mass by the number of moles; this gives the molar mass.

$$\frac{0.935 \text{ g}}{0.02186 \text{ mol}} = \frac{42.8 \text{ g}}{1 \text{ mol}} \quad \text{or} \quad 42.8 \text{ g}/\text{mol}$$

Use the general expression for the molar mass of a gas to calculate the molar mass of the gas sample in Example 9-10. Use the data given in the example.

Note the general pattern for finding the molar mass of a gas from the mass, pressure, volume, and temperature of a sample. The molar mass is the ratio of the sample mass and the number of moles.

$$\frac{\text{mass}}{n}$$

and the number of moles from the gas law is $n = PV/RT$. Thus, the molar mass can be calculated using the general expression

$$\text{molar mass} = \frac{\text{mass}}{PV/RT} \quad \text{or} \quad \frac{\text{mass} \times RT}{PV}$$

## 9-16 STOICHIOMETRY AND THE GAS LAWS

Amounts of gases are expressed in terms of the volume they occupy at a specific pressure and temperature. Consider, for example, the decomposition of potassium chlorate to give oxygen gas.

$$2KClO_3 \xrightarrow{\text{heat}} 2KCl + 3O_2$$

Some possible questions related to this equation are

1. What volume of oxygen at specific $P$ and $T$ is formed when a given mass of potassium chlorate reacts?
2. How many grams of potassium chlorate need to react to give a specified volume of oxygen at specific $P$ and $T$?

Answers to questions like these require the use of gas laws and stoichiometric reasoning. A gas in a reaction is related to other species through the combined use of the gas laws and molar ratios from the balanced equation. The gas laws relate the volume of gas at specific conditions to the number of moles of the gas. The number of moles provides an entry to stoichiometric calculations. Figure 9-7 shows a stoichiometry map illus-

**FIGURE 9-7**
Stoichiometry map showing how gas laws relate to stoichiometric calculations.

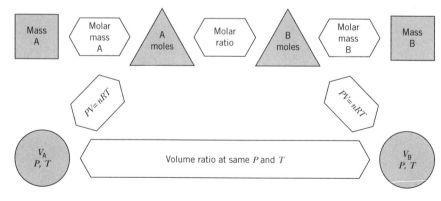

trating how the volumes of gas samples are integrated into stoichiometric calculations. Chemistry in Action 9-2 is an example of using gas laws with stoichiometry.

We can use the **molar volume, (22.4 L/mol)$_{STP}$** to relate the volume of a gas at STP to the number of moles. Alternatively, we can use the **ideal gas law** to relate the volume of a gas at any conditions to the number of moles of the gas.

---

**EXAMPLE 9-11**

How many liters of oxygen gas collected at STP can form when a 246-g sample of potassium chlorate is heated to give potassium chloride and oxygen gas? The balanced equation for the reaction is

$$2KClO_3 \xrightarrow{\text{heat}} 2KCl + 3O_2$$

The information is

$$2KClO_3 \xrightarrow{\text{heat}} 2KCl + 3O_2$$

|  |  |  |
|---|---|---|
| mass | 246 g |  |
| molar mass | 122.5 g/mol |  |
| moles | mol KClO$_3$ $\Longrightarrow$ | mol O$_2$ |
| volume |  | ? L (STP) |

Notice that we include mass, molar mass, and moles. Part of the question involves a mass-to-mole stoichiometric calculation. We include volume since the question asks for the volume of a gas from the mass of a reactant. Consulting the stoichiometry map in Figure 9-7 we see that the liters of oxygen gas produced at STP are found by

1. Finding the number of moles of KClO$_3$ from the mass of KClO$_3$ using its molar mass.
2. Using the molar ratio from the balanced equation to find the number of moles of oxygen from the number of moles of KClO$_3$.
3. Using the molar volume to find the volume of oxygen gas at STP from the number of moles of oxygen.

First, we use the molar mass of potassium chlorate to find the number of moles of KClO$_3$ [39.099 + 35.453 + 3(15.999) = 122.5].

$$246\,\cancel{g} \times \frac{1 \text{ mol KClO}_3}{122.5\,\cancel{g}}$$

Second, we multiply the moles of potassium chlorate by the molar ratio to find the moles of oxygen.

$$246\,\cancel{g} \times \frac{1 \text{ mol } \cancel{KClO_3}}{122.5\,\cancel{g}} \times \frac{3 \text{ mol O}_2}{2 \text{ mol } \cancel{KClO_3}}$$

Finally, we find the volume of oxygen gas at STP by multiplying the moles of oxygen by the molar volume.

$$L\ O_2\ \text{at STP} = 246\ \cancel{g} \times \frac{1\ \cancel{mol\ KClO_3}}{122.5\ \cancel{g}} \times \frac{3\ \cancel{mol\ O_2}}{2\ \cancel{mol\ KClO_3}} \times \frac{22.4\ L}{1\ \cancel{mol\ O_2}}$$

$$= 67.5\ L\ O_2\ \text{at STP}$$

### 9-17 VOLUME-TO-VOLUME CALCULATIONS

For reactions that involve two or more gaseous reactants or products, it is sometimes necessary to figure out how the volume of one species relates to the volume of another. Often this kind of calculation is solved quite readily if we know how the volumes of gases relate to the numbers of moles. The ideal gas law, $PV = nRT$, reveals that the volume of a gas directly relates to the number of moles if the pressure and temperature are constant; $V = nRT/P$. Furthermore, the ideal gas law shows that two gas samples having the same volume, pressure, and temperature must contain the same number of moles of gas. This idea was first suggested by Amedeo Avogadro in 1811, long before the ideal gas law was known. The relation is often stated in a form called Avogadro's law as shown in the margin and Figure 9-8.

Any samples of gases that have the same volumes at the same temperature and pressure must have the same number of moles. Another way to state this is that the number of moles of a gas relates directly to the volume. Thus, what is true for the number of moles of a gas in a reaction is also true for the number of liters of the gas. This means that it is possible to interpret a reaction involving gases at a constant temperature and pressure on a volume basis rather than a molar basis. The volumes of gases involved in a reaction relate through volume ratios that are equal to molar ratios. This is true as long as the gases are at the same temperature and pressure. Consider the reaction of hydrogen gas and chlorine gas to give hydrogen chloride gas.

$$H_2(g) + Cl_2(g) \longrightarrow 2HCl(g)$$

In this reaction, 1 mole of hydrogen reacts with 1 mole of chlorine to give 2 moles of hydrogen chloride. The ratios of the volumes of the gases involved is the same as the ratio of the number of moles. This is true as long as the gases have the same temperature and pressure.

**Volume ratios** can be expressed in the same manner as molar ratios.

$$\frac{1\ L\ H_2}{1\ L\ Cl_2} \qquad \frac{1\ L\ H_2}{2\ L\ HCl} \qquad \frac{1\ L\ Cl_2}{1\ L\ H_2}$$

$$\frac{2\ L\ HCl}{1\ L\ H_2} \qquad \frac{1\ L\ Cl_2}{2\ L\ HCl} \qquad \frac{2\ L\ HCl}{1\ L\ Cl_2}$$

**Avogadro's Law:**

*Equal volumes of gases at the same temperature and pressure contain the same number of moles.*

**FIGURE 9-8**
Avogardro's law.

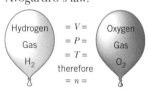

Volume ratios can be written for any reaction involving two or more gases. Of course, volume ratios do not apply to chemicals that are not in the gaseous state. These ratios are conversion factors used to relate the volume of one gas to the volume of another gas in stoichiometric calculations. They are used in a way that is similar to molar ratios. See the stoichiometry map in Figure 9-7.

---

**EXAMPLE 9-12**

How many liters of hydrogen chloride gas are produced at 780 torr and 25°C when 32 L of chlorine gas at 780 torr and 25°C react with sufficient hydrogen? The balanced equation is

$$H_2(g) \quad + \quad Cl_2(g) \longrightarrow 2HCl(g)$$
$$\text{volume} \qquad\qquad 32\ L \Longrightarrow ?\ L$$

Since the two gases are at the same conditions of temperature and pressure, the volume ratio is used as a factor.

$$L\ HCl = 32\ \cancel{L\ Cl_2} \times \frac{2\ L\ HCl}{1\ \cancel{L\ Cl_2}} = 64\ L\ HCl$$

Methane gas reacts with oxygen gas to give carbon dioxide and water. A cubic foot of methane has a volume of 28.3 L at STP. How many liters of oxygen gas at STP are needed to react with 28.3 L of methane at STP?

---

## *Gas Laws*

Charles' Law: The volume of a gas sample at constant pressure is directly proportional to the temperature. $V = kT$

Boyle's Law: The volume of a gas sample at constant temperature is inversely proportional to the pressure. $V = k/P$

Pressure-Temperature Law: The pressure of a gas sample at constant volume is directly proportional to the temperature. $P = kT$

The combined gas law is used for changes in the volume, temperature, and pressure of a gas sample.

$$\frac{V_2 P_2}{T_2} = \frac{V_1 P_1}{T_1}$$

The ideal gas law is used to find the $P$, $V$, $n$, or $T$ of a gas sample. $PV = nRT$

The universal gas constant $R$ has a value of **0.0821 L atm/K mol.**

**STP** or standard temperature and pressure are

|  |  |  |
|---|---|---|
| **0°C** | or | **273 K** |
| **760 torr** | or | **1 atm** |

The **molar volume** expresses the volume of a mole of an ideal gas at STP. **22.4 L/mol**

Dalton's Law of Partial Pressures: The total pressure of a mixture of gases is equal to the sum of the partial pressures of the components. $P_t = P_a + P_b + P_c + \ldots$

The **molar mass of a gas** is found by measuring the mass, volume, temperature, and pressure of a sample of the gas.

$$\text{Molar Mass} = (\text{mass})RT/PV$$

Avogadro's Law: Equal volumes of gases at the same temperature and pressure contain the same number of moles.

In stoichiometric calculations gas laws are used to related moles of gases to volumes at specific pressures and temperatures.

$$V_A \text{ at } P \text{ and } T \qquad V_B \text{ at } P \text{ and } T$$
$$\searrow PV = nRT \qquad \nearrow PV = nRT$$
$$\text{mass A} \longleftrightarrow \text{mol A} \longleftrightarrow \text{mol B} \longleftrightarrow \text{mass B}$$

## 9-18 PRIMARY AIR POLLUTANTS

As you know, air is a mixture of gases of which nitrogen and oxygen are the major components. The activities of any society produce waste gases. Industrial processes and the burning of trash produce certain gases, and smoke and cars and trucks produce exhaust gases. There is a difference between gases and smoke. Smoke consists of finely divided solids and liquids suspended in air. In cigarette smoke, for example, you can see these suspended solid and liquid particulates as they become dispersed in the air. You cannot see the gaseous products in the smoke even though they are present. When gases and smoke mix with the atmosphere, they can become semipermanent components. Just because the products are released into the air does not mean they are gone. When pollutants accumulate in specific geographical areas, they can cause serious air pollution.

Those gases produced by an industrial society and released into the atmosphere are called **primary air pollutants.** The five major primary pollutants are **carbon monoxide, sulfur dioxide, nitrogen oxides, volatile organic compounds,** and **particulates.** Table 9-1 describes the nature and origin of each of these primary air pollutants. Consider a burning cigarette as a source of primary air pollutants. As the cigarette smolders the incomplete combustion produces carbon monoxide gas. Any sulfur-containing impurities in the tobacco burn to form sulfur dioxide gas. The high temperature of the burning tobacco forms nitrogen oxides. The hot tobacco decomposes to produce a variety of volatile organic compounds including nicotine. Cigarette smoke contains liquid and solid particulates

***Table 9-1*** Sources of the Primary Air Pollutants

**Carbon monoxide** is formed in the incomplete combustion of fossil fuels.

$$\text{carbon-containing compounds} + O_2 \longrightarrow CO + H_2O$$

**Sulfur dioxide** is formed when sulfur-containing impurities in fossil fuels are burned.

$$\text{sulfur-containing compounds} + O_2 \longrightarrow SO_2$$

**Nitrogen oxides** are formed when nitrogen and oxygen combine at higher temperatures of fossil fuel combustion.

$$N_2 + O_2 \longrightarrow 2NO$$
$$2NO + O_2 \longrightarrow 2NO_2$$

**Volatile organic compounds** result when fossil fuels and solvents evaporate or are not completely burned.

$$\text{solvents or fossil fuels} \longrightarrow HC + VOC$$

Hydrocarbons include a variety of hydrogen–carbon compounds represented collectively as HC.

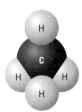

**Particulates** consist of liquid droplets and tiny pieces of solids that are formed during the combustion process.

which deposit a variety of chemicals in the lungs. In the atmosphere primary air pollutants come mainly from automobiles and trucks, electrical power plants, industrial and manufacturing processes, and solid waste incineration.

## 9-19 TEMPERATURE INVERSIONS

This topic was mentioned in Chapter 1 but let's consider this important idea in more detail. The accumulation of air pollutants within a restricted geographical area can cause an air pollution episode. Because of weather fluctuations and winds, air masses can be swept horizontally from one region to another region of the atmosphere. Horizontal air movement can disperse and dilute air pollutants.

Another means by which air mass movement can occur is the rising of warm air to higher regions of the atmosphere. When air near the surface of the earth heats, it becomes less dense, rises vertically, and is replaced by cooler air from the higher regions of the atmosphere. This vertical movement or circulation of the surface air can disperse air pollutants into higher regions of the atmosphere.

Normally the temperature of the air decreases with altitude, as illustrated in Figure 1-5. The air mass near the surface of the earth is warmer. Sometimes a cooler mass of air moves in at low altitude and underlies the warm air mass. This results in a layer of cooler air under a layer of warm air. As shown in Figure 1-5, the layer of warm air becomes trapped between the cool air mass and the cooler air of higher altitudes. This phenomenon is a **temperature inversion.** When an inversion occurs, the temperature of the air decreases with altitude until the warm air layer is reached. At this level the temperature begins to increase with altitude until the overlying cooler air is reached. Beyond this point the temperature decreases with altitude as usual. A temperature inversion prevents normal air circulation.

## 9-20  SMOG

A temperature inversion can result in a trapped and immobile air mass in which pollutants accumulate. Often, temperature inversions break up at night, but a new inversion may form the next day if the weather is favorable. Sometimes an inversion may persist for days. Temperature inversions often occur during cloudless, sunny days; which complicates air pollution episodes.

Once a temperature inversion has formed, primary air pollutants become trapped and accumulate in localized areas. If the inversion occurs on a warm, cloudless day, the primary air pollutants can react to form **secondary air pollutants.** The sunlight-induced chemical reactions that produce secondary pollutants are **photochemical reactions.** The inversion layer acts as a large reaction vessel in which photochemical reactions and subsequent reactions produce a variety of secondary pollutants known collectively as smog. Such **photochemical smog** is typical of air pollution episodes in various regions of the United States, especially in California. The hazy brown appearance of smog is a result of accumulated particulates and nitrogen dioxide but many components of smog are colorless gases. Photochemical smog is a complex mixture of chemical compounds produced by reactions between nitrogen oxides, hydrocarbons, particulates, water vapor, and oxygen.

An important example of a secondary pollutant is **ozone, $O_3$.** Ozone is not normally found in the lower atmosphere but it is formed by the following set of reactions. Nitrogen oxide, NO, is produced by the combination of nitrogen and oxygen at the high temperatures of internal combustion engines, incinerators, and power plants. Nitrogen oxide reacts with oxygen in the air to form reddish-brown nitrogen dioxide sometimes called nitric oxide.

$$2NO + O_2 \longrightarrow 2NO_2$$

Atomic oxygen forms when nitrogen dioxide is exposed to the ultraviolet (UV) light of sunshine.

$$NO_2 \xrightarrow{\text{UV}} NO + O$$

Individual atoms of oxygen are very reactive and, when they collide with molecules of oxygen adsorbed on the surface of particulates, they form ozone gas.

$$O + O_2(\text{particulates}) \longrightarrow O_3 + \text{particulates}$$

Ozone is a toxic gas that can damage plants and cause respiratory problems in humans.

## A Hole in the Ozone Layer

Ozone is an undesirable secondary air pollutant in lower regions of the atmosphere. In contrast, ozone in the upper region of the atmosphere, called the stratosphere, has an important function. The ozone in the stratosphere screens the earth from harmful ultraviolet light. Ozone forms in the stratosphere through a photochemical reaction of molecular oxygen.

$$3O_2 \xrightarrow{\text{UV}} 2O_3$$

Ozone molecules formed in the stratosphere absorb ultraviolet light. In this way ozone serves to shield the earth's surface from a significant fraction of UV light that is part of solar radiation. Ozone concentrations stay at relatively fixed levels because ultraviolet light also decomposes ozone molecules to reform oxygen molecules.

$$2O_3 \xrightarrow{\text{UV}} 3O_2$$

The formation and decomposition of ozone results in a steady-state concentration of ozone in the stratosphere. The resulting **layer or region of ozone** forms in the stratosphere at an altitude of 25 to 30 km (16 to 19 miles). This layer is sometimes called the **ozonosphere.**

Recent observations have revealed that the stratospheric ozone concentration is decreasing, especially in the north and south polar regions. This decrease in concentration has become known as the thinning of the ozone layer. Significant decreases are called holes in the layer and they appear to vary with the seasons. Holes have also been observed over some industrialized regions of the northern hemisphere as well as the poles.

A decrease in the ozone concentration can allow higher than normal amounts of UV light to reach the surface of the earth. This increase in radiation could cause damage to humans, other animals, plants, bacteria, and plankton in the ocean. Damage to humans may cause an increase in some kinds of skin cancer and an increase in the incidence of cataracts. In 1993 the World Health Organization reported that malignant melanoma, a form of skin cancer, was increasing at an annual rate of 5 to 10 percent. The increase was especially notable among fair-skinned people. However, the increase is not necessarily related to the thinning of the ozone layer. Other, possibly more likely sources are increases in outdoor activities and life-style changes among people in developed societies. Melanoma is largely preventable by the appropriate use of sunscreens and proper outdoor clothing.

Apparently the **thinning of the ozone layer** is a result of synthetic compounds containing chlorine, fluorine, and carbon known as chlorofluorocarbons, CFCs, or as Freons. CFCs have been widely used as coolants in refrigerators and air conditioners, as propellants in aerosol cans, as industrial solvents, and as foaming agents in plastic products (Styrofoam and polyurethane foams). CFCs released by industrial uses and consumer products become semipermanent components of the atmosphere. The CFC molecules slowly diffuse to the stratosphere where high-energy UV radiation causes them to decompose to form chlorine atoms.

$$Cl_2CF_2 \xrightarrow{\text{UV}} ClCF_2 + Cl \text{ (a chlorine atom)}$$

Chlorine atoms serve to catalyze the decomposition of ozone. A chlorine atom can react with an ozone molecule to decompose it.

$$Cl + O_3 \longrightarrow ClO + O_2$$

Then, the ClO molecule produced reacts with an oxygen atom to reform the chlorine atom.

$$Cl + O \longrightarrow Cl + O_2$$

The chlorine atom can go on to decompose another ozone molecule, and so on. One chlorine atom can decompose about 100,000 ozone molecules before it reacts with some other molecule and becomes inactive. The result is that CFCs can cause a decrease in the steady-state concentration of ozone.

Since the thinning of the ozone layer is caused by human-made gases, the phenomena can be controlled by phasing out their use. Most industrialized countries have begun to decrease emissions of CFCS. Unfortunately, most of the CFCs recently released will not reach the stratosphere for five or more years. There is no way to retrieve these gases so the problem could get worse in the near future. CFCs are still being used in countries throughout the world. In 1992 several European countries, the United States, and Canada agreed to ban all CFC production by the year 1997. Together these countries manufacture about 75 percent of the world's supply of CFCs. Industries are trying to develop alternatives to CFCs. The elimination of human-made, ozone-destroying gases requires cooperation between all countries of the world. Alternative chemicals will likely, at least in the short term, be more costly.

The value of $R$ can be calculated (to one digit) by using some simple data. $R$ is calculated from the volume, temperature, pressure and number of moles of a gas in a gas sample. You need a clear straw, some water, a pen or pencil, and a metric ruler.

1. Hold the straw upright and dip one end into water so that some water enters the straw.
2. Put your finger over the other end of the straw to seal it and lift it from the water.
3. Mark the level of water in the straw by using a pen or pencil; then remove your finger from the top of the straw.
4. Use the ruler to measure the height of the air sample you had in the straw and the diameter of the straw. Make the measurements in units of centimeters.
5. Calculate the volume of the air sample in cubic centimeters by assuming that the sample is the shape of a cylinder, having the measured height and diameter. Express the volume in liters.
6. Assume that the temperature of the sample is 20°C (normal room temperature) and change this temperature from Celsius to Kelvin.
7. The pressure of the air sample was slightly less than atmospheric pressure, but it is close to 1 atm. Use a pressure of 1 atm. Why was the pressure slightly less than 1 atm?
8. The density of air is about 1.2 g/L so use your sample volume to estimate the mass of air in your sample.
9. Air is mainly nitrogen and oxygen so use a molar mass of 29 g/mol to estimate the number of moles of gas in the sample. Explain why this value can be used.
10. Use the volume, temperature, pressure, and number of moles of gas in the ideal gas law to calculate a value for $R$, the gas constant.

This is a quantitative exercise so make measurements carefully. You need: two zip-lock bags, baking soda, white vinegar, and a measuring cup with metric measure.

1. Carefully measure one-half teaspoon of baking soda. Do this by obtaining one level teaspoonful and removing one-half of it with a knife. This will leave the desired amount in the spoon.
2. Carefully place all of the sample of baking soda into one of the bottom corners of a zip-lock bag. Grasp the corner of the bag in your hand to contain the sample.
3. Measure 50 mL of vinegar and pour it into the other corner of the bag. Do not let it mix with the baking soda. Push as much air as you can out of the bag and seal it. Be sure it is tightly sealed.

4. Release your grip of the baking soda and mix it with the vinegar. Allow the reaction to occur as you agitate the bag for a few minutes. Then set the bag on a flat surface so that it stands upright. Leave it for a few minutes.

5. Note the volume of the bag with the gas in it. Add water to the second bag and place it next to the first. Support the second bag with your hand and fill it with water until it appears to have a volume about the same as the first bag.

6. Carefully measure the amount of water in the second bag by pouring it into the measuring cup. You'll have more than one cupful so empty the measuring cup each time you measure a portion of the water. Record the total volume of the water in milliliters.

Use the data in this exercise to calculate the pressure of the carbon dioxide in the bag of gas. Assume that the temperature of the gas is 20°C, which is about room temperature. Record the data in a table like that shown below.

| NaHCO₃ SAMPLE | T | BAG VOLUME | VINEGAR VOLUME |
|---|---|---|---|
| 1/2 tsp is 2.0 g | 20°C | _____ mL | 50 mL |

Subtract the vinegar volume from the bag volume _____

Assume the resulting volume equals the gas volume and express the gas volume in liters _____

Express the temperature of the gas in kelvins _____

The reaction that produced the carbon dioxide was

$$NaHCO_3(s) + HC_2H_3O_2(aq) \longrightarrow CO_2(g) + NaC_2H_3O_2(aq) + H_2O$$

Using this equation and the mass of $NaHCO_3$, calculate the number of moles of carbon dioxide formed. Use the volume, the temperature, and the number of moles of carbon dioxide in the ideal gas law to calculate the pressure of carbon dioxide in the bag of gas. Express the pressure in atmospheres and in torr. Explain why it makes sense that the pressure of the carbon dioxide is greater than 1 atm.

## QUESTIONS

### Section 9-1

1. What is a vapor?

2. List five chemicals that are gases at normal earth conditions.

### Section 9-2

3. Define pressure. What is the difference between pressure and force?

### Section 9-3

4. When a diver is under water, what is the cause of the pressure he or she experiences?

5. When the diver in Question 4 is standing on the beach, what is the cause of the pressure he or she experiences?

6. Explain how a barometer works. Describe how you would construct a mercury barometer.

7. Why does a person experience shortness of breath at higher altitudes?

### Section 9-4

8. Define a torr. Define an atmosphere. Give a factor that relates these two pressure units.

9. What is the SI unit of pressure? What is a bar?

10. The barometer of Question 6 reads 31.6 inches of mercury. What is the pressure in torr? What is the pressure in atmospheres? (Hint: 25.4 mm = 1 in.)

11. Suppose that as you construct a barometer, you find that the lab is out of liquid mercury. You decide instead to use water. If the atmospheric pressure is 30 inches of mercury, how long a tube will you need for your water barometer? Express your answer in inches, feet, and meters. Mercury is 13.6 times as dense as water (e.g., a column of mercury 1 inch long corresponds to the same pressure as a column of water 13.6 inches long).

12. Change the following pressures from torr to atmospheres.

    (a) 787 torr    (b) 1093 torr    (c) 689 torr

13. Change the following pressures from atmospheres to torr.

    (a) 3.22 atm    (b) 0.965 atm    (c) 1.46 atm

### Sections 9-5 to 9-7

14. State the following gas laws in words and in the form of an equation:

    (a) Boyle's law    (b) Charles' law

    (c) pressure–temperature law

### Section 9-8

15. *A gas sample in a flexible container occupies 462 mL at 31°C and 1.46 atm. If the temperature is constant, find the volume the gas occupies when the pressure is

    (a) 1.85 atm    (b) doubled    (c) halved

    (d) 20 times its original value

16. A gas in a flexible container occupies 125 mL at 25°C and 2.10 atm. If the temperature is constant, find the volume the gas occupies when the pressure is

    (a) 1.78 atm    (b) tripled

    (c) decreased to one-fourth its original value

    (d) decreased to one-tenth its original value

17. *Suppose an air mattress has a volume of 109 L at −9.9°C and 0.931 atm on a mountain pass. You then take it to the jungle where the temperature is 41°C and the pressure is 1.08 atm. What volume does the air mattress occupy in the jungle?

18. At sea level a potato chip bag is found to occupy a volume of 491 mL at 763 torr. When the bag is taken to the mountains it is found to occupy a volume of 502 mL. What is the atmospheric pressure in the mountains if the temperature is constant?

19. A gas sample in a syringe has a volume of 43.7 mL at 757 torr. The plunger of the syringe is depressed so that the gas has a volume of 27.2 mL. What is the new pressure of the gas sample if the temperature is constant?

20. An inflated balloon has a volume of 3.7 L at 1.02 atm and 22°C. What volume will the balloon occupy at a pressure of 0.961 atm if the temperature is constant?

21. * A gas sample in a flexible container occupies 625 mL at 28°C and 1.98 atm. If the pressure remains constant, find the volume it occupies when

    (a) the Celsius temperature is doubled

    (b) the Kelvin temperature is doubled

    (c) the Kelvin temperature is decreased to one-third its original value

22. A gas in a flexible container occupies 199 mL at 27°C and 1.35 atm. If the pressure remains constant, find the volume it occupies when

    (a) the temperature is 93°C

    (b) the Kelvin temperature is tripled

    (c) the Celsius temperature is decreased to one-half its original value

23. *A gas in a rigid container occupies 475 mL at 32°C and 1.08 atm. Find the pressure when

    (a) the Celsius temperature is tripled

    (b) the Kelvin temperature is tripled

    (c) the Kelvin temperature is decreased to one-third its original value

24. A gas in a rigid container occupies 854 mL at 39°C and 0.975 atm. Find the pressure when

    (a) the Celsius temperature is doubled

    (b) the Kelvin temperature is doubled

    (c) the Kelvin temperature is decreased to one-half its original value

25. *Suppose a used aerosol can contains a gas at 0.999

atm and 22°C. When this can is heated in a fire to 596°C, what is the internal pressure? (Danger, an aerosol can may explode under these conditions. Never put an aerosol can in a fire.)

26. * Your inner ear is normally open to the atmosphere through the eustachian tube, which keeps the pressure of the inner ear equal to atmospheric pressure. Suppose that the pressure is 758 torr. An infection blocks your eustachian tube and closes your inner ear from the atmosphere. When you run a fever, the temperature of your ear increases from 36°C to 41°C. What is the pressure in your inner ear if the volume remains constant? Why do your ears sometimes pop when you change altitude?

27. A balloon of air occupies 8.5 L at 24°C and 0.999 atm. What volume will it occupy if it is placed in a freezer at −5°C while the pressure remains constant?

28. A tire is filled to a pressure of 28 psi at 21°C. After driving the pressure reads 31 psi. Assuming the volume is constant determine the new temperature of the tire in degrees Celsius.

## Section 9-9

29. *A gas in a flexible container occupies 1.00 L at 0°C and 760 torr. Find the volume when

   (a) the temperature is 110°C and the pressure is 0.756 atm
   (b) the Kelvin temperature is doubled and the pressure is doubled
   (c) the Kelvin temperature is doubled and the pressure is decreased to one-half its original value

30. A gas in a flexible container occupies 585 mL at 20°C and 0.993 atm. Find the volume when

   (a) the temperature is 80°C and the pressure is 0.800 atm
   (b) the Kelvin temperature is doubled and the pressure is doubled
   (c) the Kelvin temperature is decreased to one-half its original value and the pressure is doubled

31. *A weather balloon contains 175 L of helium gas at 28°C and 761 torr at the surface of the earth. What volume does the balloon occupy at an altitude of 30 km where the temperature is 233 K and the pressure is 11.5 torr?

32. The air in a cylinder of a diesel engine occupies 945 mL at 29°C and 1.00 atm. What is the pressure in the cylinder when the air is compressed to 67.5 mL (14:1 compression ratio) and heats to 485°C? When the fuel explodes in the compression chamber, the temperature increases to 2000°C for an instant. What is the pressure of the gases in the cylinder at this instant before expansion occurs?

33. A weather balloon contains 200 L of helium at 25°C and 758 torr on the ground. What volume does the balloon occupy at an altitude of 20 km where the temperature is 240 K and the pressure is 78.5 torr?

34. A gas sample in a flexible container occupies a volume of 299 mL at 313 K and 1.04 atm. It is heated and the pressure is changed, which causes a change in volume. What is the new temperature if the pressure changes to 1.01 atm and the new volume is 717 mL?

## Sections 9-10 to 9-13

35. State the ideal gas law both in words and as an algebraic equation. Define each term in your equation.

36. *1.00 g of argon gas occupies $4.17 \times 10^2$ mL at 200 K and 749 torr. Use these data to determine the value of the gas constant $R$ having units of mL torr/K mol.

37. One mole of a gas occupies $2.45 \times 10^4$ mL at 300 K and 764 torr. Use these data to determine the value of the gas constant $R$ having units mL torr/K mol.

38. *How many moles of acetylene gas, $C_2H_2$, are contained in a 50 gal tank at 22°C and 2.12 atm? (1 gal = 3.785 L)

39. A tank of oxygen gas contains 147 L of oxygen at 26°C and 4.79 atm pressure. How many moles of oxygen are in the tank?

40. What volume will 1.00 gram of water occupy if it is converted to steam at 100°C and 0.998 atm?

41. *A 6.25-g sample of solid carbon dioxide (dry ice) is placed in a 750 mL sealed container at 20°C. What is the pressure of the gas in the container when all the $CO_2$ has changed from solid to gas. (Danger, do not try this experiment.)

42. A small container of butane, $C_4H_{10}$, is attached to an empty 2.50 L tank at 24°C. If a 3.1 g sample of the butane is released from the container as a gas, what is the pressure of butane gas in the tank? (Hint: Find the moles from the mass of butane.)

43. *The Goodyear blimp holds $5.91 \times 10^6$ L of helium, He, at 27°C and 763 torr. What is the mass of the helium in the blimp? (Hint: This mass can be found from the number of moles.)

44. A tank contains 28.5 L of nitrogen gas at 0°C and 1.00 atm pressure. What is the mass of $N_2$ in the tank?

## Section 9-14

45. What is standard temperature and pressure, STP?

46. What is the molar volume for gases, and how can it be used to relate the volume of a gas sample to the number of moles of gas in the sample?

47. How many moles are contained in each of the following?

    (a) 68.2 L $SO_2$ at STP   (b) 31.4 L $N_2$ at STP
    (c) 579 L $CO_2$ at STP

48. What volume at STP is occupied by each of the following number of moles of gas?

    (a) 10.0 mol $NH_3$   (b) 9.47 mol Ne
    (c) 8056 mol $NO_2$

49. Determine the number of moles in each of the following gas samples.

    (a) 36.7 L of Ar at STP   (b) 480 mL of CO at STP
    (c) 951 mL of $Cl_2$ at STP

## Section 9-15

50. *A sample of air on earth is primarily a mixture of $N_2$, $O_2$, and Ar. If the partial pressures of these gases are 0.758 atm, 0.187 atm, and $5.11 \times 10^{-2}$ atm, respectively, what is the total pressure in atmospheres?

51. A sample of the atmosphere of Venus is made up of $CO_2$, $H_2O$, $O_2$, and $N_2$. If the partial pressure (on Venus) of $CO_2$ is 16.79 atm, $H_2O$ is $9.50 \times 10^{-2}$ atm, $O_2$ is $9.45 \times 10^{-2}$ atm, and $N_2$ is 1.79 atm, what is the atmospheric pressure on Venus?

52. *Disparlure is a pheromone or sex attractant produced by the gypsy moth. Male gypsy moths can detect this gas in very small amounts. If the partial pressure of disparlure in a sample of air is $1.07 \times 10^{-18}$ atm, how many moles of disparlure are there in 1.00 L of air if the temperature is 18°C? How many molecules is this?

53. *A sample of oxygen gas is prepared in the lab. You collect a 644-mL sample of gas at 27°C by displacement of water, so your oxygen sample is saturated by water vapor. Water vapor is a gas and contributes to the total pressure. If the total pressure of the sample is 756 torr and the partial pressure of the water vapor is 26.7 torr, what is the partial pressure of the oxygen gas that was produced? How many moles of oxygen molecules are in the sample?

54. A 249-mL sample of hydrogen prepared in the laboratory is collected at 23°C by the displacement of water and becomes saturated with water vapor. Since water vapor is a gas, it contributes to the pressure of the mixture. If the total pressure of the sample is 761 torr and the partial pressure of water vapor is 21.2 torr, what is the partial pressure of the hydrogen? What is the mass of $H_2$ in the sample?

## Section 9-16

55. *Halothane is a general anesthetic gas. If a 28.4-g sample occupies a volume of 3.46 L at 19°C and 0.998 atm, what is the molar mass of halothane?

56. A 0.328-g sample of a liquid is vaporized. The vapor sample is found to occupy 56.8 mL at 95°C and 0.996 atm pressure. Calculate the molar mass.

57. * Butane is used as a fuel. If a 75-g sample of butane occupies a volume of 30.5 L at 1.02 atm and 20°C, what is the molar mass of butane? Butane is 82.6% C and 17.4% H. Find the empirical formula and the molecular formula of butane.

58. Acetylene is a hydrocarbon used in welding. If a 4.35-g sample of acetylene occupies 3.70 L at 1.12 atm and 29°C, what is the molar mass? Acetylene is 92.3% C and 7.74% H. Find the empirical formula and the molecular formula of acetylene.

59. A liquid hydrocarbon is found to have the empirical formula $C_3H_6$. If a 0.842-g sample of the liquid is vaporized and the vapor occupies 303 mL at 1.07 atm and 121°C what is the molar mass of this hydrocarbon. What is its molecular formula?

60. A 1.26-g sample of ethyl ether occupies 429 mL at 40°C and 749 torr. What is the molar mass of ethyl ether?

## Section 9-17

61. *Smelling salts contain ammonium carbonate, which can decompose to form ammonia, a mild heart stimulant. The ammonium carbonate decomposes by the reaction:

$$(NH_4)_2CO_3(s) \longrightarrow 2NH_3(g) + CO_2(g) + H_2O(\ell)$$

    (a) How many milliliters of $NH_3$ at 21°C and 1.00 atm can be formed from 2.00 g of $(NH_4)_2CO_3$?

    (b) How many grams of ammonium carbonate are needed to give 555 mL of ammonia at 21°C and 0.997 atm?

(c) How many milliliters of $CO_2$ at 21°C and 0.997 atm can be formed along with 75.0 mL of $NH_3$ at 21°C and 0.997 atm?

62. *An oxyacetylene torch burns according to the following equation:

$$2C_2H_2(g) + 5O_2(g) \longrightarrow 4CO_2(g) + 2H_2O(g)$$

(a) How many liters of $C_2H_2$ at STP are needed to produce 500 mL of $CO_2$ at 625°C and 0.999 atm?

(b) How many grams of $O_2$ are required to react with 25.0 L of acetylene at 0.989 atm and 23°C?

(c) How many liters of $C_2H_2$ at STP are needed to react with 716 L of $O_2$ at STP?

63. *Bicarbonate of soda is sodium hydrogen carbonate, $NaHCO_3$. If a small amount of this compound is ingested, it reacts with stomach acid by the reaction:

$NaHCO_3(s) + H_3O^+(aq) \longrightarrow$
$$Na^+(aq) + 2H_2O + CO_2(g)$$

(Caution, do not ingest bicarbonate of soda without medical advice.)

(a) If 1.86 g of $NaHCO_3$ react to form 237 mL of $CO_2$ at 37°C, what is the pressure of the $CO_2$?

(b) Refer to part (a). Why does a person often burp after ingesting bicarbonate of soda?

(c) How many grams of $NaHCO_3$ are needed to form 350 mL of $CO_2$ at 25°C and 0.990 atm?

64. In a catalytic muffler on an automobile, carbon monoxide is combined with oxygen:

$$2CO(g) + O_2(g) \xrightarrow{\text{Pt}} 2CO_2(g)$$

(a) How many grams of CO can react with 38 L of oxygen at 275°C and 1.00 atm?

(b) How many liters of CO at 145°C and 770 torr can react with 21.4 L of oxygen at 145°C and 770 torr.

(c) How many liters of $CO_2$ at 20°C and 760 torr can be formed when 7.95 L of oxygen at 145°C and 770 torr react?

(d) How many grams of $CO_2$ are formed when 164 L of CO at 275°C and 0.325 atm react with oxygen?

## Section 9-18

65. State Avogadro's law.

66. You are handed two balloons. One contains Kr(g) and one contains $H_2$(g). They have identical pressure, volume, and temperature. Which balloon has the larger number of molecules of gas? Which has the greater mass? Explain your reasoning.

67. Ammonia is produced by the reaction:

$$3H_2(g) + N_2(g) \longrightarrow 2NH_3(g)$$

(a) How many liters of $NH_3$ at STP can be formed when 50 L of $N_2$ at STP react with hydrogen?

(b) How many liters of $H_2$ at 25°C and 1.00 atm are needed to react with 918 L of $N_2$ at 25°C and 1.00 atm?

(c) How many liters of $NH_3$ at STP can be produced when 6.37 L of $N_2$ at STP reacts with 18.1 L of $H_2$ at STP. (Hint: This is a limiting reactant problem.)

## Sections 9-19 to 9-21

68. List the five primary air pollutants.

69. What is a temperature inversion?

70. What is a secondary air pollutant?

71. What is photochemical smog and how is a temperature inversion involved in smog formation?

72. How is ozone formed as a secondary air pollutant?

73. What is the hole in the ozone layer and what is its likely cause?

## Questions to Ponder

74. Suggest some solutions to chronic air pollution. What are you willing to do to combat air pollution?

75. What can societies do to help heal the hole in the ozone layer?

76. What are your views about smoking tobacco? What are your views of secondary tobacco smoke?

# WATER

# AND

# SOLUTIONS

## 10-1 SOLUTIONS

When you mix sugar in water and stir the mixture, the sugar disappears. The sugar is not gone; it has become intimately mixed with the water. The sugar dissolves in the water and forms a solution. A solution is a homogeneous mixture of chemicals in which the fundamental particles of the chemicals interact and intermingle as defined in the margin.

Chemical compounds are characterized by fixed, unchanging composition. Samples of sugar, for instance, always contain the same elements in the same fixed proportions. A solution has a known composition but the amounts of the chemicals in the mixture can vary. Sugar solutions, for instance, range from very dilute to very concentrated in sugar content. Solutions are not compounds but unique types of mixtures. They are not gross mixtures of one substance suspended in another, such as sand in water. Instead, because of the great degree of intermingling of the particles and the forces of interaction that exist between them, the components of a solution are mixed uniformly. Filtration can separate sand suspended in water, but more involved methods are needed to separate the components of a solution. One way to get the salt out of seawater, for instance, is to evaporate the water. Seawater is boiled and the water vapor is condensed to isolate pure water. Incidently, this process requires large amounts of energy and usually is not an economical way to get pure water from seawater.

**Solution**:
*A mixture of two or more chemicals in which the particles intermingle on an atomic, molecular, or ionic level.*

**319**

Some complex solutions, such as seawater, have many components. A simple two-component solution like sugar in water is a **binary solution.** A solution forms when one chemical dissolves another chemical. The **solvent** is the dissolver and the dissolved substance is the **solute.** When sugar is dissolved in water, the water is the solvent and the sugar is the solute. The mixing of chemicals to form a solution involves the intermixing of their constituent molecules or ions. The solute particles uniformly distribute among the solvent molecules to give a homogeneous mixture.

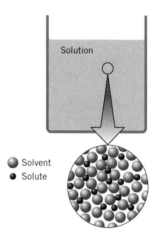

Solutions can occur as liquids, solids, or gases. Some metallic alloys are solid solutions in which one metal dissolves in another. All mixtures of gases are essentially solutions but we usually don't distinguish solvents and solutes in gas mixtures. Liquid solutions are the most common type of solution and water is by far the most common solvent. **Aqueous solutions** are those in which water is the solvent. Liquid solutions not involving water are simply nonaqueous solutions. Tincture of iodine, for example, is a nonaqueous solution of iodine dissolved in ethyl alcohol.

When a solute dissolves in a solvent the various chemical particles commingle. This commingling results from the attractive forces between the particles of solute and solvent. To understand how chemicals dissolve to make solutions we need to take a closer look at the properties of molecules and the attractive forces that exist between them.

## 10-2 ELECTRONEGATIVITY

Covalent bonds involve the sharing of electron pairs. Molecules have covalently bonded atoms. In some kinds of covalent bonds one of the atoms may attract the shared pair more strongly than the other atom. The result is unequal sharing of the electron pair. What this means is that the shared pair of electrons is pulled closer to one of the atoms than the other.

Electronegativity relates to the ability of atoms to attract shared electron pairs.

**Electronegativity** is a measure of the tendency of an atom of an element to attract electrons in a bond. Consider a molecule of hydrogen chloride as an example. Chlorine in the molecule strongly attracts electrons. As a result the electron cloud representing the bonding electrons shifts towards the chlorine giving it a slight negative charge. The charge results because electrons have negative charge.

Write the Lewis electron dot structure for a molecule of hydrogen chloride, HCl.

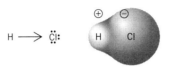

This does not mean that the electrons leave the covalent bond and form a negative chloride ion. The chlorine attracts the bonding electrons more strongly than the hydrogen. This behavior is related to electronegativity. We say that chlorine has a higher electronegativity than hydrogen. It is possible to measure the ability of various elements to attract electrons in covalent bonds by observing their bonding behaviors and the properties of the compounds they form.

A relative scale of electronegativities was first established by Linus Pauling, an American chemist. On his **electronegativity scale** a value of 4.0 was assigned to fluorine, the element of highest electronegativity. The electronegativities of the other elements are expressed relative to fluorine. Figure 10-1 shows a periodic table listing the electronegativities of the ele-

**FIGURE 10-1** The electronegatives of elements. Generally electronegativities increase from bottom to top within a group and increase from left to right within a period.

Electronegativity increases →

| | | | | | | | | | | | | | | | | | |
|---|---|---|---|---|---|---|---|---|---|---|---|---|---|---|---|---|---|
| H<br>2.1 | | | | | | | | | | | | | | | | | |
| Li<br>1.0 | Be<br>1.5 | | | | | | | | | | | B<br>2.0 | C<br>2.5 | N<br>3.0 | O<br>3.5 | F<br>4.0 | |
| Na<br>0.9 | Mg<br>1.2 | | | | | | | | | | | Al<br>1.5 | Si<br>1.8 | P<br>2.1 | S<br>2.5 | Cl<br>3.0 | |
| K<br>0.8 | Ca<br>1.0 | Sc<br>1.3 | Ti<br>1.5 | V<br>1.6 | Cr<br>1.6 | Mn<br>1.5 | Fe<br>1.8 | Co<br>1.8 | Ni<br>1.8 | Cu<br>1.9 | Zn<br>1.6 | Ga<br>1.6 | Ge<br>1.8 | As<br>2.0 | Se<br>2.4 | Br<br>2.8 | |
| Rb<br>0.8 | Sr<br>1.0 | Y<br>1.2 | Zr<br>1.4 | Nb<br>1.6 | Mo<br>1.8 | Tc<br>1.9 | Ru<br>2.2 | Rh<br>2.2 | Pd<br>2.2 | Ag<br>1.9 | Cd<br>1.7 | In<br>1.7 | Sn<br>1.8 | Sb<br>1.9 | Te<br>2.1 | I<br>2.5 | |
| Cs<br>0.7 | Ba<br>0.9 | La<br>1.1 | Hf<br>1.3 | Ta<br>1.5 | W<br>1.7 | Re<br>1.9 | Os<br>2.2 | Ir<br>2.2 | Pt<br>2.2 | Au<br>2.4 | Hg<br>1.9 | Tl<br>1.8 | Pb<br>1.8 | Bi<br>1.9 | Po<br>2.0 | At<br>2.2 | |
| Fr<br>0.7 | Ra<br>0.9 | Ac<br>1.0 | Lanthanides: 1.1–1.2<br>Actinides: 1.1–1.2 | | | | | | | | | | | | | | |

Electronegativity increases

ments. Look at the table and notice which elements have high electronegativities and which have low electronegativities. Also notice how the electronegativities vary in the table.

Generally elements having small atoms with little shielding (see Sections 6-20 and 6-21) have high electronegativities. On the other hand, those elements with larger atoms and many electrons to shield the nucleus have low electronegativities. Notice that, as might be expected, the electronegativities of the elements within a given group of representative elements tend to increase from the bottom to the top of the group. Furthermore, within a given period of representative elements, the electronegativities increase from left to right. The elements in the upper right of the periodic table, excluding the noble gases, have the greatest electronegativities.

Fluorine has the highest electronegativity (4.0) of all of the elements, while oxygen has the next highest (3.5) followed by chlorine (3.0) and nitrogen (3.0). The greater the electronegativity of an element, the greater is the tendency for atoms of that element to attract electrons when bonded. As we might expect metals generally have the lowest electronegativity values since they tend to lose electrons to form cations rather than gain electrons. Electronegativity differences between the two atoms involved in a covalent bond affect the distributions of bonding electrons in molecules.

## 10-3 POLAR BONDS

All molecules are electrically neutral. Unlike an ion a molecule does not carry a charge since the total number of electrons present equals the total number of protons. Although they are neutral, some molecules contain an electron distribution that causes them to have regions of relatively concentrated negative charge and regions of relatively concentrated positive charge. In a neutral atom, the electrons surround the nucleus, and the average center of negative charge coincides with the positive center of the nucleus. In some molecules, the average positions of electrons change because of unequal sharing of electron pairs. A hydrogen molecule is pictured in Figure 10-2. The bonding electron cloud has a greater density between the two positive nuclei. Since the two nuclei are the same, they equally share the electron pair. In such a molecule the average centers of positive and negative charge coincide; there is no separation of charges.

In a molecule of hydrogen chloride, HCl, the chlorine attracts the bonding electron pair more than does the hydrogen. The chlorine strongly attracts the shared electrons giving a partial negative charge around the chlorine and a partial positive charge around the hydrogen (see Fig. 10-3). It is important to note that the chlorine attracts the bonding electron pair but the bond between hydrogen and chlorine remains.

**FIGURE 10-2**   The centers of positive and negative charge coincide in a hydrogen molecule.

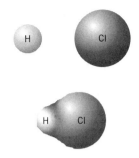

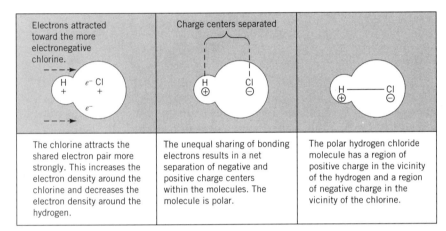

| Electrons attracted toward the more electronegative chlorine. | Charge centers separated | |
| --- | --- | --- |
| The chlorine attracts the shared electron pair more strongly. This increases the electron density around the chlorine and decreases the electron density around the hydrogen. | The unequal sharing of bonding electrons results in a net separation of negative and positive charge centers within the molecules. The molecule is polar. | The polar hydrogen chloride molecule has a region of positive charge in the vicinity of the hydrogen and a region of negative charge in the vicinity of the chlorine. |

**FIGURE 10-3** The centers of positive and negative charge are separated in a hydrogen chloride molecule, causing it to be polar.

In a molecule such as HCl, the average centers of positive and negative charge do not coincide. The molecule carries no overall charge but there is an uneven distribution of positive and negative charges. The bond between the chlorine and the hydrogen is a polar covalent bond.

A **polar covalent bond** is a covalent bond between atoms of different electronegativity. The term polar refers to opposite parts (e.g., the north and south poles of the earth or the opposite poles of a magnet). A polar bond has a negative region around the more electronegative atom and a positive region around the less electronegative atom. In general, the greater the difference in electronegativity, the greater is the polarity of the bond.

## 10-4 POLAR MOLECULES

Polar bonds in molecules affect the overall distribution of charge in the molecule. The polar bond in a simple linear molecule like hydrogen chloride results in a net separation of positive and negative charge centers within the molecule (see Fig. 10-3). The unequal distribution of charge in a molecule of hydrogen chloride makes it a polar molecule. It has a negative region or negative pole and a positive region or positive pole. A **polar molecule** is a molecule in which there is a net separation of positive and negative charge centers.

A water molecule is another good example of a polar molecule. Within the molecule, the hydrogens form polar bonds with the more electronegative oxygen. Furthermore, the water molecule happens to have an angular shape (see Section 7-18).

The polar bonds result in a slight negative charge in the region of the oxygen and in slight positive charges in the region of the hydrogens. The angular shape of the molecule results in a net separation of negative and positive charge centers.

We can easily predict whether a simple molecule is polar or nonpolar. A polar molecule must satisfy two criteria.

1. It must have one or more polar bonds.
2. It must have a shape that results in a net separation of positive and negative charge centers.

To make a prediction, first draw a dot structure for a molecule and predict its shape. Then decide if any of the bonds are polar by considering differences in electronegativities of the bonded atoms. To be polar, a molecule must have a **nonsymmetrical** shape as well as polar bonds. To judge whether a molecule is polar you can manipulate it in your imagination. Imagine what would happen if you pulled outward on the polar bonds in the molecule. If you pull equally on all polar bonds that have the same kinds of atoms, would the central atom move or would the equal pulling forces cancel? In symmetrical molecules the pulling forces would cancel and the central atom would not move. A molecule of carbon dioxide has a central carbon bonded to two more-electronegative oxygen atoms. The carbon-oxygen bonds are polar. Within the molecule, the atoms bond in a linear fashion. Imagine pulling with equal force on the oxygen atoms.

Since the molecule is linear and the two bonds are of the same type, the central carbon atom would not move. The symmetrical linear shape of the molecule results in no net separation of charge centers. The average centers of negative and positive charge happen to coincide. In other words, there is a symmetrical distribution of charge in the molecule, and it is not polar.

If a polar molecule is not symmetrical the central atom would move. A water molecule has two polar hydrogen–oxygen bonds and a bent shape. If we imagine pulling outward on these bonds the central oxygen would move. Water molecules are polar molecules

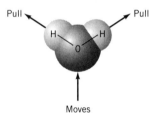

Molecules in which there is no net separation of charge are **nonpolar molecules.** There are two types of nonpolar molecules. One type includes those molecules in which the atoms have little or no differences in electronegativity and, thus, no polar bonds. Chlorine, $Cl_2$, molecules are non-

You can observe the electrical properties of polar water molecules. To do this you need a plastic knife, a cloth towel (dry), and a water faucet. Turn on the water faucet so that a very small stream of water is flowing. Rub the knife with the towel about 20 times and bring the edge of the knife near the stream of water. Assuming the knife has a negative electrostatic charge, draw a picture to show how the knife attracts the polar water molecules.

Are molecules of ammonia, NH₃, polar molecules? Explain your answer and draw a picture.

polar. Why is this true? The other type includes those molecules that have polar bonds but a symmetrical shape resulting in no net separation of charge, for example, carbon dioxide. Table 10-1 shows a few examples of polar and nonpolar molecules.

**Table 10-1** Some Polar and Nonpolar Molecules

| | Compound | | Polarity of Molecules |
|---|---|---|---|
| Nonpolar | Chlorine | | (Nonpolar bond) |
| | Sulfur trioxide | | (The molecule has polar bonds but the symmetry of the molecule makes it nonpolar.) |
| | Carbon Tetrachloride | | (The molecule has polar bonds but the symmetry of the molecule makes it nonpolar.) |
| Polar | Hydrogen chloride | | (Polar bonds with unsymmetrical molecule.) |
| | Ammonia | | (Polar bonds with unsymmetrical molecule.) |
| | Water | | (Polar bonds with unsymmetrical molecule.) |
| | Ethyl alcohol | | (Polar C—O and O—H bonds with unsymmetrical molecule.) |

# 10-5  MOLECULAR ATTRACTIONS

Imagine two molecules touching. Atoms within the molecules are held by covalent bonds. Do the molecules attract one another? Yes, intermolecular attractive forces between molecules exist and can be significant when molecules are in close proximity. The attractive forces between molecules are called **van der Waals forces,** named for the Dutch scientist Johannes van der Waals who first suggested their existence in the late 1800s. According to the kinetic molecular theory of gases, no significant attractive forces exist between gaseous particles. Gas molecules move so rapidly and are so far apart that the attractive forces are not very important. The attractive forces become important when molecules have slower speeds and are in close proximity. In liquids and solids particles have relatively slower speeds and are close together. In liquids and solids, attractive forces between particles are significant.

Polar molecules can attract one another through their oppositely charged regions.

Imagine two bar magnets. If they are far apart or move by one another quickly, the attractive forces are not noticeable. However, when the magnets are slowly brought closer they interact. If the opposite poles of the magnets match, attractive forces pull them together. On the other hand, if like poles are aligned, repulsive forces push the magnets apart. Polar molecules interact in a similar fashion, but the attractive and repulsive forces arise from the **electric poles.** The interactions are electrostatic rather than magnetic.

When polar molecules are in close proximity, the opposite poles attract one another. This is called **polar interaction** or **dipole interaction.** If the attractive forces are strong enough, the molecules become loosely attached to one another.

Although **Intermolecular forces** of attraction are not as strong as chemical bonds they can still cause molecules to attract and associate. As discussed in Chapter 7 aggregations of molecules are characteristic of the liquid and solid states. Molecules aggregate because of intermolecular attractive forces.

Intermolecular attractive forces also occur between nonpolar molecules. These **nonpolar molecular attractions** are weaker than **polar molecular attractions,** but they are important in substances consisting of nonpolar molecules. Nonpolar molecules attract when the negatively charged electrons within molecules are momentarily attracted to the positively charged nuclei of other molecules.

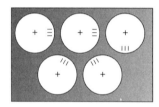

When nonpolar molecules are in close proximity, nonpolar molecular attractions cause the molecules to become loosely attached to one another. This is called **nonpolar interaction.**

## 10-6  THE HYDROGEN BOND

Water molecules are highly polar and forces of attraction between them are much stronger than most dipole-dipole attractions. The hydrogen atoms in water molecules are attracted to oxygen atoms in other water molecules ($-H\longleftrightarrow O-$). These attractions are called hydrogen bonds. A **hydrogen bond** is the force of attraction between a hydrogen of one molecule and an electronegative atom of another molecule. In other words, a hydrogen bond is an electrostatic force of attraction between the positive hydrogen atom region of one polar molecule and the negative region of another polar molecule. The hydrogen bond is a special kind of chemical bond. It is a much weaker bond than the covalent or ionic bond but much stronger than simple dipole-dipole molecular attractions. Furthermore, hydrogen bonds occur only between certain polar, hydrogen-containing molecules. Hydrogen bonding commonly occurs with molecules in which hydrogen is bonded to the electronegative elements oxygen, nitrogen, or fluorine (e.g., $H_2O$, $NH_3$, HF). Hydrogen bonding is also very important in some vital biological molecules, such as proteins and deoxyribonucleic acids (DNA) as discussed in Chapter 17.

Liquid water can be pictured as a loose collection of water molecules that are hydrogen-bonded to one another. As is the case with any liquid, the molecules of water are in constant motion within the collection. The hydrogen bonds between molecules are continually being formed and broken. The molecules move about, continually interacting with one another breaking their hydrogen bonds with some molecules while forming new bonds with other molecules. The hydrogen bond is weak enough to be broken by kinetic motion of the molecules. Consequently liquid water consists of ever-changing groupings or clusters of water molecules. With this dynamic view of water in mind, let us consider the nature of water solutions.

 **THE DISSOLVING PROCESS**

Why is water such a good solvent? Water is a liquid that has highly polar molecules and is an excellent solvent for many ionic and molecular compounds. Polar water molecules can interact with the ions of ionic compounds. That is, the electrical charge centers of polar water molecules can attract ions since they are charged particles. The positive ends of water molecules attract negative ions and the negative ends of water molecules attract positive ions.

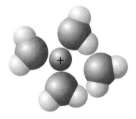

Polar water molecules can also interact with other types of polar molecules. The electric poles of water molecules attract the electric poles of other polar molecules. As a result, water can dissolve many polar molecular substances. Although water is a polar solvent, many nonpolar molecular compounds dissolve in water to some degree. For instance, ocean, river, and lake waters contain some dissolved nitrogen, oxygen, and carbon dioxide. These nonpolar compounds are, however, only slightly soluble. Such natural water solutions also contain a variety of dissolved ions. Seawater, for instance, has relatively high concentrations of dissolved ions, mostly sodium ions and chloride ions.

When we put some salt crystals in water and stir the mixture, the crystals become smaller and smaller and soon completely dissolve. Water can

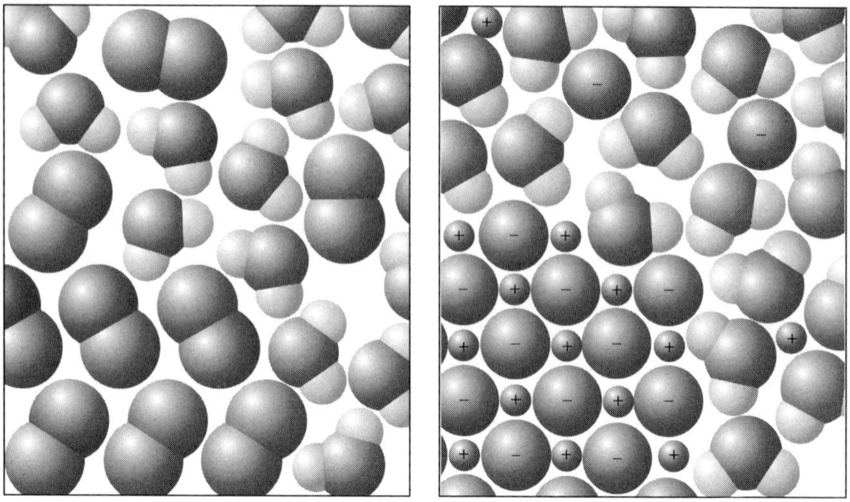

**FIGURE 10-4**    The dissolving of a solute in water. Water dissolves a solute by attracting the solute particles from the crystalline solid.

(*a*) The dissolving of a molecular solid.          (*b*) The dissolving of an ionic solid.

readily dissolve many ionic substances, since strong interactions occur between the ions and the polar water molecules. The interactions are strong enough to pull the ions from the solid and compel them to enter the solution. Figure 10-4(b) illustrates the dissolving of an ionic solid. The positively charged ends of the water molecules attract and surround negative ions until they are pulled into solution. The negatively charged ends of the water molecules attract and surround positive ions until they are pulled into solution. When an ionic compound, such as salt, dissolves in water, the process can be shown as the separation of the ions composing the compound.

$$NaCl(s) \longrightarrow Na^+(aq) + Cl^-(aq)$$

The (aq) following the formulas of the ions shows they are in aqueous solution.

Many substances that have polar molecules also readily dissolve in water. Polar water molecules interact with and attract the polar solute molecules, allowing them to dissolve as illustrated in Figure 10-4(a). Except for a few unique cases, polar molecules do not form ions when they dissolve; instead they remain as molecules dispersed among the water molecules. The dissolving of the polar molecular compound sucrose or sugar is represented as

$$C_{12}H_{22}O_{11}(s) \longrightarrow C_{12}H_{22}O_{11}(aq)$$

The (aq) following the formula shows that the molecules are in aqueous solution.

## 10-8 LIKE DISSOLVES LIKE

The properties of molecular compounds relate to the polarities of their molecules. An important principle of dissolving is "like dissolves like." This means that a polar solvent, such as water, will readily dissolve polar compounds. Polar compounds such as alcohols, ammonia, and sugar easily dissolve in water. In fact, since ionic compounds can be viewed as compounds of extreme polarity (completely separated centers of positive and negative charge), we expect many ionic compounds to dissolve in water. However, those ionic compounds in which the attractive forces between ions are quite strong do not readily dissolve in water. Generally, water is a good solvent for polar molecular compounds and many ionic compounds. Water can dissolve compounds composed of particles with charges or charge centers.

Waxy, oily, greasy, and fat-like compounds, characteristically have nonpolar molecules. Such compounds do not dissolve in water. Nonpolar compounds do not readily dissolve in water since no polar interactions occur. Solvents or paint thinners containing hydrocarbon compounds, such as hexane, readily dissolve oily and greasy materials. The hydrocarbons are nonpolar compounds so they interact with nonpolar solutes. To illustrate the differences note that water is used to clean up acrylic paints whereas paint thinners (containing nonpolar hydrocarbon solvents) are used to clean up oil-based paints. Furthermore, as expected, paint thinners and water do not dissolve in one another. **Like dissolves like** means polar dissolves polar and nonpolar dissolves nonpolar. Reversing this generalization, we can say that polar does not dissolve nonpolar and nonpolar does not dissolve polar. Incidently, latex rubber is made of nonpolar molecules. Latex rubber repels water so water or water solutions cannot pass through. Vaseline and oil-based materials have nonpolar molecules. These materials can break down latex rubber membranes found, for example, in rubber gloves and condoms. To keep latex rubber products safe don't let them contact oily materials.

You need two small glasses, water, vegetable oil, salt, and vaseline or some solid butter or margarine. This exercise may be messy so work on some newspaper or paper towel. Put less than one-fourth of a cup of water in one glass and about an equal volume of vegetable oil in the second glass. Add a sample of salt about the size of an aspirin tablet to each glass. Stir the contents of each glass and record your observations. To each glass add a sample of vaseline, butter, or margarine about the size of an aspirin tablet. Use your fingers to stir and mash the vaseline in each glass for a few minutes. Record your observations. Before cleaning up, pour the contents of one glass into the other and observe the mixture.

## 10-9 DISSOLVING AND EQUILIBRIUM

Imagine forming a salt solution in water. When the salt mixes with water, it readily dissolves to form a solution. As more salt is added, a point is reached at which no more salt dissolves and solid salt settles to the bottom of the mixture. There is a limit to the amount of salt we can dissolve in a given amount of water. Once a solution is saturated with dissolved salt no additional salt will dissolve unless the temperature is changed.

When a solute is placed in contact with the solvent, the solute particles enter the solution. The dissolved solute particles and solvent particles are

Recrystallization    Dissolving

**FIGURE 10-5**
Dynamic equilibrium
between dissolved and
undissolved solute.

To simulate dynamic equi-
librium you need a small
plastic or paper cup and a
water tap in a sink. Place a
small hole in the bottom of
the cup with a pencil or scissors. Fill the cup with water
and allow it to leak out into
the sink. Direct a stream of
water from the tap to the
cup and adjust the flow to
equal the leaking of the cup.
Adjust the water flow so
that as much water enters
the cup as leaves the cup
and the water level remains
constant. You have pro-
duced a steady-state
amount of water in the cup,
which is analogous to dy-
namic equilibrium. Water is
continually entering and
leaving the cup but there is
no net change in the
amount of water in the cup.

continually moving about in the solution. Because of this motion, some
solute particles migrate to a position between the solution and the undissolved solid and become reattached to the solid. This reformation of the
solid solute is called crystallization. Dissolving and crystallization are opposite processes. If excess solute is present, a point is reached when the rate
of dissolving equals the rate of crystallization. Figure 10-5 illustrates this
situation.

The rate of dissolving refers to the number of particles entering the solution per unit time. The rate of crystallization refers to the number of
particles leaving the solution per unit time. The competition between dissolving and crystallization and the eventual equality of their rates illustrate
a phenomenon called **dynamic equilibrium.** At equilibrium, crystallization
is taking place just as fast as dissolving. The word equilibrium is a composite of the two Latin terms, *equi* 'equal' and *librium* 'balance'. **Equilibrium**
refers to a state in which opposing forces or processes equally balance one
another.

Because equilibrium is important in chemistry, it is very important to
get a feeling for dynamic equilibrium. Picture the situation in which many
particles are dissolving at the surface of the solid. At equilibrium, just as
many particles become reattached to the surface of the solid per second as
leave the surface per second. A "standoff" is reached between dissolving
and crystallization; the concentration of the dissolved solute is fixed and
no more will dissolve. At equilibrium the concentration of the dissolved
solute remains constant.

This sort of equilibrium is not static since two opposing processes are
taking place; it is a dynamic equilibrium. Because of the equal rates, there
is no net change in the amount of dissolved solute after equilibrium is
reached. At equilibrium the particles that represent the dissolved solute
are not constant; a given particle is continually changing between the
dissolved and undissolved state. Equilibrium is reached only if the temperature is fixed, since a change in temperature may change the rate of dissolving, and before a new state of equilibrium is reached, the concentration of dissolved solute may change.

## 10-10  SATURATED SOLUTIONS AND SOLUBILITY

To prepare a solution we normally just add solute to the solvent and stir
the mixture. Typically as we add more and more of a solid solute to a solution, we reach a point at which dynamic equilibrium exists. Once a condition of equilibrium is reached no more solute dissolves and we have a saturated solution as defined in the margin of the next page. Chemisty in
Action 10-1 is about making a saturated solution of aspirin.

A saturated solution has a constant concentration of solute. An **unsaturated solution** simply contains less solute than a saturated solution. Most
solutions we use in the laboratory are unsaturated. It is possible to prepare

solutions of some solids that contain relatively large amounts of solute. Often the solubility of the solute increases with temperature. In such cases, heating a saturated solution containing some undissolved solute causes more solute to dissolve. Normally when such a solution cools, the excess solute recrystallizes as the solid. However, sometimes the result is a solution containing more solute than a saturated solution at the same temperature, a **supersaturated solution.** A supersaturated solution is in an unstable state and can change to a stable saturated solution when the excess solute recrystallizes. For example, you may have observed honey (a supersaturated solution of sugars in water) slowly crystallizing in storage. It is changing from a supersaturated solution to a saturated solution.

It is possible to dissolve 204 grams of sucrose in 100 grams of water at 20°C. Sucrose is quite soluble in water. In contrast 36 grams of sodium chloride and only $1.5 \times 10^{-4}$ gram of silver chloride dissolves in 100 grams of water at 20°C. Sodium chloride is reasonably soluble and silver chloride is essentially insoluble. For each solute there is a definite amount that can dissolve in a specific amount of solvent at given conditions. This amount is the solubility of the solute as defined in the margin. Each substance has a characteristic solubility in water at a given temperature. Solubilities normally vary with temperature. For example, the solubility of sodium chloride in water at 0°C is 35.7 g/100 g of water, and at 100°C, it is 39.7 g/100 g of water.

**Saturated Solution:** *A solution in which dissolved and undissolved solute are in dynamic equilibrium.*

**Solubility:** *The amount of a chemical that dissolves in a specific amount of solvent.*

## 10-11 MOLARITY

A teaspoon of sugar in a cup of water gives the solution a sweet taste. However, the solution made from a tiny pinch of sugar in a cup of water does not have a sweet taste. Both solutions contain sugar but they differ in concentration of the solute. **Concentration** can be expressed as the amount of solute per unit amount of solution or unit amount of solvent. In chemistry, the most useful way to express solution concentrations is in terms of the number of solute particles per unit volume of solution. Of course, the number of solute particles is the number of moles. The most common expression of concentration in chemistry is **molarity,** *M* as defined in the margin. The molarity is the ratio of moles of solute to liters of solution. To calculate the molarity of a solution we need to know the number of moles of solute in a specified volume of solution. Note that molarity uses volume of solution not volume of solvent. The number of moles of solute refers to moles of formula units of an ionic compound or moles of molecules of a molecular compound. Notice that since the number of moles of solute is related to its mass and molar mass the molarity expression could be written as

$$M = \frac{(\text{g solute}) / (\text{mol solute}/\#\text{g})}{\text{L of solution}}$$

**Molarity:** *The number of moles of solute per liter of solution.*

$$\text{molarity} = \frac{\text{moles of solute}}{\text{liters of solution}}$$

$$\text{or} \quad M = \frac{n}{V}$$

*where* n *is the number of moles and* V *is the volume of the solution in liters.*

As can be seen from the expression, it is easy to calculate the molarity of a solution if we know the mass of solute dissolved in enough solvent to produce a specific volume of solution. Suppose a solution contains 117 g of sodium chloride dissolved in enough water to form 500 mL of solution. To find the molarity of the solution, first find the number of moles of solute then divide by the volume of the solution in liters. The ratio of moles to volume is the molarity. The number of moles of sodium chloride in the solution is found using its molar mass (22.99 + 35.453 = 58.44 g/mol NaCl).

$$117\,\cancel{g} \times \frac{1\ \text{mol NaCl}}{58.44\,\cancel{g}}$$

To find the molarity divide the number of moles of solute by the volume of the solution. Note that 500 mL is 0.500 L since one liter is 1000 mL.

$$\frac{117\,\cancel{g}}{0.500\ \text{L}} \times \frac{1\ \text{mol NaCl}}{58.44\,\cancel{g}} = \frac{4.00\ \text{mol NaCl}}{1\ \text{L}}$$

Typically this concentration is expressed as 4.00 *M* NaCl and reads as 4.00–M–N–A–C–L. Alternate ways of reading this concentration are 4.00 molar NaCl or 4.00 moles NaCl per liter. It is important to note that a concentration given in molarity can always be expressed as a factor relating moles of solute to liters of solution (mol/L). As we shall see, the molarity of a solution is used as a factor in important chemical calculations.

---

**EXAMPLE 10-1**

A solution contains 225 g of glucose, $C_6H_{12}O_6$, dissolved in enough water to form 825 mL of solution. What is the molarity of the solution?

The concept involved is molarity so we need to use $M = n/V$. The data is

$$
\begin{array}{ll}
\text{solute} & C_6H_{12}O_6 \\
\text{mass solute} & 225\ \text{g} \\
\text{solution volume} & 825\ \text{mL or 0.825 L}
\end{array}
$$

The number of moles of glucose in the solution is found using its molar mass (6(12.01) + 12(1.0079) + 6(15.994) = 180.2 g/mol $C_6H_{12}O_6$).

$$225\,\cancel{g} \times \frac{1\ \text{mol } C_6H_{12}O_6}{180.2\,\cancel{g}}$$

Dividing by the volume in liters gives the molarity.

$$X\ M\ C_6H_{12}O_6 = \frac{225\,\cancel{g}}{0.825\ \text{L}} \times \frac{1\ \text{mol } C_6H_{12}O_6}{180.2\,\cancel{g}}$$

$$= \frac{1.51\ \text{mol } C_6H_{12}O_6}{1\ \text{L}} \quad \text{or} \quad 1.51\ M\ C_6H_{12}O_6$$

**10-6**

A cup has a volume of 237 mL. A teaspoon of sugar has a mass of 8.0 g. What is the molarity of sucrose, $C_{12}H_{22}O_{11}$, in a solution which has one teaspoon of sugar dissolved in water to make 237 mL of solution?

# 10-12 MOLARITY AS A UNIT FACTOR

It is convenient to store some substances as solutions. When a sample is needed, simply measure a portion of the solution. As long as the molarity of a solution is known, it is possible to dispense a specific number of moles of the solute by measuring a specific volume of the solution. For instance, in hospitals specific amounts of chemicals are given to patients by using prescribed volumes of intravenous (IV) solutions. The molarity of the solution serves as a conversion factor to find the moles of solute from the volume of solution. This is analogous to the use of density as a conversion factor. Density is the mass per unit volume of a substance and molarity is the moles per unit volume of a solution. Density relates mass and volume of a substance. **Molarity** relates moles of solute and volume of a solution.

Solution of
known molarity
$M$

$$M = \frac{n}{V}$$

When using molarity in a calculation, include the units so that molarity becomes a property factor. If a solution is 4.00 $M$ NaCl, the molarity as a factor is

$$\frac{4.00 \text{ mol NaCl}}{1 \text{ L}} \quad \text{or} \quad \frac{4.00 \text{ mol NaCl}}{1000 \text{ mL}}$$

Since 1 L is 1000 mL we can use units of mililiters in molarity if we wish. That is, we can convert 1 L to milliliters by just substituting 1000 mL for 1 L. The idea of molarity was developed to give a simple way to relate amount of solute in a solution to the volume of the solution. Molarity is used in two ways.

Measured volume
of the solution
$V$
Number of moles
of solute
$n = VM$

1.  To find the number of moles in a given volume of solution by multiplying the volume by the molarity.

$$n = VM$$

2.  To find the volume of solution needed to contain a specific number of moles of solute by multiplying the moles by the inverted molarity

$$V = n \frac{1}{M}$$

---

**EXAMPLE 10-2**

How many moles of dissolved NaOH are in 25.0 mL of a 6.00 $M$ NaOH solution?

Molarity is involved so we need to use $M = n/V$. The given information is

solute   NaOH
molarity   6.00*M*
solution volume   25.0 mL      Unknown ? mol NaOH

We can express the molarity as a factor.

$$\frac{6.00 \text{ mol NaOH}}{1 \text{ L}}$$

Find the number of moles by multiplying the volume in liters by the molarity.

To change 25.0 mL from milliliters to liters we divide by 1000 or move the decimal point three places to the left: 25.0 mL = 0.0250 L.

$$X \text{ mol NaOH} = 0.0250 \cancel{L} \times \frac{6.00 \text{ mol NaOH}}{1 \cancel{L}} = 0.150 \text{ mol NaOH}$$

Alternatively we can use the molarity with the 1 L converted to 1000 mL.

$$X \text{ mol NaOH} = 25.0 \cancel{mL} \times \frac{6.00 \text{ mol NaOH}}{1000 \cancel{mL}} = 0.150 \text{ mol NaOH}$$

---

**EXAMPLE 10-3**

How many milliliters of a 2.00 *M* NaCl solution are needed to provide 0.250 moles of NaCl?

Molarity is involved so we need to use *M = n/V*. The information is

solute   NaCl
moles solute   0.250 mol NaCl
molarity   2.00 *M*                Unknown ? mL of solution

The molarity of NaCl as a factor is

$$\frac{2.00 \text{ mol NaCl}}{1 \text{ L}}$$

To find the volume, multiply the given number of moles by the inverted molarity (1 L/2.00 mol NaCl) to give the volume in liters, which can then be converted to milliliters.

$$X \text{ mL} = 0.250 \cancel{\text{mol NaCl}} \times \frac{1 \cancel{L}}{2.00 \cancel{\text{mol NaCl}}} \times \frac{1000 \text{ mL}}{1 \cancel{L}} = 125 \text{ mL}$$

**10-7**

(a) What is the molarity of a glucose, $C_6H_{12}O_6$, solution that contains 9.01 grams of glucose in 500 mL of solution?

(b) What volume of a 0.100 *M* $C_6H_{12}O_6$ solution contains 0.0500 mole of glucose?

(c) How many moles of $C_6H_{12}O_6$ are in a 10.0-mL sample of a 0.100 *M* $C_6H_{12}O_6$ solution?

## 10-13 STANDARD SOLUTIONS

It is useful to store substances in solutions and measure volumes as needed. Solutions of known concentration are called **standard solutions.**

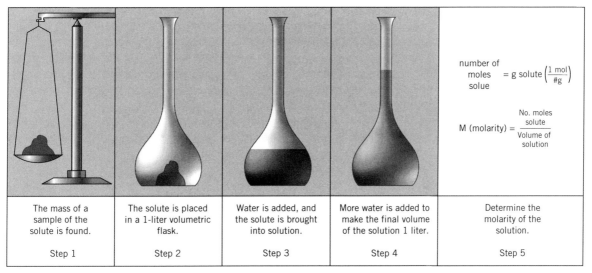

| The mass of a sample of the solute is found. | The solute is placed in a 1-liter volumetric flask. | Water is added, and the solute is brought into solution. | More water is added to make the final volume of the solution 1 liter. | Determine the molarity of the solution. |
|---|---|---|---|---|
| Step 1 | Step 2 | Step 3 | Step 4 | Step 5 |

**FIGURE 10-6** The preparation of a standard solution.

A standard solution is prepared by weighing a sample of solute and dissolving the sample in enough solvent to give a known volume of solution. A volumetric flask, calibrated to contain a known volume of liquid, is often used to prepare standard solutions. Figure 10-6 illustrates the use of the volumetric flask. Standard solutions have known molarities. To prepare a standard solution we need to know three things.

1. The formula of the solute.
2. The desired molarity of the solute.
3. The total volume of solution to be prepared.

The product of the volume and molarity of the solution will give the number of moles of solute needed to prepare the solution. The number of grams of solute needed is easily found from the number of moles using its molar mass.

**EXAMPLE 10-4**

How many grams of sodium chloride are needed to prepare 250 mL of a 0.500 $M$ solution of NaCl?

Molarity is involved so we need to use $M = n/V$. The information is

|                  |                            |                    |
|------------------|----------------------------|--------------------|
| solute           | NaCl                       |                    |
| molarity         | 0.500 $M$                  |                    |
| solution volume  | 250 mL or 0.250 L          | Unknown ? g NaCl   |

ACTIVITY 10-8

How many grams of glucose are needed to prepare 500 mL of a 0.200 $M$ $C_6H_{12}O_6$ solution? Explain how such a standard solution would be prepared.

We cannot find the mass of NaCl directly but we can find the number of moles needed by multiplying the solution volume in liters by the molarity. Note that we want to cancel liters and retain moles.

$$0.250\,\cancel{L} \times \frac{0.500 \text{ mol NaCl}}{1\,\cancel{L}}$$

Next, to find the grams of NaCl multiply the number of moles by the molar mass (58.44 g/mol NaCl). Here we want to cancel moles and end up with grams.

$$X \text{ g NaCl} = 0.250\,\cancel{L} \times \frac{0.500\,\cancel{\text{mol NaCl}}}{1\,\cancel{L}} \times \frac{58.44 \text{ g}}{1\,\cancel{\text{mol NaCl}}} = 7.31 \text{ g}$$

The standard solution is prepared by weighing 7.31 g of sodium chloride, placing it in a 250 mL volumetric flask, and dissolving it in enough water to give 250 mL of solution.

**FIGURE 10-7** The molarity of a solution is inversely proportional to the volume.

When the volume of a solution is increased (the solution is diluted), the molarity decreases. Since the number of mole is constant, dividing by a larger volume gives a smaller molarity.

When the volume of a solution is decreased (the solvent is evaporated), the molarity increases. Since the number of moles is constant, dividing by a smaller volume gives a larger molarity.

## 10-14 DILUTION

Restaurants often keep cola drinks in concentrated forms and dilute them with water when needed. Aqueous solutions are made less concentrated by diluting with water. In the laboratory it is sometimes necessary to prepare a solution of a lower molarity by dilution of a more concentrated solution.

As shown in Figure 10-7, the molarity is inversely proportional to the volume. This makes sense if you consider that adding water to a solution increases the volume, which decreases the molarity. As the volume gets larger the molarity gets smaller. Molarity is the number of moles of solute per liter of solution.

$$M = \frac{n}{V}$$

We can see that the number of moles of solute in a solution is the product of the volume and the molarity. If we have a known volume of a solution, it contains a specific number of moles.

$$n = V_c M_c$$

The subscript c refers to concentrated. Adding water to the sample dilutes it, changing the volume and, thus, the molarity. Dilution does not change the number of moles of solute, it simply makes the solution less concentrated. Thus,

$$n = V_d M_d$$

where the subscript d refers to diluted. Dilution increases the volume, so the molarity decreases. The number of moles is unchanged by dilution.

$$n = V_c M_c \qquad n = V_d M_d \qquad n = n$$

This gives the general **dilution formula:**

$$V_c M_c = V_d M_d$$

There are two ways to use this expression.

1. Given the initial molarity and volume of a concentrated solution, find the volume of the solution of lower molarity. That is, $V_c$, $M_c$, and $M_d$ are known and $V_d$ is to be calculated.
2. Find the volume of a more concentrated solution needed to prepare a specific volume of a more dilute solution. That is, $V_d$, $M_d$, and $M_c$ are known and $V_c$ is to be calculated.

Simply put, we can always calculate the fourth term in this expression if we know the three other terms.

---

**EXAMPLE 10-5**

To what volume should 10.0 mL of 15.0 $M$ aqueous ammonia be diluted to give a 2.00 $M$ solution? First, summarize the data.

| $V_c$ | $M_c$ | $V_d$ | $M_d$ |
|---|---|---|---|
| 10.0 mL | 15.0 $M$ | ? | 2.00 $M$ |

Use the relation:

$$V_c M_c = V_d M_d$$

and solve for $V_d$

$$V_d = \frac{V_c M_c}{M_d}$$

$$V_d = 10.0 \text{ mL} \times \frac{(15.0\,\cancel{M})}{(2.00\,\cancel{M})} = 75.0 \text{ mL}$$

Note that using milliliters for the volume gives the calculated answer in milliliters. To prepare the less concentrated solution 10.0 mL of the concentrated solution is diluted with water to give a new volume of 75.0 mL.

---

**EXAMPLE 10-6**

How many milliliters of 6.00 $M$ hydrochloric acid are needed to prepare 100 mL of 0.500 $M$ hydrochloric acid by dilution? First, summarize the data.

| $V_c$ | $M_c$ | $V_d$ | $M_d$ |
|-------|-------|-------|-------|
| ? | 6.00 $M$ | 100 mL | 0.500 $M$ |

Then use the relation:

$$V_c M_c = V_d M_d$$

and solve for $V_c$

$$V_c = \frac{V_d M_d}{M_c}$$

$$V_c = (100 \text{ mL}) \times \frac{(0.500 \,M)}{(6.00 \,M)} = 8.33 \text{ mL}$$

To prepare the less concentrated solution, 8.33 mL of the more concentrated solution is diluted with water to give a new volume of 100 mL.

**ACTIVITY 10-9**

(a) What volume of 5.00 $M$ NaCl is needed to prepare 500 mL of 0.500 $M$ NaCl by dilution?

(b) To what volume should 10.0 mL of 3.00 $M$ NaCl be diluted to give 1.00 $M$ NaCl?

---

## 10-15 SOLUTION STOICHIOMETRY

As we know, the molarity of a solution is a property factor used to relate the number of moles of solute to the volume of the solution, and many laboratory chemicals are stored as solutions of known molarity. These standard solutions are sometimes called **reagent solutions** and are used to carry out chemical reactions in solution. Stoichiometric questions about solution reactions involve **solution stoichiometry.** Calculations for such questions include the usual stoichiometric conversions coupled with the molarity of the reagent solutions. Stoichiometry maps were discussed in Sections 5-8 and 9-16. Molarity is also used in stoichiometric calculations and it fits into the stoichiometry map as shown in Figure 10-8.

We start with an example using moles. A solution of sodium hydroxide and a solution of acetic acid react according to the equation:

$$\text{NaOH(aq)} + \text{HC}_2\text{H}_3\text{O}_2\text{(aq)} \longrightarrow \text{H}_2\text{O} + \text{NaC}_2\text{H}_3\text{O}_2\text{(aq)}$$

How many milliliters of 6.00 $M$ NaOH solution are needed to react with a solution containing 0.300 mole of acetic acid, $\text{HC}_2\text{H}_3\text{O}_2$? Following the pattern of stoichiometry calculations we can summarize the question by making a stoichiometry table of the given data as shown on the next page.

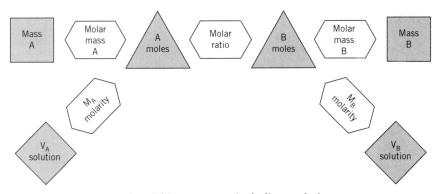

**FIGURE 10-8**   A stoichiometry map including molarity.

$$NaOH(aq) \ + \ HC_2H_3O_2(aq) \longrightarrow H_2O + NaC_2H_3O_2(aq)$$

| | | |
|---|---|---|
| moles | mol NaOH $\Longleftarrow$ | 0.300 mol $HC_2H_3O_2$ |
| molarity | 6.00 $M$ | |
| solution volume | ? mL | |

We include molarity in the table since it is the factor that relates solution volume to number of moles of solute. We want to know what volume of 6.00 $M$ NaOH contains enough moles to react with the 0.300 mole $HC_2H_3O_2$. Consult the map to follow the calculation sequence.

In the first step, find the moles of sodium hydroxide from the number of moles of $HC_2H_3O_2$ using the molar ratio from the equation.

$$0.300 \ \cancel{mol \ HC_2H_3O_2} \times \frac{1 \ mol \ NaOH}{1 \ \cancel{mol \ HC_2H_3O_2}}$$

Then, find the volume of solution using the molarity, so that the moles cancel and liters remain. The liters can then be converted to milliliters.

$$0.300 \ \cancel{mol \ HC_2H_3O_2} \times \frac{1 \ \cancel{mol \ NaOH}}{1 \ \cancel{mol \ HC_2H_3O_2}}$$

$$\times \frac{1 \ \cancel{L}}{6.00 \ \cancel{mol \ NaOH}} \times \frac{1000 \ mL}{1 \ \cancel{L}} = 50.0 \ mL$$

As was the case with typical stoichiometric problems, no calculations are necessary until we have the complete setup of the problem.

The same general approach is used to relate the number of grams of a reactant to the volume of a reagent. The molar mass is used to find the number of moles from the mass.

**EXAMPLE 10-7**

How many grams of oxalic acid, $H_2C_2O_4$, react with 25.0 mL of a 0.100 $M$ NaOH solution if the reaction is

$$2NaOH(aq) + H_2C_2O_4(aq) \longrightarrow 2H_2O + Na_2C_2O_4(aq)$$

The information is

$$2NaOH(aq) \quad + \quad H_2C_2O_4(aq) \longrightarrow 2H_2O + Na_2C_2O_4(aq)$$

|                 |                                            |
|-----------------|--------------------------------------------|
| mass            | ? g                                        |
| molar mass      | 90.04 g/mol                                |
| moles           | mol NaOH $\Longrightarrow$ mol $H_2C_2O_4$ |
| molarity        | 0.100 $M$                                  |
| solution volume | 25.0 mL or 0.0250 L                        |

Consult the stoichimetry map to follow the calculation sequence. To put this question on a stoichiometric basis, first find the number of moles of NaOH from the volume and molarity of the NaOH solution.

$$0.0250 \, \cancel{L} \times \frac{0.100 \text{ mol NaOH}}{1 \, \cancel{L}}$$

Then, find the number of moles of oxalic acid using the molar ratio from the equation.

$$0.0250 \, \cancel{L} \times \frac{0.100 \text{ mol NaOH}}{1 \, \cancel{L}} \times \frac{1 \text{ mol } H_2C_2O_4}{2 \text{ mol NaOH}}$$

Finally the number of moles of oxalic acid is changed to mass using its molar mass.

$$H_2C_2O_4: 2(1.0079) + 2(12.01) + 4(16.00) = 90.04$$

$$X \text{ g } H_2C_2O_4 = 0.0250 \, \cancel{L} \times \frac{0.100 \text{ mol NaOH}}{1 \, \cancel{L}}$$

$$\times \frac{1 \text{ mol } H_2C_2O_4}{2 \text{ mol NaOH}} \times \frac{90.04 \text{ g}}{1 \text{ mol } H_2C_2O_4} = 0.113 \text{ g}$$

## **10-16** SOME OTHER SOLUTION CONCENTRATION TERMS

### Percent-by-Mass (or Weight) Composition

The amount of each component in a solution can be expressed simply as a percent. To calculate the **percent by mass** of a solution component, divide the mass of the component by the mass of the solution and then multiply by 100.

$$\text{percent component} = \frac{\text{mass component}}{\text{mass solution}} \times 100$$

Percent compositions are sometimes used in industry and medicine to express concentration of solutions. For example, a salt solution called physiological saline is used in medicine; it must have a specific concentration of sodium chloride. Suppose a physiological saline solution is prepared by mixing 4.6 g of sodium chloride in enough water to make 500 g of solution. To find the percent of sodium chloride divide the mass of NaCl by the solution mass and multiply this fraction by 100.

$$\% \text{ NaCl} = \frac{4.6 \text{ g NaCl}}{500 \text{ g solution}} \times 100 = 0.92\% \text{ NaCl}$$

---

**EXAMPLE 10-8**

A solution consists of 20.6 g of glucose dissolved in 53.5 g of water. What is the percent by mass of glucose in the solution?

The mass of glucose is 20.6 g, and the mass of solvent is 53.5 g. The total mass of the solution containing this amount of solute is found by adding the mass of the solute to the mass of the solvent.

$$\text{mass solution} = 53.5 \text{ g} + 20.6 \text{ g} = 74.1 \text{ g}$$

To find the percent by mass of glucose divide the mass of glucose by the mass of the solution then multiply by 100.

$$\% \text{ glucose} = \frac{20.6 \text{ g glucose}}{74.1 \text{ g soution}} \times 100 = 27.8\% \text{ glucose}$$

---

## Molality

**Molality,** *m,* is another solution concentration term that is sometimes used in chemistry. It is defined as the number of moles of solute per kilogram of solvent.

$$\text{molality} = \frac{\text{number of moles of solute}}{\text{number of kilograms solvent}} \quad \text{or} \quad m = \frac{n}{\text{kg solvent}}$$

where $n$ is the number of moles of solute and the denominator is the number of kilograms of solvent. Note that molality expresses the amount of solute per kilogram of solvent not per kilogram of solution. Be careful not to confuse molality with molarity since the names are very similar. A lowercase *m* is used as an abbreviation for molality. Molality is not used often but, as we shall see, it does have some important uses.

**EXAMPLE 10-9**

A solution contains 18.0 g of glucose dissolved in 500 g of water. What is the molality of glucose in the solution?

The concept involved is molality so we need to use $m = n/kg$ solvent. The data is

> solute $C_6H_{12}O_6$     Unknown ? $m\ C_6H_{12}O_6$
> mass solute   18.0 g
> solvent mass   500 g or 0.500 kg, since there are 1000 g/kg

Since we need number of moles to find molality we use the molar mass of the solute (180.2 g/mol $C_6H_{12}O_6$). Note that the mass of solvent in grams is easily changed to kilograms by dividing by 1000 or moving the decimal point three places to the left.

First find the number of moles of glucose in the given mass using its molar mass.

$$18.0\ g \times \frac{1\ mol\ C_6H_{12}O_6}{180.2\ g}$$

Dividing this by the number of kilograms of water gives the molality.

$$X\ m\ C_6H_{12}O_6 = \frac{18.0\ g}{0.500\ kg\ H_2O} \times \frac{1\ mol\ C_6H_{12}O_6}{180.2\ g}$$

$$= \frac{0.200\ mol\ C_6H_{12}O_6}{1\ kg\ H_2O} \quad or \quad 0.200\ m\ C_6H_{12}O_6$$

**Parts per Million**

There are about $1.4 \times 10^{21}$ g of dissolved aluminum in the oceans of the world. This is indeed a large amount of aluminum but, since the total volume of the oceans is so great, the concentration of aluminum is quite low. So low, in fact, that we find it convenient to refer to the concentration in terms of **parts per million** or **ppm**. Parts per million is analogous to percent except that percent refers to parts per hundred. For comparison imagine yourself in a crowd of one million people; you'd be one part per million. One person in one hundred is one percent.

The parts per million concentration of a solute can be calculated by dividing the mass of the solute in a sample of solution by the mass of the solution and multiplying by $10^6$, which is a million.

$$\mathbf{ppm} = \frac{\mathbf{mass\ solute}}{\mathbf{mass\ solution}} \times \mathbf{10^6}$$

For dilute water solutions that have densities close to 1 g/mL the volume of the solution in milliliters can be used in place of the mass of the solution. As an example of a ppm calculation suppose a 0.50 kg sample of drinking water is found to contain 2.2 mg of fluoride ions. To calculate the ppm of $F^-$ we need the mass of solute in grams and the mass of solution in grams. 2.2 mg is $2.2 \times 10^{-3}$ g and 0.50 kg is 500 g. Thus the ppm of fluoride ion is

$$\frac{2.2 \times 10^{-3} \, \text{g} \, F^-}{500 \, \text{g}} \times 10^6 = 4.4 \text{ ppm } F^-$$

---

**EXAMPLE 10-10**

The U.S. Environmental Protection Agency (EPA) drinking water standards allow 1.00 ppm $Ba^{2+}$. If a 1.5-liter sample of drinking water is found to contain 2.00 mg of $Ba^{2+}$, does the water meet or exceed the standard?

Calculate the parts per million $Ba^{2+}$ in the water by dividing the grams of $Ba^{2+}$ by the number of milliliters of solution. Milliliters can be used in place of the number of grams for dilute solutions.

$$\frac{2.00 \times 10^{-3} \text{ g } Ba^{2+}}{1.5 \times 10^3 \text{ mL}} \times 10^6 = 1.3 \text{ ppm } Ba^{2+}$$

Based upon this result the drinking water exceeds the EPA standard for barium ions.

---

It is of interest to note that the above expression for parts per million can be altered to calculate the **parts per billion** (ppb) or even **parts per trillion** (ppt) concentration of a solute. This is done by using $10^9$ (a billion) or $10^{12}$ (a trillion) in place of $10^6$ in the expression.

## Drinking Water Standards

According to the Federal Safe Drinking Water Act the EPA requires public water systems to test for a variety of dissolved chemicals and bacteria in the water they supply. For these potential water pollutants the EPA has established **maximum contaminant levels (MCL),** which are legal standards. Any system with 15 or more connections must notify its customers of any violation of MCLs in supplied water. Nearly 15% of Americans use private water sources. The EPA recommends that these water sources be tested periodically for contaminants, especially nitrate ions and coliform bacteria.

The EPA standards include a long list of organic chemicals that could show up in drinking water from industrial and pesticide use. The standards also include testing for radioactivity and coliform bacteria. The presence of excess coliform bacteria is an indication that the water could be contaminated with sewage or other harmful bacteria. Here is a list of the MCLs for some common chemicals that occur dissolved in water in ionic forms.

| CHEMICAL | MCL (PPM) | CHEMICAL | MCL (PPM) |
| --- | --- | --- | --- |
| Aluminum | 1 | Arsenic | 0.05 |
| Barium | 1 | Cadmium | 0.01 |
| Chloride | 500 | Chromium | 0.05 |
| Copper | 1 | Fluoride | 2 |
| Iron | 0.3 | Lead | 0.015 |
| Manganese | 0.05 | Mercury | 0.002 |
| Nitrate | 45 | Selenium | 0.01 |
| Silver | 0.05 | Sulfate | 500 |
| Zinc | 5 | Total dissolved solids | 1000 |

## 10-17 ELECTROLYTES AND NONELECTROLYTES

A rule of electrical safety is never to use an electrical appliance when you are wet or are standing in water. This is to avoid an electrical shock caused by electricity flowing from the appliance through your body to the ground. You might wonder how water can conduct electricity. Pure water is not a good conductor of electricity. It is the presence of dissolved ions in water that make it an electrical conductor. Chemistry in Action 10-2 is about testing the conductivity of several solutions.

A solution is tested for electrical conductivity by using an apparatus that has electrodes connected to a source of electrical current. If the solution conducts, it completes the circuit between the electrodes and current flows in the conductivity apparatus. A solution conducts electricity by the movement of ions toward the electrodes and by chemical reactions at the electrodes (see Section 13-11).

If ions are present in an aqueous solution, the solution will conduct electricity. A substance that forms an aqueous solution that conducts electricity is called an **electrolyte.** A solution that conducts is assumed to contain ions. Soluble ionic compounds are electrolytes. Molecular compounds that dissolve in water and react to form ions are also electrolytes. Incidentally, deionized water is water that is purified by removing all ions by a special chemical process called ion exchange. Deionized water does not conduct electricity. A substance that forms an aqueous solution that

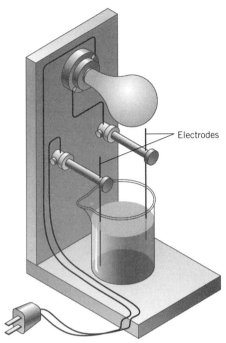

A conductivity apparatus.

does not conduct is a **nonelectrolyte.** Most soluble molecular substances (e.g., sugar and alcohol) are nonelectrolytes. Characterizing a substance as a nonelectrolyte means that it does not form ions when it dissolves.

Conductivity measurements reveal that solutions of some electrolytes are strong conductors, whereas solutions of others are quite weak conductors. According to these tests, some electrolytes are classified as strong electrolytes and others as weak electrolytes. The difference comes from the extent to which substances form ions when they dissolve in water. Substances that dissolve in water to form no products other than positive and negative ions are **strong electrolytes.** Strong electrolytes can be ionic or molecular compounds. All soluble ionic compounds are strong electrolytes since they separate into ions when they dissolve. **Weak electrolytes** are substances that dissolve in water and react with water to only a slight extent, producing relatively low concentrations of ions. Low concentration of ions account for the slight conduction of electricity.

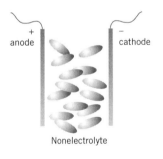

## 10-18   STRONG AND WEAK ELECTROLYTES

Suppose we had a water solution of 1 molar hydrogen chloride (1 $M$ HCl) and a water solution of 1 molar acetic acid (1 $M$ HC$_2$H$_3$O$_2$). We can as-

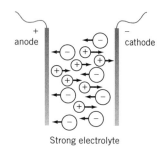

Strong electrolyte

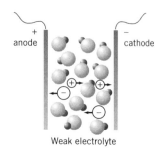

Weak electrolyte

sume they have the same amount of dissolved solute per liter of solution since they have the same molarity. Both solutions are found to conduct electricity, so we classify hydrogen chloride and acetic acid as electrolytes. However, the hydrogen chloride solution is a much stronger conductor of electricity than the acetic acid solution. Hydrogen chloride is a strong electrolyte and acetic acid is a weak electrolyte.

This difference in conductivity suggests that the acetic acid solution has a lower concentration of ions than does the hydrogen chloride solution. Since they are both electrolytes, and we are comparing solutions of the same concentration, we could expect the solutions to contain the same concentrations of ions. However, they do not. The difference in the concentration of ions is a result of the extent to which these two substances enter into chemical reaction with water to form ions.

When some molecular substances, such as hydrogen chloride, dissolve in water, they enter into a chemical reaction that completely consumes the substance to form ions. The reaction between hydrogen chloride and water is

$$HCl + H_2O \longrightarrow H_3O^+(aq) + Cl^-(aq)$$

Note that hydrogen chloride and water react to give chloride ions and $H_3O^+$ ions, called hydronium ions. We'll learn more about this important ion in Chapters 11 and 12.

Some molecular electrolytes dissolve in water and react slightly with the water, forming relatively low concentrations of ions. In other words these compounds readily dissolve in water but relatively few of the molecules react with water. For example, when acetic acid dissolves in water, the reaction is

$$HC_2H_3O_2(aq) + H_2O \rightleftharpoons H_3O^+(aq) + C_2H_3O_2^-(aq)$$

This equation has a double arrow between reactants and products to convey the fact that it is a reversible chemical reaction involving a dynamic **chemical equilibrium.** This means that the reactants react to form the products and the products in turn react to reform the reactants. The result is a mixture of reactants and products. In the case of acetic acid, a weak electrolyte, the equilibrium favors the reactant side, which means there is only a slight reaction with water. The **double arrow** showing a smaller arrow to the right and a larger arrow to the left ($\rightleftharpoons$) suggests a slight reaction toward the product side. The reaction favors the reactant side and the equilibrium mixture has relatively high concentrations of reactants and relatively low concentractions of products.

In previous discussions, we used chemical equations showing reactants and products separated by a single arrow. When it is important to convey the fact that a reaction involves chemical equilibrium, we write a double arrow. Actually all chemical reactions are theoretically reversible and involve chemical equilibrium. When a reaction favors the formation of the product and the reaction mixture contains very little of the original reactants, a single arrow is used in an equation. Furthermore, if it is not important to show that a reaction involves chemical equilibrium use a single arrow. You'll see some equations written with double arrows and some with single arrows. See Chapter 14 for a discussion of chemical equilibrium.

## 10-19  SUBSTANCES IN SOLUTION

When working with solutions, a question arises concerning which formulas should be used to represent the dissolved solutes. This can be an important point of chemical vision. An aqueous solution is always clear; we can see through it. If the dissolved solute is colorless the solution is colorless. If the dissolved solute is colored, the solution is colored. As part of our chemical vision we need a way to picture the particles found in solutions. The picture depends upon whether the solute is a strong electrolyte, weak electrolyte, or nonelectrolyte. We know that a weak electrolyte reacts somewhat with water to produce some ions. Ammonia is a weak electrolyte and it reacts with water.

$$NH_3(aq) + H_2O \rightleftharpoons NH_4^+(aq) + OH^-(aq)$$

Notice that the sizes of the arrows in the equation reveal that ammonia reacts only slightly with water. It is not convenient nor necessary to write the formulas of all of the species when we want to represent a solution of ammonia. Since $NH_3$ is the species present in the greatest concentration and the ions are present in relatively low concentrations, a solution of ammonia is represented most conveniently by the formula $NH_3(aq)$.

It is standard practice to represent a **solution of a weak electrolyte** by the formula of the most abundant species in solution. If desired this formula is followed by (aq) to show a water solution. Keep in mind that water is always present in an aqueous solution, and its presence is either implied or indicated by (aq). In the case of a weak electrolyte, the most abundant or major species in solution is the molecular form of the dissolved substance.

---

**EXAMPLE 10-11**

An aqueous solution of acetic acid, $HC_2H_3O_2$, is a weak conductor of electricity. How should an aqueous solution of acetic acid be represented? That is, how do we picture the major particles that populate the solution?

Since acetic acid is a weak electrolyte, it is best to represent a solution of acetic acid by its molecular formula: $HC_2H_3O_2(aq)$

---

Nonelectrolytes do not react with water when they dissolve; their molecules just commingle with water molecules. **Solutions of nonelectrolytes** are represented by the molecular formula followed by (aq) if desired. For instance, an aqueous solution of the nonelectrolyte ethyl alcohol is represented as $C_2H_5OH(aq)$

A chemical that is a strong electrolyte is completely ionized in solution. The major species in solution are the cations and anions produced when the substance dissolves. In the case of most ionic substances, it is a simple task to decide which ions are involved. A **solution of an ionic compound** is represented by the formulas of the cation and the anion in the compound. Picture the ions as separate rather than combined by ionic bonds. Write their formulas as separate ions followed by (aq) to indicated an aqueous solution.

---

**EXAMPLE 10-12**

How is a solution of sodium chloride, NaCl, represented?

We represent solutions of a soluble ionic compound by the formulas of its constituent cation and anion. Therefore, we show a solution of sodium chloride as $Na^+(aq) + Cl^-(aq)$

---

In the case of strong electrolytes that are molecular rather than ionic, we must know how the substance reacts with water. A **solution of a molecular strong electrolyte** is represented by the formulas of the ions produced in the reaction between water and the strong electrolyte. Hydrogen chloride gas is a strong electrolyte. When it dissolves in water it reacts according to the equation:

$$HCl(g) + H_2O \longrightarrow H_3O^+(aq) + Cl^-(aq)$$

Since hydronium ions and chloride ions are the species formed when HCl dissolves in water, we should represent an aqueous solution of HCl, called hydrochloric acid, as

$$H_3O^+(aq) + Cl^-(aq)$$

Table 10-2 lists some common electrolytes and nonelectrolytes. There are four important ideas to keep in mind when representing substances in aqueous solution.

1. Ionic compounds are shown as the formulas of the separate cation and anion.

2. Molecular compounds that are strong electrolytes are shown as the ions they form in water. (As shown in Table 10-2, the three common molecular strong electrolytes are HCl, $HNO_3$, and $H_2SO_4$.)

**Table 10-2**   Some Strong Electrolytes, Weak Electrolytes, and Nonelectrolytes

| COMPOUND NAME | FORMULA | MAJOR SPECIES IN SOLUTION |
|---|---|---|
| *Molecular Strong Electrolytes* | | |
| Sulfuric acid | $H_2SO_4$ | $H_3O^+(aq) + HSO_4^-(aq)$ |
| Hydrogen chloride | HCl | $H_3O^+(aq) + Cl^-(aq)$ |
| Nitric acid | $HNO_3$ | $H_3O^+(aq) + NO_3^-(aq)$ |
| *Soluble Ionic Compounds* | | |
| Sodium chloride | NaCl | $Na^+(aq) + Cl^-(aq)$ |
| Potassium sulfate | $K_2SO_4$ | $2K^+(aq) + SO_4^{2-}(aq)$ |
| Ammonium nitrate | $NH_4NO_3$ | $NH_4^+(aq) + NO_3^-(aq)$ |
| Sodium hydroxide | NaOH | $Na^+(aq) + OH^-(aq)$ |
| Potassium hydroxide | KOH | $K^+(aq) + OH^-(aq)$ |
| Barium hydroxide | $Ba(OH)_2$ | $Ba^{2+}(aq) + 2OH^-(aq)$ |
| Sodium acetate | $NaC_2H_3O_2$ | $Na^+(aq) + C_2H_3O_2^-(aq)$ |
| Silver nitrate | $AgNO_3$ | $Ag^+(aq) + NO_3^-(aq)$ |
| *Weak Electrolytes* | | |
| Ammonia | $NH_3$ | $NH_3(aq)$ |
| Hydrogen fluoride | HF | $HF(aq)$ |
| Hydrogen sulfide | $H_2S$ | $H_2S(aq)$ |
| Phosphoric acid | $H_3PO_4$ | $H_3PO_4(aq)$ |
| Acetic acid | $HC_2H_3O_2$ | $HC_2H_3O_2(aq)$ |
| *Nonelectrolytes* | | |
| Methyl alcohol | $CH_3OH$ | $CH_3OH(aq)$ |
| Ethyl alcohol | $C_2H_5OH$ | $C_2H_5OH(aq)$ |
| Sucrose | $C_{12}H_{22}O_{11}$ | $C_{12}H_{22}O_{11}(aq)$ |
| Hydrogen peroxide | $H_2O_2$ | $H_2O_2(aq)$ |

3. Molecular compounds that are weak electrolytes are shown by the formula of the molecule.

4. Nonelectrolytes are shown by the formula of the molecule.

 **COLLIGATIVE PROPERTIES OF SOLUTIONS**

When we mix commercial antifreeze with water the resulting solution has a freezing point that is lower than pure water. In addition its boiling point is higher than pure water. Salt mixed with ice and water in an ice-cream

Write formulas that represent the chemical particles in the following solutions.

(a) hydrogen peroxide, $H_2O_2$, a nonelectrolyte

(b) nitric acid, $HNO_3$, a strong molecular electrolyte like HCl

(c) ammonium sulfate, $(NH_4)_2SO_4$, an ionic compound

maker lowers the freezing point so that the ice cream mixture freezes. Salt in water also causes the boiling point of the solution to be higher than that of pure water. The presence of a solute in a solvent makes the properties of the solution different from those of the pure solvent. Chemistry in Action 10-3 looks at the effects of solute on freezing point. Freezing point and boiling point are properties of solutions known as colligative properties. **Colligative properties** relate to the concentration of solute particles not to the chemical identity of the solute particles. The term colligative refers to the collective effect of the solute particles. These properties are generally the same no matter what solute is involved; they depend on the concentration not on the identity of the solute particles. This is true for the boiling point as long as the solute is nonvolatile. A nonvolatile solute is one that does not evaporate or boil from the solution. Being nonvolatile is not a prerequisite for freezing point depression.

Pure water freezes at 0°C. A water solution containing 15% by mass of sodium chloride freezes at around −10°C. The solute depresses the freezing point of the solvent. When a solute dissolves in a solvent, the resulting solution typically boils at a higher temperature than the pure solvent and freezes at a lower temperature than the pure solvent. These phenomena are called **boiling point elevation** and **freezing point depression** (see Fig. 10-9). The main component of commercial antifreeze is ethylene glycol. The ethylene glycol lowers the freezing point of the water in the radiator so that it does not freeze in cold weather. The antifreeze also elevates the boiling point so that it does not boil as readily as pure water. That is why commercial antifreeze products are sold as coolants. Some plants and animals have special solutes in the solutions of their cells that protect them from freezing in cold weather.

The change in the freezing point of a solvent is directly proportional to the amount of solute in the solution. If the amount of solute is expressed in terms of molality, $m$, the change in freezing point is given by $\Delta T = K_f m$ where $\Delta T$ is the change in freezing point, $m$ is the molality of the solute, and $K_f$ is a proportionality constant called the **molal freezing point depression constant.** Each solvent has a unique value for this constant. For water the value of $K_f$ is ($1.86°C/1\ m$). This means that one mole of solute per kilogram of water depresses the freezing point by $1.86°C$. To find the freezing point depression for a water solution we simply multiply the constant by the molality of the solute. However, if the solute is a strong electrolyte we need to know the total ionic concentration since the ions act as independent particles. Consequently, the equation works best for solutes that are not strong electrolytes.

**EXAMPLE 10-13**

Calculate the freezing point of a 2.00 $m$ sucrose solution. Since the question involves the freezing point of a water solution we use $\Delta T = K_f m$.

● Solute
● Solvent

(a)

(b)

Boiling point elevation:The presence of solute molecules interferes with the ability of the solvent molecules to evaporate. Consequently, a higher temperature is needed to make the solution boil. The increase in the boiling point of a solvent caused by the addition of a nonvolatile solute is called boiling point elevation.

Freezing point depression:The presence of solute particles interferes with the ability of the solvent molecules to freeze to form crystals. Consequently, a lower temperature is needed to reach the freezing point. The decrease in the freezing point of a solvent caused by the addition of a solute is called freezing point depression. Freezing point depression is used in the making of homemade ice cream. Salt is added to an ice–water mixture to lower the freezing temperature enough to freeze the ice cream mixture.

**FIGURE 10-9**
Boiling point elevation and freezing point depression.

$$\Delta T = (1.86°C/1 \; m)(2.00 \; m) = 3.72°C$$

$\Delta T$ is the freezing point depression or decrease in freezing point. Thus, to find the value of the freezing point we need to subtract $\Delta T$ from the freezing point of pure water.

$$\text{Solution freezing point} = 0°C - 3.72°C = -3.72°C$$

The change in the boiling point of a solvent is directly proportional to the amount of solute in the solution. If the amount of solute is expressed in terms of molality, $m$, the change in boiling point is given by $\Delta T = K_b m$ where $\Delta T$ is the change in boiling point, $m$ is the molality of the solute, and $K_b$ is a proportionality constant called the **molal boiling point elevation constant.** Each solvent has a unique value for this constant. For water the value of $K_b$ is $(0.52°C/1 \; m)$. This means that one mole of solute per kilogram of water elevates the boiling point by $0.52°C$. As is the case for freezing point depression, the boiling point equation works best with

Calculate the freezing point and boiling point of a 2.00 *m* glucose solution?

solutes that are not strong electrolytes but it can be modified to be used with strong electrolyte solutions.

## 10-21 OSMOSIS

When red blood cells are placed in pure water, water enters the cells causing them to swell and eventually burst. On the other hand, when red blood cells are placed in a concentrated salt solution, water migrates out and they shrivel and shrink. These two processes result from the same phenomenon, known as osmosis. **Osmosis** is a phenomenon involving solutions separated by a membrane. The membrane (e.g., blood cell membrane) acts as a barrier between two solutions. It has the property of allowing certain types of molecules to pass through while preventing the passage of other species in solution. Such a membrane is a **semipermeable membrane,** since it is only permeable to selected species.

Semipermeable membranes that allow the passage of water but not solutes are **osmotic membranes.** Imagine a solution separated from a sample of pure water (or another solution of lower concentration of solute) by an osmotic membrane. In such a situation, the water molecules spontaneously penetrate the membrane from both directions but not at equal rates. More water molecules move from the pure solvent side of the membrane to the solution side than move from the solution side to the pure solvent side. Figure 10-10 illustrates this type of osmosis.

**FIGURE 10-10**
Osmosis and osmotic pressure.

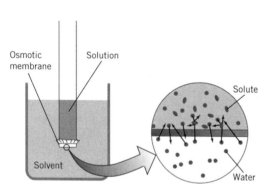

Osmosis: Enlarged view of two solutions separated by a semi-permeable membrane.

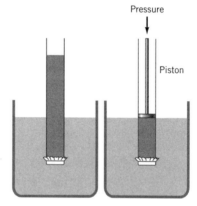

Osmotic pressure: In the osmosis process, solvent molecules pass from the less concentrated solution to the more concentrated solution. The process of osmosis can be reversed by applying pressure to the more concentrated side. The amount of pressure needed just to counteract the osmosis process is called the osmotic pressure. Pressures in excess of the osmotic pressure can cause reverse osmosis.

The migration of water through a semipermeable membrane occurs because of a **concentration gradient.** A gradient exists when the concentrations are different on either side of the membrane. The natural tendency is for the solvent to move from a region where it is present in higher concentration to a region where it is present in lower concentration. The osmotic membrane allows the water but not the solute to migrate. The net result of osmosis is the transfer of water across the membrane to the side having the lower concentration of water. Note that we are referring to the concentration of water not solute. A dilute solution has a relatively low concentration of solute and a relatively high concentration of water. A concentrated solution has a relatively high concentration of solute and a relatively low concentration of water. Thus, water moves from the dilute solution into the concentrated solution of solute.

The mechanism of osmosis is not completely understood, but it appears to occur between aqueous solutions separated by an osmotic membrane no matter what solute is involved. Osmosis is stopped by exerting an opposing pressure or force per unit area on the solution side of the membrane. This is illustrated in Figure 10-10. The applied pressure counteracts the natural flow of solvent through the membrane. Another way to view this situation is that osmosis results in a pressure being exerted on the solution of higher concentration of solute. This pressure is called **osmotic pressure.** Osmotic pressure is counteracted by applying a pressure to the side of the membrane with the higher concentration of solute. The osmotic pressure of a solution is measured by finding the opposing pressure needed to balance the pressure of osmosis. In some solutions the osmotic pressure can be quite great and may reach hundreds of atmospheres. Osmotic pressure is one of the factors involved in the uptake of water by trees and other plants.

Red blood cells are surrounded by osmotic membranes and the blood cells contain various dissolved solutes. When red blood cells are placed in a concentrated salt solution, water moves from the cells to the salt water by osmosis. The cells shrivel and shrink and the process is termed crenation. On the other hand, blood cells swell and burst in fresh water because water enters the cells by osmosis. Inward osmosis that swells and bursts the cells is called hemolysis. In general, water passes through an osmotic membrane from the side having the lower concentration of solute to the side having the higher concentration of solute.

FOCUS
10

## Solutions

Solution: A mixture of two or more chemicals in which the particles intermingle on an atomic, molecular, or ionic level.

Electronegativity: A measure of the tendency of an atom of an element to attract electrons in a bond.

Polar covalent bond: A covalent bond between atoms of different electronegativity.

Polar molecule: A molecule in which there is a net separation of positive and negative charge centers. To be polar a molecule must have one or more polar bonds and a shape that results in a net separation of positive and negative charge centers.

Nonpolar molecule: A molecule in which there is no net separation of charge. One type of nonpolar molecule is a molecule in which the atoms have little or no differences in electronegativity. Another type is a molecule that has polar bonds but a symmetrical shape resulting in no net separation of charge centers.

Like dissolves like means that a polar solvent dissolves polar solutes. A polar solvent like water readily dissolves polar molecular compounds and many ionic compounds. A nonpolar solvent dissolves nonpolar solutes. Nonpolar hydrocarbon solvents readily dissolve waxy, oily, greasy, and fat-like compounds that characteristically have nonpolar molecules.

Saturated Solution: A solution in which dissolved and undissolved solute are in dynamic equilibrium.

Solubility: The amount of a chemical that dissolves in a specific amount of solvent.

Molarity: The number of moles of solute per liter of solution

$$\textbf{molarity} = \frac{\textbf{moles of solute}}{\textbf{liters of solution}} \quad \text{or} \quad M = \frac{n}{V}$$

When a solution is diluted with solvent the molarity of the solute decreases. The dilution equation is

$$V_c M_c = V_d M_d$$

The concentration of a solute can be expressed in terms of percent by mass

$$\textbf{percent component} = \frac{\textbf{mass component}}{\textbf{mass solution}} \, \textbf{100}$$

The concentration of a solute or impurity can be expressed in terms of parts per million or ppm

$$\textbf{ppm} = \frac{\textbf{mass solute}}{\textbf{mass solution}} \, \textbf{10}^6$$

Molality: The number of moles of solute per kilogram of solvent

$$\textbf{molality} = \frac{\textbf{number of moles of solute}}{\textbf{number of kilograms solvent}} \quad \text{or} \quad m = \frac{n}{\textbf{kg solvent}}$$

Electrolytes: Compounds which form solutions that conduct electricity.

Strong electrolytes: Soluble ionic compounds and a few molecular compounds ($HCl$, $HNO_3$, $H_2SO_4$) that dissolve in water to form ions.

Weak electrolytes: Soluble molecular compounds that react slightly with water upon dissolving to form relatively low concentrations of ions.

Nonelectrolytes: Compounds which form solutions that do not conduct electricity. Soluble molecular compounds that do not form ions when dissolved in water.

Freezing point depression: When a solute is dissolved in a solvent the freezing point of the solution is lower than the freezing point of the pure solvent.

The freezing point depression equation is

$$\Delta T = K_f m$$   $\Delta T$ is the change in freezing point, $m$ is the molality of the solute and $K_f$ is the molal freezing point depression constant

For water $K_f$ is (1.86 °C/1 $m$)

Boiling point elevation—When a (nonvolatile) solute is dissolved in a solvent the boiling point of the solution is higher than the boiling point of the pure solvent.

The boiling point elevation equation is

$$\Delta T = K_b m$$   $\Delta T$ is the change in boiling point, $m$ is the molality of the solute and $K_b$ is the molal boiling point elevation constant.

For water $K_b$ is (0.52 °C/1 $m$)

You need: four aspirin tablets, three glasses, two coffee filters, cold water, and boiling hot water. Put one filter in the other to double the thickness. Instead of filters you can use three or four napkins. To make a filter from the napkins stack them, place your finger in the middle of the stack, and fold them to form a cone. Use some sticky tape to hold the napkins in the cone shape.

1. Crush two aspirin tablets and place them in a glass. Crush the other two tablets and place them in the second glass.
2. Add one-fourth of a cup of cold water to one of the glasses. Stir the mixture and record your observations.
3. Add one-fourth of a cup of boiling hot water to the second glass. Stir and record your observations. While the solution is still hot quickly continue with Step 4.
4. Hold the filter over the third glass and pour the contents of the second glass into the filter. Collect the hot liquid that passes through the filter in the third glass.
5. Let the filtered liquid cool for about 30 minutes and observe it. If you do not see crystals in the glass add a crumb of crushed aspirin as a seed crystal and observe. You can speed the cooling process by immersing the glass in a bowl or larger glass that has some ice water in it.

Describe the appearance of the aspirin crystals. You can reach in and touch them. What do you conclude about the solubility of aspirin in cold and hot water?

You need: a 9-volt battery, salt, white vinegar, medicinal hydrogen peroxide solution, tape, some aluminum foil, and a shallow container such as a very small glass, or the bottom cut from a small plastic glass.

1. Cut two small pieces of aluminum foil measuring about 4 cm long and 0.5 cm wide. Use some tape to attach the aluminum to the electrodes of the battery. Bend the strips so that they extend upwards from the top of the battery. When you use the apparatus look very closely for any activity at the surface of the aluminum electrodes.

2. Be sure that the shallow container is clean and dry. Put some hydrogen peroxide solution in the container. Carefully immerse the aluminum electrodes into the solution and record your observations.

3. Rinse the container with some vinegar and put a sample of vinegar in it. Carefully immerse the aluminum electrodes into the vinegar and record your observations.

4. Rinse the container with some water and put a sample of tap water in it. Carefully immerse the aluminum electrodes into the tap water and record your observations.

5. Add a sample of salt about the size of an aspirin tablet to the water in the container. Gently stir to dissolve the salt. Carefully immerse the aluminum electrodes into the salt solution and record your observations.

Describe the similarities and differences in the conductivities of the solutions. Explain why the solutions had the conductivities you observed.

This exercise is an appealing investigation of a property of solutions. You need: two Zip-Lock bags, milk, sugar, ice, vanilla flavoring, salt and a four-cup bowl, pan, or container. You can use soy milk in place of milk or use orange juice as the only ingredient.

1. Put ice in the bowl along with some water to make an ice–water mixture. Add more ice as it melts.

2. Place one heaping teaspoon of sugar, one-third of a cup of milk and five drops of vanilla in a Zip-Lock bag. Press out the air and seal the bag. Place this bag in a second bag, press out the air, and seal it.

3. Immerse the bags into the ice–water mixture. Make sure they are below the surface and leave for five minutes. After five minutes feel the contents of the bags and record your observations.

4. Add one-fourth of a cup of salt to the ice–water mixture in the bowl and stir the mixture. Add more ice as it melts.

5. Immerse the Zip-Lock bags into the ice–water–salt mixture. Make sure the bags are sealed and below the surface. Stir the contents of the bowl continually. Feel the contents of the bags every minute or so and let them remain in the ice–water–salt mixture for 5 to 10 minutes.

6. Remove the bags from the bowl and rinse the outside surface with water. Dispose of the contents of the inner bag in any way that seems appropriate to the occasion. Record your observations. Repeat this exercise as often as needed. Experiment with different sample sizes and various ingredients.

# QUESTIONS

## Section 10-1

1. Define the following terms.

   (a) solution

   (b) solvent

   (c) solute

## Section 10-2

2. Define electronegativity.

3. List the four most electronegative elements in descending order. Which group contains the elements of lowest electronegativity?

## Sections 10-3 and 10-4

4. What is a polar covalent bond?

5. Describe a polar molecule and give an example.

6. There are two different types or categories of nonpolar molecules. Describe each of the types and give examples.

7. Are molecules of the following compounds polar or nonpolar?

   (a) $HF$    (b) $CS_2$   (c) $CH_4$

   (d) $NCl_3$   (e) $CH_3OH$

## Sections 10-5 and 10-6

8. Explain how polar molecules attract one another. What is a hydrogen bond?

9. Sketch two water molecules and show what portions of the molecules can attract when they are in close proximity.

10. Explain how nonpolar molecules attract one another.

11. Sketch two nonpolar iodine, $I_2$, molecules and show how they can attract one another.

## Section 10-7

12. Describe what happens, in terms of the interaction between water molecules and ions, when the ionic compound potassium chloride, $KCl$, dissolves in water.

13. Why does a solid solute usually dissolve faster when it is in the form of tiny crystals rather than larger crystals? Why do most solid solutes dissolve faster when the solute and solvent are stirred together?

14. Sodium dichromate, $Na_2Cr_2O_7$, is an orange-yellow solid. The color is due to the presence of the dichromate ion, which has an orangy-yellow color. When sodium dichromate is dissolved in water, the entire solution takes on an orangy-yellow color. Give an explanation for this in terms of the dissolving process.

15. Most gases are less soluble in hot water and more soluble in cold water. Natural waters contain dissolved oxygen and nitrogen from the atmosphere.

   (a) When tap water is heated, small gas bubbles form. Give an explanation for this observation.

   (b) When natural waters become overly heated by power plant or industrial wastewaters, fish, and other aquatic life cannot survive. Give an explanation for this.

16. Most gases are more soluble at higher pressures and less soluble at lower pressures. Carbonated water is water saturated with carbon dioxide under pressure. Why does some of the carbon dioxide gas bubble off when a bottle of carbonated beverage is opened?

## Section 10-8

17. What does "like dissolves like" mean?

18. Explain why water is such a good solvent for ionic compounds.

19. Suppose you were asked to find a good solvent for a compound that has nonpolar molecules. Do you think water would be the best solvent? Why?

## Sections 10-9 and 10-10

20. What is meant by the term dynamic equilibrium?

21. Define the following terms.

    (a) saturated solution   (b) solubility

    (c) supersaturated solution

    (d) unsaturated solution

22. A bottle on the laboratory shelf is marked "saturated NaCl." The bottle contains a clear solution on top of some white crystals. Describe the nature of this solution in terms of dissolving, recrystallization, and dynamic equilibrium.

## Section 10-11

23. Give a definition for molarity. How is the symbol $M$ interpreted?

24. A solution in the laboratory has a label marked 2.00 $M$ KNO$_3$. Explain its meaning.

25. *A sample of wine vinegar is found to contain 18.9 g of acetic acid, HC$_2$H$_3$O$_2$ in 341 mL of the solution. Calculate the molarity of acetic acid in the vinegar.

26. A 325-mL sample of beer is found to contain 9.75 grams of ethyl alcohol, C$_2$H$_6$O. Calculate the molarity of ethyl alcohol in the beer.

27. *You analyze a cup of coffee and find that it contains 88 mg of caffeine, C$_8$H$_{10}$N$_4$O$_2$. What is the molarity of caffeine in your coffee if one cup contains 237 mL of coffee?

28. A 49.5-mL sample of medicinal hydrogen peroxide solution is found to contain 1.49 grams of H$_2$O$_2$. What is the molarity of H$_2$O$_2$ in the solution?

29. Following a dose of an antibiotic, $1.18 \times 10^6$ units of penicillin G (C$_{16}$H$_{18}$O$_4$N$_2$S) are recovered from 165 mL of the patient's urine. What is the molarity of penicillin G in the urine if $10^6$ units of penicillin equals 0.600 g?

30. A sodium chloride solution is prepared by adding 1.50 moles of NaCl to 1.00 L of water. Explain why this solution is not 1.50 $M$ in NaCl.

31. *A solution is prepared by dissolving 20.5 mL of glacial acetic acid, HC$_2$H$_3$O$_2$, in enough water to produce 115 mL of solution. If the density of glacial acetic acid is 1.049 g/mL, determine the molarity of the acetic acid solution. Glacial acetic acid is a common name for pure acetic acid.

32. A solution is prepared by dissolving 28.4 mL of ethyl alcohol, C$_2$H$_6$O, in enough water to produce 0.275 L of solution. If the density of pure ethyl alcohol is 0.790 g/mL, determine the molarity of the ethyl alcohol solution.

33. *A solution is prepared by dissolving 19.5 g of Na$_2$SO$_4$ in enough water to form 785 mL of solution. When the solute dissolves it forms ions.

$$Na_2SO_4(s) \longrightarrow 2Na^+(aq) + SO_4^{2-}(aq)$$

    (a) What is the molarity of the SO$_4^{2-}$ in the solution?

    (b) What is the molarity of the Na$^+$ in the solution?

34. Iron(III) sulfate, Fe$_2$(SO$_4$)$_3$, dissolves in water and forms ions according to the equation:

$$Fe_2(SO_4)_3(s) \longrightarrow 2Fe^{3+}(aq) + 3SO_4^{2-}(aq)$$

If 36.2 g of iron(III) sulfate are dissolved in 350 mL of solution, find

    (a) the molarity of the sulfate ion, SO$_4^{2-}$, in solution

    (b) the molarity of the iron(III) ion, Fe$^{3+}$, in solution

## Sections 10-12 and 10-13

35. *How many milliliters of a 0.215 $M$ potassium nitrate solution are needed to provide 0.0323 mole of dissolved KNO$_3$?

36. How many milliliters of a 0.789 $M$ acetic acid solution are needed to provide 0.475 mole of HC$_2$H$_3$O$_2$?

37. *A solution is 0.363 $M$ in H$_2$O$_2$. How many milliliters of this solution will contain 1.75 g of H$_2$O$_2$?

38. A MgCl$_2$ solution is 1.50 $M$. How many milliliters of this solution will contain 16.3 g of MgCl$_2$?

39. What is a standard solution? Describe how one is prepared.

40. *Determine the number of grams of solute needed to prepare the following solutions.

    (a) 300 mL of a 0.925 $M$ KOH solution

    (b) 520 mL of a 0.487 $M$ NaNO$_3$ solution

    (c) 1.17 L of a 0.125 $M$ C$_2$H$_5$OH (ethyl alcohol) solution

(d) 60 mL of a 0.180 $M$ $C_6H_{12}O_6$ (glucose) solution

(e) 725.0 mL of a 2.12 $M$ $Li_2CO_3$ solution

41. Determine the number of grams of solute needed to prepare the following solutions.

(a) 235 mL of a $4.95 \times 10^{-2}$ $M$ $HC_2H_3O_2$ (acetic acid) solution

(b) 8.30 L of a 0.396 $M$ $FeCl_3$ solution

(c) 740 mL of a 0.200 $M$ NaOH solution

(d) 3.15 L of a 0.500 $M$ $C_2H_5OH$ (ethanol) solution

(e) 910 mL of a $6.75 \times 10^{-3}$ $M$ $K_2CO_3$ solution

42. *Plasma is the fluid portion of blood. The concentration of acetylsalicylic acid (aspirin), $C_9H_8O_4$, in your plasma is found to be $2.99 \times 10^{-4}$ $M$ after you take two aspirin. If the volume of your plasma is 5.85 L, how many grams of aspirin are in your blood?

43. A person can just taste the sweetness in $2.0 \times 10^{-2}$ $M$ sucrose, or table sugar, $C_{12}H_{22}O_{11}$. How many grams of sucrose are contained in a cup of tea (250 mL) which is $2.0 \times 10^{-2}$ $M$ in sucrose?

44. *A cleaning solution is 9.75 $M$ in ammonia, $NH_3$. How many grams of ammonia are contained in 1.00 quarts of the cleaning solution? (1 qt = 0.946 L)

## Section 10-14

45. What proportionality exists between the volume and molarity of a solution? What happens to the molarity of a water solution when the volume is increased by diluting the solution with water?

46. *To what volume should 25.0 mL of a 1.37 $M$ KI solution be diluted so that the solution is 0.375 $M$?

47. To what volume should 150 mL of 5.0 $M$ acetic acid be diluted to give a 0.100 $M$ solution?

48. *You need 0.250 $M$ hydrochloric acid for an experiment but all you can find on the shelf is 6.00 $M$ hydrochloric acid. If you need 100 mL of 0.250 $M$ acid, how many milliliters of the 6.00 $M$ acid will you need to prepare it?

49. How many milliliters of 12 $M$ $NH_3$ are needed to prepare 75 mL of 1.0 $M$ $NH_3$ by dilution?

50. *Potassium chloride, KCl, is used to treat people with hypokalemia (low blood potassium levels). Potassium Chloride Injection, USP, is a 3.00 $M$ solution of KCl. You must dilute this to 0.075 $M$ before giving it to a patient. To prepare 50 mL of the dilute solution, what volume of the more concentrated KCl solution do you have to use?

51. How many milliliters of 6.00 $M$ hydrochloric acid are needed to prepare 10.0 L of 0.100 $M$ hydrochloric acid by dilution?

52. *How many milliliters of a 6.00 $M$ sulfuric acid solution are needed to prepare 1.2 L of a 0.050 $M$ sulfuric acid solution by dilution?

53. How many milliliters of 12.00 $M$ nitric acid are needed to prepare 220 mL of 0.150 $M$ nitric acid solution by dilution?

54. *An $FeCl_3$ solution is 2.50 $M$. Find the molarity of the following.

(a) a 10.0-mL sample diluted with water to a volume of 40.0 mL

(b) a 75.0-mL sample allowed to evaporate to a volume of 40.0 mL

(c) A 1.25-L sample in which an additional 25.0 g of $FeCl_3$ is dissolved (assume no change in volume)

55. An NaCl solution is 2.00 $M$. Find the molarity of the following.

(a) a 10.0-mL sample diluted with water to a volume of 30.0 mL

(b) a 100-mL sample allowed to evaporate to a volume of 60.0 mL

(c) A 1.10 L sample in which an additional 10.0 g of NaCl is dissolved (assume no change in volume)

## Section 10-16

56. *A cup of tea is 12.5% by mass sugar, $C_{12}H_{22}O_{11}$. If a cup of tea contains 300 g of solution, how many grams of sugar did you add?

57. A solution is prepared by dissolving 0.727 g of potassium chloride in 90.0 g of water. What is the percent by mass of KCl in the solution?

58. *A 3.0% by mass solution of hydrogen peroxide, $H_2O_2$, is sold in drugstores as a mild antiseptic. If a bottle of hydrogen peroxide contains 448 g of solution, how many grams of $H_2O_2$ does it contain?

59. A physiological saline solution contains 0.92% NaCl by mass. How many grams of NaCl are in 320 grams of this solution?

60. *Ocean water contains 1.31 g of magnesium ions, $Mg^{2+}$, per kilogram of ocean water. Calculate the parts per million concentration of $Mg^{2+}$ ions in ocean water.

61. Ocean water contains 0.401 g of calcium ions, $Ca^{2+}$,

per kilogram of ocean water. Calculate the parts per million concentration of $Ca^{2+}$ ions in ocean water.

62. *Ocean water contains $1.39 \times 10^{-4}$ ppm silver ions, $Ag^+$. If the volume of the ocean is $1.5 \times 10^9$ km$^3$, and the density is 1.005 g/cm$^3$, calculate how many grams of $Ag^+$ are contained in the ocean.

63. Bromine can be extracted from ocean water by reacting bromide ions, $Br^-$, with chlorine.

$$Cl_2(aq) + 2Br^-(aq) \longrightarrow Br_2(\ell) + 2Cl^-(aq)$$

If ocean water contains $6.8 \times 10^{-2}$ g of bromide ions, $Br^-$, per kilogram, how many kilograms of water must be processed to obtain 1 metric ton, 1000 kg, of bromine, $Br_2$?

64. *200 ppm $CO_2$ dissolved in water is lethal to fish. If a 2.0-L sample of Mississippi River water contains 0.68 g of dissolved $CO_2$, is the water likely to be lethal to fish?

65. The Environmental Protection Agency standards for drinking water allow an MCL of 0.01 ppm $Cd^{2+}$. If a 1.25-L sample of drinking water contains 0.014 mg of $Cd^{2+}$, does the water meet or exceed the maximum allowable concentration?

66. *Medicinal hydrogen peroxide solution (3.0%) contains 3.0 g of $H_2O_2$ in 97 g of water. What is the molality of hydrogen peroxide in the solution.

67. A solution contains 35.0 g of glucose in 100 g of water. What is the molality of glucose in the solution?

## Section 10-15

68. *How many milliliters of 1.00 $M$ hydrochloric acid are needed to dissolve 15.00 g of aluminum metal in acid, if the reaction is

$$2Al(s) + 6HCl(aq) \longrightarrow 2AlCl_3(aq) + 3H_2(g)$$

69. How many grams of calcium metal can react with 118 mL of 0.375 $M$ hydrochloric acid if the reaction is

$$Ca(s) + 2HCl(aq) \longrightarrow CaCl_2(aq) + H_2(g)$$

70. *How many liters of carbon dioxide gas measured at STP can form when 40.0 mL of 12.00 $M$ HCl reacts with calcium carbonate by the reaction:

$$CaCO_3(s) + 2HCl(aq) \longrightarrow$$
$$CaCl_2(aq) + CO_2(g) + H_2O(\ell)$$

71. How many liters of carbon dioxide gas measured at 21°C and 1.00 atm can form when 225 mL of 1.00 $M$ nitric acid react with calcium carbonate by the reaction:

$$CaCO_3(s) + 2HNO_3(aq) \longrightarrow$$
$$Ca(NO_3)_2(aq) + CO_2(g) + H_2O(\ell)$$

## Sections 10-17 and 10-18

72. What are electrolytes and nonelectrolytes?

73. What is meant by the statement that a solution contains ions?

74. Is it possible for a solution to contain only positive ions and no negative ions? Why?

75. What simple test can be used to determine whether or not a solution contains ions?

76. Why is it dangerous to handle electrical equipment while you are immersed in water or standing in water?

77. What is the difference between strong electrolytes and weak electrolytes?

78. Why would you expect a solution of a soluble ionic compound to conduct electricity?

## Section 10-19

79. *Give the formulas of the ions that are present in aqueous solutions of the following ionic compounds.
    (a) sodium sulfate, $Na_2SO_4$
    (b) potassium dichromate, $K_2Cr_2O_7$
    (c) sodium fluoride, NaF
    (d) lithium nitrate, $LiNO_3$
    (e) ammonium chloride, $NH_4Cl$
    (f) silver nitrate, $AgNO_3$

80. Give the formulas of the ions that are present in aqueous solutions of the following ionic compounds.
    (a) ammonium sulfate, $(NH_4)_2SO_4$
    (b) potassium permanganate, $KMnO_4$
    (c) lithium carbonate, $Li_2CO_3$
    (d) calcium nitrate, $Ca(NO_3)_2$
    (e) sodium chromate, $Na_2CrO_4$
    (f) sodium phosphate, $Na_3PO_4$

81. *Give the formulas of the major species that are found in solutions of the following compounds.
    (a) nitric acid, $HNO_3$ (strong electrolyte-like HCl)
    (b) ethyl alcohol, $C_2H_5OH$ (nonelectrolyte)
    (c) hydrogen sulfide, $H_2S$ (weak electrolyte)
    (d) sucrose, $C_{12}H_{22}O_{11}$ (nonelectrolyte)

(e)  hydrogen fluoride, HF (weak electrolyte)

(f)  copper(II) sulfate, $CuSO_4$ (strong electrolyte)

(g)  hydrogen peroxide, $H_2O_2$ (nonelectrolyte)

(h)  calcium chloride, $CaCl_2$ (strong electrolyte)

(i)  ammonia, $NH_3$ (weak electrolyte)

82. Give the formulas of the major species that are found in solutions of the following compounds.

(a)  glucose, $C_6H_{12}O_6$ (nonelectrolyte)

(b)  potassium chloride, KCl (strong electrolyte)

(c)  methyl alcohol, $CH_3OH$ (nonelectrolyte)

(d)  magnesium sulfate, $MgSO_4$ (strong electrolyte)

(e)  phosphoric acid, $H_3PO_4$ (weak electrolyte)

(f)  ethylene glycol, $C_2H_6O_2$ (nonelectrolyte)

(g)  sodium hydrogen carbonate, $NaHCO_3$ (strong electrolyte)

(h)  silver nitrate, $AgNO_3$ (strong electrolyte)

(i)  oxalic acid, $H_2C_2O_4$ (weak electrolyte)

## Sections 10-20 and 10-21

83. What are colligative properties of solutions?

84. How are the freezing and boiling points of a solvent affected by the presence of a solute?

85. Ethylene glycol is added to the radiator system of an automobile as an antifreeze and coolant. How does ethylene glycol function as an antifreeze and coolant?

86. To make homemade ice cream it is necessary to cool the mixture below 0°C. How does an ice and salt mixture provide the required cooling?

87. *Calculate the freezing point of a 3.0% hydrogen peroxide solution. See Question 66.

88. Calculate the freezing point of a solution that contains 50.0 g of sucrose dissolved in 200 g of water?

89. *Calculate the boiling point of a solution that contains 50.0 g of sucrose dissolved in 200 g of water?

90. The freezing point of a glucose solution is −3.2°C. In this solution how many grams of glucose are dissolved in each 100 g of water?

91. Describe the phenomenon of osmosis.

92. Cellular membranes in plants and animals are osmotic membranes. Explain the following in terms of osmosis.

(a)  Sliced fruit will form its own juice when sugar is added.

(b)  Drinking salt water can cause dehydration of the body and sometimes death.

(c)  Solutions injected into the body intravenously must have the same ionic concentrations as blood; they are called isotonic solutions.

(d)  Prunes swell when they are placed in water.

(e)  Cucumbers shrink when they are placed in a concentrated salt solution to make pickles.

C H A P T E R

*11*

# CHEMICAL

# REACTIONS

# REVEALED

## 11-1 EVIDENCE FOR REACTIONS

Chemical processes are dynamic. The kinetic molecular view of matter pictures atoms, molecules, and ions in motion. Chemical reactions are fundamental to chemistry. Reactions involve chemical particles moving about and colliding with one another. Collisions between particles result in the breaking of chemical bonds and formation of new bonds. Chemical reactions are active processes in which chemicals transform. Some transformations are noticeably dramatic, for instance, the explosion of nitroglycerine in dynamite is a fast and furious reaction. An **explosion** is a very fast chemical reaction that releases energy, some of which takes the form of heat and light. In an explosion, chemical products expand rapidly at speeds of many thousands of meters per second. An explosion scatters bits of solid materials and produces a shock wave in the surrounding air. This is why a loud sound accompanies an explosion. In contrast, some reactions are slow but relentless. Iron in the environment slowly rusts; it reacts with oxygen and transforms into a compound, the compound from which humans extracted the iron in the first place. It can take years for an iron sample to rust completely. See Section 13-8 for more about rust.

How can a chemical reaction be witnessed? When charcoal burns in air we notice that it glows, releases heat, and seems to ultimately disappear. The white ash remaining contains noncombustible compounds formed

A C T I V I T Y

11-1

The equation for the explosion of nitroglycerine is

$$4C_3H_5(NO_3)_3(\ell) \longrightarrow 6N_2(g) + 10H_2O(g) + 12CO_2(g) + O_2(g)$$

The instant a 100-g sample of nitroglycerine explodes assume the gases formed have a temperature of 500°C and occupy a volume of 0.250 L. What is the pressure of this mixture of gases under these conditions?

**365**

from impurities in the charcoal. From a chemical view a reaction involves atoms changing associations. Atoms cannot be seen but they can be symbolically represented in equations. The following equation represents the fact that charcoal burns in air to form carbon dioxide gas:

$$C + O_2 \longrightarrow CO_2 + \text{energy}$$

We cannot see atoms react but we can witness reactions by noting physical changes in the chemicals as they are transformed. Some possible changes are:

a gas forms or disappears

a solid forms or disappears

a liquid forms or disappears

a color change occurs

energy is released as heat and sometimes as light (exothermic reactions)

energy is absorbed as heat from the surroundings (endothermic reactions)

Hard-water scale can be removed from a teakettle by pouring vinegar in the kettle and allowing it to sit overnight. In the morning the kettle can be rinsed and scrubbed. Write an equation for the reaction of acetic acid with solid calcium carbonate to give aqueous calcium acetate, carbon dioxide gas, and water.

It is important to realize that these kinds of changes can occur in processes that are not chemical reactions. Can you give some examples of such changes in which no chemical reaction occurs? Physical changes are suggestions that a chemical reaction has occurred but not proof.

Consider a few examples of reactions in which changes do occur. When magnesium metal reacts with oxygen in air the metal disappears and a white, powdery solid is produced. The reaction is accompanied by the noticeable release of heat and light. When iron metal with a metallic sheen rusts, the solid metal is transformed into a reddish (rusty) solid. When we treat hard-water scale in a teakettle (solid calcium carbonate) with vinegar, the solid slowly disappears and bubbles of gas form. When nitroglycerine explodes the liquid chemical rapidly disappears and gases form along with a very noticeable release of energy.

## 11-2 RATES OF REACTIONS

When the reactants involved in a reaction come into contact, a reaction may occur spontaneously. In many cases the reaction has to be initiated with some source of energy. We need to heat the reactants to encourage the reaction. Charcoal, for example, has to be heated before it reacts spontaneously with oxygen in air. Some reactions are fast and some are slow. In airbags, used as safety devices in automobiles, a fast chemical reaction is required but one that is only initiated by an impact. An airbag contains the unique chemical called sodium azide, $NaN_3$, as a solid. Upon impact an electrical source initiates its decomposition to form sodium and nitrogen gas.

$$2NaN_3 \longrightarrow 2Na + 3N_2$$

This reaction occurs very rapidly and the nitrogen gas expands to fill the bag within 0.04 second. The process only works because the reaction occurs at a very fast rate.

The speed at which a reaction takes place can be expressed as the **rate of the reaction.** The rate of a reaction depends on several factors, especially on the exact way in which the reactants interact to give products. Since a reaction involves the breaking and forming of chemical bonds, contact between the species making up the reactants is quite important. That is, for the reaction to proceed, the species making up the reactants must come in contact through collisions. This idea is called the **collision theory** of chemical reactions. According to the theory, the chemical particles of the reactants must collide with sufficient energy to break chemical bonds. Collisions can produce intermediate particles in which old bonds break and new bonds form ultimately giving the products of the reaction.

Imagine a chemical reaction starting as the reactants mix (see Fig. 11-1). The atoms, molecules, or ions that compose the reactants intermingle and collide with one another. Collisions are very important in chemical reactions, since the contact of colliding species provides a way in which bonds can be broken and new bonds formed. Collisions alone do not cause a reaction. A specific set of species must collide with sufficient energy and proper spatial orientation for a reaction to occur. Nevertheless, with many collisions, enough are successful to ultimately give the products of the reaction. How fast the process occurs is expressed as the rate of the reaction. For most reactions, the rate is measured by observing how fast a reactant is used or how fast a product is formed. Rates are expressed as grams per unit time or moles per unit time. Time units are normally seconds or minutes but in some reactions they are hours or days.

## 11-3  FACTORS AFFECTING REACTION RATES

We can illustrate some factors that affect the rate of a reaction using a familiar example. The burning of wood includes the reaction of the carbohydrates (cellulose) of wood with the oxygen in the air. As it begins to burn and releases heat, the wood becomes hotter, and it burns at a faster rate. The rate of the reaction increases with temperature. Another way to get the wood to burn faster is to break it up into smaller pieces. The state of subdivision of the wood increases the reaction rate. If we limit the air supply, the wood does not burn as quickly and may just smolder. Lowering the concentration of oxygen decreases the reaction rate. On the other hand, if we fan the fire, increasing the air supply, it burns more rapidly.

Chemical Reactions

Lower temperature

Higher temperature

*(a)* An increase in temperature increases the speed of the particles. This causes more frequent and more successful collisions, thereby increasing the rate of the reaction.

Larger pieces of a reactant

Many smaller pieces of a reactant

*(b)* An increase in the state of subdivision provides for more contact between reactants and thus more frequent collisions. This increases the rate of the reaction.

Lower concentration

Higher concentration

*(c)* An increase in the concentration increases the number of particles in a given volume. This increases the frequency of collisions, and thus the rate of the reaction increases.

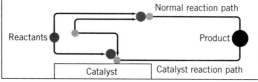

Reactants

Normal reaction path

Product

Catalyst    Catalyst reaction path

*(d)* A catalyst provides an alternate path for a reaction, which allows the reaction to occur at a faster rate than it does when following the uncatalyzed path.

**FIGURE 11-1**
Factors affecting reaction rates.

Increasing the concentration of oxygen increases the reaction rate. Using Figure 11-1 we consider in more detail temperature, subdivision, and concentration. The effect of catalysts on reactions is discussed in Section 11-4.

**Temperature**

An increase in the temperature of a reaction system always causes the chemical particles to move more rapidly. Thus, heating the reactants allows more frequent collisions and an increased reaction rate. More important, heating produces more high-energy particles, which react more readily. In other words, collisions of these high-energy particles are more likely to form products, and the rate of the reaction increases accordingly. See Figure 11-1 (a).

As an example consider the reaction of hydrogen gas and nitrogen gas to form ammonia gas. For a room temperature mixture of hydrogen gas and nitrogen gas the rate of the reaction is so slow that no detectable reaction occurs. Even with an iron catalyst the reaction at room temperature is very slow. We need to heat the reaction system to about 425°C so the rate is high enough to form significant amounts of ammonia in a reasonable time.

$$N_2(g) + 3H_2(g) \xrightarrow[425°C]{Fe} 2NH_3(g)$$

## Subdivision of Reactants

Starch is the main component of wheat grain. Oxygen in the air reacts spontaneously with starch, but the rate of the reaction of oxygen with the individual grains of wheat is so low that it is not noticed. On the other hand, there have been cases in which empty grain elevators filled with tiny pieces of finely divided starch have exploded violently, destroying the buildings. The rate of the reaction of starch with oxygen increases dramatically when the starch is present as very tiny pieces called grain dust.

When solids and liquids are involved in a reaction, the subdivision of the materials influences the rate. This makes sense if we realize that tinier pieces have greater **surface area** so they provide much greater contact of the reactants and more collisions. To increase the rate of a reaction divide the reactants into smaller parts and mix them to increase contact. Solids can be broken into pieces or powdered and liquids can be sprayed to form tiny droplets called aerosols. See Figure 11-1(b).

Objective: Observe the effect of temperature on a reaction rate.

You need: two Zip-loc bags, a measuring cup, two cups (or beakers), white vinegar, two Tums tablets, cold water, and boiling hot water.

1. To each Zip-loc bag add one-fourth of a cup of vinegar but do not seal the bags.

2. Fill one of the cups three-quarters full of cold water and the other cup three-quarters full of boiling hot water.

3. Place a Tums tablet in each bag, seal the bags then place one bag in the cold water and one bag in the hot water.

4. Observe the contents of the bags as they sit in the water for at least five minutes. Every few minutes you can lift a bag from the water to observe it. Feel the contents and then put it back in. Record your observations.

Objective: Observe the effect of subdivision on a reaction rate.

The main chemical component of wood and cotton is cellulose. You need: a short length of pencil wood with the paint removed and a small fluffy wad of cotton. Stretch the cotton to make a loose, fluffy wad. You can get a small wad of cotton from a cotton swab (Q-tip). Strike a match and hold the pencil wood in the flame. Place the cotton wad on a flat surface and ignite it with a match. Record your observations.

## Concentration

Acetylene, $C_2H_2$, is a hydrocarbon fuel used in oxyacetylene welding torches. Acetylene gas burns readily in air, giving off heat.

$$2C_2H_2(g) + 5O_2(g) \longrightarrow 4CO_2(g) + 2H_2O(g) + energy$$

Mixing acetylene gas and pure oxygen gas increases the rate of the reaction in an oxyacetylene welding torch. The rate increases so much that the flame of the torch produces a temperature high enough to melt metals. The rate of reaction in the torch increases because the concentration of oxygen is increased. Concentration refers to the number of particles in a given volume. Air is only about 21% oxygen, so pure oxygen is more concentrated in oxygen than air. When the concentration increases, the frequency of collisions is increased and the reaction speeds up. In other words, higher concentrations of reactants result in more collisions.

Generally, the rate of a reaction involving gases or chemicals dissolved in water is changed by changing the number of reactant particles in a given volume, that is, by changing the concentration of one or more reactants. Increasing the concentration increases the reaction rate, and decreasing the concentration decreases the reaction rate. See Figure 11-1(c). For species dissolved in aqueous solution, the concentration or molarity increases by dissolving more of the species. For gaseous reactants, the concentration is increased by using a purer form of the gas or by compressing

Objective: Observe the effect of changes in concentration on reaction rate.

You need: a measuring cup with metric measure, three small glasses, white vinegar, baking soda, a teaspoon, and water. Assume that one teaspoon is approximately 5 mL.

1. Add two teaspoons of vinegar to the measuring cup and dilute it with water to give a total volume of 100 mL. Stir to mix and pour it into one of the glasses.

2. Rinse the measuring cup. Add two teaspoons of the newly prepared vinegar–water solution to the measuring cup and dilute it with water to 100 mL. Stir to mix and pour the solution into an empty glass. Remove two teaspoons of this solution to make it the same volume as the first solution.

3. Measure 90 mL of pure white vinegar in the measuring cup and pour it into an empty glass.

4. Position the glasses next to one another and put one level teaspoon of baking soda into each of the three cups of vinegar solution. Record your observations.

Assume that a teaspoon has a volume of 5 mL and that the pure vinegar is 0.83 *M* $HC_2H_3O_2$. Calculate the molarity of acetic acid in each of the two diluted vinegar solutions before the baking soda was added.

the gas. Compressing a gas results in more molecules per unit volume. For instance, the industrial production of ethyl alcohol, $C_2H_5OH$, involves the reaction of ethylene, $C_2H_4$, and water. To raise the rate of the reaction to a productive level, the reaction is carried out in a closed system at 400°C and very high pressures.

$$C_2H_4(g) + H_2O(g) \xrightarrow[\text{pressure}]{400°C} C_2H_5OH(g)$$

---

### Controlled Explosions in Rocket Engines

The launching of rockets and space shuttles requires large amounts of energy. The energy comes from fast chemical reactions. Rockets that propel space shuttles into space are called booster rockets; they use solid chemicals that react and quickly provide large amounts of energy. Solid booster rockets can contain powdered aluminum as a fuel and solid ammonium perchlorate as an oxidizer. The reaction produces temperatures of around 5700°C and products that are expelled from the rocket nozzle with great velocities to give the lifting thrust. The equation for the reaction taking place in such rockets is

$$3NH_4ClO_4(s) + 3Al(s) \longrightarrow AlCl_3(s) + Al_2O_3(s) + 6H_2O(g) + 3NO(g) + \text{energy}$$

A space shuttle craft has its own rocket engines that use different chemicals than used in booster rockets. The shuttle carries tanks of liquid hydrogen and liquid oxygen. These chemicals are fed to the rocket engines and ignited to form water. The hydrogen is the fuel and the oxygen is the oxidizer. This very exothermic reaction provides energy for the shuttle engines.

---

### 11-4  CATALYSTS

Catalysts are very important in controlling the rates of reactions. Many industrial and biological processes need catalysts for efficiency; the reactions are too slow without them. Catalysts are fascinating chemicals. The reaction between hydrogen and nitrogen gas to form ammonia does not occur at a significant rate even at high pressures and high temperatures. However, if some finely divided iron is added to the reaction vessel, the rate of the reaction increases dramatically and the reaction becomes productive. The iron is not a reactant and does not chemically change in the reaction. It is a catalyst. Chemistry in Action 11-1 investigates the effect of a variety of catalysts on the decomposition of hydrogen peroxide in water.

A **catalyst** is a chemical that increases the speed or rate of a chemical reaction without undergoing permanent chemical change. The $H_2O_2$ in an aqueous solution of hydrogen peroxide does not noticeably decompose at room temperature. A small amount of manganese(IV) oxide, $MnO_2$, catalyst causes the $H_2O_2$ to decompose rapidly at room tempera-

**ACTIVITY 11-6**

Some rocket engines on space shuttles use the reaction of hydrogen and oxygen as a source of energy. Write an equation for the reaction of hydrogen and oxygen to give water. Include energy on the side of the equation to show that the reaction is exothermic.

ture. Since the catalyst is not a reactant or product we show it above the arrow in the equation.

$$2H_2O_2(aq) \xrightarrow{\text{MnO}_2} 2H_2O(\ell) + O_2(g)$$

If you place some medicinal hydrogen peroxide disinfectant on a cut, a different catalyst, called an enzyme, present in blood and tissue, causes the rapid decomposition of the $H_2O_2$. The oxygen gas formed causes the bubbling and fizzing.

Specific catalysts are found to catalyze specific reactions but some reactions have no known catalyst. A catalyst for the synthesis of ammonia was sought for many years. Tests were run using many chemicals. In time, finely divided iron was found to work as a catalyst and it is now used in the industrial synthesis of ammonia. Generally, there are three types of catalysts.

### Heterogeneous Catalysts

Heterogeneous catalysts are surface or contact catalysts; they are typically metals, metal oxides, or other solids immersed in a gaseous or liquid mixture of the reactants. They function by facilitating the reaction on the surface of the solid. A common application of a heterogeneous catalyst is the catalytic converter on automobiles. One of the air pollutants in the exhaust of the internal combustion engine is carbon monoxide, CO. The carbon monoxide is converted to carbon dioxide by mixing the exhaust gases with air and passing this mixture over a bed of finely divided platinum. The reaction is

$$2CO(g) + O_2(g) \xrightarrow{\text{Pt}} 2CO_2(g)$$

The platinum catalyst allows the reaction to occur rapidly at the exhaust temperatures. The catalytic converter also catalyzes the reaction of unburned hydrocarbon gasoline with oxygen to give carbon dioxide and water.

Another heterogeneous catalyst is rhodium. Gaseous methyl alcohol or methanol, $CH_3OH$, and carbon monoxide, CO, do not noticeably react. When solid rhodium is placed in the reaction vessel, a rapid reaction occurs forming acetic acid.

$$CH_3OH(g) + CO(g) \xrightarrow{\text{Rh}} HC_2H_3O_2(g)$$

The rhodium serves as a contact catalyst. It accelerates the reaction so that it can be used for the commercial manufacture of acetic acid. Without the catalyst the reaction is too slow.

ACTIVITY 11-7

Unburned hexane, $C_6H_{14}$, in automobile exhaust reacts with oxygen in the catalytic converter. Write a balanced equation for the burning of hexane and show the Pt as a catalyst.

## Homogeneous Catalysts

Homogeneous catalysts function by being intimately mixed with the reactants in gaseous mixtures or liquid solutions. In other words, the catalyst is in the same physical state as the reactants; unlike a heterogeneous catalyst, it does not facilitate surface reactions. For example, when a compound containing iron(III) ions dissolves in a water solution of hydrogen peroxide, the $H_2O_2$ decomposes quickly.

$$2H_2O_2(aq) \xrightarrow{Fe^{3+}} 2H_2O + O_2(g)$$

The iron(III) ion catalyst is not a solid but rather is a dissolved ion, homogeneous with the solution. Another example of a homogeneous catalyst is the catalyst in commercial epoxy glue. When a solution containing the catalyst is mixed with the glue reactants, the catalyst encourages a reaction that forms the hardened products. Without the catalyst the reaction is very slow and the glue is not useful.

## Enzyme Catalysts

Many reactions that take place in living systems would occur too slowly if they were not catalyzed. Plants and animals contain biological catalysts to speed up and control the many chemical reactions involved in life and growth. Biological catalysts are known as **enzymes.** Literally thousands of enzymes are involved in life processes. These enzymes allow a variety of specific reactions to occur at useful rates. In our bodies enzymes catalyze reactions that would otherwise occur very slowly at body temperatures. For example, the human body produces many specialized digestive enzymes to aid the digestion of food molecules. Without digestive enzymes foods would not break down fast enough to be absorbed by the body. For more about enzymes see Chapter 17.

How a catalyst works may seem a bit mysterious. In some reactions it is not known in detail how the catalyst actually functions. However, it is known that the catalyst provides an alternate path or sequence of intermediate reactions that are faster than the uncatalyzed reaction. That is, the catalyst provides another way for the reaction to occur but gives the same products. See Figure 11-1(d). Catalyzed reactions can occur up to 10 times faster than the uncatalyzed reaction. Under specific conditions, the reaction follows the faster catalyzed route rather than the uncatalyzed route. In the case of rhodium used in the manufacture of acetic acid, it appears that methanol molecules are attracted to the surface of the solid rhodium. A given methanol molecule arranges on the surface so that when a carbon monoxide molecule collides with it acetic acid forms. This is why rhodium is a heterogeneous surface catalyst.

An **inhibitor** is a chemical that slows or decreases the rate of a chemical reaction. Very often an inhibitor slows a reaction by preventing the function of a catalyst. Sometimes inhibitors are desirable in reactions and sometimes they are not. For example, leaded gasoline, in an automobile with a catalytic converter, can poison the catalyst in the converter. The lead inhibits the catalyst by coating it with lead and lead compounds. Some antibacterial drugs function by inhibiting specific enzymes in bacteria, preventing the growth and multiplication of the bacteria. Some anticancer drugs inhibit enzymes to prevent the rapid growth and multiplication of cells. On the other hand, some inhibitors act as general poisons. Mercury and lead compounds in the body can inhibit a variety of enzymes, causing the body to malfunction. The result is heavy-metal poisoning that can cause permanent damage or death.

### 11-5 OXIDATION NUMBERS

In chemical reactions elements can change the ways in which they are chemically combined. The way in which an element is combined is described by an oxidation number. For example, in ionic compounds fluorine occurs as the fluoride ion, $F^-$, and in molecular compounds it occurs as part of a single covalent bond. An oxidation number of $-1$ is given to fluorine to describe its chemical behavior. An **oxidation number** or the **oxidation state** of an element shows the kind of simple ion it forms or the kind of covalent bonds it forms. From another view, an oxidation number is the charge an element has in a simple ion or the hypothetical charge it would have in a covalent bond if the shared electrons are assigned to the more electronegative elements (see Section 10-2 for a discussion of electronegativity). For example, the oxidation number of oxygen in the oxide ion, $O^{2-}$, is $-2$. In the compound water if we assign the shared electrons to the more electronegative oxygen, it would have a hypothetical charge of $2-$, since it would have two more electrons than a neutral oxygen atom. Each hydrogen would have a hypothetical charge of $1+$, since each would have one electron less than a neutral hydrogen atom.

In water, oxygen has an oxidation number of $-2$ and each hydrogen has an oxidation number of $+1$. In molecular hydrogen, $H_2$, the shared electrons are assigned equally to the two hydrogen atoms. Neither hydrogen is viewed as having an excess or deficiency of electrons, so the oxidation number of each hydrogen is zero.

An element may have different oxidation numbers in different compounds. For instance, sulfur forms the sulfide ion, $S^{2-}$, and can form two covalent bonds with less electronegative elements (e.g., $H_2S$) and, thus is assigned the $-2$ oxidation number. Sulfur also forms two, four, and sometimes six covalent bonds with elements of higher electronegativity, for which sulfur is assigned oxidation numbers of $+2$, $+4$, and $+6$ respectively. Note that oxidation numbers are numbers preceded by signs. The sign used with an oxidation number relates to the charge of a simple ion formed by an element or to its electronegativity relative to elements with which it covalently bonds. An oxidation number for an element consists of a number with a positive or negative sign. If an element is more electronegative than the element with which it bonds, it is given a negative oxidation number. The less electronegative element is given a positive oxidation number.

As we shall see, oxidation numbers are useful in describing certain kinds of reactions. To find an oxidation number of an element in an ion or compound we follow oxidation number rules. These rules are based on patterns that elements follow when they enter into chemical combinations.

1. The oxidation number or oxidation state of an element in the natural form uncombined with other elements is zero, 0. For example, the oxidation state of oxygen in $O_2$ is 0, the oxidation state of chlorine in $Cl_2$ is 0, and the oxidation state of sodium in metallic sodium, Na, is 0.

2. The oxidation number of an element in a simple ion is given by the charge of the ion. For example, the oxidation number of sodium in $Na^+$ is $+1$, and the oxidation number of sulfur in $S^{2-}$ is $-2$.

3. Hydrogen has an oxidation number of $+1$ when it is covalently bonded to nonmetals. When it combines with metals, in binary ionic compounds called hydrides, it has an oxidation number of $-1$.

4. Oxygen has an oxidation number of $-2$ in the vast majority of its compounds. Oxygen has a $-1$ oxidation number in compounds called peroxides, such as hydrogen peroxide, $H_2O_2$.

5. The algebraic sum of the oxidation numbers of the elements in a compound equals zero. This means that if we add all the positive oxidation numbers and all the negative oxidation numbers the sum must be zero. For example, in water, $H_2O$, each of the two hydrogen atoms is in the $+1$ oxidation state, and the oxygen is in the $-2$ oxidation state. The sum of the oxidation numbers ($+1$, $+1$, and $-2$) equals zero.

$$H_2 \qquad O$$
$$2(+1) + (-2) = +2 - 2 = 0$$

6. The sum of the oxidation numbers of the elements in a polyatomic ion equals the charge of the ion. For example, in the hydroxide ion, $OH^-$, the hydrogen is in the $+1$ oxidation state and the oxygen is in the $-2$

oxidation state. The sum of the oxidation numbers corresponds to the charge of the ion, which is 1−.

$$O \quad H^-$$
$$(-2) + (+1) = -1$$

7. Metals normally have positive oxidation numbers in compounds. Nonmetals have negative oxidation numbers when combined with metals. However, nonmetals can have positive oxidation numbers when combined with more electronegative nonmetals. The more electronegative elements always take on the negative oxidation numbers. Table 11-1 is a periodic table that shows the oxidation numbers of most ele-

*Table 11-1*  The Oxidation Numbers of Elements

Oxidation numbers of most of the elements. The most common oxidation numbers are in bold.

| Period | 1 (I A) | 2 (II A) | 3 (III B) | 4 (IV B) | 5 (V B) | 6 (VI B) | 7 (VIII B) | 8 | 9 (VIII) | 10 | 11 (I B) | 12 (II B) | 13 (III A) | 14 (IV A) | 15 (V A) | 16 (VI A) | 17 (VII A) | 18 (NOBLE GASES) |
|---|---|---|---|---|---|---|---|---|---|---|---|---|---|---|---|---|---|---|
| 1 | 1 H **+1** −1 | | | | | | | | | | | | | | | | | 2 He |
| 2 | 3 Li **+1** | 4 Be **+2** | | | | | | | | | | | 5 B **+3** | 6 C **+4** +2 −4 | 7 N **+5** +4 +3 +2 +1 −1 −2 −3 | 8 O +2 −1 **−2** | 9 F **−1** | 10 Ne |
| 3 | 11 Na **+1** | 12 Mg **+2** | | | | | | | | | | | 13 Al **+3** | 14 Si **+4** −4 | 15 P **+5** +3 −3 | 16 S **+6** +4 +2 −2 | 17 Cl **+7** +6 +5 +4 +3 +1 −1 | 18 Ar |
| 4 | 19 K **+1** | 20 Ca **+2** | 21 Sc **+3** | 22 Ti **+4** +3 +2 | 23 V **+5** +4 +3 +2 | 24 Cr **+6** +5 +4 **+3** +2 | 25 Mn **+7** +6 **+4** +3 +2 | 26 Fe **+3** **+2** | 27 Co +3 **+2** | 28 Ni **+2** +3 | 29 Cu **+2** +1 | 30 Zn **+2** | 31 Ga **+3** | 32 Ge **+4** −4 | 33 As +5 **+3** −3 | 34 Se **+6** +4 −2 | 35 Br **+7** +5 +1 −1 | 36 Kr **+4** +2 |
| 5 | 37 Rb **+1** | 38 Sr **+2** | 39 Y **+3** | 40 Zr **+4** | 41 Nb **+5** +4 | 42 Mo **+6** +4 +3 | 43 Tc **+7** +6 +4 | 44 Ru **+8** +6 +4 +3 | 45 Rh **+3** +2 | 46 Pd **+4** +2 | 47 Ag **+1** | 48 Cd **+2** | 49 In **+3** | 50 Sn **+4** +2 | 51 Sb **+5** +3 −3 | 52 Te **+6** +4 −2 | 53 I **+7** +5 +1 −1 | 54 Xe **+6** +4 +2 |
| 6 | 55 Cs **+1** | 56 Ba **+2** | 57 La **+3** | 72 Hf **+4** | 73 Ta **+5** | 74 W **+6** +4 | 75 Re **+7** +6 +4 | 76 Os **+8** +4 | 77 Ir **+4** +3 | 78 Pt **+4** +2 | 79 Au **+3** +1 | 80 Hg **+2** +1 | 81 Tl **+3** +1 | 82 Pb **+4** +2 | 83 Bi **+5** +3 | 84 Po **+2** | 85 At **−1** | 86 Rn |
| 7 | 87 Fr **+1** | 88 Ra **+2** | 89 Ac **+3** | | | | | | | | | | | | | | | |

Transition Metals

ments. Notice that the most likely negative oxidation number for group 16 or VIA elements is −2 and for group 17 or VIIA elements is −1.

We can figure out the oxidation numbers elements in a compound or a polyatomic ion by assigning the expected oxidation numbers to the other elements then applying Rule 5 or Rule 6. Hydrogen is typically +1 and oxygen is typically −2. For example, in the compound $SO_3$ the expected oxidation number of oxygen is −2. Therefore, the oxidation number of sulfur must be +6 to match the total of the negative oxidation numbers $(3(-2) = -6)$.

---

**EXAMPLE 11-1**

What is the oxidation number of nitrogen in ammonia, $NH_3$? Assigning an oxidation number of +1 to hydrogen and using Rule 5 gives

$$\begin{array}{cc} N & H_3 \\ x + 3(+1) &= 0 \\ x + (+3) &= 0 \end{array}$$

Therefore, $x = 0 - 3$ or nitrogen has an oxidation number of −3.

---

**EXAMPLE 11-2**

What is the oxidation number of sulfur in the hydrogen sulfate ion, $HSO_4^-$?

The oxidation number of sulfur in the hydrogen sulfate ion, $HSO_4^-$, is found by assigning a +1 oxidation number to hydrogen and a −2 oxidation number to oxygen. The oxidation number of sulfur must have a value that results in the overall ionic charge of −1. So, applying Rule 6, we have

$$\begin{array}{cccc} H & S & O_4^- & \text{Charge} \\ (+1) + x + 4(-2) &=& -1 \\ (+1) + x + (-8) &=& -1 \\ x = -1 + 8 - 1 &=& +6 \end{array}$$

The oxidation number of S is +6.

---

**EXAMPLE 11-3**

What is the oxidation number of chromium in the dichromate ion, $Cr_2O_7^{2-}$? Assigning an oxidation number of −2 to oxygen and using Rule 6 gives

$$\begin{array}{ccc} Cr_2 & O_7^{2-} & \text{Charge} \\ 2x + 7(-2) &=& (-2) \\ 2x + (-14) &=& (-2) \\ 2x = -2 + 14 &=& +12 \\ x &=& +6 \end{array}$$

What is the oxidation number of carbon in the hydrogen carbonate ion, $HCO_3^-$? What are the oxidation numbers of sulfur in S, $SO_2$, $H_2SO_3$, and $SO_3$?

Each chromium has an oxidation number of +6.

It is common practice to state the oxidation numbers of elements in compounds and ions on a per atom basis. Thus, as shown above, we divide +12 by 2 to determine that each chromium atom has an oxidation number of +6.

## 11-6 ELECTRON TRANSFER REACTIONS

Chemical bonds involve electrons and chemical reactions involve the breaking and forming of chemical bonds. A very common type of chemical reaction involves the transfer of electrons between atoms. However, it is important to note that not all reactions involve electron transfer. A simple example of an electron transfer is the reaction between iron and sulfur to form iron(II) sulfide (see Figure 11-2).

$$\cdot Fe\cdot \quad \overset{2+}{\cdot \ddot{S}\colon} \longrightarrow Fe \quad \overset{2-}{\ddot{S}\colon}$$

**FIGURE 11-2** The reaction between iron and sulfur.

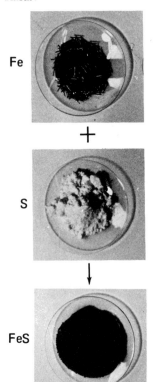

Fe

+

S

↓

FeS

Note that electrons transfer or move from the iron to the sulfur giving the iron(II) cation and the sulfide anion. The resulting cations and anions form ionic bonds. An important fact about electron transfer reactions is that the oxidation numbers of the elements change during the reaction. In the above reaction the oxidation number of iron changes from 0 to +2 and the oxidation number of sulfur changes from 0 to −2. Electron transfer reactions are also called **oxidation–reduction** reactions. **Oxidation** is defined as the loss of electrons or an increase in oxidation number. **Reduction** is defined as the gain of electrons or a decrease in oxidation number. Since one element loses electrons and another gains them, oxidation and reduction happen simultaneously. In the above reaction iron undergoes oxidation and sulfur undergoes reduction. There is more to the story of oxidation and reduction. Chapter 13 discusses this topic in detail.

Metals tend to form cations and nonmetals tend to form anions. Reactions between metals and nonmetals are examples of simple electron transfer reactions. When sodium reacts with chlorine electron transfer occurs.

$$Cl_2 \longrightarrow 2\;\cdot\ddot{C}l\colon \qquad Na\cdot\!\!\frown\!\!\cdot\ddot{C}l\colon \qquad Na^+ \colon\!\ddot{C}l\colon^-$$
$$Na\cdot\!\!\frown\!\!\cdot\ddot{C}l\colon \longrightarrow Na^+ \colon\!\ddot{C}l\colon^-$$

The term oxidation originally came from the fact that many elements react with oxygen to form oxides. As discussed in Chapter 13 oxidation now has a more involved meaning. Electron transfer does not always involve the formation of ions. In some reactions an element changes its state of combination by bonding to an element with a higher electronegativity. For example, consider the reaction of carbon with oxygen.

$$C + O_2 \longrightarrow CO_2$$

The oxidation number of carbon changes from 0 to +4 and the oxidation number of oxygen changes from 0 to −2. Carbon and oxygen become covalently bonded. Carbon has been oxidized and oxygen has been reduced even though the electrons have not completely transferred from one atom to another. If a reaction involves electron transfer then the oxidation numbers of some elements change. Thus, if you have an equation for a reaction it is easy to recognize the reaction as an electron transfer. Just look for oxidation number changes. However, keep in mind that not all of the elements in a reaction need to change oxidation number.

Show the reaction of Mg and $F_2$ as an electron transfer process.

## 11-7   SOME TYPES OF ELECTRON TRANSFER REACTIONS

We can classify simple electron transfer reactions into specific categories or types, and we can confirm that they are electron transfer reactions by noting the oxidation numbers of the elements involved.

Deduce the oxidation numbers of C, H, and O in the reactants and products of the following reaction:

$$C + H_2O \longrightarrow CO + H_2$$

### Synthesis or Combination Reactions

Synthesis or combination reactions involve the synthesis or formation of a compound from simpler chemicals.

1. Forming compounds from elements

$$3H_2 + N_2 \longrightarrow 2NH_3 \quad \text{ammonia synthesis}$$

$$S + O_2 \longrightarrow SO_2 \quad \text{sulfuric acid synthesis first step}$$

2. Forming a compound from an element and a compound

$$2SO_2 + O_2 \xrightarrow{V_2O_5} 2SO_3 \quad \text{sulfuric acid synthesis second step}$$

$$2CO + O_2 \xrightarrow{Pt} 2CO_2 \quad \text{catalytic conversion in automobiles}$$

### Decomposition Reactions

Decomposition reactions involve a compound decomposing into simpler chemicals.

1. Forming elements from compounds

$$2H_2O \xrightarrow{\text{electrolysis}} 2H_2 + O_2 \quad \text{electrical decomposition of water}$$

$$2HgO \xrightarrow{\text{heat}} 2Hg + O_2 \quad \text{thermal decomposition of mercury(II) oxide}$$

2. Forming an element and compound from a compound

$$2H_2O_2 \xrightarrow{\text{enzyme}} 2H_2O + O_2 \quad \text{enzyme-catalyzed decomposition of hydrogen peroxide}$$

## Displacement Reactions

Displacement reactions involve one element exchanging for another.

$$HgS + O_2 \xrightarrow{\text{heat}} Hg + SO_2 \quad \text{makes mercury from mercury ore}$$

$$SiO_2 + 2C \xrightarrow{\text{heat}} Si + 2CO \quad \text{makes silicon from silica sand}$$

$$6Na + Fe_2O_3 \longrightarrow 3Na_2O + 2Fe \quad \text{used in airbags to compound Na formed when } NaN_3 \text{ explodes}$$

$$C + H_2O \xrightarrow{\text{heat}} CO + H_2 \quad \text{synthesis of methanol first step}$$

## Combustion

Combustion reactions are also called burning. A common type of combustion is the reaction of organic (carbon-containing) compounds with oxygen. The complete combustion of carbon–hydrogen or carbon–hydrogen–oxygen compounds always produces carbon dioxide and water.

$$C_2H_5OH + 3O_2 \longrightarrow 2CO_2 + 3H_2O \quad \text{combustion of ethanol}$$

$$2C_8H_{18} + 25O_2 \longrightarrow 16CO_2 + 18H_2O \quad \text{combustion of octane, a major component of gasoline}$$

Sometimes combustion occurs with insufficient oxygen, which results in the formation of carbon monoxide instead of carbon dioxide. The following equation shows the incomplete combustion of octane:

$$2C_8H_{18} + 17O_2 \longrightarrow 16CO + 18H_2O$$

In the internal combustion engine most of the gasoline undergoes complete combustion to give carbon dioxide and water but a small amount undergoes incomplete combustion, which results in some carbon monoxide in the exhaust. Carbon monoxide gas is a deadly poison.

## 11-8  NET-IONIC EQUATIONS

Water solutions are convenient for storing some chemicals and good for performing some reactions. Particles of a chemical dissolved in water can intermix, move about the solution, and collide with any other dissolved particles. In solution, **ionic compounds** and **strong molecular electrolytes** form ions, which are sometimes involved in chemical reactions. To write an equation for a reaction in solution we have to know the formulas of the reactants and products. The formulas can come from experimental observations and, sometimes, by prediction based upon information about the solution. Recall from Section 10-19 that it is possible to know the nature of the solute particles in a solution. To do this we need to know whether a solute is a strong electrolyte, a weak electrolyte, or a nonelectrolyte.

Let's look at an example of an equation for a reaction that occurs in solution. When a solution of the ionic compound calcium chloride, $CaCl_2$, mixes with a solution of the ionic compound sodium carbonate, $Na_2CO_3$, the insoluble solid $CaCO_3$ forms. The solutions of the ionic compounds are represented by the separate ions they form upon dissolving.

calcium chloride $\qquad Ca^{2+}(aq) + 2Cl^-(aq)$

sodium carbonate $\qquad 2Na^+(aq) + CO_3^{2-}(aq)$

The product of the reaction is $CaCO_3$ so the equation for the reaction is

$$Ca^{2+}(aq) + 2Cl^-(aq) + 2Na^+(aq) + CO_3^{2-}(aq) \longrightarrow$$
$$CaCO_3(s) + 2Cl^-(aq) + 2Na^+(aq)$$

Using your chemical vision of solutions, imagine what happens when the two solutions mix. The ions in one solution mix with the ions of the other solution. The ions move about and interact with one another and with water molecules. The $Ca^{2+}$ and the $CO_3^{2-}$ form ionic bonds that are stronger than the attraction of water molecules resulting in the formation of solid $CaCO_3$. Notice that the $Na^+$ and $Cl^-$ do not react in any way. They remain dissolved and distributed among the water molecules of the solution. In the above equation these ions appear unchanged on both sides of the arrow.

When reactions occur in water some particles react and others do not react. Ions that do not react are known as **spectator ions** because they just "watch" the reaction. You might wonder why they are present in the first place. All solutions are electrically neutral having as much positive as negative charge. It is not possible to have a solution of pure cation or pure anion. Spectator ions are ions that do not take part in a reaction but are present to maintain equal amounts of negative and positive charge in the solution.

Often equations representing reactions in solution do not include the spectator ions. Such equations for solution reactions are called net-ionic equations. A **net-ionic equation** shows the particles that react and the particles formed and does not include any spectator ions. Deleting the spectator ions gives the net-ionic equation for the reaction of solutions of calcium chloride and sodium carbonate:

$$Ca^{2+}(aq) + CO_3^{2-}(aq) \longrightarrow CaCO_3(s)$$

This equation is chemically balanced and charge balanced. **Charge balance** means that an equation involving ions must have the same net amount of positive and negative charge on both sides of the arrow. A reaction cannot produce a net change in electrical charge; it cannot produce excess positive charge or negative charge. Why is this true?

In net-ionic equations the (aq) notation is used to emphasize the reaction is taking place in a water solution. For convenience the (aq) can be omitted. To write a net-ionic equation for a reaction use the following pattern:

1. Write the formulas of all major particles present in the solutions.
2. Write the formulas of the product or products of the reaction.
3. Delete any particles that are not involved in the reaction; normally these are spectator ions.
4. Balance the equation by making sure it is chemically balanced and charge balanced.

---

**EXAMPLE 11-4**

A solution of the ionic compound NaOH is added to a solution of the weak electrolyte acetic acid, $HC_2H_3O_2$. Write a net-ionic equation for the reaction that produces acetate ion, $C_2H_3O_2^-$, and water.

First, write the formulas of the particles that are present in the solutions that mix. Show ionic compounds as separate ions and weak electrolytes by their molecular formulas.

$$Na^+ + OH^- + HC_2H_3O_2$$

Second, write the formulas of the products.

$$Na^+ + OH^- + HC_2H_3O_2 \longrightarrow C_2H_3O_2^- + H_2O$$

Third, delete any particles that do not react. The sodium ion does not react in this case.

$$OH^- + HC_2H_3O_2 \longrightarrow C_2H_3O_2^- + H_2O$$

Fourth, balance the net-ionic equation, both chemically and by charge. In this case the equation balances chemically and has the same net charge (1–) on both sides of the arrow. It's balanced.

---

**EXAMPLE 11-5**

A water solution of hydrochloric acid (remember that this strong electrolyte is represented a $H_3O^+$ and $Cl^-$ as shown in Table 10-2) is mixed with a solution of potassium carbonate, $K_2CO_3$. Write a net-ionic equation for the reaction that forms $CO_2$ gas and water.

First, write the formulas of the particles that are present in the solutions. Hydrochloric acid is a strong electrolyte and potassium carbonate is an ionic compound.

$$H_3O^+ + Cl^- + 2K^+ + CO_3^{2-}$$

Second, write the formulas of the products.

$$H_3O^+ + Cl^- + 2K^+ + CO_3^{2-} \longrightarrow CO_2 + H_2O$$

Third, delete any particles that do not react. In this case $Cl^-$ and $K^+$.

$$H_3O^+ + CO_3^{2-} \longrightarrow CO_2 + H_2O$$

Fourth, balance the equation both chemically and by charge. To balance the 2– charge of $CO_3^{2-}$ we need to place a coefficient of 2 in front of $H_3O^+$. To balance the equation chemically, we need a coefficient of 3 for water to give 6 hydrogens and 5 oxygens on each side.

$$2H_3O^+ + CO_3^{2-} \longrightarrow CO_2 + 3H_2O$$

When a solution of calcium chloride, $CaCl_2$ and a solution of $Na_3PO_4$ are mixed a reaction occurs that forms solid $Ca_3(PO_4)_2$. Write a balanced net-ionic equation for the reaction.

## 11-9 PRECIPITATION REACTIONS

With your chemical vision you can imagine an ionic compound in solution as separate cations and anions. The ions are in continuous motion and migrate about the solution momentarily interacting with one another. As long as the compound corresponding to the cation and anion is soluble, and enough water is present, the ions remain in solution. In such cases the attractions between oppositely charged ions are not great enough to cause the ions to bond to form a crystalline ionic solid. If any of the mixed ions are the ions of an **insoluble ionic compound,** the ions bond to form crystals and the insoluble solid separates from the solution; this is illustrated in Figure 11-3. A reaction of ions in solution to form a solid compound is called an **ion combination** or a **precipitation reaction. Precipitation** reactions are not electron transfer reactions and do not involve changes in oxidation numbers. **Precipitate** is the term used to name the insoluble solid that separates from the solution.

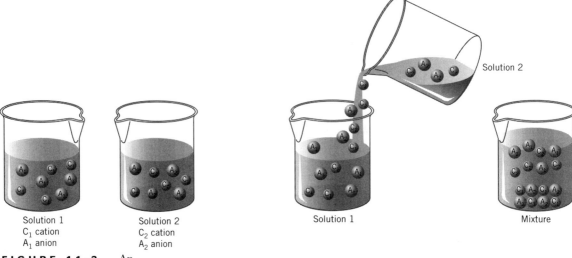

**FIGURE 11-3**   An ion combination reaction.

Solution 1
$C_1$ cation
$A_1$ anion

Solution 2
$C_2$ cation
$A_2$ anion

Solution 1

Solution 2

Mixture

A good example of a precipitation reaction is the reaction of calcium and carbonate ion discussed in the previous section.

$$Ca^{2+}(aq) + CO_3^{2-}(aq) \longrightarrow CaCO_3(s)$$

When any solutions containing these ions mix and the concentrations of the ions are high enough solid calcium carbonate precipitates. Check the oxidation numbers of the elements to confirm that they do not change in the reaction. **Hard water** is a term used for natural waters that contain relatively high concentrations of calcium ions and other ions. The use of hard water sometimes results in the precipitation of calcium carbonate to form scale in hot-water pipes and teakettles. Some natural limestone deposits, such as the White Cliffs of Dover in England, are precipitated calcium carbonate. The stalactites and stalagmites in limestone caves are calcium carbonate deposits.

Some mineral deposits were formed by natural precipitation of ionic compounds. In nature, precipitation may occur quite slowly forming large crystals. In the laboratory, however, where the concentrations of ions are relatively high, precipitation usually occurs very quickly, producing finely divided crystals. Precipitates form as soon as the solutions mix then slowly settle from the mixture. Precipitates have a variety of colors and appearances. They may be flocculent (fluffy), gelatinous (jelly-like), coagulated (curdy or lumpy), or crystalline (powdery or silky) in appearance.

## 11-10 SOLUBILITY RULES FOR IONIC COMPOUNDS

It is possible to predict whether a precipitation reaction will occur when solutions of ionic compounds are mixed. To make a prediction you need to know which ionic compounds are soluble and which are insoluble. Ionic compounds vary in solubility ranging from **very soluble** through **soluble** and **slightly soluble** to **insoluble.**

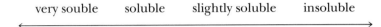

An ionic compound is termed **soluble** if enough of it can be dissolved in water to make a solution that is at least 0.1 $M$. We can consider an ionic compound that is not soluble enough to make a 0.1 $M$ solution to be slightly soluble or **insoluble.** Actually, every compound dissolves to some extent. Many, however, are quite insoluble. An example of a very insoluble compound is barium sulfate, $BaSO_4$. Barium ions are toxic so it is not a good idea to ingest them. Before being X-rayed in intestinal areas, it is standard practice to have patients drink a barium sulfate slurry (a mixture of the compound with water that is not a solution). The compound is so insoluble it passes through the intestinal tract without dissolving in any significant amount. Barium is a strong absorber of X rays so it helps in the visualization of the intestinal passages.

The solubilities of many common ionic compounds in water are summarized in solubility rules based upon experimental observations. If it is possible to prepare at least a 0.1 $M$ solution of a compound it is soluble. Typical ionic compounds contain metallic cations in combination with nonmetal anions or polyatomic anions. For our purposes the solubility rules given on the next page apply. Keep in mind that the rules do not include all ionic compounds and there may be some exceptions to the general rules. Refer to the rules or the summary given in Table 11-2 when you want to check the solubility of an ionic compound.

*Table 11-2*    Solubility List of Common Ionic Compounds[a]

| ANION | CATIONS THAT FORM PRECIPITATE | CATIONS THAT DO NOT FORM PRECIPITATE |
|---|---|---|
| $NO_3^-$ | None | All other common |
| $C_2H_3O_2^-$ | None | All other common |
| $SO_4^{2-}$ | $Ca^{2+}$, $Sr^{2+}$, $Ba^{2+}$, $Pb^{2+}$ | Most other common |
| $Cl^-$, $Br^-$, $I^-$ | $Ag^+$, $Hg_2^{2+}$, $Pb^{2+}$ | Most other common |
| $OH^-$ | Most | $Na^+$, $K^+$, $Ba^{2+}$ |
| $F^-$ | $Mg^{2+}$, $Ca^{2+}$, $Sr^{2+}$, $Ba^{2+}$, $Pb^{2+}$ | Most other common |
| $CO_3^{2-}$, $CrO_4^{2-}$, $PO_4^{3-}$ | Most | $NH_4^+$, group IA |

[a]Almost all ionic compounds of $Na^+$, $K^+$, and $NH_4^+$ are soluble.

1. Nearly all ionic compounds containing sodium ions, $Na^+$, potassium ions, $K^+$, or ammonium ions, $NH_4^+$, are **soluble.**

2. Nearly all compounds containing nitrate ions, $NO_3^-$ or acetate ions, $C_2H_3O_2^-$, are **soluble.**

3. All compounds containing sulfate ions, $SO_4^{2-}$, are **soluble,** except

| calcium sulfate | $CaSO_4$ |
| strontium sulfate | $SrSO_4$ |
| barium sulfate | $BaSO_4$ |
| lead(II) sulfate | $PbSO_4$ |

4. All ionic compounds containing chloride ions, $Cl^-$, bromide ions, $Br^-$, or iodide ions, $I^-$, are **soluble,** except silver, mercury(I), and lead(II) compounds of these ions.

| $AgCl$ | $Hg_2Cl_2$ | $PbCl_2$ |
| $AgBr$ | $Hg_2Br_2$ | $PbBr_2$ |
| $AgI$ | $Hg_2I_2$ | $PbI_2$ |

5. All ionic compounds containing fluoride ions, $F^-$, are **soluble,** except

| calcium fluoride | $CaF_2$ |
| strontium fluoride | $SrF_2$ |
| barium fluoride | $BaF_2$ |
| lead(II) fluoride | $PbF_2$ |

6. All common ionic compounds containing hydroxide ions, $OH^-$, are **insoluble,** except

| sodium hydroxide | $NaOH$ |
| potassium hydroxide | $KOH$ |
| barium hydroxide | $Ba(OH)_2$ |

7. Most ionic compounds containing carbonate ions, $CO_3^{2-}$, chromate ions, $CrO_4^{2-}$, or phosphate ions, $PO_4^{3-}$, are **insoluble** except for group IA and ammonium ion compounds.

## 11-11 PREDICTING PRECIPITATION REACTIONS

It is possible to predict whether a precipitation reaction will occur when solutions of ionic compounds are mixed. This is done by considering all the ions that are mixed and deciding whether any set of cations and anions forms an insoluble compound. Of course, the solubility rules are used to decide which compounds are insoluble. When a solution of one chemical containing cations and anions is mixed with a solution of another chemical containing cations and anions there are two possible ion combinations (see Fig. 11-4).

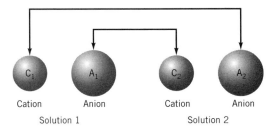

**FIGURE 11-4**
Possible cation and anion combinations.

$$\text{cation}_1 + \text{anion}_1 + \text{cation}_2 + \text{anion}_2$$

If either combination gives an insoluble compound, a precipitation reaction is predicted and a net-ionic equation can be written for the reaction. Note that in some cases it is possible to form two different precipitates.

---

**EXAMPLE 11-6**

Write a net-ionic equation for any reaction that occurs when a solution of silver nitrate, $AgNO_3$, is mixed with a solution of sodium chloride, NaCl.

Both solutes are ionic compounds so we can show them as separate cations and anions.

$$Ag^+ + NO_3^- + Na^+ + Cl^-$$

Consider the two possible combinations. By Solubility Rule 2 all nitrates are soluble so sodium nitrate will not precipitate. By Solubility Rule 4 silver chloride is insoluble. Thus, we predict that silver ions react with chloride ions to give silver chloride.

$$Ag^+ + Cl^- \longrightarrow AgCl$$

The net-ionic equation is balanced chemically and by charge.

---

**EXAMPLE 11-7**

Write a net-ionic equation for any reaction that occurs when a solution of aluminum nitrate, $Al(NO_3)_3$, is mixed with a solution of NaOH.

The ions in the solutions of these ionic compounds are

$$Al^{3+} + NO_3^- + Na^+ + OH^-$$

All nitrates are soluble so sodium nitrate does not precipitate. Aluminum hydroxide is insoluble so the equation for the precipitation reaction is

$$Al^{3+} + 3OH^- \longrightarrow Al(OH)_3$$

To balance the charge, we use three $OH^-$ for one $Al^{3+}$.

---

**EXAMPLE 11-8**

Write a net-ionic equation for any reaction that occurs when a solution of potassium acetate, $KC_2H_3O_2$, is mixed with a solution of sodium chloride.

The ions involved are

$$K^+ + C_2H_3O_2^- + Na^+ + Cl^-$$

By Solubility Rule 2, all acetates are soluble and by Solubility Rule 4, KCl is soluble. Thus neither combination gives a precipitate and we predict no reaction.

---

**EXAMPLE 11-9**

Write a net-ionic equation for any reaction that occurs when a solution of barium hydroxide, $Ba(OH)_2$, is mixed with a solution of zinc sulfate, $ZnSO_4$.

The ions in the solutions are

$$Ba^{2+} + OH^- + Zn^{2+} + SO_4^{2-}$$

By Solubility Rule 3 barium sulfate is insoluble and by Solubility Rule 5 zinc hydroxide is insoluble. Here two ion combinations occur. The equation showing both reactions is

$$Ba^{2+} + 2OH^- + Zn^{2+} + SO_4^{2-} \longrightarrow Zn(OH)_2 + BaSO_4$$

Or we could show each precipitation in a separate equation.

$$Zn^{2+} + 2OH^- \longrightarrow Zn(OH)_2$$
$$Ba^{2+} + SO_4^{2-} \longrightarrow BaSO_4$$

**ACTIVITY 11-12**

Write the net-ionic equation for any reaction that occurs when a solution of lead(II) nitrate, $Pb(NO_3)_2$, is mixed with a solution of potassium chloride, KCl.

---

## Calcium in the Body

Calcium is an important nutrient mineral in our diets. Calcium ions serve as one of the body electrolytes and it play a part in controlling the heartbeat. About 1.5% to 2.0% of body weight is calcium, most of which occurs in bones and teeth. A normal diet supplies sufficient calcium. Dairy products are a particularly rich source of calcium. Some postmenopausal women take calcium supplements to help prevent deterioration of bones. Bones and tooth enamel are made rigid and strong by calcium carbonate, $CaCO_3$, and hydroxyapatite, $Ca_5(PO_4)_3OH$. Tooth enamel is made more resistant to decay when a fluoride ion replaces the hydroxide ion in hydroxyapatite.

$$Ca_5(PO_4)_3OH + F^- \longrightarrow Ca_5(PO_4)_3F + OH^-$$

Fluoride ions form a stronger bond and harden the enamel making decay less likely. Compounds containing fluoride ions are added to some toothpastes to "fight" cavities. Fluoride ions are also present in most public water supplies, either as a natural mineral or added in small amounts as a dental prophylactic. Too much fluoride ion is undesirable because it can cause tooth enamel to become mottled (blotched with brown spots).

A calcium imbalance in the body can result in the formation of kidney stones. Most kidney stones are precipitates of insoluble calcium compounds such as calcium carbonate, calcium phosphate, and calcium oxalate. When small crystals of insoluble calcium compounds form stones in the kidneys, they can be very painful. Most of them pass from the body naturally or after medication. Sometimes they are treated by lithotripsy, which uses sound waves to break up the stones for easier passage.

**ACTIVITY 11-13**

(a) Using the information that 1.5% to 2.0% of body weight is calcium, calculate the number of kilograms of calcium in your body.

(b) Read the label of a toothpaste container and give the name or the formula of the active ingredient containing fluoride ions.

**FIGURE 11-5**    An acid-base reaction.

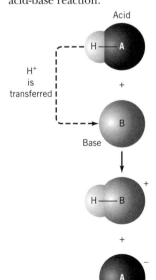

## 11-12 ACID–BASE REACTIONS

Another important kind of reaction that occurs in solutions is illustrated by the dissolving of hydrogen chloride in water to make hydrochloric acid.

$$HCl(g) + H_2O \longrightarrow H_3O^+(aq) + Cl^-(aq)$$

This reaction is neither an ion combination reaction nor an oxidation reaction; instead it involves the movement or transfer of a hydrogen ion ($H^+$) from HCl to $H_2O$. This kind of reaction is called an **acid–base reaction,** in which a covalent bond involving hydrogen is broken to give a hydrogen ion that forms a new covalent bond (see Fig. 11-5). The result is that a hydrogen ion, $H^+$, moves or transfers from one chemical to another.

A chemical that can lose or give a hydrogen ion in a reaction is called an **acid.** A chemical that can gain or bond with a hydrogen ion in a reaction is termed a **base.** Thus, reactions in which hydrogen ions transfer are acid–base reactions. Common acid–base reactions involve an acid reacting with a base to give water. Two examples are a hydrochloric acid solution reacting with a potassium hydroxide solution

$$H_3O^+ + OH^- \longrightarrow 2H_2O$$

and an acetic acid solution reacting in a sodium hydroxide solution:

$$HC_2H_3O_2 + OH^- \longrightarrow C_2H_3O_2^- + H_2O$$

In these reactions the acid and base neutralize one another to form water; such reactions are sometimes called **neutralization reactions.** The equations are net-ionic equations and do not show the spectator ions. Acid–base reactions are important in many industrial and biological processes; they are discussed in more detail in Chapter 12.

**Look closely at the two neutralization reactions given as examples. Draw pictures to show that these reactions are hydrogen ion transfer reactions. In each reaction identify the acid and base in the reactants.**

###  POLYMERIZATION AND PLASTICS

**Polymerization** is another interesting kind of reaction. In these reactions certain kinds of molecules react under specific conditions in the presence of a catalyst so that individual molecules add to one another to make very large molecules. In such a reaction hundreds or thousands of molecules link together to form very large macromolecules called **polymers.** The original molecule used to prepare the polymer is called the **monomer** (see Fig. 11-6).

**Plastics** are human-made polymers. The term plastic comes from the fact that these materials are flexible and can be easily cast and molded. Many natural organic substances such as cotton, wool, silk, and rubber are also polymers. Sections 16-2 and 16-4 discusses important biological molecules that are polymers. Chemistry in Action 11-2 compares samples of various plastics.

In an equation for a **polymerization reaction,** it is not possible to give the exact formula of the polymer, since the individual molecules, made up of hundreds or thousands of monomer units, vary in chain length. The polymerization of ethylene, $C_2H_2$ can be represented as

**Objective: Simulate the formation of a simple polymer.**

**You need eight paper clips of the same size. Link the paper clips together to make a chain. Link them in a regular way so that the larger part of each clip is connected to the smaller part of the next clip. Note that an individual clip represents a repeating unit in your paper clip "polymer." Sketch a picture of the repeating unit of your polymer chain.**

$$nCH_2{=}CH_2 \xrightarrow[\text{heat pressure}]{\text{catalyst}} H-\underset{\underset{H}{|}}{\overset{\overset{H}{|}}{C}}-\underset{\underset{H}{|}}{\overset{\overset{H}{|}}{C}}-\underset{\underset{H}{|}}{\overset{\overset{H}{|}}{C}}-\underset{\underset{H}{|}}{\overset{\overset{H}{|}}{C}}-\underset{\underset{H}{|}}{\overset{\overset{H}{|}}{C}}-\underset{\underset{H}{|}}{\overset{\overset{H}{|}}{C}}-\underset{\underset{H}{|}}{\overset{\overset{H}{|}}{C}}-\cdots$$

but this is not convenient. Notice, however, that the polymer can be represented by the repeating sequence shown on the bottom of the next page.

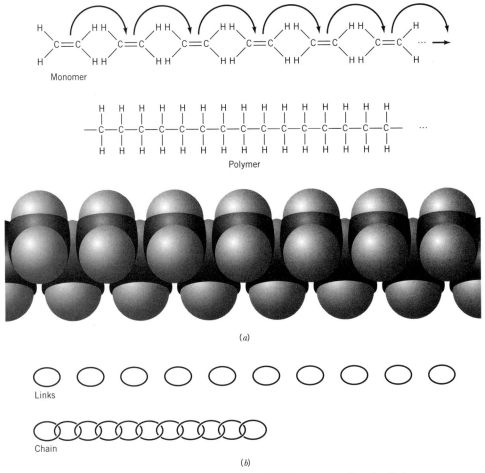

**FIGURE 11-6** A polymerization reaction. (a) Monomer molecules link to form large molecules called polymers. (b) The formation of a polymer can be likened to the formation of a chain. Monomer units link, one after another, to form a chain-like polymer molecule. Polymerization reactions are sometimes called chain reactions and the bonded atoms in the polymer molecules are called the polymer chain.

This sequence repeats along the polymer except at the ends, but the ends are a very small part of the entire polymer. Polymers are represented by the characteristic repeating sequence. The polymer of ethylene is shown as

$$\left(\begin{array}{c} H \quad H \\ | \quad | \\ -C-C- \\ | \quad | \\ H \quad H \end{array}\right)_n$$

where $n$ is some large number representing the hundreds or thousands of monomer units comprising the polymer. Using this representation, the equation for the polymerization of ethylene to form **polyethylene** is

$$n CH_2{=}CH_2 \xrightarrow[\text{heat pressure}]{\text{catalyst}} \left(\begin{array}{c} H \quad H \\ | \quad | \\ -C-C- \\ | \quad | \\ H \quad H \end{array}\right)_n$$

(monomer)                          (polymer)

Polyethylene is a waxy solid used to coat paper milk cartons and to make plastic bottles, toys, and other products. **Teflon** is a chemically inert plastic used for nonstick coatings on machines, tools, and cooking utensils. It is formed by this reaction:

$$n CF_2{=}CF_2 \longrightarrow \left(\begin{array}{c} F \quad F \\ | \quad | \\ -C-C- \\ | \quad | \\ F \quad F \end{array}\right)_n$$

(tetrafluoroethylene)        (polytetrafluoroethylene
                              or Teflon)

**ACTIVITY 11-16**

Make a sketch of the "polymerization reaction" that formed your paper clip polymer in Activity 11-15. Show the reaction in a form similar to the polyethylene reaction.

Part of a
teflon polymer

Another type of polymerization reaction is called condensation copolymerization. **Copolymers** have two kinds of monomers that typically join as alternating units along the polymer chain. **Polyesters** are plastics made from the condensation reaction of an organic acid monomer and an alcohol monomer.

$$\text{etc. } HO-\overset{\overset{\displaystyle O}{\|}}{C}-C_6H_4-\overset{\overset{\displaystyle O}{\|}}{C}-\fbox{OH} \quad \fbox{H}-OCH_2CH_2O-\fbox{H} \quad \fbox{HO}-\overset{\overset{\displaystyle O}{\|}}{C}-C_6H_4-\overset{\overset{\displaystyle O}{\|}}{C}-OH \text{ etc.}$$

ethylene
glycol

terephthalic
acid

↓

$$\left(-O-CH_2-CH_2-O-\overset{\overset{\displaystyle O}{\|}}{C}-C_6H_4-\overset{\overset{\displaystyle O}{\|}}{C}-\right)_n$$

(Dacron, a polyester)

Polyesters, such as Dacron, are used in synthetic fibers. **Nylons** are copolymers made from the condensation of organic compounds called amines and organic acids.

$$\text{etc. } H-\overset{\overset{\displaystyle H}{|}}{N}-(CH_2)_6-\overset{\overset{\displaystyle H}{|}}{N}-\fbox{H} \quad \fbox{HO}-\overset{\overset{\displaystyle O}{\|}}{C}-(CH_2)_4-\overset{\overset{\displaystyle O}{\|}}{C}-\fbox{OH} \quad \fbox{H}-\overset{\overset{\displaystyle H}{|}}{N}-(CH_2)_6-\overset{\overset{\displaystyle H}{|}}{N}-H \text{ etc.}$$

↓

$$\left(-\overset{\overset{\displaystyle O}{\|}}{C}-(CH_2)_4-\overset{\overset{\displaystyle O}{\|}}{C}-\overset{\overset{\displaystyle H}{|}}{N}-(CH_2)_6-\overset{\overset{\displaystyle H}{|}}{N}-\right)_n$$

(a nylon)

Large quantities of polymers are produced worldwide in the forms of molded plastics, films, fibers, and a variety of synthetic rubbers. Table 11-3 lists some of the types and uses of plastics. Synthetic fibers are in wide use in carpets, clothing, and other fabrics. Approximately 40 billion pounds of plastics are manufactured in the United States annually. This amounts to about 150 pounds of plastics per person per year.

**ACTIVITY 11-18**

Inventory the room you are in and list all of the items that you think contain plastics, synthetic fibers, or synthetic rubbers.

**ACTIVITY 11-17**

To simulate the formation of a copolymer from two different monomers, you need eight paper clips of the same size. Link the paper clips together to make a chain. Link them so that the larger parts of every set of two clips are connected. Note that a set of two clips in sequence represents a repeating unit in your paper clip polymer. Sketch a picture of the repeating unit of your polymer chain. Sketch the general form of the polymerization reaction that formed your paper clip polymer.

Recycling
coding symbols for plastics

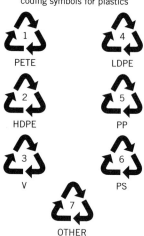

| | |
|---|---|
| 1 PETE | Polyethylene terephthalate |
| 2 HDPE | High-density polyethylene |
| 3 V | Vinyl or polyvinylchloride |
| 4 LDPE | Low-density polyethylene |
| 5 PP | Polypropylene |
| 6 PS | Polystyrene |
| 7 OTHER | All other plastics |

*Table 11-3*   Polymers: Plastics, Resins, and Rubbers

| NAME | USE |
|---|---|
| Polyethylene | Electrical insulation, packaging material (sandwich bags, plastic wrap), molded toys and utensils, milk carton coatings |
| Polyethylene terephthalate | Clear plastic bottles |
| Polypropylene | Molded containers, bottles, hospital utensils (sterilizable), washing machine parts, automobile interior parts |
| Polyvinyl chloride | Electrical insulation, toys, garden hoses, automobile seat covers, washable wallpaper, packaging material, bottles (shampoo), "patent leather," vinyl flooring, water pipes |
| Polytetrafluoroethylene | Teflon, electrical insulator, chemically inert material, nonstick coatings (pots and tools) |
| Polyvinyl acetate | Paints, adhesives for textiles, paper, and wood, sizing for textiles |
| Polyvinyl alcohol | Emulsifiers in cosmetics, water-soluble packaging materials |
| Polymethyl methacrylate (acrylic) | Plexiglas, Lucite, paints, light fixtures, signs, airplane windows, helicopter bubbles, dentures |
| Polystyrene | Styrofoam, packaging material, bottle caps, refrigerator interiors, toys, containers, kitchen utensils, foam insulation |
| Polyvinyldiene chloride | Saran wrap, packing material, coating material |
| Nylon (polyamides) | Machine parts (gears, cams), boil-in-the-bag containers, brushes, surgical sutures, fishing lines, carpets, stockings, clothing |
| Cellulose acetate | Photographic film, toothbrushes, combs |
| Phenol–formaldehyde resins | Bakelite, electrical insulators |
| Melamine–formaldehyde | Dishes, buttons, Melmac, Formica |
| Polyester resins | Molding compounds, Fiberglas resins, Dacron, Mylar film, synthetic fibers, paints, permanent-press clothing |
| Epoxide resins | Epoxy glues, coatings |
| Polyurethanes | Foams for packaging, insulation, furniture |
| Silicones | Polishes, lubricants, high-temperature rubber |
| *cis*-Polyisoprene | Natural rubber or latex rubber |
| Styrene–butadiene rubber | Synthetic rubber |
| Polychloroprene | Neoprene synthetic rubber |

## *Chemical Reactions*

Evidence for chemical reactions:

1. A gas forms or disappears. When an Alka-Seltzer tablet dissolves a gas bubbles from the solution.
2. A solid forms or disappears. When a candle burns the solid wax disappears.
3. A liquid forms or disappears. When ethyl alcohol burns the liquid alcohol disappears.
4. A color change occurs. When sucrose decomposes the white sugar changes to a black solid.
5. Energy is released as heat and sometimes as light (exothermic reactions). When a candle burns heat and light are released.
6. Energy is absorbed as heat from the surroundings (endothermic reactions). When baking soda reacts with vinegar heat is absorbed and the temperature of the solution drops.

Reaction Rate: The speed of a reaction as measured by the decrease in the concentration or amount of a reactant or the increase in the concentration or amount of a product.

Factors that affect reaction rates:

1. Temperature: Reaction rates are increased by heating. Hot wood burns faster than cooler wood.
2. State of Subdivision: For solids and liquids reaction rates are increased by increasing the state of subdivision of a reactant. Wood burns faster when it is subdivided into smaller pieces of solid.
3. Concentration: Reaction rates are increased by increasing the concentration of a reactant. Acetylene reacts faster when mixed with pure oxygen gas compared to its reaction in air.
4. Catalyst: A chemical that increases the speed or rate of a chemical reaction without undergoing permanent chemical change. The chemicals in the mantle of a camping lantern catalyze the reaction of the lantern fuel with oxygen in the air.

Oxidation number or oxidation state: A signed number that reflects the state of chemical combination of an element.

Oxidation number rules:

1. The oxidation number of an element in the natural form uncombined with other elements is zero, 0. The oxidation number of iron in Fe is 0 and the oxidation number of iodine in $I_2$ is 0.
2. The oxidation number of an element in a simple ion is given by the charge of the ion. The oxidation number of calcium in $Ca^{2+}$ is +2 and the oxidation number of chlorine in $Cl^-$ is −1.
3. Hydrogen normally has an oxidation number of +1 in its compounds.
4. Oxygen normally has an oxidation number of −2 in its compounds. However, oxygen has a −1 oxidation number in compounds called peroxides, such as hydrogen peroxide, $H_2O_2$.
5. The algebraic sum of the oxidation numbers of the elements in a compound equals zero. In $CaCl_2$ the sum of the oxidation numbers equals zero, $(+2) + 2(-1) = 0$.
6. The sum of the oxidation numbers of the elements in a polyatomic ion equals the charge of the ion. For hydronium ion, $H_3O^+$, the sum of the oxidation numbers is $3(+1) + (-2) = +1$.

7. Metals normally have positive oxidation numbers in compounds. Nonmetals have negative oxidation numbers when combined with metals. However, nonmetals can have positive oxidation numbers when combined with more electronegative nonmetals. The more electronegative elements always take on the negative oxidation numbers related to their groups in the periodic table.

Common types of chemical reactions:

1. Electron transfer or redox reactions—reactions in which elements change oxidation number. Typically one element increases oxidation number and another element decreases oxidation number as a result of electron transfer. $2Fe + 3Cl_2 \longrightarrow 2FeCl_3$

2. Precipitation or ion combination reactions—reactions in which cations and anions in solutions combine to form an insoluble ionic compound. $Fe^{3+}(aq) + 3OH^-(aq) \longrightarrow Fe(OH)_3(s)$

3. Acid-base reactions—reactions in which an acid reacts with a base and a hydrogen ion transfers from the acid to the base. $HF(aq) + OH^-(aq) \longrightarrow H_2O + F^-(aq)$

4. Polymerization reactions—reactions in which monomer molecules react to form macromolecules called polymers. monomer $\longrightarrow$ polymer     styrene $\longrightarrow$ polystyrene

Hydrogen peroxide in water solution can decompose to give water and oxygen gas.

$$2H_2O_2(aq) \longrightarrow 2H_2O(\ell) + O_2(g)$$

The rate of the reaction can be observed by noting how fast the gas is released. In this exercise you are to observe the effect of various catalysts on the reaction. You need: medicinal hydrogen peroxide solution, a tablespoon, some small glasses, a shiny penny, a dull penny, a leaf from a green plant, commercial bleach, some small rusty iron object (if available).

In each part of this exercise place three tablespoons of hydrogen peroxide solution in a glass, add the chemical to be used as a catalyst, and observe the process for a few minutes. After each test you can rinse out the glass with water so that it can be used for the next test or use several glasses.

1. Put a shiny penny and a dull penny in a glass of hydrogen peroxide. Compare them.
2. Crush a leaf from a green plant and put it in a glass of hydrogen peroxide.
3. Put two drops of commercial bleach in a glass of hydrogen peroxide.
4. Put a small rusty object in a glass of hydrogen peroxide.

Summarize your observations by making a table listing the chemical tested and the catalytic effect noted. Describe the effects as none, very slight, noticeable, or great.

You need: water, vegetable oil, one-fourth of a cup of saturated salt solution, seven steel paper clips, matches, and three glasses

1. Collect six samples of plastic materials. Try to find samples of apparently different clear plastics, although colored plastics will do. You need three small samples of each plastic no larger than about 1 cm by 1 cm. Also obtain three small samples of Styrofoam plastic, such as a Styrofoam cup or egg carton.

2. You are going to observe some properties of the plastics. Make a table having entries for appearance, density, and effect of heat. Work in a well-ventilated area and avoid breathing any fumes when heating plastics.

3. Manipulate a sample of each plastic with your fingers to note whether it is stretchable, flexible, or rigid.

4. To one glass add one-fourth of a cup of water. To the second glass add one-fourth of a cup of saturated salt solution. To the third glass add one-fourth of a cup of vegetable oil.

5. Test the plastics one at a time by dropping a piece into the water. Record whether it floats or sinks. If the sample floats in water, add a piece to vegetable oil and record whether it sinks or floats. On the other hand, if the sample sinks in water, add a piece to the saturated salt solution and record whether it floats or sinks. Most plastics repel water so you may need to push the samples below the surface with your finger.

6. The density of vegetable oil is 0.8 g/mL, the density of water is 1 g/mL, and the density of saturated salt solution is 1.2 g/mL. Use these densities to describe the densities of each plastic sample based upon your observations in Step 5. Styrofoam contains pockets of gas, used to make the foam, so its density is relatively low.

7. Test the effect of heat on each plastic. To make a test, straighten the larger end of a paper clip to make a device for holding a piece of plastic. A piece of plastic can be placed in the springy part of the smaller loop of the clip. Put a piece of plastic in the clip and hold a lighted match under the sample. The sample may melt or burn so keep the match under it until you observe the effect of heat. Record your observations in the data table. **CAUTION: Work in a well-ventilated area and be careful not to let hot plastic touch your skin.**

# QUESTIONS

### Section 11-1

1. Describe some factors that serve as evidence for chemical reactions.

2. Imagine a burning candle. Describe any evidence for chemical reactions that occur when a candle burns.

3. What evidence is there for the metabolism of carbohydrates in our bodies?

### Section 11-2

4. What is meant by the rate of a chemical reaction?

5. What is a chemical explosion?

6. How does an airbag work?

## Section 11-3

7. Tell how each of the following can alter the rate of a chemical reaction.

(a) temperature    (b) state of subdivision

(c) concentration

8. Suppose you want to burn some coal in air quickly. Explain how each of the following factors can be changed to increase the rate of the combustion of coal.

(a) temperature

(b) state of subdivision

(c) concentration

9. Explain how the burning of wood is encouraged by each of the following. Indicate the factor that affects the rate in each case.

(a) A match is used to start the fire.

(b) Kindling is used to get the fire started.

(c) Air is blown on a freshly lit fire.

(d) The firewood is loosely stacked with numerous air spaces.

10. What is a catalyst and how does a catalyst alter the rate of a chemical reaction?

11. *In each of the following cases, explain what happens in terms of the factor or factors that affect reaction rates.

(a) Flour dust in a flour mill can cause an explosion.

(b) Nitrogen and oxygen in the air combine in the high-temperature operation of the internal combustion engine.

$$N_2 + O_2 \longrightarrow 2NO$$

(c) Increasing pressure in the synthesis of ammonia speeds up the reaction.

$$N_2 + 3H_2 \longrightarrow 2NH_3$$

(d) Microorganisms can make ammonia from $N_2$ at low temperature and pressure.

(e) Fish become very active and sometimes die in heated water.

(f) Smoking close to a hospital oxygen mask can cause a fire.

(g) Wood shavings are easier to ignite than larger pieces of wood.

(h) A lighted candle will extinguish when burned in a closed container.

12. In each of the following cases, explain what happens in terms of the factor or factors that affect reaction rates.

(a) Swimmers become very lethargic and sometimes die in extremely cold water.

(b) Decreasing the air supply in a Bunsen burner gives a "cooler" flame.

(c) A person can become weak and lose energy at high altitudes.

(d) A safety match ignites when it is rubbed on the striking surface.

(e) A piece of aluminum metal, when coated with mercury, will quickly react with oxygen in the air to form aluminum oxide.

(f) Coal dust in a coal mine can cause an explosion.

(g) A wad of cotton will burn more rapidly than a piece of wood even though they both are composed of cellulose.

(h) Methyl alcohol can be made from carbon monoxide and hydrogen by carrying out the following reaction at high pressures in the presence of a catalyst.

$$CO + 2H_2 \longrightarrow CH_3OH$$

13. Describe the chemical reactions used in rockets.

## Section 11-4

14. What is a catalyst?

15. Describe or define the following:

(a) heterogeneous catalyst

(b) homogeneous catalyst    (c) enzyme

16. Explain the function of a catalytic converter on an automobile.

17. Generally how does a catalyst work?

18. What is an inhibitor?

## Section 11-5

19. What is an oxidation number?

20. If a compound contains two elements, which element would be expected to have a positive oxidation number and which would be expected to have a negative oxidation number?

21. When an element is in the form of a simple ion what is the oxidation number of the element?

22. What is the oxidation number of an element when it is uncombined with other elements? What are the expected oxidation numbers of oxygen and hydrogen in their compounds?

23. What is true about the sum of the oxidation numbers of the elements in a compound?

24. What is true about the sum of the oxidation numbers of the elements that compose a polyatomic ion?

25. *Determine the oxidation number of the element other than hydrogen and oxygen in the following. Use the expected oxidation numbers for hydrogen and oxygen.

(a) $NO_3^-$         (l) $CO_2$
(b) $NO$             (m) $C_2O_4^{2-}$
(c) $NO_2$           (n) $CH_2O$
(d) $MnO_4^-$        (o) $Ag_2O$
(e) $MnO_4^{2-}$     (p) $AsO_3^{3-}$
(f) $SO_2$           (q) $AsO_4^{3-}$
(g) $SO_4^{2-}$      (r) $NH_3$
(h) $Cr_2O_7^{2-}$   (s) $SiO_3^{2-}$
(i) $H_2C_2O_4$      (t) $Al(OH)_4^-$
(j) $ClO_3^-$        (u) $H_2O_2$ (use +1 for H
(k) $MnO_2$              and deduce O)

26. Determine the oxidation number of the element other than hydrogen and oxygen in the following. Use the expected oxidation numbers for hydrogen and oxygen.

(a) $NH_4^+$        (e) $Cl_2O$
(b) $IO_3^-$        (f) $CH_4$
(c) $CrO_4^{2-}$    (g) $I_2$
(d) $H_2SO_4$       (h) $HNO_3$
(i) $MgO$           (n) $N_2O_4$
(j) $HSO_4^-$       (o) $P_4O_{10}$
(k) $HCO_3^-$       (p) $BrO_3^-$
(l) $PO_4^{3-}$     (q) $CO$
(m) $N_2H_4$        (r) $CaH_2$

## Section 11-6

27. What are electron transfer reactions?

28. Write equations for the reactions of the following elements. List the oxidation number for each element on both sides of the equation.

(a) $Ca + Br_2$   (b) $Mg + O_2$   (c) $K + F_2$
(d) $S + O_2$ to give $SO_2$

29. Write equations for the reactions of the following elements. List the oxidation number for each element on both sides of the equation.

(a) $Mg + N_2$   (b) $Ba + S$   (c) $Na + I_2$
(d) $N_2 + O_2$ to give $NO$

## Section 11-7

30. Classify the following electron transfer reactions as synthesis (combination), decomposition, displacement, or combustion. Give the oxidation numbers of the elements in the reactants and products.

(a) $2NCl_3 \longrightarrow N_2 + 3Cl_2$
(b) $H_2 + Cl_2 \longrightarrow 2HCl$
(c) $C_2H_4 + 3O_2 \longrightarrow 2CO_2 + 2H_2O$
(d) $3MnO_2 + 4Al \longrightarrow 2Al_2O_3 + 3Mn$

31. Classify the following electron transfer reactions as synthesis (combination), decomposition, displacement, or combustion. Give the oxidation numbers of the elements in the reactants and products.

(a) $2NaN_3 \longrightarrow 2Na + 3N_2$
(b) $2C_6H_6 + 15O_2 \longrightarrow 12CO_2 + 6H_2O$
(c) $2C + O_2 \longrightarrow 2CO$
(d) $CH_4 + Cl_2 \longrightarrow CH_3Cl + HCl$

## Section 11-8

32. What is a net-ionic equation? Give an example.

33. What are spectator ions?

34. *Write balanced net-ionic equations for the following reactions.

(a) Some zinc metal, Zn, is added to a solution containing dissolved copper(II) chloride and produces copper metal, Cu, and $Zn^{2+}$(aq).

(b) $Cl_2$(aq) reacts with a solution containing $Br^-$ (aq) and produces $Br_2$(aq) and $Cl^-$(aq).

(c) A solid piece of zinc is added to a hydrochloric acid solution and produces hydrogen gas, water, and zinc ion, $Zn^{2+}$(aq).

(d) A solution of $K_2CrO_4$ is added to a solution of $BaCl_2$ and produces solid $BaCrO_4$.

(e) A solution of NaOH is added to a solution of $Na_2HPO_4$ and produces $H_2O$ and $PO_4^{3-}$(aq).

(f) Solid $CaCO_3$ is added to a solution of hydrochloric acid and produces $CO_2$, $H_2O$, and $Ca^{2+}$(aq). [Hint: $CaCO_3$(s) is one reactant and hydrochloric acid, the other reactant, is $H_3O^+(aq)$ and $Cl^-(aq)$.]

35. Write balanced net-ionic equations for the following reactions.

(a) $Cl_2(aq)$ reacts with a solution containing $I^-(aq)$ to give $Cl^-(aq)$ and $I_2$.

(b) Some copper, Cu, metal is added to a solution containing dissolved $AgNO_3$; a reaction occurs and produces Ag metal and $Cu^{2+}(aq)$.

(c) A solution of KI is added to a solution of $Pb(NO_3)_2$ and solid $PbI_2$ is formed.

(d) A solution of KOH is added to a solution of $NaHCO_3$, producing $H_2O$ and $CO_3{}^{2-}(aq)$.

(e) A piece of solid Na metal reacts with water to form hydrogen gas, hydroxide ion, $OH^-$, and sodium ion, $Na^+$.

(f) A solution of the weak electrolyte oxalic acid, $H_2C_2O_4$, is mixed with a solution of NaOH; in the reaction water and oxalate ion, $C_2O_4{}^{2-}$, are formed.

## Section 11-9

36. What is a precipitation reaction? Give an example.

## Sections 11-10 and 11-11

37. What are the differences between soluble, slightly soluble, and insoluble ionic compounds?

38. Give an example of an insoluble ionic compound.

39. *Write balanced net-ionic equations for any precipitation reactions that occur when the following solutions are mixed. Use the general approach to predicting precipitation reactions and refer to the solubility rules in Section 11-10.

(a) mercury(I) nitrate solution and sodium bromide solution

(b) potassium fluoride solution and barium nitrate solution

(c) lithium chloride solution and magnesium chloride solution

(d) ammonium nitrate solution and sodium nitrate solution

(e) potassium hydroxide solution and calcium chloride solution

(f) barium nitrate solution and sodium phosphate solution

(g) lead(II) acetate solution and sodium iodide solution

40. Write balanced net-ionic equations for any precipitation reactions that occur when the following solutions are mixed. Use the general approach to predicting precipitation reactions and refer to the solubility list given in Section 11-10.

(a) lead(II) nitrate solution and potassium chloride solution

(b) sodium fluoride solution and calcium nitrate solution

(c) ammonium chloride solution and sodium chloride solution

(d) sodium hydroxide solution and magnesium chloride solution

(e) barium nitrate solution and sodium sulfate solution

(f) sodium chloride solution and zinc chloride solution

(g) silver nitrate solution and potassium bromide solution

41. *Predict and give balanced net-ionic equations for any precipitation reactions that occur when the following solutions are mixed. Base the predictions on the solubility list given in Section 11-10.

(a) $MgCl_2$ solution and $Ba(OH)_2$ solution

(b) $Ca(NO_3)_2$ solution and NaF solution

(c) $Pb(NO_3)_2$ solution and $Na_3PO_4$ solution

(d) $FeCl_3$ solution and $Pb(NO_3)_2$ solution

(e) KOH solution and NaCl solution

(f) $AgNO_3$ solution and NaI solution

(g) $FeCl_3$ solution and $Ba(OH)_2$ solution

(h) $Ba(OH)_2$ solution and $CdSO_4$ solution

(i) NaCl solution and $(NH_4)_2SO_4$ solution

(j) $Pb(NO_3)_2$ solution and $K_2CrO_4$ solution

(k) $Mg(C_2H_3O_2)_2$ solution and NaF solution

42. What is the role of calcium in the human body?

## Section 11-12

43. What is an acid–base reaction?

44. What is a neutralization reaction?

45. Which of the following reactions are acid–base reactions?

(a) $C + O_2 \longrightarrow CO_2$

(b) $H_3O^+ + CN^- \longrightarrow HCN + H_2O$

(c) $HF + NH_3 \longrightarrow NH_4^+ + F^-$

(d) $H_3PO_4 + OH^- \longrightarrow H_2PO_4^- + H_2O$

(e) $CH_4 + 2O_2 \longrightarrow 2H_2O + CO_2$

46. For each of the acid–base reactions of Question 45 give the formula of the acid and the base.

### Section 11-13

47. Using the terms monomer and polymer, explain polymerization reactions.

48. What is a plastic? Give some examples.

49. What is a copolymer? Give an example.

50. Write an equation for the polymerization reaction in which ethylene, $CH_2{=}CH_2$, forms polyethylene.

51. *The average mass of a polyethylene macromolecule is approximately $5.0 \times 10^4$ u. Determine the number of ethylene units in such a macromolecule. (Hint: There are 28 u per ethylene unit.)

52. The compound styrene

$$CH_2{=}C\begin{smallmatrix} H \\[4pt] \\ C_6H_5 \end{smallmatrix}$$

polymerizes to form polystyrene. Give an equation for this reaction.

53. The compound methyl methacrylate

$$CH_2{=}C\begin{smallmatrix} CH_3 \\[4pt] \\ CO_2CH_3 \end{smallmatrix}$$

polymerizes to form polymethyl methacrylate, usually called Plexiglass, Lucite, or acrylic plastic. Give an equation for this reaction.

54. The compound vinyl chloride

$$CH_2{=}C\begin{smallmatrix} Cl \\[4pt] \\ H \end{smallmatrix}$$

polymerizes to form polyvinyl chloride, PVC. Give an equation for this reaction.

### Questions to Ponder

55. Should dual airbags be required on all new cars? Argue your case.

56. Are plastics good or bad? Explain your answer.

57. It is possible to add trace elements to chemicals that are used to make explosives. Such chemical tagging could be used to trace the manufacturer of the chemical. Do you agree that a federal law should be passed to require the tagging of all chemicals used to make explosives? Argue your case.

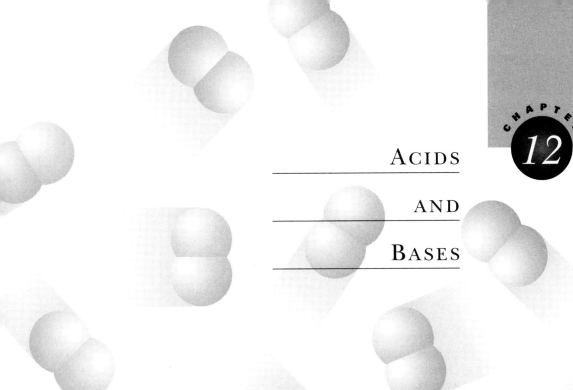

# ACIDS
# AND
# BASES

## 12-1 THE NATURE OF ACIDS AND BASES

Vinegar has a distinct and characteristic sour taste. Vinegar is a water solution of acetic acid and it serves as a good example of an acid in aqueous solution. Taste buds located on the sides of our tongues respond to sour tastes. Citric acid in lemon juice tastes sour, as does ascorbic acid or vitamin C. You've undoubtedly tasted other sour foods. The chemical structures of the molecules of acetic acid, citric acid, and vitamin C are quite different, but they do share certain chemical properties. In fact, a unique chemical property of these acids, as well as other acids, accounts for the sour taste of acid solutions. In times past, chemists often tasted chemicals to test their properties. This practice resulted in many cases of sickness and death. It is not good to taste or smell laboratory chemicals indiscriminately since many are toxic. It is interesting that the properties of some chemicals have been discovered in modern times by accidental tasting. The hallucinogenic drug LSD was discovered by a chemists who accidently ingested some of it and experienced the startling effects. Similarly, the artificial sweetener aspartame (Nutrasweet) was discovered when a chemist accidently tasted it.

It is possible to classify compounds according to similarities in chemical properties, that is, similarities in the kinds of chemical reactions they undergo. Chemists noted long ago that certain substances, now known as

**403**

You need some white vinegar, pieces of red and blue litmus paper, and two shiny zinc–copper pennies. Dip pieces of red and blue litmus paper into the vinegar and observe the colors. Use the tip of a knife or nail file to scratch the surface of one of the pennies to expose some of the silvery zinc metal. Make several large scratches. Place this penny and the other penny on a flat surface and cover each penny with a few drops of vinegar. Observe them for five minutes. Copper metal will not react with acetic acid but zinc metal does react. Write a net-ionic equation for the reaction of zinc metal with $HC_2H_3O_2$ to give $Zn^{2+}$, $C_2H_3O_2^-$, and hydrogen gas.

acids, were characterized by a set of common properties. The word acid comes from the Latin *acidus* 'sour'.

Here are some experimental observations on solutions of **acids.**

1. Acids taste sour in diluted solutions and in concentrated solutions cause skin burns by dissolving tissue.
2. Acids change the color of litmus, a vegetable dye, from blue to red.
3. Acids chemically react with and neutralize the effect of chemical bases (neutralization reactions).
4. Acids react with certain metals to dissolve them and give metal ions and hydrogen gas.

Another class of compounds, called bases, are in a sense the chemical opposites of acids.

Here are some experimental observations on solutions of **bases.**

1. Bases taste bitter and feel slippery in diluted solutions and in concentrated solutions cause skin burns by dissolving tissue.
2. Bases change the color of litmus from red to blue.
3. Bases chemically react with and neutralize the effect of acids.

Some of the first known bases were produced by strongly heating substances isolated from water extracts of wood ashes. Compounds obtained in this manner were viewed as the basis, or base, from which other compounds could be made. Thus, they came to be known as bases.

You need red and blue litmus paper, milk of magnesia ($Mg(OH)_2$), vinegar, and a teaspoon. Wet your index finger and thumb then rub them together. Put one or two drops of milk of magnesia on your fingers then rub them together. Dip red and blue litmus paper in a sample of milk of magnesia. Wipe of any excess chemical and observe the colors. Note the distinctive odor of vinegar. Put two small drops of milk of magnesia in a teaspoon, nearly fill the spoon with vinegar, and stir the contents with a toothpick, a stick, or your finger. Note the odor of the mixture.

## 12-2 Definitions for Acids and Bases

As a deeper understanding of the nature of chemicals developed, it became apparent that there were many substances that could be classified as acids or bases. It is now possible to give specific chemical definitions for these classes of compounds. Several definitions are possible and we consider the most useful of them.

In 1884, a Swedish chemist, Svante Arrhenius, proposed the first significant definitions of acid and base. He defined an acid as a chemical that forms hydrogen ions ($H^+$) in water solution and a base as a chemical that forms hydroxide ions ($OH^-$) in water solution. In the Arrhenius view, the behavior of hydrogen chloride as an acid is represented as

$$HCl \longrightarrow H^+ + Cl^-$$

and the behavior of sodium hydroxide as a base is represented as

$$NaOH \longrightarrow Na^+ + OH^-$$

The Arrhenius theory established the fact that hydrogen ions and hydroxide ions are related to acid and base behavior in water. The theory, however, was not general enough to include all chemicals that behave as acids and bases. More general definitions come from the **Brønsted–Lowry** theory of acids and bases. Definitions are given in the margin.

A **proton, ($H^+$),** is a positively charged hydrogen ion, a bare hydrogen nucleus, that is, a hydrogen atom without an electron. A lone proton is formed by the breaking of a covalent bond involving hydrogen and some other element so that the shared electron pair remains with the other element. In this discussion hydrogen ions are called protons. The loss of a proton by an acid can be represented generally as

$$H{-}A \longrightarrow H^+ + :A^-$$

where $H^+$ is the proton, A is an element that covalently bonds to hydrogen, and : represents the electrons of the covalent bond.

A proton is a unique chemical particle; it is a hydrogen ion but, unlike other ions, it has no electrons. When an acid loses a proton a base gains it. This means that free protons are not produced by acids; instead they are passed from acids to bases, consequently a **base** is a species that can form a covalent bond with a proton. A base gains or accepts a proton by forming a covalent bond with it. Before continuing your study of acids and bases you should review the nomenclature of acids given in Sections 8-7 and 8-8.

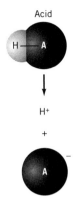

**Acid:**
*A chemical species that can lose hydrogen ions or protons in a chemical reaction; a proton donor.*

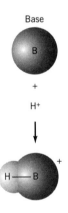

**Base:**
*A chemical species that can gain hydrogen ions or protons in a chemical reaction; a proton acceptor.*

## 12-3   ACID–BASE REACTIONS

Acid–base reactions were mentioned in Section 11-12. Let's look at one of these reactions in more detail. A solution of acetic acid has the typical properties of an acid. A solution of sodium hydroxide has the typical properties of a base. When these two solutions are mixed the acid and base neutralize each others' properties in a chemical reaction. An acid reacts

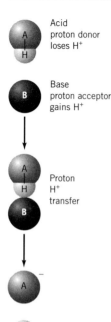

Acid
proton donor
loses H⁺

Base
proton acceptor
gains H⁺

Proton
H⁺
transfer

**FIGURE 12-1** An acid–base reaction.

with a base in a proton transfer or acid–base reaction (see Fig. 12-1). As an example of an acid–base reaction, consider the reaction between acetic acid and the hydroxide ion present in a sodium hydroxide solution. The reaction occurs when a solution of NaOH is mixed with a solution of $HC_2H_3O_2$. The net-ionic equation (see Section 11-8) for the reaction is

$$HC_2H_3O_2(aq) + OH^-(aq) \xrightarrow{\text{proton transfer}} H_2O + C_2H_3O_2^-(aq)$$

Look closely at the formulas and note that an $H^+$ transfers from $HC_2H_3O_2$ to $OH^-$ to give $H_2O$ and $C_3H_3O_2^-$.

Any acid–base reaction can be generally represented as

$$\underset{\text{acid}}{H-A} + \underset{\text{base}}{B} \xrightarrow{\text{proton transfer}} \underset{\text{acid}}{H-B} + \underset{\text{base}}{A}$$

In such a chemical reaction an acid loses a proton to a species that is capable of bonding to the proton more readily than the original acid. Thus, we say that an acid, H—A, loses a proton to form a base, A; and a base, B, gains a proton to form an acid, H—B.

In an acid–base reaction there is an acid as a reactant and an acid as a product. The acid that is a reactant is said to be a stronger acid than the acid formed as a product. A **stronger acid** has a greater tendency to lose a proton. Furthermore, in any acid–base reaction there is a base as a reactant and a base as a product. The base that is a reactant is said to be a stronger base than the base formed as a product. A **stronger base** has a greater tendency to gain a proton.

| H—A | + | B | ⇌ | H—B | + | A |
|---|---|---|---|---|---|---|
| weaker proton binder (the stronger acid) | | stronger proton acceptor (the stronger base) | | greater proton binder (the weaker acid) | | weaker proton acceptor (the weaker base) |

In simple terms, when an acid–base reaction occurs, the stronger acid and stronger base form a weaker acid and a weaker base.

Typically acid–base reactions are reversible and involve chemical equilibrium so double arrows are often, but not always, shown in the equations. When the larger arrow is shown to the right (⇌) the acid–base reaction is extensive in the direction of the products. When the larger arrow is directed to the left (⇌) this means that the reaction is not extensive towards the products but only occurs to a slight extent. (See Section 10-18).

## 12-4 IDENTIFYING ACIDS AND BASES

Every Brønsted–Lowry acid has a corresponding base formed by the loss of a proton, and each base has a corresponding acid formed by the gain of a proton. An acid and the base it forms by the loss of a proton are a **conjugate acid–base pair.** Likewise, a base and the acid it forms by gaining a proton are a **conjugate acid–base pair.** In the acid–base reaction between acetic acid and hydroxide ions, referred to in the previous section, $HC_2H_3O_2$ and $C_2H_3O_2^-$ are a conjugate acid–base pair and $H_2O$ and $OH^-$ are a conjugate acid–base pair.

$$\overbrace{HC_2H_3O_2(aq) + OH^-(aq)}^{\text{conjugate pair}} \rightleftharpoons H_2O + C_2H_3O_2^-(aq)$$
(conjugate pair)

The double arrow indicates that the reaction is extensive towards the product side. In any acid–base reaction there are two sets of conjugate acid–base pairs. Each pair is related by the loss or gain of a proton.

When an acid loses a proton it is converted to its conjugate base. The base that gains the proton is, in turn, converted to its conjugate acid (see Fig. 12-1). Acid–base reactions are common types of reactions in water solutions. In any Brønsted–Lowry acid–base reaction it is possible to label each acid and base. The acetic acid–hydroxide ion reaction can be labeled as

$$\begin{array}{cccc} \text{acid 1} & \text{base 2} & \text{acid 2} & \text{base 1} \\ HC_2H_3O_2(aq) & + \; OH^-(aq) & \rightleftharpoons \quad H_2O & + \quad C_2H_3O_2^-(aq) \\ \text{stronger} & \text{stronger} & \text{weaker} & \text{weaker} \\ \text{acid} & \text{base} & \text{acid} & \text{base} \end{array}$$

The labels mean that acid 1 pairs with base 1 and base 2 pairs with acid 2. In the reaction the stronger acid, acetic acid, reacts with the stronger base, the hydroxide ion, to form the weaker acid, water, and the weaker base, the acetate ion. Be careful with this terminology. The stronger an acid the greater its ability to lose a proton; the stronger a base the greater its ability to gain a proton. The term stronger means more chemically reactive; stronger acids and stronger bases react to form weaker acids and weaker bases.

---

**EXAMPLE 12-1**

Label the conjugate acid–base pairs in the following reaction and indicate which acid is the stronger acid and which base is the stronger base.

$$HC_2H_3O_2(aq) + NH_3(aq) \rightleftharpoons NH_4^+(aq) + C_2H_3O_2^-(aq)$$

On the left, we pick the species that loses an $H^+$ as the acid and the species that gains an $H^+$ as the base. On the right, the acid is the conjugate acid of the reactant base and the base is the conjugate base of the reactant acid. The stronger acid and base are those that more readily lose and gain protons. Thus, according to the size of the double arrow, the stronger acid and base are on the left.

| acid 1 | | base 2 | | acid 2 | | base 1 |
|--------|--|--------|--|--------|--|--------|
| $HC_2H_3O_2(aq)$ | + | $NH_3(aq)$ | $\rightleftharpoons$ | $NH_4^+(aq)$ | + | $C_2H_3O_2^-(aq)$ |
| stronger acid | | stronger base | | weaker acid | | weaker base |

---

### EXAMPLE 12-2

Indicate which species are conjugate acid–base pairs in the reaction:

$$H_3O^+(aq) + F^-(aq) \rightleftharpoons HF(aq) + H_2O$$

**CAUTION: Hydrofluoric acid is very dangerous.**

The conjugate acid–base pairs are related by the loss and gain of an $H^+$. So we look for the acid as the reactant that loses an $H^+$ to form its conjugate base as a product. The base is the reactant that gains an $H^+$ to form its conjugate acid as a product.

| | ACIDS | BASES |
|--|-------|-------|
| Conjugate pair | $H_3O^+$ | $H_2O$ |
| Conjugate pair | HF | $F^-$ |

**12-3**

For the following acid base reaction label the conjugate acid–base pairs (see Example 12-1) and make a table of the acids and bases (see Example 12-2).

$$H_2SO_4 + H_2O \longrightarrow$$
$$H_3O^+(aq) + HSO_4^-(aq)$$

---

## 12-5 ACIDIC AND BASIC SOLUTIONS

Many common acids are stored as solutions made by dissolving them in water. When the acid hydrogen chloride is dissolved in water, the following acid–base reaction occurs:

$$HCl + H_2O \longrightarrow H_3O^+(aq) + Cl^-(aq)$$

| | | hydronium | chloride |
|--|--|-----------|----------|
| acid | base | ion | ion |

Hydrogen chloride is an example of an acid that is a strong electrolyte since it reacts almost completely with water to form hydronium ions and chloride ions. Note that since the reaction is essentially complete, no double arrow is used in the equation. In other words, HCl in water is 100% ionized.

Many bases are also stored as water solutions. When ammonia gas is mixed with water, the following equilibrium reaction occurs.

$$NH_3 + H_2O \rightleftharpoons NH_4^+(aq) + OH^-(aq)$$

| | | ammonium | hydroxide |
|--|--|----------|-----------|
| base | acid | ion | ion |

Ammonia is quite soluble in water but reacts to only a slight extent, which means that ammonia is the weaker base and $H_2O$ is the weaker acid in the

reaction. The relative sizes of the arrows in the equation show that the reaction does not occur extensively to the right. Note that water behaves as
an acid in this reaction but as a base in the reaction with hydrogen chloride. Water can be an acid in some reactions and a base in other reactions;
this behavior is compatible with its structure.

$\overset{..}{\underset{H \quad H}{O}}$     gain of a proton forms     $\overset{..}{\underset{H \; \underset{H}{H}}{O^+}}$     hydronium ion

$\overset{..}{\underset{H \quad H}{O}}$     loss of a proton forms     $:\overset{..}{\underset{..}{O}}—H^-$     hydroxide ion

Whether water acts as an acid or a base depends upon the chemical
species that react with it. A species that is a stronger acid than water will
donate a proton to water. A species that is a stronger base than water will
accept a proton from water.

An animal species that can exist both on land and in water is an am-
phibian. A chemical species that can act as an acid or a base is said to be
amphiprotic. Notice the word amphiprotic is related to the word amphib-
ian and an amphiprotic species can play two roles. An **amphiprotic** species
is one that can gain a proton or lose a proton, depending upon the other
chemicals with which it is mixed. Since water is amphiprotic, some proton
transfer occurs between individual water molecules in any sample of pure
water or any water solution. That is, water molecules react with one another to a slight extent as shown by the equilibrium equation:

$$H_2O + H_2O \rightleftharpoons H_3O^+ + OH^-$$

This reversible reaction proceeds only slightly to the right, producing, in
pure water, equal but relatively low concentrations of hydronium and hydroxide ions. This means that a sample of water contains mostly water
molecules along with very minor concentrations of $H_3O^+$ and $OH^-$. Pure
water or water solutions that contain neither acids nor bases have equal
concentrations of hydronium and hydroxide ions. Such solutions are
called **neutral solutions.**

When an acid is mixed with water it reacts with water to form hydronium ions. If we represent an acid by the general formula HA, the equation for the reaction of an acid with water is

$$HA + H_2O \rightleftharpoons H_3O^+ + A^-$$

A water solution that has a concentration of hydronium ions that is
greater than that in pure water is called an **acidic solution.** When an acid
dissolves in water, an acidic solution results. In highly acidic solutions the
concentrations of hydronium ions are relatively high. Whereas in slightly
acidic solutions the concentrations of hydronium ions are relatively low
but still higher than pure water.

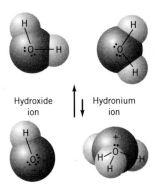

Hydroxide ion

Hydronium ion

A base mixed with water reacts to form some hydroxide ions. If we use B as a general formula for a base the equation for the reaction of a base with water is

$$B + H_2O \rightleftharpoons BH^+ + OH^-$$

A water solution that has a concentration of hydroxide ions greater than in pure water is called a **basic solution.** Dissolved in water, bases form basic solutions also known as **alkaline solutions.** In highly basic solutions the concentrations of hydroxide ions are relatively high. Whereas in slightly basic solutions the concentrations of hydroxide ions are relatively low but still higher than pure water.

## 12-6 AN ACID–BASE TABLE

We know that water solutions of hydrogen chloride and acetic acid are acidic, but solutions of these acids differ in one very important way. As discussed in Section 10-18, hydrogen chloride is a strong electrolyte and acetic acid is a weak electrolyte. The reaction of each of these acids with water reveals how they differ.

$$HCl + H_2O \longrightarrow H_3O^+(aq) + Cl^-(aq)$$

$$HC_2H_3O_2 + H_2O \rightleftharpoons H_3O^+(aq) + C_2H_3O_2^-(aq)$$

Hydrogen chloride reacts completely with water to form hydronium ions and chloride ions, that is, it is 100% ionized in solution. Hydrogen chloride is an example of a strong acid. Other common strong acids are nitric acid, $HNO_3$, and sulfuric acid, $H_2SO_4$. Recall that these are also the three common strong molecular electrolytes. The term **strong acid** is used because an acid of this type mixed with water reacts completely with water to form hydronium ions and their corresponding anions (the conjugate base of the acid).

In contrast to the strong acid hydrogen chloride, acetic acid has only a slight tendency to react with water to form hydronium ions and acetate ions. Acids that react with water only to a slight extent to form relatively low concentrations of hydronium ions and the corresponding conjugate base are called **weak acids.** Many acids are known to be weak acids. When weak acid solutions are compared, it is found that weak acids vary in their ability to react with water; that is, some have a greater tendency to lose protons to water than others. The strength of a weak acid relative to other weak acids can be determined by experiment. In fact, it is possible to rank or list acids according to their strengths.

Some common acids ranked in order of decreasing strength are shown in Table 12-1. The stronger the acid, the higher it ranks in the table. Of course, the three common strong acids are listed at the top, but even

*Table 12-1*   Acid–Base Table of Common Brønsted–Lowry Acids and Bases

| | ACIDS | | BASES | | |
|---|---|---|---|---|---|
| *Strong Acids*<br>React completely<br>with water to<br>form $H_3O^+$ and<br>conjugate base | Sulfuric acid | $H_2SO_4$ | $HSO_4^-$ | Hydrogen<br>sulfate ion | *Very Weak Bases*<br>Will not react<br>with $H_3O^+$ to<br>form conjugate<br>acid |
| | Hydrogen chloride | HCl | $Cl^-$ | Chloride ion | |
| | Nitric acid | $HNO_3$ | $NO_3^-$ | Nitrate ion | |
| *Weak Acids*<br>Do not react<br>extensively with<br>water | Hydronium ion | $H_3O^+$ | $H_2O$ | Water | *Weak Bases* |
| | Oxalic acid | $H_2C_2O_4$ | $HC_2O_4^-$ | Hydrogen<br>oxalate ion | |
| | Hydrogen sulfate ion | $HSO_4^-$ | $SO_4^{2-}$ | Sulfate ion | |
| | Phosphoric acid | $H_3PO_4$ | $H_2PO_4^-$ | Dihydrogen<br>phosphate ion | |
| | Hydrogen fluoride | HF | $F^-$ | Fluoride ion | |
| | Hydrogen oxalate<br>ion | $HC_2O_4^-$ | $C_2O_4^{2-}$ | Oxalate ion | |
| | Acetic acid | $HC_2H_3O_2$ | $C_2H_3O_2^-$ | Acetate ion | |
| | Carbon dioxide (aq) | $(CO_2 + H_2O)$ | $HCO_3^-$ | Hydrogen<br>carbonate ion | |
| | Hydrogen sulfide | $H_2S$ | $HS^-$ | Hydrogen<br>sulfide ion | |
| | Dihydrogen<br>phosphate ion | $H_2PO_4^-$ | $HPO_4^{2-}$ | Hydrogen<br>phosphate ion | |
| | Hydrogen sulfite ion | $HSO_3^-$ | $SO_3^{2-}$ | Sulfite ion | |
| | Ammonium ion | $NH_4^+$ | $NH_3$ | Ammonia | |
| | Hydrogen cyanide | HCN | $CN^-$ | Cyanide ion | |
| | Hydrogen carbonate<br>ion | $HCO_3^-$ | $CO_3^{2-}$ | Carbonate ion | |
| | Hydrogen phosphate<br>ion | $HPO_4^{2-}$ | $PO_4^{3-}$ | Phosphate ion | |
| | Hydrogen sulfide ion | $HS^-$ | $S^{2-}$ | Sulfide ion | |
| | Water | $H_2O$ | $OH^-$ | Hydroxide<br>ion | *Strong Base* |

(Left margin, top-to-bottom arrow: Decreasing strength)

(Right margin, bottom-to-top arrow: Increasing base strength)

among the weak acids there is variation in strength. Be careful to distinguish between "strong acid" and "stronger acid." Strong acid has a definite meaning. All strong acids are 100% ionized in water.

Any acid has a corresponding conjugate base which it can form by the loss of a proton. The conjugate base of each acid is included in Table 12-1. The bases are listed in the right-hand column. The conjugate bases are also listed according to relative strength. However, they are listed with the weakest at the top and the strongest at the bottom. This makes sense because the stronger an acid the weaker its conjugate base, and the weaker an acid the stronger its conjugate base.

Look closely at Table 12-1. Note that water is listed as a very weak acid at the bottom of the list of acids and as a weak base near the top of the list of bases. Any of the strong acids, when mixed with water, reacts to give hydronium ions, $H_3O^+$, and the conjugate base. Thus, the **hydronium ion** is the strongest acid that can exist in water.

Weak acids include ionic species as well as molecular substances. For example, $H_2S$ is a molecular weak acid and the hydrogen sulfide ion, $HS^-$, is an even weaker acid. All weak acids react only slightly with water, so the major species in a weak acid solution is the acid itself. For instance, a solution of hydrogen sulfide contains $H_2S$ as the major species since it only reacts to a slight extent. When carbon dioxide is dissolved in water to give carbonated water, the solution is acidic. Since $CO_2$ has no proton, we can explain its acidic nature as a close association and reaction with water. The equation for the reaction of carbon dioxide with water to give an acidic solution is

$$CO_2 + 2H_2O \rightleftharpoons H_3O^+ + HCO_3^-$$

As a consequence of this behavior, carbon dioxide is included in the table as $CO_2 + H_2O$. You may see this written as $H_2CO_3$, called carbonic acid, but there is little evidence for the existence of such a compound. If it does form, it quickly reacts with water or separates into carbon dioxide and water.

Returning to Table 12-1, note that most common bases are negatively charged ionic species. The only exceptions are water and ammonia, $NH_3$. Water and ammonia are molecular bases. Other molecular bases are known but are not included in the table. In a water solution of any base the base is the major species. For example, in a solution of ammonia, $NH_3(aq)$ is the major species.

$$\underset{\text{major}}{NH_3} + H_2O \rightleftharpoons \underset{\text{minor}}{NH_4^+(aq) + OH^-(aq)}$$

And in a solution of NaOH, $Na^+$ and $OH^-$ are the major species.

$$NaOH(s) \longrightarrow \underset{\text{major}}{Na^+(aq) + OH^-(aq)}$$

## 12-7 COMMON ACID AND BASE SOLUTIONS

In the laboratory, you may encounter water solutions of strong acids such as hydrochloric acid, sulfuric acid, or nitric acid. These acids are also widely used in industry for various chemical processes. Strong acid solutions contain hydronium ions and the conjugate base (see Table 12-1). To represent the major species in these solutions, we write the formulas of both ions.

| | |
|---|---|
| hydrochloric acid solution | $H_3O^+(aq) + Cl^-(aq)$ |
| sulfuric acid solution | $H_3O^+(aq) + HSO_4^-(aq)$ |
| nitric acid solution | $H_3O^+(aq) + NO_3^-(aq)$ |

A solution of a strong acid always contains the hydronium ion as a major ion. **Solutions of strong acids** are good sources of hydronium ions.

A weak acid does not react extensively with water. Thus, a solution of a molecular weak acid contains the unreacted acid as the major species. A solution of a **molecular weak acid** is represented by the formula of the acid. For example, a water solution of hydrogen fluoride, HF, is represented as

$$HF(aq)$$

and a water solution of phosphoric acid, $H_3PO_4$, is represented as

$$H_3PO_4(aq)$$

Likewise, weak acids that are ionic species are represented by the formula of the ion. But how do we prepare solutions containing these ions? Recall from Section 10-19 that a soluble ionic compound is represented in solution by the formulas of the ions comprising the compound. If we want a solution containing hydrogen sulfate ions, for example, we dissolve $NaHSO_4$ in water. The solution is represented as

$$Na^+(aq) + HSO_4^-(aq)$$

Incidentally, this compound is the main ingredient in some toilet bowl cleaners. On the label of such a product it may be listed as sodium bisulfate, a common name for sodium hydrogen sulfate. In a solution of this compound the sodium ion is present to balance the charge, but the acid in solution is the hydrogen sulfate ion.

Solutions of bases are similarly represented by the major species in solution. A solution of a **molecular base** is represented by the formula of the base. Thus, a water solution of ammonia is represented as

$$NH_3(aq)$$

Note in Table 12-1 that many of the ionic bases are ions that you learned about in Chapters 7 and 8. As ions, these bases occur in many ionic compounds. Solutions of these bases are formed by dissolving soluble ionic compounds. For example, a solution of potassium carbonate, $K_2CO_3$, is represented as shown on the next page.

How would you prepare a water solution containing the weak acid ammonium ion, $NH_4^+$?

How would you prepare a solution containing the base phosphate ion, $PO_4^{3-}$?

Make a table with columns for strong acids, weak acids, weak bases, and strong bases. Give the formula for one example of each of these. Also give the formulas of the major species found in a solution of each of your examples.

$$2K^+(aq) + CO_3^{2-}(aq)$$

The carbonate ion is the base in the solution.

As Table 12-1 reveals, the hydroxide ion, $OH^-$, is the strongest base that can exist in water. Note its location in the table. A solution containing hydroxide ions can be prepared by dissolving a soluble ionic compound containing the hydroxide ion. The common soluble hydroxide-ion-containing compounds, are the strong bases sodium hydroxide, NaOH, potassium hydroxide, KOH, and barium hydroxide, $Ba(OH)_2$. In aqueous solutions, these compounds are represented as the hydroxide ion and the cation.

| | |
|---|---|
| sodium hydroxide solution | $Na^+(aq) + OH^-(aq)$ |
| potassium hydroxide solution | $K^+(aq) + OH^-(aq)$ |
| barium hydroxide solution | $Ba^{2+}(aq) + 2OH^-(aq)$ |

Sodium hydroxide solutions are the most common sources of hydroxide ions for use in laboratory work, and sodium hydroxide is by far the most important industrial hydroxide.

### 12-8 NEUTRALIZATION REACTIONS

Acid–base reactions can be truly amazing. A concentrated solution of hydrochloric acid can be quite dangerous; it can dissolve metals and cause skin burns. A concentrated solution of sodium hydroxide is also very dangerous. However, when these solutions are mixed in the correct proportions, the result is a neutral salt solution, and the dangerous properties are no longer present. **DANGER: Never mix concentrated acid and base solutions, since they react with vigor and heat.** Most acid–base reactions are not this dramatic, but in all acid–base reactions the original acid and base are altered chemically.

When we mix a solution containing an acid with a solution containing a base, an acid–base reaction may occur. A solution of a strong acid contains hydronium ions, potent proton donors. A solution of a strong base contains hydroxide ions, potent proton acceptors. When a solution of hydrochloric acid is mixed with a solution of sodium hydroxide, the ions in the solutions are

$$H_3O^+(aq) + Cl^-(aq) \quad \text{and} \quad Na^+(aq) + OH^-(aq)$$

Mixing the solutions brings all the ions into contact.

$$H_3O^+(aq) + Cl^-(aq) + Na^+(aq) + OH^-(aq)$$

The hydronium ion reacts with the hydroxide ion as shown by the net ionic equation shown on the next page.

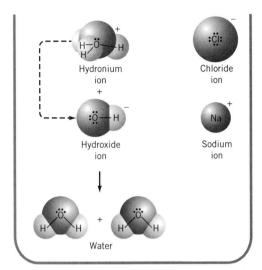

**FIGURE 12-2** The neutralization reaction between hydrochloric acid and sodium hydroxide. The hydronium ions and hydroxide ions react and the sodium ions and chloride ions are spectator ions.

$$H_3O^+(aq) + OH^-(aq) \rightleftharpoons 2H_2O$$

The sodium ion and the chloride ion are spectator ions and do not react. Thus, the acid–base reaction results in a solution of sodium ions and chloride ions, as shown in Figure 12-2.

The reaction between a solution of a soluble hydroxide and a solution of a strong acid always produces water. When the solutions are added in the correct proportions so that enough $H_3O^+$ is present to react with all of the $OH^-$, the resulting solution is neither acidic nor basic but is neutral. These reactions are typical neutralization reactions.

## 12-9 SOME FACTS ABOUT ACIDS AND BASES

One reason acid solutions have to be handled with care is that they can cause burns to the skin. Acids can dissolve and destroy tissue. Wart removers are typically solutions of weak acids. Solutions having relatively high concentrations of hydronium ions are especially dangerous. Strong acid solutions always have higher concentrations of hydronium ions than weak acid solutions of comparable molarities. For example, a 6 $M$ hydrochloric acid solution contains 6 moles of hydronium ions per liter, while a 6 $M$ acetic acid solution contains only about $1 \times 10^{-2}$ moles of hy-

WEAR YOUR
SAFETY GOGGLES

*"Do as you ought'er
always add acid
to water"*

Anonymous

DANGER

SULFURIC
ACID

DANGER

HYDROCHLORIC
ACID

DANGER

ACID

dronium ions per liter. Even though all acid solutions should be handled carefully, great care should be used with strong acid solutions. **Always wear safety goggles when working with acids** since your eyes are much more sensitive than your skin.

In the laboratory you may encounter solutions of the common strong acids: sulfuric, nitric, and hydrochloric. Always read the labels of bottles containing these acids very carefully. Sometimes, acid solutions are designated as concentrated or dilute. The molarities of typical **concentrated and dilute acid solutions** are listed in Table 12-2. Whatever its concentration, an acid solution should be handled with respect. Especially avoid adding water to concentrated acids, since the heat released upon mixing may cause the water to boil and splatter the acid.

The only strong acid that you may find in a hardware or pool supply store is hydrochloric acid. It is sold as **muriatic acid** and is used to adjust the acidity of swimming pools and as a cleaning agent. It should be used with gloves and safety goggles. Sulfuric acid is used in lead storage batteries in automobiles. Be careful not to spill battery acid on your skin or clothing. Concentrated sulfuric acid is especially dangerous since it not only causes skin burns but it can absorb water from skin tissue and undergo other reactions with body chemicals. If you happen to spill acid on yourself or your clothing, rinse it off with large quantities of water. You can treat acid on the skin or clothing with a paste made from baking soda (sodium bicarbonate). Never try to neutralize an acid burn with a solution of a strong base because the mixture will generate heat and make the burn worse. Since sulfuric acid is dangerous never watch a battery that is being charged or jump started. Batteries have been known to explode while being charged so just stand clear and don't look.

Basic solutions with high concentrations of hydroxide ion are also potentially dangerous. These solutions can react with and dissolve skin, so handle them with care. As Table 12-1 shows, the hydroxide ion is the strongest base in water; solutions of sodium hydroxide, NaOH, and potassium hydroxide, KOH, are the most common sources of hydroxide ions. The common name for sodium hydroxide is lye or caustic soda and potassium hydroxide is sometimes called caustic potash. The term caustic is well chosen, since solutions of these compounds are quite caustic and will dissolve proteins, vegetable oils, and animal fats. In fact, solutions of these bases are the main

*Table 12-2*    Molarities of Concentrated and Dilute Laboratory Acids

| FORMULA | ACID | MOLARITY (M) | |
| --- | --- | --- | --- |
| | | CONCENTRATED | DILUTE |
| $HCl(aq)$ | Hydrochloric | 12 | 6 |
| $HNO_3(aq)$ | Nitric | 16 | 6 |
| $H_2SO_4(aq)$ | Sulfuric | 18 | 3 |

components of commercial drain cleaners; they dissolve hair, fatty grease deposits, and other organic matter. Needless to say, these solutions should be handled with great care. Strong base solutions can cause severe eye damage so, just as was emphasized with strong acids, **always wear eye covering or goggles when working with solutions of strong bases.**

Some basic solutions are commonly used as household cleaners, since the hydroxide ion helps to dissolve fats and vegetable oils, it "cuts grease." Note the position of ammonia in Table 12-1. Ammonia is used in a variety of household cleaning solutions. Sometimes $Na_3PO_4$ (commonly called trisodium phosphate or TSP) and $Na_2CO_3$ (commonly called washing soda) solutions are used for cleaning purposes. Note the positions of the phosphate ion and the carbonate ion are in Table 12-1. TSP is strong enough that it can injure your skin and eyes. Use rubber gloves and goggles when working with TSP.

As you know, common baking soda or bicarbonate of soda is sodium hydrogen carbonate, $NaHCO_3$. The hydrogen carbonate ion, $HCO_3^-$ (sometimes called the bicarbonate ion), is amphiprotic. Look for $HCO_3^-$ in the acid and base sides of Table 12-1. The hydrogen carbonate ion can be used to react with the hydronium ion or the hydroxide ion. So sodium hydrogen carbonate solid, or as a solution, can be used for treatment of acid and base spills. Solutions of $NaHCO_3$, or powdered $NaHCO_3$, are usually available in the laboratory for emergencies.

Occasionally, the product formed when a solution of a strong acid is added to a solution of a weak base is only slightly soluble. An example of this is the reaction that occurs between a strong acid solution and a solution containing the carbonate ion.

$$2H_3O^+ + CO_3^{2-} \rightleftharpoons 3H_2O + CO_2(g)$$

During this reaction slightly soluble carbon dioxide bubbles out of solution and provides a test for the presence of carbonate ions. If a strong acid solution is added to a test solution and carbon dioxide is formed, carbonate ions or hydrogen carbonate ions are indicated.

There are three other bases in Table 12-1 that form insoluble gases when mixed with strong acids. **CAUTION: All these gases are very dangerous and toxic.** The bases are the fluoride ion, the sulfide ion, and the cyanide ion. They react with the hydronium ion as follows.

$$H_3O^+(aq) + F^-(aq) \rightleftharpoons H_2O + HF(g)$$
dangerous poisonous gas

$$2H_3O^+(aq) + S^{2-}(aq) \rightleftharpoons 2H_2O + H_2S(g)$$
poisonous gas with a
rotten-egg odor

$$H_3O^+(aq) + CN^-(aq) \rightleftharpoons H_2O + HCN(g)$$
very dangerous poisonous gas
with an almond-like odor

To illustrate the carbonate test, white vinegar can be used as an acid. In the laboratory a strong acid solution is used. Place some baking soda in a cup or glass and add a few drops of white vinegar. Place some eggshell, a calcium tablet, or a piece of chalk in a cup or glass and add a teaspoon of vinegar. Use Table 12-1 to write equations that show why the test works for either $CO_3^{2-}$ or $HCO_3^-$?

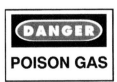

**Never** add acid solutions to solutions containing fluoride ions, sulfide ions, or cyanide ions since poison gases are formed. **Never** unnecessarily smell a solution.

## 12-10 STOICHIOMETRY AND ACID–BASE REACTIONS

By law, a commercial vinegar solution must have at least a set minimum concentration of acetic acid. The label on commercial vinegar may read 5% acidity, which corresponds to an acetic acid molarity of about 0.8 $M$ $HC_2H_3O_2$. If a product does not have the minimum concentration of acetic acid, it cannot be called vinegar. Suppose we had a sample of commercial vinegar and wanted to know the concentration of acetic acid in the sample. A solution of a base is added to the acid solution to supply a base to react with the acid. If the molarity of the base solution is known and we measure the volume of base solution needed to supply enough base to completely react with the acid, the amount of acid can be determined by stoichiometric calculations. To accomplish such an analysis we start with a carefully measured volume of the acetic acid solution. Then we add a sodium hydroxide solution of known concentration to react with the acetic acid. Since we know the equation for the reaction taking place and the volume and concentration of the sodium hydroxide solution used, the amount of acetic acid in the vinegar can be calculated. The reaction between acetic acid and the hydroxide ion is

$$HC_2H_3O_2 + OH^- \rightleftharpoons C_2H_3O_2^- + H_2O$$

Let us assume, for example, that 33.3 mL of a 0.500 $M$ solution of $OH^-$ is required to react with the acetic acid in a 20.0-mL sample of the commercial vinegar. We want to calculate the molarity of the acid solution using these data. This is actually a solution stoichiometry question of the type discussed in Section 10-15.

|  | $HC_2H_3O_2$ | + | $OH^- \rightleftharpoons C_2H_3O_2^- + H_2O$ |
|---|---|---|---|
| moles | mol $HC_2H_3O_2$ | $\Longleftarrow$ | mol $OH^-$ |
| molarity | ? $M$ | | 0.500 $M$ |
| solution volume | 20.0 mL or | | 33.3 mL or |
| | 0.0200 L | | 0.0333 L |

We find the number of moles of hydroxide ion involved in the reaction by using $n = VM$.

$$0.0333\ \cancel{L} \times \frac{0.500\ \text{mol OH}^-}{1\ \cancel{L}}$$

Next from the balanced equation, we see that the molar ratio between the hydroxide ion and acetic acid is

$$\frac{1 \text{ mol HC}_2\text{H}_3\text{O}_2}{1 \text{ mol OH}^-}$$

We multiply the number of moles of hydroxide ion by this ratio to give the number of moles of acetic acid involved in the reaction.

$$0.0333 \,\cancel{L} \times \frac{0.500 \,\cancel{\text{mol OH}^-}}{1 \,\cancel{L}} \times \frac{1 \text{ mol HC}_2\text{H}_3\text{O}_2}{1 \,\cancel{\text{mol OH}^-}}$$

Finally, the concentration of the acetic acid in moles per liter is obtained by dividing the number of moles by the initial volume of the acid sample in liters.

$$0.0333 \,\cancel{L} \times \frac{0.500 \,\cancel{\text{mol OH}^-}}{1 \,\cancel{L}} \times \frac{1 \text{ mol HC}_2\text{H}_3\text{O}_2}{1 \,\cancel{\text{mol OH}^-}} \times \frac{1}{0.0200 \text{ L}}$$

$$= \frac{0.833 \text{ mol HC}_2\text{H}_3\text{O}_2}{1 \text{ L}} \quad \text{or} \quad 0.833 \, M \,\text{HC}_2\text{H}_3\text{O}_2$$

The solution meets the vinegar standard.

## 12-11  TITRATIONS

Let's consider the actual laboratory techniques used to determine the molarity of an acetic acid solution. Such a determination is accomplished by using a solution sample of carefully measured volume. A solution of known concentration of sodium hydroxide is added to the sample of acid in the proper proportion so that the number of moles of acetic acid can be found by stoichiometric calculations. Mixing measured amounts of solutions for purposes of determining the concentration of one solution or the number of grams of some species in solution is known as a **titration.**

Titration is a very important experimental method used for the quantitative analysis of solutions. In practice, the addition and measurement of the volume of the solution of known concentration is performed using a **buret,** pictured in Figure 12-3. A titration is performed by filling a buret with the known solution, called the **titrant,** and placing a sample of the solution to be titrated in a flask. The titrant is then slowly delivered to the flask until the necessary amount has been added. The point at which the necessary volume of titrant has been added is termed the **end point** of the titration.

The end point of a titration is usually detected by placing a small amount of a substance called an **indicator** in the reaction flask. The indicator is chosen so that it reacts with the titrant when the end point is reached. The idea is that as soon as all of the titrated species has reacted

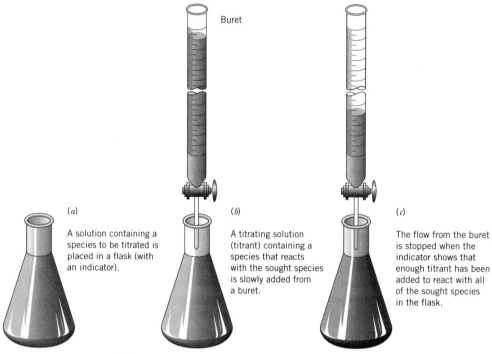

Buret

(a)

A solution containing a
species to be titrated is
placed in a flask (with
an indicator).

(b)

A titrating solution
(titrant) containing a
species that reacts
with the sought species
is slowly added from
a buret.

(c)

The flow from the buret
is stopped when the
indicator shows that
enough titrant has been
added to react with all
of the sought species
in the flask.

**FIGURE 12-3**    The
titration process using a
buret.

with the titrant, any additional amount of titrant reacts with the indicator.
Usually, the reaction of the indicator and titrant produces a colored product that can be seen. **Phenolphthalein** is an indicator that is pink in basic
solutions and colorless in acidic solutions. Some indicators are colored to
begin with and produce a different colored product, so the color change
indicates the end point. Once the end point is found, the volume of
titrant used to reach the end point can be determined from the buret. See
Figure 10-8 for a stoichiometry map to find the mass of a titrated species.

## 12-12 TITRATION CALCULATIONS

When a titration is done using a buret, the main piece of experimental
data is the volume of titrant used to reach the end point in the titration.
The product of the volume and concentration of the titrant gives the
number of moles of the titrant. In turn, the number of moles of the
titrated species is found using the appropriate molar ratio from the balanced equation for the titration reaction. Chemistry in Action 12-1 is a teaspoon titration; the experiment uses a teaspoon instead of a buret.

If we want to determine the number of grams of some species in a solution by titration the experimental data needed from the titration are

volume of titrant used in titration $= V_t$      molarity of titrant $= M_t$

We use these data and the molar ratio to find the number of moles of the titrated species. Then find the mass of this species from the number of moles using its molar mass. The general calculation sequence is:

$$V_t M_t \longrightarrow \begin{array}{c} \text{moles} \\ \text{titrant} \end{array} \xrightarrow[\substack{\text{molar ratio} \\ \text{from equation}}]{} \begin{array}{c} \text{moles} \\ \text{titrated} \\ \text{species} \end{array} \xrightarrow[\substack{\text{molar mass} \\ \text{of species}}]{} \begin{array}{c} \text{grams} \\ \text{species} \end{array}$$

---

**EXAMPLE 12-3**

A solution of ammonia requires 32.5 mL of a 0.227 $M$ hydrochloric acid solution to titrate it to the end point. Calculate the number of grams of ammonia in the original solution if the reaction is

$$H_3O^+ + NH_3 \rightleftharpoons NH_4^+ + H_2O$$

The information is

|  | titrant | | unknown | | |
|---|---|---|---|---|---|
|  | $H_3O^+$ | $+$ | $NH_3$ | $\rightleftharpoons$ | $NH_4^+ + H_2O$ |
| mass |  |  | ? g | | |
| molar mass |  |  | 17.03 g/mol | | |
| moles | mol $H_3O^+$ | $\Longrightarrow$ | mol $NH_3$ | | |
| molarity | 0.227 $M$ | | | | |
| volume | 32.5 mL or | | | | |
|  | 0.0325 L | | | | |

We find the number of moles of hydronium ion used in the titration from the volume and the molarity of the hydrochloric acid titrant. Then, find the moles of ammonia in the titrated solution from the moles of hydronium ion. Finally, find the grams of ammonia from the number of moles and its molar mass.

$$X \text{ g } NH_3 = 0.0325 \cancel{L} \times \frac{0.227 \cancel{\text{mol } H_3O^+}}{1 \cancel{L}} \times \frac{1 \cancel{\text{mol } NH_3}}{1 \cancel{\text{mol } H_3O^+}} \times \frac{17.03 \text{ g}}{1 \cancel{\text{mol } NH_3}}$$

$$= 1.26 \times 10^{-1} \text{ g}$$

---

A solution of an acid can be titrated to the end point with a base to find the molarity of the acid, $M_a$. In such a titration it is necessary to measure the volume of the original acid sample before it is titrated. The experimental data needed from the titration are

volume of base needed in the titration $\quad\quad V_b$
molarity of base $\quad\quad M_b$
volume of the acid solution of unknown molarity $\quad V_a$

Figure 12-4 shows that in general the calculation setup to find the molarity of the acid in a titration is

$$M_a = M_b \frac{V_b}{V_a} \left(\begin{array}{c} \text{molar} \\ \text{ratio} \end{array}\right)$$

$$M_a = M_b \frac{V_b}{V_a} \left(\begin{array}{c} \text{Molar} \\ \text{ratio} \end{array}\right)$$

When the molar ratio is one to one the relation simplifies to

$$M_a = M_b \frac{V_b}{V_a}$$

The relation can be rearranged to give

$$M_a V_a = M_b V_b$$

**FIGURE 12-4**
Calculations for an acid–base titration.

---

**EXAMPLE 12-4**

It is found that 37.2 mL of a 0.575 $M$ sodium hydroxide solution is needed to titrate a 25.0-mL sample of a solution of oxalic acid, $H_2C_2O_4$ to the end point. What is the molarity of the oxalic acid solution? The equation for the reaction is

|  | unknown | titrant |  |
|---|---|---|---|
|  | $H_2C_2O_4$ | $+$ 2OH$^-$ | $\rightleftharpoons$ 2H$_2$O + C$_2$O$_4{}^{2-}$ |
| moles | mol $H_2C_2O_4$ $\Longleftarrow$ | mol OH$^-$ | |
| molarity | ? $M_a$ | 0.575 $M_b$ | |
| volumes | $V_a$ = 25.0 mL | $V_b$ = 37.2 mL | |
|  | or 0.0250 L | or 0.0372 L | |

The data includes the volume of titrant, the molarity of the titrant, the volume of the solution that is titrated, and the molar ratio from the balanced equation. In this case the molar ratio is 1:2. Using these data gives

$$\frac{0.575 \text{ mol OH}^-}{1 \text{ L}} \times \frac{0.0372 \text{ L}}{0.0250 \text{ L}} \times \frac{1 \text{ mol } H_2C_2O_4}{2 \text{ mol OH}^-}$$

$$= \frac{0.428 \text{ mol } H_2C_2O_4}{1 \text{ L}} \quad \text{or} \quad 0.428 \text{ } M \text{ } H_2C_2O.$$

Notice how the calculation fits the general setup shown above. Incidentally, solution volumes in milliliters could just as easily be used and give the same result. Why?

---

In an acid–base titration in which there is a 1:1 molar ratio of acid to base the general calculation setup is simply

$$M_a = M_b \frac{V_b}{V_a}$$

which can be rearranged to give an expression to find any of the terms

$$M_aV_a = M_bV_b$$

**EXAMPLE 12-5**

A sodium hydroxide solution is used to titrate a hydrochloric acid solution. The neutralization reaction for this titration is

$$H_3O^+ + OH^- \rightleftharpoons 2H_2O$$

If 23.54 mL of a 0.1025 $M$ solution of NaOH is required to titrate a 20.00-mL sample of hydrochloric acid to the end point, what is the molarity of the hydrochloric acid? The data are

volume of base used       $= V_b = 23.54$ mL

molarity of base          $= M_b = 0.1025$ $M$

volume of acid titrated   $= V_a = 20.00$ mL

molarity of acid          $= M_a = ?$

Note that the molarity of the NaOH is also the molarity of the $OH^-$. The molar ratio is 1:1 so the molarity of the unknown acid is simply

$$M_a = M_b \frac{V_b}{V_a} = 0.1025 \ M \times \frac{23.54 \ \text{mL}}{20.00 \ \text{mL}} = 0.1206 \ M \ H_3O^+$$

## 12-13 $K_W$ THE WATER CONSTANT

Acidic solutions are characterized by the presence of the hydronium ion. In fact, it is the hydronium ion that is responsible for the similar acidic properties of such solutions. Some solutions have high concentrations of hydronium ions and some have low concentrations. As mentioned in Section 12-5, in any sample of water or a water solution, the hydronium ion and the hydroxide ion are in equilibrium with water.

$$H_2O + H_2O \rightleftharpoons H_3O^+ + OH^-$$

In pure water or a neutral aqueous solution the concentrations of hydronium and hydroxide ions are equal. In a neutral solution at 25°C, the concentrations of each of these two ions is $1.0 \times 10^{-7}$ $M$. These facts can be expressed as

$$[H_3O^+] = [OH^-] = 1.0 \times 10^{-7} \ M$$

where the square brackets [   ] denote concentration in moles per liter of the species enclosed in the bracket. In pure water or any water solution the concentrations of hydronium and hydroxide ions are fixed. However, if the concentration of one of these ions increases the other decreases in proportion. That is, if a solution becomes more acidic because of an increase in the hydronium ion concentration it becomes proportionally less basic since the hydroxide ion concentration decreases. On the other hand, if a solution becomes more basic because of an increase in the hydroxide ion concentration it becomes proportionally less acidic since the hydronium ion concentration decreases. The relation between the concentrations of these ions is

$$[H_3O^+] = K_w/[OH^-]$$

This inverse proportion between the hydronium and hydroxide ion concentrations is analogous to other inverse proportions, such as Boyle's law, $V = k/P$. The hydronium ion–hydroxide ion expression can be rearranged to show the product of the two concentrations is equal to the proportionality constant.

$$K_w = [H_3O^+][OH^-]$$

This proportion indicates that if $[OH^-]$ increases, then $[H_3O^+]$ decreases and vice versa. The proportionality constant represented by $K_w$ (read K–W) is called the ion product of water. Its value is determined by experiment. The typical value for $K_w$ is $1.0 \times 10^{-14}$. This means that in pure water or water solutions the product of the concentrations of hydronium and hydroxide ions has a constant value of $1.0 \times 10^{-14}$ at 25°C.

When an acid or a base is dissolved in water the hydronium ion–hydroxide ion equilibrium is affected. An acid in solution reacts with water to increase the concentration of hydronium ions. As the hydronium ion concentration increases, the hydroxide ion concentration decreases proportionally. An acidic solution is characterized by a concentration of hydronium ions that is greater than $1.0 \times 10^{-7}$ M and a concentration of hydroxide ions that is less than $1.0 \times 10^{-7}$ M.

A base in solution increases the concentration of hydroxide ions. The increased hydroxide ion concentration results in a lower concentration of hydronium ions. They vary in inverse proportion. A basic solution is characterized by a concentration of hydroxide ions that is greater than $1.0 \times 10^{-7}$ M and a concentration of hydronium ions that is less than $1.0 \times 10^{-7}$ M.

| | |
|---|---|
| 1.0 M | $10^0$ M |
| 0.1 M | $10^{-1}$ M |
| 0.01 M | $10^{-2}$ M |
| 0.001 M | $10^{-3}$ M |
| 0.000 1 M | $10^{-4}$ M |
| 0.000 01 M | $10^{-5}$ M |
| 0.000 001 M | $10^{-6}$ M |
| 0.000 000 1 M | $10^{-7}$ M |
| 0.000 000 01 M | $10^{-8}$ M |
| 0.000 000 001 M | $10^{-9}$ M |
| 0.000 000 000 1 M | $10^{-10}$ M |
| 0.000 000 000 01 M | $10^{-11}$ M |
| 0.000 000 000 001 M | $10^{-12}$ M |
| 0.000 000 000 000 1 M | $10^{-13}$ M |
| 0.000 000 000 000 01 M | $10^{-14}$ M |

## 12-14 PH AND POH

The concentration of hydronium ion in solutions can vary greatly. Some solutions are highly acidic; they have relatively high concentrations of hydronium ion. A 1 M hydrochloric acid solution, for instance, has a hydronium ion concentration of 1 M. Some solutions are highly basic; they have relatively high concentrations of hydroxide ions and correspondingly low concentrations of hydronium ions. A 1 M sodium hydroxide solution, for instance, has a hydroxide ion concentration of 1 $M$ and a hydronium ion concentration of $1 \times 10^{-14}$ $M$. In pure water and neutral solutions the concentration of hydronium ions is $1.0 \times 10^{-7}$ $M$. Common solutions of acids and bases have hydronium ion concentrations ranging from around 1 $M$ to $1 \times 10^{-14}$ $M$ and corresponding hydroxide ion concentrations ranging from around $1 \times 10^{-14}$ $M$ to 1 $M$. Let's see how the concentrations of these ions vary in inverse proportion. For instance, if a solution is 1.0 $M$ $H_3O^+$ then the $OH^-$ concentration is $1.0 \times 10^{-14}$ $M$. The reason is provided by the water constant.

$$K_w = [H_3O^+][OH^-] = 1.0 \times 10^{-14}$$

therefore

$$[OH^-] = K_w/H_3O^+]$$

If $[H_3O^+] = 1.0$ $M$   and    $K_w = 1.0 \times 10^{-14}$

then

$$[OH^-] = (1.0 \times 10^{-14})/(1.0)$$
$$= 1.0 \times 10^{-14} \ M$$

The hydroxide concentration is $1.0 \times 10^{-14}$ $M$

The concentration of hydronium ion in a solution can be expressed in terms of molarity. However, since the possible concentrations cover a very wide range, a special scale has been devised to express hydronium ion concentrations in aqueous solutions. This is called the pH (read P–H) scale. Similarly various hydroxide ion concentrations in solutions can be expressed on a pOH (read P–O–H) scale. Definitions are in the margin.

**pH:**
*The negative logarithm of the hydronium ion concentration. pH = –log[$H_3O^+$] where [$H_3O^+$] represents the molarity of the hydronium ion.*

**pOH:**
*The negative logarithm of the hydroxide ion concentration. pOH = –log[$OH^-$] where [$OH^-$] represents the molarity of the hydroxide ion.*

---

## Logarithms

A number can be expressed in terms of 10 raised to a power. Simple examples are 100 expressed as $10^2$ and 0.001 expressed as $10^{-3}$. Any number can be expressed as a power of 10, for example, 2 as a power of 10 is $10^{0.3}$ and 9.5 as a power of 10 is $10^{0.98}$. The base 10 logarithm of a number is the power to which 10 is raised to give the number. In general any number, $y$, can be expressed as

$$y = 10^x$$

The log of the number $y$ is $x$. Thus, the log of $10^2$ is 2, the log of $10^{-3}$ is –3, the log of 2 is 0.3 and the log of 9.5 is 0.98. The numerical value of the logarithm of a number can be found using a scientific calculator. To find the logarithm of a number using a calculator with a log key, just key in the number and press the log key. If you have a calculator with a log key practice by finding the logarithms of 100, 0.001, 2, and 9.5.

The definitions of pH and pOH allow for the expression of ion concentrations as powers of 10 rather than normal numbers. In fact, the term pH can be viewed as power (p) of the hydronium ion concentration and pOH as the power of the hydroxide ion concentration. pH and pOH allow concentrations to be expressed on a logarithmic scale which makes them simpler. However, since the logarithms of hydronium and hydroxide ion concentrations are typically negative (i.e., $\log 10^{-7} = -7$), the definitions of pH and pOH include the negative logarithm. That means the logarithm has its sign changed or multiplied by $-1$ during the calculation of pH, for example, $-\log 10^{-7} = -(-7) = 7$. By this convention the pH or pOH of a solution is normally positive. pH is a convenient way to express the acidity of a solution and pOH is a convenient way to express the basicity of a solution. The pH scale reduces the range of potential hydronium ion concentrations to a scale ranging from about 0 to 14. It is possible, however, to have pH values lower than 0 and higher than 14.

$$pH = -\log[H_3O^+]$$

| $[H_3O^+]$ | $10^0$ | $10^{-1}$ | $10^{-2}$ | $10^{-3}$ | $10^{-4}$ | $10^{-5}$ | $10^{-6}$ | $10^{-7}$ | $10^{-8}$ | $10^{-9}$ | $10^{-10}$ | $10^{-11}$ | $10^{-12}$ | $10^{-13}$ | $10^{-14}$ |
|---|---|---|---|---|---|---|---|---|---|---|---|---|---|---|---|
| pH | 0 | 1 | 2 | 3 | 4 | 5 | 6 | 7 | 8 | 9 | 10 | 11 | 12 | 13 | 14 |

Note that $1\ M = 10^0 M$ since $10^0 = 1$. The pOH scale reduces the range of potential hydroxide ion concentrations to a scale ranging from about 14 to 0.

| $[OH^-]$ | $10^{-14}$ | $10^{-13}$ | $10^{-12}$ | $10^{-11}$ | $10^{-10}$ | $10^{-9}$ | $10^{-8}$ | $10^{-7}$ | $10^{-6}$ | $10^{-5}$ | $10^{-4}$ | $10^{-3}$ | $10^{-2}$ | $10^{-1}$ | $10^0$ |
|---|---|---|---|---|---|---|---|---|---|---|---|---|---|---|---|
| pOH | 14 | 13 | 12 | 11 | 10 | 9 | 8 | 7 | 6 | 5 | 4 | 3 | 2 | 1 | 0 |

For a given solution the sum of its pH and pOH is 14.00. This useful fact can be expressed as

$$pH + pOH = 14.00$$

This expression is used to easily calculate the pOH from the pH of a solution or the pH from the pOH. For example, if we know that a solution has a pH of 9.0, then the pOH must be $14.00 - 9.0 = 5.0$.

## 12-15 PH AND POH CALCULATIONS

If we know the hydronium ion concentration in a solution the pH is easily calculated using the relation $pH = -\log[H_3O^+]$. If we know the hydroxide ion concentration in a solution the pOH is easily calculated using $pOH = -\log[OH^-]$. If we know the pH we can find the pOH by subtracting from 14.00 or vice versa to get the pH from the pOH. Note, in the pH scale given above, that there is a 10-fold difference in acidity between subsequent pH units. This means that a given pH value corresponds to a hydronium ion concentration that is 10 times greater than the next higher pH value on the scale. At pH 4 $[H_3O^+]$ is 10 times greater than at pH 5, 100 times greater than at pH 6, 1000 times greater than at pH 7, and so on. A solution of pH 0 has a hydronium concentration that is 100 trillion times more than a solution of pH 14. Keep in mind that higher values of pH correspond to lower concentrations of hydronium ions. Higher pH values correspond to solutions of lower acidity. This is a consequence of the negative logarithm in the definition of pH. It is interesting that an acidic solution having a hydronium ion concentration greater than 1 $M$ has a negative pH. For instance, the pH of a solution having a hydronium ion concentration of 10 $M$ is $-1$; pH $= -\log[H_3O^+]$ and $-\log(10) = -1$. However, such highly concentrated solutions are not common. Why are these solutions very dangerous to handle?

For solutions with $[H_3O^+]$ ranging from 1 $M$ to $10^{-14}$ $M$, the pH ranges from 0 to 14. Solutions of lower pH are more acidic than solutions of higher pH. Solutions with pH less than 7 are **acidic,** whereas solutions with pH greater than 7 are **basic.** Basic solutions are sometimes called alkaline solutions. A neutral solution at 25°C has a pH of 7 and pOH of 7. Notice that solutions with lower pOH are more basic than solutions of higher pOH.

| | highly acidic | | | | slightly acidic | | | neutral | | slightly basic | | | | highly basic |
|---|---|---|---|---|---|---|---|---|---|---|---|---|---|---|
| pH | 0 | 1 | 2 | 3 | 4 | 5 | 6 | 7 | 8 | 9 | 10 | 11 | 12 | 13 | 14 |
| pOH | 14 | 13 | 12 | 11 | 10 | 9 | 8 | 7 | 6 | 5 | 4 | 3 | 2 | 1 | 0 |

The pH is used as a convenient means of expressing the hydronium ion concentration in a solution. If $[H_3O^+]$ is a whole-number power of 10 the pH is

$$[H_3O^+] = 10^{-pH}$$

This means that the pH is the numerical value of the power of 10 with its sign changed. Thus, a solution with $[H_3O^+] = 10^{-14}$ $M$ has a pH of 14, and a solution with $[H_3O^+] = 10^{-9}$ $M$ has a pH of 9. Similarly, if $[OH^-]$ is an integral power of 10, the pOH is easily determined as

$$[OH^-] = 10^{-pOH}$$

If $[OH^-] = 10^{-8}$ the pOH is 8 and if $[OH^-] = 10^{-2}$ the pOH is 2. See Chemistry in Action 12-2 for an exercise in observing the pH of various solutions.

When the hydronium ion concentration is not an integral power of 10, the concentration is used in the definition of pH and the logarithm must be determined. For example, a solution with $[H_3O^+] = 2 \times 10^{-3}$ $M$ has a pH of

$$pH = -\log(2 \times 10^{-3})$$

A hand calculator that allows the input of exponential numbers and has a base 10 logarithm key (log) can be used to calculate pH. To do this, first key in the hydronium ion concentration. Then find the log using the logarithm key and change the sign of the result to give the negative log. See Appendix 2 for more on the use of a calculator for pH calculations. In this case the calculation sequence is 2 (EE) 3 (+/−) (log) (+/−) = 2.70. . . = 2.7. Round to one digit past the decimal since the data had only one digit. When reporting a pH the digits on the left of the decimal correspond to the exponent and do not count as significant digits. Only those digits to the right of the decimal point are significant digits.

---

**EXAMPLE 12-6**

What is the pH of a solution that has a hydronium ion concentration of $9.5 \times 10^{-9}$ *M*? What is the pOH of the solution? Is the solution acidic or basic?

To find the pH using a hand calculator, just key in the hydronium ion concentration, push the log key, and change the sign (for the negative log).

$$9.5 \times 10^{-9} \text{ (log) (+/−) } 8.0222. . .$$

or 8.02 to two significant digits. The pOH of this solution is

$$14.00 - 8.02 = 5.98$$

The pH is above 7 and the pOH is below 7 so the solution is basic.

---

**ACTIVITY 12-8**

What are the pOH and pH of a solution that has $[OH^-] = 3.3 \times 10^{-3}$ *M*? Is the solution acidic or basic?

---

Hydronium ions are present in all water solutions. Gastric juices, citrus fruit juices, soft drinks, vinegar, urine, milk, and some natural fresh waters are acidic. Pure water is neutral, whereas blood is slightly basic. Table 12-3 lists the pH of some common solutions.

## 12-16 ESTIMATING THE pH OR pOH OF SOLUTIONS

The approximate pH of a solution can be found without using a calculator. Suppose we had a sample of lemon juice having a hydronium ion concentration of $3.0 \times 10^{-3}$ *M*. The pH is found using the concentration in the pH definition.

$$pH = -\log(3.0 \times 10^{-3})$$

To estimate, we round the concentration down to the next lower power and up to the next higher power: $1 \times 10^{-3}$ and $10 \times 10^{-3}$ which equals $1 \times 10^{-2}$. In this way we know that the pH is between 2 and 3 so it must be 2 point something or 2.?. A simple pattern for obtaining an estimate of the

*Table 12-3*   pH Values of Some Common Solutions

| PH | | |
|---|---|---|
| 0 | 1 *M* Hydrochloric acid | 0 |
| 1 | Gastric juices | 1.0–3.0 |
| 2 | Lemon juice | 2.2–2.4 |
| | Vinegar | 2.4–3.4 |
| 3 | Wines | 2.8–3.8 |
| 4 | Carbonated water (saturated) | 3.8 |
| 5 | Black coffee | 4.8–5.2 |
| 6 | Milk | 6.3–6.6 |
| | Saliva | 6.5–7.5 |
| 7 | Pure water | 7.0 |
| 8 | Blood plasma | 7.3–7.5 |
| | 0.1 *M* Sodium hydrogen carbonate | 8.4 |
| 9 | Soap solutions | 8.5–10.0 |
| 10 | Saturated magnesium hydroxide | 10.5 |
| 11 | | |
| 12 | 1 *M* Aqueous ammonia | 11.6 |
| | Saturated calcium hydroxide (limewater) | 12.4 |
| 13 | | |
| 14 | 1 *M* Sodium hydroxide | 14 |

pH or pOH of a solution is

1. If the ion concentration is one with a whole number power such as $1 \times 10^{-n}$ or $1.0 \times 10^{-n}$ the pH or pOH value is $n.0$ or $n.00$
2. Otherwise, use the power in the $\times 10$ part of the concentration, change its sign and subtract one.
3. The approximate pH or pOH value is this number point something.

As an example let's estimate the pH of a soap solution having a hydronium ion concentration of $8.5 \times 10^{-10}$. Subtracting 1 from 10 suggests that the pH is 9 point something. Since, the sum of pH and pOH is 14, the pOH must be 4 point something. Of course, you can calculate the precise pH with a calculator. However, you can use the quick estimation method to decide whether a solution is acidic or basic and to double-check your calculation of the precise pH or pOH.

ACTIVITY
12-9

(a) Estimate the pH of a sample of wine that has a hydronium ion concentration of $1.8 \times 10^{-4}$ *M*. Is it acidic or basic?

(b) Estimate the pOH of a solution of KOH that has a hydroxide ion concentration of $5.0 \times 10^{-3}$ *M*. What is the approximate pH of the KOH solution?

# *Acids, Bases, pH, and pOH*

### BRØNSTED-LOWRY ACIDS AND BASES

Acid: A chemical species that can lose hydrogen ions or protons in a chemical reaction. A proton donor.

Base: A chemical species that can gain hydrogen ions or protons in a chemical reaction.
A proton acceptor.

A proton transfer or acid-base reaction can be represented as

$$\text{H—A} \ + \ \text{B} \ \xrightarrow{\text{proton transfer}} \ \text{H—B} \ + \ \text{A}$$
$$\text{acid}_1 \qquad \text{base}_2 \qquad\qquad\qquad \text{acid}_2 \quad\ \text{base}_1$$

Conjugate acid-base pair: An acid and the base that it forms by the loss of a proton or a base and the acid it forms by gaining a proton.

Acidic solution: A water solution that has a concentration of hydronium ion that is greater than that in pure water. When an acid dissolves in water, an acidic solution results.

Basic solution: A water solution that has a concentration of hydroxide ion that is greater than that in pure water. When a base dissolves in water, a basic solution results.

A solution of a strong acid always contains hydronium ion as a major ion. Three common strong acids are

hydrochloric acid solution      $H_3O^+(aq) + Cl^-(aq)$

sulfuric acid solution      $H_3O^+(aq) + HSO_4^-(aq)$

nitric acid solution      $H_3O^+(aq) + NO_3^-(aq)$

A weak acid does not react extensively with water and a solution of a molecular weak acid contains the unreacted acid as the major species. A solution of a molecular weak acid is represented by the formula of the acid and an ionic weak acid is represented by its formula.

A solution of a weak base is represented by the formula of the base.

The common strong bases are sodium hydroxide, NaOH, potassium hydroxide, KOH, and barium hydroxide, $Ba(OH)_2$. In aqueous solutions, these compounds are represented as hydroxide ion and the cation.

Titration: A method of chemical analysis that involves mixing measured amounts of solutions for purposes of determining the concentration of one solution or number of grams of some species in solution.

End point: The point at which the necessary volume of titrant has been added in a titration.

The molarity of an acid solution can be found by titrating a known volume with a base solution of known molarity. The titration equation is given on the next page.

$$M_a = M_b \frac{V_b}{V_a} \frac{\text{molar}}{\text{ratio}}$$

The molar ratio comes from the balanced equation for the reation.

When the molar ratio in an acid-base titration is one to one the titration equation is simply

$$M_a = M_b \frac{V_b}{V_a} \quad \text{or} \quad M_a V_a = M_b V_b$$

The product of the concentrations of hydronium ion and hydroxide ion in water solutions is a constant: $K_w = [H_3O^+][OH^-]$

pH is the negative logarithm of the hydronium ion concentration.

**pH $= -$ log[H$_3$O$^+$]** where [H$_3$O$^+$] represents the molarity of hydronium ion.

pOH is the negative logarithm of the hydroxide ion concentration.

**pOH $= -$ log[OH$^-$]** where [OH$^-$] represents the molarity of hydroxide ion.

|  |  |  |  | neutral |  |  |  |
|---|---|---|---|---|---|---|---|
| highly acidic |  | slightly acidic |  | slightly basic |  |  | highly basic |

| pH | 0 | 1 | 2 | 3 | 4 | 5 | 6 | 7 | 8 | 9 | 10 | 11 | 12 | 13 | 14 |
|---|---|---|---|---|---|---|---|---|---|---|---|---|---|---|---|
| pOH | 14 | 13 | 12 | 11 | 10 | 9 | 8 | 7 | 6 | 5 | 4 | 3 | 2 | 1 | 0 |

The sum of the pH and pOH is 14.00. **pH + pOH = 14.00**

To find the pH from the hydronium ion concentration use **pH $= -$ log[H$_3$O$^+$]**

To find the pOH from the hydroxide ion concentration use **pOH $= -$ log[OH$^-$]**

To estimate the pH or pOH of a solution using the hydronium ion or hydroxide ion concentration

1. If the ion concentration is $1 \times 10^{-n}$ or $1.0 \times 10^{-n}$ the value is $n.0$ or $n.00$
2. Otherwise, use the power in the $\times 10$ part of the concentration, change its sign and subtract one.
3. The estimated value is this number point something.

To find the hydronium ion concentration from the pH use **[H$_3$O$^+$] = 10$^{-pH}$** (See Appendix 2.)

To find the hydroxide ion concentration from the pOH use **[OH$^-$] = 10$^{-pOH}$**

In this exercise you are to do a teaspoon titration. This is a quantitative exercise so measure level teaspoons carefully. You need: commercial milk of magnesia (regular strength not extra strength), white vinegar, two glasses, two teaspoons, a kitchen knife or flat blade, and a laxative tablet containing phenolphthalein (if available).

1. Shake the container of milk of magnesia vigorously and pour a teaspoonful into a clean teaspoon. Since the material is thick you can get a level teaspoonful by spreading it with a knife and using the flat part of the knife to scrape off any excess material from the top of the spoon.

2. Place the milk of magnesia and the spoon in a clean glass.

3. If you have a laxative tablet, crush it and add a sample less than the size of a grain of rice to the glass of milk of magnesia, otherwise go on to Step 4.

4. Pour some vinegar into the second glass so that you can easily remove teaspoon samples.

5. Use another teaspoon to put a level teaspoon of vinegar into the milk of magnesia and stir for a few minutes with the spoon that is in the glass. Note the slight pink color of the phenolphthalein.

6. Each time you add a level teaspoon of vinegar stir the mixture for a few minutes and note the appearance. Add the vinegar one teaspoonful at a time until the mixture is no longer milky and is no longer pink. If you are not using phenolphthalein add the vinegar until the mixture is no longer milky. Be sure to count the number of teaspoons of vinegar used in the titration.

The active ingredient in milk of magnesia is solid $Mg(OH)_2$, and vinegar contains $HC_2H_3O_2$. Write a balanced equation for the acid–base reaction between these chemicals. The label of commercial milk of magnesia claims that a teaspoon contains 400 mg of $Mg(OH)_2$. Use your titration data to verify this claim. Assume that one teaspoon has a volume of 5 mL and vinegar is 0.83 $M$ $HC_2H_3O_2$.

The pigmented juice of red cabbage contains chemicals that act as acid–base indicators. In this exercise you are going to test cabbage juice as an indicator. You need: one cup of chopped red cabbage, a pan to heat the cabbage in, several glasses, a teaspoon, a saturated solution of baking soda, milk of magnesia, white vinegar, laundry detergent, seltzer water (seltzer water is a simple solution of carbon dioxide in water; tonic water and soda water contain other chemicals in addition to carbon dioxide), four aspirin tablets or vitamin C tablets, and a sample of saliva. You can generate a saliva sample by chewing a rubber band or wax. Note: Do not use an aluminum pan.

1. Chop enough red cabbage to provide a packed cupful. Place the cabbage in a pan along with one cup of water. Bring the cabbage to boil and let it simmer for a few minutes. Remove the pan from the heat and let it cool. After it cools pour the liquid from the pan into a glass. This juice is your acid–base indicator solution. It needs to be fresh so use it as soon as possible.

2. Prepare a data table and record the color of the indicator in each test. If you have enough glasses keep the samples for color comparison. Unless instructed otherwise use 100 mL samples of solu-

tions. Do not let the solutions mix or contaminate one another. Furthermore, do not contaminate your supply of indicator solution.

3. Place a sample of vinegar in a glass, add a teaspoon of indicator, and stir.

4. Place a sample of seltzer water in a glass, add a teaspoon of indicator, and stir.

5. Place a sample of saturated baking soda solution in a glass, add a teaspoon of indicator, and stir.

6. Put a teaspoon of laundry detergent in a glass containing 100 mL of water, add a teaspoon of indicator, and stir.

7. Put a few drops of milk of magnesia in a glass, add a teaspoon of water and a teaspoon of indicator, and stir.

8. Put a small amount of saliva in a glass, add a teaspoon of indicator, and stir.

9. Crush four aspirin or vitamin C tablets. Put them in a glass containing 100 mL of water and stir for a few minutes then add a teaspoon of indicator and stir.

10. If you happen to have some household ammonia, test a sample with the indicator. If you happen to have some lye (danger), put a sample in 100 mL of water and test it with the indicator. Considering that the pH of the solutions increases in the order of vinegar (2.5), seltzer water (4), saliva (7), saturated baking soda (8), milk of magnesia (10), and laundry detergent (11) make a list of the various colors of the indicator corresponding to the solutions and their pH values. Using the color of the aspirin or vitamin C solution, which solution does it most closely match in acidity? Do the same for the ammonia or lye solution. **CAUTION:** Lye is very caustic. Use safety goggles.

11. To observe how the pH changes during a titration, place one-fourth of a cup of vinegar in a glass, add two teaspoonfuls of indicator and stir. Add a teaspoonful of milk of magnesia to the glass and stir for a minute. You don't have to measure the teaspoonful carefully. Add another teaspoonful of milk of magnesia and stir. Continue adding teaspoonfuls of milk of magnesia until you see a dramatic color change. You will not see a dramatic change until you reach the end point, when all of the acid has been titrated.

## QUESTIONS

### Section 12-1

1. List the common properties of water solutions of acids.

2. List the common properties of water solutions of bases.

### Section 12-2

3. Give the Brønsted–Lowry definitions for an acid and a base.

4. What is a proton in reference to acid–base reactions?

### Section 12-3

5. Describe an acid–base reaction and give an example.

6. What is the meaning of a double arrow in an acid–base reaction?

### Section 12-4

7. What is a conjugate acid–base pair? Give examples.

8. *Complete the following table by supplying the conjugate acid or base as needed.

| ACID | BASE |
|------|------|
| $HNO_2$ | ____ |
| ____ | $H_2O$ |
| $HSO_4^-$ | ____ |
| ____ | $NH_3$ |
| $H_2O$ | ____ |

9. For each of the following acid–base reactions pick out the acids and the bases and indicate the conjugate acid–base pairs.

    (a) $HNO_3 + H_2O \longrightarrow H_3O^+(aq) + NO_3^-(aq)$

    (b) $OH^-(aq) + NH_4^+(aq) \rightleftharpoons NH_3(aq) + H_2O$

    (c) $H_3O^+(aq) + F^-(aq) \rightleftharpoons HF(aq) + H_2O$

10. For each of the following acid–base reactions pick out the acids and the bases and indicate the conjugate acid–base pairs.

    (a) $H_2SO_4 + H_2O \longrightarrow H_3O^+(aq) + HSO_4^-(aq)$

    (b) $OH^-(aq) + H_3PO_4 \rightleftharpoons H_2O + H_2PO_4^-(aq)$

    (c) $H_3O^+(aq) + OH^-(aq) \rightleftharpoons H_2O + H_2O$

## Section 12-5

11. What is an acidic solution compared to a neutral solution? What is a basic solution compared to a neutral solution?

12. What does it mean when a species is said to be amphiprotic?

13. Water is amphiprotic. Write the electron dot structure for water and explain how it can behave as an acid or a base.

## Section 12-6

14. Explain the structure of Table 12-1.

15. Explain the difference between a weak acid and a strong acid.

## Section 12-7

16. What is a strong acid and what are the major species in a solution of a strong acid? Give an example.

17. What is a weak acid and what are the major species in a solution of a weak acid? Give an example.

18. What are the major species in a solution of a strong base? What are the major species in a solution of a weak base?

19. According to Table 12-1 what is the strongest acid that can exist in water?

20. According to Table 12-1 what is the strongest base that can exist in water?

## Section 12-8

21. What is a neutralization reaction?

22. Give the net-ionic equation for the neutralization reaction that occurs when a solution of nitric acid is added to a solution of sodium hydroxide.

23. Give the net-ionic equation for the neutralization reaction that occurs when a solution of hydrochloric acid is mixed with a solution of potassium hydroxide.

## Section 12-9

24. Why are some acid and base solutions potentially dangerous?

25. Which bases in Table 12-1 produce poison gases when mixed with a solution of a strong acid? **CAUTION: Never mix compounds or solutions containing these ions with acids.**

26. Why do 1.0 $M$ solutions of hydrochloric acid and acetic acid differ in their concentrations of the hydronium ion?

27. Which acid is used as an electrolyte in lead storage batteries?

28. Why is it not a good idea to watch a battery as it is being jump started?

29. Why is baking soda, $NaHCO_3$, useful for the treatment of acid spills.

## Sections 12-10 to 12-12

30. What is a titration?

31. Define the following terms with respect to a titration.

    (a) buret

    (b) titrant

    (c) indicator

    (d) end point

32. *A vinegar sample is titrated with a sodium hydroxide solution to the end point. Using the equation and the experimental data given below, calculate the molarity of the acetic acid in the vinegar.

$$HC_2H_3O_2(aq) + OH^-(aq) \rightleftharpoons H_2O + C_2H_3O_2^-(aq)$$

| *M* NAOH | VOLUME NAOH USED | VOLUME VINEGAR TITRATED |
|---|---|---|
| 0.515 *M* | 21.15 mL | 10.0 mL |

33. *Calculate the molarity of a hydrochloric acid solution if 27.5 mL of a 0.675 *M* KOH solution are needed to titrate a 20.0-mL sample of the acid solution to the end point. The reaction involved is

$$H_3O^+(aq) + OH^-(aq) \rightleftharpoons H_2O + H_2O$$

34. *A sample of sodium hydrogen phosphate, $Na_2HPO_4$, is dissolved in water and titrated with a sodium hydroxide solution to the end point. If 19.4 mL of a 0.789 *M* NaOH solution was required to titrate the sample, calculate the mass in grams of $Na_2HPO_4$ in the sample. The reaction is

$$HPO_4{}^{2-}(aq) + OH^-(aq) \rightleftharpoons H_2O + PO_4{}^{3-}(aq)$$

35. A sample of a cleaning solution containing ammonia was titrated with a hydrochloric acid solution. Using the equation and titration data given below, calculate the molarity of the ammonia in the cleaning solution.

$$H_3O^+(aq) + NH_3(aq) \rightleftharpoons H_2O + NH_4{}^+(aq)$$

| MOLARITY *M* HYDROCHLORIC ACID | VOLUME ACID USED | VOLUME NH₃ SOLUTION TITRATED |
|---|---|---|
| 1.10 *M* | 9.28 mL | 10.50 mL |

36. A vinegar sample is titrated with a sodium hydroxide solution. Using the reaction and experimental data given below, calculate the molarity of acetic acid in the vinegar.

$$HC_2H_3O_2(aq) + OH^-(aq) \rightleftharpoons H_2O + C_2H_3O_2{}^-(aq)$$

| *M* NAOH | VOLUME NAOH USED | VOLUME VINEGAR TITRATED |
|---|---|---|
| 0.768 *M* | 21.2 mL | 25.0 mL |

37. A sample of an ammonia solution is titrated with 0.100 *M* hydrochloric acid. It requires 38.5 mL of the acid to titrate the ammonia. If the reaction is

$$H_3O^+(aq) + NH_3(aq) \rightleftharpoons NH_4{}^+(aq) + H_2O$$

calculate the number of grams of $NH_3$ in the sample.

38. Calculate the molarity of a KOH solution if 48.52 mL of a 0.165 *M* hydrochloric acid solution are required to titrate 20.0 mL of the base solution. The reaction is

$$H_3O^+(aq) + OH^-(aq) \rightleftharpoons 2H_2O$$

## Section 12-13

39. What is $K_w$ and what does it represent?

40. How is the ion product of water useful?

## Section 12-14

41. Give a definition of pH. Give a definition of pOH.

42. What are acidic and basic solutions in terms of the hydronium ion concentration and the hydroxide ion concentration? What are they in terms of pH and pOH?

43. What are the concentrations of the hydronium ion and the hydroxide ion in pure water or a neutral solution at 25°C? Explain why they are equal.

44. Draw a number line from 0 to 14 to represent pH values. Indicate on the line the regions corresponding to acidic solutions and basic solutions. Draw an arrow pointing to the pH of neutral solutions. List pOH values corresponding to pH values on your line.

## Sections 12-15 and 12-16

45. *Estimate and calculate the pH of the following solutions and indicate whether a solution is acidic or basic.

(a) an ammonia solution that has $8.9 \times 10^{-9}$ *M* $H_3O^+$

(b) a sample of cerebrospinal fluid that has $2.7 \times 10^{-8}$ *M* $H_3O^+$

(c) a sample of urine that has $6.9 \times 10^{-7}$ *M* $H_3O^+$

(d) a citric acid solution that has $9.9 \times 10^{-4}$ *M* $H_3O^+$

(e) a 1.0 *M* NaOH solution that has $1.0 \times 10^{-14}$ *M* $H_3O^+$

(f) a 0.50 *M* hydrochloric acid solution, a strong acid

(g) a sample of gastric juice that has $3.2 \times 10^{-2}$ $M$ $H_3O^+$

(h) a sample of rain water that has $1.1 \times 10^{-7}$ $M$ $H_3O^+$

(i) a sample of acid rain water that has $2.3 \times 10^{-4}$ $M$ $H_3O^+$

(j) a sample of carbonated water that has $2.8 \times 10^{-4}$ $M$ $H_3O^+$

(k) a sample of blood plasma that has $4.0 \times 10^{-8}$ $M$ $H_3O^+$

46. Estimate and calculate the pH of the following solutions and indicate whether a solution is acidic or basic.

(a) an ammonia solution that has $8.4 \times 10^{-9}$ $M$ $H_3O^+$

(b) a sample of saliva that has $3.5 \times 10^{-7}$ $M H_3O^+$

(c) a sample of rain that has $1.8 \times 10^{-7}$ $M H_3O^+$

(d) an apple juice that has $9.3 \times 10^{-4}$ $M H_3O^+$

(e) a sample of milk that has $2.3 \times 10^{-7}$ $M H_3O^+$

(f) a 0.10 $M$ NaOH solution that has $1.0 \times 10^{-13}$ $M$ $H_3O^+$

(g) a sample of wine that has $3.8 \times 10^{-3}$ $M H_3O^+$

(h) a sample of coffee that has $6.2 \times 10^{-5}$ $M H_3O^+$

(i) a solution of soap that has $9.7 \times 10^{-8}$ $M H_3O^+$

(j) a sample of lemon juice that has $2.4 \times 10^{-2}$ $M$ $H_3O^+$

(k) a sample of hydrochloric acid that has $5.5$ $M$ $H_3O^+$

47. *Estimate and calculate the pOH of the following solutions and indicate whether a solution is acidic or basic.

(a) an ammonia solution that has $2.3 \times 10^{-5}$ $M$ $OH^-$

(b) a sample of cerebrospinal fluid that has $6.2 \times 10^{-6}$ $M OH^-$

(c) a sample of soap solution that has $6.9 \times 10^{-6}$ $M$ $OH^-$

(d) a citric acid solution that has $1.0 \times 10^{-11}$ $M OH^-$

(e) a 1.0 $M$ NaOH solution

(f) a 0.50 $M$ hydrochloric acid solution, a strong acid

(g) a sample of bile juice that has $3.2 \times 10^{-6}$ $M OH^-$

(h) a sample of alkaline lake water that has $9.2 \times 10^{-5}$ $M OH^-$

(i) a sample of acid rainwater that has $7.7 \times 10^{-11}$ $M OH^-$

(j) a sample of white wine that has $2.1 \times 10^{-10}$ $M$ $OH^-$

(k) a sample of blood plasma that has $6.0 \times 10^{-7}$ $M$ $OH^-$

48. Estimate and calculate the pOH of the following solutions and indicate whether the solution is acidic or basic.

(a) an ammonia solution that has $2.0 \times 10^{-6}$ $M OH^-$

(b) a sample of saliva that has $3.5 \times 10^{-7}$ $M OH^-$

(c) a sample of rain that has $1.9 \times 10^{-8}$ $M OH^-$

(d) an apple juice that has $7.0 \times 10^{-12}$ $M OH^-$

(e) a sample of milk that has $8.1 \times 10^{-8}$ $M OH^-$

(f) a 0.10 $M$ NaOH solution

(g) a sample of beer that has $3.6 \times 10^{-9}$ $M OH^-$

(h) a sample of coffee that has $4.0 \times 10^{-10}$ $M OH^-$

(i) a solution of soap that has $8.5 \times 10^{-6}$ $M OH^-$

(j) a sample of lemon juice that has $7.6 \times 10^{-13}$ $M$ $OH^-$

(k) a 0.5 $M$ solution of hydrochloric acid

49. Estimate and calculate the pOH of the solutions in parts (a) to (e) of Question 46.

50. Estimate and calculate the pH of the solutions in parts (a) to (e) of Question 47.

# OXIDATION

## AND

## REDUCTION

CHAPTER

13

## 13-1 ELECTRON TRANSFER REACTIONS

When you turn on a flashlight or a hand calculator, electrical energy is instantly available to operate the device. Batteries supply electrical energy through electron transfer reactions. Electron transfer reactions in batteries are exothermic reactions, in which most of the energy released takes the form of electrical energy.

When a piece of zinc metal is added to a solution containing copper(II) ions, the following reaction occurs spontaneously:

$$Zn(s) + Cu^{2+}(aq) \longrightarrow Zn^{2+}(aq) + Cu(s)$$

In this reaction zinc metal changes to zinc ions and copper(II) ions to copper metal. Note how the oxidation number of zinc changes and how the oxidation number of copper changes. Review Section 11-5. Chemistry in Action 13-1 shows a way to observe the reaction between zinc metal and copper(II) ions.

The zinc–copper reaction can be carried out in another interesting way. A piece of zinc (the zinc electrode) is placed in a solution containing zinc ions and a piece of copper (the copper electrode) is placed in a solution containing copper(II) ions. Of course, negative ions are also present in these solutions, but it is the metal ions that are involved in the chemical reaction. A conducting bridge, called a salt bridge, connects the two solutions as shown in Figure 13-1. A salt bridge is a glass tube that contains a

**437**

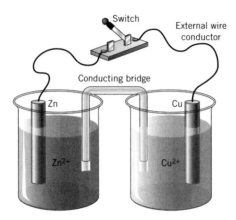

**FIGURE 13-1**
Galvanic or voltaic cell.

solution of an ionic compound (an electrolyte). The bridge provides a way to have electrical contact between the two solutions without actually mixing them. The conducting bridge contains electrolyte ions that move within the bridge from one solution to the other providing electrical contact.

Since the two solutions are not in direct contact, no reaction occurs. However, when the electrodes are connected to one another by a metal wire, an electron transfer reaction occurs. The electrons flow through the wire from one side of the system to the other. The reaction at the zinc electrode (the anode) is

$$Zn(s) \longrightarrow Zn^{2+}(aq) + 2e^-$$

The lost electrons flow through the wire to the copper electrode (the cathode). The copper (II) ions gain electrons at the copper electrode.

$$Cu^{2+}(aq) + 2e^- \longrightarrow Cu(s)$$

As soon as the electrodes are connected by a wire, the reactions occur spontaneously, and electrons flow through the wire as electrical current. The current flow is a source of useful electrical energy. This setup is an example of a **voltaic cell** or **galvanic cell.** A voltaic cell is an apparatus in which electrode reactions are separated by a salt bridge. When the electrodes are connected by an external conductor, electrical work is produced as electrons flow. **Batteries** of all sorts are voltaic cells specially packaged to provide electrical energy from electron transfer reactions (see Section 13-8). When the electrodes in a battery are connected by a metallic conductor, electricity flows as the reaction takes place.

In the copper–zinc cell described above, zinc metal changes to zinc ions and copper(II) ions change to copper metal. To get an equation for the overall reaction add the two electrode reactions.

$$Zn(s) \longrightarrow Zn^{2+}(aq) + 2e^-$$
$$Cu^{2+}(aq) + 2e^- \longrightarrow Cu(s)$$

$$\overline{Zn(s) + Cu^{2+}(aq) + 2e^- \longrightarrow Cu(s) + Zn^{2+}(aq) + 2e^-}$$

The electrons lost by the zinc are gained by the copper ions; after canceling the electrons on both sides, the equation becomes

$$Zn(s) + Cu^{2+}(aq) \longrightarrow Cu(s) + Zn^{2+}(aq)$$

Notice that this is the same equation for the spontaneous reaction that occurs when zinc metal is placed directly in a solution containing copper(II) ions. Electron transfer reactions typically occur when reactants are in direct contact. However, it is possible to carry out some electron transfer reactions by separating the reaction into two halves in a voltaic cell.

## 13-2  OXIDATION AND REDUCTION DEFINED

Consider the electron transfer reaction between sodium metal and chlorine gas.

$$Na(s) + Cl_2(g) \longrightarrow 2NaCl(s)$$

In an electron transfer reaction, elements change their chemical combinations. As a result they change oxidation number. Sodium has an oxidation number of 0 in metallic sodium and an oxidation number of +1 in NaCl. Chlorine has an oxidation number of 0 in chlorine gas and oxidation number of −1 in NaCl. In the reaction sodium has undergone oxidation and chlorine has undergone reduction as defined in the margin and Figure 13-2.

Electron transfer reactions are oxidation–reduction reactions or redox reactions. You can recognize them by noting whether changes have occurred in the oxidation numbers of any elements involved in the reaction. Since oxidation and reduction occur simultaneously, a redox reaction must involve an increase in the oxidation number of one element and a decrease for another element.

Another example of an electron transfer reaction is the reaction that occurs when sodium metal is placed in water.

$$2Na(s) + 2H_2O \longrightarrow 2Na^+(aq) + 2OH^-(aq) + H_2(g)$$

In this reaction sodium changes from elemental sodium to a charged sodium ion and hydrogen combined with oxygen in water changes to hydrogen gas. Sodium is not a spectator ion because it changes from a solid metal to an ion in the reaction. The hydrogen covalently bonded to oxygen in water becomes covalently bonded to itself in hydrogen gas. Look closely at the oxidation numbers of the elements involved in the reaction. The oxidation number of sodium has changed from 0 in sodium metal to +1 in the sodium ion. Sodium must have lost electrons.

**Oxidation:**

*Loss of electrons; the increase in oxidation number of an element.*

**Reduction:**

*Gain of electrons; the decrease in oxidation number of an element.*

**FIGURE 13-2**
Oxidation is an increase in oxidation number and reduction is a decrease in oxidation number.

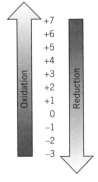

$$\overset{0}{\vphantom{2}}2Na + 2H_2O \longrightarrow \overset{+1}{\vphantom{2}}2Na^+ + 2OH^- + H_2$$

The oxidation number of hydrogen has changed from +1 in water to 0 in hydrogen gas. Hydrogen must have gained electrons.

$$\overset{+1}{\vphantom{2}}2Na + 2H_2O \longrightarrow 2Na^+ + 2OH^- + \overset{0}{H_2}$$

The oxidation number of oxygen is −2 in water, on the left, and −2 in the hydroxide ion, on the right. Oxygen has not changed oxidation number, so it has not lost or gained electrons and two of the hydrogen atoms have not changed oxidation number. In the reaction, sodium has undergone oxidation and hydrogen has undergone reduction.

## 13-3  HALF-REACTIONS

For convenience as well as clarification, the equation for a redox reaction is often separated into two parts. One part involves the loss of electrons (oxidation) and the other involves the gain of electrons (reduction). For example, the reaction between sodium metal and water is represented as

$$Na\ (s) \longrightarrow Na^+(aq) + e^- \qquad \text{oxidation; loss of electrons}$$
$$2e^- + 2H_2O \longrightarrow 2OH^-(aq) + H_2(g) \quad \text{reduction; gain of electrons}$$

Such representations of a redox reaction are called **half-reactions** or **half-equations.** One half-reaction represents the oxidation half of the overall reaction and the other represents the reduction half of the overall reaction. Of course, one half-reaction does not occur without the other half-reaction taking place simultaneously. A complete redox reaction is a combination of the two half-reactions.

In a redox reaction, the number of electrons lost must balance the electrons gained, since no electrons can be created or destroyed in a reaction. Electrons are lost by one element and gained by another. To incorporate this into the complete reaction, the half-reactions often must be adjusted by multiplying one or both by the appropriate numbers to balance the electrons lost with the electrons gained. The oxidation half of the sodium–water reaction given above is multiplied by 2, two electrons lost balance the two electrons gained in the reduction half.

$$2Na \longrightarrow 2Na^+ + 2e^-$$
$$2e^- + 2H_2O \longrightarrow 2OH^- + H_2$$

After balancing the electrons, the two half-reactions can be added to give the complete redox equation.

$$2Na + 2e^- + 2H_2O \longrightarrow 2Na^+ + 2e^- + 2OH^- + H_2$$

The electrons appearing on both sides are canceled to give

$$2Na + 2H_2O \longrightarrow 2Na^+ + 2OH^- + H_2$$

Many redox reactions are not easily balanced using the typical trial and error methods. Systematic methods used to balance half-equations and redox equations for more complicated reactions are discussed in Appendix 4.

## 13-4 OXIDIZING AND REDUCING AGENTS

In a redox reaction one species acts as an electron donor. It loses electrons and is oxidized since oxidation is loss of electrons. Sodium metal in the sodium–water reaction is an electron donor. The electrons lost by the electron donor reduce another species. An **electron donor** that causes another species to be reduced is called a reducing agent. A **reducing agent** is a species that reduces another species and is itself oxidized in the process.

The species that gains electrons in a redox reaction is an electron acceptor. By accepting electrons it is reduced since gain of electrons is reduction. Water in the sodium–water reaction accepts or gains electrons from sodium metal. The **electron acceptor** is called an oxidizing agent because it oxidizes another species. An **oxidizing agent** oxidizes another species, and is itself reduced in the process.

It is usually easy to identify the reducing agent in a redox reaction. It is that species having an element that undergoes an increase in oxidation number. The element changes from a lower to a higher oxidation number. In the sodium–water reaction, sodium is the reducing agent and its oxidation number changes from 0 to + 1.

$$\underset{\substack{\text{reducing agent;} \\ \text{reduces hydrogen} \\ \text{in water}}}{\overset{0}{2Na}} + 2H_2O \xrightarrow{\text{reduction}} \overset{+1}{2Na^+} + H_2 + 2OH^-$$

Also it is usually just as easy to identify the oxidizing agent since it is that species having an element that undergoes a decrease in oxidation number. The element changes from a higher to a lower oxidation number. In the sodium–water reaction, water is the oxidizing agent and the oxidation number of hydrogen changes from +1 to 0.

$$2Na + \overset{+1}{2H_2O} \xrightarrow{\text{oxidation}} 2Na^+ + \overset{0}{H_2} + 2OH^-$$
$$\underset{\substack{\text{oxidizing agent;} \\ \text{oxidizes sodium}}}{}$$

Each redox reaction must have an oxidizing agent and a reducing agent. We identify the oxidizing and reducing agents by noting the oxidation number changes of elements in the reaction.

**EXAMPLE 13-1**

Label the oxidizing and reducing agent in the following redox reaction:

$$S + O_2 \longrightarrow SO_2$$

The sulfur undergoes an increase in oxidation number from 0 to +4. It is oxidized, so oxygen is the oxidizing agent. The oxygen changes oxidation number from 0 to –2. It is reduced, so sulfur is the reducing agent.

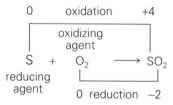

**EXAMPLE 13-2**

Label the oxidizing and reducing agents in the following redox reaction:

$$5Fe^{2+} + MnO_4^- + 8H^+ \longrightarrow 5Fe^{3+} + Mn^{2+} + 4H_2O$$

Look for any change in ion charge or change in combination. The iron changes its oxidation number from +2 to +3; it is oxidized. The manganese changes its oxidation number from +7 in $MnO_4^-$ to +2 in $Mn^{2+}$; it is reduced. Hydrogen and oxygen do not change their oxidation numbers in the reaction. Hydrogen is +1 and oxygen is –2 on both sides of the equation. The iron is oxidized by the permanganate ion, so $MnO_4^-$ is the oxidizing agent. Note that the manganese in the permanganate ion changes oxidation number but the entire ion is viewed as the oxidizing agent. The permanganate ion is reduced by the iron(II) ion, so $Fe^{2+}$ is the reducing agent.

+2        oxidation        +3

oxidizing agent

$$5Fe^{2+} + MnO_4^- + 8H^+ \longrightarrow 5Fe^{3+} + Mn^{2+} + 4H_2O$$

reducing agent    +7     reduction     +2

The hydrogen ion, $H^+$, in this reaction is neither an oxidizing nor a reducing agent. It is involved in the reaction as a carrier of charge and appears in the balanced equation. As you learned in Chapter 12, the $H^+$ exists in water as hydronium ion, $H_3O^+$. When dealing with redox reactions it is more convenient to use $H^+$ instead of $H_3O^+$. In any case, hydrogen ion participates in many redox reactions as a charge carrier. This means that it is involved in the reaction as a carrier of positive charge to help balance the charge.

ACTIVITY 13-1

Write the oxidation and reduction half-reactions and label the oxidizing and reducing agents in the following redox reaction:

$$2H^+ + Zn \longrightarrow Zn^{2+} + H_2$$

## 13-5 MEASURING STRENGTHS OF OXIDIZING AND REDUCING AGENTS

When oxidizing and reducing agents are mixed, a redox reaction may occur. Oxidizing and reducing agents vary in their ability to react with each other. How extensively oxidizing and reducing agents react depends on the strength of the substance with which they are mixed.

When an oxidizing agent reacts, it is reduced and forms a potential reducing agent. For example, chlorine, $Cl_2$, can act as an oxidizing agent, and the chloride ion, $Cl^-$, can act as a reducing agent.

$$\text{oxidizing agent} \quad Cl_2 + 2e^- \longrightarrow 2Cl^-$$
$$\text{reducing agent} \quad 2Cl^- \longrightarrow Cl_2 + 2e^-$$

When a reducing agent reacts, it is oxidized and forms a potential oxidizing agent. For example, zinc metal is a reducing agent and zinc ion formed by the oxidation of zinc is a potential oxidizing agent.

$$\text{reducing agent} \quad Zn \longrightarrow Zn^{2+} + 2e^-$$
$$\text{oxidizing agent} \quad Zn^{2+} + 2e^- \longrightarrow Zn$$

Oxidation and reduction processes are potentially reversible. It is possible to couple or associate each oxidizing agent with its reduced form that is a reducing agent. Such pairs are called **redox couples.** Chlorine, $Cl_2$, and the chloride ion, $Cl^-$ are a redox couple and the zinc ion, $Zn^{2+}$, and zinc metal are a redox couple.

The stronger the oxidizing agent the greater its tendency to oxidize reducing agents. In comparison, the reduced form of a stronger oxidizing agent is a weaker reducing agent. Fluorine, $F_2$, is a very strong oxidizing agent, and its reduced form, the fluoride ion, $F^-$, is a very weak reducing agent. The stronger a reducing agent the greater its tendency to reduce oxidizing agents. In comparison, the oxidized form of a stronger reducing agent is a weaker oxidizing agent. Sodium metal is a very strong reducing agent, and its oxidized form, sodium ion, $Na^+$, is a very weak oxidizing agent.

How is it possible to measure the strength of an oxidizing agent or a reducing agent? One way is to incorporate the agent into a galvanic cell. In a cell, two half-reactions can be physically separated and compared. Recall from Section 13-1 that when the two electrodes in contact with the half-reactions are connected by an external circuit, a spontaneous electron transfer can occur. As the reaction occurs, the oxidizing agent gains electrons and the reducing agent loses electrons. A measure of the tendency for two half-reactions to occur is found by placing a voltmeter in the circuit. A voltmeter is an electronic device that measures the electromotive force between two locations in an electrical circuit. Electromotive force is a force that acts when there is a difference in electrical charge between two points. If a voltmeter, is attached to the terminals of a battery or the electrodes of a cell the voltage capability can be measured. The greater the voltage reading the greater the tendency for spontaneous electron transfer to occur. By convention, the strength of an oxidizing agent is measured by comparing it, in a galvanic cell, to the hydrogen ion–hydrogen half-reaction.

$$2H^+(aq) + 2e^- \longrightarrow H_2(g)$$

The electrode for this half-reaction is referred to as the **hydrogen electrode** when it is part of a galvanic cell (see Fig. 13-3). To make measurements of cell voltages a specific half-reaction it is set up as part of the galvanic cell. The voltage between this half of the cell and the hydrogen electrode is measured using a voltmeter. By convention, this is done under the conditions that each dissolved species has a concentration of 1 mol/L and gases are at 1 atm pressure.

By convention, if the oxidizing agent gains electrons and the hydrogen electrode loses electrons, the half-reaction being tested has a positive cell voltage. For example, a fluorine–fluoride half-reaction gives a voltage of +2.87 volts when coupled with the hydrogen electrode.

$$F_2(g) + 2e^- \longrightarrow 2F^-(aq) \quad (\text{+2.87 volts})$$

A half-reaction involving acidified permanganate gives a voltage of +1.49 volts.

$$MnO_4^-(aq) + 8H^+(aq) + 5e^- \longrightarrow Mn^{2+} + 4H_2O \quad (\text{+1.49 volts})$$

This means fluorine is a stronger oxidizing agent than acidified permanganate ion.

A variety of oxidizing agents have been measured against the hydrogen electrode and ranked according to their strengths. When a half-cell involv-

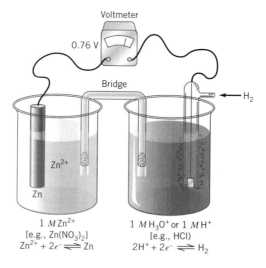

**FIGURE 13-3** A galvanic cell including a hydrogen electrode and a voltmeter.

ing zinc ions and zinc metal is tested against hydrogen, the zinc ions do not gain electrons; instead the zinc metal loses electrons and the hydrogen ions, in the hydrogen electrode, gain electrons. (See Fig. 13-3.) When this happens, the measured voltage associated with the half-reaction is given a negative sign.

$$Zn^{2+}(aq) + 2e^- \longrightarrow Zn(s) \quad (-0.76 \text{ volt})$$

The negative voltage shows that the zinc ion is a weaker oxidizing agent than the hydrogen ion. Furthermore, it also means that the zinc ion is a weaker oxidizing agent than any agent having a positive voltage.

Table 13-1 is a brief **redox table** which ranks some common oxidizing agents according to strength. They are ranked in order of decreasing cell voltages. Since each oxidizing agent has a corresponding reducing agent, the reducing agents are also listed according to strength. Stronger oxidizing agents have weaker reducing agents in the couple, so the reducing agents in the table are listed according to increasing strength. That is, the table starts with weaker reducing agents at the top and the strengths increase down the table. Note that the completely balanced half-reaction is given for each oxidizing agent in the table. Also note that some oxidizing agents have the hydrogen ion included in their half-reactions. The hydrogen ion is not acting as an acid but is acting to balance the charge in the reaction. If the solution is not acidic, the reaction does not occur.

*Table 13-1*   A Redox Table Including Some Common Oxidizing and Reducing Agents

| | OXIDIZING AGENT | | REDUCING AGENT | | CELL VOLTAGE |
|---|---|---|---|---|---|
| Fluorine | $F_2(g)$ | | $+ 2e^- \rightleftharpoons$        $2F^-(aq)$ | Fluoride ion | +2.87 |
| Hydrogen peroxide | $H_2O_2(aq)$ | + | $2H^+(aq) + 2e^- \rightleftharpoons$        $2H_2O$ | Water | +1.78 |
| Permanganate ion | $MnO_4^-(aq)$ | + | $8H^+(aq) + 5e^- \rightleftharpoons 4H_2O + 2Mn^{2+}(aq)$ | Manganese(II) ion | +1.49 |
| Chlorine | $Cl_2(g)$ | | $+ 2e^- \rightleftharpoons$        $2Cl^-(aq)$ | Chloride ion | +1.36 |
| Dichromate ion | $Cr_2O_7^{2-}(aq)$ | + | $14H^+(aq) + 6e^- \rightleftharpoons 7H_2O + 2Cr^{3+}(aq)$ | Chromium(III) ion | +1.33 |
| Bromine | $Br_2(\ell)$ | | $+ 2e^- \rightleftharpoons$        $2Br^-(aq)$ | Bromide ion | +1.07 |
| Nitrate ion | $NO_3^-(aq)$ | + | $4H^+(aq) + 3e^- \rightleftharpoons 2H_2O + NO(g)$ | Nitrogen oxide | +0.96 |
| Iron(III) ion | $Fe^{3+}(aq)$ | | $+ 1e^- \rightleftharpoons$        $Fe^{2+}(aq)$ | Iron(II) ion | +0.77 |
| Iodine | $I_2(s)$ | | $+ 2e^- \rightleftharpoons$        $2I^-(aq)$ | Iodide ion | +0.54 |
| Carbon dioxide | $2CO_2(g)$ | + | $2H^+(aq) + 2e^- \rightleftharpoons$        $H_2C_2O_4(aq)$ | Oxalic acid | +0.49 |

## 13-6 USING A REDOX TABLE

Redox reactions can be predicted using Table 13-1. A prediction is based on the fact that when an oxidizing agent and reducing agent mix, a reaction occurs if a weaker oxidizing agent and weaker reducing agent form as products. The agents in the table are listed according to strength. Generally, any oxidizing agent reacts with any reducing agent below it in the table. In contrast, no oxidizing agent reacts with a reducing agent lying above it in the table.

$$\left.\begin{array}{l}\text{stronger}\\\text{oxidizing}\\\text{agent}\end{array}\right\} \longrightarrow \left\{\begin{array}{l}\text{weaker}\\\text{reducing}\\\text{agent}\end{array}\right.$$

$$\left.\begin{array}{l}\text{weaker}\\\text{oxidizing}\\\text{agent}\end{array}\right\} \longleftarrow \left\{\begin{array}{l}\text{stronger}\\\text{reducing}\\\text{agent}\end{array}\right.$$

In any reaction predicted from the table the oxidizing agent reacts with the reducing agent to form the products given in the table. For example, suppose a solution containing iron(III) ions, $Fe^{3+}$, mixes with a solution containing iodide ions, $I^-$. In Table 13-1 the iron(III) ion, $Fe^{3+}$, is found as an oxidizing agent, and the iodide ion, $I^-$, is found as a reducing agent. Since the oxidizing agent is above the reducing agent, the predicted reaction is

$$Fe^{3+} + e^- \longrightarrow Fe^{2+} \qquad \text{reduction}$$
$$2I^- \longrightarrow I_2 + 2e^- \quad \text{oxidation}$$

Note that the oxidation part is the reverse of the reaction in the table. To write a complete net-ionic equation for the redox reaction, copy from the table the half-reaction of the oxidizing agent and the reversed half-reaction of the reducing agent. Then multiply each half-reaction by the appropriate factors so that the number of electrons gained equals the number of electrons lost. Start with the two half-reactions.

$$Fe^{3+} + e^- \longrightarrow Fe^{2+}$$
$$2I^- \longrightarrow I_2 + 2e^-$$

Multiplying the iron half-reaction by 2 and adding the half-reactions gives the final equation.

$$\begin{array}{l}2Fe^{3+} + 2e^- \longrightarrow 2Fe^{2+}\\\underline{\qquad\qquad 2I^- \longrightarrow I_2 + 2e^-}\\2Fe^{3+} + 2I^- \longrightarrow 2Fe^{2+} + I_2\end{array}$$

Chemistry in Action 13-2 is about the oxidation of iodide ions to make elemental iodine.

**EXAMPLE 13-3**

What reaction will take place when an acidic solution containing dichromate ions, $Cr_2O_7^{2-}$, is mixed with a solution containing bromide ions, $Br^-$? The acidic solution supplies hydrogen ions.

The possible reactants are $Cr_2O_7^{2-}$, $H^+$, and $Br^-$. Using Table 13-1, we note the position of the acidified dichromate ion is above the bromide ion.

$$14H^+ + Cr_2O_7^{2-} + 6e^- \longrightarrow 2Cr^{3+} + 7H_2O$$
$$Br_2 + 2e^- \longleftarrow 2Br^-$$

Therefore, we write the reverse of the half-reaction of the reducing agent, $Br^-$.

$$2Br^- \longrightarrow Br_2 + 2e^-$$

Then we multiply by 3 to balance the electrons. The final equation is obtained by adding the two half-reactions

$$14H^+ + Cr_2O_7^{2-} + 6e^- \longrightarrow 2Cr^{3+} + 7H_2O$$
$$\underline{6Br^- \longrightarrow 3Br_2 + 6e^-}$$
$$14H^+ + Cr_2O_7^{2-} + 6Br^- \longrightarrow 3Br_2 + 2Cr^{3+} + 7H_2O$$

**EXAMPLE 13-4**

What reaction will take place when an acidic solution containing dichromate ions, $Cr_2O_7^{2-}$, mixes with a solution containing chloride ions, $Cl^-$?

The possible reactants are $Cr_2O_7^{2-}$, $H^+$, and $Cl^-$.

Noting the positions of the acidified dichromate ion and the chloride ion in Table 13-1, we would predict no reaction, since the couple with $Cl^-$ is above the couple with $Cr_2O_7^{2-}$ in the table.

$$Cl_2 + 2e^- \longrightarrow 2Cl^-    \text{No reaction}$$
$$14H^+ + Cr_2O_7^{2-} + 6e^- \longrightarrow 2Cr^{3+} + 7H_2O$$

## 13-7   THE ACTIVITY SERIES OR ELECTROMOTIVE SERIES OF METALS

Table 13-2 shows another redox table that includes some common metals, water, and the hydrogen ion. This table is often called the **electromotive series** of metals or the **activity series** of metals because it lists common metals according to their strength as reducing agents. Note that metals, as reducing agents, are listed on the right in the table. Stronger reducing agents occur at the bottom of the list and they decrease in strength from

***Table 13-2*** The Electromotive Series of Metals or the Activity Series of Metals

| | OXIDIZING AGENT | REDUCING AGENT | | CELL VOLTAGE | |
|---|---|---|---|---|---|
| Silver ion | $Ag^+(aq)$ $+ 1e^- \rightleftharpoons Ag(s)$ | Silver | +0.80 | |
| Mercury(II) ion | $Hg^{2+}(aq)$ $+ 2e^- \rightleftharpoons Hg(\ell)$ | Mercury | +0.79 | |
| Iron(III) ion | $Fe^{3+}(aq)$ $+ 1e^- \rightleftharpoons Fe^{2+}(aq)$ | Iron(II) ion | +0.77 | |
| Copper(II) ion | $Cu^{2+}(aq)$ $+ 2e^- \rightleftharpoons Cu(s)$ | Copper | +0.34 | |
| Hydrogen ion | $2H^+(aq)$ $+ 2e^- \rightleftharpoons H_2(g)$ | Hydrogen | 0.00 | |
| Lead(II) ion | $Pb^{2+}(aq)$ $+ 2e^- \rightleftharpoons Pb(s)$ | Lead | −0.13 | |
| Nickel(II) ion | $Ni^{2+}(aq)$ $+ 2e^- \rightleftharpoons Ni(s)$ | Nickel | −0.25 | These metals dissolve in acid solutions |
| Cobalt(II) ion | $Co^{2+}(aq)$ $+ 2e^- \rightleftharpoons Co(s)$ | Cobalt | −0.28 | |
| Cadmium ion | $Cd^{2+}(aq)$ $+ 2e^- \rightleftharpoons Cd(s)$ | Cadmium | −0.40 | |
| Iron(II) ion | $Fe^{2+}(aq)$ $+ 2e^- \rightleftharpoons Fe(s)$ | Iron | −0.41 | |
| Zinc ion | $Zn^{2+}(aq)$ $+ 2e^- \rightleftharpoons Zn(s)$ | Zinc | −0.76 | |
| Water | $2H_2O$ $+ 2e^- \rightleftharpoons H_2(g) + 2OH^-(aq)$ | Hydrogen | −0.83 | |
| Aluminum ion | $Al^{3+}(aq)$ $+ 3e^- \rightleftharpoons Al(s)$ | Aluminum | −1.67 | |
| Magnesium ion | $Mg^{2+}(aq)$ $+ 2e^- \rightleftharpoons Mg(s)$ | Magnesium | −2.38 | |
| Sodium ion | $Na^+(aq)$ $+ 1e^- \rightleftharpoons Na(s)$ | Sodium | −2.71 | These metals react spontaneously in water |
| Calcium ion | $Ca^{2+}(aq)$ $+ 2e^- \rightleftharpoons Ca(s)$ | Calcium | −2.76 | |
| Potassium ion | $K^+(aq)$ $+ 1e^- \rightleftharpoons K(s)$ | Potassium | −2.92 | |

the bottom to the top of the table. You can use this table to predict reactions in the same manner as Table 13-1. That is, any oxidizing agent reacts with any reducing agent below it in the table and will not react with a reducing agent above it.

Metal ions are listed in the table as oxidizing agents; metals are listed as reducing agents. The table shows which metals and metal ions react with one another. A specific metal ion reacts with any metal that appears below it in the table. When zinc metal is placed in a solution containing copper(II) ions, $Cu^{2+}$, a reaction occurs. In the table, the oxidizing agent $Cu^{2+}$ is above the couple involving the reducing agent, Zn.

$$Cu^{2+} + 2e^- \longrightarrow Cu \quad \text{reduction}$$
$$Zn^{2+} + 2e^- \longleftarrow Zn \quad \text{oxidation}$$

Thus, the predicted reaction is zinc metal reacting with copper(II) ion to give copper metal and zinc ion. The copper half equation is taken from the table and added to the reverse of the zinc half equation:

$$Cu^{2+} + 2e^- \longrightarrow Cu \qquad \text{reduction}$$
$$\underline{\phantom{Zn + Cu^{2+}}Zn \longrightarrow Zn^{2+} + 2e^- \quad \text{oxidation}}$$
$$Zn + Cu^{2+} \longrightarrow Cu + Zn^{2+}$$

Recall that this was the spontaneous electron transfer reaction mentioned in Section 13-1.

Table 13-2 lists metals according to their chemical activity as reducing agents. The more active metals are at the bottom of the table and the less active metals are at the top of the table. An active metal is a strong reducing agent so it is easily oxidized. Less active metals are weaker reducing agents and are less easily oxidized. It is not uncommon to find naturally occurring copper metal since this metal has a low activity. Active metals, such as sodium and potassium, are never found in nature. Furthermore, samples of these metals must be carefully stored since they are very reactive. They are usually stored in containers of kerosene since they do not readily react with hydrocarbons. Samples of metals of intermediate activity, such as iron, tend to oxidize slowly when exposed to the environment. Over time pure iron, for example, readily rusts and corrodes in the environment. See the end of this section for more about rust.

Use Table 13-2 as an activity series of metals to predict reactions that may occur between various metals and metal ions in solution. Generally, any metal ion reacts with any metal below it to oxidize the metal to its ion and reduce the original ion to its metallic form. More active metals are those that are strong reducing agents at the bottom of the table. These active metals readily form ions. The activity series reveals some practical information. The Statue of Liberty is made of copper metal. It has a green color due to the oxidation of copper to form various copper compounds. These compounds form a coating on the copper which prevents it from further oxidation. Alas, the statue was built on an iron frame. According to the activity series any solution of copper ions that contacts iron reacts with the iron to give iron(II) ions. This caused the iron frame to corrode and lose its strength. Note the position of iron and copper in Table 13-2.

$$Cu^{2+} + 2e^- \longrightarrow Cu$$
$$Fe^{2+} + 2e^- \longleftarrow Fe$$

The equation for the reaction that occurs between copper(II) ions and iron is found by reversing the iron half-reaction and adding the two half-reactions

$$Cu^{2+} + Fe \longrightarrow Fe^{2+} + Cu$$

Occurring over a period of many years, this reaction caused the corrosion and decay of the iron frame inside the statue. That is why extensive repairs were needed in 1986.

**ACTIVITY 13-2**

(a) Which metals in Table 13-2 do you think are sometimes found in nature as the free metals?

(b) Which metals in Table 13-2 cannot exist as metals in water?

**ACTIVITY 13-3**

Repeat Chemistry in Action 13-1 but, in place of the zinc-exposed penny, place a steel paper clip in the solution of the copper(II) compound. Record your observations. Explain your observations using the terms oxidation and reduction.

---

**EXAMPLE 13-5**

What reaction occurs when a piece of nickel metal is added to a solution of mercury(II) nitrate?

First we write the formulas of the species that are mixed. Remember that we represent samples of metals by their symbols and we represent solutions of ionic compounds as the separate cation and anion.

$$Ni(s) + Hg^{2+} + NO_3^-$$

The mercury(II) ion is above nickel in the redox table, so we would predict the following reaction obtained by adding the half-equations in the table.

$$Ni(s) + Hg^{2+} \longrightarrow Hg(\ell) + Ni^{2+}$$

---

**13-4**

What reaction occurs when solid copper metal is added to a solution of silver nitrate, $AgNO_3$?

Table 13-2 also shows why some metals dissolve in acidic solutions and some do not. The hydrogen ion occurs in the table as an oxidizing agent. Note the location of the hydrogen ion in Table 13-2. When any of the metals below $H^+$ in the table are added to an acidic solution, a reaction occurs producing $H_2$ gas and the ion of the metal. These metals dissolve in acidic solutions. For example, note the relative positions of the hydrogen ion and magnesium metal in the table.

$$2H^+ + 2e^- \longrightarrow H_2(g)$$
$$Mg^{2+} + 2e^- \longleftarrow Mg$$

When magnesium metal is added to a solution of hydrochloric acid (a strong acid containing $H_3O^+$ for which we use $H^+$) a reaction occurs. To write the overall equation reverse the half-equation for the reducing agent and add the two half-equations.

$$2H^+ + Mg \longrightarrow Mg^{2+} + H_2(g)$$

**DANGER: Never mix metals at the bottom of the table with acid solutions. They can react violently causing an explosion.**

---

**EXAMPLE 13-6**

What reaction occurs when a piece of aluminum metal is added to a hydrochloric acid solution?

First write the formulas of the species that are mixed. Use $H^+$ and $Cl^-$ for hydrochloric acid.

$$Al(s) + H^+ + Cl^-$$

Hydrogen ion is above aluminum in the redox table. Thus, we expect the following reaction to take place. The equation is obtained by adding the two half-equations after each is multiplied by numbers to balance the electrons lost and gained.

$$2Al(s) + 6H^+ \longrightarrow 2Al^{3+} + 3H_2(g)$$

Notice that $H_2O$ also appears in Table 13-2 as an oxidizing agent. Some more active metals readily react with water. We would expect that any metal below $H_2O$ in the table would react with $H_2O$ to form $H_2$ and the metal ion. However, Mg reacts only in very hot water, and Al does not react at normal temperatures because an impervious layer of aluminum oxide coats aluminum samples. That is why aluminum is so corrosion-resistant and keeps its shine. The other three metals below $H_2O$ in the table react vigorously with water. For instance, when potassium metal is placed in water, a very exothermic reaction occurs. We get the equation for the reaction by reversing the potassium half-reaction, multiplying it by 2, and adding it to the water half-reaction.

$$2H_2O + 2e^- \longrightarrow 2OH^- + H_2(g)$$
$$2K(s) \longrightarrow 2K^+ + 2e^-$$
$$\overline{2K(s) + 2H_2O \longrightarrow 2K^+ + 2OH^- + H_2(g)}$$

(a) Give the equation for the reaction that occurs when a piece of solid cobalt metal is mixed with a solution of hydrochloric acid.

(b) Give the equation for the reaction that occurs when a piece of calcium metal is added to water. [Hint: One of the products is solid $Ca(OH)_2$.]

## Rust

Corrosion refers to the reaction of metals with atmospheric oxygen. Iron metal corrodes to form rust. **Rusting** of iron occurs when iron combines with oxygen in a moist environment.

$$4Fe(s) + 3O_2 \xrightarrow{\ H_2O\ } 2Fe_2O_3(s) \text{ (rust)}$$

Many common metals corrode to form a protective surface layer of metal oxide. The layer prevents further reaction with oxygen and the metals retain their metallic luster. Examples of such metals are chromium, nickel, tin, and aluminum. However, in a moist environment, in the presence of sea spray or acidic chemicals, most metals are consumed by corrosion. In contrast to other metals, iron oxides easily flake from the surface of iron exposing fresh metal surfaces. Stainless steels are alloys containing nickel and chromium which help to protect the metal surfaces and make them corrosion-resistant. Rain, made acidic when acidic air pollutants dissolve, is responsible for significant rusting of steel bridges, buildings, and railroad tracks. Sea sprays and the salts used in winter deicing of roads promote the rusting of steel in automobiles.

One way to protect iron and steel structures is to cover them with paint. Of course, if the paint peels, rusting can occur. Some large bridges, such as the Golden Gate Bridge in California, are painted repeatedly to inhibit corrosion. Iron coated with a layer of zinc metal is called galvanized iron. Consult Table 13-2 and explain how zinc in contact with the iron prevents the formation of $Fe^{2+}$ and, therefore, rusting.

**Cathodic protection** is a technique used to protect steel in water heaters, buried fuel tanks, underground pipes, and hulls of ships. A metal that is a stronger reducing agent than iron is attached, by wire, to the iron object. Pieces of magnesium metal are used for corrosion protection of tanks, pipes, and water heaters. The magnesium metal is more easily oxidized than the iron so it corrodes in preference to the iron. However, since the magnesium metal is consumed by oxidation it has to be replaced occasionally. Nevertheless, cathodic protection saves money by preventing corrosion of expensive metal installations.

## 13-8 BATTERIES AND FUEL CELLS

In Section 13-1 we saw that it was possible to obtain useful electrical energy from a spontaneous redox reaction. This is done by separating the oxidizing and reducing agents in half-cell compartments. The compartments are connected by a barrier containing a conducting electrolyte to provide electrical contact between the compartments. The redox reaction would occur spontaneously if we mixed the contents of the two compartments. This, however, would not give any useful electrical work. To create a potential source of useful electrical energy a conducting electrode is placed in each half-cell.

When the electrodes are connected by a wire, the half-reactions occur in each compartment, and electrons flow through the connecting wire. As oxidation and reduction occur, electrons flow from the anode, where oxidation takes place, through the wire to the cathode, where reduction occurs. Of course, the flow of electrical current can be used to do electrical work (light a bulb, run a motor, etc.). Much of the energy released in the reaction is converted to electrical energy but some is lost as heat. This is why batteries sometimes get hot as they are used.

A device in which the chemical energy of a redox reaction is converted to electrical energy is a **battery** or **electrical cell.** The terminals of a battery are the electrode contacts. The oxidizing agent and reducing agent are contained in the cell and separated by an electrolyte solution. The cell has the potential of doing electrical work. When we connect the electrodes by an external conductor, we get useful electrical work. Opening the circuit stops the electron flow and the reaction.

The strength of the cell depends on the relative strengths of the oxidizing and reducing agents. The strength can be expressed in terms of the potential difference or voltage between the electrodes. Higher voltages are obtained with stronger oxidizing and reducing agents or by connecting cells in series. In the practical construction of a battery, the active

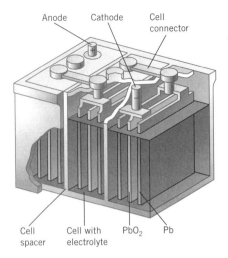

Anode    Cathode    Cell
                    connector

Cell      Cell with    PbO$_2$    Pb
spacer    electrolyte

**FIGURE 13-4**  A
three-cell lead storage
battery.

components cannot react with the structural and electrolyte materials used.

A common example of a battery is the **lead storage battery** used in automobiles (see Fig. 13-4). A lead storage battery consists of cells using an anode coated with lead metal and a cathode coated with lead(IV) oxide, PbO$_2$. The electrodes are immersed in a sulfuric acid electrolyte solution. A 6-volt battery has three of these cells in series, and a 12-volt battery has six. The discharge reaction for such a battery is

$$Pb(s) + PbO_2(s) + 2SO_4{}^{2-}(aq) + 4H^+(aq) \longrightarrow 2PbSO_4 + 2H_2O + energy$$

During the discharge, the lead anode is oxidized. This oxidation half-reaction is

$$Pb(s) + SO_4{}^{2-}(aq) \longrightarrow PbSO_4(s) + 2e^-$$
anode

Simultaneously, the lead(IV) oxide cathode is reduced as shown by the reduction half-reaction:

$$PbO_2 + 4H^+(aq) + SO_4{}^{2-}(aq) + 2e^- \longrightarrow PbSO_4(s) + 2H_2O$$
cathode

A lead storage battery runs down when the lead sulfate accumulates at the electrodes. A lead storage battery can be recharged by pumping electrons into the battery in the direction opposite to the discharge direction. This reverses the electrode reactions and changes the electrodes back into lead and lead(IV) oxide. Reversing the discharge reaction gives the charging reaction.

Ideally, the alternator on a car should keep a battery charged, and it should last indefinitely. But, in time, the electrode materials begin to flake

ACTIVITY
*13-6*

Give the equation for the reaction that occurs when a lead storage battery is charged.

Brass cap

Zinc outer anode

Graphite cathode

Moist paste of $NH_4Cl$, $MnO_2$, carbon

A typical dry-cell battery

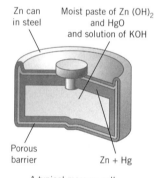

Zn can in steel

Moist paste of Zn (OH)$_2$ and HgO and solution of KOH

Porous barrier

Zn + Hg

A typical mercury cell

**FIGURE 13-5** A typical dry-cell battery and a mercury-cell battery.

off, and the internal parts loosen, causing the battery to fail. Common dry-cell batteries used in flashlights and other devices are not rechargeable. A dry cell runs down when the oxidizing and reducing agents are consumed. Figure 13-5 shows the components of a typical **dry cell.** They are called dry cells because the electrolyte material is damp but is not a liquid like the electrolytes used in lead batteries. Common dry cells use zinc as a reducing agent and manganese(IV) oxide as an oxidizing agent. Nickel–cadmium dry cells are similar except they can be recharged and reused. A mercury cell, also shown in Figure 13-5, is a small battery used in watches and calculators. Mercury cells use zinc as an oxidizing agent and mercury(II) oxide as a reducing agent. It is important to realize that all these types of batteries contain toxic and heavy metals (lead, manganese, cadmium, and mercury). Consequently you should recycle batteries when possible.

A battery stops working when the reactants are consumed. A cell in which the oxidizing and reducing agents are continually supplied and the products continually removed is a **fuel cell.** Fuel cells are the subject of a great deal of research, and may in the future be available as sources of industrial and domestic electricity. Special fuel cells are used as electrical sources in spacecraft. As shown in Figure 13-6, a typical fuel cell in a spacecraft involves the reaction of hydrogen gas and oxygen gas. The electrodes are nickel metal in contact with a potassium hydroxide solution. The hydrogen and oxygen fuels are fed into the cells continuously to produce electricity. Hydrogen gas reacts at the anode.

$$\text{oxidation} \quad H_2(g) + 2OH^-(aq) \longrightarrow 2H_2O + 2e^-$$

And oxygen gas reacts at the cathode.

$$\text{reduction} \quad 4e^- + O_2(g) + 2H_2O \longrightarrow 4OH^-(aq)$$

The net discharge reaction, obtained by combining the half-reactions, is the combination of hydrogen and oxygen to form water.

$$2H_2(g) + O_2(g) \longrightarrow 2H_2O + \text{electrical energy}$$

Fuel cells are used in spacecraft to run electrical equipment and instruments; the water produced is used as needed.

## 13-9 ELECTROLYSIS

The normal spontaneous discharge reaction of a lead battery is reversed by pumping electrons into the cell in the reverse direction of discharge. In a car this is done by the alternator, but another electron source such as a generator or even a stronger battery could be used. As long as the electron source can supply enough potential difference or voltage, electrode reactions occur that would not normally occur. Electrical energy is used as a source of energy to cause a redox reaction to occur.

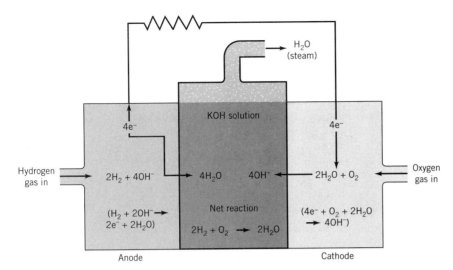

**FIGURE 13-6**   A hydrogen–oxygen fuel cell.

Imagine placing a piece of metal, carrying a negative electrical charge, into a solution containing cations and anions. The negatively charged metal attracts the positive ions in the solution. Thus, the cations migrate toward the metal and form a layer of positive ions around the metal. If a positively charged piece of metal is placed in the solution, the anions are attracted to the metal and form a layer of negative ions around the metal.

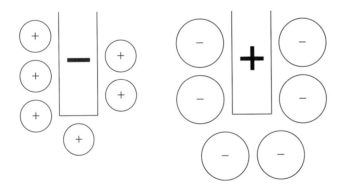

How would it be possible to obtain the two charged pieces of metal? This is done by connecting the two pieces of metal with wires to the two terminals of a battery or a generator, as shown in Figure 13-7. The battery or generator serves as a device for pumping electrons from one piece of metal to the other. This ease of movement of electrons in metals comes from the unique structure of metals. Since one piece of metal connected to the battery has an excess of electrons, it is negatively charged. In contrast, the piece of metal that is deficient in electrons is positively charged.

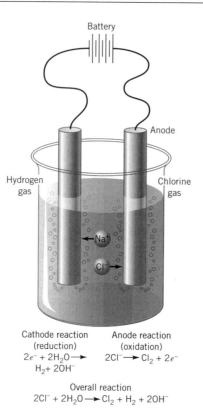

**FIGURE 13-7**
Electrolysis of a sodium
chloride solution.

The electron pump or battery serves to keep a voltage or potential difference between the two pieces of metal. If the two pieces of metal connected to the electron pump are immersed in a solution containing ions, the cations move toward the negative piece of metal and the anions toward the positive piece of metal. The pieces of metal act as electrodes. The negative electrode that attracts the cations is called the **cathode,** and the positive electrode that attracts the anions is called the **anode.** This is why positive ions and negative ions are called cations and anions.

An interesting phenomenon occurs if the potential difference between the electrodes is great enough. The battery or generator provides a driving force that causes electron loss and gain. At the cathode surface some species gain electrons. At the anode some species lose electrons. A redox reaction is induced using an external source of electrical energy. That is, since the battery or generator pumps electrons, a situation arises in which one species in solution loses electrons at the anode and another gains electrons at the cathode. In this manner, a complete electrical circuit is set up in which electrons move to the cathode, where they are gained by one species in solution. Simultaneously, another species in solution loses electrons to the anode to provide more electrons to be pumped to the cathode. This is essentially the reverse of a battery, which supplies electrical energy by a spontaneous redox reaction. The process in which electricity passes through a solution and a redox reaction occurs is called **electrolysis.** The term electrolysis literally means breaking down by electricity. Electrolysis can also occur when electricity passes though a molten sample of an ionic compound that contains separate cations and anions.

## 13-10 ELECTROLYSIS OF SODIUM CHLORIDE

The electrolysis of a water solution of sodium chloride, as illustrated in Figure 13-7, serves as a good example of industrial electrolysis. In this process an external source of electricity serves as an electron pump. When the electrodes are connected to the source of electricity the anode attracts chloride ions, where they lose electrons and form molecular chlorine. This half-reaction is

$$2Cl^-(aq) \longrightarrow Cl_2(g) + 2e^- \quad \text{occurs at the anode}$$

Since molecular chlorine is diatomic, two chloride ions must react to produce one chlorine molecule.

A half-reaction in which a species loses or gains electrons at an electrode and changes to a new species is called an **electrode reaction.** An electrode reaction, such as that given above, occurs only when a simultaneous reaction involving the gain of electrons takes place at the other electrode.

In this example, we would expect the sodium ions to migrate toward the cathode. However, sodium ions do not gain electrons in aqueous solution to form metal because water is more easily reduced. Refer to Table 13-2 and note that solid sodium metal cannot exist in water. Water gains electrons according to the electrode reaction

$$\text{occurs at cathode} \quad 2e^- + 2H_2O \longrightarrow H_2(g) + 2OH^-(aq)$$

The electrode reactions occur simultaneously when the solution of sodium chloride is subjected to electrolysis. For every two electrons gained at the cathode, two electrons are lost at the anode. Notice that this electrode reaction produces the hydroxide ion, a base. This is the electrode reaction you may have used in electrical conductivity tests. (See Activity 13-7.) The hydroxide ion changes phenolphthalein from colorless to pink. The electrode reactions are actual chemical reactions involving electron transfer. The overall reaction corresponding to the electron transfer that occurs during the electrolysis of the sodium chloride is

$$2Cl^-(aq) + 2H_2O \xrightarrow{\text{electrolysis}} Cl_2(g) + H_2(g) + 2OH^-(aq)$$

This is a simple exercise to observe the electrolysis of a sodium chloride solution. You need a 9-volt battery, a napkin or a piece of paper towel, salt, water, a laxative tablet containing phenolphthalein, and a piece of white paper.

1. Place the white paper on a flat, nonporous surface and put the napkin or paper towel on top of it.

2. Place a small sample of salt on the paper towel. Use a sample about the size of an aspirin tablet.

3. Crush the laxative tablet and sprinkle the powdered part of the tablet on top of the salt.

4. Put a small amount of water on top of the solids so that most of the salt dissolves.

5. Turn the battery upside down and touch the wet paper with the terminals of the battery. Hold the battery in place for a moment. Record your observations.

Refer to the discussion of the electrolysis of a sodium chloride solution. Which terminal of the battery was the cathode and which was the anode? Explain your answer.

The sodium ion is a spectator ion in the reaction. When a rather concentrated solution of sodium chloride is subjected to electrolysis, chlorine gas is produced at the anode and hydrogen gas at the cathode. Furthermore, for every two chloride ions that react, two hydroxide ions form; this balances the charge. The result is a solution of sodium hydroxide.

In the United States over 10 million tons of chlorine are manufactured each year by the chlor-alkali process using an electrolysis reaction similar to that described above. The production of some metals, such as aluminum, and electroplating, such as silver plating and chromium plating, also use electrolysis. Table 13-3 lists some common examples of electrolysis. Chemistry in Action 13-3 is about electroplating some metals.

***Table 13-3***   Examples of Electrolysis

### Electrolytic Production of Aluminum

Aluminum oxide from bauxite ore is dissolved in the molten mineral cryolite. The melt is subjected to electrolysis to form aluminum metal.

$$2Al_2O_3 + 3C \xrightarrow[\text{cryolite}]{\text{molten}} 3CO_2 + 4Al$$
$$Al^{3+} + 3e^- \longrightarrow Al \quad \text{cathode}$$
$$C + 2O^{2-} \longrightarrow CO_2 + 4e^- \quad \text{anode}$$

### Electrolytic Production of Magnesium

Magnesium chloride obtained from seawater is melted and subjected to electrolysis to give magnesium metal.

$$MgCl_2 \xrightarrow{\text{molten}} Mg + Cl_2$$
$$Mg^{2+} + 2e^- \longrightarrow Mg \quad \text{cathode}$$
$$2Cl^- \longrightarrow Cl_2 + 2e^- \quad \text{anode}$$

### Electrolysis of Water

The electrolysis of water gives hydrogen and oxygen gases. Some nonelectrolyzable ions must be present in water for electrolysis.

$$2H_2O \longrightarrow 2H_2 + O_2$$
$$2H_2O + 2e^- \longrightarrow H_2 + 2OH^- \quad \text{cathode}$$
$$2H_2O \longrightarrow 4e^- + O_2 + 4H^+ \quad \text{anode}$$

### Electroplating of Silver

In silver plating by electrolysis, a pure silver anode is used with a base metal cathode. During electrolysis, silver is plated on the cathode.

$$Ag(pure) \longrightarrow Ag(plate)$$
$$Ag^+ + 1e^- \longrightarrow Ag(plate) \quad \text{cathode}$$
$$Ag \longrightarrow Ag^+ + 1e^- \quad \text{anode}$$

### Electroplating of Tin

Tin plating of "tin" cans involves the electrolytic transfer of tin from a pure tin anode to a steel sheet. The tin-plated steel is used to make cans.

$$Sn(pure) \longrightarrow Sn(plate)$$
$$Sn^{2+} + 2e^- \longrightarrow Sn(plate) \quad \text{cathode}$$
$$Sn \longrightarrow Sn^{2+} + 2e^- \quad \text{anode}$$

# Aluminum

The largest-scale industrial electrolysis in the world is used in the making of aluminum metal from its ore. Aluminum is the most abundant metal in nature and is found in various aluminum silicate minerals, rocks, and soils. Aluminum is a low-density, corrosion-resistant metal used in a variety of structural materials and products, including cooking utensils, high-tension electrical lines, aluminum cans, and aluminum foil. Structural alloys of aluminum, containing small amounts of silicon and other metals, are used in buildings, window frames, airplanes, trailers, trucks, and automobiles.

Aluminum is found as a compound in the aluminum ore called **bauxite.** Once it is mined, bauxite is treated to form aluminum oxide. The industrial Hall process uses aluminum oxide as a raw material. More than 5 million tons of aluminum are produced in the United States each year by this process, developed in 1886 by Charles Hall, an American chemist. In the Hall process, aluminum oxide is dissolved in a compound called cryolite, $Na_3AlF_6$, at 800°C to 1000°C. Carbon electrodes are immersed in the molten mixture and electricity is passed through the melt. See Table 13-3 for the electrode reactions. Electrolysis reduces the aluminum ions to aluminum metal at the cathode. The carbon is oxidized to carbon dioxide at the anode. The overall reaction that takes place is

$$2Al_2O_3 + 3C \xrightarrow{\text{electrolysis}} 4Al + 3CO_2$$

At the high temperature of the process the aluminum forms as a liquid which is tapped from the electrolysis vessel then cast as a solid. See the figure in Section 5-8.

The refining of aluminum requires large amounts of electricity. As a result, aluminum producing plants are situated near sources of cheap electricity such as hydroelectric dams. Not only is energy used in the electrolysis of aluminum ore but energy is also expended in the mining, transporting, and purifying of the ore. More energy is needed to forge the metal into various materials. The energy used to make one aluminum can and use it once is estimated to be about 7000 kJ or 7000 Btu. In contrast, 65 percent less energy is used to make a new can from recycled aluminum metal. Recycling aluminum not only saves bauxite but also energy.

How many tons of aluminum oxide are needed to make 5 million tons of aluminum by the Hall process? How many tons of carbon are needed to make 5 million tons of aluminum? (Hint: You can use tons in the mass-to-mass stoichiometric calculation without changing to grams. The gram units in the molar masses will cancel leaving tons for the answer.)

## *Oxidation and Reduction*

Oxidation: Loss of electrons; the increase in oxidation number of an element.

**LOX     ION**

Reduction: Gain of electrons; the decrease in oxidation number of an element.

**GER     DON**

Oxidation-reduction reactions or redox reactions: Electron transfer reactions are characterized by changes in the oxidation numbers of some elements involved in the reaction. A redox reaction involves an increase in the oxidation number of one element and a decrease for another element.

Reducing agent: An electron donor that causes another species to be reduced. A species that reduces another species and is itself oxidized in the process. It is that species having an element that undergoes an increase in oxidation number, changing from a lower to a higher oxidation number.

Oxidizing agent: An electron acceptor that causes another species to be oxidized. A species that oxidizes another species and is itself reduced in the process. It is that species having an element that undergoes a decrease in oxidation number, changing from a higher to a lower oxidation number.

Redox Table: A list of oxidizing agents ranked according to decreasing strength and the corresponding reducing agents ranked according to increasing strength.

Generally, any oxidizing agent reacts with any reducing agent below it in a redox table. In contrast, no oxidizing agent reacts with a reducing agent lying above it in the table.

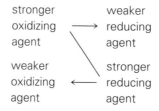

In a reaction predicted from a redox table the oxidizing agent reacts with the reducing agent to form the products given in the table. To write a complete net-ionic equation for a predicted redox reaction:

1. Copy the half reaction of the oxidizing agent from the table as it appears.
2. Copy the half reaction of the reducing agent but reverse it from the way it appears in the table.
3. Multiply each half reaction by the proper numbers to give the same number of electrons lost and as are gained, and add them to give the final equation.

Electromotive series of metals or the activity series of metals: A redox table that lists common metals according to their strength as reducing agents and metal ions according to their strengths as oxidizing agents. Any oxidizing agent reacts with any reducing agent below it in the table and will not react with a reducing agent above it. The series reveals which metals and metal ions react with one another. More active metals are those that are strong reducing agents and readily form metal ions.

Battery or electrical cell: A device in which the chemical energy of a redox reaction is converted to electrical energy. The terminals of a battery are the electrode contacts, and the oxidizing agent and reducing agent are contained in the cell and separated by an electrolyte solution.

Fuel cell: An electrical cell in which the oxidizing and reducing agents are continuously supplied and the products are continuously removed.

Electrolysis: The process in which electricity is passed through a solution and causes a redox reaction.

To observe the reaction between zinc metal and copper(II) ions use a sample of zinc metal and a solution of a copper(II) compound. You need: a small glass, a shiny penny, 10 dull pennies (not shiny), white vinegar, and a small amount of baking soda.

1. Use sandpaper or rough concrete to scrape the edge of the shiny copper–zinc penny to expose fresh zinc. Rotate the penny and scrape it so that all of the copper metal is removed from the edge. Also scrape or scratch the top and bottom to expose as much zinc as you can.
2. Place two tablespoons of vinegar in a glass and add 10 dull pennies to the glass. Stir the pennies a few times as you let them sit for five minutes.
3. Remove the pennies from the vinegar and mix a portion of baking soda about the size of one third of an aspirin tablet with the vinegar. It is important not to add too much baking soda.
4. Place the exposed zinc penny in the glass of solution and let it sit for five minutes or more.
5. Remove the penny, rinse it, and record your observations.

Often when copper metal quickly deposits on another metal it has a black appearance rather than the familiar copper color.

Iodine is an important trace element in the body. Various compounds containing the iodide ion supply iodine in the diet. Iodized salt is table salt that contains a small amount of potassium iodide as a nutrient supplement. Iodide ions are colorless so we can't see them in salt. In this exercise you are going to make some elemental iodine from the iodide in iodized salt. You need: a box of commercial iodized salt, two glasses, a zip-lock bag, water, hydrogen peroxide solution (medicinal hydrogen peroxide), white vinegar, four aspirin tablets, commercial mineral oil, and two coffee filters. Instead of filters you can shape four napkins into a cone which can be used as a filter. Instead of mineral oil use two tablespoons of charcoal lighter, Energine, or lighter fluid. **CAUTION: These fluids are flammable and should be used in a well-ventilated space.**

1. Place half a cup of iodized salt in a glass and add about half a cup of water. Stir the mixture for a few minutes.

2. Place one coffee filter paper inside the other to make a double filter. Pass the salt solution through the filter and collect the liquid in another glass.

3. Pour one-fourth of a cup of hydrogen peroxide into the glass of salt solution and add about one-eighth of a cup of vinegar. Stir the solution, let it stand for several minutes and record your observations.

4. When iodine and starch are mixed they form a purple color. This color can be used as a test for iodine or a test for starch. Place four crushed aspirin tablets in a glass and add two teaspoons of the yellow solution from Step 3. Aspirin tablets contain a small amount of starch used to bind the aspirin particles. Also use some of the iodine solution to test some copier paper and some bread or cracker for starch. Record your observations.

5. You can extract elemental iodine from the yellow solution. Put half a cup of the yellow solution into a zip-lock bag along with six tablespoons of mineral oil. Iodine is more soluble in the oil than it is in water. Seal the bag and gently agitate the liquids for a few minutes. Record your observations.

Write a balanced net-ionic equation for the reaction between hydrogen peroxide and iodide ions. The vinegar provides some hydrogen ions for the reaction (see Table 13-1). Considering the polarities of the solvents why would you expect elemental iodine to be more soluble in mineral oil than in water?

In this exercise you are going to electroplate some metals. You need: a 9-volt battery, three shiny pennies, magic tape, aluminum foil, salt (to make a small amount of saturated salt solution), calamine lotion, white vinegar, a Pepto Bismol tablet, and mercurochrome (the disinfectant type from a drugstore), paper towel or napkin, and a flat dish or flat nonporous, nonmetallic surface. Calamine lotion containing zinc oxide is a source of zinc, Pepto Bismol is a source of bismuth, and mercurochrome is a source of mercury **(CAUTION: Mercury can cause heavy metal poisoning so do not use too much of it.)**

1. Tear or cut a 3 inch by 1 inch piece of paper towel or napkin. Fold it in half twice to give a 1 inch by 1 inch length. Place the folded paper on a flat dish or surface.

2. Fold some aluminum foil into a half inch square so that it has thickness about the same as a penny.

3. Place four drops of calamine lotion in a tablespoon. Add some vinegar and stir the mixture. Pour the mixture onto one end of the flat paper towel.

4. Pour saturated salt solution on the other end of the paper towel and be sure that it is soaked with solution.

5. Sit a shiny penny on top of the calamine lotion and sit the folded aluminum foil on the paper near the penny. Invert a 9-volt battery and balance it so that the large terminal is touching the penny and the small terminal is touching the aluminum. If needed, use a 30 cm length of tape to hold the battery in an inverted position. Keep the battery in place for about 5 minutes.

6. Remove the battery, pick up the penny, and dry it. Record your observations.

7. Use a new piece of paper and repeat steps 1 to 6. In place of the calamine lotion and vinegar mixture use a crushed Pepto Bismol tablet.

8. Use a new piece of paper and repeat steps 1 to 6. In place of the calamine lotion and vinegar mixture pour enough mercurochrome onto the paper to wet it thoroughly. When you finish dry the penny and rub it with your fingers. **(CAUTION: Mercurochrome stains are very hard to remove from clothing, but they will wear off your skin.)**

## QUESTIONS

### Section 13-1

1. What is an electron transfer reaction?

2. What is a voltaic or galvanic cell?

### Sections 13-2 to 13-4

3. Give definitions for oxidation and reduction.

4. Why do oxidation and reduction always occur simultaneously?

5. Describe the following terms.
   (a) half-reactions    (b) oxidizing agent
   (c) reducing agent    (d) redox reaction

6. *Tell whether or not each of the following reactions is an oxidation–reduction reaction. For each of the reactions that is oxidation–reduction, identify the oxidizing agent and the reducing agent. (Hint: If no elements change oxidation number in a reaction then the reaction is not a redox reaction.)
   (a) $CH_4 + 2O_2 \longrightarrow 2H_2O + CO_2$
   (b) $Ag^+ + Cl^- \longrightarrow AgCl$
   (c) $I_2O_5 + 5CO \longrightarrow I_2 + 5CO_2$
   (d) $NH_3 + HCl \longrightarrow NH_4Cl$
   (e) $H_3O^+ + F^- \longrightarrow HF + H_2O$

   (f) $SO_2 + NO_2 \longrightarrow SO_3 + NO$
   (g) $2Cl_2 + CH_4 \longrightarrow 4HCl + C$
   (h) $Zn^{2+} + 2OH^- \longrightarrow Zn(OH)_2$
   (i) $Ni + 2H^+ \longrightarrow Ni^{2+} + H_2$

7. *For each of the redox reactions in Question 6, give the oxidation numbers of the elements that change oxidation number. For example, in (a), C changes from −4 to +4 and O changes from 0 to −2.

8. Tell whether or not each of the following reactions is an oxidation–reduction reaction. For each of the reactions that is oxidation–reduction, identify the oxidizing agent and the reducing agent. (Hint: If no elements change oxidation number in a reaction then the reaction is not a redox reaction.)
   (a) $Ca + 2H_2O \longrightarrow Ca(OH)_2 + H_2$
   (b) $SO_3 + H_2O \longrightarrow H_2SO_4$
   (c) $2CO + O_2 \longrightarrow 2CO_2$
   (d) $Pb^{2+} + CrO_4^{2-} \longrightarrow PbCrO_4$
   (e) $Cl_2 + 2I^- \longrightarrow I_2 + 2Cl^-$
   (f) $Al^{3+} + 3OH^- \longrightarrow Al(OH)_3$
   (g) $4Cl_2 + CH_4 \longrightarrow CCl_4 + 4HCl$
   (h) $Zn + 2H^+ \longrightarrow Zn^{2+} + H_2$
   (i) $2H_2 + O_2 \longrightarrow 2H_2O$

9. For each of the redox reactions in Question 8, give the oxidation numbers of the elements that change oxidation number. For example, in (a), Ca changes from 0 to +2 and two of the H's change from +1 to 0.

10. In each of the following industrial and biological redox reactions identify the oxidizing agent and the reducing agent.

    (a) $N_2 + 3H_2 \longrightarrow 2NH_3$

    (b) $2H_2 + CO \longrightarrow CH_3OH$

    (c) $CH_4 + 2H_2O \longrightarrow CO_2 + 4H_2$

    (d) $Fe_2O_3 + 3CO \longrightarrow 2Fe + 3CO_2$

    (e) $6CO_2 + 6H_2O \longrightarrow C_6H_{12}O_6 + 6O_2$

    (f) $C_6H_{12}O_6 + 2O_2 \longrightarrow 2HC_2H_3O_2 + 2CO_2 + 2H_2O$

## Section 13-5

11. What is a redox couple?

12. *In the reaction:

$$Zn + 2H^+ \longrightarrow Zn^{2+} + H_2$$

Zn and $Zn^{2+}$ are a redox couple; $H^+$ and $H_2$ are a redox couple. In each of the following redox reactions identify the two sets of redox couples.

    (a) $CH_4 + 2H_2O \longrightarrow CO_2 + 4H_2$

    (b) $Fe_2O_3 + 3CO \longrightarrow 2Fe + 3CO_2$

    (c) $Ca + 2H_2O \longrightarrow Ca(OH)_2 + H_2$

    (d) $Cl_2 + 2I^- \longrightarrow I_2 + 2Cl^-$

    (e) $2Ag^+ + Zn \longrightarrow 2Ag + Zn^{2+}$

13. How is it possible to measure the strength of an oxidizing agent?

14. In a table of oxidizing and reducing agents in which the oxidizing agents are listed according to decreasing strength why are the corresponding reducing agents listed according to increasing strength?

## Section 13-6

15. *Using Table 13-1 as a guide, give balanced net-ionic equations for any redox reactions that occur when the following solutions or substances are mixed. The term acidified means that acid has been added to a solution to supply hydrogen ions.

    (a) An NaBr solution is added to an acidified solution of hydrogen peroxide, $H_2O_2$.

    (b) An NaI solution is added to an acidified solution of $Na_2Cr_2O_7$.

    (c) An $FeCl_2$ solution is added to an acidified solution of $KMnO_4$.

    (d) An oxalic acid solution, $H_2C_2O_4$, is added to an acidified solution of $H_2O_2$.

    (e) An $Fe(NO_3)_3$ solution is added to a KI solution.

16. Using Table 13-1 as a guide, give balanced net-ionic equations for any redox reactions that occur when the following solutions or substances are mixed. The term acidified means that acid has been added to a solution to supply hydrogen ions.

    (a) Chlorine, $Cl_2$, is added to an NaBr solution.

    (b) Iodine, $I_2$, is added to an NaCl solution.

    (c) An oxalic acid solution, $H_2C_2O_4$, is added to an acidified solution of $Na_2Cr_2O_7$.

    (d) A nitric acid solution is added to a solution of $Fe(NO_3)_2$.

17. The elemental form of a halogen will replace the halide ion of any halogen that occurs below it in the periodic table. Use Table 13-1 to explain this statement. Make a short redox table that includes all the halogens as oxidizing agents and all the halide ions as reducing agents. Where in the table would you fit fluorine and the fluoride ion?

18. Ocean water contains low concentrations of bromide ions, $Br^-$. Elemental bromine can be obtained from ocean water by adding chlorine, $Cl_2$. Refer to Table 13-1 and give a balanced net-ionic equation to show the reaction for this process.

19. In Chemistry in Action 13-2 the reaction does not occur without vinegar. Use Table 13-1 to explain why.

## Section 13-7

20. *Using Table 13-2 as a guide, give balanced net-ionic equations for any redox reactions that occur when the following solutions or substances are mixed.

    (a) A small piece of potassium metal is added to water.

    (b) A piece of zinc is added to water.

    (c) A copper penny is added to a $Hg(NO_3)_2$ solution.

    (d) A piece of zinc is added to a $Pb(C_2H_3O_2)_2$ solution.

    (e) A solution of $NiCl_2$ is mixed with a solution of $Zn(NO_3)_2$.

21. Using Table 13-2 as a guide, give balanced net-ionic equations for any redox reactions that occur when the following solutions or substances are mixed.

    (a) A small piece of calcium metal is added to a hydrochloric acid solution. **CAUTION: Do not try this experiment.**

    (b) A piece of iron metal is added to a solution of $Cu(NO_3)_2$.

    (c) A piece of aluminum metal is added to a solution of $AgNO_3$.

    (d) A piece of silver metal is added to a hydrochloric acid solution.

22. Corrosion or rusting of iron occurs when oxygen oxidizes iron in moist environments to form ionic iron compounds. Use Table 13-2 to explain the following statements.

    (a) Galvanized iron is zinc-coated iron; coating iron with zinc inhibits rusting.

    (b) Iron pipes buried in the ground can be protected from corrosion by connecting a sheet of magnesium metal to the pipeline.

    (c) Aluminum screws should not be used with iron materials.

23. Copper metal is obtained from low-grade ore by dissolving the copper compounds in the ores then adding scrap iron. Use Table 13-2 to explain how this works and give an equation.

24. Use Table 13-2 to explain why highly acidic foods should not be wrapped in aluminum foil.

25. When silver metal becomes tarnished it is coated with a compound of silver. The tarnish can be removed by wrapping a silver object in an aluminum foil and placing it in a pan of salty water. Use Table 13-2 to explain this. (Hint: The silver compound in the tarnish contains $Ag^+$ ions which react; the salt is not involved in any chemical reaction.)

26. An alloy of silver that contains tin and some mercury can be used for tooth fillings. This alloy does not dissolve in the foods we eat. Use Table 13-2 to explain why silver and mercury are good metals for tooth fillings.

27. Use Table 13-2 to explain why the metals silver, mercury, and copper are sometimes found in nature as uncombined metals whereas other metals, especially, sodium, calcium and potassium, are never found in nature as uncombined metals.

28. Acidic air pollutants are absorbed in rain to make acid rain. Use Table 13-2 to explain why acid rains encourage the decomposition and rusting of iron bridges and other iron structures.

29. Use Table 13-2 to explain why it can be dangerous for sodium or potassium metal to come in contact with water.

30. It is sometimes said that metals "dissolve" in acids. Give an explanation of this statement using Table 13-2.

31. What is rust and how does it occur?

32. How can rust be prevented or inhibited?

### Section 13-8

33. What is a battery and how does a battery function. What are the anode and cathode of a battery?

34. Describe a lead storage battery and give the overall discharge reaction. How is a lead storage battery recharged and what happens when it is recharged?

35. Describe a fuel cell. What advantage does a fuel cell have over a battery?

### Sections 13-9 and 13-10

36. What is electrolysis?

37. What reaction occurs when a concentrated solution of sodium chloride is subjected to electrolysis?

38. List some examples of the use of electrolysis.

39. Describe the industrial process for making aluminum metal.

40. Why is the recycling of aluminum important?

### Questions to Ponder

41. Since some batteries contain toxic metals, do you believe that batteries should have deposits that can be redeemed when they are recycled? Argue your case.

42. Aluminum refining plants use relatively large amounts of electricity. Often, such plants pay discount prices for electricity. State whether or not you believe aluminum refining should be subsidized with low electricity rates? Explain your views.

# CHEMICAL

# EQUILIBRIUM

## 14-1 CHEMICAL EQUILIBRIUM

When reactants are mixed, they react to give products at a certain rate. See Section 11-2 for a discussion of reaction rates. In many reactions, as the reactants form products the products react to reform the reactants. When this occurs, the reaction is said to be **reversible** and is viewed as a system involving **chemical equilibrium.** To get a picture or feeling for chemical equilibrium consider the apparatus shown in Figure 14-1. The figure shows two containers of water each equipped with a pump that work to pump the water over the barrier. Imagine that we start with the re- actant side full and the product side empty. As the water is pumped to the product side it is pumped back to the reactant side. As long as both pumps are working where does the water go? If both pumps are working, the system would soon reach an equilibrium-like steady state in which there would be some water on the reactant side and some on the product side. The water would continue to be pumped but a "standoff" would be reached in which the rate of the forward process would equal the rate of the reverse process. A double arrow could be used to represent the bal- anced steady state. If the pumps are of equal strength the amounts of wa- ter on the reactant and product side would be the same. This case could be represented by double arrows of equal size ($\rightleftharpoons$). If the forward pump is stronger than the reverse pump a steady state would still be attained.

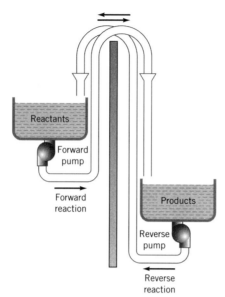

**FIGURE 14-1** Two containers are separated by a barrier. Each container has a pump that pumps water over the barrier. The pumps operate in opposite directions. When the rate of forward pumping equals the rate of reverse pumping, the system is in equilibrium.

However, in such a case the amount of water on the product side would be greater than the amount on the reactant side. This case could be represented by a double arrow indicating that the system favors the product side ($\rightleftharpoons$). In contrast, if the reverse pump is stronger than the forward pump a steady state would still be attained. In this case, however, the amount of water on the reactant side would be greater than the amount on the product side. This case could be represented by a double arrow indicating that the system favors the reactant side ($\rightleftharpoons$). Chemical reactions, of course, do not involve pumps but can involve a forward reaction and a reverse reaction. Forward and reverse reactions are opposing processes that are occurring at the same time but in opposite directions.

## 14-2 EQUILIBRIUM SYSTEMS

To illustrate the idea of a reversible chemical reaction involving chemical equilibrium, consider the following reaction:

$$N_2O_4(g) \rightleftharpoons 2NO_2(g)$$
colorless     reddish-brown

**Dynamic Chemical Equilibrium:**

*The state that exists when the rate of a forward reaction equals the rate of a reverse reaction.*

This reversible reaction includes two opposing reactions: colorless dinitrogen tetroxide, $N_2O_4$, decomposes to form reddish-brown nitrogen dioxide, $NO_2$, and nitrogen dioxide combines to form dinitrogen tetroxide.

The two reactions in this example occur in opposition. Dynamic chemical equilibrium results when two opposing reactions occur resulting in a mixture of reactants and products in which the relative amounts are fixed. Equilibrium is reached when the rates of the two opposing reactions become equal. Both reactions occur, but since equilibrium exists the net result is a continuous cyclic situation in which the reactant gives the products and the products react to give the original reactant. At equilibrium no further change in the system is observed.

A reversible reaction has forward and reverse directions. Typically, the **forward reaction** is written from left to right in the equation and the **reverse reaction** is written from right to left. Of course, the designation of the forward and reverse reaction is arbitrary and is used only as an aid in discussing the reaction. In the reversible reaction used in this example the forward reaction, the decomposition of dinitrogen tetroxide, competes with the reverse reaction, the combination of nitrogen dioxide. Since both reactions are taking place, the net result is a standoff that produces a mixture of reactants and products.

Normally it is not possible to see a chemical reaction come to equilibrium as it occurs, but in this equilibrium the $NO_2$ has a reddish-brown color and $N_2O_4$ is colorless. Suppose we place some pure dinitrogen tetroxide in a sealed glass vessel as illustrated in Figure 14-2. The dinitrogen tetroxide would react to give nitrogen dioxide. As the nitrogen dioxide forms it would react to give dinitrogen tetroxide. As equilibrium is approached the mixture would change from colorless to a slight reddish-brown color corresponding to the equilibrium mixture of the two gases. How soon equilibrium is reached depends upon the rates of the reactions

**FIGURE 14-2**   Idealized equilibrium between dinitrogen tetroxide and nitrogen dioxide. A pure sample of dinitrogen tetroxide produces, in time, an equilibrium mixture of $N_2O_4$ and $NO_2$ in which the concentrations of the reactant and product remain constant. The ratio of the square of the $NO_2$ concentration to the $N_2O_4$ concentration equals a constant value at a fixed temperature. See Figure 14-3.

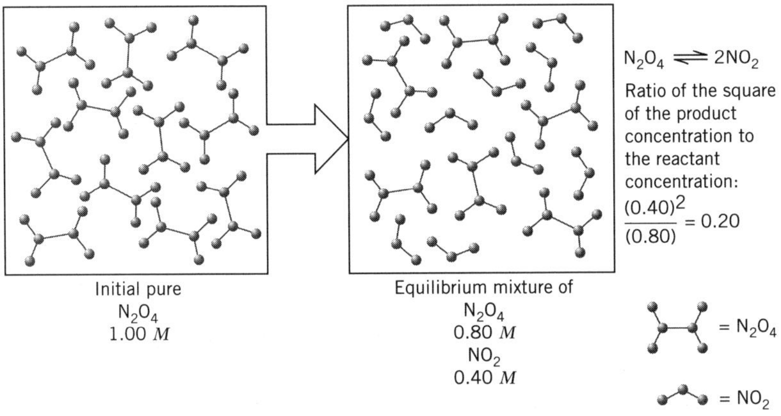

Initial pure
$N_2O_4$
1.00 M

Equilibrium mixture of
$N_2O_4$
0.80 M
$NO_2$
0.40 M

$N_2O_4 \rightleftharpoons 2NO_2$

Ratio of the square of the product concentration to the reactant concentration:

$$\frac{(0.40)^2}{(0.80)} = 0.20$$

= $N_2O_4$

= $NO_2$

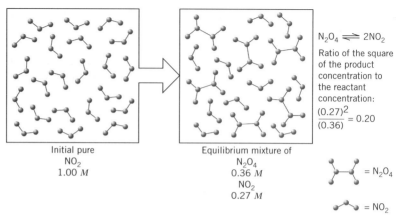

**FIGURE 14-3** Idealized equilibrium between dinitrogen tetroxide and nitrogen dioxide. A pure sample of nitrogen dioxide produces, in time, an equilibrium mixture of $N_2O_4$ and $NO_2$ in which the concentrations of the reactant and product remain constant. The ratio of the square of the $NO_2$ concentration to the $N_2O_4$ concentration equals a constant value at a fixed temperature. See Figure 14-2.

involved. At equilibrium, the mixture takes on a uniform color because it contains a specific concentration of each gas.

If we started with some pure nitrogen dioxide it would react to form dinitrogen tetroxide and, as the dinitrogen tetroxide is formed it would react to give nitrogen dioxide. This is illustrated in Figure 14-3. As equilibrium is approached, the original reddish-brown color of the nitrogen dioxide would fade to a lighter shade corresponding to the equilibrium mixture of the two gases. Again at equilibrium, the mixture takes on a uniform color since it contains specific concentrations of each gas. Recall from Section 10-9 that dynamic equilibrium in a saturated solution results in a fixed concentration of dissolved solute. As shown in Figures 14-2 and 14-3 the nitrogen dioxide–dinitrogen oxide equilibrium reaction produces fixed concentrations of both gases. These fixed concentrations are called **equilibrium concentrations.** The constant equilibrium concentrations are reflected in the fact that at equilibrium the ratio of the concentrations of the products and reactant is a constant value.

## 14-3 EQUILIBRIUM CONSTANTS

When the dinitrogen tetroxide–nitrogen dioxide reaction reaches equilibrium the result is a steady-state mixture of reactants and products. The forward and reverse reactions continue to occur at equilibrium but the reactions balance out to give constant concentrations of both nitrogen dioxide and dinitrogen tetroxide.

Any reaction system involving chemical equilibrium is characterized by the fact that at equilibrium there is a ratio of the concentrations of the products to the reactants that is equal to a constant numerical value. This ratio is called the equilibrium constant. An equilibrium system represented by the general equation

$$a\,A + b\,B \rightleftharpoons c\,C + d\,D$$

can be described by an **equilibrium constant expression** of the general form

$$K_{eq} = \frac{[C]^c[D]^d}{[A]^a[B]^b}$$

where $K_{eq}$ is the equilibrium constant and the square brackets refer to the concentrations of the products and reactants in moles per liter. Each concentration is raised to a power corresponding to its coefficient in the equilibrium equation. To find the numerical value of $K_{eq}$ for an equilibrium system, it is necessary to experimentally measure the concentration of the products and reactants at equilibrium at a constant temperature. The ratio of the molar concentrations raised to the appropriate powers gives the numerical value of $K_{eq}$ for the equilibrium system at the specified temperature.

As an example let us write an equilibrium constant expression for the reaction

$$N_2O_4(g) \rightleftharpoons 2NO_2(g)$$

The equation has a coefficient of 2 for the nitrogen dioxide. In a sense, there are two products.

$$N_2O_4(g) \rightleftharpoons NO_2(g) + NO_2(g)$$

We include the concentration of both products and the reactant in the equilibrium constant expression

$$K_{eq} = \frac{[NO_2][NO_2]}{[N_2O_4]}$$

Note that $[NO_2][NO_2]$ is the same as $[NO_2]^2$. So the algebraically equivalent form is

$$K_{eq} = \frac{[NO_2]^2}{[N_2O_4]}$$

The above expression reveals that at equilibrium the ratio of the square of the concentration of nitrogen dioxide to the concentration of dinitrogen tetroxide is a constant. See Figs. 14-2 and 14-3.

The reactants and products in an equilibrium system may be liquids,

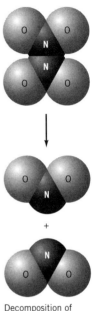

Decomposition of dinitrogen tetroxide to form nitrogen dioxide

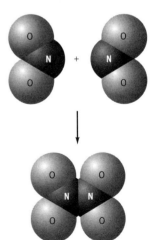

Combination of nitrogen dioxide to form dinitrogen tetroxide

solids, gases, or species in solution. It is important to note that only gases and substances in solution can have variable concentrations in equilibrium reactions. Any solid or liquid in a reaction will have a constant concentration—a fixed number of moles per liter. This is because the concentration of a solid or a liquid is directly related to its density. Since the density is constant then the concentration is constant. Thus, **gases and solutes in solution are included in equilibrium constant expressions** and liquids and solids are not included. Recall from Section 12-13 that liquid water reacts with itself.

$$2H_2O \rightleftharpoons H_3O^+(aq) + OH^-(aq)$$

The equilibrium constant expression for this reaction is

$$K_w = [H_3O^+][OH^-]$$

Note that the equilibrium constant for this reaction does not include water since water is a liquid.

The value of the equilibrium constant for a reaction is calculated from experimental measurements of the concentrations of the products. The equilibrium constant is found experimentally by measuring the molar concentration of each species at equilibrium. At 25°C the pH of pure water is 7.00 corresponding to a hydronium ion concentration of $1.0 \times 10^{-7}$ *M*. In pure water the concentrations of hydronium and hydroxide ions are equal so the value of the equilibrium constant for water is

$$K_w = [H_3O^+][OH^-] = (1.0 \times 10^{-7})(1.0 \times 10^{-7}) = \mathbf{1.0 \times 10^{-14}}$$

---

**EXAMPLE 14-1**

The dinitrogen tetroxide–nitrogen dioxide reaction may be carried out at a variety of temperatures. At any of these temperatures the equilibrium equation for the reaction is

$$N_2O_4(g) \rightleftharpoons 2NO_2(g)$$

Analysis of the equilibrium mixture for the reaction at 100°C gives the following concentration data in moles per liter.

$$[N_2O_4] = 0.36 \qquad [NO_2] = 0.27$$

The equilibrium constant expression for the reaction is

$$K_{eq} = \frac{[NO_2]^2}{[N_2O_4]}$$

The value of the equilibrium constant is found simply by substituting the equilibrium concentrations into the expression.

Equilibrium mixture

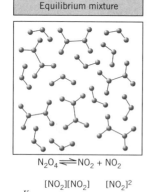

$$N_2O_4 \rightleftharpoons NO_2 + NO_2$$

$$K_{eq} = \frac{[NO_2][NO_2]}{[N_2O_4]} = \frac{[NO_2]^2}{[N_2O_4]}$$

$$K_{eq} = \frac{[NO_2]^2}{[N_2O_4]} = \frac{(0.27)^2}{(0.36)} = 0.20$$

Note that the value of the equilibrium constant in the above example is close to 1. It is neither a very large nor a very small value. An equilibrium constant value that is near 1, say between 0.1 and 10, shows that the equilibrium mixture contains similar concentrations of reactants and products. Such a value reveals that the equilibrium reaction does not significantly favor one side of the reaction or the other. On the other hand, an equilibrium constant value that is much smaller than 0.1 indicates that the equilibrium largely favors the reactants and a value much greater than 10 indicates that the equilibrium largely favors the products. For instance, $K_w = 1 \times 10^{-14}$ for the reaction of water molecules and this indicates the reaction only occurs to a very slight extent. Further examples of equilibrium reactions are given in Table 14-1.

## 14-4 THE EXTENT OF A REACTION

As discussed in the previous section, it is possible to calculate the value of an equilibrium constant using experimentally measured concentrations of

*Table 14-1*   Some Equilibrium Reactions

| | |
|---|---|
| Formation of nitrogen oxide gas from nitrogen gas and oxygen gas at 25°C | $N_2 + O_2 \rightleftharpoons 2NO$ |
| Formation of nitrogen dioxide gas from nitrogen oxide gas and oxygen gas at 25°C | $2NO + O_2 \rightleftharpoons 2NO_2$ |
| Formation of ammonia gas from nitrogen and hydrogen gas at 425°C | $N_2 + 3H_2 \rightleftharpoons 2NH_3$ |
| Reaction of ammonia and water at 25°C | $NH_3(aq) + H_2O \rightleftharpoons NH_4^+(aq) + OH^-(aq)$ |
| Reaction of hydrogen fluoride and water at 25°C | $HF(aq) + H_2O \rightleftharpoons H_3O^+(aq) + F^-(aq)$ |
| Dissolving of silver chloride in water at 25°C | $AgCl(s) \rightleftharpoons Ag^+(aq) + Cl^-(aq)$ |
| Formation of calcium carbonate from its aqueous ions at 25°C | $Ca^{2+}(aq) + CO_3^{2-}(aq) \rightleftharpoons CaCO_3(s)$ |

the reactants and products. The numerical value of $K_{eq}$ reflects the extent of the reaction. A relatively large value of $K_{eq}$ indicates that the reaction produces a higher concentration of products than reactants. An equation for an equilibrium favoring the products shows a larger arrow toward the product side.

$$a\,A + b\,B \rightleftharpoons c\,C + d\,D$$

A relatively small value of $K_{eq}$ indicates that the concentration of reactants is higher than that of the products. The equilibrium favors the reactant side. An equation for this equilibrium state gives the larger arrow toward the reactants.

$$a\,A + b\,B \rightleftharpoons c\,C + d\,D$$

When the equilibrium concentrations of reactants and products are close to one another, $K_{eq}$ has a value close to 1. In such a case the two arrows would be about the same size.

$$a\,A + b\,B \rightleftharpoons c\,C + d\,D$$

Note that the equations given in Table 14-1 have arrows that reflect the extent of the reactions. When it is not necessary to show the extent of a reaction or the favored direction is not known, equal arrows are used ($\rightleftharpoons$).

A temperature change causes the equilibrium to shift, producing a new ratio of product-to-reactant concentrations. At a new temperature, the new equilibrium is reflected by a different value of the equilibrium constant. A measured value for an equilibrium constant applies at a specific temperature.

### 14-5 EQUILIBRIUM CONSTANT EXPRESSIONS

By convention an equilibrium constant expression is written with the concentrations of the products in the numerator and the concentrations of the reactants in the denominator. To deduce an **equilibrium constant expression** for an equilibrium reaction, we will use the following pattern. This pattern applies to the equilibrium reactions with which we will be concerned. There are a few reactions for which these guidelines do not apply but we will not be concerned with them.

1. Write an equilibrium equation and note whether the species involved are gases, solids, liquids, or in aqueous solution.
2. Enclose the formulas of all products that are gases or aqueous species in square brackets for the numerator of the expression. If a species has a coefficient greater than 1, raise its concentration to the power given by the coefficient.

3. Enclose the formulas of all reactants that are gases or aqueous species in square brackets for the denominator of the expression using the coefficients as powers.

4. This ratio of the product concentrations to the reactant concentrations, each raised to a power corresponding to its coefficient in the balanced equation, is equal to the constant, $K_{eq}$.

---

**EXAMPLE 14-2**

The equilibrium equation for the reaction of nitrogen and hydrogen gas to give ammonia is

$$N_2(g) + 3H_2(g) \rightleftharpoons 2NH_3(g)$$

What is the equilibrium constant expression?

Since all of the components are gases their concentrations are included in the expression and each is raised to the power corresponding to its coefficient.

$$K_{eq} = \frac{[NH_3]^2}{[N_2][H_2]^3}$$

---

**EXAMPLE 14-3**

The equilibrium equation for the reaction of nitrogen and oxygen gas to produce nitrogen oxide gas is

$$N_2(g) + O_2(g) \rightleftharpoons 2NO(g)$$

What is the equilibrium constant expression?

Since all the components are gases, their concentrations are included in the expression and each is raised to the power corresponding to its coefficient.

$$K_{eq} = \frac{[NO]^2}{[N_2][O_2]}$$

---

**EXAMPLE 14-4**

The equilibrium equation for the dissolving of calcium carbonate is

$$CaCO_3(s) \rightleftharpoons Ca^{2+}(aq) + CO_3^{2-}(aq)$$

What is the equilibrium constant expression?

$CaCO_3$ is a solid and its concentration cannot change, so it does not appear in the expression. But, the aqueous ionic species do appear.

$$K_{eq} = [Ca^{2+}][CO_3^{2-}]$$

Write equilibrium constant expressions for the following equilibrium reactions:

(a) $PbCl_2(s) \rightleftharpoons$
$$Pb^{2+}(aq) + 2Cl^-(aq)$$

(b) $CaCO_3(s) \rightleftharpoons$
$$CaO(s) + CO_2(g)$$

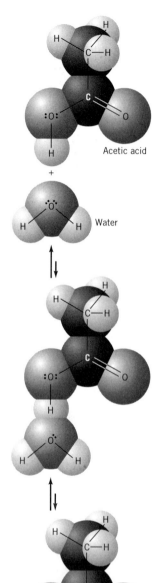

Acetic acid

+

Water

Acetate ion

+

Hydronium
ion

### 14-6 EQUILIBRIUM OF ACETIC ACID

An example of an equilibrium reaction in water is the reversible reaction between acetic acid and water.

$$HC_2H_3O_2 + H_2O \rightleftharpoons H_3O^+(aq) + C_2H_3O_2^-(aq)$$

When pure acetic acid is added to water the $HC_2H_3O_2$ reacts with $H_2O$ to form $H_3O^+$ (hydronium ion) and $C_2H_3O_2^-$ (acetate ion). When these two ions begin to accumulate, they react with one another to reform the acetic acid and water. In time, the rates of the two reactions become equal and chemical equilibrium is reached. At equilibrium acetic acid and water are continually reacting to give hydronium ions and acetate ions, and the ions are continually reacting to give the original reactants. The equilibrium constant expression for the reaction is

$$K_{eq} = \frac{[H_3O^+][C_2H_3O_2^-]}{[H_2O][HC_2H_3O_2]}$$

Note that since water is a liquid and since very little water is used in the reaction, the concentration of water is constant. This can be shown as

$$K_{eq}[H_2O] = \frac{[H_3O^+][C_2H_3O_2^-]}{[HC_2H_3O_2]} = K_a$$

This kind of equilibrium constant for a weak acid like acetic acid is represented by the symbol $K_a$, where the *a* stands for acid. The equilibrium reaction in the case of acetic acid produces relatively low concentrations of ions as shown by a $K_a$ value at 25°C of $1.8 \times 10^{-5}$. A solution of acetic acid is a mixture of acetic acid molecules and water molecules with relatively low concentrations of hydronium and acetate ions. The ion concentrations are low and the molecular acetic acid has by far the highest concentration of all the species in the solution.

A state of dynamic equilibrium in a reaction is denoted by use of a double arrow in the equation for the reaction. Keep in mind that the relative sizes of the arrows show which species exist at greater concentrations at equilibrium. The larger arrow points in the direction of the species present at greater concentrations. The equilibrium is said to favor the direction of the larger arrow. The reversible reaction between acetic acid and water (see the equation given above) favors the formation of water and acetic acid. Another way of stating this is that water and acetic acid enter into a relatively slight chemical reaction in which relatively low concentrations of hydronium and acetate ions are formed. This is why a solution of acetic acid is a weak electrolyte and acetic acid is called a weak acid.

### 14-7 EQUILIBRIUM CONSTANTS FOR WEAK ACIDS

When a weak acid is in water it reacts with water in a reversible chemical reaction (see Section 12-6). The **equilibrium reaction of any weak acid,**

**HA,** in water is represented by

$$HA + H_2O \rightleftharpoons H_3O^+ + A^-$$

where HA is the acid and $A^-$ is the conjugate base. Equilibrium is reached when the rates of the forward and reverse reactions are equal. Freshly prepared solutions of acids quickly reach equilibrium. At equilibrium, the solution contains constant concentrations of the various species. The concentration of water is essentially constant since very little is used in the reaction. The actual values of the concentrations of the other species depend on the acidic properties of the particular acid involved. Remember that weak acids vary in strength.

For a weak acid equilibrium system, the ratio of the concentrations of the products to the concentrations of the reactants is constant. This situation can be represented by the general **equilibrium constant expression for any weak acid.**

$$K_a = \frac{[H_3O^+][A^-]}{[HA]}$$

where $K_a$ is called the **acid ionization constant** or the **acid constant** for a weak acid.

An acid constant expression states that the product of the molar concentrations of the hydronium ion and the conjugate base of the acid divided by the molar concentration of the weak acid is constant. This constant has the same numerical value for any solution of a given acid at a specific temperature. Each weak acid has a unique $K_a$ value.

Acid ionization constants can be determined experimentally by preparing a solution of a weak acid and measuring the equilibrium concentrations of the hydronium ion, the weak acid and its conjugate base. The concentration values are used in the equilibrium expression to calculate the numerical value of $K_a$.

ACTIVITY
**14-2**

Use the formulas for formic acid, $HCO_2H$, and its conjugate base, $HCO_2^-$, and show how they fit the general weak acid equilibrium equation given above.

---

**EXAMPLE 14-5**

Calculate the value of the acid ionization constant, $K_a$, for acetic acid if a solution of acetic acid is experimentally found to have the following equilibrium concentrations:

| $[HC_2H_3O_2]$ | $[H_3O^+]$ | $[C_2H_3O_2^-]$ |
|---|---|---|
| 0.10 $M$ | 1.34 × 10$^{-3}$ $M$ | 1.34 × 10$^{-3}$ $M$ |

$K_a$ is found by use of the equilibrium expression for the reaction as written from the equilibrium equation. The equilibrium equation for acetic acid in water is

$$HC_2H_3O_2 + H_2O \rightleftharpoons H_3O^+ + C_2H_3O_2^-$$

The acid constant expression is the ratio of the product concentrations to reactant concentration.

When lactic acid, $HC_3H_5O_3$, dissolves in water, it reacts with water to give hydronium ions and lactate ions, $C_3H_5O_3^-$. Write an equation showing this reaction as an equilibrium and write an acid constant expression from the equation. A solution of the acid is found to contain 0.025 $M$ $HC_3H_5O_3$, $1.8 \times 10^{-3}$ $M$ $H_3O^+$ and $1.8 \times 10^{-3}$ $M$ $C_3H_5O_3^-$. Calculate the value of the acid ionization constant for lactic acid.

$$K_a = \frac{[H_3O^+][C_2H_3O_2^-]}{[HC_2H_3O_2]}$$

By substituting the known concentrations into this expression it is possible to calculate the numerical value for $K_a$.

$$K_a = \frac{[H_3O^+][C_2H_3O_2^-]}{[HC_2H_3O_2]} = \frac{(1.34 \times 10^{-3})(1.34 \times 10^{-3})}{(0.10)} = 1.8 \times 10^{-5}$$

The numerical value of $K_a$ reflects the extent of the reaction of the acid with water, which reflects the strength of the acid. The stronger a weak acid the larger its $K_a$ value. Of course, this would be true because higher concentrations of hydronium ion and conjugate base relative to the concentration of unreacted weak acid would correspond to a larger value of $K_a$. $K_a$ values for weak acids are always less than one so the double arrows used for the reactions of weak acids in water are often written as $\rightleftharpoons$. The $K_a$ values for some common acids are given in Table 14-2. If acids are listed according to decreasing $K_a$ values the result is an acid–base table like Table 12-1.

## 14-8 LE CHÂTELIER'S PRINCIPLE

Before exploring some practical applications of equilibrium we need to consider the behavior of equilibrium systems. When a chemical system is

***Table 14-2*** $K_a$ Values for Some Weak Acids

| ACID | FORMULA | $K_a$ |
|---|---|---|
| Oxalic acid | $H_2C_2O_4$ | $5.9 \times 10^{-2}$ |
| Hydrogen sulfate ion | $HSO_4^-$ | $1.0 \times 10^{-2}$ |
| Phosphoric acid | $H_3PO_4$ | $7.1 \times 10^{-3}$ |
| Hydrogen fluoride | HF | $6.9 \times 10^{-4}$ |
| Hydrogen oxalate ion | $HC_2O_4^-$ | $5.2 \times 10^{-5}$ |
| Acetic acid | $HC_2H_3O_2$ | $1.8 \times 10^{-5}$ |
| Carbon dioxide + water | $CO_2 + H_2O$ | $4.4 \times 10^{-7}$ |
| Dihydrogen phosphate ion | $H_2PO_4^-$ | $6.2 \times 10^{-8}$ |
| Hydrogen cyanide | HCN | $5.8 \times 10^{-10}$ |
| Hydrogen carbonate ion | $HCO_3^-$ | $4.7 \times 10^{-11}$ |
| Hydrogen phosphate ion | $HPO_4^{2-}$ | $4.5 \times 10^{-13}$ |

at equilibrium, it will remain in this state indefinitely unless the equilibrium is disturbed in some manner. An equilibrium will shift in one direction or another when some factor upsets the equilibrium. A general principle regarding this situation is called Le Châtelier's principle in honor of the French chemist who first stated it in 1888. See margin.

An equilibrium system is in a state of balance. When something is done to upset the balance the equilibrium shifts in a way that allows the attainment of a new state of balance. To illustrate Le Châtelier's principle, the equilibrium reaction involved in the formation of ammonia from hydrogen and nitrogen can be used (see Fig. 14-4).

$$N_2(g) + 3H_2(g) \rightleftharpoons 2NH_3(g) + energy$$

At equilibrium the equilibrium concentrations of the reactants and products are fixed unless something is done to change them.

What factors affect an equilibrium system?

1. A change in the concentration of a species that is a reactant or product.
2. A change in temperature of the equilibrium mixture.
3. A change in the pressure of an equilibrium mixture involving one or more gases.

A change in the amount of a solid or liquid that is a reactant or product in an equilibrium system does not affect the equilibrium. This is because the concentration of a solid or liquid is not changed by changing the amount. A simple example to illustrate this point is a saturated solution of sodium chloride. In a saturated solution a dynamic equilibrium exists between dissolved and undissolved solute. If more solid sodium chloride is added to a saturated solution no more salt will dissolve. In other words, the equilibrium is not affected by adding the solid and it is not possible to dissolve more salt by adding solid salt.

**Le Châtelier's Principle:**
*When a factor affecting an equilibrium system is changed, the equilibrium will shift in a direction that tends to counteract the change.*

## Concentration

When a system is at equilibrium, the concentration of each species is constant. The **concentration** of a gaseous or dissolved species can be changed by adding more of it to the system or by removing some of it from the system. When this is done, the equilibrium shifts in the direction that tends to diminish an increase in concentration or replenish a decrease in concentration. In other words, the equilibrium shifts away from an increase and towards a decrease.

As illustrated in Figure 14-4(a), if more hydrogen or nitrogen is added to the equilibrium system, the equilibrium will shift toward the ammonia side. This shift tends to diminish the increased concentration of nitrogen or hydrogen and maintain the equilibrium balance of the relative concen-

(a) The effect of a concentration change on equilibrium

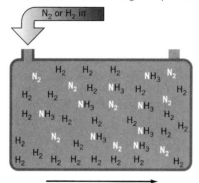

$$K_{eq} = \frac{[NH_3]^2}{[N_2][H_2]^3}$$

An increase in either of these concentrations causes a shift in the equilibrium toward the ammonia, which increases the concentration of $NH_3$ and tends to maintain a constant ratio of concentrations.

$$N_2 + 3H_2 \rightleftharpoons 2NH_3 + energy$$

An increase in the concentration of $N_2$ or $H_2$ causes the equilibrium to shift in the $NH_3$ direction.

(b) The effect of a pressure change on an equilibrium involving gases

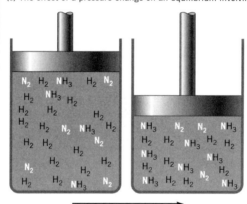

$$K_{eq} = \frac{[NH_3]^2}{[N_2][H_2]^3}$$

An increase in pressure essentially increases all the concentrations of the gases. The equilibrium shifts toward the ammonia, which increases the concentration of $NH_3$ and tends to maintain a constant ratio of concentrations.

$$N_2 + 3H_2 \rightleftharpoons 2NH_3 + energy$$

When the pressure is increased, the equilibrium shifts toward the side having the lesser number of moles of gas, which tends to counteract the pressure change.

(c) The effect of a temperature change on equilibrium

Temperature Increased

$$N_2 + 3H_2 \rightleftharpoons 2NH_3 + energy$$

Temperature Decreased

At equilibrium an energy balance exists. When the temperature is increased energy is added to the equilibrium system. The equilibrium shifts toward the nitrogen-hydrogen side, and a new energy balance is established at the higher temperature. When the temperature is decreased, the equilibrium shifts toward the ammonia and a new energy balance is established at the lower temperature.

**FIGURE 14-4** Idealized representation of how an equilibrium system is affected by concentration, pressure, and temperature.

trations of the species. If the concentration of ammonia is decreased by removal of ammonia, the equilibrium will shift toward the ammonia side, which tends to replenish the decrease and maintain the equilibrium balance of concentrations. This is true since the ratio must remain constant if the system is in equilibrium.

## Pressure

What happens to an equilibrium system involving gases when the total pressure of the system is changed? Recall from Chapter 9 that the pressure of a gas sample, even one containing a mixture of gases, is given by $P = nRT/V$. One way to increase the pressure is to decrease the total volume of the system by compressing it. The ideal gas law shows that the pressure of a gas mixture depends on the total number of molecules, that is, the number of moles of gas. The equilibrium system can counteract an increase in pressure by shifting in the direction that gives the lesser number of moles or molecules of gas. This shift allows the concentrations of the gases to change so that the ratio of products to reactants remains constant.

As illustrated in Figure 14-4(b), the ammonia reaction involves 1 mole of nitrogen and 3 moles of hydrogen forming 2 moles of ammonia. An increase in pressure would cause the equilibrium to shift toward the ammonia side, since this would produce fewer moles of gas and reduce the pressure. On the other hand, a decrease in the total pressure of the equilibrium system would cause the equilibrium to shift in the direction that forms the greater number of moles of gas. A decrease in the total pressure of the ammonia equilibrium would cause a shift toward the nitrogen–hydrogen side. A pressure change does not change the equilibrium constant but can change the relative amounts of reactants and products. In general, a decrease in pressure causes a shift in equilibrium to make more moles of gas and an increases in pressure causes a shift to make fewer moles of gas.

## Temperature

An equilibrium system involves the production and consumption of energy. The ammonia equilibrium is exothermic in the forward direction and endothermic in the reverse direction.

$$\text{exothermic direction} \longrightarrow$$
$$N_2(g) + 3H_2(g) \rightleftharpoons 2NH_3(g) + \text{energy}$$
$$\longleftarrow \text{endothermic direction}$$

At equilibrium, the same amount of energy produced in the forward reaction is used in the reverse reaction. An energy balance exists at equilibrium, and no heat is lost or gained by the system.

Make a saturated solution of $NaHCO_3$ by adding six heaping teaspoons of baking soda to one cup of water. Stir thoroughly. Let the solution stand in a glass for about 10 minutes. Pour into a glass about one-fourth of a cup of the clear solution from the top of the saturated baking soda solution. Add a teaspoon of solid NaCl to this glass and stir to dissolve the salt. Let the glass sit and observe it. Write an equilibrium equation showing solid $NaHCO_3$ in equilibrium with sodium ions and hydrogen carbonate ions. Use Le Châtelier's principle to explain what happened when sodium chloride was dissolved in the saturated solution of baking soda.

Carbonated beverages contain dissolved carbon dioxide gas. The equilibrium for the carbon dioxide can be represented as

$$CO_2(g) \rightleftharpoons CO_2(aq)$$

Use this equation and Le Châtelier's principle to explain why carbon dioxide gas bubbles out of solution when a can or bottle of beverage is opened.

**ACTIVITY 14-6**

The dissolved carbon dioxide in carbonated water can be viewed as involving the following equilibrium:

$$CO_2(g) \rightleftharpoons CO_2(aq)$$

You need a 12 by 4-inch piece of aluminum foil, some carbonated water, such as mineral water or soda, and some matches. Use the foil to shape a spoon with a handle. Pour some carbonated water into the spoon. Light a match and hold it under the liquid in the spoon. Record your observations. Use your observations to decide which side of the above equilibrium equation should have the word energy. Explain your reasoning.

As shown in Figure 14-4(c), a change in temperature of the reaction system causes a shift in equilibrium. An increase in temperature increases the energy of the system. The equilibrium shifts in the direction that diminishes the increase in energy and tends to maintain the energy balance. In other words, an **increase in temperature** will cause the equilibrium to shift in the direction of the endothermic reaction. When the ammonia equilibrium system is heated, the equilibrium shifts in the nitrogen–hydrogen direction.

A **decrease in temperature** of an equilibrium system will cause the equilibrium to shift in the direction of the exothermic reaction. This tends to replenish the decrease in energy. When the ammonia equilibrium system is cooled, the equilibrium shifts in the direction that increases the yield of ammonia. It is interesting to note, however, that this is not done in the industrial manufacture of ammonia because it does not allow ammonia to be produced fast enough. Recall from Section 11-3 that reaction rates are increased by increasing the temperature. The nitrogen–hydrogen reaction is very slow at room temperature even when a catalyst is present. Cooling the reaction system to shift the equilibrium makes the reaction occur even more slowly. Consequently, to make the reaction occur fast enough, it is run at elevated temperatures. This is an example where increasing the rate of a reaction by heating is more important than shifting the equilibrium by cooling. In other words, it does not do any good to try to shift the equilibrium if the reaction is so slow that equilibrium cannot be readily attained. But in many chemical reactions lower temperatures or elevated temperatures can be used to shift the equilibrium in the direction of desirable products. Chemistry in Action 14-1 investigates the effect of temperature on an iodine–starch equilibrium.

### Catalyst

Recall that a catalyst is a species that increases the rate of a reaction and is not chemically changed. A catalyst will not cause equilibrium to shift because it has the same effect on the rates of the forward and reverse reactions. Putting a catalyst in a system already in equilibrium has no effect. However, a catalyst does cause a system to reach equilibrium more quickly since it will speed both the forward and reverse reactions. In fact, in many industrial reactions a catalyst is used to allow the reaction to reach equilibrium quickly. Biological catalysts or enzymes allow many reactions in the body to quickly reach equilibrium.

## The Bends

The **bends** or **decompression illness** can affect deep-sea divers and scuba divers. Air contains oxygen and nitrogen. The oxygen has a physiological function but nitrogen is physiologically inert. Air that you breathe mixes with blood in the lungs and the blood becomes saturated with nitrogen gas. The equilibrium reaction is

$$N_2(g) \rightleftharpoons N_2(aq)$$

Higher air pressures correspond to increased concentrations of nitrogen. At higher pressures nitrogen is more soluble.

As divers descend in water they are subjected to increased pressures. Diving to 34 feet is equivalent to one atmosphere of pressure in excess of the prevailing atmospheric pressure. The increase in pressure causes more nitrogen than normal to dissolve in the blood. Incidentally, if too much nitrogen dissolves in the blood a diver may experience nitrogen narcosis, a dangerous effect very similar to excess alcohol in the blood. If a diver rises too quickly the decrease in pressure causes nitrogen gas to be released into the blood. The released gas forms bubbles in the tissue and blood vessels. The result is extreme pain in the joints and muscles and loss of bladder and rectal muscle control. If not treated, the bends can result in coma and death. It is possible to treat a bends victim in a pressure chamber. The pressure inside the chamber is increased causing the nitrogen bubbles to dissolve in the blood. The pressure is then slowly decreased to allow time for the nitrogen to be exhaled from the lungs. Proper diving procedures to avoid the bends include slowly ascending to provide a gradual decrease in pressure. It is interesting that astronauts who space walk in soft space suits also risk the bends. Why? To prevent the bends resulting from a space walk an astronaut prebreathes a mixture of oxygen and noble gases to rid the blood of nitrogen.

## 14-9 HYDRONIUM IONS IN WEAK ACID SOLUTIONS

As discussed in Chapter 12 solutions of weak acids have typical acidic properties that reflect the presence of hydronium ions. The hydronium ion concentration in a solution of a weak acid can be found from the acid ionization constant expression for the acid. A monoprotic weak acid in solution exists in equilibrium with water.

$$HA + H_2O \rightleftharpoons H_3O^+ + A^-$$

In solutions of typical weak acids we can assume that the concentration of $H_3O^+$ and the concentration of $A^-$ are equal since they are formed in equal amounts by the reaction.

$$[H_3O^+] = [A^-]$$

Thus, we can substitute $[H_3O^+]$ for $[A^-]$ in the acid ionization constant expression.

$$K_a = \frac{[H_3O^+][A^-]}{[HA]} = \frac{[H_3O^+][H_3O^+]}{[HA]} = \frac{[H_3O^+]^2}{[HA]} = K_a$$

Write an equilibrium equation and an equilibrium constant expression for the dissolving of nitrogen gas in water and use it, along with Le Châtelier's principle, to explain why nitrogen is more soluble at higher pressures and less soluble at lower pressures.

Solving this expression for $[H_3O^+]^2$ gives

$$[H_3O^+]^2 = K_a[HA]$$

and taking the square root of both sides gives a general equation for finding the **hydronium ion concentration in a solution of any weak acid.**

$$\left[H_3O^+\right] = \sqrt{K_a[HA]}$$

$$[H_3O^+] = \sqrt{K_a[HA]}$$

To find the hydronium ion concentration we need to know the concentration of the weak acid and its $K_a$ value. This general equation does not apply to low concentrations of weak acids that have relatively large $K_a$ values.

---

**EXAMPLE 14-6**

Determine the concentration of hydronium ions in a 0.83 M solution of acetic acid. $K_a$ for acetic acid is $1.8 \times 10^{-5}$.

Given: 0.83 M $HC_2H_3O_2$ and $K_a = 1.8 \times 10^{-5}$
Want: $[H_3O^+]$

The square root of the product of the $K_a$ and the concentration of the weak acid gives the hydronium ion concentration.

$$[H_3O^+] = \sqrt{(1.8 \times 10^{-5})(0.83)} = \sqrt{1.5 \times 10^{-5}} = 3.9 \times 10^{-3}$$

A calculator can be used to find square roots.

---

**ACTIVITY 14-8**

Estimate and calculate the pH of the acetic acid solution discussed in Example 14-6.

Recall from Section 12-14 that the pH of a solution is the negative logarithm of the hydronium ion concentration. If we want to find the pH of a weak acid solution of known concentration we can find the hydronium ion concentration then use it to calculate the pH.

---

**EXAMPLE 14-7**

What is the pH of a 1.0 M formic acid, $HCO_2H$, solution if the $K_a$ of formic acid is $1.9 \times 10^{-4}$?

First find the $[H_3O^+]$ from the square root of the product of the $K_a$ and the concentration of the acid.

$$[H_3O^+] = \sqrt{(1.9 \times 10^{-4})(1.0)} = \sqrt{1.9 \times 10^{-4}} = 1.4 \times 10^{-2} .$$

Next use the hydronium ion concentration to find the pH.

$$pH = -\log[H_3O^+] = -\log(1.4 \times 10^{-2})$$

The pH by estimation is 1 point something. The pH found with a calculator is

$$1.9 \times 10^{-2} \quad \boxed{\log} \quad \boxed{+/-} \quad 1.72$$

## 14-10 BUFFER SOLUTIONS

Pure water has a pH of 7.00. When a small amount of acid is added to water, the pH decreases; the solution becomes acidic. When a small amount of base is added to pure water, the pH increases; the solution becomes basic. Blood serum, the fluid base of blood, behaves differently than water. Blood serum has a pH of 7.4. When a small amount of a strong acid or strong base is added to serum, the pH changes only a slight amount, if at all. The difference between serum and pure water is that serum is a buffer solution. A **buffer solution** is a solution that resists a change in pH on addition of small amounts of an acid or base. Serum is a naturally occurring buffer system and it contains more than one buffering agent. It is possible to prepare a simple buffer solution in the laboratory by mixing the appropriate chemicals.

A **simple buffer solution** contains a mixture of a weak acid and the conjugate base of that acid at approximately equal concentrations. Some complex buffer solutions are mixtures of more than one acid and base. Blood contains several buffer systems. One of the buffer systems in blood involves equilibrium between the dihydrogen phosphate ion, $H_2PO_4^-$, and its conjugate base, the hydrogen phosphate ion, $HPO_4^{2-}$.

$$\underset{\text{conjugate acid}}{H_2PO_4^-} + H_2O \rightleftharpoons \underset{\text{conjugate base}}{HPO_4^{2-}} + H_3O^+$$

This equilibrium, along with other buffer systems in blood, serves to maintain a fairly constant pH. The relatively constant pH results from the tendency of the equilibrium concentrations of the species involved in the buffer system to be maintained by a shift in the equilibrium as predicted by Le Châtelier's principle. That is, when some acid or base is added to a buffer the equilibrium shifts in the direction that tends to absorb the change in acidity. In a buffer system a balance exists between the weak acid and its conjugate base. If the balance is upset by changing the hydronium ion concentration the system shifts to counteract the change. The term buffer comes from the idea that the shift in equilibrium tends to counteract or cushion a change in the hydronium ion concentration. A buffer system is a "shock absorber" in the sense that it absorbs or adjusts to an increase or decrease in hydronium ion concentration.

The function of the dihydrogen phosphate ion–hydrogen phosphate ion buffer is illustrated in Figure 14-5. If some $OH^-$ is added to such a solution, it will react with some $H_3O^+$ and be neutralized. The decrease in $H_3O^+$ concentration causes a shift in the equilibrium toward the right; the $H_2PO_4^-$ reacts to replenish most of the hydronium ions that were used.

If some $H_3O^+$ ions are added to the solution, a right to left reaction will occur between the $H_3O^+$ ions and the $HPO_4^{2-}$ ions to produce additional $H_2O$ and $H_2PO_4^-$. This shift removes most of the added $H_3O^+$ ions from the system. Thus, because of the equilibrium between $HPO_4^{2-}$ and $H_2PO_4^-$, the $H_3O^+$ concentration or pH will be maintained at almost a constant

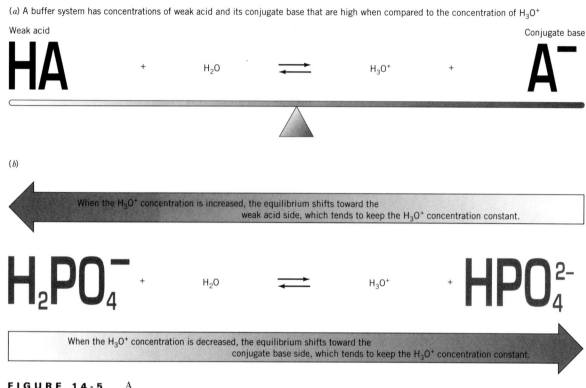

(*a*) A buffer system has concentrations of weak acid and its conjugate base that are high when compared to the concentration of $H_3O^+$

Weak acid                                                                                                    Conjugate base

$$HA \quad + \quad H_2O \quad \rightleftharpoons \quad H_3O^+ \quad + \quad A^-$$

(*b*)

When the $H_3O^+$ concentration is increased, the equilibrium shifts toward the weak acid side, which tends to keep the $H_3O^+$ concentration constant.

$$H_2PO_4^- \quad + \quad H_2O \quad \rightleftharpoons \quad H_3O^+ \quad + \quad HPO_4^{2-}$$

When the $H_3O^+$ concentration is decreased, the equilibrium shifts toward the conjugate base side, which tends to keep the $H_3O^+$ concentration constant.

**FIGURE 14-5**   A buffer system.

level when small amounts of acid or base are added to the system. However, a buffer system has a limited capacity to resist changes in $H_3O^+$ concentration, and if too much acid or base is added, the capacity of the buffer can be exceeded. Nevertheless, the dihydrogen phosphate–hydrogen phosphate ion buffer system in blood helps to maintain the hydronium ion concentration of blood within critical limits.

Buffers are used in chemistry when we want a solution that will have a specific pH and will resist a change in pH on the addition of small amounts of acid or base. Buffer solutions that maintain a specific pH can be prepared by adding the appropriate acid and its conjugate base in specific proportions. This is often done by mixing an ionic compound containing the conjugate base with a solution of the weak acid. For example, a buffer solution can be prepared by adding sodium acetate to an acetic acid solution. The sodium acetate dissolves to form acetate ions, the conjugate base of acetic acid. Chemistry in Action 14-2 investigates a buffer solution.

## 14-11 FINDING THE pH OF BUFFER SOLUTIONS

The acid ionization constant expression for an acid can be used to determine the hydronium ion concentration of a buffer solution that contains a known concentration of a weak acid and its conjugate base. A $K_a$ value for the acid is also needed in the calculation. Tabulations of $K_a$ values for various acids are available in chemical reference books. As an example, the determination of the hydronium ion concentration in a buffer solution that is 2.0 $M$ in acetic acid and 1.5 $M$ in acetate ion is shown here. The equilibrium equation is

$$HC_2H_3O_2(aq) + H_2O \rightleftharpoons H_3O^+(aq) + C_2H_3O_2^-(aq)$$

and the acid ionization constant expression is

$$K_a = \frac{[H_3O^+][C_2H_3O_2^-]}{[HC_2H_3O_2]} = 1.8 \times 10^{-5}$$

It is given that $[C_2H_3O_2^-] = 1.5$ and $[HC_2H_3O_2] = 2.0$ and we want $[H_3O^+]$. The expression is first solved for $[H_3O^+]$ and the given values are used to find the value of $[H_3O^+]$.

$$[H_3O^+] = 1.8 \times 10^{-5} \times \frac{[HC_2H_3O_2]}{[C_2H_3O_2^-]}$$

$$[H_3O^+] = 1.8 \times 10^{-5} \times \frac{(2.0)}{(1.5)} = 2.4 \times 10^{-5}$$

Using this $[H_3O^+]$ value we find that the buffer has a pH of 4.62.

In general, the **hydronium ion concentration for any buffer solution** of a weak acid and its conjugate base

$$HA + H_2O \rightleftharpoons H_3O^+ + A^-$$

is given by

$$\mathbf{[H_3O^+]} = \frac{K_a\mathbf{[HA]}}{\mathbf{[A^-]}} \quad \begin{array}{l} \text{conjugate acid} \\ \text{conjugate base} \end{array} \qquad\qquad [H_3O^+] = K_a\frac{[HA]}{[A^-]}$$

where $K_a$ is the known acid constant, [HA] is the concentration of the weak acid, and $[A^-]$ is the concentration of its conjugate base. If desired the pH of a buffer can be calculated using the value of the hydronium ion concentration.

---

**EXAMPLE 14-8**

What is the pH of a buffer solution consisting of a 1.0 $M$ solution of ammonium ions in equilibrium with 2.0 $M$ ammonia? The $K_a$ of $NH_4^+$ is $5.6 \times 10^{-10}$.

First, we consider the equilibrium reaction of the acid by showing the reaction of the acid and water to give the hydronium ion and the conjugate base of the acid. In this case the ammonium ion is the acid and ammonia is its conjugate base.

$$\begin{array}{cccc} NH_4^+ + H_2O & \rightleftharpoons & NH_3 & + H_3O^+ \\ \text{acid} & & \text{base} & \end{array}$$

The hydronium ion concentration of the buffer can be found, given the following data:

| $[NH_4^+]$ | $[NH_3]$ | $K_a$ |
|---|---|---|
| 1.0 M | 2.0 M | $5.6 \times 10^{-10}$ |

$$[H_3O^+] = \frac{K_a[NH_4^+]}{[NH_3]} = 5.6 \times 10^{-10} \times \frac{(1.0)}{(2.0)}$$

$$[H_3O^+] = 3.0 \times 10^{-10}$$

Finally, the pH can be found using $pH = -\log[H_3O^+]$. The pH is 9 point something. The pH found with a calculator is $3.0 \times 10^{-10}$ (log) (+/-) 9.522. . . or 9.52.

**A buffer solution contains 0.25 M $H_2PO_4^-$ and 0.16 M $HPO_4^{2-}$. The $K_a$ of $H_2PO_4^-$ is $6.23 \times 10^{-8}$. What is the pH of the buffer solution?**

## Chemical Equilibrium

Chemical equilibrium: The state of a reversible chemical reaction when the rate of the forward reaction equals the rate of the reverse reaction.

A reaction system involving chemical equilibrium is characterized by the fact that at equilibrium the ratio of the concentrations of the products to the reactants is equal to a constant value. This ratio is called the equilibrium constant.

An equilibrium system represented by the general equation

$$a A + b B \rightleftharpoons c C + d D$$

is described by an equilibrium constant expression of the general form:

$$K_{eq} = \frac{[C]^c[D]^d}{[A]^a[B]^b}$$

where $K_{eq}$ is the equilibrium constant, and the square brackets refer to the concentrations of the products and reactants in moles per liter. Each concentration is raised to a power corresponding to its coefficient in the equilibrium equation.

Steps used to derive the equilibrium constant expression for an equilibrium system are:

1. Write an equilibrium equation and note whether the species involved are gases, solids, liquids, or in aqueous solution.
2. Enclose the formulas of all products that are gases or aqueous species in square brackets for the numerator of the expression. If any of these species has a coefficient greater than 1, show its concentration raised to a power given by the coefficient.
3. Enclose the formulas of all reactants that are gases or aqueous species in square brackets for the denominator of the expression using the coefficients as powers.

4. The ratio of the product concentrations to the reactant concentrations, each raised to a power corresponding to its coefficient in the balanced equation, is equal to the constant, $K_{eq}$.

The relative sizes of double arrows in equilibrium equations relate to the relative numerical values of $K_{eq}$.

relatively small $K_{eq}$ value          $K_{eq}$ value in the range of one          relatively large $K_{eq}$ value
$\rightleftharpoons$                                  $\rightleftharpoons$                                     $\rightleftharpoons$

Le Châtelier's Principle: When a factor affecting an equilibrium system is changed, the equilibrium will shift in a direction that tends to counteract the change.

Factors that effect an equilibrium system are:

1. A change in the concentration of some gaseous or dissolved species that is a reactant or product. Equilibrium shifts away from an increase in concentration or towards a decrease in concentration.
2. A change in temperature of the equilibrium mixture. When a mixture is heated, the equilibrium shifts in the endothermic direction, and when a mixture is cooled, the equilibrium shifts in the exothermic direction.
3. A change in the pressure of an equilibrium mixture involving one or more gases. When the pressure is increased, the equilibrium shifts towards the side having fewer moles of gas, and when the pressure is decreased, the equilibrium shifts towards the side having more moles of gas.

The equilibrium reaction of any weak acid, HA, in water is represented by the general equation:

$$HA + H_2O \rightleftharpoons H_3O^+ + A^-$$

where HA is the acid and $A^-$ is its conjugate base. For a weak acid equilibrium system, the ratio of the concentrations of the products to the concentrations of the reactants is constant.

$$K_a = \frac{[H_3O^+][A^-]}{[HA]}$$

where $K_a$ is called the acid ionization constant or the acid constant for a weak acid.

The general equation for finding the hydronium ion concentration in a solution of any weak acid is

$$[H_3O^+] = \sqrt{K_a\,[HA]}$$

To find the hydronium ion concentration multiply the concentration of the weak acid by its $K_a$ value and take the square root of the product.

Buffer solution: A solution that resists a change in pH on addition of an acid or base. A simple buffer solution contains a mixture of a weak acid and the conjugate base of that acid at approximately equal concentrations.

The hydronium ion concentration for any buffer solution of a weak acid and its conjugate base is found using

$$[H_3O^+] = \frac{K_a[HA]}{[A^-]}$$

where $K_a$ is the known acid constant, [HA] is the concentration of the weak acid, and $[A^-]$ is the concentration of its conjugate base.

You need: a box of iodized salt, a filter or some napkins arranged as a filter, hydrogen peroxide solution, white vinegar, cornstarch (or flour), two zip-lock bags, three glasses, ice, and hot water.

1. Follow the directions in Chemistry in Action 13-2 to prepare the iodine solution from salt. Use the yellow iodine solution as instructed below.

2. Add one teaspoon of cornstarch to half a cup of very hot water and stir the mixture for a few minutes. Let the mixture cool.

3. To each of the zip-lock bags add three tablespoons of the starch mixture and two teaspoons of the yellow iodine solution. Squeeze the air out of the bags and seal them. Use one of the bags for color comparison.

4. Make an ice–water mixture in one glass and add hot water—but not boiling hot— to the other glass.

5. Dip one of the zip-lock bags into the hot water, let it sit for a minute or so and observe it. Compare its color to the contents of the other bag.

6. Move the bag from the hot water to the ice water. Let it sit for a minute or so and observe it. Compare its color to the contents of the other bag. Repeat the process of dipping the bag in hot then cold water a few more times. Add more hot water or ice as needed.

The iodine forms a complex compound with the starch that has a distinct color. The equilibrium involved can be represented as

$$I_2 + starch \rightleftharpoons starch–I_2 \text{ (purple color)}$$

Use this equation and Le Châtelier's principle to explain your observations. Based upon your observations, write the word energy on the appropriate side of the above equation.

You need: a cup of chopped red cabbage, a pan with a lid, milk of magnesia, white vinegar, lemon juice, and three glasses. Prepare red cabbage juice indicator as directed in Chemistry in Action 12-2.

The buffer solution you are going to make is an acetic acid–acetate ion buffer. Write an equation for the reaction of acetic acid and water in equilibrium with hydronium ions and acetate ions.

1. Put two teaspoons of lemon juice in a glass and add two teaspoons of indicator. Use this glass for color comparison.

2. Put a few drops of milk of magnesia in a glass along with three teaspoons of water. Add two teaspoons of indicator and stir. Use this glass for color comparison.

3. To prepare the acetic acid–acetate ion buffer add two level teaspoons of milk of magnesia to a glass followed by two tablespoons of vinegar. Then add two more tablespoons of vinegar and stir to mix. Add two teaspoons of indicator and note the color. Note teaspoons and tablespoons.

4. Test the resistance of the buffer solution to a change in pH by adding acid and base to it. Add one teaspoon of lemon juice (acid) and stir. Add a second teaspoon of lemon juice and stir. Add two drops of milk of magnesia (base) and stir. Add two more drops of milk of magnesia and stir. Record your observations.

5. Assume that your acetic acid–acetate ion buffer contains 0.3 $M$ $HC_2H_3O_2$ and 0.4 $M$ $C_2H_3O_2^-$. Calculate the pH of your buffer solution using $K_a = 1.8 \times 10^{-5}$ and the method discussed in Section 14-11.

## QUESTIONS

### Sections 14-1 to 14-5

1. What are a reversible chemical reaction and chemical equilibrium.

2. How are double arrows used to express the extent of an equilibrium reaction?

3. What is an equilibrium constant? How is it determined? How does the value of a constant indicate the direction and extent of an equilibrium reaction?

4. *Experimental observation of the following equilibrium system gives the equilibrium concentrations shown. Give the acid ionization constant expression and calculate the numerical value of $K_a$.

$$HC_2H_3O_2 + H_2O \rightleftharpoons H_3O^+ + C_2H_3O_2^-$$

| | $[HC_2H_3O_2]$ | $[H_3O^+]$ | $[C_2H_3O_2^-]$ |
|---|---|---|---|
| equilibrium concentrations in mol/L | 2.12 | $6.2 \times 10^{-3}$ | $6.2 \times 10^{-3}$ |

5. Experimental observation of the following equilibrium system gives the equilibrium concentrations shown. Give the equilibrium constant expression and calculate the numerical value of $K_{eq}$.

$$2HI(g) \rightleftharpoons H_2(g) + I_2(g)$$

| | $[HI]$ | $[H_2]$ | $[I_2]$ |
|---|---|---|---|
| equilibrium concentrations in mol/L | 0.20 | 0.64 | 0.64 |

6. *Write equilibrium constant expressions for the following equilibrium reactions.

(a) $2HCl(g) \rightleftharpoons H_2(g) + Cl_2(g)$

(b) $2SO_2(g) + O_2(g) \rightleftharpoons 2SO_3(g)$

(c) $3O_2(g) \rightleftharpoons 2O_3(g)$

(d) $HCN(aq) + H_2O \rightleftharpoons H_3O^+(aq) + CN^-(aq)$

(e) $NH_3(aq) + H_2O \rightleftharpoons NH_4^+(aq) + OH^-(aq)$

7. Write equilibrium constant expressions for the following equilibrium reactions.

(a) $C(s) + CO_2(g) \rightleftharpoons 2CO(g)$

(b) $HNO_2(aq) + H_2O \rightleftharpoons H_3O^+(aq) + NO_2^-(aq)$

(c) $CH_4(g) + H_2O(g) \rightleftharpoons CO(g) + 3H_2(g)$

(d) $CO_2(aq) + 2H_2O \rightleftharpoons H_3O^+(aq) + HCO_3^-(aq)$

(e) $2NO(g) + Br_2(g) \rightleftharpoons 2NOBr(g)$

8. Explain the meaning of a double arrow in an equation and how arrows are used to convey information about the extent of a reaction.

9. *For each of the following equilibrium reactions, give an appropriate equilibrium equation. Use the value of the equilibrium constant as a guide for writing a larger arrow and a smaller arrow in the appropriate directions.

(a) Ethylene, $C_2H_4$, gas, and steam in equilibrium with gaseous ethyl alcohol. Note that water is a gas in the reaction and is included in the equilibrium constant expression.

$$K_{eq} = \frac{[C_2H_5OH]}{[C_2H_4][H_2O]} = 8.2 \times 10^3$$

(b) Carbon monoxide gas and steam in equilibrium with hydrogen gas and carbon dioxide gas.

$$K_{eq} = \frac{[H_2][CO_2]}{[CO][H_2O]} = 3.4 \times 10^3$$

(c) Carbon dioxide and water in equilibrium with hydrogen carbonate ions and hydronium ions in your kidneys.

$$K_a = \frac{[H_3O^+][HCO_3^-]}{[CO_2]} = 4.3 \times 10^{-7}$$

(d) Hydrogen fluoride in solution and water are in equilibrium with hydronium ions and fluoride ions.

$$K_a = \frac{[H_3O^+][F^-]}{[HF]} = 7.0 \times 10^{-4}$$

10. For each of the following equilibrium reactions, give an appropriate equilibrium equation. Use the value of the equilibrium constant as a guide for writing a larger arrow and a smaller arrow in the appropriate directions.

(a) Phosphorus pentachloride gas in equilibrium with phosphorus trichloride gas and chlorine gas.

$$K_{eq} = \frac{[PCl_3][Cl_2]}{[PCl_5]} = 1.8$$

(b) Hydrogen gas and chlorine gas in equilibrium with hydrogen chloride gas.

$$K_{eq} = \frac{[HCl]^2}{[H_2][Cl_2]} = 2.5 \times 10^4$$

(c) Solid $NH_4HS$ in equilibrium with ammonia gas and hydrogen sulfide gas.

$$K_{eq} = [NH_3][H_2S] = 1.8 \times 10^{-4}$$

(d) The weak base ammonia and water in equilibrium with ammonium ions and hydroxide ions.

$$K_b = \frac{[NH_4^+][OH^-]}{[NH_3]} = 1.8 \times 10^{-5}$$

## Section 14-6

11. Write the equation for the equilibrium reaction of propionic acid $CH_3CH_2CO_2H$ in water and give the equilibrium constant expression for the reaction.

12. An acetic acid solution contains 2.0 $M$ $HC_2H_3O_2$, $3.6 \times 10^{-5}$ $M$ $H_3O^+$, and $3.6 \times 10^{-5}$ $M$ $C_2H_3O_2^-$. Using these data, calculate the value of the acid ionization constant for acetic acid in water. Based on this value why is acetic acid classified as a weak acid?

## Section 14-7

13. *Write acid constant expressions for the following weak acids in water.

(a) $H_2PO_4^-(aq) + H_2O \rightleftharpoons H_3O^+(aq) + HPO_4^{2-}(aq)$

(b) $HCO_3^-(aq) + H_2O \rightleftharpoons H_3O^+(aq) + CO_3^{2-}(aq)$

(c) $NH_4^+(aq) + H_2O \rightleftharpoons H_3O^+(aq) + NH_3(aq)$

14. *Calculate the value of the acid constant, $K_a$, for each of the following solutions having the equilibrium concentrations shown.

(a) $HSO_4^-(aq) + H_2O \rightleftharpoons H_3O^+(aq) + SO_4^{2-}(aq)$
    0.25 $M$            $5.5 \times 10^{-2}$ $M$   $5.5 \times 10^{-2}$ $M$

(b) $HCO_3^-(aq) + H_2O \rightleftharpoons H_3O^+(aq) + CO_3^{2-}(aq)$
    1.15 $M$            $8.1 \times 10^{-6}$ $M$   $8.1 \times 10^{-6}$ $M$

(c) $NH_4^+(aq) + H_2O \rightleftharpoons H_3O^+(aq) + NH_3(aq)$
    3.75 $M$            $4.6 \times 10^{-5}$ $M$   $4.6 \times 10^{-5}$ $M$

15. Using the values of the acid constants calculated in Question 14, rank the three weak acids from strongest to weakest. How does this ranking correspond to the positions of these acids in Table 12-1?

16. Write equilibrium constant expressions for the following weak acids in water.

(a) $HF(aq) + H_2O \rightleftharpoons H_3O^+(aq) + F^-(aq)$

(b) $HC_2O_4^-(aq) + H_2O \rightleftharpoons H_3O^+(aq) + C_2O_4^{2-}(aq)$

(c) $H_3PO_4(aq) + H_2O \rightleftharpoons H_3O^+(aq) + H_2PO_4^-(aq)$

17. Calculate the value of the acid constant, $K_a$, for each of the following solutions having the equilibrium concentrations shown.

(a) $HF(aq) + H_2O \rightleftharpoons H_3O^+(aq) + F^-(aq)$
    0.50 $M$            $1.32 \times 10^{-2}$ $M$   $1.32 \times 10^{-2}$ $M$

(b) $HC_2O_4^-(aq) + H_2O \rightleftharpoons H_3O^+(aq) + C_2O_4^{2-}(aq)$
   2.00 *M*                    $1.01 \times 10^{-2}$   $1.01 \times 10^{-2}$

(c) $H_3PO_4(aq) + H_2O \rightleftharpoons H_3O^+(aq) + H_2PO_4^-(aq)$
   0.876 M                    $2.5 \times 10^{-2}$   $2.5 \times 10^{-2}$

18. Using the values of the acid constants calculated in Question 17, rank the three acids from the strongest to the weakest.

### Section 14-8

19. Give a statement of Le Châtelier's principle.

20. *Using Le Châtelier's principle, predict any shift in equilibrium in the reaction

$$energy + H_2(g) + I_2(g) \rightleftharpoons 2HI(g)$$

when the following occur.

   (a) the equilibrium system is cooled
   (b) extra $H_2$ is added to the system
   (c) a catalyst is added to the system
   (d) the pressure of the equilibrium system is decreased
   (e) the concentration of HI is increased

21. Using Le Châtelier's principle, predict any shift in equilibrium in the reaction

$$CO(g) + H_2O(g) \rightleftharpoons CO_2(g) + H_2(g) + energy$$

when the following occur.

   (a) the temperature of the equilibrium system is increased
   (b) the concentration of $H_2O(g)$ is increased
   (c) a catalyst is added to the system
   (d) the equilibrium system is cooled
   (e) the pressure of the equilibrium system is decreased

22. *Using Le Châtelier's principle, predict any shift in equilibrium in the reaction

$$N_2(g) + 3H_2(g) \rightleftharpoons 2NH_3(g) + energy$$

when the following occur.

   (a) the temperature of the equilibrium system is increased
   (b) the concentration of $H_2$ is increased
   (c) the concentration of $N_2$ is decreased
   (d) the temperature of the equilibrium system is decreased

   (e) the pressure on the equilibrium system is increased

23. Using Le Châtelier's principle, predict any shift in equilibrium in the reaction

$$energy + C(s) + H_2O(g) \rightleftharpoons CO(g) + H_2(g)$$

when the following occur.

   (a) the temperature of the equilibrium system is increased
   (b) the concentration of $H_2$ is increased
   (c) a catalyst is added to the system
   (d) the concentration of CO is decreased
   (e) the pressure of the equilibrium system is increased

24. *Using Le Châtelier's principle, predict any shift in equilibrium in each of the following systems.

   (a) $energy + H_2O(s) \rightleftharpoons H_2O(\ell)$, when the system is cooled
   (b) $2CO(g) + O_2(g) \rightleftharpoons 2CO_2(g)$, when a catalyst is added
   (c) $Mg(OH)_2(s) \rightleftharpoons Mg^{2+}(aq) + 2OH^-(aq)$, when extra $OH^-$ from sodium hydroxide is added
   (d) $CO_2(aq) + 2H_2O \rightleftharpoons H_3O^+(aq) + HCO_3^-(aq)$, when $H_3O^+$ from hydrochloric acid is added
   (e) $CaCO_3(s) \rightleftharpoons CaO(s) + CO_2(g)$, when the pressure is decreased
   (f) $2H_2O(g) + energy \rightleftharpoons 2H_2(g) + O_2(g)$, when the temperature is decreased and the pressure is increased
   (g) $Ag^+(aq) + Cl^-(aq) \rightleftharpoons AgCl(s)$, when $Cl^-$ from sodium chloride is added

25. Using Le Châtelier's principle, predict any shift in equilibrium in each of the following systems.

   (a) $energy + 2N_2O(g) \rightleftharpoons 2N_2(g) + O_2(g)$, when the system is cooled
   (b) $2SO_2(g) + O_2(g) \rightleftharpoons 2SO_3(g)$, when a catalyst is added
   (c) $CaCO_3(s) \rightleftharpoons Ca^{2+}(aq) + CO_3^{2-}(aq)$, when extra $Ca^{2+}$ from calcium chloride is added
   (d) $NH_3(aq) + H_2O \rightleftharpoons NH_4^+(aq) + OH^-(aq)$, when $OH^-$ from sodium hydroxide is added
   (e) $H_2(g) + CO_2(g) \rightleftharpoons H_2O(g) + CO(g)$, when the pressure is decreased

(f) $2NO(g) + Br_2(g) \rightleftharpoons 2NOBr(g) + energy$, when the temperature is decreased and the pressure is increased

(g) $Pb^{2+}(aq) + 2Cl^-(aq) \rightleftharpoons PbCl_2(s)$, when $Cl^-$ from sodium chloride is added

26. Dissolved carbon dioxide, $CO_2$, and water are in equilibrium with hydrogen carbonate ions, $HCO_3^-$, and hydronium ions, $H_3O^+$, in your blood. Write an equilibrium equation. Using Le Châtelier's principle, explain what happens to the concentration of hydronium ions in your blood when you become anxious and start to hyperventilate, or breathe rapidly, which removes carbon dioxide from your body. What happens to the pH of blood as a result of hyperventilation?

27. In the internal combustion engine a gasoline–air mixture is compressed in the piston chamber and burned producing a high temperature. Nitrogen oxide, NO, an air pollutant, is formed in the internal combustion engine by the reaction:

$$energy + N_2(g) + O_2(g) \rightleftharpoons 2NO(g)$$

According to Le Châtelier's principle, what condition in the internal combustion engine causes a shift in this equilibrium toward the formation of NO?

28. Hemoglobin (represented symbolically as Hb) is a complex protein found in red blood cells. Hemoglobin is involved in oxygen transport in the body according to the reaction:

$$Hb + O_2 \rightleftharpoons HbO_2$$

Using Le Châtelier's principle, explain how hemoglobin transports oxygen from the lungs, where there is a high concentration of oxygen, to the cells fed by blood capillaries, where there is a low concentration of oxygen.

29. Explain the bends or decompression illness using Le Châtelier's principle.

## Section 14-9

30. *Find the hydronium ion concentration of a 0.10 $M$ solution of $NaHSO_4$. $K_a = 1.2 \times 10^{-2}$ for the weak acid $HSO_4^-$. What is the pH of this solution?

31. Find the hydronium ion concentration of a 2.25 $M$ solution of HF. $K_a = 3.5 \times 10^{-4}$ for HF. What is the pH of this solution?

32. *Formic acid is a weak acid, $K_a = 1.8 \times 10^{-4}$. What is the hydronium ion concentration in a 0.315 $M$ solu-

tion of formic acid? What is the pH of this solution?

33. Vinegar, an acetic acid solution, typically has a concentration of 0.83 $M$ $HC_2H_3O_2$. If $K_a = 1.8 \times 10^{-5}$ for acetic acid what is the hydronium ion concentration in vinegar? What is the pH of vinegar?

## Sections 14-10 and 14-11

34. Describe a simple buffer solution and give an example to explain how a buffer system functions.

35. Use the following general equation for an acid in equilibrium with its conjugate base to explain how a buffer solution functions upon the addition of small amounts of acid or base.

$$HA + H_2O \rightleftharpoons H_3O^+ + A^-$$

36. *Calculate the hydronium ion concentration of a buffer solution containing 1.25 $M$ acetic acid, $HC_2H_3O_2$, and 0.75 $M$ sodium acetate, $NaC_2H_3O_2$. ($K_a = 1.8 \times 10^{-5}$ for $HC_2H_3O_2$)

37. *Calculate the hydronium ion concentration of a buffer solution containing 2.5 $M$ hydrofluoric acid, HF, and 1.8 $M$ potassium fluoride, KF. ($K_a = 3.5 \times 10^{-4}$ for HF) **CAUTION: Hydrofluoric acid solutions are very dangerous and should be used with utmost care and ventilation.**

38. *Calculate the pH of a buffer solution containing 1.5 $M$ sodium dihydrogen phosphate, $NaH_2PO_4$, and 3.0 $M$ sodium hydrogen phosphate, $Na_2HPO_4$. The equilibrium is

$$H_2PO_4^-(aq) + H_2O \rightleftharpoons H_3O^+(aq) + HPO_4^{2-}(aq)$$

and $K_a = 6.23 \times 10^{-8}$ for $H_2PO_4^-$, which is the acid species in the buffer.

39. *Calculate the pH of a buffer solution containing 1.0 $M$ $HCO_2H$ and 0.92 $M$ $HCO_2^-$. ($K_a = 1.9 \times 10^{-4}$ for $HCO_2H$)

40. Calculate the hydronium ion concentration of a buffer solution containing 0.75 $M$ acetic acid, $HC_2H_3O_2$, and 0.50 $M$ sodium acetate, $NaC_2H_3O_2$. ($K_a = 1.8 \times 10^{-5}$ for $HC_2H_3O_2$)

41. Calculate the hydronium ion concentration of a buffer solution containing 2.3 $M$ $HCO_2H$ and 1.4 $M$ $HCO_2^-$. ($K_a = 1.9 \times 10^{-4}$ for $HCO_2H$)

42. Calculate the pH of a buffer solution containing 1.6 $M$ sodium dihydrogen phosphate, $NaH_2PO_4$, and 1.20 $M$ sodium hydrogen phosphate, $Na_2HPO_4$. The equilibrium is

$$H_2PO_4^-(aq) + H_2O \rightleftharpoons H_3O^+(aq) + HPO_4^{2-}(aq)$$

and $K_a = 6.23 \times 10^{-8}$ for $H_2PO_4^-$, which is the acid in the buffer.

43. Calculate the pH of a buffer solution containing 1.80 *M* sodium oxalate, $Na_2C_2O_4$, and 3.10 *M* sodium hydrogen oxalate, $NaHC_2O_4$. The equilibrium is

$$HC_2O_4^-(aq) + H_2O \rightleftharpoons H_3O^+(aq) + C_2O_4^{2-}(aq)$$

and $K_a = 5.2 \times 10^{-5}$ for $HC_2O_4^-$, which is the acid species in the buffer.

44. A buffer solution contains acetic acid and acetate ion. If the pH of the buffer is 4.74 what is the value of the ratio of acetate ion concentration to the acetic acid concentration? $K_a = 1.8 \times 10^{-5}$

# NUCLEAR

# CHEMISTRY

## 15-1 ATOMIC NUCLEI

Chemistry deals with the structure and behavior of atoms. Nuclear chemistry deals with the structures and behaviors of atomic nuclei—the tiny, massive centers of atoms. For purposes of discussion we can view an atomic nucleus as a clump or aggregate of protons and neutrons. **Protons** and **neutrons** as nuclear particles are called **nucleons.** As we shall see, other particles such as beta particles, alpha particles, and gamma rays, are associated with the properties of nuclei. Table 15-1 has a summary of the symbols and properties of these nuclear particles.

All atoms contain nuclei surrounded by electrons. When discussing the properties of nuclei, we will view them without consideration of the electrons. However, keep in mind that except for a few important cases, nuclei are always parts of atoms not separate particles. The nuclei of the atoms of a given element always have the same number of protons; this is the **atomic number** of the element. Recall that **isotopes** are atoms of a given element that contain different numbers of neutrons. Most elements have atoms with more than one combination of neutrons and protons. The types of nuclei found in the isotopes of the various elements are called **nuclides.** Nuclide is a general term that refers to the various types of nuclei that are known. Numerous nuclides occur in nature and others are made by methods discussed in Section 15-11.

An atomic nucleus is very small relative to the diameter of an atom. Place a dime on the ground and imagine it to be the nucleus of a hydrogen atom. Walk about one-half a kilometer or three-tenths of a mile from the dime and imagine that you are at the outer limits of the electron space of the hydrogen atom.

**Table 15-1** Common Nuclear Particles

| NAME | MASS (u) | CHARGE | SYMBOL |
|---|---|---|---|
| Proton | 1.007825 | 1+ | $_1^1\text{H}, p$ |
| Neutron | 1.008665 | 0 | $_0^1 n, n$ |
| Electron | 0.000549 | 1− | $_{-1}^0 e, e^-$ |
| Alpha particle | 4.00260 | 2+ | $_2^4\text{He}, \alpha$ |
| Gamma ray | 0 | 0 | $\gamma, h\nu$ |

A nuclide is represented by a special symbol. The atomic number gives the number of protons in a nuclide. The **nucleon number** or **mass number** of a nuclide is defined as the sum of the number of protons and neutrons. It is deduced by adding the number of neutrons to the number of protons. The general symbol for a nuclide is

$$_Z^A X$$

where X is the symbol of the element corresponding to the nuclide, $A$ is the nucleon number, and $Z$ is the atomic number. A few examples of nuclides are

$$_1^1\text{H} \qquad _2^1\text{H} \qquad _6^{12}\text{C} \qquad _8^{16}\text{O} \qquad _{92}^{238}\text{U}$$

Read a nuclide symbol as the element name followed by the nucleon number (i.e., $_1^1\text{H}$, hydrogen one; $_1^2\text{H}$, hydrogen two; $_6^{12}\text{C}$, carbon twelve; $_8^{16}\text{O}$, oxygen sixteen; $_{92}^{238}\text{U}$, uranium two-thirty-eight). A nuclide is sometimes referred to by the element name followed by the nucleon number. For instance, the above nuclides are hydrogen-1, hydrogen-2, carbon-12, oxygen-16 and uranium-238.

There are hundreds of known nuclides and tabulations of all known nuclides are available in reference books. Table 15-2 lists the common nuclides of a few elements. If the nucleon number of a nuclide is known, the number of neutrons in a nuclide is easily found by subtracting the atomic number from the nucleon number: $A - Z$.

**ACTIVITY 15-2**

Make a list of all the isotopes in Table 15-2 and indicate the number of protons and neutrons in each isotope. In addition, include the neutron to proton ratio for each isotope which is found by dividing the number of neutrons by the number of protons.

**Table 15-2** Common Nuclides of Some Elements

| | | | |
|---|---|---|---|
| Hydrogen | $_1^1\text{H}$ | $_1^2\text{H}$ | $_1^3\text{H}$ |
| Helium | $_2^3\text{He}$ | $_2^4\text{He}$ | |
| Carbon | $_6^{12}\text{C}$ | $_6^{13}\text{C}$ | $_6^{14}\text{C}$ |
| Oxygen | $_8^{16}\text{O}$ | $_8^{17}\text{O}$ | $_8^{18}\text{O}$ |
| Uranium | $_{92}^{234}\text{U}$ | $_{92}^{235}\text{U}$ | $_{92}^{238}\text{U}$ |

## 15-2 RADIOACTIVITY

Most naturally occurring nuclides are stable and retain their structure indefinitely. Some nuclides, however, are not stable and spontaneously come apart or decay over time. Unstable nuclides are radioactive. For example, the nucleus of a uranium-238 atom can spontaneously break apart to form a helium-4 nucleus (two protons and two neutrons) and a new nucleus containing all of the remaining protons and neutrons.

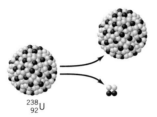

$^{238}_{92}U$

In a collection of radioactive nuclei such spontaneous decay does not occur all at once. Instead it continues over time until eventually all of the nuclei in the collection decay.

**Radioactivity** is the spontaneous decay of a nucleus to form another nucleus and a nuclear particle. Some nuclides are radioactive and some are not. Radioactivity relates to the ratio of the number of neutrons to protons in a nucleus. Generally the ratio of neutrons to protons in stable nuclides ranges from a minimum of 1 to 1 to a maximum of about 1.6 to 1. A nuclide that has too low a **neutron-to-proton ratio** is likely to be radioactive. Furthermore, a nuclide with too high a neutron-to-proton ratio is also likely to be radioactive. All nuclides with more than 83 protons are radioactive. This means that all isotopes of elements beyond bismuth are radioactive. Apparently all nuclei with more than 83 protons spontaneously decay no matter how many neutrons are present. Look at a periodic table to see which elements follow bismuth, atomic number 83.

## 15-3 TYPES OF RADIOACTIVE DECAY

In a collection of radioactive nuclei the nuclei decay by splitting off nuclear particles and forming new nuclei. The nuclear particles produced during radioactive decay are ejected from the original nucleus with large amounts of kinetic energy. These particles radiate or move out in all directions from the collection of nuclei. These energetic nuclear particles make radioactive substances dangerous, but sometimes, in certain controlled situations, they can be useful.

There are specific types or modes of radioactive decay. The three most common types of decay are described below. Other less common types are known but they are not relevant to our discussion.

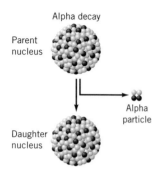

Alpha decay

Parent nucleus

Alpha particle

Daughter nucleus

## Alpha Particle Decay

A nucleus, called the **parent,** decays by emitting a high-speed helium-4 nucleus known as an **alpha particle (α).** The loss of an alpha particle forms a new nucleus, called the daughter. The **daughter nucleus** has a nucleon number that is four less than the parent nucleus and an atomic number that is two less. Alpha decay is represented by the general nuclear equation

$$\underset{\substack{\text{parent}\\\text{nucleus}}}{{}^{A}_{Z}X} \longrightarrow \underset{\substack{\text{daughter}\\\text{nucleus}}}{{}^{A-4}_{Z-2}Y} + \underset{\substack{\text{alpha}\\\text{particle}}}{{}^{4}_{2}He}$$

where X is the parent element and Y is the newly formed daughter element. Some examples of alpha decay are

Uranium-238 decays to form thorium-234:

$$^{238}_{92}U \longrightarrow {}^{234}_{90}Th + {}^{4}_{2}He$$

Radium-226 decays to form radon-222:

$$^{226}_{88}Ra \longrightarrow {}^{222}_{86}Rn + {}^{4}_{2}He$$

The numerous alpha particles emitted from a sample of a radioactive substance are called **alpha radiation.** The alpha particles radiate from the sample in all directions.

A nuclear equation is not the same as a chemical equation but does represent the balance or conservation of nucleons characteristic of that nuclear process. In a **nuclear equation** the sums of the nucleon numbers on each side of the arrow are equal, and the sums of the atomic numbers are equal.

ACTIVITY
15-3

In each of the two nuclear equations given here confirm that the number of nucleons is conserved by adding the mass numbers and atomic numbers on each side of the arrow. What is the difference between the parent and daughter in each case. Locate the parent and daughter in the periodic table.

## Beta Particle Decay

Some nuclei decay by emitting a high-speed electron called a **beta particle (β).** The daughter nucleus formed by beta decay has the same nucleon number as the original nucleus and an atomic number that is one greater.

$$\underset{\substack{\text{parent}\\\text{nucleus}}}{{}^{A}_{Z}X} \longrightarrow \underset{\substack{\text{daughter}\\\text{nucleus}}}{{}^{A}_{Z+1}Y} + \underset{\substack{\text{beta}\\\text{particle}}}{{}^{0}_{-1}e}$$

Beta particles are electrons that come from nuclei. During beta decay it appears that a neutron in the nucleus decays to form a proton and an electron. The electron is ejected as a beta particle and the proton is absorbed by the nucleus. Thus beta decay of a nucleus decreases the number of neutrons by one and increases the number of protons by one. Some examples of beta decay are

Plutonium-241 decays to give americium-241:

$$^{241}_{94}Pu \longrightarrow {}^{241}_{95}Am + {}^{0}_{-1}e$$

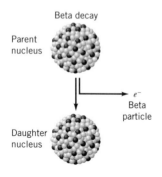

Beta decay

Parent nucleus

$e^-$
Beta particle

Daughter nucleus

Bismuth-210 decays to give polonium-210:

$$^{210}_{83}\text{Bi} \longrightarrow {}^{210}_{84}\text{Po} + {}^{0}_{-1}e$$

The numerous beta particles emitted from a sample of a radioactive substance are called **beta radiation.** The beta particles radiate from the sample in all directions.

### Gamma Ray Emission

Sometimes the daughter nucleus formed in an alpha decay or a beta decay is produced in an energetically excited state. It is in a higher than normal energy state. This excited energy state is not the normal state of the nucleus and it quickly drops to a lower energy state. In the process the nucleus emits a form of radiant energy called a **gamma ray ($\gamma$).** This is similar to the emission of radiant energy by excited electrons in energized atoms as discussed in Chapter 6. The difference is that excited nuclei emit much-higher-energy radiation than excited atoms. Gamma rays have energies similar to but higher than X rays. The emission of gamma radiation by a radioactive substance can accompany the alpha radiation or beta radiation produced by the decay of nuclides of the substance.

In summary, the three major types of radiation are alpha radiation, beta radiation, and gamma radiation. Nuclides that are **alpha emitters** produce alpha radiation ($\alpha$). Some alpha emitters also produce gamma radiation ($\alpha$, $\gamma$). Nuclides that are **beta emitters** produce beta radiation ($\beta$). Some beta emitters also involve gamma radiation ($\beta$, $\gamma$). Some examples are:

Thorium 230 decays by alpha and gamma emission:

$$^{230}_{90}\text{Th} \longrightarrow {}^{226}_{88}\text{Ra} + {}^{4}_{2}\text{He} + \gamma$$
$$\text{thorium-230} \qquad\qquad\qquad\qquad \text{alpha} \qquad \text{gamma}$$

Lead 210 decays by beta and gamma emission:

$$^{210}_{82}\text{Pb} \longrightarrow {}^{210}_{83}\text{Bi} + {}^{0}_{-1}e + \gamma$$
$$\text{lead-210} \qquad\qquad\qquad\qquad \text{beta} \qquad \text{gamma}$$

### 15-4 COMPLETING NUCLEAR EQUATIONS

The fact that the nucleon numbers and atomic numbers are equal in a nuclear equation can be useful. If a parent nucleus is known to decay by a specific decay mode, we can predict the daughter. It is important to note, however, that a beta particle has a negative charge. When a beta particle occurs in a nuclear equation its atomic number is treated as though it is $-1$. For example, hydrogen-3, the radioactive isotope of hydrogen, decays by beta particle, ${}^{0}_{-1}e$, emission. What is the daughter produced? The incomplete nuclear equation is given on the next page.

$$_1^3H \longrightarrow _{-1}^0e + ?$$

The product must have a nucleon number of 3 to equal the 3 on the left and its atomic number must be 2, since 2 − 1 on the right would equal 1 on the left. The equation becomes

$$_1^3H \longrightarrow _{-1}^0e + _2^3$$

Consulting a list of elements for atomic number 2 we see that the product must be an isotope of helium having the symbol $_2^3He$.

$$_1^3H \longrightarrow _{-1}^0e + _2^3He$$

---

**EXAMPLE 15-1**

Uranium-235 decays by alpha emission. Consulting a list of elements for the atomic number of uranium and using $_2^4He$ for an alpha particle, the incomplete nuclear equation is

$$_{92}^{235}U \longrightarrow _2^4He + ?$$

Since the sum of the nucleon numbers of the products must be equal to the nucleon number of the parent, the daughter product has a nucleon number of 231 (235 − 4). Furthermore, since the sum of the atomic numbers must be equal on each side, the daughter has an atomic number of 90 (92 − 2). Consulting a list of elements for atomic number 90, the daughter product must be an isotope of thorium having the symbol $_{90}^{231}Th$. Thus, the complete nuclear equation is

$$_{92}^{235}U \longrightarrow _2^4He + _{90}^{231}Th$$

---

**ACTIVITY 15-4**

Complete the following incomplete nuclear equations:

(a) $_{94}^{241}Pu \longrightarrow _{95}^{241}Am + $ ?

(b) $_{88}^{226}Ra \longrightarrow$ ? $ + _2^4He$

---

**15-5 HALF-LIFE**

Radioactive decay is random and spontaneous. It is not possible to stop it or prevent it. Such decay does follow a pattern that depends upon the total number of radioactive nuclei in the sample. Radioactive decay is somewhat similar to the popping of corn. Imagine a collection of hot corn kernels. They begin to pop and continue to pop over time until the last few kernels pop. Of course, the nuclei of a radioactive isotope are not heated like popcorn. They spontaneously decay and continue to decay over a time until they all have decayed. When popping corn, you may have noticed that at first many kernels pop, then the frequency of pops decreases as the number of kernels decreases. In a radioactive sample the frequency of decays is great at first, then the activity decreases as the number of nuclei decreases. The rate of decay is proportional to the number of nuclei, so the frequency is great at the beginning then tapers off.

A sample of a radioactive element will continue to decay over time until all the nuclei have decayed. For some nuclides this will happen within seconds, minutes, hours, or days. For other nuclides it may require months, years, or thousands of years for total decay. The rate of decay is a characteristic of a given radioactive nuclide. A common way to express the rate of decay of a radioactive element is in terms of its half-life.

**Half-life** is the time required for the decay of one-half of a sample of a radioactive substance. For instance, if we start with 10 grams of radon-222, after four days 5 grams will remain, after 8 days 2.5 grams will remain, and so on. The amount of radon-222 decreases by one-half every four days so the half-life is 4 days. Figure 15-1 illustrates the idea of a half-life. The half-lives of nuclides vary widely, ranging from fractions of seconds to billions of years. Each radioactive nuclide has a characteristic half-life. Table 15-3 lists a few typical half-lives.

The half-lives of nuclides are measured by experiment. The radioactivity of a sample of an element is found by periodically measuring the activity (in terms of the number of decays per unit time) of a sample over an appropriate period. As illustrated in Figure 15-2, the half-life can be found from a plot of the activity versus time. As a sample decays the activity is great when there are numerous radioactive nuclei and the activity decreases as the number of radioactive nuclei decreases. The decrease in activity with time follows a pattern as illustrated in Figure 15-2. The half-life is deduced from the pattern. Incidentally, special experimental methods are needed to measure half-lives that are very long or very short. Chemistry in Action 15-1 is about measuring the half-life of water leaking from a small hole pierced in a bottle.

Refer to your plot of the experimental data for the water bottle exercise in Chemistry in Action 15-1. What is the experimental half-life of your leaking bottle. In this case the half-life is the time required to drain one-half of the initial amount of water from the bottle.

---

## Carbon-14 Dating

Carbon-14 is a radioactive isotope of carbon having a half-life of 5680 years. It is formed in the atmosphere by a nuclear process involving atmospheric nitrogen. This process is a consequence of cosmic radiation striking nitrogen in the earth's atmosphere. Cosmic radiation contains high-energy alpha particles and protons that continually enter the earth's atmosphere from outer space (from the cosmos). As a result carbon-14 is continually formed by cosmic radiation and since it decays over time there is a constant steady-state concentration of carbon-14 in the atmosphere. The carbon-14 combines with oxygen to make some radioactive carbon dioxide in the atmosphere. It appears that the concentration of radioactive carbon dioxide in the atmosphere has been essentially constant for thousands of years. Since plants use carbon dioxide in photosynthesis, all plants contain small amounts of carbon-14 distributed among the carbon-containing compounds. Moreover, since animals eat plants, all animals also contain small amounts of carbon-14 incorporated in carbon-containing compounds. In fact, all living plants and animals have a steady-state amount of carbon-14 since as some decays more is assimilated by photosynthesis or using plants as food. This fact has been experimentally confirmed by measuring the amount of carbon-14 per gram of recently living plant and animal tissue.

Once a plant or animal dies no additional carbon-14 is added by life processes. The carbon-14 present at death decays and slowly decreases the amount of carbon-14 in the dead organism. Carbon-14 dating is a method of estimating the age of an ancient organic object by measuring the amount of carbon-14 per gram of the object and comparing this to the amount present in a living organisms. For instance if the carbon-14 content of an object is found to be one-half the content in living organisms the object must be about 5580 years old, which is one half-life for carbon-14. The carbon-14 content of an object is found by carefully measuring the activity of the sample due to this radioactive isotope.

Carbon-14 dating has been compared to dates obtained from tree rings and other objects of known age and has been found to be reasonably accurate. Dating in this way has been a useful tool for estimating the age of many samples of organic matter found in archaeological sites and objects of historical interest. Some examples of the age of objects found by carbon-14 dating are tabulated here.

|  | AGE (YEARS) |
| --- | --- |
| Charcoal from Lascaux cave, France | 15,516 |
| Hair from ancient Egyptian burial site | 5,744 |
| "Ice man" found in the Alps | 5,300 |
| Wood from an Egyptian tomb | 2,190 |
| Dead Sea Scrolls (Book of Isaiah) | 1,917 |
| Ancient corncobs from Texas site | 1,553 |
| Indian House Pit in Oregon | 430 |

## 15-6 RADIOACTIVITY LEVELS

When working with a radioactive sample, it is important to know how radioactive it is or how "hot" the sample is. The **becquerel, Bq,** is the unit of radioactivity corresponding to one radioactive decay per second. This unit is named in honor of Antoine Becquerel, the French physicist who first discovered nuclear radiation. The curie is another unit used to express intensity of radioactivity. A **curie** is a unit of activity equal to 37 billion decays per second, or $3.7 \times 10^{10}$ Bq. It is named in honor of Marie Curie, the Polish chemist who was a pioneer in research on radioactivity. She also was the winner of two Nobel Prizes in science and was the only person to have a daughter who also won the Nobel Prize.

The actual amount in grams of a radioactive material that has an activity of 1 curie is related to the rate of decay of a radioactive element. For example, 1 gram of radium-226 will produce 1 curie of radiation. A curie

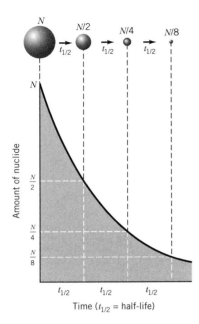

FIGURE 15-1 To understand half-life, start with a given amount of radioactive nuclide, $N$. One-half the amount remains after one half-life has passed. One-fourth remains after two half-lives have passed. The halving continues until the sample has completely decayed.

is a relatively large amount of radioactivity. Normally, for safety, samples with **millicurie** ($10^{-3}$ curie) and **microcurie** ($10^{-6}$ curie) activities are used in laboratory work. Radioactive materials must be used with care and appropriate shielding.

As discussed in Section 15-10 nuclear radiation is dangerous to living systems. Special units are used to refer to radiation doses as they relate to humans. One **rad** (radiation absorbed dose) is the radiation dose that

Carefully take the cover off of a battery-operated smoke detector. Read the label and record any information about the amount of radioactive material used in the detector. The isotope americium-241, used in smoke detectors, is an alpha emitter. Write a balanced nuclear equation showing its alpha decay and identify the daughter isotope. **Caution: Never dismantle or take apart the pieces of a smoke detector.**

***Table 15-3*** Some Typical Half-Lives

| NUCLIDE | SYMBOL | HALF-LIFE |
|---|---|---|
| Uranium-238 | $^{238}_{92}U$ | 4.5 billion years |
| Plutonium-239 | $^{239}_{94}Pu$ | 24,400 years |
| Carbon-14 | $^{14}_{6}C$ | 5680 years |
| Americium-241 | $^{241}_{95}Am$ | 458 years |
| Cobalt-60 | $^{60}_{27}Co$ | 5.3 years |
| Iodine-125 | $^{125}_{53}I$ | 60 days |
| Radon-222 | $^{222}_{86}Rn$ | 3.8 days |
| Iodine-123 | $^{123}_{53}I$ | 13.3 hours |
| Californium-242 | $^{242}_{98}Cf$ | 3.7 minutes |
| Lawrencium-256 | $^{256}_{103}Lr$ | 8 seconds |
| Polonium-214 | $^{214}_{84}Po$ | 0.00016 second |

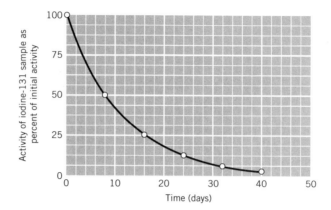

**FIGURE 15-2** A plot of the activity of an iodine-131 sample versus time.

transfers $1 \times 10^{-5}$ J of energy per gram of tissue. The effect of radiation depends on the type of radiation as well as its energy. The rem (radiation equivalent human) is used as a unit of radiation dose. A **rem** value is determined by measuring the dose in rad and multiplying by factors related to the kind of radiation (i.e., alpha, beta, gamma, X rays or neutrons). It is estimated that the minimum lethal dose for humans (50% would die in 30

**FIGURE 15-3** Uranium-238 radioactive decay series. The half-lives of the nuclides are shown along with the mode of decay (y = years, d = days, h = hours, m = minutes, and s = seconds).

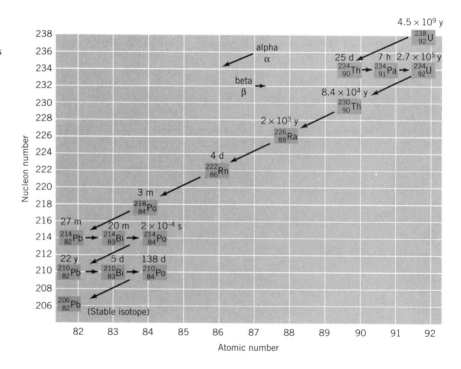

days) is 590 rem. Fortunately, such high doses of radiation are rare. Unfortunately, many victims of the use of atomic bombs in the Japanese cities of Hiroshima and Nagasaki and victims of the Chernobyl nuclear accident (see Section 15-21) were exposed to lethal rem doses and died of radiation exposure. A typical American is exposed to less than 0.2 rem in an entire year. This exposure is mainly from low-level environmental sources. All humans are constantly exposed to nuclear radiation from cosmic radiation, naturally occurring isotopes, radioactive materials that have entered the environment from nuclear power plants, and fallout from nuclear weapons testing. This low-level but relatively constant environmental radiation is called **background radiation.**

## 15-7 RADIOACTIVE ELEMENTS

Nearly 330 different nuclides are found in nature. These nuclides represent the naturally occurring isotopes of the elements. Some elements have only one naturally occurring nuclide (e.g., fluorine: $^{19}_{9}F$), whereas other elements have many natural nuclides (e.g., tin: $^{112}_{50}Sn$, $^{114}_{50}Sn$, $^{115}_{50}Sn$, $^{116}_{50}Sn$, $^{117}_{50}Sn$, $^{118}_{50}Sn$, $^{119}_{50}Sn$, $^{120}_{50}Sn$, $^{122}_{50}Sn$, $^{124}_{50}Sn$). Most natural nuclides are stable and not radioactive. Recall that all the isotopes of the 26 elements of atomic number greater than bismuth are radioactive. A few naturally occurring nuclides of the other elements, such as $^{204}_{82}Pb$ and $^{40}_{19}K$, are radioactive.

Radioactive nuclides are unstable and undergo decay to form other radioactive nuclides or stable nuclides. This means that the daughter of a decay is often radioactive, although in some cases the daughter is stable. Studies of the naturally occurring elements have revealed that all radioactive nuclides of atomic number greater than 82 belong to one of three **decay chains or series.** In other words, these elements decay by alpha or beta emission according to definite patterns. Figure 15-3 shows the radioactive decay series that starts with uranium-238, the most common uranium isotope. In this series, uranium-238 decays into thorium-234, which, in time, decays into another radioactive daughter, and so on, until stable lead-206 forms. Other natural radioactive nuclides belong to other series.

Knowledge of a radioactive decay series provides a way to estimate the age of the earth. Assuming that uranium-238 formed when the earth formed, a sample of undisturbed rock containing uranium-238 would then contain a certain amount of lead-206, the end product of the uranium-238 decay chain. The amount of lead-206 would depend on how long the uranium had been decaying. Since the half-life of uranium-238 is $4.5 \times 10^9$ years, a measurement of the amounts of uranium-238 and lead-206 suggests the age of the rock. In one half-life, 1 g of $^{238}_{92}U$ would decay to about 0.4 g of $^{206}_{82}Pb$, and 0.5 g of $^{238}_{92}U$ would remain. By using this and similar methods of rock dating an estimation of the age of the earth is 4.5 billion years.

### 15-8 IONIZING RADIATION

The emission of radiation is a characteristic of radioactivity. The emitted radiation is used to detect and identify radioactive substances, however, the radiation can be very dangerous to living organisms. The particles given off during alpha or beta decay are high-speed, high-energy particles. These particles are ejected from the decaying nucleus and travel outward into the material surrounding the radioactive substance. Alpha or beta particles lose their energy by interacting with the atoms and molecules of matter. They collide with atoms or molecules causing them to lose electrons, as illustrated in Figure 15-4. In other words, these particles lose their energies by causing atoms and molecules to ionize. The electron and positive ion produced by such an interaction are called an **ion pair.**

An alpha or beta particle passing through matter causes the formation of numerous ion pairs. The more massive alpha particles ($_2^4$He) are less penetrating than beta particles ($_{-1}^0 e$), but they produce more ion pairs. A typical alpha particle travels about 6 cm in the air and produces about 40,000 ion pairs, whereas a typical beta particle travels 1000 cm in the air and produces only about 2000 ion pairs. When these particles pass through materials more dense than air, they still produce ion pairs but travel shorter distances. Gamma radiation consists of electromagnetic radiation similar to X rays. Gamma rays are not massive like alpha and beta

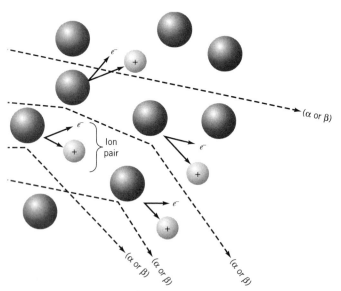

**FIGURE 15-4**    The formation of ion pairs occurs when ionizing radiation passes through matter. Some of the atoms and molecules are ionized, forming ion pairs. One ion pair consists of a separate electron and positive ion.

particles, but they also interact with matter to form ion pairs. Gamma rays are much more penetrating than alpha or beta radiation and can pass through dense samples of matter.

Because radiation causes the formation of ion pairs in the matter through which it passes, it is termed **ionizing radiation.** Other less common types of nuclear decay processes produce radiation in the form of X rays, and some produce high-energy, free-moving neutrons. These types of radiation also form ion pairs when they penetrate matter. **X rays** are less penetrating than gamma rays and **free neutrons** are an extremely penetrating form of radiation.

## 15-9 DETECTION OF RADIATION

We cannot see, hear, or feel nuclear radiation. It is interesting, however, that some very radioactive samples have elevated temperatures because of the energy released by radioactive decay. Used fuel rods from nuclear reactors, for example, are stored in water to keep them from getting too hot. Methods of detecting radiation are based on the ionizing effects of radiation. A typical device used to detect radiation is the **Geiger–Müller tube,** as illustrated in Figure 15-5. The tube consists of a metal cylinder with a thin plastic window on the end. The tube is filled with a special gas and has a thin metal rod in the center. Using an external source of power, a voltage difference is maintained between the rod and the cylinder. When a particle of ionizing radiation enters the tube, some ion pairs form. The electrons formed in the ionization are strongly attracted to the positively charged center rod. They move toward the rod with great speed, and in the process, they act essentially like beta particles and produce more ion pairs. This forms a relatively large number of electrons, which quickly flow toward the center rod.

The flow of electrons is equivalent to a small current passing through the tube. It stimulates an electronic counter, connected to the tube, which recognizes the current flow and registers it as an ionizing event caused by a particle of radiation. Each time a particle enters the tube and causes the electron "avalanche," the counter will tally the event. In this way, the Geiger–Müller tube and the counter provide a way to measure the **activity** of a radioactive substance. Silicon based solid state detectors are sometimes used as alternatives to Geiger-Müller tubes.

The number of decays per second or the number of decays per minute associated with a radioactive sample is measured with a radiation counter. This is one way in which the half-life of a radioactive nuclide is found. Radiation counters are also used to detect the presence of radioactive elements and to find the amount of a radioactive element in a sample. Geiger counters are very useful, since nuclear radiation cannot be seen or felt. Geiger counters provide a way to "see and hear" radioactivity. In 1993 a small sample of cesium-137 was lost on a California freeway. The sample was located using a radiation detector.

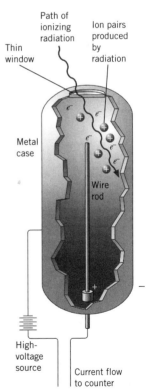

**FIGURE 15-5** The operating principle of a Geiger–Müller tube.

CAUTION

RADIOACTIVE
MATERIAL

The
International
symbol for
radiation

## 15-10 RADIATION DANGER

Why is radiation dangerous? Gamma rays, X rays, and high-speed alpha and beta particles can penetrate living tissue and cause damage in varying degrees. Billions upon billions of high-speed particles and the accompanying gamma radiation can be produced by a few milligrams of a radioactive substance. The radiation moves outward in all directions and penetrates any matter surrounding the substance. The radiation interacts with the matter and loses its energy by forming numerous ion pairs. Ionizing radiation can cause damage to biological materials since ion pairs within tissue cells can form altered and often highly reactive chemicals that react with other cell chemicals. These reactions can disrupt normal cell chemistry and can cause permanent cell damage. Sometimes ionizing radiation damages cell chromosomes. Chromosome damage may cause the cell to malfunction or die. Occasionally damage to chromosomes can cause cells to become cancerous, which results in tumor formation. Too much radiation can induce cancer.

Ionizing radiation can damage developing embryos and cause mutations in cells of the gonads (i.e., egg or sperm cells). Mutations result from damage to the genes of the chromosomes. The abnormal chromosomes are passed on to any children born from the cells. **Gene mutation** in these children can result in sickness, cancer, and even death. Of course, the amount of damage to cells caused by radiation depends upon the amount of exposure that occurs. Obviously **low doses** of radiation are not as dangerous as **high doses** and **short-term exposure** is less dangerous than **long-term exposure.** When you get a medical X ray this is a short-term exposure. X-ray technicians, on the other hand, have to protect themselves from long-term exposure. Obviously you should avoid any unnecessary exposure to sources of radiation. Exposure to high levels of radiation may cause **radiation sickness** with symptoms of nausea, vomiting, and weakness. This is followed by a period characterized by fatigue, weight loss, fever, diarrhea, internal bleeding, and loss of hair. If too much damage is done to the body, a person can die of radiation sickness. Many people exposed to radiation during the World War II nuclear bomb attacks on the Japanese cities of Hiroshima and Nagasaki died of radiation sickness. Many more died from the explosions and fires caused by the bombs.

Alpha particles cannot penetrate skin and are not dangerous externally. However, if you eat or inhale an alpha emitter, it can cause severe damage inside the body. Beta particles can penetrate the skin to some extent and may cause skin burns. Normal clothing serves to shield against exposure to beta particles. Gamma radiation, X rays, and neutrons can penetrate the body and thus are the most harmful type of external radiation. It is important to guard against undue exposure to these forms of penetrating radiation.

Since radioactive substances can be dangerous they must be handled with special equipment and stored in protective, shielded containers. For

Why should pregnant women avoid X rays especially in the early stages of pregnancy? Call your local hospital or medical clinic and ask about their X-ray policy for pregnant women.

example, radioactive wastes at nuclear power plants are stored in thick concrete-lined pools of water. How long a radioactive sample needs to be stored depends upon the half-lives of the nuclides present. After about **10 half-lives** have passed the radioactivity of most nuclides will have dropped to a negligibly low level. Consequently, a general guideline is that after 10 half-lives it is possible to dispose of a sample of a radioactive element. For example, the half-life of iodine-131 is 8.1 days, so a sample of this element needs to be stored for more than 81 days before disposal. Of course, the safe disposal of a radioactive sample also depends upon the type of element involved and the amounts of these elements. As discussed in Section 15-19 radioactive wastes from nuclear reactors must be stored for thousands of years before they are safe. Chemistry in Action 15-2 uses the successive halving of a piece of paper to illustrate the reduction in activity during consecutive half-lives.

Suppose you have 1000 grams of a radioactive element. Calculate the number of grams that remain after each half-life for 10 half-lives. Also calculate the percentage of the original sample that remains after each half-life. To calculate simply divide by 2 ten times. For instance, after the first half-life 500 grams remain as 1000/2 = 500. This is 50% of the original sample because (500/1000) × 100% = 50%.

## Radon in the Home

Rocks and soils contain small amounts of uranium and other radioactive elements that are daughters in the uranium decay series. These isotopes are partly responsible for background radiation and they normally represent no hazard. However, one of the daughters of uranium, **radon-222,** is a gas and as a gas it can diffuse through soils and porous and cracked rocks. Furthermore, it can dissolve somewhat in water and accumulate in water wells. As a result, radon can accumulate in homes and buildings by diffusing through cracks and openings in foundations and through water pipes.

Radon-222 is an alpha emitter with a half-life of 3.8 days. It can potentially be harmful if inhaled. More significantly it decays to give daughters that are various radioactive isotopes of polonium, bismuth, and lead. These radon daughters are solids and small amounts can settle out in the environment and become attached to dust and smoke. When inhaled these isotopes can lodge in the lungs and decay to produce alpha, beta, and gamma radiation. In this way radon represents a lung cancer risk and is considered to be the leading cause of lung cancer among nonsmoking Americans.

Dangerous levels of radon are not common in most homes but can be significant in certain regions. Radon levels are highest in geological areas with granite and black shale covered by porous soils and where there are relatively high levels of uranium. Areas that have been found by the U.S. Environmental Protection Agency to be of particular risk are eastern Pennsylvania, eastern Washington, Colorado, New Mexico, western Florida, and parts of Maine, New Jersey, and New York.

Radon levels can be measured using commercially available test kits or by the use of special instruments. Typically, indoor levels of radon are less than 1 picocurie per liter of air. A **picocurie** is $10^{-12}$ curie or a trillionth of a curie. If radon levels exceed 4 picocuries per liter some remedial action is appropriate such as patching foundation cracks, capping sump pumps, and increasing ventilation with fresh air.

**ACTIVITY 15-9**

Radon-222 is an alpha emitter. Write a nuclear equation to show the alpha decay of radon and identify the daughter. Write a nuclear equation for the alpha decay of the daughter and identify its daughter.

## 15-11 NUCLEAR TRANSMUTATIONS

A high-speed nuclear particle can, under certain conditions, collide with a nucleus and cause a nuclear reaction that produces a different nucleus. **Transmutation** is the name given to such a nuclear reaction. In a transmutation, starting nuclei are changed to new nuclei. Some transmutations occur naturally, but most are done in nuclear laboratories using special techniques and equipment. An example of a transmutation is

$$^{14}_{7}\text{N} \quad + \quad ^{1}_{0}\text{n} \quad \longrightarrow \quad ^{14}_{6}\text{C} \quad + \quad ^{1}_{1}\text{H}$$

target    projectile    product nucleus    particle

In this reaction, the neutron is the projectile, and the $^{14}_{7}\text{N}$ is the target nucleus. A transmutation involves a collision of a **projectile particle** with the **target nucleus** and results in new combinations of neutrons and protons as products.

Nuclear scientists have used nuclear transmutations as a means of preparing artificial nuclides. These are **human-made isotopes.** Nearly all of the naturally occurring nuclides have been used as targets, and a variety of nuclear particles, such as protons ($_1^1H$), deuterons ($_1^2H$), neutrons ($_0^1n$), alpha particles ($_2^4He$), and beta particles ($_{-1}^0e$) have been used as projectiles. The projectiles are given relatively high kinetic energies before being projected at the target nuclei. Particle accelerating devices such as cyclotrons, linear accelerators, synchrotrons, and nuclear reactors produce these high-energy projectiles. Special equipment is needed to produce human-made isotopes. All of the elements having atomic numbers greater than uranium (92) do not occur in nature. These transuranium elements (93 to 109) have been synthesized by nuclear scientists.

## 15-12  NUCLEAR FISSION

Some nuclear decay processes produce neutrons as products. The neutrons produced by these processes serve as projectiles in other nuclear transmutations. In 1938 the physicists Otto Hahn and Lise Meitner made a startling report about their research on the use of neutrons as projectiles with uranium target nuclei. This research showed that the isotope uranium-235 undergoes a wholly different kind of transmutation. Upon the capture of a neutron the uranium-235 nucleus forms an unstable nucleus that splits into two nuclei of smaller size plus a few neutrons. **Nuclear fission** is the name given to this unique type of transmutation. The fission of uranium-235 is represented by the general nuclear equation:

$$_{92}^{235}U + _0^1n \longrightarrow _{36}^{85}Kr + _{56}^{148}Ba + 3_0^1n + energy$$

This example is just one of many possible fission reactions of uranium-235. When uranium-235 undergoes fission the unstable nucleus can split in a variety of ways. Therefore, fission produces a mixture of **fission product nuclei.** As a simple analogy, if you tried to break many pieces of chalk in half not all pieces would be the same size. You would have an assortment of chalk "fission products." A specific fission event forms two fission product nuclei and some neutrons. The possible product nuclei of the fission of uranium-235 range in nucleon number from about 70 to 165. Most nuclei formed by fission are radioactive and some are stable. Thus, the collection of fission products or collection of elements produced by fission is highly radioactive.

Only a few kinds of isotopes undergo fission. These are the **fissile isotopes.** The most important fissile isotopes are **uranium-235** and **plutonium-239.** Uranium-235 is a rare natural isotope of uranium. It makes up

ACTIVITY
**15-10**

To simulate fission you need some commercial bubble-making solution and two wire or plastic bubble-making loops. Make a bubble and catch it between the two loops. Gently pull the bubble apart by moving the loops and watch it undergo fission. Bubbles around 1 inch in diameter work best. Note the sizes of the two bubbles formed by fission. Repeat the making and fission of bubbles several times and note the varying sizes of the fission fragments.

about 0.7% of natural uranium. The most common isotope of uranium is uranium-238. Plutonium-239 does not occur in nature but it is made in nuclear reactors as discussed in Section 15-22.

There are two important characteristics of fission. First, as indicated in the example of uranium-235, fission always gives neutrons as products. In fact, normally two or more neutrons form for every one neutron used to initiate a fission. Second, when fission occurs, more stable product nuclei form from the less stable parent nucleus. This change from less stable to more stable nuclei produces a large amount of energy. This energy represents **nuclear energy** that accompanies fission. The complete fission of 1 gram of uranium-235 can produce about 80 billion joules of energy. This means that the complete fission of 1 kg of uranium-235 produces as much energy as the burning of about 2600 tons of coal or 14,000 barrels of crude oil. The energy of fission is the basis for the production of atomic energy or nuclear energy.

## 15-13 ENERGY FROM FISSION

During fission, smaller more stable nuclei form when energy is released. What is the source of this nuclear energy? Consider a typical fission of uranium-235 and the masses of the nuclei and particles involved.

$$^{235}_{92}U \ + \ ^{1}_{0}n \ \longrightarrow \ ^{94}_{38}Sr \ + \ ^{139}_{54}Xe \ + \ 3\,^{1}_{0}n$$

relative
masses    235.0439    1.0087        93.9154        138.9178    3(1.0087)

Adding the masses on each side of the equation gives 236.0526 on the left and 235.8593 on the right. In nuclear fission loss of mass occurs as it changes to energy. This is the basis of nuclear energy. In the above fission process, when 235.0439 grams of uranium-235 undergo fission the loss in mass is the difference between the initial and final masses (236.0526 g − 235.8593 g = 0.1933 g). In a sense, mass is a form of energy. Energy is released in nuclear transmutations and nuclear decays because mass is changed to energy. The law of conservation of matter states that mass is conserved in normal chemical processes. Nuclear decay and transmutations are not normal chemical processes. Mass is lost during these nuclear processes and the loss manifests as energy.

The relation between mass and energy is given by the famous formula first stated by Albert Einstein.

$$E = mc^2$$

where $m$ is the mass in kg and $c$ is the speed of light in m/s. The speed of light refers to the speed at which radiant energy passes through empty space. The formula is used to calculate amounts of energy produced by nuclear processes.

**EXAMPLE 15-2**

In the fission of a 0.235-kg sample of uranium-235, $1.9 \times 10^{-4}$ kg of mass is converted to energy. If the speed of light is $3.00 \times 10^{8}$ m/s and 1 J = 1 kg m$^2$/s$^2$, how many joules of energy are produced by the fission?

$$E = mc^2$$

To obtain the energy in joules we need to use the mass in kilograms and the speed of light in meters per second.

$$E = mc^2 = (1.9 \times 10^{-4} \text{ kg}) (3.00 \times 10^8 \text{ m/s})^2 = (1.9 \times 10^{-4} \text{ kg}) (9.00 \times 10^{16} \text{ m}^2/\text{s}^2) = 1.7 \times 10^{13} \text{ kg m}^2/\text{s}^2 \text{ or } 1.7 \times 10^{13} \text{ J}$$

To put this number in perspective note that it is equivalent to the energy released by burning about 20 million moles of methane gas.

What is the mass in grams of 20 million moles of methane? How many tons is this mass?

## 15-14 FISSION CHAIN REACTIONS

Neutrons are always products of the fission of uranium-235. The neutrons formed during fission can cause other uranium-235 nuclei to undergo fission. Each fission produces more neutrons, which in turn can cause further fission. In small samples of uranium-235 many neutrons are ejected into the surroundings. The fission that does occur in small samples is under control. As shown in Figure 15-6, if a sample contains enough ura-

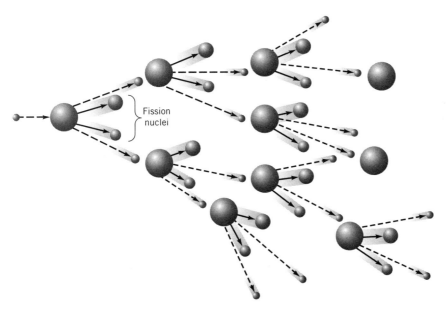

Fission nuclei

**FIGURE 15-6** In a nuclear chain reaction the neutrons produced by a fission event can cause fission of other nuclei. Since each fission event typically produces twice the number of neutrons needed for one fission, a nuclear chain reaction can occur within a large collection of fissionable nuclei.

Schematic for
a fission bomb

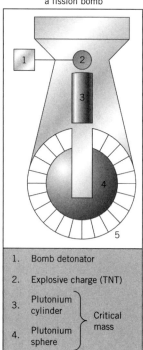

1.  Bomb detonator

2.  Explosive charge (TNT)

3.  Plutonium
    cylinder          } Critical
                        mass
4.  Plutonium
    sphere

5.  Neutron reflector

nium-235 nuclei, a **fission chain reaction** can occur. In a chain reaction one fission step gives neutrons that cause further fission steps, these can cause still others, and so on. As long as, on the average, one of the neutrons formed by a specific fission causes another fission, a fission chain reaction becomes self-propagating and self-sustaining.

The minimum amount of fissionable material in which a fission chain reaction can be self-sustaining is the **critical mass.** The critical mass of pure uranium-235 is less than 1 kg when the uranium sample is spherical for maximum fission efficiency. It is also important to note that not all neutrons produced in fission will cause further fission. Some of the very-high-energy neutrons will escape from the sample and others will not cause fission until they lose some kinetic energy. Neutrons of lower kinetic energy, called **thermal neutrons,** have a greater chance of causing further fission events.

When a fission chain reaction becomes self-sustaining, it is said to be **critical** and it will continue as long as fissionable nuclei are present. Under certain conditions a fission chain reaction becomes uncontrollable and produces a tremendous number of fissions in a short period. This type of **supercritical** condition can result in an explosion and a release of large amounts of energy. This is the basis for the construction of **atomic bombs** or **fission bombs** (see margin).

Fission chain reactions are also used in nuclear power plants. In a nuclear reactor the fission occurs at a subcritical or critical level. It is controlled by use of special neutron absorbing materials within the reactor. The nuclear fuel components in a nuclear reactor are designed so that fission in the reactor cannot become supercritical; a reactor cannot explode like a nuclear bomb. However, the components of a reactor can rise to very high temperatures as they absorb fission energy.

## 15-15 NUCLEAR REACTORS

A **nuclear reactor** is a device in which nuclear fission is carried out in a controlled fashion. A **power reactor** converts the energy of fission into electrical energy. Figure 15-7 shows the fundamental components of a nuclear reactor. The **core** of the reactor contains **nuclear fuel rods.** The fuel rods contain fissionable isotopes as enriched uranium.

Recall that natural uranium contains about 0.7% uranium-235. This means that there is about one uranium-235 atom for every 140 uranium-238 atoms. Special enrichment processes increase the percentage of uranium-235 in uranium samples to a level of about 3%. At this level self-sustaining fission in the uranium is possible. Nuclear weapons need uranium enriched to more than 90% uranium-235.

Nuclear fuel pellets are made from the 3% enriched uranium and they

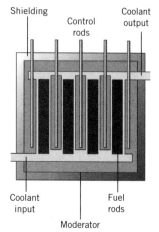

**FIGURE 15-7** The components of a nuclear reactor.

are encased in steel or zirconium alloy rods. The pellets, made of a compound of uranium called uranium oxide, have cylindrical shapes about 1 inch long and 1/2 inch in diameter. The pellets are stacked on one another to make a fuel rod about 1.5 m long. Fuel rods are suspended in the center or core of the reactor. A typical reactor has around 40,000 fuel rods.

The **reactor core** also contains a substance called a **moderator** that serves to slow neutrons produced by fission. Fission neutrons normally have such high energies they do not cause enough further fission events. The moderator slows the neutrons to produce more thermal neutrons for further fission. Also located in the core of the reactor are **control rods.** These rods, made of cadmium or boron steel, absorb neutrons. By moving the rods within the core it is possible to control the number of neutrons. When the control rods are partially removed, the number of neutrons increases to a level at which fission becomes self-sustaining. The reactor becomes critical. Lowering the rods into the core decreases fission to a subcritical level.

A **coolant** circulates within the core to absorb the energy of fission as heat energy. The heated coolant is the energy source used to make electricity. A simple analogy is the coolant in an automobile engine. The coolant absorbs the heat in the engine, is cooled by a radiator, and returned to the engine. In a reactor, the hot coolant is used to make electricity and is cooled before returning to the reactor. Typical power reactors in the United States use water as a coolant and as a moderator. These water-using reactors are sometimes called **light water reactors (LWR).**

Shielding material surrounds the core of the reactor to prevent radiation leakage. Typical **shielding** consists of steel plates and thick layers of concrete. In a power reactor, a thick steel pressure vessel contains the reactor core. A power reactor is housed in a containment structure for safety. The containment structure has concrete walls 3 to 4 feet thick and a steel liner. Typically an **emergency coolant** supply is available in a power reactor. If the primary coolant supply is interrupted in any way, the emergency core cooling system (ECCS) is activated.

## 15-16 REACTORS IN USE

Around 200 nuclear power reactors are in use worldwide. In the United States there are nearly 100 commercially licensed power reactors and 15 or 20 under construction. Nuclear power provides about 16% of the electricity generated in the United States. Many reactors were ordered by electrical utilities in the mid-1960s and early 1970s. However, beginning in the mid-1970s most reactor orders were canceled and there have been no new orders since 1978. The decrease in orders and construction of reactors suggests that possibly only around 100 power reactors will be in use in the

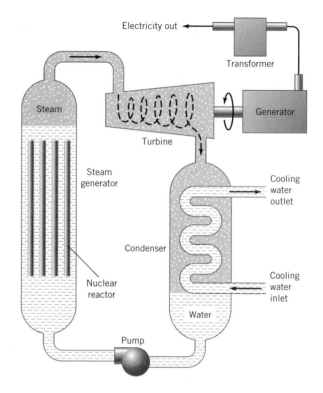

Electricity out ← Transformer

Generator

Steam

Turbine

Steam generator

Nuclear reactor

Condenser

Cooling water outlet

Cooling water inlet

Water

Pump

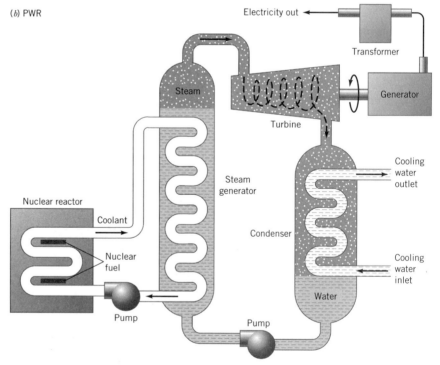

Electricity out ←

Transformer

Generator

Steam

Turbine

Steam generator

Nuclear reactor

Coolant

Nuclear fuel

Condenser

Cooling water outlet

Cooling water inlet

Water

Pump

Pump

United States by the year 2000. The useful lifetime of a power reactor is around 30 years so many reactors in use today will be shut down within about 10 years.

Two types of power reactors are used in the United States. The **boiling water reactor (BWR)** design uses enriched uranium fuel. The core is in a pressurized vessel in which steam forms by the heat of fission (see Fig. 15-8). The steam drives a turbine that generates electricity. In this design, the primary coolant converts to steam to drive the turbine and is condensed to liquid water using external cooling water.

The **pressurized water reactor** or **PWR** also uses enriched uranium fuel and water as the primary coolant (see Fig. 15-8). These reactors run under high pressure so that the primary coolant water does not form steam. The water becomes superheated water with a temperature of about 300°C. The superheated water heats a secondary supply of water so that it becomes steam. The steam runs the turbine. Steam in the secondary water loop is condensed back to liquid water using external cooling water.

Incidentally a typical nuclear power plant is about 30% efficient. This means that 30% of the energy of fission is converted to electricity and the remainder is waste heat. This waste heat is released to the atmosphere and warms the external cooling water. Conventional fossil fuel power plants have greater efficiencies than nuclear power plants.

## 15-17  FUSION

**Fusion** is another type of transmutation process that can produce energy. At high temperatures, it is possible for some lighter nuclei to fuse to form heavier nuclei. Fusion can produce enormous amounts of energy. In fact, the energy produced by the sun comes from fusion.

Nuclear scientists found that the temperatures produced during fission, known as **thermonuclear temperatures,** could cause certain fusion reactions. These fusion reactions involve hydrogen and lithium nuclides. The destructive power of a hydrogen bomb is a result of a very rapid self-sustaining fusion reaction. A hydrogen bomb uses the heat of a uranium or plutonium fission bomb to provide the thermonuclear temperatures needed for fusion. The fusionable isotopes hydrogen-2, deuterium, and hydrogen-3, tritium, are contained in a bomb along with lithium-6. The thermonuclear temperature induces the fusion of the hydrogen isotopes as shown on the next page.

**FIGURE 15-8**    See opposite page. Light water nuclear power plants: (a) boiling water reactor, BWR (b) pressurized water reactor, PWR.

Schematic for
a hydrogen bomb

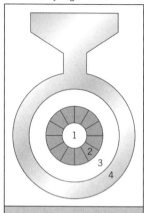

1. $_1^2H$ and $_1^3H$ as lithium
   compounds

2. Plutonium fission bomb

3. $_3^6Li$ $_1^2H$ (lithium deuteride)

4. Metal case

$$_1^2H + {}_1^3H \longrightarrow {}_2^4He + {}_0^1n + \text{energy}$$

$$\text{⊖} + \text{⊖⊖} \longrightarrow \text{⊖⊖} + \text{●} + \text{energy}$$

The neutron formed in this process interacts with lithium-6 to give more tritium.

$$_3^6Li + {}_0^1n \longrightarrow {}_2^4He + {}_1^3H$$

The tritium can then fuse with more deuterium to give a self-sustaining and explosively fast chain reaction that occurs in a fraction of a second.

The fusion process can produce larger amounts of energy for a given amount of material than the fission process. Uncontrolled thermonuclear fusion releases a large amount of energy and accounts for the tremendous destructive power of thermonuclear bombs. Since these **thermonuclear bombs** use hydrogen as fuel they are sometimes called **hydrogen bombs** or **H-bombs.** And since they require a fission bomb to trigger the fusion process they are also called **fission–fusion bombs.** No way has yet been found to carry out a controlled fusion process so that the energy released can be conveniently used. However, controlled fusion is the subject of much research, and someday fusion reactors may be developed to provide vast amounts of low-cost energy using common fusionable isotopes.

## Nuclear Chemistry

The general symbol for a nuclide is

$$_Z^AX$$

where X is the symbol of the element corresponding to the nuclide, A is the nucleon number, and Z is the atomic number.

Radioactivity: The spontaneous decay of a nucleus to form another nucleus and a nuclear particle.

Alpha Particle Decay: A nucleus, called the parent, decays by emitting a high-speed helium 4 nucleus known as an alpha particle ($\alpha$). The loss of an alpha particle forms a new nucleus, called the daughter. The daughter nucleus has a nucleon number that is four less than the parent nucleus and an atomic number that is two less.

| $_Z^AX$ | $\longrightarrow$ | $_{Z-2}^{A-4}Y$ | $+$ | $_2^4He$ |
|---|---|---|---|---|
| Parent nucleus | | Daughter nucleus | | Alpha particle |

Beta Particle Decay: The decay of a nucleus by emission of a high-speed electron called a beta particle (β). The daughter nucleus formed by beta particle decay has the same nucleon number as the parent nucleus and an atomic number that is one greater

$$_{Z}^{A}X \longrightarrow \quad _{Z+1}^{A}Y \quad + \quad _{-1}^{0}e$$

$$\text{Parent} \qquad\qquad \text{Daughter} \qquad\qquad \text{Beta}$$
$$\text{nucleus} \qquad\qquad \text{nucleus} \qquad\qquad \text{particle}$$

Gamma Ray Emission: In some cases the daughter nucleus formed in an alpha decay or a beta decay is produced in an energetically excited state or a higher than normal energy state. This excited energy state is not the normal state of the nucleus, and it quickly drops to a lower energy state. In the process the nucleus emits a form of radiant energy called a gamma radiation (γ). Gamma rays have energies similar to x rays.

Half-life: The time required for the decay of one-half of a sample of a radioactive substance.

Becquerel, Bq: The unit of radioactivity corresponding to one radioactive decay per second.

Curie, Ci: A unit of activity equal to 37 billion decays per second, or $3.7 \times 10^{10}$ Bq.

Rad (radiation absorbed dose): The radiation dose that transfers $1 \times 10^{-5}$ J of energy per gram of tissue. The effect of radiation depends on the type of radiation as well as its energy.

Rem (radiation equivalent human): A unit of radiation dose. A rem value is determined by measuring the dose in rads and multiplying by factors related to the kind of radiation (i.e., alpha, beta, gamma, x rays, or neutrons).

Background radiation: Nuclear radiation from cosmic radiation, naturally occurring isotopes, radioactive materials that have entered the environment from nuclear power plants, and fallout from nuclear weapons testing.

Nuclear transmutation: A process in which a high-speed nuclear particle collides with a nucleus resulting in a nuclear reaction that produces a different nucleus. The high-speed particle is called the projectile, and the nucleus it collides with is called the target.

Nuclear fission: A nuclear process in which a neutron projectile is captured by target nucleus like uranium 235 and plutonium 239. Upon the capture of a neutron an unstable nucleus is formed that splits into two nuclei of roughly equal size plus a few neutrons accompanied by the release of energy.

Energy released in nuclear processes comes from the conversion of mass to energy. The relation between mass and energy is given by the famous formula first stated by Albert Einstein:

$$E = mc^2$$

where $m$ is the mass in kilograms (kg) and $c$ is the speed of light in meters per second (m/s). The speed of light refers to the speed at which radiant energy passes through empty space.

Nuclear power reactor: A device in which nuclear fission is carried out in a controlled fashion and the energy of fission is converted to electrical energy.

Nuclear fusion: A type of high-temperature nuclear transmutation in which some lighter nuclei fuse to form heavier nuclei accompanied by the release of energy.

## 15-18 PROS AND CONS OF NUCLEAR ENERGY

There are several advantages and disadvantages to the wide-scale use of fission reactors for electrical power. Power reactors can provide large amounts of energy and thus can replace conventional fossil fuel sources. At this time, power reactors are a feasible alternative to diminishing supplies of fossil fuels.

Since the fission process is not a combustion process, power reactors do not give off large amounts of waste gases that can become air pollutants. However, some highly radioactive gases, such as krypton-85, form in relatively small amounts during fission. These gases are different from normal air pollutants but are an especially dangerous kind of pollutant with which we have to contend. In fact, the use of power reactors introduces several new kinds of pollution problems.

Impurities in the reactor coolant can become radioactive when exposed to the reactor core. These are **low-level radioactive wastes** and must be disposed of by sealing in special containers that are placed in semipermanent storage. Some low-level radioactive wastes leak into the secondary cooling waters. When this water returns to the source, the radioactivity enters the environment. This kind of radioactive leakage must be monitored continuously.

Figure 15-9 illustrates how radioactive substances enter the food chain and are transmitted to humans. The inhalation of radioactive materials can induce lung cancer. Some radioactive isotopes remain in the body because of their chemical similarity to common body chemicals. **Cesium-137,** a beta emitter of 30-year half-life, is chemically similar to potassium and can become incorporated in all the cells of the body. **Strontium-90,** a beta

**FIGURE 15-9**
Transmission of radioactive substances to humans.

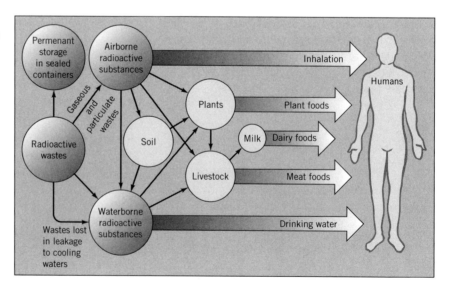

emitter of 28-year half-life, is chemically similar to calcium. Since calcium is a main constituent of bones and teeth, strontium-90 can accumulate in these regions of the body. The result is that the body is continually exposed to radiation from this nuclide.

## 15-19 RADIOACTIVE WASTES

When fission occurs, the fission products produced accumulate in the used or spent fuel rods of the reactor. Periodically **spent rods** must be removed from a reactor and replaced. The fission products in the rods are extremely radioactive and hazardous. They include a variety of nuclides produced as fission products and some plutonium-239 produced from the uranium-238 in the fuel rods. Materials from spent fuel rods are called **high-level radioactive wastes.** Such wastes must be kept in special containers for safety reasons. In 1980 an estimated 6,700 metric tons (1 metric ton = 1000 kg) were in storage in the United States and an estimated minimum of 73,000 metric tons will be in storage by the year 2000. These wastes will represent around 40 billion curies of radiation. Spent fuel rod wastes are currently stored on nuclear reactor sites in continually cooled pools of water. Ideally, such wastes should be processed to remove the uranium and plutonium. In the mid-1990s no spent fuel rod processing facilities were in use in the United States. Even if these elements are removed, the remaining wastes must be safely stored in isolated locations for thousands of years. In 1994 no national high-level storage sites were available in the United States. How high-level nuclear wastes are stored and where they are to be stored is one of the most challenging dilemmas of the use of nuclear energy.

Another aspect of nuclear energy is the problem of **waste heat.** Nuclear power plants are currently less efficient in energy conversion than fossil fuel plants. A conventional nuclear power plant (meeting current safety standards) produces 50% more waste heat than a comparable fossil fuel plant. More efficient nuclear plants are in development. Nevertheless, if nuclear energy begins to substantially replace fossil fuel energy, waste heat will be of greater concern.

## 15-20 REACTOR SAFETY

The fuel rods in the reactor core are the source of heat energy in a power reactor. The primary coolant has the important function of absorbing this heat. If the **primary coolant** supply is disrupted, the temperature of the core can rise dramatically. Theoretically the temperature of the core could reach the level of **core meltdown.** At meltdown the fuel rods would melt and form a high-temperature mass that would maintain high temperatures due to continued fission within it. This mass would melt anything that tried to contain it. Within days it would melt through the core vessel and the concrete containment shielding below it. Then it would sink into the earth and probably be dispersed in a steam explosion when it reached the water table. The resulting explosion could scatter fission products over the environment near the plant site and downwind from the site. The possible penetration into the earth by a melted core is sometimes called the **China Syndrome.** The China Syndrome did not occur during the Chernobyl accident in 1989. However, very hot masses of highly radioactive core material did accumulate beneath the destroyed reactor. See Section 15-21 for a discussion of the Chernobyl accident.

Reactor safety features help prevent a core meltdown and the accidental escape of fission products from the reactor. Nuclear power plants have a series of barriers between the nuclear fuel and the outside environment. A core meltdown is a highly unlikely event and no such disaster has occurred. The probability of such a meltdown accident is extremely low, but if one occurred, the results could be disastrous. In 1979, a partial core meltdown occurred in the **Three Mile Island reactor** located in Pennsylvania. The accident resulted from a coolant flow cutoff. Fortunately, a total core meltdown did not occur.

Core meltdown in a reactor could release radioactive materials into water and air in the surrounding environment. Winds or waterways could spread this material. The major problem with this type of pollution is that it is extremely dangerous to life and cannot be seen, tasted, or smelled. It takes special instruments to detect radiation. Furthermore, some radioactive isotopes have long half-lives, which means that a contaminated area would need evacuation and would be unusable for habitation or agriculture for many years. A partial core melting could result in the permanent shutdown of the plant. Furthermore, cleanup and safety operations can be very costly.

 ## REACTOR ACCIDENTS

The most serious nuclear reactor accident in the United States occurred in the **Three Mile Island reactor** located in Pennsylvania only 12 miles from the metropolitan area of Harrisburg. On March 28, 1979 a combination of equipment failure and human operator error resulted in the partial meltdown of the reactor. Fortunately, total meltdown did not occur and the various containment features of the reactor worked to prevent the release of significant amounts of radioactive materials into the environment. Nonetheless, some radioactive material was released. The Three Mile Island incident made it clear that serious nuclear reactor accidents can occur. The damaged nuclear reactor had to be abandoned and cleanup costs have reached hundreds of millions of dollars.

On April 26, 1986 a nuclear power reactor located near a city called Chernobyl in the Ukraine exploded, killing and hospitalizing numerous people and spreading radioactive materials in the local areas in the former USSR and other parts of Europe. The **Chernobyl reactor** used graphite not water as a moderator. The fuel in the reactor was about 75% used as it was being shut down for maintenance. During the shutdown, plant operators planned an experiment with the electrical generator associated with the reactor. To carry out the experiment they unwisely turned off the **ECCS** and withdrew nearly all the control rods as the reactor was being powered down. They also switched the primary coolant pumps from external electricity to electricity still being produced by the reactor and disabled the automatic shutdown mechanism of the reactor. This was all done on the assumption that since the reactor was being shut down none of these safety features would be needed. Unfortunately because of a design flaw in this type of reactor and the lack of operational safety features, the reactor experienced an increased rate of fission as it was being shut down. The high temperature caused by this power surge resulted in the destruction of the reactor.

The fuel rods reached a very high temperature and ruptured, spilling very hot material into the cooling water. The steam pressure became so great that it blew off the 1000-ton lid of the reactor destroying the reactor and the reactor building. It is thought that a second explosion quickly followed the first. This explosion occurred from the hydrogen gas that was formed when steam reacted with the zirconium metal used in the reactor. In any case, the explosion resulted in a plume of debris that rose some 1200 m or three-quarters of a mile into the atmosphere. Following this the hot graphite of the reactor began to burn producing a fire that lasted 11 days.

The explosion and fire spread radioactive materials from the fission products in the fuel rods. The area in a 30-km radius around the reactor was highly contaminated. In fact, 135,000 people were evacuated from this area and will never be allowed to return. Lesser amounts of radioactive materials were deposited as fallout from the explosion and fire in various regions of the republics of the former USSR and Europe. This disaster represents the single largest release of radioactive materials into the environment. It is estimated that from 50 to 180 million curies of radiation were released. Much of the radioactive material has dissipated since the isotopes had short half-lives. The most significant isotopes that were released include iodine-131 with an 8-day half-life. This isotope was absorbed into the thyroid glands of many people and has caused an increase in thyroid cancer. The other significant isotope is cesium-137 with a half-life of 30 years. This cesium that has been deposited in the soil can become part of the food chain.

The accident caused the direct death of 31 people and over 200 were hospitalized with radiation poisoning. Millions of people were affected with low levels of radiation and thousands of people were relocated from contaminated lands. It is difficult to predict the effect of the accident since nothing of this magnitude has occurred in the past. However, it is expected that radiation from the accident will result in at least 40,000 excess cancer deaths in the future. These cancers will occur mainly in the republics of the former USSR and neighboring regions of Europe. Needless to say one lesson of Chernobyl is that serious nuclear accidents can happen.

## 15-22 PLUTONIUM

A final concern over the large-scale use of nuclear energy involves plutonium. **Plutonium-239,** produced in nuclear reactors, has a half-life of about 24,000 years. A nuclear accident involving plutonium could contaminate an area for a very long time. The chance of such a disaster means that the safety factors in nuclear reactors must be carefully developed and continuously reviewed. However, earthquakes, human error of plant operators, and even the chance of sabotage are unpredictable factors.

Plutonium is present in spent fuel rods. It can be extracted from spent fuel in special fuel processing plants. In 1994 the only operating processing plant is located in the United Kingdom. It is important to realize that the dismantling of nuclear weapons in the United States and Russia is resulting in tons of excess plutonium that have to be safely stored indefinitely. The critical mass of plutonium-239 is about 5 kg. It is conceivable that a crude atomic bomb can be constructed from this amount of plutonium. Furthermore, plutonium can be used to make atomic bombs and hydrogen bombs by any country having the necessary technology.

In the United States, it has been a tradition to have a separation of nuclear power reactors and the military use of plutonium. That is, the plutonium produced in commercial nuclear reactors is not used to construct nuclear weapons. Fortunately, most countries do not have the technology to separate plutonium from nuclear reactor wastes. However, any country that develops this technology could make significant amounts of plutonium for the manufacture of nuclear weapons.

The desirable aspects and possible necessity, contrasted with the many disadvantages of nuclear power, present a significant environmental and social dilemma. We will continue to have to make decisions on the use of nuclear power use in the future.

You need: a 2-L clear plastic bottle with a cap, a paper clip, matches, a pen or pencil, a ruler, some magic tape, and a stopwatch or watch with a second hand

1. At the top of the bottle is a cap and at the bottom is a taper. Make a tiny hole in the plastic bottle just above the taper. Straighten a paper clip, heat one end with a match, and use the hot end to make the hole in the bottle. The hole should be no larger than 2 mm in diameter.

2. Place a length of tape on the bottle starting next to the hole and extending up the bottle. Use a marker to mark a line on the tape next to the hole and another line one-half inch above the hole. Continue marking lines on the side of the bottle one-half inch apart until you reach the top of the bottle. Only mark the part of the bottle that is straight and stop making marks where the bottle changes shape near the cap. Number the lines starting at zero next to the hole and increasing by one up the bottle.

3. The bottle needs to be located on a flat surface and the hole pointed towards a sink or some large container to catch the water.

4. Fill the bottle with water and put the cap on it. To observe how the rate of leaking ("decay") depends upon the amount of water remove the cap and note the decrease in the leak rate as the amount of water decreases. Also note the size of the leaking water stream.

5. Collect quantitative data on the leak rate as follows. Fill the bottle with water and put the cap on it. Be sure to fill it above the highest marked line. Be ready to start timing. Remove the cap and when the water level reaches the first mark start timing the leak. This is time zero. Record the time when the water level reaches each marked line. Continue recording the time until the water level reaches the mark one inch above the hole. The process can be repeated if you want to collect another set of data.

6. Your data consists of the relative amounts of water and elapsed time in seconds. The line corresponding to time zero is the total initial amount of water and is represented by the highest line on the bottle. Make a table of the data; as the time increases the height of each line (in inches) should decrease. Your last recorded time should be when the water level reaches the one inch line.

7. To picture the decay rate of the leaking bottle plot the relative amounts of water (the half-inch lines) versus the recorded times in seconds. The relative amounts will range from zero to the highest numbered half-inch line. The leak rate follows a slight curve, so connect the points of your graph with a curved smooth line. From your plot you can note that the leak rate slows down as the amount of water decreases. The curve levels off with time following an exponential pattern.

You need: an 8.5 inch by 11 inch piece of paper, scissors, and a metric ruler

1. Measure the length and height of the paper in units of millimeters and calculate the area of the paper in square millimeters.

2. Carefully cut the paper in half. Cut one of the halves in half. Continue cutting the halves until you have cut the paper in half 10 times.

3. Measure the length and height of the last remaining piece of paper in units of millimeters and calculate its area in square millimeters.

4. Divide the small area by the initial area of the sheet of paper and multiply by 100 to express the remaining area as a percent of the initial area. Record your results.

## QUESTIONS

### Section 15-1

1. Describe the structure of a typical atomic nucleus.

2. What are isotopes? What are nuclides?

3. Define the terms atomic number and nucleon number. How can the nucleon number and atomic number of a nuclide be used to find the number of neutrons in the nuclide.

4. Give the nuclide symbols used to represent the following nuclides. Refer to a list of elements for the necessary atomic numbers.

   (a) carbon-14   (e) potassium-40
   (b) oxygen-16   (f) plutonium-239
   (c) uranium-238   (g) iodine-131
   (d) radium-226   (h) cobalt-60

5. Determine the number of neutrons in the nuclei of each of the nuclides in Question 4.

6. Give the names of the following nuclides and determine the number of neutrons in nuclei of each nuclide.

   (a) $^{235}_{92}U$   (e) $^{241}_{95}Am$
   (b) $^{51}_{24}Cr$   (f) $^{218}_{84}Po$
   (c) $^{3}_{1}H$   (g) $^{222}_{86}Rn$
   (d) $^{249}_{98}Cf$   (h) $^{90}_{38}Sr$

### Section 15-2

7. The isotopes of some elements are radioactive. What does the term radioactive mean?

### Section 15-3

8. (a) What is alpha decay? What happens to the nucleon number and atomic number of a nuclide that undergoes alpha decay?

   (b) What is beta decay? What happens to the nucleon number and atomic number of a nuclide that undergoes beta decay?

   (c) What is gamma ray emission?

9. What is the structure and charge of an alpha particle?

10. What is the structure and charge of a beta particle? How is it thought that a negatively charged beta particle can come from a nucleus?

11. What does the term radiation mean in reference to radioactive decay? What are the three common types of nuclear radiation?

### Section 15-4

12. *Write nuclear equations for the alpha decay of the following nuclides.

   (a) $^{239}_{94}Pu$   (d) $^{241}_{95}Am$
   (b) $^{214}_{84}Po$   (e) $^{207}_{84}Po$
   (c) $^{238}_{92}U$   (f) $^{210}_{82}Pb$

13. Write nuclear equations for the alpha decay of the following nuclides.

   (a) $^{208}_{87}Fr$   (d) $^{252}_{99}Es$
   (b) $^{242}_{94}Pu$   (e) $^{243}_{96}Cm$
   (c) $^{251}_{98}Cf$   (f) $^{230}_{90}Th$

14. *Write nuclear equations for the beta decay of the following nuclides.

   (a) $^{197}_{78}Pt$   (d) $^{214}_{82}Pb$
   (b) $^{14}_{6}C$   (e) $^{239}_{93}Np$
   (c) $^{82}_{35}Br$   (f) $^{60}_{27}Co$

15. Write nuclear equations for the beta decay of the following nuclides.

   (a) $^{211}_{82}Pb$   (d) $^{32}_{15}P$
   (b) $^{28}_{12}Mg$   (e) $^{42}_{19}K$
   (c) $^{140}_{56}Ba$   (f) $^{20}_{8}O$

16. *Complete the following nuclear equations by supplying the missing nuclide or particle.

   (a) _____ $\longrightarrow$ $^{28}_{14}Si + ^{0}_{-1}e$
   (b) $^{226}_{88}Ra \longrightarrow$ _____ $+ ^{4}_{2}He$
   (c) $^{234}_{91}Pa \longrightarrow ^{234}_{92}U +$ _____
   (d) _____ $\longrightarrow ^{234}_{90}Th + ^{4}_{2}He$

17. Complete the following nuclear equations by supplying the missing nuclide or particle.

   (a) $^{237}_{93}Np \longrightarrow ^{4}_{2}He +$ _____
   (b) $^{210}_{84}Po \longrightarrow$ _____ $+ ^{206}_{82}Pb$
   (c) $^{210}_{82}Pb \longrightarrow$ _____ $+ ^{210}_{83}Bi$
   (d) $^{60}_{27}Co \longrightarrow ^{0}_{-1}e +$ _____

### Section 15-5

18. What is meant by the term half-life?

19. *Radon-222, an alpha emitter, has a half-life of 3.82 days.

   (a) Give a nuclear equation for the alpha decay of radon-222.

   (b) Radon is a gas. Why is it a dangerous alpha emitter when it is in the atmosphere?

   (c) If a radon-222 sample had a mass of 0.50 g,

what would the mass of the sample be after 11.5 days?

(d) If we started with 1,000,000 radon-222 atoms, how many would remain after 10 half-lives had passed? (Hint: Use a calculator and divide by 2 for each half-life).

20. Plutonium-239 is an alpha emitter with a half-life of 24,000 years. It is absorbed into the bones when it enters the body. It does not occur naturally but is produced as a by-product of nuclear reactors.

(a) Give a nuclear equation for the alpha decay of plutonium-239.

(b) Why is plutonium an especially dangerous nuclide if it enters the environment in significant amounts?

(c) A dose of 1 μg (1 μg = $10^{-6}$ g) of plutonium-239 in the human body can cause serious radiation poisoning or death. How many plutonium atoms are in a 1.0-μg sample? (Hint: The nucleon number can be used as a molar mass.)

(d) A rule of thumb for radioactive materials is that 10 half-lives are needed to decrease the amount of a radioactive material to an "acceptable" level. If a part of the environment became contaminated with plutonium-239, how many years would have to pass before the acceptable level of contamination had been reached?

21. * Iodine-131, a beta emitter, has a half-life of about 8 days. It is used medically to diagnose thyroid disease and treat thyroid cancer. The body uses iodine to form thyroxine, a thyroid hormone, so most iodine in the body concentrates in the thyroid.

(a) Give a nuclear equation for the beta decay of iodine-131.

(b) If we start with 10 μg (1 μg = $10^{-6}$ g) of iodine-131, how many grams are left after 24 days? How many after 80 days? (Hint: Divide the amount by 2 for each half-life. Use a calculator.)

(c) Why is iodine-131 dangerous when it is released by a nuclear power plant during an accident?

(d) A precautionary measure for people in the vicinity of a nuclear accident is to take massive doses of nonradioactive iodine (in the form of iodide ion). How would this help prevent the dangerous effects of iodine-131?

## Section 15-6

22. What is a becquerel? What is a curie? Some home smoke detectors contain 1 μCi of americium-241, an alpha emitter. How many nuclear disintegrations occur per second in this amount of the nuclide? **CAUTION: Never dismantle a smoke detector or let children touch or play with a smoke detector. For information about the disposal or an old detector contact the manufacturer or your local fire department.**

23. What is a rad? What is a rem? What is background radiation?

## Section 15-7

24. Which of the elements are naturally radioactive and have no stable isotopes?

25. What is a radioactive decay series? The half-life of radon-222 is 3.82 days yet small amounts are found in nature. Why?

26. Why are uranium ores not only a source of uranium but also radium? How is the uranium and lead content of certain rocks used to estimate their age?

27. Helium gas is found trapped in porous rock deposits. How do you suppose that this helium was formed?

## Section 15-8

28. Describe what happens when radiation passes through matter. Why is it called ionizing radiation?

29. Why is nuclear radiation dangerous to humans and other living organisms? Which type of radiation is most dangerous?

## Section 15-9

30. How does a Geiger–Müller tube function?

31. How can the activity of a radioactive nuclide be used to determine its half-life?

## Section 15-10

32. What is radiation sickness?

33. How are cancer, genetic damage, and mutation associated with nuclear radiation?

## Section 15-11

34. Define the following terms.

(a) nuclear transmutation

(b) target nuclei

(c) projectile

(d) human-made isotopes

35. *Complete the following equations representing nuclear transmutations.

(a) $^{59}_{26}\text{Fe} + ^{1}_{0}n \longrightarrow ^{60}_{27}\text{Co} + \underline{\hspace{1cm}}$

(b) $^{246}_{96}\text{Cm} + ^{12}_{6}\text{C} \longrightarrow \underline{\hspace{1cm}} + 4^{1}_{0}n$

(c) $^{27}_{13}\text{Al} + ^{4}_{2}\text{He} \longrightarrow \underline{\hspace{1cm}} + ^{1}_{0}n$

(d) $\underline{\hspace{1cm}} + ^{0}_{-1}e \longrightarrow ^{32}_{15}\text{P}$

(e) $^{2}_{1}\text{H} + \underline{\hspace{1cm}} \longrightarrow ^{4}_{2}\text{He} + ^{1}_{0}n$

(f) $\underline{\hspace{1cm}} + ^{1}_{0}n \longrightarrow ^{3}_{1}\text{H} + ^{4}_{2}\text{He}$

(g) $^{16}_{8}\text{O} + ^{1}_{0}n \longrightarrow \underline{\hspace{1cm}} + ^{4}_{2}\text{He}$

36. Complete the following equations representing nuclear transmutations.

(a) $^{15}_{7}\text{N} + \underline{\hspace{1cm}} \longrightarrow ^{12}_{6}\text{C} + ^{4}_{2}\text{He}$

(b) $^{24}_{12}\text{Mg} + ^{2}_{1}\text{H} \longrightarrow ^{25}_{12}\text{Mg} + \underline{\hspace{1cm}}$

(c) $^{6}_{3}\text{Li} + ^{1}_{0}n \longrightarrow ^{4}_{2}\text{He} + \underline{\hspace{1cm}}$

(d) $^{9}_{4}\text{Be} + ^{4}_{2}\text{He} \longrightarrow ^{12}_{6}\text{C} + \underline{\hspace{1cm}}$

(e) $^{238}_{92}\text{U} + ^{16}_{8}\text{O} \longrightarrow \underline{\hspace{1cm}} + 5^{1}_{0}n$

(f) $^{3}_{1}\text{H} + ^{2}_{1}\text{H} \longrightarrow ^{4}_{2}\text{He} + \underline{\hspace{1cm}}$

(g) $\underline{\hspace{1cm}} + ^{1}_{1}\text{H} \longrightarrow ^{15}_{7}\text{N} + ^{4}_{2}\text{He}$

## Section 15-12

37. Describe nuclear fission and a nuclear chain reaction.

38. Which common nuclides are fissionable? Which is naturally occurring and which is human-made?

39. What is meant by the critical mass of a fissionable nuclide?

## Section 15-13

40. *Using the equation $E = mc^2$, calculate the energy released when 7.0160 grams of lithium-7 and 1.0078 g of hydrogen-1 undergo the following fusion reaction:

$$^{7}_{3}\text{Li} + ^{1}_{1}\text{H} \longrightarrow 2^{4}_{2}\text{He}$$

relative
masses    7.0160          1.0078          2(4.0026)

41. Using the equation $E = mc^2$, calculate the energy re-

leased when 226.0254 grams of radium-226 undergoes alpha decay.

$$^{226}_{88}\text{Ra} \longrightarrow ^{222}_{86}\text{Rn} + ^{4}_{2}\text{He}$$

relative
masses    226.0254          222.0154          4.0026

## Section 15-15

42. List and describe the five basic components of a nuclear reactor. What is a LWR?

43. What is the composition of the fuel rods used in a nuclear power reactor.

## Section 15-16

44. Describe a nuclear power plant designed to produce electricity. Explain the differences between a boiling water reactor (BWR) and a pressurized water reactor (PWR).

45. Why is it not possible to have a nuclear explosion in a nuclear power reactor.

## Section 15-17

46. Describe nuclear fusion. How do fusion and fission differ?

## Sections 15-18 to 15-22

47. Explain or give a description of each of the following.

(a) low-level radioactive wastes

(b) fission product wastes

(c) waste heat

(d) the "China Syndrome" or core meltdown

48. List some of the pros and cons of nuclear power plants.

49. What is plutonium-239? How can this isotope be used and why is it dangerous?

50. Suggest some solutions to long-term disposal of high-level nuclear wastes.

51. Do you feel that nuclear energy should be an important source in the United States? Argue your case.

## CHAPTER 16

# ORGANIC
# CHEMISTRY

## 16-1 SYNTHETIC ORGANIC CHEMICALS

For centuries chemists have isolated and investigated chemical compounds found in nature. They classified them into two general groups: organic and inorganic. Inorganic compounds were those isolated from mineral sources. Organic compounds were those isolated from plant and animal sources. The word organic comes from the word for a living system, organism. Upon investigation it was found that the one thing organic compounds had in common was that they all were carbon-containing compounds. Today, synthetic and natural carbon-containing compounds are considered to be **organic compounds.** Except for the carbonates and cyanides, inorganic compounds do not contain carbon.

Much effort has been expended in investigating the composition of compounds found in nature. When chemists began experimenting with organic compounds they soon realized they could synthesize some compounds that were exactly the same as naturally occurring substances. Moreover, they discovered they could synthesize numerous carbon-containing compounds that were not found in nature. These achievements contributed to the development of a vast technology for the production of synthetic chemicals.

Several decades ago **synthetic organic** substances were not common in consumer products. Today such human-made substances surround us and are used in plastics, synthetic fibers, medicines, pesticides, synthetic rubbers, detergents, cosmetics, explosives, food additives, and food preservatives. Synthetic organic chemicals are prepared from naturally occurring compounds. The most common source of these raw materials is petroleum. Numerous industrial chemical processes are used to convert natural petroleum products into useful synthetic chemicals.

## 16-2  CARBON CHEMISTRY

The branch of chemistry called **organic chemistry** is devoted to the study of the compounds of carbon. The most important property of carbon atoms is their ability to form chemical bonds with one another and with a variety of other elements. Carbon usually forms four covalent bonds in organic compounds. This tetravalency, or tendency to form four bonds, correlates with the fact that carbon has four valence electrons.

$$\cdot \overset{\displaystyle \cdot}{\underset{\displaystyle \cdot}{C}} \cdot$$

When a carbon atom forms covalent single bonds with four other atoms, the four pairs of electrons are **tetrahedrally** distributed about the carbon atom, as shown in the margin of the next page. For convenience, the four possible covalent single bonds of carbon are represented on the flat page as

$$-\overset{\displaystyle |}{\underset{\displaystyle |}{C}}-$$

in which the lines represent the four single bonds of carbon. Other atoms are bonded to carbon through these bonds. The four bonds of carbon are actually oriented in three-dimensional space in a tetrahedral fashion. As you know, it is easier to use a flat projection of the surface of the earth as a map. In a similar fashion it is more convenient to use flat projections of carbon bonds even though they distort the actual bond angles.

One of the reasons there are so many organic compounds is that carbon atoms can form strong bonds with other carbon atoms as well as other nonmetals. This means that it is possible to have chains of carbon atoms

ACTIVITY
16-1

Appendix 5 has some simple patterns that can be used to make flat models for molecules of organic compounds. Trace these patterns and cut them out. They can be used to help visualize molecules. For instance, lie two of the four-bond-carbon patterns side by side to represent two carbons bonded by a single covalent bond. Imagine hydrogens to be bonded on the remaining open bonds when you use these patterns.

bonded to one another. As an analogy imagine carbon atoms and the atoms they bond with to be nature's "lego kit." As with any construction kit there are many interesting combinations; to start with, carbon atoms may be linked by single covalent bonds.

$$-\overset{|}{\underset{|}{C}}-\overset{|}{\underset{|}{C}}-$$

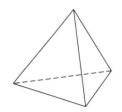

As illustrated in Figure 16-1, carbon atoms can bond together in a variety of ways much like the links of a chain. In fact, the bonded carbon sequence in an organic compound is called the **carbon chain.** There are literally billions of possible sequences of bonded carbon atoms. This accounts for the fact that there are millions of known organic compounds. Fortunately these compounds can be studied as groups of compounds having similar structures. Organic chemistry involves the study of the various groups of compounds. This chapter describes 12 of the most common groups of organic compounds.

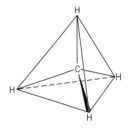

## 16-3 BONDING IN ORGANIC COMPOUNDS

Organic compounds involve bonded carbon sequences or chains in which one or more of the carbons are bonded to atoms of other elements. Hydrogen is the element most often bonded to carbon in organic compounds. Hydrogen is the element most often bonded to carbon in organic compounds. A hydrogen atom can share one pair of electrons with a carbon atom.

○ Hydrogen
● Carbon

$$-\overset{|}{\underset{|}{C}}-H$$

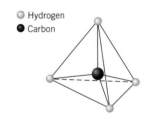

The formula above does not represent a compound but shows how a carbon–hydrogen bond can be represented. In a compound all the bonds of carbon must be used. The compound methane, $CH_4$, can be represented as

$$H-\overset{\overset{\displaystyle H}{|}}{\underset{\underset{\displaystyle H}{|}}{C}}-H$$

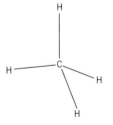

This formula, which shows the sequence of bonded atoms, is a structural formula. The formula $CH_4$, which indicates only the type and number of atoms in a molecule, is a **molecular formula. Structural formulas** are used to show which atoms are bonded to one another in a molecule. Structural formulas are somewhat like Lewis structures except they do not show the

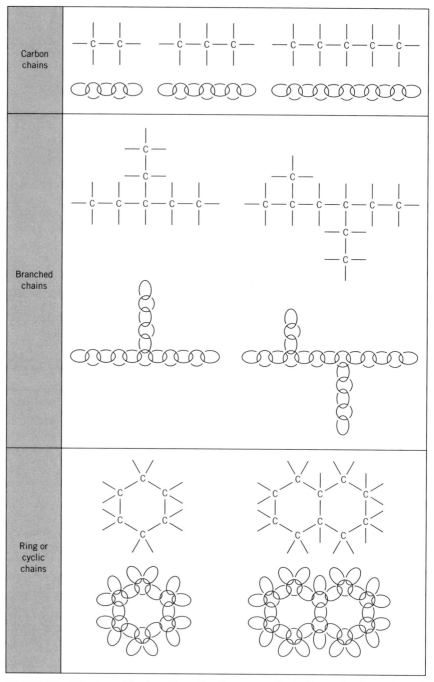

**FIGURE 16-1** The bonding of carbon atoms is similar to the formation of chains from individual links. As shown, carbon chains can be linear, branched, and in cyclic or ring forms.

nonbonding electrons. Molecular formulas are used to indicate the composition of molecules of a compound and structural formulas are used to emphasize the bonding sequence in a molecule. The molecular formula for butane is $C_4H_{10}$. It has the following structural formula:

$$H-\underset{\underset{H}{|}}{\overset{\overset{H}{|}}{C}}-\underset{\underset{H}{|}}{\overset{\overset{H}{|}}{C}}-\underset{\underset{H}{|}}{\overset{\overset{H}{|}}{C}}-\underset{\underset{H}{|}}{\overset{\overset{H}{|}}{C}}-H$$

Use four of the four-bond-carbon patterns to make a model of butane.

Instead of writing the complete structural formula of a compound like butane, the compound may be represented by a **condensed structural formula.** A condensed formula reveals the bonding sequence without showing all the bonds. For example, the condensed formula for butane is

$$CH_3CH_2CH_2CH_3$$

Compare this condensed formula to the structural formula. In the condensed formula the atoms follow the carbon to which they are bonded. A condensed formula is interpreted as indicating that the carbons are bonded to one another in sequence and that each carbon is bonded to the hydrogens or other atoms that surround it in the formula. Condensed formulas are more convenient to write than full structural formulas.

Two carbon atoms are capable of sharing two pairs of electrons with one another to form a **double covalent bond** or a **double bond.**

$$\overset{\diagdown}{\underset{\diagup}{}}C=C\overset{\diagup}{\underset{\diagdown}{}}$$

Connect two double-bond-carbon patterns to make a model of ethylene. Look up ethylene in a large dictionary or encyclopedia and list some of its chemical uses.

The double bond uses two bonds for each carbon, leaving two other positions for bonds to other atoms. For example, the compound ethylene, $C_2H_4$, includes a double bond and four carbon-hydrogen single bonds.

$$\underset{H}{\overset{H}{\diagdown}}C=C\underset{H}{\overset{H}{\diagup}} \quad \text{or} \quad CH_2\!=\!CH_2$$

In some compounds, two carbon atoms share three pairs of electrons to form a **triple covalent bond** or simply a **triple bond.** When two carbon atoms are bonded by a triple bond, each carbon can form one other bond with a different atom. For example, acetylene, $C_2H_2$, involves a triple bond and two single bonds to hydrogen atoms.

$$H-C\equiv C-H \quad \text{or} \quad HC\equiv CH$$

Triple bonds are less common than single bonds and double bonds. The kinds of bonds carbon forms with oxygen, nitrogen, and the halogens are shown in Table 16-1.

**Table 16-1** Typical Bonds Between Elements and Carbon

| ELEMENT BONDING TO CARBON | EXAMPLES |
|---|---|

**Oxygen**

Single Bond

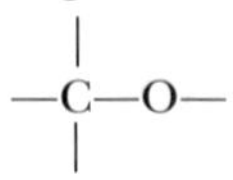

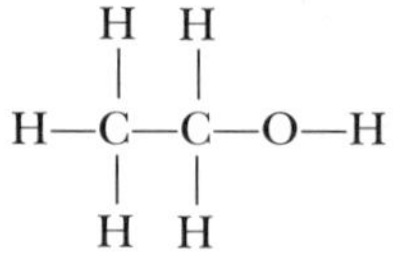

$$-\overset{|}{\underset{|}{C}}-O-$$

$$\begin{array}{c} \text{H} \quad \text{H} \\ | \quad | \\ \text{H}-\overset{|}{\underset{|}{\text{C}}}-\overset{|}{\underset{|}{\text{C}}}-\text{O}-\text{H} \\ | \quad | \\ \text{H} \quad \text{H} \end{array}$$
ethyl alcohol

Double bond

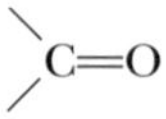

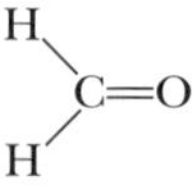

$$\diagdown_{\diagup}C=O$$

$$\begin{array}{c} \text{H}\diagdown \\ \qquad C=O \\ \text{H}\diagup \end{array}$$
formaldehyde

**Nitrogen**

Single bond

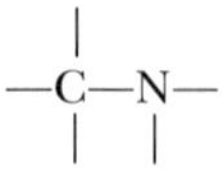

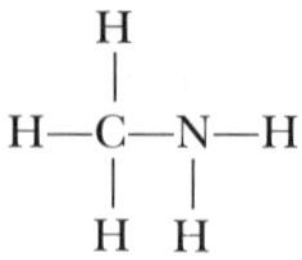

$$-\overset{|}{\underset{|}{C}}-\overset{|}{N}-$$

$$\begin{array}{c} \text{H} \\ | \\ \text{H}-\overset{|}{\underset{|}{\text{C}}}-\overset{|}{\text{N}}-\text{H} \\ | \quad | \\ \text{H} \quad \text{H} \end{array}$$
methylamine

Double bond

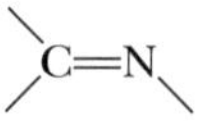

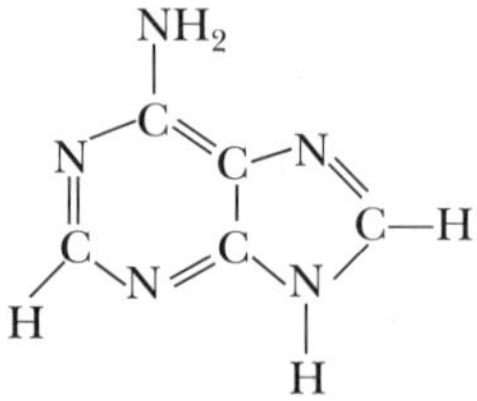

$$\diagdown_{\diagup}C=N\diagdown$$

adenine

**Halogens**

Single bond

Chlorine $-\overset{|}{\underset{|}{C}}-Cl$

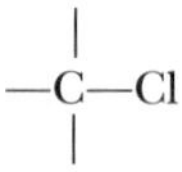

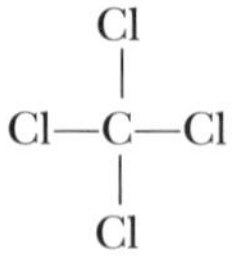

$$\begin{array}{c} \text{Cl} \\ | \\ \text{Cl}-\overset{|}{\underset{|}{\text{C}}}-\text{Cl} \quad \text{carbon tetrachloride} \\ | \\ \text{Cl} \end{array}$$

Bromine $-\overset{|}{\underset{|}{C}}-Br$

$$\begin{array}{c} \text{H} \\ | \\ \text{H}-\overset{|}{\underset{|}{\text{C}}}-\text{Br} \quad \text{methyl bromide} \\ | \\ \text{H} \end{array}$$

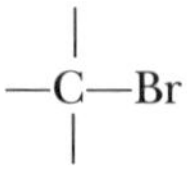

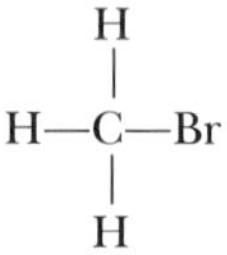

Fluorine $-\overset{|}{\underset{|}{C}}-F$

$$\begin{array}{c} \text{Cl} \\ | \\ \text{F}-\overset{|}{\underset{|}{\text{C}}}-\text{F} \\ | \\ \text{Cl} \end{array}$$
dichlorodifluoromethane (Freon 12)

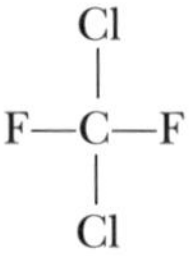

Iodine $-\overset{|}{\underset{|}{C}}-I$

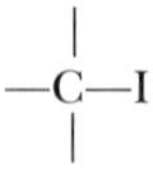

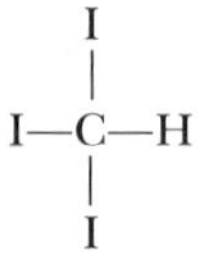

$$\begin{array}{c} \text{I} \\ | \\ \text{I}-\overset{|}{\underset{|}{\text{C}}}-\text{H} \quad \text{iodoform} \\ | \\ \text{I} \end{array}$$

## 16-4 ALKANES

Organic compounds containing only carbon and hydrogen are called **hydrocarbons.** Hydrocarbons that include only single-bonded carbons and no double or triple bonds are called **saturated hydrocarbons.** It is useful to study some of these hydrocarbons, since they serve as a basis for the structure of a large number of organic compounds. There are many possible saturated hydrocarbons; the simplest is methane, $CH_4$, next comes ethane, $C_2H_6$.

$$CH_3CH_3$$

Third is propane, $C_3H_8$.

$$CH_3CH_2CH_3$$

Notice that these compounds differ by $—CH_2—$ units.

$$CH_4 \qquad CH_3—CH_2—H \qquad CH_3—CH_2—CH_3$$

The structural formulas of a large number of hydrocarbons differ by $—CH_2—$ units. Any group of compounds in which the members differ in this manner is called a **homologous series.** The saturated hydrocarbons comprise a homologous series corresponding to the general formula $C_nH_{2n+2}$ where $n$ can be any integer. For example, $CH_4$, $n = 1$; $C_2H_6$, $n = 2$; $C_3H_8$, $n = 3$. These compounds are called the **alkanes.** Eight of the most common alkanes are listed in Table 16-2.

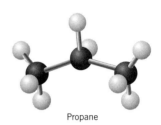

Propane

ACTIVITY 16-4

Use the general formula for the alkanes to determine the molecular formula for butane ($n = 4$), octane ($n = 8$), and decane ($n = 10$).

## 16-5 ALKYL GROUPS

A large number of organic compounds can be considered to be derived from the substitution of one or more hydrogens of alkanes by another atom or group of atoms. For instance, the compound methyl alcohol

$$H—\overset{\displaystyle H}{\underset{\displaystyle H}{C}}—OH$$

is viewed as the substitution of an H of methane with an OH. Another view of this compound is that it is made up of portions or parts that are special groups of covalently bonded atoms. In this view, the above compound is composed of a

$$H—\overset{\displaystyle H}{\underset{\displaystyle H}{C}}— \quad \text{group bonded to a } —OH \text{ group}$$

***Table 16-2***    Names and Formulas of Some Alkanes

| NAME | MOLECULAR FORMULA | STRUCTURAL FORMULA | CONDENSED STRUCTURAL FORMULA[a] |
|------|-------------------|--------------------|-------------------------------|
| Methane | $CH_4$ | | $CH_4$ |
| Ethane | $C_2H_6$ | | $CH_3CH_3$ |
| Propane | $C_3H_8$ | | $CH_3CH_2CH_3$ |
| Butane | $C_4H_{10}$ | | $CH_3CH_2CH_2CH_3$ or $CH_3(CH_2)_2CH_3$ |
| Pentane | $C_5H_{12}$ | | $CH_3CH_2CH_2CH_2CH_3$ or $CH_3(CH_2)_3CH_3$ |
| Hexane | $C_6H_{14}$ | | $CH_3CH_2CH_2CH_2CH_2CH_3$ or $CH_3(CH_2)_4CH_3$ |
| Heptane | $C_7H_{16}$ | | $CH_3CH_2CH_2CH_2CH_2CH_2CH_3$ or $CH_3(CH_2)_5CH_3$ |
| Octane | $C_8H_{18}$ | | $CH_3CH_2CH_2CH_2CH_2CH_2CH_2CH_3$ or $CH_3(CH_2)_6CH_3$ |

[a]The $(CH_2)_n$ notation refers to $n$ $CH_2$ units in a row.

A **group** is a sequence of bonded atoms with an open bond position that can be bonded to another atom or group. In methyl alcohol a $CH_3$— group and an —OH group are bonded.

Groups derived from alkanes can be considered to be formed by removing one hydrogen to leave an open bonding position. Groups do not exist by themselves; they are parts of compounds. The formula of a group is written with an open bonding position. Any two groups can bond through these open positions. Similar to kit constructions, chemical groups can be attached to make a more complex structure.

The groups formed from the alkanes are called **alkyl groups.** They are named by using the name of the alkane from which they are derived, with the -ane ending changed to -yl. Some examples are

Use your paper models. Link a four-bond-carbon pattern and an —OH pattern to make a model of methyl alcohol.

$$
\begin{array}{cccc}
& \text{H} & & \text{H} \quad \text{H} \\
& | & & | \quad\;\; | \\
\text{H}\!-\!\text{C}\!- & & \text{H}\!-\!\text{C}\!-\!\text{C}\!- & \\
& | & & | \quad\;\; | \\
& \text{H} & & \text{H} \quad \text{H}
\end{array}
$$

methyl group   ethyl group   propyl group   isopropyl group

These groups can be more easily written as the condensed formulas $CH_3-$ (methyl), $CH_3CH_2-$ (ethyl), $CH_3CH_2CH_2-$ (propyl), and $(CH_3)_2CH-$ (isopropyl). Table 16-3 lists some common alkyl groups. The general symbol $R-$ is used to represent any alkyl group. A few examples of substituted compounds containing alkyl groups are

Use three four-bond-carbon patterns and an —OH pattern to make a model of propyl alcohol and isopropyl alcohol.

| | |
|---|---|
| $CH_3OH$ | methyl alcohol |
| $CH_3CH_2Cl$ | ethyl chloride |
| $(CH_3)_2CHOH$ | isopropyl alcohol |

*Table 16-3*   Names and Formulas of Some Alkyl Groups

| FORMULA | GROUP NAME |
|---|---|
| $CH_3-$ | Methyl |
| $CH_3CH_2-$ | Ethyl |
| $CH_3CH_2CH_2-$ | Propyl |
| $CH_3CH- \quad$ or $\quad (CH_3)_2CH-$ | Isopropyl |
| $CH_3CH_2CH_2CH_2-$ | Butyl |
| $CH_3CHCH_2- \quad$ or $\quad (CH_3)_2CHCH_2-$ | Isobutyl |
| $CH_3-C- \quad$ or $\quad (CH_3)_3C-$ | *tert*-Butyl or tertiary Butyl |
| $R-$ | General symbol for any alkyl group |

### 16-6 Alkenes

Hydrocarbons that have a double bond between any two carbons are called **unsaturated hydrocarbons** or **olefins.** The simplest olefin is ethylene.

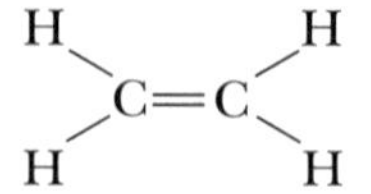

$$\begin{array}{ccc} H & & H \\ \diagdown & & \diagup \\ & C = C & \\ \diagup & & \diagdown \\ H & & H \end{array}$$

Compounds formed from ethylene by adding $—CH_2—$ units are members of the homologous series called the **alkenes.** The general formula for alkenes is $\mathbf{C_n H_{2n}}$ ($CH_2{=}CH_2$, $n = 2$; $CH_3CH{=}CH_2$, $n = 3$, etc.).

As discussed in Section 11-3, alkenes are used to make plastics. Some common alkenes and alkenes substituted with other groups are shown below.

$$CH_2{=}CH_2$$

Ethylene, a natural fruit hormone used to ripen fruit.

$$CH_3—CH{=}CH_2 \quad \text{or} \quad CH_3CH{=}CH_2$$

Propylene or propene, used to make polypropylene plastic.

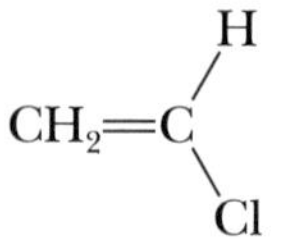

$$\begin{array}{c} H \\ \diagup \\ CH_2{=}C \quad \text{or} \quad CH_2{=}CHCl \\ \diagdown \\ Cl \end{array}$$

Chloroethylene or vinyl chloride, used to make PVC plastic.

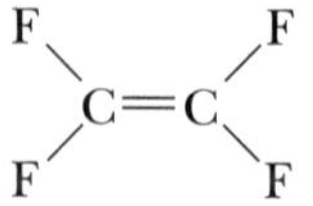

$$\begin{array}{ccc} F & & F \\ \diagdown & & \diagup \\ & C = C & \quad \text{or} \quad F_2C{=}CF_2 \\ \diagup & & \diagdown \\ F & & F \end{array}$$

Tetrafluoroethylene, used to make Teflon plastic.

Whenever a double bond occurs in a sequence of carbons, it is called an **unsaturated linkage.** Molecules of **polyunsaturated vegetable oils** involve carbon sequences with unsaturated linkages. However, vegetable oils are not classified as alkenes, since they contain oxygen as well as carbon and hydrogen. Vegetable oils are classified as lipids and are discussed in Section 17-4.

Compounds with double bonds are generally quite chemically reactive in comparison to alkanes. The typical reactions involving alkenes are **addition reactions** in which the double bond becomes a single bond and other atoms bond to the carbons of the original double bond. An example of an addition reaction involving an alkene and iodine is

$$\underset{H}{\overset{H}{\diagdown}} C = C \underset{H}{\overset{H}{\diagup}} \;+\; I_2 \;\longrightarrow\; I-\underset{\underset{H}{|}}{\overset{\overset{H}{|}}{C}}-\underset{\underset{H}{|}}{\overset{\overset{H}{|}}{C}}-I \quad \text{1,2-diiodoethane}$$

Use your patterns to make models of ethylene and 1,2-diiodoethane for comparison. Use circles to represent iodines.

##  ALCOHOLS

The alkanes are saturated hydrocarbons, and the alkenes are hydrocarbons with a double bond. In a similar fashion other kinds of organic compounds have unique structural features. Organic compounds can be classified and grouped according to structural characteristics that sets them apart from other compounds.

Compounds containing an **—OH (hydroxy group)** bonded to an **R—(alkyl group)** are called **alcohols (R—OH).** The hydroxy group is characteristic of alcohols. The —OH gives these compounds their characteristic properties and is it called the functional group for alcohols. Organic compounds can be classified by their **functional groups.** A few typical alcohols are described below.

$$\textbf{CH}_3\textbf{OH}$$
**methyl alcohol**
**or methanol**

Methyl alcohol, also called wood alcohol, is used in the manufacture of numerous chemical products such as formaldehyde, jet fuel, and antifreeze. It is very poisonous and can cause blindness.

$$\overset{\textbf{OH}}{\overset{|}{\textbf{CH}_3\textbf{CHCH}_3}} \quad \text{or} \quad \textbf{(CH}_3\textbf{)}_2\textbf{CHOH}$$
**isopropyl alcohol**

Isopropyl alcohol is used as a disinfectant and as rubbing alcohol.

$$\textbf{CH}_3\textbf{CH}_2\textbf{OH} \quad \text{or} \quad \textbf{C}_2\textbf{H}_5\textbf{OH}$$
**ethyl alcohol**
**or ethanol**

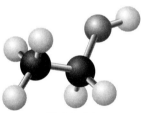

Ethyl alcohol

Millions of tons of ethyl alcohol are produced annually in the United States. It is used in numerous manufacturing processes and in the preparation of alcoholic beverages. One method used to make ethyl alcohol is the fermentation of sugars. All forms of alcoholic beverages contain ethyl alcohol and it is the physiological active ingredient. Most alcohols are poisonous and relatively large doses of ethyl alcohol can be deadly. Denatured ethyl alcohol contains ingredients like methanol or benzene to make it unfit for drinking.

To observe fermentation of a sugar you need sucrose, a ¼-oz package of baker's dry yeast, a zip-lock bag, and a bowl or a cup.

1. Dissolve one tablespoon of sugar in one-fourth of a cup of water and place the solution in a zip-lock bag.

2. Add a package of yeast to the bag, squeeze out the air, and seal the bag.

3. Put the bag in a bowl or cup of warm water. Let the bag sit for 15 to 30 minutes. The fermentation of sugars includes glucose reacting to give ethyl alcohol and carbon dioxide gas. Write a balanced equation for this reaction. In this exercise what evidence suggests that fermentation took place?

Alcohols that have more than one hydroxy group attached to a carbon sequence are called **polyhydroxy alcohols.** Two important polyhydroxy alcohols are described below.

$$\text{HO—CH}_2\text{—CH}_2\text{—OH} \quad \text{or} \quad \text{CH}_2\text{OHCH}_2\text{OH}$$
**ethylene glycol**

Ethylene glycol is used as an antifreeze and engine coolant.

$$\begin{array}{ccc} \text{OH} & \text{OH} & \text{OH} \\ | & | & | \\ \text{CH}_2 & \text{—CH—} & \text{CH}_3 \end{array} \quad \text{or} \quad \text{CH}_2\text{OHCHOHCH}_2\text{OH}$$
**glycerol or glycerine**

Glycerol, a common lubricant, is used in the manufacture of plastics, drugs, cosmetics, inks, food products, and nitroglycerine, an explosive.

## 16-8 ISOMERS

One of the reasons there is such a variety of organic compounds is that carbon can bond with itself and other elements in numerous ways. Consider a compound with the formula $C_5H_{12}$. A possible structure corresponding to this formula is the straight chain form

$$\begin{array}{ccccc} \text{H} & \text{H} & \text{H} & \text{H} & \text{H} \\ | & | & | & | & | \\ \text{H—C—} & \text{C—} & \text{C—} & \text{C—} & \text{C—H} \\ | & | & | & | & | \\ \text{H} & \text{H} & \text{H} & \text{H} & \text{H} \end{array} \quad \text{or} \quad \text{CH}_3\text{CH}_2\text{CH}_2\text{CH}_2\text{CH}_3$$

It is possible to have branched carbon sequences with the same molecular formula.

Use your patterns to make models of ethylene glycol and glycerine.

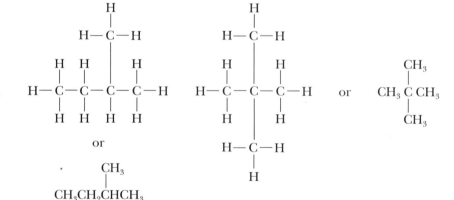

Count the carbons and hydrogens to see that these three compounds have the same molecular formula. The three formulas represent different compounds. These compounds do not have the same structure, but they do have the same molecular formula, $C_5H_{12}$. Compounds with the same molecular formula but different structural formulas are called **isomers.** Isomerism is quite common in organic chemistry and greatly increases the number of possible compounds.

As another example of isomerism, consider the formula $C_2H_6O$. Two structures corresponding to this formula are

$$\underset{\text{ethyl alcohol}}{H-\overset{\displaystyle H}{\underset{\displaystyle H}{C}}-\overset{\displaystyle H}{\underset{\displaystyle H}{C}}-O-H} \quad \text{and} \quad \underset{\text{dimethyl ether}}{H-\overset{\displaystyle H}{\underset{\displaystyle H}{C}}-O-\overset{\displaystyle H}{\underset{\displaystyle H}{C}}-H}$$

Even though they have the same molecular formula, these two structures represent totally different compounds with different chemical properties. For example, ethyl alcohol is a liquid that boils at 78.5°C and dimethyl ether is a gas at room temperature.

Since the same molecular formula can in many cases represent more than one substance, compounds cannot be classified by their molecular formulas. However, they can be classified according to characteristic bonding sequences in their structural formulas, the functional group. For instance, the compounds

$$H-\overset{\displaystyle H}{\underset{\displaystyle H}{C}}-OH \quad \text{and} \quad H-\overset{\displaystyle H}{\underset{\displaystyle H}{C}}-\overset{\displaystyle H}{\underset{\displaystyle H}{C}}-OH$$

both include an alkyl group bonded to an —OH group. Thus, both of these compounds are classified as alcohols.

Use your patterns to make models of the two alcohols shown here.

## 16-9   ALDEHYDES AND ACIDS

Alcohols that have the hydroxy group attached to a carbon, which is in turn bonded to only one other carbon ($R—CH_2OH$), are called **primary alcohols.** When a primary alcohol reacts with certain oxidizing agents, the $—CH_2OH$ grouping can be oxidized to an aldehyde group.

$$RCH_2OH + \text{oxidizing agent} \longrightarrow \underset{\text{aldehyde group}}{R-C\overset{\displaystyle O}{\diagdown}_H} \quad \text{or} \quad RCHO$$

Compounds that contain an aldehyde group and correspond to the formula above are called aldehydes. The **functional group** of **aldehydes** is **—CHO.** Some examples are shown on the next page.

Acetaldehyde

$$H - C \overset{O}{\underset{H}{\big<}} \quad \text{or} \quad \text{HCHO formaldehyde}$$

$$CH_3 - C \overset{O}{\underset{H}{\big<}} \quad \text{or} \quad CH_3CHO \text{ acetaldehyde}$$

ACTIVITY 16-11

Use your patterns to make models of formaldehyde and acetaldehyde.

Formaldehyde is used for the manufacture of plastics such as Formica. A water solution of formaldehyde is called formalin. It is used as a disinfectant and to preserve biological specimens. Acetaldehyde is used in the manufacture of plastics and in some medicines.

Acetic acid, butyric acid, citric acid, and benzoic acid are all acids and they also belong to a special group of organic acids known as **carboxylic acids.** Aldehydes react with certain oxidizing agents to oxidize the aldehyde group to a group called a carboxyl group.

$$R - C \overset{O}{\underset{H}{\big<}} \quad + \text{ oxidizing agent} \longrightarrow R - C \overset{O}{\underset{O-H}{\big<}}$$

$$\text{or} \quad \text{RCOOH} \quad \text{or} \quad \text{RCO}_2\text{H}$$

Organic compounds that include a carboxyl group and correspond to the general formula above are called the **carboxylic acids.** The functional group is **—COOH** or **—CO₂H**. The formulas of some carboxylic acids are shown below; they are all weak acids. Formic acid is an active ingredient in ant stings, acetic acid is the active ingredient in vinegar and butyric and propionic acids are two of the odorous chemicals in sweat.

$$H - C \overset{O}{\underset{OH}{\big<}} \quad \text{or} \quad \text{HCOOH formic acid}$$

$$CH_3C \overset{O}{\underset{OH}{\big<}} \quad \text{or} \quad CH_3COOH \text{ acetic acid}$$

$$CH_3CH_2C \overset{O}{\underset{OH}{\big<}} \quad \text{or} \quad CH_3CH_2COOH \text{ propionic acid}$$

$$CH_3CH_2CH_2C \overset{O}{\underset{OH}{\big<}} \quad \text{or} \quad CH_3CH_2CH_2COOH \text{ butyric acid}$$

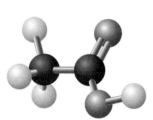

Acetic acid

ACTIVITY 16-12

Use your patterns to make a model of acetic acid.

## 16-10 ESTERS

A carboxylic acid can react with an alcohol to form a product called an ester and water. This is called a **condensation reaction.**

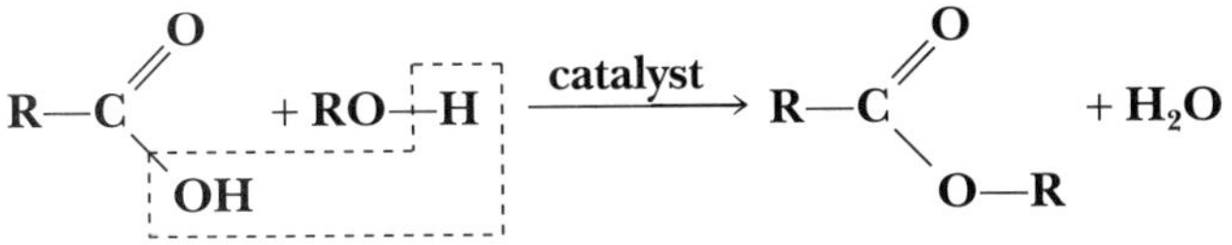

$$R-C\overset{O}{\underset{OH}{{\diagup}}} + RO{-}H \xrightarrow{\text{catalyst}} R-C\overset{O}{\underset{O-R}{{\diagup}}} + H_2O$$

The RO— grouping of the alcohol replaces the —OH of the acid. For example,

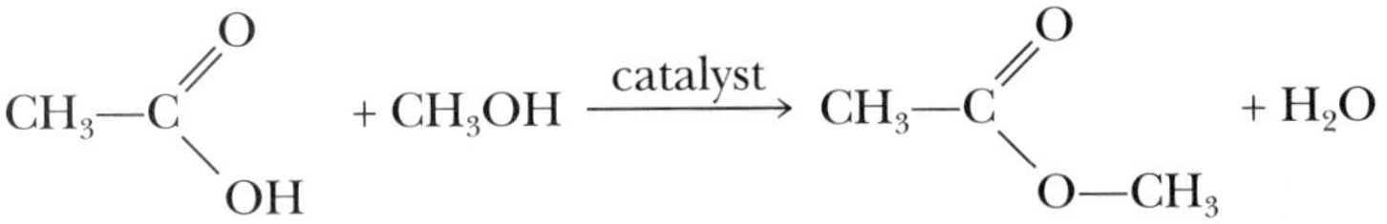

$$\underset{\text{acetic acid}}{CH_3-C\overset{O}{\underset{OH}{{\diagup}}}} + \underset{\text{methyl alcohol}}{CH_3OH} \xrightarrow{\text{catalyst}} \underset{\text{methyl acetate}}{CH_3-C\overset{O}{\underset{O-CH_3}{{\diagup}}}} + H_2O$$

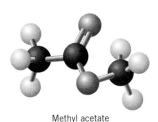

Methyl acetate

Note that the general formula for **esters** including the functional group is R—COO—R. Many esters occur naturally and are responsible for some of the tastes and odors of fruits. Table 16-4 lists some typical esters of this type. Chemistry in Action 16-1 shows you how to make isopropyl acetate.

Esters can also be formed between noncarboxylic acids and alcohols. It is possible to have **phosphate esters** derived from phosphoric acid where one or more of the —OH groups are replaced by —OR groups.

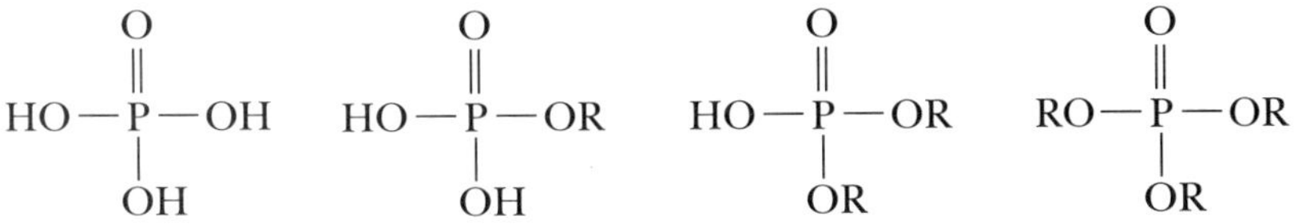

$$\underset{\underset{OH}{\big|}}{HO-\overset{\overset{O}{\|}}{P}-OH} \qquad \underset{\underset{OH}{\big|}}{HO-\overset{\overset{O}{\|}}{P}-OR} \qquad \underset{\underset{OR}{\big|}}{HO-\overset{\overset{O}{\|}}{P}-OR} \qquad \underset{\underset{OR}{\big|}}{RO-\overset{\overset{O}{\|}}{P}-OR}$$

An example is

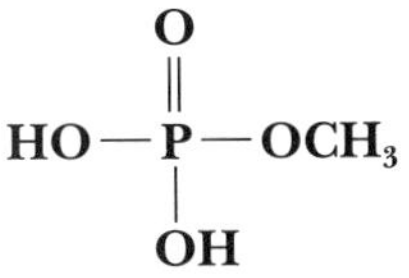

$$\underset{\underset{OH}{\big|}}{HO-\overset{\overset{O}{\|}}{P}-OCH_3}$$

**methyl dihydrogen phosphate**

Methyl dihydrogen phosphate is used as a gasoline additive to control the ignition process. Phosphate esters are important in nucleic acids as discussed in section 17-8.

## 16-11 AMINES

The structural formula of the base ammonia is

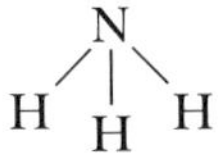

$$\underset{H\ \underset{H}{|}\ H}{N}$$

**Table 16-4**   Some Esters

| NAME | STRUCTURE | SOURCE OR FLAVOR |
|------|-----------|------------------|
| Ethyl formate | $CH_3CH_2-O-\overset{\overset{\textstyle O}{\|}}{C}-H$ | Rum |
| Isobutyl formate | $CH_3\overset{\overset{\textstyle CH_3}{\|}}{C}HCH_2-O-\overset{\overset{\textstyle O}{\|}}{C}-H$ | Raspberries |
| Ethyl acetate | $CH_3CH_2-O-\overset{\overset{\textstyle O}{\|}}{C}CH_3$ | Used in lacquers |
| Pentyl acetate (amyl acetate) | $CH_3(CH_2)_4-O-\overset{\overset{\textstyle O}{\|}}{C}CH_3$ | Bananas |
| Isopentyl acetate (isoamyl acetate) | $CH_3\overset{\overset{\textstyle CH_3}{\|}}{C}HCH_2CH_2-O-\overset{\overset{\textstyle O}{\|}}{C}CH_3$ | Pears |
| Octyl acetate | $CH_3(CH_2)_7-O-\overset{\overset{\textstyle O}{\|}}{C}CH_3$ | Oranges |
| Ethyl butyrate | $CH_3CH_2-O-\overset{\overset{\textstyle O}{\|}}{C}CH_2CH_2CH_3$ | Pineapples |
| Pentyl butyrate | $CH_3(CH_2)_4-O-\overset{\overset{\textstyle O}{\|}}{C}CH_2CH_2CH_3$ | Apricots |
| "Waxes" | $CH_3(CH_2)_n-O-\overset{\overset{\textstyle O}{\|}}{C}(CH_2)_nCH_3$ | Carnauba wax[a] Beeswax[b] Spermaceti[c] |

[a]Where $n$ represents (23–33) $CH_2$ groups.
[b]Where $n$ represents (25–27) $CH_2$ groups.
[c]Where $n$ represents (14–15) $CH_2$ groups.

Organic compounds in which one or more H atoms of ammonia are replaced by an alkyl group, —R, are called amines.

$$R-N\overset{\textstyle H}{\underset{\textstyle H}{<}} \quad \text{or} \quad R-NH_2 \quad \text{or} \quad RNH_2$$

The **—NH₂** group is called the **amino group.** The **amines** are **organic bases** because of their similarity to ammonia. Amines with one alkyl group are called primary amines. Amines with two alkyl groups, which may be different, are called secondary amines, and those with three alkyl groups are called tertiary amines.

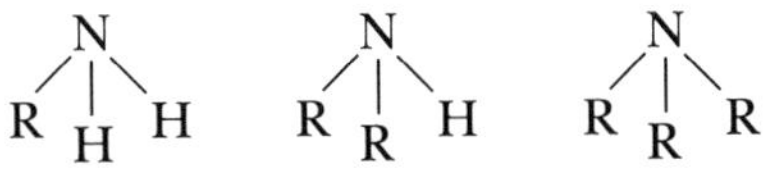

RNH₂
primary
amine

R₂NH
secondary
amine

R₃N
tertiary
amine

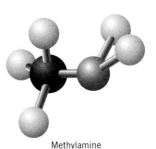

Methylamine

Examples are

**primary**        **CH₃CH₂NH₂**
               **ethylamine**

**secondary**      **CH₃CH₂—N—H**
                              |
                          **CH₃**
               **ethylmethylamine**

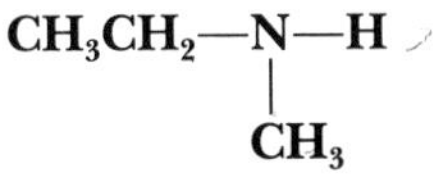

**tertiary**       **CH₃—N—CH₃**
                          |
                       **CH₃**
               **trimethylamine**

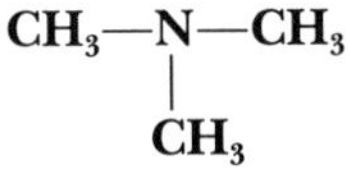

Use the three-bond-nitrogen pattern and others to make models of ethylamine and ethylmethylamine.

## 16-12 AMIDES

Under certain conditions, ammonia or some amine can react with a carboxylic acid in a condensation reaction that is quite similar to the formation of an ester.

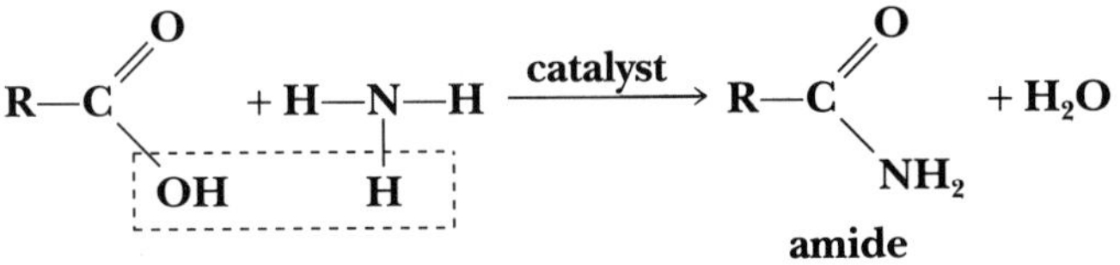

$$R—C\overset{O}{\diagdown}\underset{OH}{} + H—N—H\underset{H}{} \xrightarrow{\text{catalyst}} R—C\overset{O}{\diagdown}\underset{NH_2}{} + H_2O$$

amide

The product of such a reaction is called an **amide.** This bonding of an amine to an acid is called an **amide linkage.** The general formula for simple amides is **R—CO—NH₂.** As we shall see in Section 17-5, amide linkages are involved in the formation of proteins. Two examples of amides are

Use your patterns to make a model of acetamide.

CH₃—C—NH₂ acetamide        CH₃CH₂CH₂C—NH₂ butyramide

## 16-13 AMINO ACIDS

Amino acids are important in biochemistry and nutrition. As discussed in Section 17-5 amino acids are the building blocks of proteins found in living organisms. The common **amino acids** are carboxylic acids with an —$NH_2$ group bonded to the carbon next to the carboxyl carbon. The —$NH_2$ group is called an **amino group.**

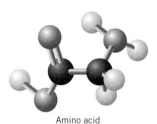

Amino acid

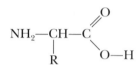

They are called amino acids since they have both an amino group and a carboxyl group. In an amino acid the —**R** group may be an **alkyl group** (e.g., —$CH_3$) or a substituted alkyl group (e.g., —$CH_2OH$). The common amino acids differ only in the structure of the —R group. The 21 most important amino acids that are found in nature have been given common names. For example, the simplest amino acid is glycine in which the —R group is simply a hydrogen.

*16-15*

Use your patterns to make a model of glycine.

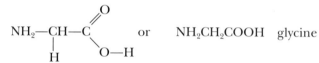

The names and structures of some other amino acids are given in Table 16-5.

## 16-14 CYCLIC COMPOUNDS AND BENZENE

In some hydrocarbons, the carbon sequence is one in which the carbons form a ring. These hydrocarbons are called **cyclic** or **ring compounds** (see Fig. 16-1). If each carbon of the ring has two hydrogens bonded to it, the compound is called a cycloalkane. For example, cyclohexane can be represented by the formula

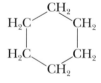

And cyclopentane is

Cyclopentane

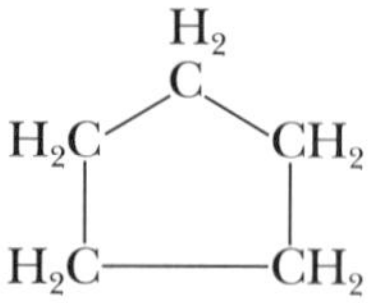

**Table 16-5** Common Amino Acids Including the Essential Amino Acids (the Groups Shown Are Bonded to $NH_2CHCOOH$)

| No. | Name | Group | Symbol |
|---|---|---|---|
| 1. | Glycine | —H | Gly[a] |
| 2. | Alanine | $-CH_3$ | Ala |
| 3. | Serine | $-CH_2OH$ | Ser |
| 4. | Cysteine | $-CH_2SH$ | Cys |
| 5. | Cystine | $-CH_2-S-S-CH_2-$ | Cys—S—S—Cys[b] |
| 6. | Threonine[c] | $-CH-CH_3$ with OH | Thr |
| 7. | Valine[c] | $CH_3-CH-CH_3$ | Val |
| 8. | Leucine[c] | $-CH_2-CH-CH_3$ with $CH_3$ | Leu |
| 9. | Isoleucine[c] | $-CH(CH_3)(CH_2-CH_3)$ | Ile |
| 10. | Methionine[c] | $-CH_2-CH_2-S-CH_3$ | Met |
| 11. | Aspartic acid | $-CH_2CO_2H$ | Asp |
| 12. | Glutamic acid | $-CH_2-CH_2-CO_2H$ | Glu |
| 13. | Lysine[c] | $-CH_2-CH_2-CH_2-CH_2-NH_2$ | Lys |
| 14. | Arginine | $-CH_2-CH_2-CH_2-NHC(NH)NH_2$ | Arg |
| 15. | Phenylalanine[c] | $-CH_2-C_6H_5$ | Phe |
| 16. | Tyrosine | $-CH_2-C_6H_4-OH$ | Tyr |
| 17. | Tryptophan[c] | $-CH_2$(indole) | Trp |
| 18. | Histidine | $-CH_2$(imidazole) | His |
| 19. | Proline | (pyrrolidine ring, $H_2C-CH_2$, $H_2C$, $N-H$, $CHCO_2H$) | Pro |
| 20. | Hydroxyproline | (HOHC—CH_2, $H_2C$, $N-H$, $CHCO_2H$) | Hyp |

[a]These are the official IUPAC symbols for amino acids.
[b]Cystine involves two cysteine units joined by disulfide linkage (—S—S—).
[c]Essential amino acids needed in the diet of humans.

Often, for convenience, cyclic compounds are represented by a geometrical figure corresponding to the carbon sequence.

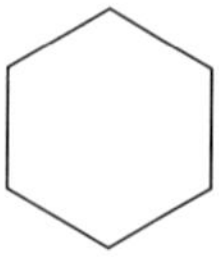

represents 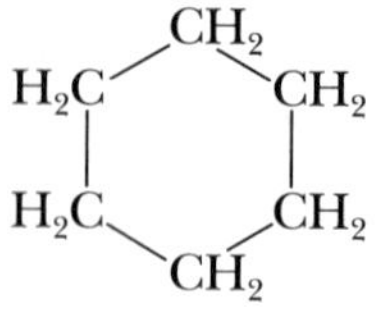 cyclohexane

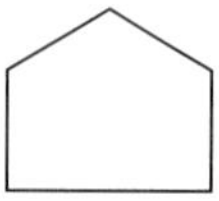

represents 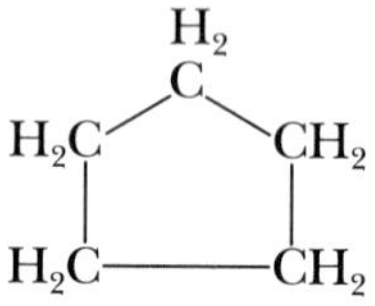 cyclopentane

In these figures each corner represents a $CH_2$ group. **Substituted ring compounds** can be represented using the figure with the group that replaced an —H connected to the figure. For instance, the alcohol cyclohexanol, $C_6H_{11}OH$, can be shown as

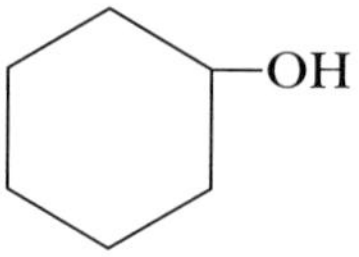

—OH

**Benzene, $C_6H_6$,** is an important ring compound in organic chemistry. This compound can be represented by the structural formulas

Use six of your double-bond-carbon patterns to make a model of benzene. What is the shape of your model?

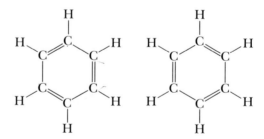

Benzene is an unusually stable ring compound, and many important substances are formed by substituting other groups for hydrogens on the benzene ring. Notice that the structural formulas given above indicate that benzene has alternating single and double bonds around the ring. Either of the two arrangements shown could be used to represent benzene. Actually, these two formulas are different. One shows the double bonds in one possible arrangement, and the other shows the double bonds in another arrangement. However, studies have revealed that all the carbon–carbon bonds in benzene are the same. Thus, neither of the structures represents the actual bonding electron distribution in the benzene ring. The actual structure is some where between these two extremes.

How should we represent benzene? One possible way is to draw the structure to indicate that the carbon–carbon bonds are neither completely single nor completely double bonds.

However, since it is not convenient to write this structure every time we want to represent benzene, it is possible to use a geometrical figure, as was done with the cyclic alkanes. The common symbol used to represent the formula of benzene is

Sometimes benzene is represented by similar symbols which show the alternating bonds.

or

Benzene is used as a starting material to prepare many useful compounds. Compounds involving a benzene ring belong to a class of compounds called **aromatic** compounds. Substituted benzenes are represented by showing the substituted group attached to the ring. Some common aromatic compounds are listed in Table 16-6. For example, the structure of aspirin or acetylsalicylic acid is

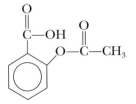

Use your patterns to make a model of aspirin.

**Table 16-6**  Some Typical Aromatic Compounds

| Formula | Name | Comment |
|---|---|---|
| $CH_3$ (benzene ring) | Toluene | Used to prepare TNT and as a solvent |
| CHO (benzene ring) | Benzaldehyde | Flavoring agent, "oil of almond" |
| Cl (benzene ring) Cl | *para*-Dichlorobenzene | Mothballs |
| $CH_3$ with $NO_2$, $NO_2$, $NO_2$ (benzene ring) | 2,4,6-Trinitrotoluene | TNT, an explosive |
| $\overset{O}{\overset{\|}{C}}-O-CH_3$, OH (benzene ring) | Methyl salicylate | Oil of wintergreen; used in rubbing ointments |
| $\overset{O}{\overset{\|}{C}}-OH$, $O-\overset{O}{\overset{\|}{C}}-CH_3$ (benzene ring) | Acetylsalicylic acid | Aspirin; analgesic (pain reliever) |
| OH, $OCH_3$, CHO (benzene ring) | Vanillin | Vanilla flavoring |
| $CO_2H$ (benzene ring) $CO_2H$ | Terephthalic acid | Used in the manufacture of polyesters such as Dacron, Fortrel, and Mylar |

## 16-15 POLYCYCLIC AND HETEROCYCLIC COMPOUNDS

Compounds in which two or more rings mutually share carbons are called **polycyclic** (many ring) **compounds.** Polycyclic compounds, such as cholesterol, are common in biochemistry.

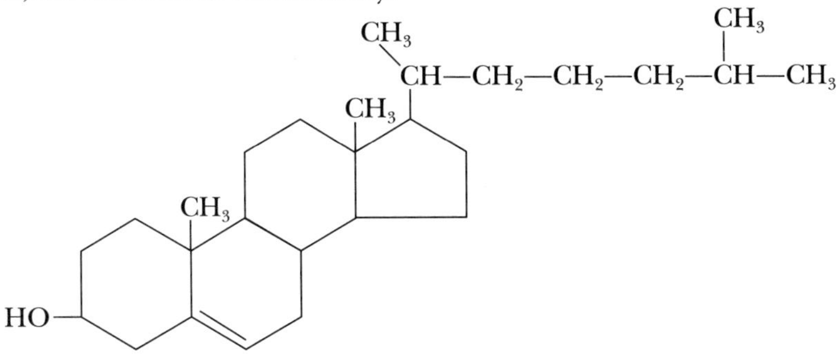

cholesterol

Two unique cyclic compounds are shown below.

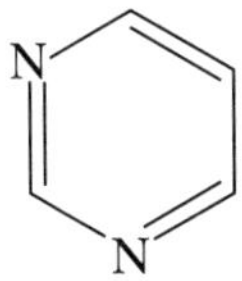

pyrimidine 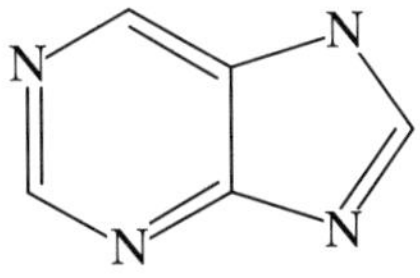            purine

Such cyclic compounds that include purine and pyrimidine have atoms other than carbon in their rings. Cyclic compounds that have carbon and another atom in their ring or rings are called **heterocyclic compounds.** Some other important heterocyclic compounds are

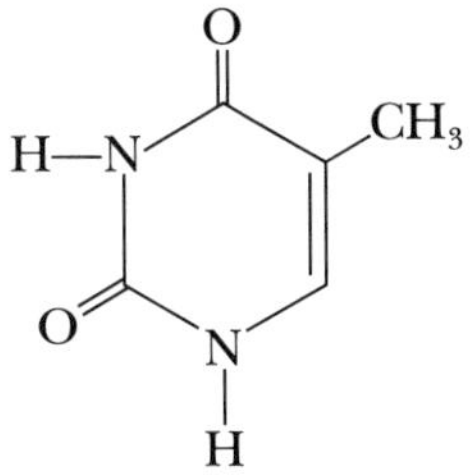

thymine 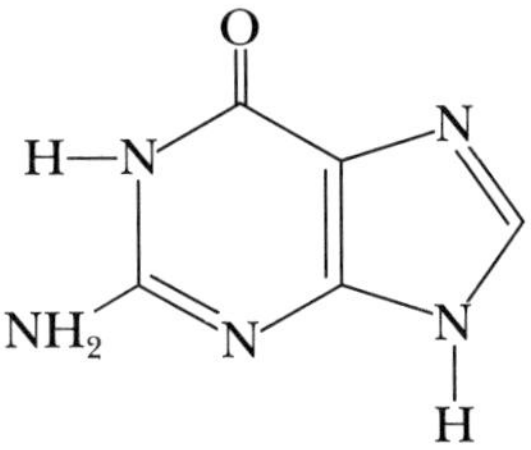            guanine

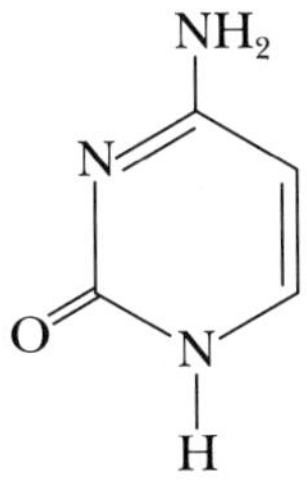

cytosine 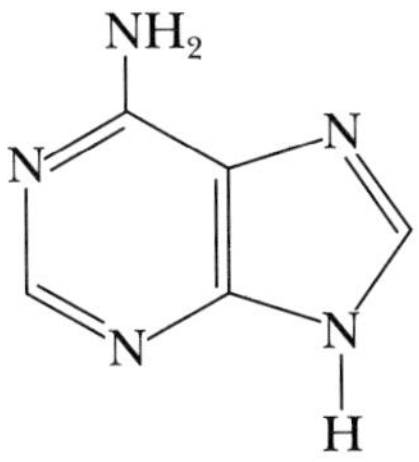            adenine 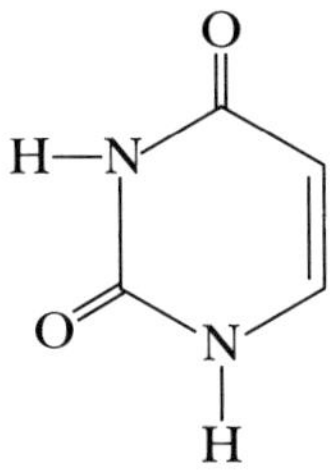            uracil

These five compounds are the heterocyclic bases or heterocyclic amines, discussed in Section 17-8, involved in biologically important ribonucleic acids (RNA) and deoxyribonucleic acids (DNA).

Some important classes of organic compounds and some common names of organic compounds have been discussed in this chapter. Organic chemistry covers many other compounds, chemical reactions, and systems of nomenclature. The organic chemistry considered in this chapter is summarized in Focus 16.

## Organic Compounds

| NAME | GENERAL FORMULA | FUNCTIONAL GROUP | EXAMPLE |
|---|---|---|---|
| Alkanes | $C_nH_{2n+2}$ | R— | $CH_3CH_2CH_3$ propane |
| Alkenes | $C_nH_{2n}$ | $\diagdown C{=}C \diagup$ | $CH_3CH{=}CH_2$ propene |
| Alcohols | R—OH | $-\overset{\displaystyle \mid}{\underset{\displaystyle \mid}{C}}-OH$ | $CH_3CH_2CH_2OH$ propanol |
| Aldehydes | R—CHO | $-\overset{\displaystyle O}{\overset{\displaystyle \parallel}{C}}-H$ | $CH_3CHO$ acetaldehyde |
| Acids | $R-CO_2H$ | $-\overset{\displaystyle O}{\overset{\displaystyle \parallel}{C}}-OH$ | $CH_3CO_2H$ acetic acid |
| Esters | $R-CO_2-R$ | $-\overset{\displaystyle O}{\overset{\displaystyle \parallel}{C}}-O-\overset{\displaystyle \mid}{\underset{\displaystyle \mid}{C}}-$ | $CH_3CO_2CH_2CH_3$ ethyl acetate |
| Amines | $RNH_2,\ R_2NH,\ R_3N$ | $-\overset{\displaystyle \mid}{\underset{\displaystyle \mid}{C}}-\overset{\displaystyle \mid}{N}-$ | $CH_3NH_2$ methylamine |
| Amides | $RCONH_2$ | $-\overset{\displaystyle O}{\overset{\displaystyle \parallel}{C}}-\overset{\displaystyle \mid}{N}-$ | $CH_3CONH_2$ acetamide |
| Amino acids | $NH_2CHRCO_2H$ | $-\overset{\displaystyle \mid}{\underset{\displaystyle \mid}{N}}-\overset{\displaystyle \mid}{\underset{\displaystyle \mid}{C}}-\overset{\displaystyle O}{\overset{\displaystyle \parallel}{C}}-OH$ | $NH_2CH_2CO_2H$ glycine |

| NAME | GENERAL FORMULA | FUNCTIONAL GROUP | EXAMPLE |
|------|-----------------|------------------|---------|
| Cyclic alkanes | $C_nH_{2n}$ | | cyclohexane |
| Aromatics | Substituted benzene ring | | Cl chlorobenzene |
| Heterocyclics | Element other than carbon in ring | | pyridine |

You need: a zip-lock bag, white vinegar, rubbing alcohol (70% isopropyl alcohol), a bowl and boiling hot water.

1. Place half a cup of vinegar in a zip-lock bag. Note the odor of acetic acid in vinegar.
2. Add two teaspoons of rubbing alcohol to the bag, expel the air and seal the bag. Cautiously note the odor of isopropyl alcohol.
3. Put the zip-lock bag in a bowl and carefully pour boiling hot water over it. Let the bag sit in the water for 10 minutes.
4. With caution, open the bag and carefully smell the contents. The ester is safe but still do not breathe too much of it. Describe the odor.
5. The ester you made was isopropyl acetate. Write the structural formula of this ester and use your patterns to make a model of it.

## QUESTIONS

### Section 16-1

1. Give a definition for organic chemistry.
2. What are synthetic organic chemicals?

### Sections 16-2 and 16-3

3. Describe the following and give a structure for each.
   (a) the tetravalency of carbon
   (b) a carbon chain
   (c) a carbon–carbon double bond
   (d) a carbon–carbon triple bond
4. Describe how carbon may form bonds with the following atoms and give an example of each.
   (a) hydrogen
   (b) oxygen
   (c) nitrogen
   (d) a halogen
5. What is the difference between a molecular formula and a structural formula?

### Section 16-4

6. Describe the group of compounds known as saturated hydrocarbons or alkanes. Draw the structure of an alkane with three carbons.
7. Alkanes have the general formula $C_nH_{2n+2}$. Deduce formulas for the following alkanes.
   (a) heptane, $n = 7$
   (b) dodecane, $n = 12$
   (c) nonane, $n = 9$
   (d) decane, $n = 10$
8. Give the formulas of the following alkanes.
   (a) methane
   (b) octane
   (c) pentane
   (d) ethane
   (e) propane
   (f) hexane

### Section 16-5

9. What is an alkyl group?
10. Give the name of each of the following alkyl groups.

(a) $-CH_3$

(d) $-\overset{\displaystyle CH_3}{\underset{\displaystyle CH_3}{C}}-CH_3$

(b) $CH_3CH_2CH_2CH_2-$   (e) $CH_3\overset{\displaystyle CH_3}{CH}CH_2-$

(c) $(CH_3)_2CH-$   (f) $-CH_2CH_3$

11. Give the structural formula of each of the following alkyl groups.
   (a) isobutyl group
   (b) methyl group
   (c) isopropyl group
   (d) tertiary butyl group
   (e) propyl group
   (f) ethyl group

12. Explain what R-group means.

13. Pick out the alkyl group or groups in the following formulas and give their names.

(a) $CH_3-\overset{\displaystyle O}{\overset{\|}{C}}-CH_3$

(b) $(CH_3)_2CH-O-CH_3$

(c) $CH_3CH_2CH_2OH$

(d) $CH_3CH_2-\overset{\displaystyle O}{\overset{\|}{C}}-O-CH_2CH_3$

### Section 16-6

14. What is the distinguishing characteristic of a compound that is classed as an alkene? Give the name and structural formula of a typical alkene.

### Section 16-8

15. What are structural isomers? Which of the following sets of compounds are isomers?

(a) $CH_3-\overset{\displaystyle CH_3}{\underset{\displaystyle CH_3}{C}}-CH_3$ and $CH_3CH_2CH_2CH_2CH_3$

(b) $CH_3-O-CH_3$ and $CH_3CH_2OH$

(c) $CH_3\overset{\displaystyle CH_3}{CH}CH_2CH_3$ and $CH_3CH_2CH_2CH_3$

## Section 16-7

16. What is an alcohol? Give an example of the formula of a primary alcohol.

17. What is the structure of the polyhydroxy alcohol glycerol or glycerine?

## Sections 16-9 and 16-10

18. Describe the chemical nature and give an example of each of the following classes of organic compounds.

    (a) aldehyde

    (b) carboxylic acid

    (c) ester

19. *Give the formula of the esters formed by the condensation reaction between the following pairs of carboxylic acid and alcohol.

$$
\text{(a)}\quad \underset{\displaystyle}{HC}\!-\!OH \quad\text{and}\quad CH_3OH
$$
$$
\text{(with } O \text{ double bonded to } HC)
$$

$$
\text{(b)}\quad CH_3CH_2C\!-\!OH \quad\text{and}\quad (CH_3)_2CHOH
$$
$$
\text{(with } O \text{ double bonded to } C)
$$

20. *Give the formulas of the carboxylic acid and alcohol that react to give the following esters.

(a) *n*-octyl acetate, $CH_3(CH_2)_7\!-\!O\!-\!CCH_3$ (with O double bonded to C)

(b) *n*-pentyl butyrate

$$CH_3(CH_2)_4\!-\!O\!-\!CCH_2CH_2CH_3$$ (with O double bonded to C)

(c) ethyl acetate, $CH_3CH_2\!-\!O\!-\!CCH_3$ (with O double bonded to C)

(d) ethyl dihydrogen phosphate
$CH_3CH_2\!-\!OPO_3H_2$

## Section 16-11

21. Define the following.

    (a) an amine

    (b) a primary amine

    (c) a secondary amine

    (d) a tertiary amine

## Section 16-12

22. What is an amide and which two compounds can react in a condensation reaction to form an amide?

## Section 16-13

23. Describe the chemical nature of amino acids.

24. *Amino acids can react with one another to form amide linkages. The carboxylic acid group of one amino acid reacts with the amino group of another. Give the formula of the compound with the amide linkage formed by the condensation reaction of the amino acids glycine and alanine in which glycine contributes the carboxyl group.

$$
\underset{\text{glycine}}{NH_2CH_2\!-\!\overset{\displaystyle O}{\overset{\|}{C}}\!-\!OH} \qquad \underset{\underset{\text{alanine}}{CH_3}}{H\!-\!NHCH\!-\!\overset{\displaystyle O}{\overset{\|}{C}}\!-\!OH}
$$

25. Refer to Question 24. Since amino acids have both carboxyl groups and amino groups, any two amino acids can link in two ways. Give the formula of the compound formed by the amide linkage of alanine and glycine in which alanine contributes the carboxyl group and glycine the amino group.

## Section 16-14

26. What is a cyclic hydrocarbon? Give an example.

27. Draw the symbolic formula used to represent benzene?

28. What is an aromatic compound? Give an example.

## Section 16-15

29. What is a heterocyclic compound?

## Sections 16-4 to 16-15

30. *Classify each of the following compounds as alkane, alkene, alcohol, aldehyde, carboxylic acid, ester, amine, amide, amino acid, aromatic, or heterocyclic. Some may fall into more than one class.

    (a) $CH_3CH_2CH_2OH$

    (b) $CH_3CH_2\overset{\displaystyle O}{\overset{\|}{C}}\!-\!OH$

    (c) (benzene ring)$\overset{\displaystyle O}{\overset{\|}{C}}\!-\!O\!-\!CH_3$

    (d) $CH_3CH_2CH_2CH_3$

(e) 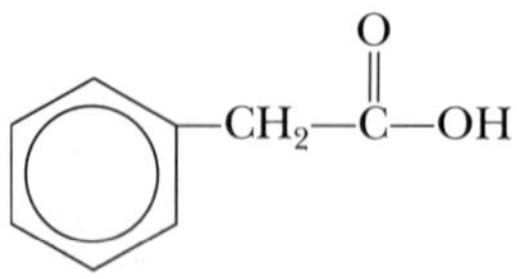 Benzene ring—$CH_2$—$\overset{\displaystyle O}{\overset{\|}{C}}$—OH

(f) 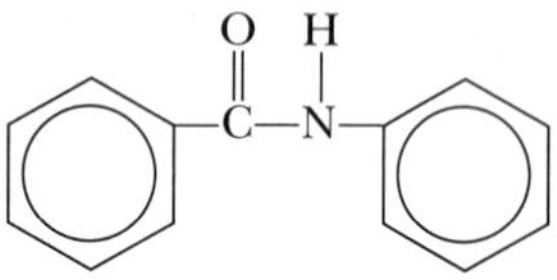 Benzene ring—$\overset{\displaystyle O}{\overset{\|}{C}}$—$\overset{\displaystyle H}{\overset{|}{N}}$—Benzene ring

(g) 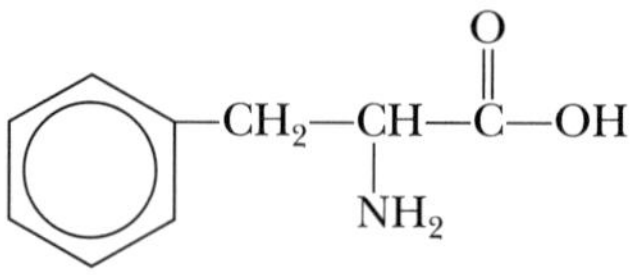 Benzene ring—$CH_2$—$\underset{\displaystyle NH_2}{\overset{\displaystyle O}{\overset{|}{CH}}}$—$\overset{\displaystyle O}{\overset{\|}{C}}$—OH

(h) Pyrrole ring with N—H

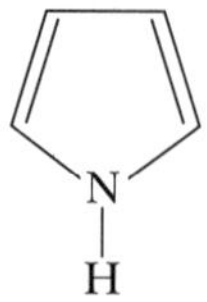

(i)  $CH_3NH_2$

(j)  $CH_3CH{=}CH_2$

(k) 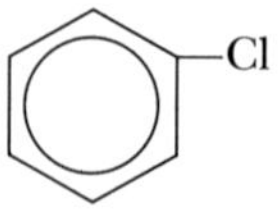 Benzene ring—Cl

(l) 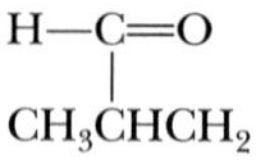  H—C=O
    $CH_3CHCH_2$

(m) Benzene ring—$\overset{\displaystyle O}{\overset{\|}{C}}$—OH

(n) $CH_3$—$\underset{\displaystyle CH_3}{\overset{\displaystyle |}{N}}$—$CH_3$

(o) HO—$\overset{\displaystyle O}{\overset{\|}{C}}$—$\underset{\displaystyle CH_3}{\overset{\displaystyle |}{CH}}$—$NH_2$

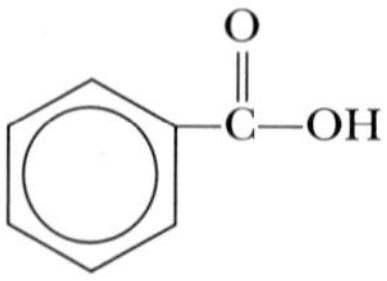

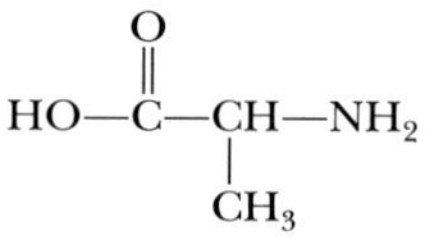

(p) 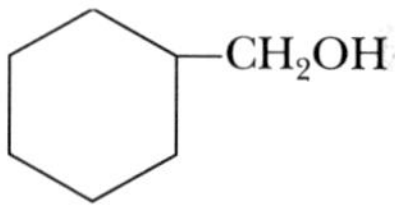 Cyclohexane ring—$CH_2OH$

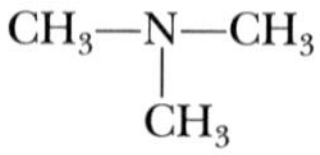

# THE

# CHEMISTRY

# OF LIFE:

# BIOCHEMISTRY

## 17-1 BIOCHEMISTRY

A living organism is an enormously complex mixture of various kinds of chemical compounds that act in concert to maintain life. Considering the great diversity of plants and animals, it is amazing that many of the same chemical compounds and chemical processes are shared by all life forms. Most biochemical compounds are organic. Combined hydrogen, oxygen, carbon, and nitrogen make up about 99% of the atoms of living organisms. Calcium, chlorine, magnesium, phosphorus, iron, potassium, sodium, and sulfur in the form of ions and in compound make up most of the remaining 1%. Several other elements are present in small amounts as trace elements that are also essential to life processes.

Animal and plant life contain a wide variety of organic and inorganic compounds. **Biochemistry** is the study of compounds involved in biological processes and the changes they undergo during life processes. The field of biochemistry is one of the frontiers of science. Many exciting discoveries are being made by biochemists. Such discoveries range from the establishment of a theory to explain the chemistry of heredity to the development of drugs and vaccines to treat diseases. Biochemists and molecular biologists have developed methods to alter the genes of bacteria and yeasts so they can be used to manufacture useful proteins. These methods

**559**

are the basis for the exciting field of **biotechnology.** The list of products being manufactured or considered for manufacture by biotechnological methods includes insulin, interferon (used to treat infections and inhibit tumor growth), the enzyme urokinase (used to dissolve blood clots), rennin (used in cheese making) and cellulase (used to make glucose sugar from plant cellulose). It would be impossible for humans to manufacture these compounds without modern methods of biotechnology. However, the synthesis of many chemicals by plants and animals cannot be duplicated in the laboratory. It is not possible, for instance, for humans to make carbohydrates from carbon dioxide and water but green plants carry out this synthesis everyday.

This chapter is a brief introduction to biochemistry. Some groups of compounds fundamental to biochemistry are described. In addition, the chemical aspects of several interesting biochemical processes are presented.

## 17-2 CARBOHYDRATES

**Carbohydrates** or **sugars** are produced by green plants during photosynthesis. Carbohydrates contain only carbon, hydrogen, and oxygen. They occur in a variety of sizes, ranging from simple sugars called **monosaccharides** to polymeric molecules called **polysaccharides.** Most of the common monosaccharides are **polyhydroxy aldehydes,** carbon chains with an aldehyde functional group on one end and a hydroxy group on each of the other carbons (see Sections 16-7 and 16-9). Among the most important monosaccharides are **ribose** and **glucose,** also known as dextrose.

$$
\begin{array}{cc}
& \text{O} \\
& \| \\
\text{O} & \text{C—H} \\
\| & | \\
\text{C—H} & \text{H—C—OH} \\
| & | \\
\text{H—C—OH} & \text{HO—C—H} \\
| & | \\
\text{H—C—OH} & \text{H—C—OH} \\
| & | \\
\text{H—C—OH} & \text{H—C—OH} \\
| & | \\
\text{CH}_2\text{—OH} & \text{CH}_2\text{—OH} \\
\text{D-ribose} & \text{D-glucose} \\
\text{(ribose)} & \text{(glucose)}
\end{array}
$$

These structures represent the open-chain forms of these sugars. For reference, the carbons are numbered starting from the aldehyde group. For example, the carbons in glucose are numbered as

Glucose actually occurs predominantly as a **heterocyclic ring** compound. The ring is normally formed by intramolecular reaction between the aldehyde group on C-1 and the —OH group on C-5. When glucose forms a ring compound, two isomers are possible. The two ring forms of **glucose** can be pictured as

α-glucose ⇌ carbon 1 ⇌ β-glucose

The —OH on C-1 is above the ring in the β-form and below in the α-form. These cyclic forms can be more conveniently written as a heterocyclic rings using a geometrical figure like those used to represent cycloalkanes. Thus, the two forms of glucose can be represented as

β-glucose
C-1—OH above ring

α-glucose
C-1—OH below ring

The biochemically important cyclic form of **ribose** involves a five-atom ring represented by the formula shown on the next page.

You can use paper models to picture the difference between α-glucose and β-glucose. On a clean sheet of paper draw two separate hexagons and draw small rectangular areas extending from the six corners of each hexagon at 120° angles. Use scissors to cut out the hexagons with the rectangular shapes attached. To model β-glucose hold a hexagon flat, tear off the rectangle on the upper right, then move clockwise to the next rectangle and bend it to point up. This is C-1. Continue clockwise, bend the next rectangle so it points down, and alternate the rectangles up and down around the hexagon. To model α-glucose hold the other hexagon flat, tear off the rectangle on the upper right, then move clockwise to the next rectangle and bend it to point down. Continue clockwise, bend the next rectangle so it also points down, and alternate the other rectangles up and down around the hexagon. Number the carbons clockwise starting with C-1. The oxygen of the ring is the corner without a rectangle. C-6 extends from C-5. Make sketches of your models.

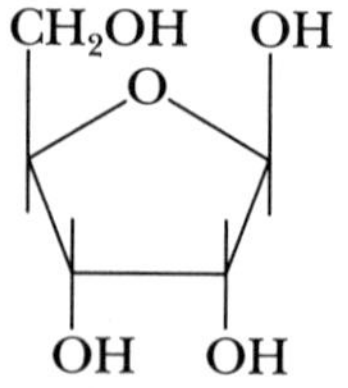

β-ribose

Another cyclic sugar of biochemical importance has the ribose structure with the —OH group on C-2 replaced by a hydrogen. Because the oxygen of the —OH group is not present, this compound is called **2-deoxyribose.**

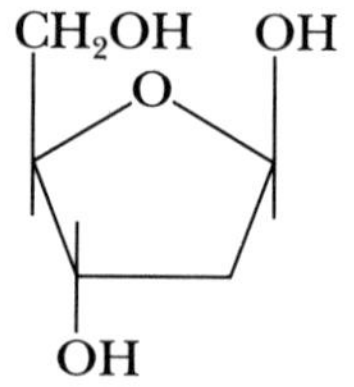

β-2-deoxyribose

As discussed in Section 17-8, ribose and deoxyribose are involved in the structures of DNA and RNA.

## 17-3 DISACCHARIDES AND POLYSACCHARIDES

Glucose combined in cellulose, starch, and sugars or as the free monosaccharide is undoubtedly the most abundant organic compound in nature. It is an important food source for humans that is found in plants, fruits, and vegetables and is present in the bloodstream of certain animals including humans. Two α-glucose molecules can link together through specific —OH groups to form the compound **maltose.**

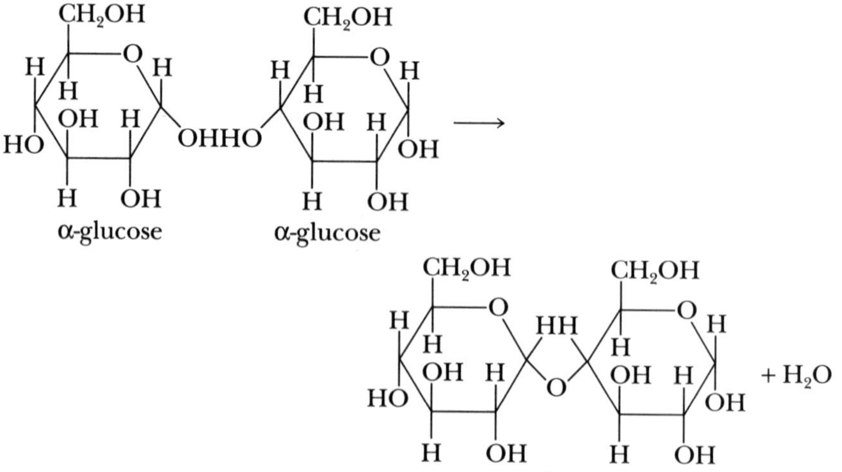

α-glucose          α-glucose

$+ H_2O$

maltose

The two α-glucose units are linked through C-1 and C-4 by what is called an **α-linkage** or **α-glycoside linkage.** Maltose is an example of a sugar called a **disaccharide** in which two monosaccharide units are joined.

Another disaccharide of interest is **sucrose** or common **table sugar,** which involves α-glucose linked to the β-form of the five-membered ring of the monosaccharide called **fructose.** Sucrose is a common chemical made by plants. It is especially abundant in sugarcane and sugar beet, which are its major commercial sources. Honey is sugar solution that has been collected by bees and chemically decomposed by them into glucose and fructose. Fructose has a slightly sweeter taste than sucrose so honey tastes sweeter than table sugar. The formulas below show sucrose and its components.

**To help picture how the glucose units link in maltose, you need two α-glucose models like those you made in Activity 17-1. Hold one model with the fingers of your left hand so that the oxygen is on the upper right. Hold the other model in a similar way with the fingers of your right hand. Bring the models together so that they touch at OH groups of C-1 on the left and C-4 on the right. This represents the α-linkage in maltose.**

α-glucose     β-fructose

sucrose

Many glucose units can link together to form polymeric molecules known as **polysaccharides.** Polysaccharides that include long chains of hundreds or thousands of glucose units are found in plants and animals. They are **complex carbohydrates. Starch** is a polysaccharide that occurs in the seeds and roots of many plants (see Fig. 17-1). It can be digested and is an important human food source. Actually, there are two forms of starch that occur in plants. **Amylose** consists of long chains of α-glucose units. **Amylopectin** consists of branched chains of α-glucose units. **Glycogen** is a polysaccharide found in animals. The structure of glycogen is similar to that of amylopectin. In our bodies the liver and muscles remove excess glucose from the bloodstream and convert it into glycogen. Glycogen serves as an energy source since it can be quickly broken down by bodily processes to form glucose. Metabolism of the glucose provides energy.

To picture the difference between amylose and amylopectin make a chain of 10 paper clips to represent part of an amylose molecule. To represent amylopectin attach a shorter chain of 4 clips at clip number 4 of the longer chain and a chain of 3 clips at clip number 8 of the longer chain. One of the two forms of starch is more soluble in water than the other. Which form do you think is more soluble and why?

**FIGURE 17-1**
Amylose and amylopectin
starch.

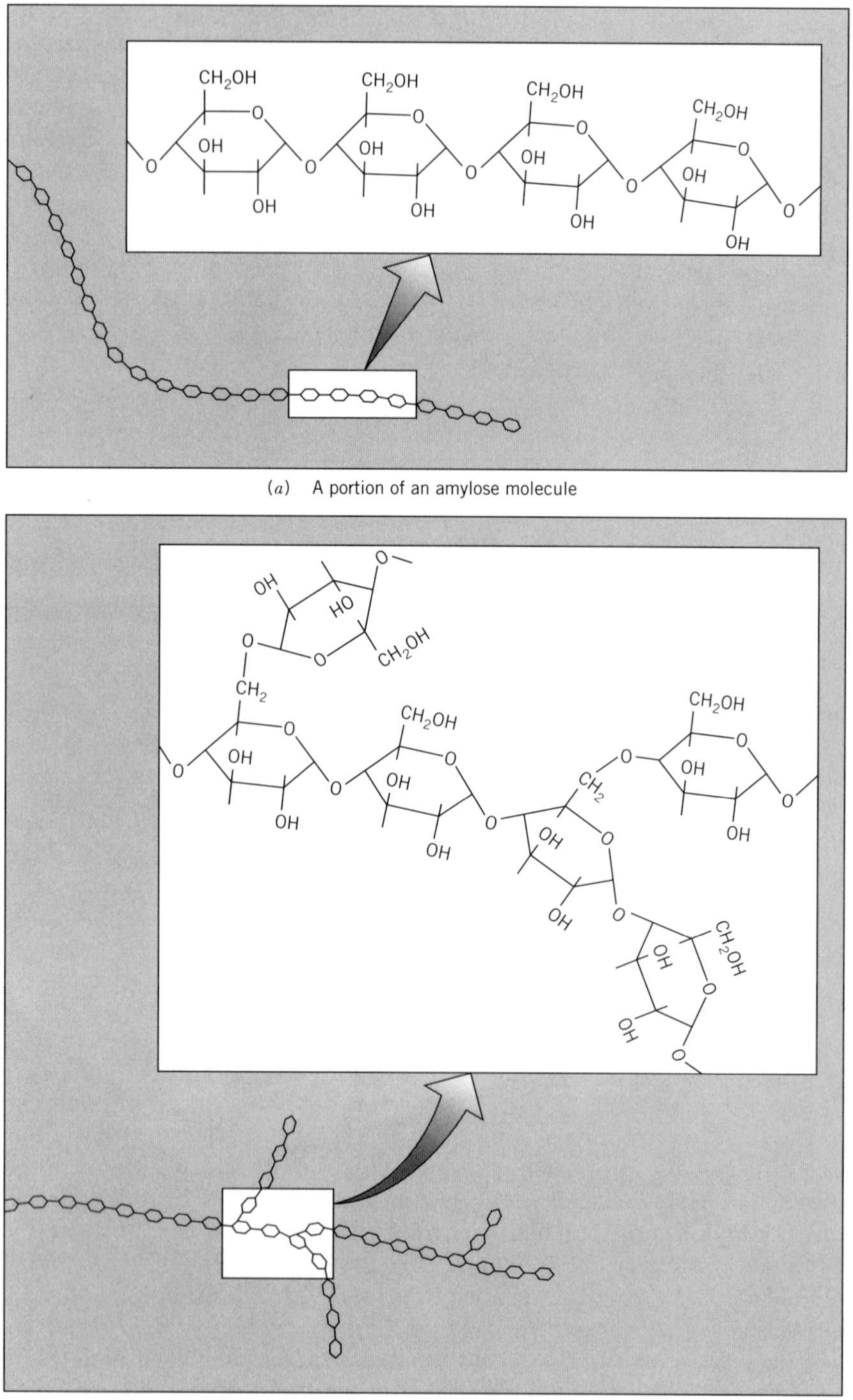

(*a*)   A portion of an amylose molecule

(*b*)   A portion of an amylopectin molecule

Another important polysaccharide, **cellulose,** makes up the structural material of plants. Cellulose consists of long chains of β-glucose units. It cannot be digested by humans, but it can be broken down by certain microorganisms. Cellulose is obtained in large amounts from trees and the cotton plant. It is used in the manufacture of paper, cotton cloth, cellophane, cellulose nitrate (used in paints and explosives), cellulose acetate, and rayon (synthetic fibers used to make textiles). The structural difference between starch and cellulose is simply that starch consists of α-glucose molecules bonded to form polymer molecules, and cellulose consists of β-glucose molecules bonded to form polymer molecules. We can digest starch but not cellulose. Complex carbohydrates are starches and cellulose is a type of dietary fiber sometimes called roughage.

## 17-4 LIPIDS

When the cells of living organisms are crushed and mixed with solvents, certain chemical components can be dissolved. Some cellular components are water soluble, whereas others are sparingly soluble in water. Those sparingly soluble in water are quite soluble in nonpolar organic solvents such as carbon tetrachloride. Those biochemical compounds that can be extracted from crushed cells by organic solvents are called **lipids.** Lipids represent a wide variety of compounds but only the **simple lipids** and **steroids** will be considered here.

**Simple lipids** are **esters of glycerol** and long-chain carboxylic acids. These esters, sometimes called **triglycerides,** can be represented by the general formula given below.

$$
\begin{array}{c}
\quad\quad\ \ \overset{\displaystyle O}{\underset{\displaystyle \|}{\ }} \\
R\!-\!C\!-\!O\!-\!CH_2 \\
\quad\quad\ \ \overset{\displaystyle O}{\underset{\displaystyle \|}{\ }}\quad\ | \\
R'\!-\!C\!-\!O\!-\!CH_2 \\
\quad\quad\ \ \overset{\displaystyle O}{\underset{\displaystyle \|}{\ }}\quad\ | \\
R''\!-\!C\!-\!O\!-\!CH_2
\end{array}
$$

where R, R′, and R″ correspond to various hydrocarbon groups. The esters of glycerol are called **fats** or **oils.** The difference between fats and oils is that fats are solid whereas oils are liquid at room temperature. This is the difference between meat fats, butter, and solid margarine compared to bottled vegetable oils. Fats are found mainly in animals, and oils are found mainly in plants. Fats and oils are a major food energy source. In our bodies, fats are stored in **adipose tissue,** which serves as a protective covering for certain parts of the body. Fats are also a reserve energy source for the body.

To observe a fatty acid you need a bar of soap (real soap such as Ivory soap, not some synthetic detergent bar), a glass, some white vinegar, and water. Soap is an ionic compound containing sodium ions and fatty acid ions. Symbolically a soap can be represented as NaFA. Soap solutions have fatty acid ions (FA⁻) that are the conjugate bases of fatty acids. This is why soap solutions are basic.

1. Put one small shaving from a bar of soap into a glass; don't use too much.

2. Add two tablespoons of water and mix the soap into the water with your fingers. Note how it feels.

3. Rinse your hands with water and note the effect of soap as you rinse.

4. Add one-fourth of a cup of vinegar to the glass and observe. Mix the material in the glass with your fingers and note how it feels.

5. Rinse your hands with water and note that fatty acids are different from soap.

Write an equation to show how a fatty acid anion, FA⁻, reacts with water slightly to give a basic solution. Write another equation to show how a fatty acid, HFA, is formed when acetic acid is added to a soap solution.

Glycerol esters can be catalytically broken down or hydrolyzed into glycerol and the constituent acids, called **fatty acids.**

$$
\begin{array}{l}
\underset{\text{O}}{R-\overset{\text{O}}{\overset{\|}{C}}-O-CH_2} \\
\underset{\text{O}}{R'-\overset{\text{O}}{\overset{\|}{C}}-O-CH} + 3H_2O \\
R''-\overset{\text{O}}{\overset{\|}{C}}-O-CH_2
\end{array}
\xrightarrow{\text{enzymes}}
\begin{array}{l}
R-\overset{\text{O}}{\overset{\|}{C}}-OH \\
+ \\
R'-\overset{\text{O}}{\overset{\|}{C}}-OH \\
+ \\
R''-\overset{\text{O}}{\overset{\|}{C}}-OH
\end{array}
+
\begin{array}{l}
CH_2-OH \\
CH-OH \\
CH_2-OH \\
\text{glycerol}
\end{array}
$$

fatty acids

Table 17-1 lists some typical fatty acids obtained by hydrolysis of simple lipids. Fatty acids with no carbon–carbon double bonds are called **saturated fatty acids,** those with one carbon–carbon double bond are **monounsaturated fatty acids** and those with two or more carbon–carbon double bonds are **polyunsaturated fatty acids. Partially hydrogenated vegetable oils** are oils that have been reacted with hydrogen to saturate some of the

*Table 17-1*   Typical Fatty Acids

| NAME | FORMULA |
|---|---|
| *Saturated Fatty Acids* | |
| Myristic acid | $CH_3(CH_2)_{12}\overset{O}{\overset{\diagup\!\!\diagup}{C}}-OH$ |
| Palmitic acid | $CH_3(CH_2)_{14}\overset{O}{\overset{\diagup\!\!\diagup}{C}}-OH$ |
| Stearic acid | $CH_3(CH_2)_{16}\overset{O}{\overset{\diagup\!\!\diagup}{C}}-OH$ |
| *Monounsaturated Fatty Acid* | |
| Oleic acid | $CH_3(CH_2)_7CH{=}CH(CH_2)_7\overset{O}{\overset{\diagup\!\!\diagup}{C}}{\diagdown}_{OH}$ |
| *Polyunsaturated Fatty Acids* | |
| Linoleic acid | $CH_3(CH_2)_4CH{=}CHCH_2CH{=}CH(CH_2)_7\overset{O}{\overset{\diagup\!\!\diagup}{C}}{\diagdown}_{OH}$ |
| Linolenic acid | $CH_3CH_2CH{=}CHCH_2CH{=}CHCH_2CH{=}CH(CH_2)_7\overset{O}{\overset{\diagup\!\!\diagup}{C}}{\diagdown}_{OH}$ |

double bonds. Partially hydrogenated vegetable oils contain some saturated fats.

Fats and oils include common products such as butter, lard, olive oil, peanut oil, corn oil, canola oil, and safflower oil. Table 17-2 lists the compositions of some common fats and oils. Those fats with high percentages of saturated fatty acids are called saturated fats. Those oils with high percentages of unsaturated fatty acids are called polyunsaturated oils. Some oils have high percentages of monounsaturated fatty acids. Chemistry in Action 17-1 is about testing vegetable oils for the presence of unsaturated fats.

**Table 17-2**  Percent Fatty Acid Composition of Some Common Fats and Oils

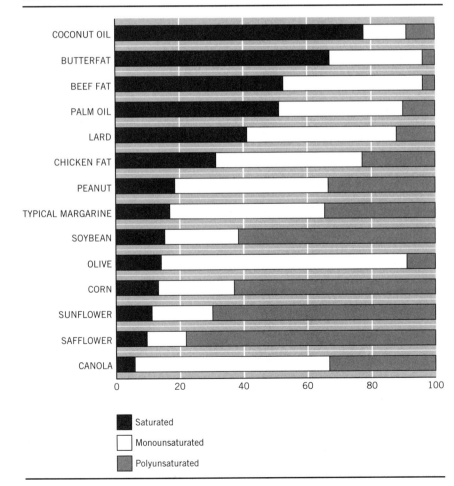

- ■ Saturated
- □ Monounsaturated
- ▨ Polyunsaturated

Use the patterns from Chapter 16 to make an eight-carbon saturated fatty acid including seven carbons and a carboxyl group on the end. Count carbons starting from the carboxyl carbon. Add two of the double-bond carbon patterns between C-5 and C-6 to make a ten-carbon monounsaturated fatty acid. Sketch the structure. Add two more double-bond carbon patterns between C-3 and C-4 to make a twelve-carbon polyunsaturated fatty acid. Sketch the structure.

Read the labels of one or more kinds of vegetable oil and record any information concerning the contents.

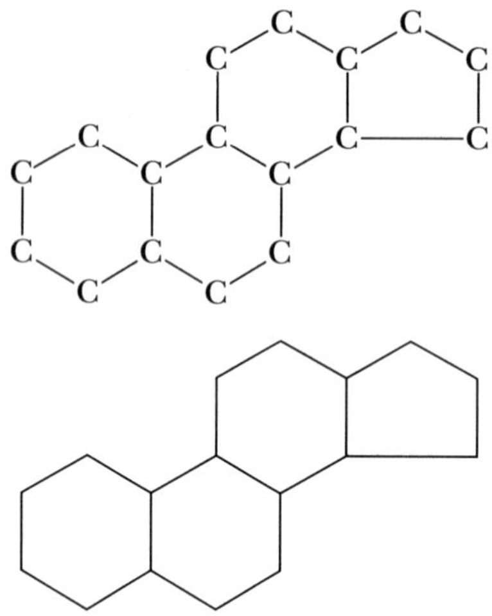

**FIGURE 17-2** The carbon sequence in steroids.

Steroids comprise another important class of lipids. Steroids have quite different chemical structures than simple lipids. Most **steroids** have the basic polycyclic ring structure shown in Figure 17-2. A wide variety of steroids are found in the body, and most have very important physiological activities. Consider the structure and activities of some steroids.

Cholesterol is the most prevalent steroid in the body. The formula of **cholesterol** is

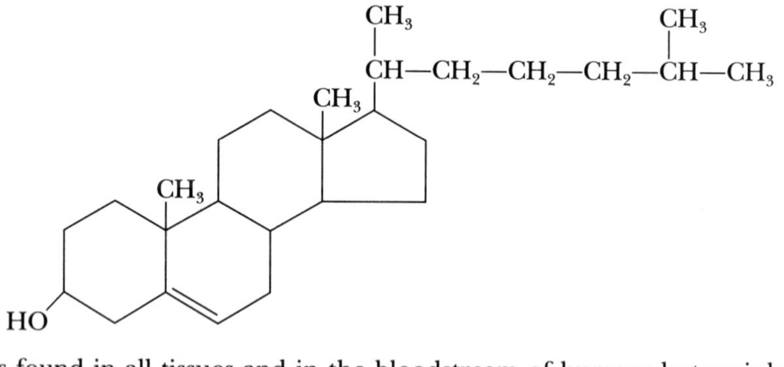

It is found in all tissues and in the bloodstream of humans but mainly in the brain, spinal cord, and nerve tissue. Cholesterol is the starting material for many essential body chemicals such as vitamin D and sex hormones. It is also a component of all cell membranes. Cholesterol is found in meats and other animal products. It is never found in vegetable products.

The male and female **sex hormones** are also steroids. **Hormones** are chemicals produced in the body that carry messages; they stimulate specific bodily processes. A variety of hormones are involved in the development and activities of human sex glands. It is known that these hormones chemically stimulate sexual processes, but the specifics of such functions are not known. The structure of the male hormones testosterone and androsterone produced in the testes are shown below.

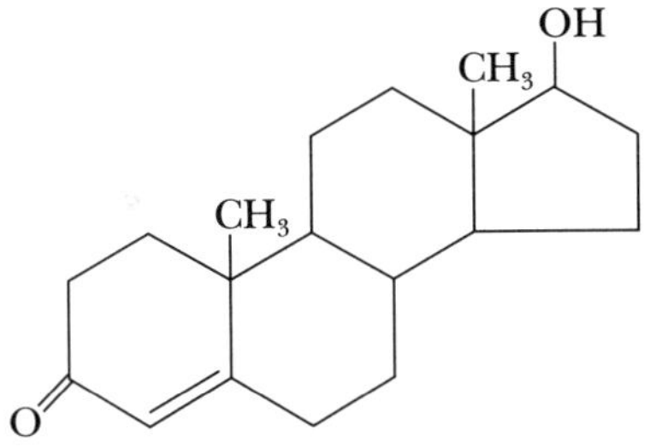

testosterone 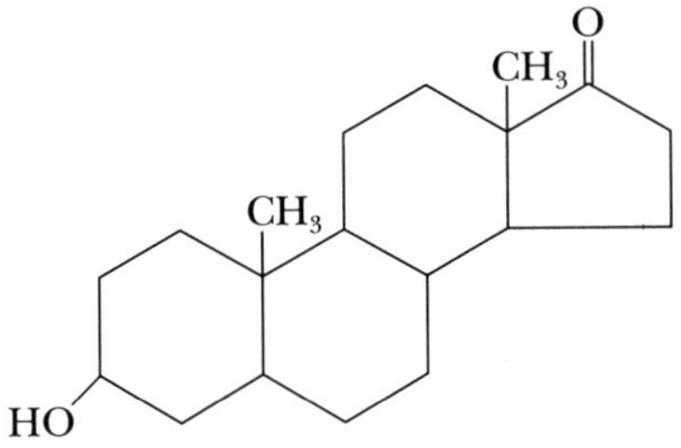                androsterone

The female hormones control the menstrual cycle and maintain the fetus during pregnancy. The structures of two important female sex hormones estrone and progesterone are shown below.

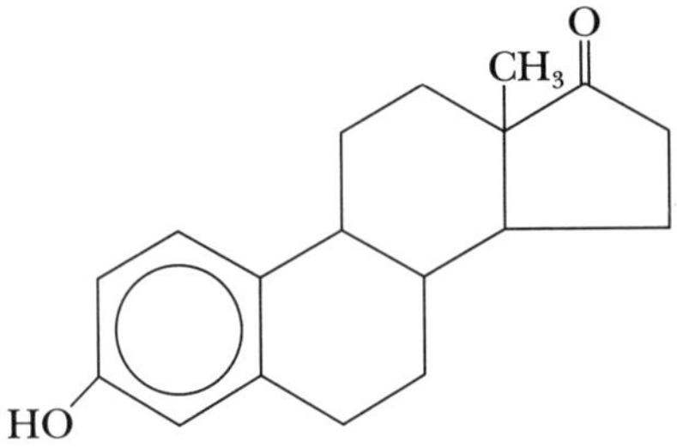

estrone 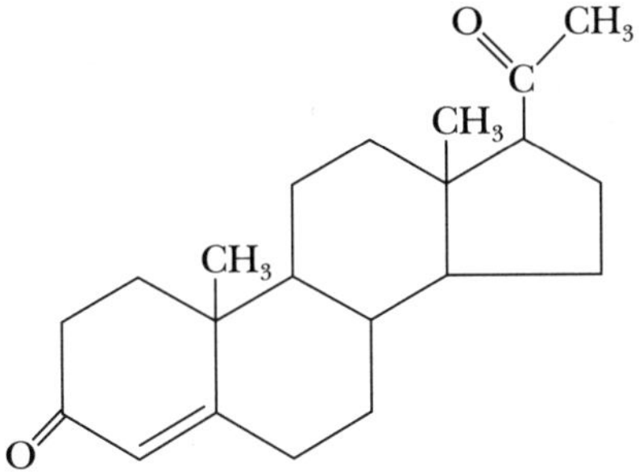                progesterone

Notice the similarities in the structures of male and female hormones. Slight differences in the molecular structures of these hormones give them totally different functions.

## 17-5 PROTEINS

Proteins are polymeric materials found in living organisms. They serve as structural materials in the body and are fundamental to many of the processes of life. **Proteins** are polymers synthesized from **amino acid monomers** in the cells of plants and animals. Proteins from animals and plants are an important food, since they provide amino acids that are essential to the body in the production of needed proteins. Table 16-5 lists amino acids and those that are essential in our diet are noted. The **essential amino acids** cannot be manufactured in our cells, so they must be part of our diets. When proteins are digested, they are broken down by digestive enzymes into their constituent amino acids. The amino acids then become available to the cells for protein synthesis.

Recall that amino acids are both amines and carboxylic acids (see Section 16-13). It is possible for the carboxyl group of one amino acid to condense with the amino group of another to form an amide linkage between the two original amino acid units (see Section 16-12.) For example,

$$\underset{\text{glycine}}{NH_2CH_2\overset{\displaystyle O}{\overset{\displaystyle \|}{C}}\!-\!OH} + \underset{\text{alanine}}{H\!-\!NH\underset{\displaystyle CH_3}{\overset{\displaystyle |}{CH}}\overset{\displaystyle O}{\overset{\displaystyle \|}{C}}OH} \longrightarrow \underset{\text{glycylalanine}}{NH_2CH_2\overset{\displaystyle O}{\overset{\displaystyle \|}{C}}\!-\!NH\underset{\displaystyle CH_3}{\overset{\displaystyle |}{CH}}\overset{\displaystyle O}{\overset{\displaystyle \|}{C}}OH} + H_2O$$

Compounds of two or more amino acids linked by amide linkages are called peptides. The characteristic bond of peptides is

$$-\overset{\displaystyle O}{\overset{\displaystyle \|}{C}}-\underset{\displaystyle |}{N}-$$

These bonds are called **peptide** bonds or **peptide linkages.** Table 16-5 lists the official abbreviations for amino acids used in peptide sequences. Thus, glycylalanine can be written as

Gly-Ala

With this notation, it is assumed that the amino acid on the left has contributed the —OH in the formation of the peptide bond and the amino acid on the right has contributed an —H of an —NH$_2$ group. Glycylalanine is a **dipeptide.**

Three amino acids linked by peptide bonds constitute a **tripeptide.** Using three amino acids, such as glycine, Gly, alanine, Ala, and valine, Val, it is possible to form six different tripeptides.

| | |
|---|---|
| Gly-Ala-Val | Gly-Val-Ala |
| Val-Gly-Ala | Val-Ala-Gly |
| Ala-Val-Gly | Ala-Gly-Val |

As the number of amino acid units increases, the number of possible combinations increases greatly.

**Polypeptides** involving hundreds or thousands of amide linked amino acid units are **proteins.** Proteins serve as structural components of cells, skin, muscles, bone interior, and nerves; as enzymes and hormones; and in many other important functions of the body. Table 17-3 lists some of the functions of proteins. In spite of their diversity, all proteins are polypeptides. Analyses reveal that all proteins in humans are made from about 20 different amino acids.

The sequences in which the amino acids are linked have been determined for many proteins. This requires an extensive amount of laboratory

ACTIVITY
17-7

Use the patterns from Chapter 16 to make a model of glycylalanine.

ACTIVITY
17-8

Peptides can contain more than one unit of the same amino acid. Using abbreviations, give the formulas for the tripeptides that can be formed from two units of glycine, Gly, and one unit of valine, Val.

*Table 17-3*  Functions of Proteins in the Body

| TYPE OF PROTEIN | FUNCTION |
|---|---|
| *Structural proteins (insoluble in water)* | |
| Collagens | Found in connective tissue |
| Elastins | Found in tendons and arteries |
| Myosins | Found in muscle tissue |
| Keratins | Found in hair and nails |
| | |
| *Globular proteins (can be dispersed in water solutions)* | |
| Albumins | Found in blood |
| Globulins | Involved in oxygen transport in the body (hemoglobin) and in body defense against disease (gamma globulin) |
| | |
| *Conjugated proteins (complexes of proteins linked to other molecules)* | |
| Nucleoproteins | Protein–nucleic acid complexes |
| Lipoproteins | Protein–lipid complexes |
| Phosphoproteins | Protein–phosphorus compound complexes |
| Chromoproteins | Protein–pigment complexes (e.g., hemoglobin) |
| | |
| Enzymes (may be conjugated with nonprotein coenzymes) | |
| | |
| Hormones (not all hormones are proteins) | Chemical messengers |
| | |
| Antibodies | Involved in body defense against disease |

work and analysis. The amino acid sequence involved in a protein is called the **primary structure** of the protein. The primary structure of beef insulin is shown in Figure 17-3. It consists of 51 amino acid units composing two chains. The chains are bonded by three **disulfide linkages** (—S—S—) that are characteristic of the amino acid cystine (see Table 16-5 for the formula of cystine).

The primary structure of a protein gives no indication of the three-dimensional arrangement of the protein. The protein chain often takes on a shape that is determined by hydrogen bonding and other forces of attraction between the amino acid groups along its length. A hydrogen bond is the electrostatic attraction between a hydrogen atom in a molecule and an electronegative atom (O or N) in another molecule or another part of the same molecule. The interaction of amino acid groups along the protein

**FIGURE 17-3** The primary structure of beef insulin is 51 amino acid units in two chains joined by disulfide linkages characteristic of the amino acid cystine. The figure shows the amino acid sequence but does not convey the three-dimensional shape of a protein. Insulins from other animals have similar primary structures.

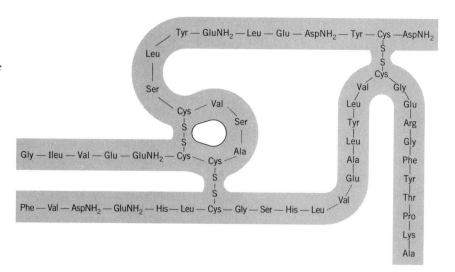

You can simulate a molecular helix using a wire twist tie. You need a new and flat twist tie about 4 inches long. Make a helix by wrapping the twist tie around the length of a pencil. Wrap it to make about four helical loops. Slip the helix off the pencil and stretch it to a length of about 2.5 inches. Sketch a picture of the helix model.

chain causes it to take on repeating patterns such as coiling and bending. These patterns of interaction are called the **secondary structure** of the protein. A common secondary structure for proteins is the **α-helix** in which the protein chain is coiled in a three-dimensional helical shape. The coiling causes the chain to take on a tubular arrangement somewhat like a coiled spring. The α-helix is illustrated in Figure 17-4. Other secondary structures are also known.

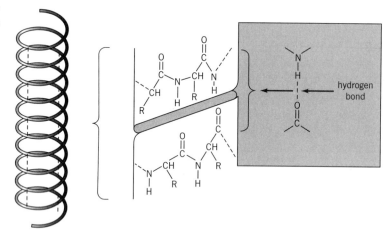

**FIGURE 17-4** The secondary structure of some proteins is a coil called the α-helix. Coiling gives the proteins a three-dimensional tubular shape. Under drastic conditions, such as heating, the secondary structure can be disrupted giving a random shape. The disruption of the secondary structure is called denaturation of the protein.

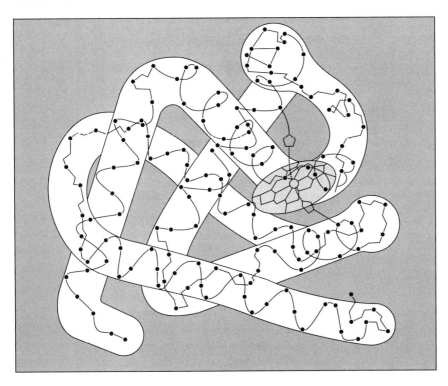

**FIGURE 17-5** The tertiary structure of myoglobin, a protein involved in the storage and transport of oxygen in muscle. Around 70% of the amino acid sequence is in the α-helix shape. The disk-shaped portion is called the heme unit.

The protein helices and other chain forms actually exist in some folded or twisted configuration. The overall three-dimensional shape is called the **tertiary structure** of the protein. Although tertiary structures are complex, some have been experimentally determined. The tertiary structure of the protein myoglobin is illustrated in Figure 17-5. Myoglobin stores and transports oxygen in muscle tissue and is related to the more complex blood protein, hemoglobin. Chemistry in Action 17-2 is about the differences between egg white and egg yolk.

## 17-6 ENZYMES

The numerous chemical reactions that occur in the body are referred to as metabolic reactions and the overall process is called metabolism. Metabolic reactions include digestion and the reactions in which certain food molecules are utilized by the body for energy. Other metabolic reactions involve the breakdown of body cells and the formation of new cells. Almost all the thousands of metabolic reactions occurring in our bodies require specific biological catalysts. These catalysts are synthesized by the body. A biological catalyst is called an **enzyme.**

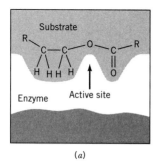

(a)

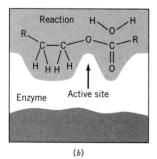

(b)

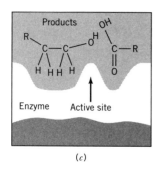

(c)

**FIGURE 17-6**    The lock-and-key theory of enzyme activity. (a) The three-dimensional shape of the enzyme in the vicinity of the active site is arranged to accommodate the substrate molecule. The active site is the portion of the enzyme that catalyzes the reaction of the substrate. (b) The substrate is held in position by the enzyme while the reaction of the substrate occurs. (c) The product or products from the substrate leave the enzyme, and the enzyme is available for another substrate.

Enzymes are large, polymeric molecules of varying complexity. All enzymes contain protein. Some enzymes are composed entirely of protein. Many enzymes consist of a protein portion called the **apoenzyme** and a nonprotein portion called the **coenzyme** or **cofactor.** These enzymes cannot function unless the apoenzyme and cofactor are joined. Biochemical reactions need specific enzymes that must be present at the correct time and place. It is thought that an enzyme functions by interacting with one of the reactants; this reactant is called the **substrate.** A given enzyme will act as a catalyst for only one kind of reaction. But how can an enzyme distinguish one reactant from another?

The commonly accepted theory of enzyme behavior is the **lock-and-key theory** illustrated in Figure 17-6. According to this theory, the enzyme has a definite three-dimensional structure that precisely fits around a part of the substrate molecule. Only a specific kind of substrate can fit into a given enzyme. Once the enzyme and substrate form an aggregate, the substrate is exposed for the reaction. Recall that a catalyst increases the rate of a reaction. Enzymes allow metabolic reactions to occur readily at body temperatures. Without enzymes these reactions would not occur fast enough to maintain life processes. After an enzyme-catalyzed reaction occurs, the product moves away from the enzyme, leaving it unchanged and available to catalyze the reaction of another substrate molecule. Chemistry in Action 17-3 is about the catalytic effect of various substances on the decomposition of hydrogen peroxide.

Certain metabolic processes involve a long sequence of enzyme-catalyzed reactions. Each enzyme must be present at the correct time for the process to occur. If any of the required enzymes are missing, the process is disrupted. Since such metabolic processes are vital to life, the absence of an enzyme can lead to illness and even death.

Some **hereditary conditions** and **diseases** result from an inability of the body to produce specific enzymes. For instance, many people have lactose intolerance caused by the body not making the digestive enzyme called lactase. For these persons the lactose in dairy products cannot be digested. The absence of certain vitamins and minerals in the diet can lead to enzyme deficiencies. A normal diet typically supplies sufficient vitamins and minerals. Most vitamins and minerals act as coenzymes or are used by the body to form important coenzymes. Several potent poisons, such as cyanide, lead, and mercury, bind to and inhibit the functioning of critical enzymes. A sufficient dose of cyanide ions results in death in a matter of seconds by interfering with an enzyme involved in energy metabolism. The absence of the enzyme causes death as a result of the body's inability to use oxygen.

## 17-7 BODY CELLS

Plants and animals consist of a large number of microscopic cells held together by structural materials. Cells differ in size and shape, depending on their function. However, all cells have several subcellular components in common. A typical animal **cell** is illustrated in Figure 17-7. Cells contain water and there is a large variety of chemicals dissolved in the water. Important bodily processes occur within cells. The cell is surrounded by a

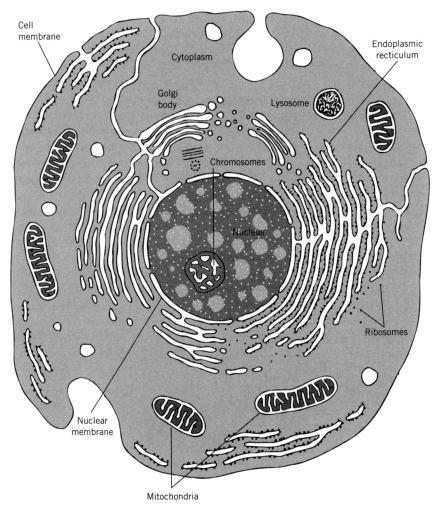

**FIGURE 17-7**
Components of a typical animal cell.

cell membrane that allows passage of selected molecules and ions in and out of the cell. The cytoplasm of the cell which consists of water-based cellular fluids contains various cellular components.

The nucleus of the cell, surrounded by a nuclear membrane, contains the chromosomes that control the hereditary characteristics of the cell. The chromosomes initiate cell division and protein synthesis. The mitochondria are membrane-contained components that are the centers of energy production for the cell. The endoplasmic reticula are areas within the cell containing the ribosomes where protein synthesis occurs. Golgi bodies are associated with protein distribution pathways in the cell. The lysosomes of the cell are involved with breakdown and degradation of cellular components when the cell membrane is ruptured.

## 17-8 NUCLEIC ACIDS: CHEMICALS OF HEREDITY

The chromosomes of the cell nucleus contain genes which serve as carriers of hereditary information. It has been found that the polymeric substance deoxyribonucleic acid (DNA) is the fundamental constituent of genes. Since genes carry hereditary information, an understanding of the structure of DNA gives great insight into hereditary processes.

DNA belongs to a class of polymeric substances called nucleic acids. Nucleic acids can be broken down into monomer units called nucleotides. The nucleotides can be further broken down into phosphoric acid, some heterocyclic amines or heterocyclic organic bases, and either deoxyribose or ribose sugar. This is illustrated in Figure 17-8. Nucleic acids that contain deoxyribose are called deoxyribonucleic acids or DNA and those that contain ribose are called ribonucleic acids or RNA.

The nucleotide monomer units of nucleic acids are composed of a phosphoric acid unit, a ribose or deoxyribose unit, and a unit of adenine, guanine, cytosine, and thymine in DNA or uracil instead of thymine in RNA. The structures of these heterocyclic amines or bases are shown in Figure 17-8. The formation of a typical nucleotide is illustrated below.

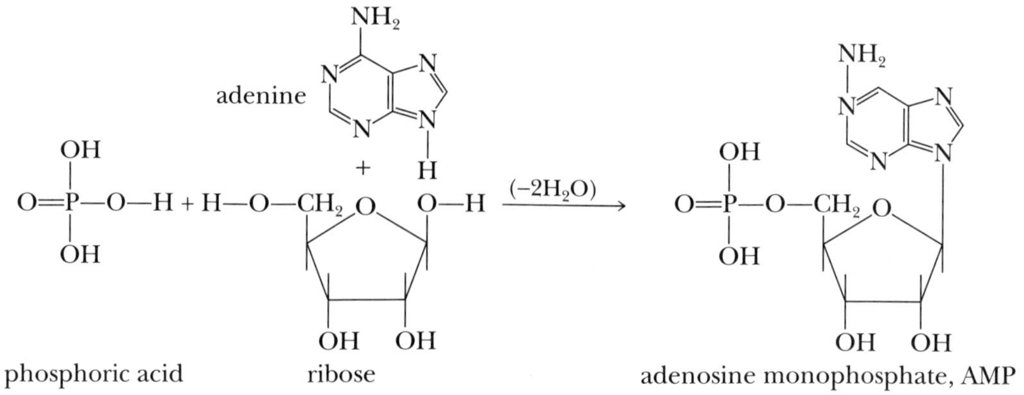

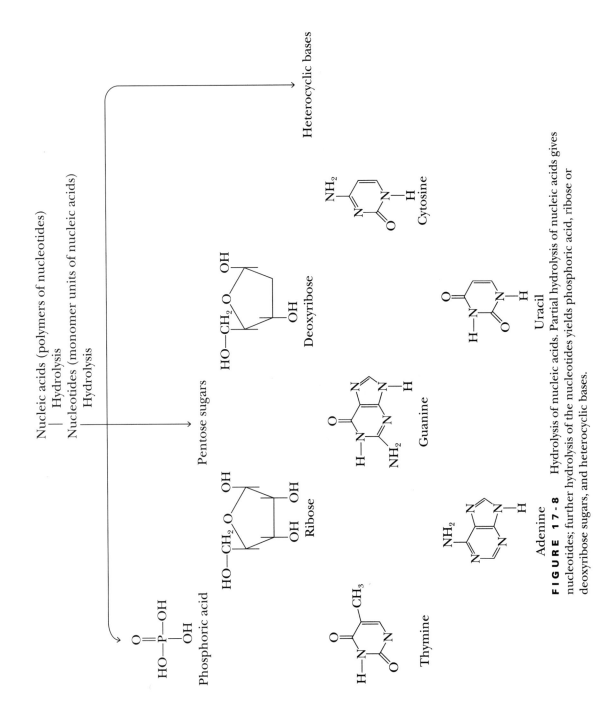

Nucleic acids (polymers of nucleotides)
| Hydrolysis
Nucleotides (monomer units of nucleic acids)
| Hydrolysis

Phosphoric acid

Pentose sugars

Ribose

Deoxyribose

Heterocyclic bases

Adenine

Guanine

Cytosine

Thymine

Uracil

**FIGURE 17-8**   Hydrolysis of nucleic acids. Partial hydrolysis of nucleic acids gives nucleotides; further hydrolysis of the nucleotides yields phosphoric acid, ribose or deoxyribose sugars, and heterocyclic bases.

Figure 17-9 shows how the nucleotide units link to form the **nucleic acid polymer,** a sequence of alternating phosphate units and sugar units. Protruding from this sequence are the heterocyclic amines. A shorthand representation of nucleic acids is given in Figure 17-10.

The **double-helix structure of DNA** proposed by Francis Crick and James Watson is one of the most significant theories of contemporary science. In 1953 Watson and Crick visualized a secondary structure of DNA that ex-

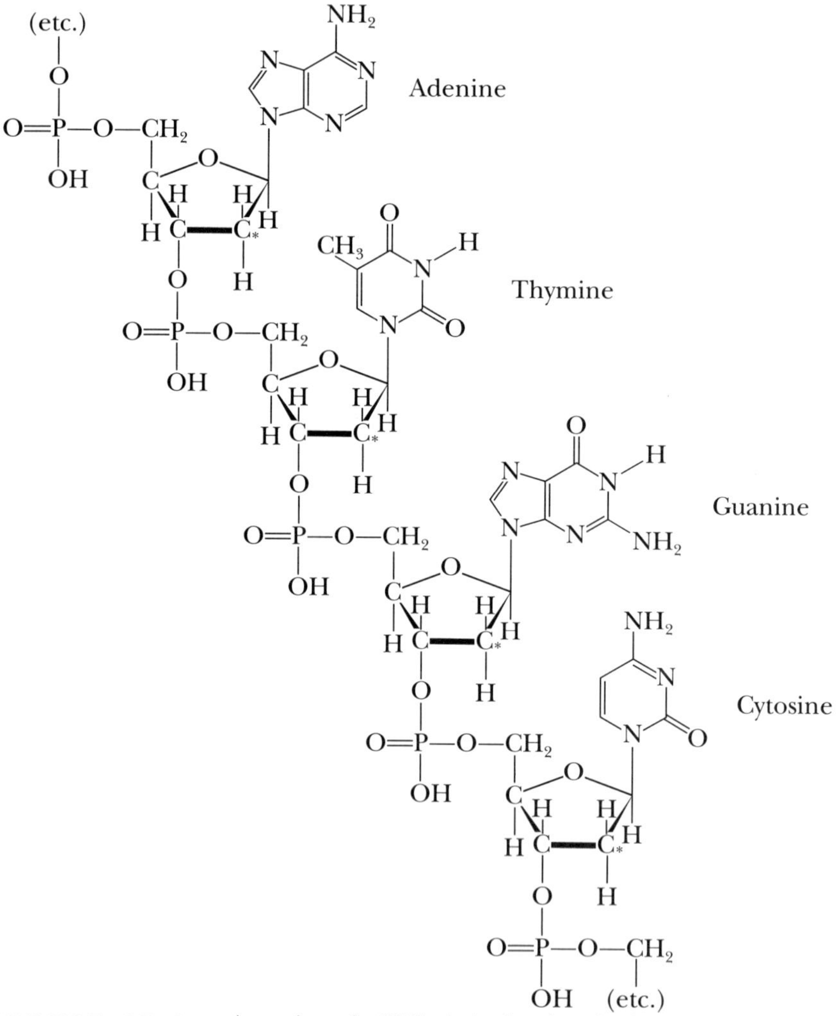

**FIGURE 17-9** A portion of a DNA chain showing the four common heterocyclic amines. Note that alternating units of deoxyribose are linked through phosphate ester bridges and the amines protrude from the sugar–phosphate polymer chain. The primary structure of RNA is the same as DNA except ribose replaces deoxyribose and uracil replaces thymine.

etc.
|
O
‖
O=P—O—Sugar unit
|
OH

Amine
|
O
‖
O=P—O—Sugar unit
|
OH

Amine'
|
O
‖
O=P—O—Sugar unit
|
OH

Amine''
|
O
|
etc.

etc.
|
P
C

P
C

P
C

P
C

etc.

= Sugar unit

= Amines

P  = Phosphate

**FIGURE  17-10**
Shorthand representation
of a nucleic acid chain.
The chain consists of
alternating sequences of
phosphate and sugar units
from which the
heterocyclic amines
extend.

plains how it functions in the cell. Two DNA strands intertwine in the form
of two helices by hydrogen bonding between thymine and adenine units
and guanine and cytosine units on opposite strands. Figure 17-11 illustrates
this kind of hydrogen bonding and Figure 17-12 illustrates a model of the
double helix of DNA. Studies have shown that cellular DNA predominantly
occurs in the double-helix form but other forms are also known.

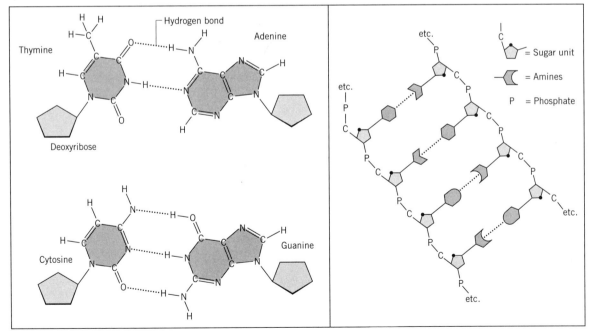

**FIGURE  17-11**    DNA strands in the double-helix form. Two strands of
DNA are held together by attraction bytween thymine and adenine and between
cytosine and guanine.

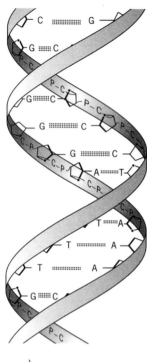

C = Deoxyribose

−P− = Phosphate ester bridge

A = Adenine

T = Thymine

G = Guanine

C = Cytosine

▦ = Hydrogen bond

**F I G U R E  1 7 - 1 2**
The double-helix form of DNA.

You can simulate a double helix using wire twist ties. You need six new and flat twist ties about 4 inches long.

1. Use two ties for the legs of a ladder and the other four for the rungs. Do this by folding the rung ties in half and wrapping them around the longer lengths. Space the rungs equally along the legs. Attach the rungs so that the flat edges of the longer ties are parallel to one another.

2. Once you have made the ladder gently twist it to arrange the ladder legs into a double-helix.

DNA in the genes serves as a starting point for the biological processes carried out in the cell. Genetic information is contained in the genes by specific sequences of heterocyclic bases on the DNA strands. In other words, the genetic information of a cell is contained as various sequences of these heterocyclic bases that are present along the DNA strands that comprise the **genes** of the **chromosomes.** The term **genetic code** refers to hereditary information that can be transmitted to new cells and to new generations. The pattern of the code is that **triplets** or **sequences of three heterocyclic bases** represent specific units of the code.

## 17-9 DNA AND THE GENETIC CODE

All cells in our bodies carry our personal genetic code. The vast majority of cells that structure our bodies are called **somatic cells** from the Greek *soma* 'body'. Certain cells located in the gonads are capable of being formed into sperm or eggs. These are called **gamete** or **germ cells** from alterations of the Latin *gignere* 'to beget'. Human somatic cells contain 46 chromosomes, but the gamete cells contain only 23. At sexual maturity, the gamete cells become active, enabling the female to ovulate and the male to produce sperm.

Human **sperm** consists of a protein sheath surrounding the DNA-structured male chromosomes. The **female egg** carries the female chromosomes. At conception, the egg is united with the sperm, and the chromosomes pair up to form a new cell called a **zygote**—a fertilized egg. The zygote multiplies by cell replication as the development of a new individual takes place. This developing being carries within its cells the genetic codes donated by the mother and the father.

As development proceeds, the cells rapidly increase in number and they take on different forms and functions of specialized parts of the body. Some cells structure the muscles, some structure the internal organs, and so on. The new individual takes on the inherited characteristics of its parents. Sometimes, one characteristic will dominate over another. For instance, the brown-eye trait of the mother may dominate over the blue-eye trait of the father, and the offspring will have brown eyes. The inherited traits come from the unique combination of the chromosomes of the parents.

All the cellular specialization and inherited characteristics are determined by the genetic code contained in the genes. As a result of various factors, including selective inherited characteristics, conditions of growth, and the environment, the new individual does not turn out to be an exact physical replica of the mother or the father. Nevertheless, it is through the genetic code that humans propagate a physical pattern of themselves to new generations.

New body cells are formed by cell division. Cell division, called **mitosis,** involves the formation of two cells from one. The new cells must each carry a set of genes containing the genetic code. In the initial phase of mitosis, the two sets of genes are formed within the cell. The double-helix strands of DNA partially separate, providing a pattern of heterocyclic base sequences upon which the cell can replicate the genes. Once replicate sets of genes are present, they migrate to different regions of the cell nucleus, and as cell division occurs, each new cell carries a set of genes.

RNA contains codes for the cellular synthesis of enzymes and other proteins. And DNA provides a pattern for RNA synthesis. Body enzymes are continually renewed so **enzyme synthesis** is a very important cellular process. The amino acids to make new enzymes come from the breakdown of old protein and from protein-containing foods. A variety of amino acids, called the **amino acid pool,** is normally available in a cell. That is why a diet should supply proteins with the essential amino acids.

Specific enzymes are synthesized in a cell according to the genetic code of the genes. These enzymes catalyze specific reactions taking place in the cell. The genetic code carries information that produces a variety of specific enzymes, resulting in the entire organism assuming various inherited characteristics. Numerous enzymes are also produced to allow the normal cellular processes to occur.

## 17-10 ENZYME SYNTHESIS

Enzyme and protein synthesis are complex cellular processes. The basic steps, summarized in Figure 17-13, begin with **RNA synthesis** in the nucleus, using DNA as a pattern. Four major types of RNA are produced. **Ribosomal RNA** migrates from the nucleus and becomes incorporated in the **ribosomes,** where protein synthesis occurs. **Heterogeneous nuclear RNA** is produced according to patterns contained on DNA segments of

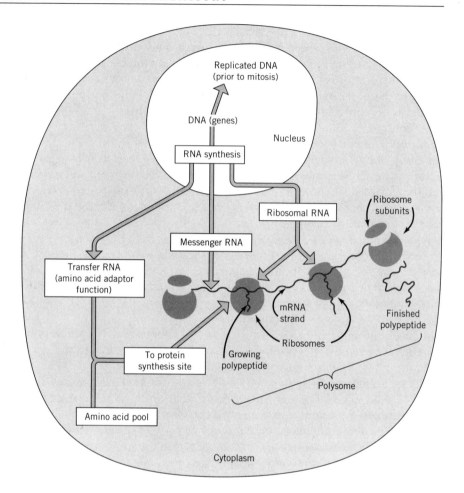

**FIGURE 17-13**
Relationships of DNA to various RNAs and protein synthesis.

the genes. Heterogeneous nuclear RNA is somehow condensed, by exclusion of unneeded portions, into **messenger RNA (mRNA).** Messenger RNA (mRNA) contains the codes for protein synthesis—triplet sequences of bases. Each mRNA base is the complement of its corresponding base in the DNA pattern, for example, guanine in the DNA pattern matches cytosine in the mRNA and adenine in the DNA pattern matches uracil in the mRNA. The mRNA migrates from the nucleus and becomes attached to one or more ribosomes. The point of attachment of the ribosomes and mRNA is the region at which protein synthesis occurs.

Various **transfer RNAs (tRNAs)** migrate from the nucleus and become attached to specific amino acids from the amino acid pool. There are at least 20 kinds of tRNA, one for each of the amino acids. The tRNAs, with the amino acids attached, migrate to the ribosomes on the mRNA. A sequence of three bases in the mRNA that serves as a code for a specific amino acid is called a **codon.** Initially a specific codon of the mRNA chain is attached to the ribosome and the ribosome can move through various codons of the mRNA chain during protein synthesis. The attachment of the ribosome and mRNA chain makes it possible for the ribosome to catalytically activate a specific codon on the mRNA chain.

The **active codon** calls for a specific amino acid. As a result the tRNA with this specific amino acid fits into the activated portion of the mRNA, and the amino acid it carries breaks off to become part of a growing chain. The ribosome then moves along the mRNA to activate another codon. The tRNA–amino acid pair corresponding to this new codon moves into the ribosome, and the amino acid becomes linked to the previous amino acid on the protein chain.

At each movement of the ribosome, a new amino acid is added to the growing protein chain. Finally, after hundreds or thousands of amino acids have been added, the ribosome reaches a codon on the mRNA that signals termination, and the newly formed protein breaks off. Actually the newly formed protein is attached to a chain of sugars and the complex is called a **glycoprotein.** The glycoproteins migrate to the Golgi bodies of the cell where they are chemically altered to form proteins. The proteins are then directed to the appropriate part of the cell in which they are to be used.

In this manner, the genetic code carried as codons by the mRNA from the DNA of the genes directs the synthesis of a specific protein used by the cell. Human cells contain enough DNA to contain codes for millions of different proteins. It is thought that only about 10% of the codes are active, but this represents hundreds of thousands of proteins to regulate human growth, development, and metabolism.

Many diseases are attributed to **genetic defects** in which the genetic code of an individual is abnormal. Two examples of genetic diseases are galactosemia, the failure to metabolize galactose and sickle cell anemia, the production of abnormal red blood cells. The abnormal code can cause a genetic disease to be passed on to the next generation, depending on whether or not the offspring develops the characteristic related to the defect. Even if the characteristic does not dominate in the offspring, it will still be present in the cells and can be passed on to the next generation.

The chemistry of genetics and enzyme behavior is an important field of research in biochemistry. This research has provided methods of early detection of many genetic diseases. Medical treatments for a few diseases have been quite successful. For instance, a diabetic cannot produce the enzyme insulin that regulates the glucose blood level but can lead a normal life as long as they take a daily dose of insulin. Human insulin can be manufactured by biotechnology processes. Biotechnology may someday provide methods of treatment, control, and possibly even cures for a variety of genetic diseases.

## *Biochemistry*

Biochemistry: The study of compounds involved in biological processes and the changes they undergo during life processes.

Foods contain carbohydrates, lipids (fats) and proteins along with water, minerals and vitamins.

Carbohydrates or sugars: Carbon, hydrogen and oxygen compounds produced by green plants during photosynthesis. They occur in a variety of sizes, ranging from simple sugars called monosaccharides to polymer molecules called polysaccharides. Most of the common monosaccharides are polyhydroxy aldehydes.

Monosaccharides or simple sugars: glucose, fructose, galactose

Disaccharides: Sugars composed of two monosaccharide units

$$\text{sucrose} \longrightarrow \text{glucose} + \text{fructose}$$
$$\text{lactose} \longrightarrow \text{glucose} + \text{galactose}$$
$$\text{maltose} \longrightarrow \text{glucose} + \text{glucose}$$

Polysaccharides or Complex Carbohydrates: Polymeric carbohydrates composed of hundreds and thousands of glucose unit monomers.

$$\text{starch} \longrightarrow \alpha\text{-glucose units} \qquad \text{energy source digestible by humans}$$
$$\text{cellulose} \longrightarrow \beta\text{-glucose units} \qquad \text{fiber or roughage not digestible by humans}$$

Simple lipids: Esters of glycerol and long-chain carboxylic acids called fatty acids.

Saturated fatty acids: Fatty acids with no carbon-carbon double bond.

Monounsaturated fatty acids: Fatty acids with one double carbon-carbon bond.

Polyunsaturated fatty acids: Fatty acids with two or more double carbon-carbon bonds.

Steroids: A kind of lipid found in the human body including cholesterol and the sex hormones.

Proteins: Polymeric materials found in living organisms that serve as structural components of cells, skin, muscles, bone interior and nerves; as enzymes and hormones; and in many other important functions of the body. Proteins are polypeptides involving hundreds or thousands of amino acid units. All proteins in humans are made up of about 20 different amino acids. The essential amino acids cannot be manufactured in our cells, so they must come from the foods we eat. When proteins are digested, they are broken down by digestive enzymes into the constituent amino acids. The amino acids then become available to the cells for protein synthesis.

Primary structure of a protein: The sequences in which the amino acids are linked in a protein molecule.

Secondary structure of a protein: The coiling and bent shapes of the chain of amino acids in a molecule of a protein caused by the interaction of amino acid groups along the chain.

Tertiary structure of a protein: The overall three-dimensional shape of a protein molecule.

Enzymes: Biological catalysts that are polymeric molecules of varying complexity. All enzymes contain protein. Some enzymes are composed entirely of protein. Many enzymes consist of a protein portion called the apoenzyme and a nonprotein portion called the coenzyme or cofactor. These enzymes cannot function unless the apoenzyme and cofactor are joined.

Chromosomes of the cell nucleus contain genes that serve as carriers of hereditary information. It has been found that the polymeric substance deoxyribonucleic acid (DNA) is the fundamental constituent of genes. DNA belongs to a class of polymeric substances called nucleic acids.

Nucleic acids: Polymeric substances that can be broken down into monomer units called nucleotides. The nucleotides can be further broken down into phosphoric acid, some heterocyclic amines or heterocyclic organic bases, and either deoxyribose or ribose sugar. Nucleic acids that contain deoxyribose are called deoxyribonucleic acids or DNA and those that contain ribose are called ribonucleic acids or RNA.

Double-helix structure of DNA: The secondary structure of DNA as proposed by Francis Crick and James Watson in 1953. DNA involves two molecular strands that intertwine in the form of two helices. The intertwining is accommodated by hydrogen bonding between thymine and adenine units and guanine and cytosine units on opposite strands.

Enzyme and Protein Synthesis: A cellular process in which the DNA of the genes directs the manufacture of proteins as they are needed. The DNA of the genes serves as a pattern on which four kinds of RNA are synthesized. Ribosomal RNA migrates from the nucleus and prepares the ribosomes for enzyme synthesis. Heterogeneous nuclear RNA is produced according to patterns contained on DNA segments on the genes. Messenger RNA is formed from the heterogeneous RNA. Messenger RNA (mRNA) contains the genetic code that serves as a pattern for the protein synthesis. The mRNA migrates from the nucleus and becomes attached to one or more ribosomes. The activated point of attachment of mRNA on a ribosome is the region at which protein synthesis occurs.

A third variety of RNA, called transfer RNA (tRNA), migrates from the nucleus and becomes attached to amino acids present in the cell. There are at least 20 kinds of tRNA, one kind for each different kind of amino acid. The tRNA-amino acid pairs migrate to the mRNA on the ribosomes. The part of the mRNA genetic code attached to the ribosome is activated and calls for a specified amino acid. The proper tRNA with this amino acid fits into the activated site, and the amino acid it carries breaks off to become a part of the protein chain. The ribosome moves along the mRNA to activate the next portion of the genetic code that calls for another amino acid. The proper tRNA delivers this amino acid to the protein chain. The movement of the ribosome continues until the complete protein is formed. The protein breaks off and becomes available to the cell. It is thought that there is enough active genetic code in human cells to synthesize around 700,000 different enzymes and proteins to regulate human growth, development and metabolism.

It is possible to test for unsaturated fats by mixing them with chemicals that react with the double bond. Elemental iodine can be used since it has a distinct pink or purple color depending upon its concentration. When it reacts with unsaturated fats the elemental color disappears. In this exercise you are going to test an unsaturated fat using iodine. You need: a box of commercial iodized salt, four or more glasses, Zip-Loc bag, water, hydrogen peroxide solution (medicinal hydrogen peroxide), white vinegar, commercial mineral oil (or a small sample of charcoal lighter, Energine, or lighter fluid), at least one kind of liquid vegetable oil and two coffee filters (or four napkins shaped into a cone that can be used as a filter). Work in a well-ventilated place.

1. Following the directions of Chemistry in Action 13-2, make some elemental iodine and extract it with mineral oil.
2. Pour the mineral oil–water mixture from the Zip-Loc bag into a glass. Be sure to transfer all the mineral oil containing the pink iodine.
3. Carefully add water to the glass to fill it so that the mineral oil level is near the top rim of the glass. Use a teaspoon or eyedropper to transfer the iodine–mineral oil to a small container or glass. Try not to transfer any water with the mineral oil.
4. Use the iodine to test for unsaturated fats as follows. Place four teaspoons of fresh mineral oil in a glass and four teaspoons of vegetable oil in another glass.
5. Use a teaspoon or eyedropper to add a few drops of the iodine–mineral oil to the mineral oil glass. Use this for color comparison.
6. Use a teaspoon or eyedropper to add a few drops of the iodine–mineral oil to the vegetable oil glass. Record your observations. Continue adding drops of the iodine–mineral oil to the vegetable oil several more times.
7. If you have another kind of vegetable oil test it with the iodine–mineral oil using the method in Step 6.

Eggs are a source of protein. Egg white contains the protein egg albumin and egg yolk contains other proteins. The yolk also contains fats and cholesterol. The bonds that hold proteins in their tertiary and secondary structures can be broken relatively easily. These bonds can be disrupted using alcohol, concentrated salt solution, heat, or mechanical agitation. Breaking these bonds denatures a protein and can cause it to coagulate. Recall how milk protein was coagulated in Chemistry in Action 3-1. Normally it is not possible to reverse the denaturization of a protein. "All the king's horses and all the king's men couldn't put Humpty together again." In this exercise you are going to denature some egg protein. You need: an egg, five glasses, salt, rubbing alcohol (70% isopropyl alcohol), boiling hot water, two 5 cm by 10 cm pieces of aluminum foil, and matches.

1. Break an egg and separate the yolk and white. Place the yolk in one glass and the white in another. Describe the appearance of each part.
2. Stir the white for a moment and divide it into three portions. Place a portion in each of three glasses. Keep a small amount in the original glass for use in Step 7.
3. Add two teaspoons of rubbing alcohol to a glass of egg white and stir. Let it sit for a few minutes and record your observations.

4. Add two teaspoons of saturated salt solution to a glass of egg white and stir. Let it sit for a few minutes and record your observations.

5. Add one fourth of a cup of boiling water to a glass of egg white and stir. Let it sit for a few minutes and record your observations. (This reaction is the basis for egg flower soup.)

6. Work in a ventilated area. Use your finger to make a slight indentation on the end of a piece of aluminum foil. Add a small amount of egg yolk (5 to 10 drops) to the indentation. Light matches and hold them under the part of the spoon containing the yolk. Use as many matches as needed to coagulate the yolk into a solid. The yolk may burn around the edges but be sure that it becomes solid. To note the presence of fats in the yolk peel back some of the solidified yolk and bend the foil to expose a piece of solid yolk. Hold a match to this piece of yolk and notice how it burns.

7. Work in a ventilated area. Use your finger to make a slight indentation on the end of the second piece of aluminum foil. Add a small amount of egg white (5 to 10 drops) to the indentation. Light matches and hold them under the part of the foil containing the white. Use as many matches as needed to coagulate the egg white into a solid. The white may burn around the edges but be sure that it becomes solid. To compare the burning of egg white to egg yolk peel back some of the white and bend the foil to expose a piece of solid white. Hold a match to this piece of white and notice how it burns.

Hydrogen peroxide, $H_2O_2$, is an oxidizing agent and can cause undesirable reactions in living cells. To prevent its accumulation all living cells contain an enzyme that catalyzes the decomposition of hydrogen peroxide. This enzyme is called catalase and it is interesting that it is found in all cells. In this exercise you are going to test for catalase. You need: commercial hydrogen peroxide solution, seven glasses, a variety of six or seven fruits, vegetables, and green leaves, and some packaged baker's yeast, if available.

1. Put individual samples of fruits, vegetables, or green leaves in separate glasses. Use samples about the size of small marbles and cut them into pieces or squeeze them to break the cell membranes. If you have some baker's yeast put a sample of it in a glass.

2. Place four tablespoons or one eighth of a cup of hydrogen peroxide solution in each glass. Let the mixtures sit for a few minutes and record your observations.

Write a balanced equation for the catalase-induced decomposition of hydrogen peroxide to give oxygen and water. Include catalase over the arrow. What do you think happens to the oxygen formed in a cell by the decomposition of hydrogen peroxide?

## QUESTIONS

### Section 17-1

1.  What is biochemistry?

2.  What four elements make up 99% of all atoms of living things?

3.  List four major types of molecules that are synthesized by living organisms.

### Sections 17-2 and 17-3

4.  Describe each of the following biological molecules and name an example of each.
    (a) carbohydrate
    (b) monosaccharide
    (c) disaccharide
    (d) polysaccharide

5.  To what uses do living organisms put carbohydrates?

6.  Give the monosaccharide units that make up the following carbohydrates:
    (a) maltose
    (b) starch
    (c) sucrose
    (d) cellulose

7.  What is the difference between cellulose and starch? Describe this difference in terms of α- and β-glucose units.

8.  Describe the two molecular forms of starch and give their common names.

9.  What is glycogen?

### Section 17-4

10. What are simple lipids? Draw the structure of a simple lipid as a triglyceride.

11. What are fatty acids. Draw the structure of a typical fatty acid.

12. How are lipids used by your body?

13. What are fats and oils?

14. When a cooking oil label says that the oil is high in polyunsaturates, what does this mean?

15. To what class of biochemicals does cholesterol belong?

16. What are steroids? Give an example.

17. What is the general chemical nature of human sex hormones?

18. Mestranol is a contraceptive agent. It has the structure of

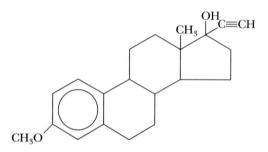

To what class of biological compounds does mestranol belong? The structure of mestranol is similar to which sex hormone?

### Section 17-5

19. Indicate the chemical nature or chemical composition of the following.
    (a) peptides
    (b) tripeptide
    (c) peptide bond
    (d) polypeptide

20. *Give the six possible tripeptides that can be formed from one molecule of each of the three amino acids threonine, valine, and leucine. Use the abbreviations for the amino acids (Thr, Val, and Leu).

21. The artificial sweetener aspartame, known as Nutrasweet®, is a compound derived from the dipeptide aspartyl phenylalanine. The structures of the amino acids aspartic acid and phenylalanine are given in Table 16-5. Draw the structure of the dipeptide used in aspartame with a peptide linkage between aspartic acid and phenylalanine. Aspartame is an ester made from the dipeptide aspartyl phenylalanine and methyl alcohol. Using your structure for aspartyl phenylalanine draw the structure of this ester.

22. Met-enkephalin is a naturally occurring morphine-like substance found in your brain. It is a pentapeptide, with the sequence tryptophan-glycine-glycine-phenylalanine-methionine. The structures of the amino acids are given below. Draw the structural formula of the pentapeptide with peptide linkages between the amino acid units. (Hint: There are two

glycine units. The tryptophan will have a free —$NH_2$ group; the methionine will have a free —COOH group.

$$NH_2-\overset{\overset{\displaystyle O}{\|}}{C}HC-OH$$
$$CH_2$$
$$C_8H_6N$$
tryptophan

$$NH_2-CH_2\overset{\overset{\displaystyle O}{\|}}{C}-OH$$
glycine

$$NH_2-\overset{\overset{\displaystyle O}{\|}}{C}HC-OH$$
$$CH_2$$
$$C_6H_5$$
phenylalanine

$$NH_2-\overset{\overset{\displaystyle O}{\|}}{C}HC-OH$$
$$CH_2$$
$$C_2H_5S$$
methionine

23. Describe the structure of a protein as a polypeptide. List some of the functions of proteins in the body.

24. Why is it important to have a variety of amino acid-containing proteins in your diet?

25. What is an α-helix in the structures of proteins?

26. *One turn of an α-helix is 3.6 amino acid units, or $5.41 \times 10^{-8}$ cm long. If your hair grows at a rate of 0.500 inch per month, how many amino acid units are you adding to a hair in a day?

27. What are the differences between the primary, secondary, and tertiary structures of a protein?

## Section 17-6

28. Define the following terms.
    (a) metabolism
    (b) enzyme
    (c) apoenzyme
    (d) coenzyme

29. What are the functions of enzymes in your body?

30. Describe the lock-and-key theory of enzyme behavior.

## Section 17-7

31. List the basic components of a typical animal cell.

32. What is the function of a lysosome in a cell?

## Section 17-8

33. Give the chemical names for RNA and DNA.

34. Describe the primary structures of the nucleic acids RNA and DNA in terms of their chemical building units. What are the differences between RNA and DNA?

35. Where is DNA located in cells?

36. Give a description of the double-helix theory of the structure of DNA. Make a rough sketch of a double helix.

## Section 17-9

37. What is meant by the genetic code? What is the code for?

38. How does inheritance occur from a chemical view?

39. Define the following terms.
    (a) genes
    (b) somatic cells
    (c) germ or gamete cells
    (d) zygote

## Section 17-10

40. What is a codon?

41. Describe the DNA-controlled synthesis of proteins, including the role of ribosomes, mRNA, and tRNA.

42. What is a genetic defect or a genetic disease? How can a genetic disease be inherited?

43. Cancer is a disease in which cells divide rapidly and uncontrollably. Chlorambucil is a drug used in cancer therapy that reacts with DNA molecules. Explain why damaging DNA molecules with chlorambucil might control cancer.

44. Why is hair loss often associated with chemotherapy used for cancer treatment?

# SOME
# USEFUL
# CONVERSION
# FACTORS

## LENGTH

1 inch (in.) = 2.54 centimeters (cm)     $\dfrac{2.54\ \text{cm}}{1\ \text{in.}}$

1 foot (ft) = 0.3048 meter (m)           $\dfrac{0.3048\ \text{m}}{1\ \text{ft}}$

1 yard (yd) = 0.9144 meter (m)           $\dfrac{0.9144\ \text{m}}{1\ \text{yd}}$

1 mile (mi) = 1.609 kilometers (km)      $\dfrac{1.609\ \text{km}}{1\ \text{mi}}$

## VOLUME

1 cup = 236.6 mL                         $\dfrac{236.6\ \text{mL}}{1\ \text{cup}}$

1 pint = 0.4731 liter (L)                $\dfrac{0.4731\ \text{L}}{1\ \text{pint}}$

1 quart = 0.9461 liter (L)               $\dfrac{0.9461\ \text{L}}{1\ \text{quart}}$

1 gallon (gal) = 3.785 liters (L)        $\dfrac{3.785\ \text{L}}{1\ \text{gal}}$

## MASS OR WEIGHT

1 pound (lb) = 453.6 grams (g)    $\dfrac{453.6 \text{ g}}{1 \text{ lb}}$

1 ounce (oz) = 28.35 grams (g)    $\dfrac{28.35 \text{ g}}{1 \text{ oz}}$

1 pound (lb) = 0.4536 kilogram (kg)    $\dfrac{0.4536 \text{ kg}}{1 \text{ lb}}$

## SPECIAL FACTORS

1 nanometer (nm) = $10^{-9}$ meter    $\dfrac{10^{-9} \text{ m}}{1 \text{ nm}}$

1 liter (L) = $10^3$ cubic centimeters ($cm^3$)    $\dfrac{10^3 \text{ cm}^3}{1 \text{ L}}$

1 milliliter (mL) = 1 cubic centimeter ($cm^3$)    $\dfrac{1 \text{ cm}^3}{1 \text{ mL}}$

1 atmosphere (atm) = 760 torr    $\dfrac{760 \text{ torr}}{1 \text{ atm}}$

1 torr = 1 millimeter Hg (1 mm Hg)    $\dfrac{1 \text{ mm Hg}}{1 \text{ torr}}$

1 atmosphere (atm) = 1.013 bar    $\dfrac{1.013 \text{ bar}}{1 \text{ atm}}$

1 calorie = 4.184 joules (J)    $\dfrac{4.184 \text{ J}}{1 \text{ cal}}$

Any of the above factors can be used in the inverted form if desired. For example, the factor relating kilograms to pounds is

$$\frac{0.4536 \text{ kg}}{1 \text{ lb}}$$

And the factor relating pounds to kilograms is

$$\frac{1 \text{ lb}}{0.4536 \text{ kg}}$$

Alternatively, the inverse of a factor can be calculated by dividing the denominator into 1. Thus, the factor relating pounds to kilograms is found by dividing 0.4536 into 1.

$$\frac{1 \text{ lb}}{0.4536 \text{ kg}} = \frac{2.205 \text{ lb}}{1 \text{ kg}}$$

Another example is the relation between kilometers and miles. The factor is expressed as

$$\frac{1 \text{ mi}}{1.609 \text{ km}}$$

Dividing 1.609 into 1 the factor becomes

$$\frac{0.6215 \text{ mi}}{1 \text{ km}}$$

To relate areas that involve square length and volumes that involve cubic length, the various length factors can be used. For example, to relate square centimeters ($cm^2$) to square inches (in.$^2$) we square the centimeter-to-inch factor.

$$\left(\frac{2.54 \text{ cm}}{1 \text{ in.}}\right)^2 = \frac{6.452 \text{ cm}^2}{1 \text{ in.}^2}$$

Another example is the factor relating cubic centimeters to cubic meters, which is found by cubing the centimeter-to-meter factor.

$$\left(\frac{100 \text{ cm}}{1 \text{ m}}\right)^3 = \left(\frac{10^2 \text{ cm}}{1 \text{ m}}\right)^3 = \frac{10^6 \text{ cm}^3}{1 \text{ m}^3}$$

## PRACTICE QUESTIONS

Calculate a factor that relates feet to meters.

Calculate a factor that relates quarts to liters.

Calculate a factor that relates $km^2$ to $mi^2$.

Calculate a factor that relates cubic centimeters to cubic inches.

Calculate a factor that relates cubic millimeters to cubic meters.

# USING A HAND CALCULATOR IN CHEMISTRY

A hand calculator can be quite useful in chemical computations. A calculator will not solve problems for you! It is convenient, however, to use a calculator to carry out the arithmetic needed to obtain a numerical answer. Normally, when solving a problem, it is a good idea to work out the complete setup before actually doing any numerical calculations. Of course, you might have to do a few preliminary calculations to obtain the values of some factors involved in the setup. Once the problem has been set up, the calculator serves as a convenient and efficient tool to help with the arithmetical part of the calculation.

In addition to the $+$, $-$, $\times$, $\div$ and $=$ or enter keys, the most useful keys on a scientific calculator are

| | |
|---|---|
| $\pm$ or CHS | used to change the sign of a number. It makes a positive number negative and a negative number positive. |
| EXP or EE | used to input exponential numbers. The key represents the $\times 10$ part of an exponential number. For practice input $2.68 \times 10^{-5}$ on your calculator. |
| log and $10^x$ | used to calculate logarithms and inverse logarithms. See the discussion given below for practice using these keys. |
| $x^2$ and $\sqrt{\phantom{x}}$ | used to square numbers and find square roots. |

Practice using your calculator by checking worked-out examples in the text. Some examples are given in this appendix for practice. If the setup of a problem involves a series of multiplications and/or divisions, it can be done as a sequence on the calculator. This is called a chain calculation. For example:

$$285 \text{ mL} \times \frac{780 \text{ torr}}{760 \text{ torr}} \times \frac{583 \text{ K}}{273 \text{ K}} = ? \text{ mL}$$

Using an algebraic calculator having an equal key, a possible calculating sequence is

$$285 \; (\times) \; 780 \; (\div) \; 760 \; (\times) \; 583 \; (\div) \; 273 \; (=)$$

Other calculation sequences are possible, since any series of multiplications or divisions would suffice. The advantage of the above sequence is that each number is used as it is encountered, working left to right. This helps avoid the accidental omission of a number.

Calculators usually do not make mistakes (unless the battery is low), but fingers can make mistakes. It is good practice to double-check a calculation to be sure that a keying error was not made. It is also good practice to check the value of a calculated answer to see if its value makes sense compared to the numbers used in the calculation. That is, if a calculated answer seems too small or too large, you may have made a calculation error.

Try the above calculation on your calculator. Using a calculator the display reads 624.64286 on an 8-digit model or 624.6428571 on a 10-digit model. The number of significant digits in a calculation involving measurements depends on the number of digits in the measurements. (See Sections 2-7 and 2-8.) A calculator does not make decisions on the number of significant digits. The user of a calculator has to judge the number of digits needed and round off the result from the calculator display. For instance, suppose we drove a car 350 miles and used 21 gallons of gas. The mileage in miles per gallon can be found by using a calculator.

$$\frac{350 \text{ mi}}{21 \text{ gal}} = 16.666667 \text{ mi/gal}$$

This is not the correct answer, since we knew the gallons to only two digits. The calculator gave us far too many digits, but it does not understand significant digits. So we round off the answer to two digits: 17 mi/gal.

Imagine that we very carefully measured our mileage and found that 21.0 gallons were used in traveling 357.0 miles. Here the three digits in the gallons should give us a three-digit mileage. On many calculator models, the answer given by the calculator is

$$\frac{357.0 \text{ mi}}{21.0 \text{ gal}} = 17. \text{ mi/gal}$$

What has happened? Has the calculator deprived us of our additional significant digit? Again, the calculator does not deal with significant digits.

Anytime a string of zeros occurs after the decimal point, most calculators drop them and give only the remaining nonzero digits. A zero can be a significant digit, as it is in this calculation. Since we know that the answer should have three significant digits, we just add a zero to the result obtained from the display: 17.0 mi/gal. This situation arises only when a calculated answer has a string of zeros to the right of the decimal point. Some calculators will actually display the string of zeros, but we still have to decide on the correct number of significant digits.

## MOLAR MASS COMPUTATION    (See Section 4-9)

To find the molar mass of a compound like sucrose, $C_{12}H_{22}O_{11}$, from the formula, the molar masses of the elements are multiplied by the corresponding subscripts and the products are added. If we want a three- or four-digit molar mass, we usually carry more digits and then round off after we have found the answer. The calculation of the molar mass of sucrose to three digits is:

C   12(12.01)

H   22(1.008)       The answer to three digits is 342.

O   11(16.00)

The following are examples of the calculation sequence on various types of calculators. Try the calculation with your calculator.

An algebraic calculator with memory:

12 $\otimes$ 12.01 $\circleddash$ $\boxed{M+}$     22 $\otimes$ 1.008 = $\boxed{M+}$     11 $\otimes$ 16.00 $\circleddash$ $\boxed{M+}$

$\boxed{MR}$ (memory recall)

Be sure that memory is cleared or set to zero before calculating.

A scientific calculator with algebraic operating system:

12 $\otimes$ 12.01 $\oplus$ 22 $\otimes$ 1.008 $\oplus$ 11 $\otimes$ 16.00 $\circleddash$

If you have no memory on your calculator, write down each product as it is determined and then add them up.

### pH AND pOH CALCULATIONS   (See Sections 12-14 and 12-15)

The pH and pOH of a solution are defined as:

$$pH = -\log[H_3O^+] \qquad pOH = -\log[OH^-]$$

To find the pH from the hydronium ion concentration or the pOH from the hydroxide ion concentration, a calculator with a log key is needed. The calculation sequence is

$[H_3O^+]$ (log) (±)    round off display for answer

$[OH^-]$ (log) (±)    round off display for answer

(±) refers to change sign key, which is the (CHS) key on some calculators.

The hydronium ion concentration corresponding to a given pH or the hydroxide ion corresponding to a given pOH is found by changing the sign of the pH or pOH value and then finding 10 raised to the power of the negated pH or negated pOH.

$$[H_3O^+] = 10^{-pH} \qquad [OH^-] = 10^{-pOH}$$

The calculation sequences on a calculator with a $10^x$ key is (substitute pOH for pH as needed):

pH (±) ($10^x$)    round off display for answer

On a scientific calculator with algebraic operating system the sequence is:

pH (±) (INV) (log)    round off display for answer

For practice find the pH corresponding to $[H_3O^+] = 1 \times 10^{-8}$. Then reverse the process to find the $[H_3O^+]$ corresponding to pH = 8.

### CALCULATIONS WITH EXPONENTIAL NUMBERS

Sometimes calculations involve exponential numbers. If you have a calculator that accepts exponential numbers, just key in the numbers in sequence as a chain calculation. On such calculators the exponent key is usually marked **EE**, **EEX**, or **EXP** and represents $\times 10$. For simple numbers like $10^3$, use 1 (EE) 3 or simple key in 1000. For practice try the following calculations to see if you obtain the answer given:

$$\left(3.62 \times 10^{-5}\right) \times \frac{5.23 \times 10^4}{8.47 \times 10^8} = 2.24 \times 10^{-9}$$

# USING
# THE VSEPR
# THEORY

It is possible to predict the shape of any molecule in which the central atom is surrounded by four pairs of valence electrons. We do this by using the Lewis structure and deciding how many bonding electron pairs are involved. Some molecules have more than four pairs of electrons around the central atom, and some have fewer than four pairs. The shapes discussed here apply only to those molecules with four pairs of electrons, which is a very common situation. Other cases are not part of this discussion. To predict a shape:

1. Write the Lewis structure of the molecule or polyatomic ion.
2. Count the number of bonding pairs and nonbonding pairs around the central atom.
3. Decide which case fits the structure.
   (a) Four bonding pairs—tetrahedral
   (b) Three bonding pairs and one nonbonding pair—triangular pyramid
   (c) Two bonding pairs and two nonbonding pairs—angular or bent
   (d) Two atoms only—always linear
4. If the structure does not fit any of these cases see whether it is one of the special cases involving multiple bonds.

(a) Two double bonds or one single bond and one triple bond—linear

(b) One double bond and two single bonds—triangular

---

**EXAMPLE A3-1**

What is the shape of sulfur dichloride, $SCl_2$, which has the following Lewis structure?

$$:\overset{..}{S}-\overset{..}{\underset{..}{C}l}:$$
$$|$$
$$:\overset{..}{\underset{..}{C}l}:$$

The central sulfur has four pairs of valence electrons. Since there are two bonding pairs and two nonbonding pairs, we predict its shape to be angular.

$$\overset{..}{\underset{\diagup\,\diagdown}{S}}$$
$$:\overset{..}{\underset{..}{C}l}:\quad :\overset{..}{\underset{..}{C}l}:$$

---

**EXAMPLE A3-2**

What is a likely shape for a molecule of chloroform, $CHCl_3$? The Lewis structure is

$$\overset{\textstyle H}{\underset{\textstyle :\overset{..}{\underset{..}{C}l}:}{:\overset{..}{\underset{..}{C}l}-\overset{|}{\underset{|}{C}}-\overset{..}{\underset{..}{C}l}:}}$$

Thus, a likely shape would be tetrahedral since there are four bonding pairs.

$$\overset{\textstyle H}{\underset{\textstyle :\overset{..}{\underset{..}{C}l}:}{\overset{|}{C}}}$$
$$:\overset{..}{\underset{..}{C}l}:\quad\ :\overset{..}{\underset{..}{C}l}:$$

---

**EXAMPLE A3-3**

What is the likely shape of a hydrogen chloride, HCl, molecule?

$$H-\overset{..}{\underset{..}{C}l}:$$

Any diatomic molecule is linear.

---

**ACTIVITY A3-1**

Find the likely shapes of the following molecules:

(a) HCN

(b) $CS_2$

(c) $PBr_3$

(d) $CF_4$

---

The VSEPR theory is also used to rationalize the shapes of some polyatomic ions. For instance, the Lewis structure of the hydronium ion is

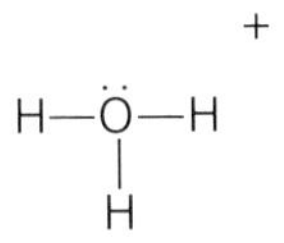

$$+$$
$$H-\overset{..}{O}-H$$
$$|$$
$$H$$

The shape would be triangular pyramidal since there are three bonding pairs and one nonbonding pair.

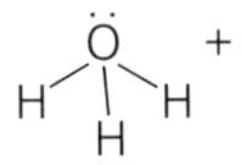

$$\overset{..}{O} \quad +$$
$$H\diagup \; | \; \diagdown H$$
$$H$$

---

**EXAMPLE A3-4**

What is the likely shape of an ammonium ion, $NH_4^+$? The Lewis structure is

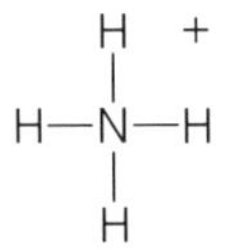

$$H \quad +$$
$$|$$
$$H-N-H$$
$$|$$
$$H$$

Since there are four bonding pairs, the shape of the ion would be tetrahedral.

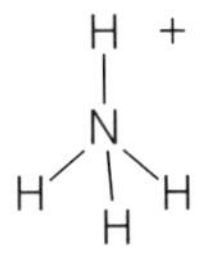

$$H \quad +$$
$$|$$
$$N$$
$$H\diagup \; | \; \diagdown H$$
$$H$$

---

The VSEPR theory may be used to rationalize the shapes of molecules or ions having more than one central atom. We do this by applying the theory to each atom that has a central position to which other atoms bond. For example, a molecule of methyl alcohol has a carbon bonded to three hydrogens and an oxygen. The oxygen is also bonded to another hydrogen.

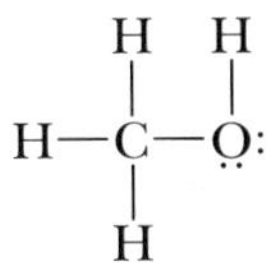

$$H \quad H$$
$$| \quad \; |$$
$$H-C-\overset{..}{O}:$$
$$|$$
$$H$$

The carbon and the oxygen are both viewed as central atoms to which other atoms bond. To apply the VSEPR theory we just count the number of electron pairs around each center and describe the shape around each center. We count the shared pair between the central atoms as a pair for both of them. In methyl alcohol the atoms are arranged around the carbon in a tetrahedral shape since it has four bonding pairs of electrons.

What is the likely shape of a molecule of acetic acid? Hint: There are three central atoms and the Lewis structure is

H  :Ö:
|    ||
H—C—C
|    |
H  :O—H

The atoms are arranged around the oxygen in an angular shape since it has two bonding pairs and two nonbonding pairs of electrons. Thus, the shape of methyl alcohol is an angular atomic arrangement attached to a tetrahedral arrangement.

H          H
 \        /
  H—C—O:
 /
H

Note that since methyl alcohol has one hydrogen bonded to oxygen, it is sometimes represented by the formula $CH_3OH$ rather than $CH_4O$.

---

**EXAMPLE A3-5**

Describe the shape of a molecule of acetaldehyde that has the following Lewis structure:

H  :O:
|    ||
H—C—C
|    |
H  H

What is the likely shape of a molecule of ethyl alcohol that is sometimes represented by the formula $CH_3CH_2OH$? (Hint: The molecule has three central atoms.)

Both carbons are central atoms. The first carbon has four bonding pairs that correspond to a tetrahedral shape. The second carbon has two single bonds and a double bond that correspond to a triangular shape. Thus, the molecule has a tetrahedral portion attached to a triangular portion.

H          :O.
 \         ‖
  H—C—C
 /         \
H           H

---

## PRACTICE QUESTIONS

What are the shapes of the following molecules? In each molecule the bonding sequence is indicated by the molecular formula.

(a) isopropyl alcohol, $CH_3CHOHCH_3$ (the three carbons are bonded in sequence and the central carbon is bonded to the oxygen that is bonded to a hydrogen)

(b) acetone, $CH_3COCH_3$ (the three carbons are bonded in sequence and the oxygen is double bonded to the middle carbon)

(c) urea, $NH_2CONH_2$ (two nitrogens are bonded to the carbon and the oxygen is double bonded to the carbon

(d) butane, $CH_3CH_2CH_2CH_3$ (the four carbons are bonded in sequence)

# BALANCING

# REDOX

# EQUATIONS

The equations of some redox reactions can be balanced using the simple trial-and-error methods discussed in Chapter 4. Many redox reactions, however, cannot be easily balanced using these methods. Redox reactions are electron transfer reactions; the number of electrons lost must equal the number of electrons gained during the reaction. This can serve as a basis for balancing redox equations.

Several different methods are used to balance redox equations. A common and useful method is known as the oxidation number method. (Review the topic of oxidation numbers in Section 11-5.) This method is especially useful for redox reactions occurring in aqueous solutions.

As an example we'll use the reaction of dichromate ions, $Cr_2O_7^{2-}$, and iron(II) ions, $Fe^{2+}$, in aqueous acidic solution. When a solution containing the pale green iron(II) ion is added to a solution containing the bright orange dichromate ion and a small amount of some strong acid, a reaction occurs and the color of the mixture changes to a shade of green. The products of the redox reaction are the green chromium(III) ion, $Cr^{3+}$, and the yellow iron(III) ion, $Fe^{3+}$.

Initially, the reaction can be represented by an unbalanced equation showing the reactants and products.

$$Cr_2O_7^{2-} + Fe^{2+} \longrightarrow Fe^{3+} + Cr^{3+}$$

**AP4-1**

The fact that this reaction takes place in an acidic solution means that some strong acid is present to provide hydronium ions or simply hydrogen ions, $H^+$, as charge carriers. We will use $H^+$ rather than $H_3O^+$ in balancing equations since it makes balancing easier. Some redox reactions occur only in aqueous acidic solutions. In contrast, some redox reactions require the presence of hydroxide ions so they occur only in basic solutions. We shall see how these ions are involved as we learn to balance redox reactions.

Here are the steps involved in balancing a redox equation.

1. Separate the reactants and products into two half-reactions for the elements that change oxidation number. Write the skeletal equations for each half-reaction.

$$Cr_2O_7^{2-} \longrightarrow Cr^{3+}$$
$$Fe^{2+} \longrightarrow Fe^{3+}$$

They are called **skeletal equations** since they give only the major reactant and product; other species may be added as the equations are balanced. Note that the skeletal equation for each half-reaction includes related species. One half-reaction includes the chromium-containing species and the other half-reaction includes the iron-containing species.

2. Work with one half-reaction at a time. Deduce the oxidation numbers of the **element that changes oxidation number;** the oxidation number rules are used for this.

   In this example the dichromate ion half-reaction is balanced first. Using the oxidation number rules the oxidation numbers of chromium are deduced. Remember to find the oxidation number per atom.

$$\overset{+6}{Cr_2}O_7^{2-} \longrightarrow \overset{+3}{Cr^{3+}}$$

3. Temporarily balance the number of atoms of the element changing oxidation number.

   Since the dichromate ion on the left has two chromium atoms, we need a coefficient of 2 for $Cr^{3+}$. This is called a **temporary coefficient** since it may become larger in the completely balanced equation.

$$\overset{+6}{Cr_2}O_7^{2-} \longrightarrow \overset{+3}{2Cr^{3+}}$$

4. Find the **total oxidation number change** by determining the change per atom and multiplying by the total number of atoms that change oxidation number. Also, decide whether electrons are lost or gained. An **increase in oxidation number is loss of electrons** and **a decrease in oxidation number is gain of electrons.**

   The chromium changes from +6 to +3, which is a 3 electron change. **A decrease in oxidation number or a gain of electrons is reduction.**

Since there are two chromiums changing the total oxidation number change corresponds to 6 electrons gained.

$$2(3e^-) = 6e^- \text{ gained}$$

$$\boxed{\begin{array}{cc} +6 & +3 \end{array}}$$
$$Cr_2O_7{}^{2-} \longrightarrow 2Cr^{3+}$$

5. **Add the electrons lost or gained** to the half-equation. Lost electrons go on the product side and gained electrons on the reactant side. It is important to put the electrons on the correct side of the equation. The 6 electrons gained are written on the reactant side of the half-reaction.

$$6e^- + Cr_2O_7{}^{2-} \longrightarrow 2Cr^{3+}$$

6. Count the charges on both sides, and **balance the charges** in the equation by adding $H^+$ or $OH^-$ to either side. If the reaction occurs in acidic solution, use $H^+$; if it occurs in basic solution, use $OH^-$; if it occurs in a neutral solution, use $H^+$ or $OH^-$ on the right as needed. That is, in a neutral solution, if negative charges are needed for balancing, use $OH^-$ and, if positive charges are needed, use $H^+$. Whether the reaction solution is acidic, basic, or neutral must be known in order to balance the equation.

In the example, the solution is acidic, so $H^+$ can be used to balance the charge. To balance the charge, count the total charge on both sides. Electrons carry negative charges and ions carry positive or negative charges. Look for the charges in the superscript positions in ions. The idea of a charge balance comes from the fact that a reaction cannot consume or produce charge. The same amount of positive and/or negative charge must occur on both sides of the arrow. Positive and negative charge on one side of the arrow cancel one another.

$$6e^- + Cr_2O_7{}^{2-} \longrightarrow 2Cr^{3+}$$
$$\begin{array}{ccc} 6- & 2- & 2(3+) \\ & 8- & 6+ \end{array}$$

Note that the total charge is found on each side by multiplying the charge of any ion by its coefficient and adding any separate charges. In the example there is a total of 8– on the left and 6+ on the right. The only way to balance the charge with hydrogen ion, $H^+$, is to add 14 hydrogen ions on the left. Since negative and positive charges on the same side cancel, this gives 6+ on both sides. In other words, since we are adding positive $H^+$ to balance the charge we need to add enough on the left to cancel the 8– charges and have 6+ charges to balance the 6+ on the right.

$$14H^+ + 6e^- + Cr_2O_7{}^{2-} \longrightarrow 2Cr^{3+}$$

7. In the final step, **balance the hydrogens and oxygens** by adding the appropriate number of $H_2O$ to either side. Count the number of hydrogens on each side then add the necessary number of waters.

In the example, the equation has 14 $H^+$, which contain 14 combined hydrogens on the left, and no hydrogens on the right. Thus, 7 waters are needed to supply the 14 hydrogens ($2 \times 7$), on the right. Adding 7 waters on the right gives the completely balanced half-reaction

$$14H^+ + Cr_2O_7^{2-} + 6e^- \longrightarrow 2Cr^{3+} + 7H_2O$$

To be sure that the half-reaction is balanced, check the chemical and charge balance. As can be seen from the equation, there are 14 hydrogens, 7 oxygens, and 2 chromiums on each side. There is a total of 6+ on each side. When tallying charge remember that positive and negative charges on the same side of the arrow cancel. Thus, the 8− charge on the left cancels 8+ of the 14+ charge on the left leaving a total charge of 6+ on each side.

The steps for balancing a redox half-reaction are summarized below. Any half-reaction can be balanced by following these steps. For success always follow the order of steps as listed. Of course, not all steps may be required in the balancing of the equations of some half-reactions.

Let's apply the same balancing steps to the iron half-reaction.

### Steps for Balancing Redox Half-Equations

1. **S**    *Skeletal half-equations*
   Separate the reactants and products into two skeletal half-reactions involving the elements that are changing oxidation number.

2. **O**    *Oxidation numbers*
   Find the oxidation numbers of the elements that are changing.

3. **T**    *Temporary coefficients*
   Balance the number of atoms of the elements that are changing oxidation number.

4. **T**    *Total oxidation number changes*
   Find the total oxidation number change for the elements that are changing oxidation number.

5. **E**    *Electrons added*
   Add electrons lost or gained.

6. **C**    *Charge balance*
   Balance charge by adding $H^+$ in acidic solutions or $OH^-$ in basic solutions or either on the right if neutral.

7. **H**    *Hydrogens and oxygen balance*
   Balance the hydrogens and oxygens by adding water on either side.

Skeletal equation:    $Fe^{2+} \longrightarrow Fe^{3+}$

Oxidation numbers:    $\overset{+2}{Fe^{2+}} \longrightarrow \overset{+3}{Fe^{3+}}$

Temporary coefficients:    No temporary coefficients are needed

Total oxidation number change:

$$\underset{Fe^{2+} \longrightarrow Fe^{3+}}{\boxed{\overset{1e^- \text{ lost}}{+2 \qquad +3}}}$$

Electrons added:    $Fe^{2+} \longrightarrow Fe^{3+} + 1e^-$    The lost electron is shown as a product.

Charge balance:    Checking the charge, we see 2+ on the left and 3+ and 1– for a total of 2+ on the right. So the half-equation is balanced with a total of 2+ on each side.

Since no hydrogens or oxygens are involved in this half-reaction, it is completely balanced. In this case, the half-reaction was found to be balanced after Step 5. Once a half-reaction is balanced, there is no need to continue.

The completely balanced redox equation is obtained by adding the two half-reactions. First, the half-reactions are multiplied by appropriate numbers to be sure that the total number of electrons lost equals the total number gained. Look at the two half-reactions in the example.

$$14H^+ + Cr_2O_7^{2-} + 6e^- \longrightarrow 2Cr^{3+} + 7H_2O$$
$$Fe^{2+} \longrightarrow Fe^{3+} + 1e^-$$

All the terms in the iron half-reaction are multiplied by 6 to have 6 electrons lost to balance the 6 electrons gained in the dichromate half-equation. Then, the half-reaction can be added and the electrons canceled.

$$14H^+ + Cr_2O_7^{2-} + 6e^- \longrightarrow 2Cr^{3+} + 7H_2O$$
$$\underline{6Fe^{2+} \longrightarrow 6Fe^{3+} + 6e^-}$$
$$14H^+ + Cr_2O_7^{2-} + 6Fe^{2+} \longrightarrow 6Fe^{3+} + 2Cr^{3+} + 7H_2O$$

To be sure the equation is balanced, check the chemical and charge balance. As can be seen from the equation, there are 14 hydrogens, 7 oxygens, 2 chromiums, and 6 irons on each side. There is a total of 24+ on each side.

$$(14+) + (2-) + 6(2+) = 24+ \qquad \text{and} \qquad 6(3+) + 2(3+) = 24+$$

Sometimes, redox reactions are written using the hydronium ion rather than the hydrogen ion. An $H^+$ occurs in water as $H_3O^+$. A hydronium ion is viewed as a hydrogen ion bonded to water. To change a balanced redox equation to include hydronium ions change the hydrogen ions to hydronium ions and add the corresponding number of waters to the other side for balance. In the above example the 14 $H^+$ are changed to 14 $H_3O^+$ and 14 $H_2O$ are added to the right to give a total of 21 $H_2O$.

$$14H_3O^+ + Cr_2O_7^{2-} + 6Fe^{2+} \longrightarrow 6Fe^{3+} + 2Cr^{3+} + 21H_2O$$

---

**EXAMPLE A4-1**

Sodium metal reacts with water to form sodium ions and hydrogen gas. Give the balanced equation for this reaction if the skeletal equation is

$$Na + H_2O \longrightarrow Na^+ + H_2$$

Separate the reaction into two half-equations and balance each half. Separate by noting the species that are obviously related.

$$Na \longrightarrow Na^+$$
$$H_2O \longrightarrow H_2$$

First, balance the sodium half.

$$\overset{\text{1}e^- \text{ lost}}{\boxed{\underset{Na}{0} \quad \quad \underset{Na^+}{+1}}}$$
$$Na \longrightarrow Na^+$$

Adding the lost electron on the product side gives

$$Na \longrightarrow Na^+ + 1e^-$$

Checking the charge, we see that the half-reaction is balanced.

Next, work with the second half-reaction.

$$\overset{2(1e^-) = 2e^- \text{ gained}}{\boxed{\underset{H_2O}{+1} \quad \quad \underset{H_2}{0}}}$$
$$H_2O \longrightarrow H_2$$

No temporary coefficients are needed, and since two hydrogens are changing from +1 to 0, this represents 2 electrons gained.

$$H_2O + 2e^- \longrightarrow H_2$$

Since the reaction solution begins as neutral, the charge is balanced by adding $OH^-$ or $H^+$ as a product. In other words, the solution is neither acidic nor basic at the beginning. Two hydroxide ions on the right balance the charge of 2– on the left.

$$H_2O + 2e^- \longrightarrow H_2 + 2OH^-$$

The hydrogens and oxygens are balanced by adding water. Since there are 4 hydrogens on the right and only 2 on the left, another water is needed on the left to give a total of 4 hydrogens on the left.

$$2H_2O + 2e^- \longrightarrow H_2 + 2OH^-$$

Checking the chemical balance, we see 4H and 2O on each side. The charge is balanced since there is 2– on each side.

The complete equation is obtained from the two half-reactions after multiplying the sodium half-reaction by 2 to balance the electrons lost with the electrons gained.

$$2Na \longrightarrow 2Na^+ + 2e^-$$
$$\underline{2H_2O + 2e^- \longrightarrow H_2 + 2OH^-}$$
$$2Na + 2H_2O \longrightarrow 2Na^+ + 2OH^- + H_2$$

---

**EXAMPLE A4-2**

Give the balanced equation for the reaction corresponding to the skeletal equation given below.

$$MnO_4^- + CHO_2^- \longrightarrow MnO_2 + CO_3^{2-} \qquad \text{(basic solution)}$$

Separate into two half-equations; one for the manganese-containing species and one for the carbon-containing species.

$$MnO_4^- \longrightarrow MnO_2$$
$$CHO_2^- \longrightarrow CO_3^{2-}$$

First, balance the permanganate ion half-reaction.

$$3e^- \text{ gained}$$
$$\boxed{+7 \qquad +4}$$
$$MnO_4^- \longrightarrow MnO_2$$

Adding the electrons gained gives

$$MnO_4^- + 3e^- \longrightarrow MnO_2$$

To balance the charge we note that the solution is basic, so hydroxide ions, $OH^-$, are used. Four hydroxide ions are needed on the right to balance the 4– charge $(1- + 3(-) = 4-)$ on the left.

$$MnO_4^- + 3e^- \longrightarrow MnO_2 + 4OH^-$$

The hydrogen and oxygen are balanced by adding water. We need 2 $H_2O$ on the left to balance the 4 H in the $OH^-$ on the right.

$$2H_2O + MnO_4^- + 3e^- \longrightarrow MnO_2 + 4OH^-$$

Next, balance the carbon half-reaction.

Skeletal equation: $\quad CHO_2^- \longrightarrow CO_3^{2-}$

Temporary coefficients: No temporary coefficients are needed

$$2e^- \text{ lost}$$
$$\boxed{+2 \qquad +4}$$
Total oxidation number change: $\quad CHO_2^- \longrightarrow CO_3^{2-}$

Electrons lost: $\quad CHO_2^- \longrightarrow CO_3^{2-} + 2e^-$ The lost electrons are shown as a product.

Balance the charge with hydroxide ions. Remember, it is a basic solution. Three $OH^-$ are needed on the left to give a total charge of 4– on each side.

$$3OH^- + CHO_2^- \longrightarrow CO_3^{2-} + 2e^-$$

Balance the hydrogens and oxygens by adding water. Two $H_2O$ are needed on the right to balance the 4 H on the left. Note that 3 H are in the $3OH^-$ and 1 H is in the $CHO_2^-$.

$$3OH^- + CHO_2^- \longrightarrow CO_3^{2-} + 2e^- + 2H_2O$$

The completely balanced equation is obtained by adding the two half-reactions after each is multiplied by a factor to adjust the electrons lost to equal the electrons gained. The permanganate half-reaction is multiplied by 2 and the carbon half-reaction by 3.

$$2(2H_2O + MnO_4^- + 3e^- \longrightarrow MnO_2 + 4OH^-)$$
$$3(3OH^- + CHO_2^- \longrightarrow CO_3^{2-} + 2e^- + 2H_2O)$$
$$\underline{\begin{array}{l} 4H_2O + 2MnO_4^- + 6e^- \longrightarrow 2MnO_2 + 8OH^- \\ 9OH^- + 3CHO_2^- \longrightarrow 3CO_3^{2-} + 6e^- + 6H_2O \end{array}}$$
$$4H_2O + 2MnO_4^- + 9OH^- + 3CHO_2^- \longrightarrow 3CO_3^{2-} + 2MnO_2 + 8OH^- + 6H_2O$$

Notice that water and hydroxide ions occur on both sides of the equation. When balancing a redox equation you may note that some species appear on both sides. This occurs because each half is balanced separately. Just cancel the appropriate number of species to give a properly balanced over-all equation. In this case, 8 $OH^-$ on the right cancel 8 on the left leaving one $OH^-$ on the left. Four $H_2O$ on the left cancel 4 $H_2O$ on the right leaving 2 $H_2O$ on the right. Thus the completely balanced equation is

$$2MnO_4^- + 3CHO_2^- + OH^- \longrightarrow 3CO_3^{2-} + 2MnO_2 + 2H_2O$$

A check of the chemical and charge balance shows the equation to be balanced.

## BALANCING WHOLE REDOX EQUATIONS

Many redox equations can be balanced by incorporating both the oxidation and reduction steps as a whole equation. A slight variation of the oxidation number method is used.

Here are the steps for balancing a whole redox equation using the earlier example of dichromate ions and iron(II) ions reacting in acid solution.

1. Start with the **skeletal equation** including the reactants and products.

$$Cr_2O_7^{2-} + Fe^{2+} \longrightarrow Fe^{3+} + Cr^{3+}$$

2. Deduce the **oxidation numbers** of the elements that are changing oxidation number.

$$\overset{+6}{\phantom{C}}\quad\overset{+2}{\phantom{F}}\quad\overset{+3}{\phantom{F}}\quad\overset{+3}{\phantom{C}}$$
$$Cr_2O_7{}^{2-} + Fe^{2+} \longrightarrow Fe^{3+} + Cr^{3+}$$

3. **Temporarily balance** the number of atoms of any element changing oxidation number. Since the dichromate ion on the left has two chromiums, we need a coefficient of 2 for $Cr^{3+}$.

$$\overset{+6}{\phantom{C}}\quad\overset{+2}{\phantom{F}}\quad\overset{+3}{\phantom{F}}\quad\overset{+3}{\phantom{C}}$$
$$Cr_2O_7{}^{2-} + Fe^{2+} \longrightarrow Fe^{3+} + 2Cr^{3+}$$

4. Find the **total change in oxidation number** for each element that changes; determine the change per atom and multiply by the total number of atoms that change oxidation number.

$$2(3e^-) = 6e^- \text{ gained}$$

$$\overset{+6}{\phantom{C}}\quad\overset{+2}{\phantom{F}}\quad\overset{+3}{\phantom{F}}\quad\overset{+3}{\phantom{C}}$$
$$Cr_2O_7{}^{2-} + Fe^{2+} \longrightarrow Fe^{3+} + 2Cr^{3+}$$
$$1e^- \text{ lost}$$

5. **Balance the electrons lost and gained** by multiplying each half by the appropriate number. This part is the major difference between this method and the half-reaction method. No electrons need to be written in the equation. To balance the 6 electrons gained multiply the iron half-reaction by 6 to make 6 electrons gained.

$$Cr_2O_7{}^{2-} + 6Fe^{2+} \longrightarrow 6Fe^{3+} + 2Cr^{3+}$$

6. **Balance the charge** by adding $H^+$ if acidic, $OH^-$ if basic or, if neutral, either $H^+$ or $OH^-$ on the right. In the example, the solution is acidic, so $H^+$ can be used to balance the charge.

$$Cr_2O_7{}^{2-} + 6Fe^{2+} \longrightarrow 6Fe^{3+} + 2Cr^{3+}$$
$$(2-) \; 6(2+) = 12+ \qquad 6(3+) = 18+ \quad 2(3+) = 6+$$
$$(2-) + (12+) = 10+ \qquad (18+) + (6+) = 24+$$

There is a total of 10+ on the left and 24+ on the right. The only way to balance the charge with hydrogen ions, $H^+$, is to add 14 hydrogen ions on the left.

$$14H^+ + Cr_2O_7{}^{2-} + 6Fe^{2+} \longrightarrow 6Fe^{3+} + 2Cr^{3+}$$

7. In the final step, **balance the hydrogens and oxygens** by adding the appropriate number of $H_2O$ to either side.

In the example, the equation has 14 $H^+$, which contain 14 combined hydrogens on the left and no hydrogens on the right. Thus, 7 waters are needed on the right to give 14 hydrogens.

$$14H^+ + Cr_2O_7{}^{2-} + 6Fe^{2+} \longrightarrow 6Fe^{3+} + 2Cr^{3+} + 7H_2O$$

To be sure the equation is balanced, check the chemical and charge balance. There are 14 hydrogens, 7 oxygens, 6 irons and 2 chromiums on each side. There is a total of 24+ on each side.

---

**EXAMPLE A4-4**

Balance the following redox skeletal equation:

Skeletal equation:   $MnO_4^- + CHO_2^- \longrightarrow CO_3^{2-} + MnO_2$     (basic)

$\phantom{Skeletal equation: MnO_4^-}$+7$\phantom{xx}$+2$\phantom{xxxxx}$+4$\phantom{xx}$+4

Oxidation numbers:   $MnO_4^- + CHO_2^- \longrightarrow CO_3^{2-} + MnO_2$     (basic)

Temporary coefficients:   No temporary coefficients are needed

Total oxidation number change:

$$\overbrace{\qquad\qquad\qquad}^{3e^-}$$

$$\boxed{+7 \qquad +2 \qquad\qquad +4 \qquad +4}$$
$$MnO_4^- + CHO_2^- \longrightarrow CO_3^{2-} + MnO_2 \text{ (basic)}$$
$$\underbrace{\qquad\qquad}_{2e^-}$$

Multiply each half-reaction appropriately to balance the electrons lost and gained. Multiply the Mn half-reaction by 2 and the C half-reaction by 3.

Tally the charges:    $2MnO_4^- + 3CHO_2^- \longrightarrow 3CO_3^{2-} + 2MnO_2$

$\phantom{Tally the charges: 2M}$(2–) + (3–) = 5–$\phantom{xxxxxx}$3(2–) = 6–

Balance the charge using $H^+$ or $OH^-$. Since the reaction conditions are basic, balance the charge by adding one $OH^-$ on the left.

Tally the H:   $2MnO_4^- + 3CHO_2^- + OH^- \longrightarrow 3CO_3^{2-} + 2MnO_2$

$\phantom{Tally the H: 2M}$3 H + 1 H = 4 H$\phantom{xxxxxxxx}$no H

Balance the H and O by adding $H_2O$. Two $H_2O$ are needed on the right.

$$2MnO_4^- + 3CHO_2^- + OH^- \longrightarrow 3CO_3^{2-} + 2MnO_2 + 2H_2O$$

Check the chemical and charge balance. The equation is completely balanced.

---

# PRACTICE QUESTIONS

Give balanced equations for the following redox reactions that take place in acidic solutions. Use the half-reaction or whole-equation method for balancing.

(a)  $ClO_3^- + I_3^- \longrightarrow Cl^- + I_2$

(Hint: Each I changes oxidation number by 2/3 from –1/3 to 0.)

(b)  $Ba + H^+ \longrightarrow Ba^{2+} + H_2$

(c)  $Cr^{3+} + BiO_3^- \longrightarrow Cr_2O_7^{2-} + Bi^{3+}$

(d) $Ag + NO_3^- \longrightarrow Ag^+ + NO_2$

(e) $H_2C_2O_4 + HNO_2 \longrightarrow CO_2 + NO$

(f) $I_2 + ClO^- \longrightarrow IO_3^- + Cl^-$

(g) $Br^- + H_2O_2 \longrightarrow Br_2 + H_2O$

(Hint: The oxidation number of O in $H_2O_2$ is $-1$.)

(h) $XeO_3 + I_3^- \longrightarrow Xe + I_2$

(Hint: Each I changes oxidation number by 2/3 from $-1/3$ to 0.)

(i) $MnO_4^- + H_2C_2O_4 \longrightarrow Mn^{2+} + CO_2$

Give balanced equations for the following redox reactions that take place in basic solutions. Use the half-reaction or whole-equation method for balancing.

(a) $BrO_3^- + F_2 \longrightarrow BrO_4^- + F^-$

(b) $N_2H_4 + Cu(OH)_2 \longrightarrow N_2 + Cu$

(Hint: H in $N_2H_4$ has the expected oxidation number of +1 and does not change in the reaction.)

(c) $SO_3^{2-} + ClO_3^- \longrightarrow SO_4^{2-} + Cl^-$

(d) $NO_3^- + PbO \longrightarrow PbO_2 + NO_2^-$

(e) $2Cl_2 \longrightarrow Cl^- + ClO_3^-$

(Hint: Use one $Cl_2$ in each half-reaction or use one $Cl_2$ for each half of the whole equation.

(f) $IO_3^- + Cr(OH)_4^- \longrightarrow I^- + CrO_4^{2-}$

(Hint: The oxidation number of Cr in $Cr(OH)_4^-$ is +3.)

(g) $Bi_2O_3 + ClO^- \longrightarrow BiO_3^- + Cl^-$

(h) $ClO_3^- + N_2H_4 \longrightarrow NO + Cl^-$

(Hint: H in $N_2H_4$ has the expected oxidation number of +1 and does not change in the reaction.)

(i) $Fe(OH)_2 + O_2 \longrightarrow Fe(OH)_3 + OH^-$

# ORGANIC

# PATTERNS

## INSTRUCTIONS

Trace or photocopy these patterns for organic compounds. Cut them out and use them to model compounds as discussed in Chapter 16. When modeling a compound, assume hydrogens are bonded to open bonds not used for other kinds of atoms.

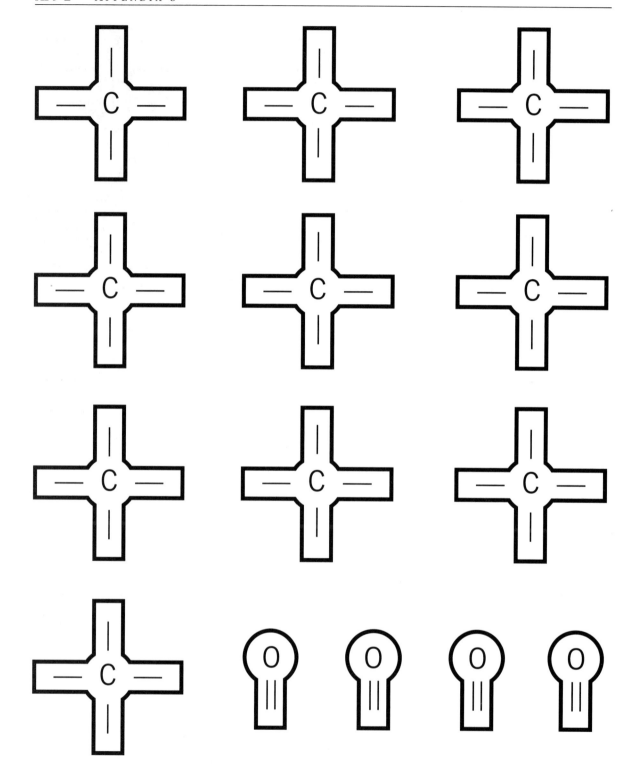

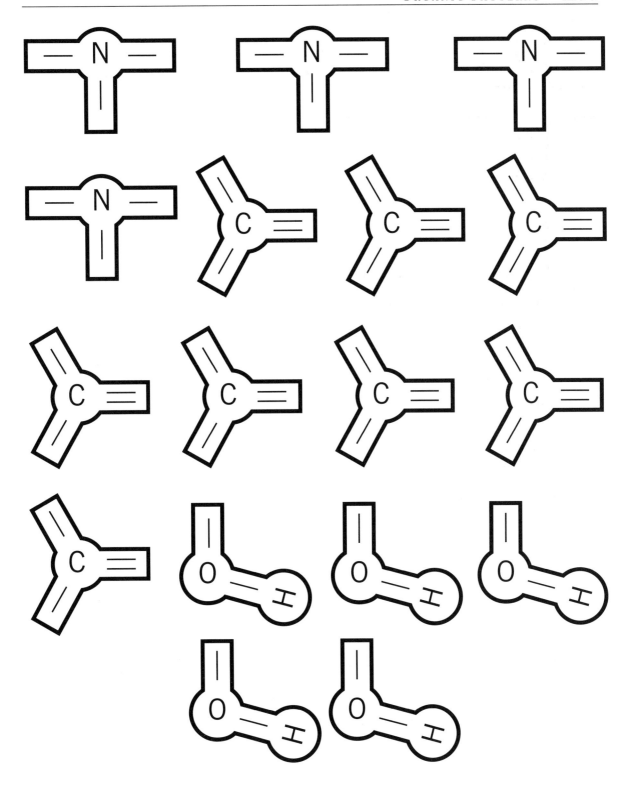

# MATERIALS FOR
## ACTIVITIES
## & CHEMISTRY
## IN ACTION
## EXERCISES

The materials and household chemicals needed for each chapter are listed here. The materials listed under the heading of general items are those that can be used throughout the exercises. Of course, you do not need all of these items at the same time and they can be shared by groups. Your instructor may assign specific exercises. Specific items needed in small amounts are listed for the exercises in each chapter. (GS = grocery store or supermarket, DS = drug store, SS = stationary store, office supply store, or bookstore.)

## GENERAL ITEMS

1. If you clean, rinse, and dry the glasses as you use them, you'll only need about 10 glasses. Eight- or ten-ounce, clear plastic glasses will work best.
2. Standard-size steel paper clips (SS)
3. Magic (Scotch) tape, 3/4-inch or 1/2-inch type (SS)
4. White vinegar (GS)
5. Fifteen shiny pennies and 12 dull pennies (Friends or Relatives).
6. Aluminum foil (GS)
7. Granulated table sugar (GS or coffee shop)
8. Boxes of iodized salt (GS)

9. Food colors (GS)

10. Baking soda (GS)

11. Liquid household bleach kept in a labeled and tightly sealed container (GS) **CAUTION: This is a dangerous solution so handle it with care.**

12. Balloons (SS, GS, DS)

13. 9-volt alkaline batteries (GS, SS, DS)

14. Bottles of 3% hydrogen peroxide solution (GS, DS)

15. An eyedropper (GS, DS, or get one from an old medicine bottle)

16. Tums tablets (GS, DS)

17. Laxative tablets containing phenolphthalein (GS, DS, read label to be sure)

18. Dry baker's yeast (GS)

19. Bottle of mineral oil (GS, DS)

20. Books of matches (GS)

21. Zip-lock bags (sandwich bag size) (GS)

22. Coffee filters, number 4 size (GS)

23. A measuring cup with standard and metric measure, one tablespoon, and two teaspoons (GS) (You could calibrate your own plastic cup in the laboratory.)

24. Bottle of regular milk of magnesia (GS, DS)

25. Aspirin tablets (GS, DS)

26. A sample of powdered sulfur the size of two aspirin tablets (see instructor)

27. Straight pins (GS, SS)

28. Small metric ruler (SS, DS or from instructor)

29. A plastic knife (see takeout restaurant)

30. Vegetable oil (GS)

31. Calamine lotion (GS, DS)

32. Pepto-Bismol tablet (GS, DS)

33. Mercurochrome (DS)

34. Twist ties about 4 inches long (GS)


## CHAPTER 1   CHEMISTRY

### Activities

1-2:   shiny penny

1-7:   balloon, vanilla extract or cologne, eyedropper or drinking straw

## Chemistry in Action

1-1:   white vinegar, baking soda, matches, a small piece of plastic wrap, household liquid bleach or dry baker's yeast, 3% hydrogen peroxide solution, toothpicks, paper towel/newspaper, two small glasses (plastic or glass)

1-2:   food coloring, two glasses, hot and cold water

# CHAPTER 2    MEASUREMENT IN CHEMISTRY

## Activity

2-11: one glass, cold water, butter/margarine, a small piece of meat (hot dog, salami, turkey and/or chicken)

## Chemistry in Action

2-1:   a half-gallon carton of milk, tape measure/ruler, bathroom, laboratory or kitchen scale

2-2:   two small glasses, water, vegetable (or mineral) oil, pieces of ice

# CHAPTER 3    ATOMS

## Activities

3-1:   9-volt battery, aluminum foil, tape, glass, tap water

3-4:   paper, scissors, balloon or clear plastic/Styrofoam cup, toothpick

3-10: white paper, knife, pencil lead, match

3-11: penny

3-12: penny, glass, white vinegar, file or sandpaper

3-13: steel paper clip (not plastic)

## Chemistry in Action

3-1:   two small glasses, half a cup of milk, white vinegar, paper towel/napkins, coffee filter or paper towel and tape

3-2:   aluminum foil, granulated sugar, matches/candle

3-3:   shiny penny, zip-lock bag, glass, boiling hot water, a small sample of sulfur powder

3-4:   two shiny pennies, paper, liquid bleach, aspirin tablet, eyedropper or Q-tip

## Chapter 4 FORMULAS and EQUATIONS
### Chemistry in Action

4-1: zip-lock bag, teaspoon, baking soda, white vinegar

4-2: white napkin/tissue, two steel paper clips, white vinegar, liquid bleach, flat plate or shallow container

4-3: small glass, white vinegar, Tums tablet or small bits of eggshell

## Chapter 5 CHEMICAL STOICHIOMETRY
### Activities

5-8: two paper matches

### Chemistry in Action

5-1: four zip-lock bags, baking soda, white vinegar

5-2: zip-lock bags, sugar, aspirin tablet, matches, steel wool

## Chapter 6 ATOMIC STRUCTURE and the PERIODIC TABLE
### Chemistry in Action

6-1: two small glasses/100 mL beakers, water, blue and yellow food colorings, matches

6-2: 9-volt alkaline battery, twist tie (paper coated), steel paper clips/staples

## Chapter 7 CHEMICAL BONDS
### Activities

7-8: four small balloons or four straight pins or toothpicks, eraser (removed from pencil end) or ball of clay

### Chemistry in Action

7-1: granulated sugar, table salt, sheets of white paper, plastic knife, dry cloth towel, matches, aluminum foil, water (distilled is best), tape, 9-volt alkaline battery

## CHAPTER 9    GASES AND GAS LAWS

### Activities

9-1:    empty juice or water bottle, water, stove

### Chemistry in Action

9-1:    clear plastic straw, water, metric rule, pen/pencil.

9-2:    two zip-lock bags, baking soda, white vinegar, metric measuring cup, water

## CHAPTER 10    WATER AND SOLUTIONS

### Activities

10-2:    plastic knife, dry cloth towel, water faucet

10-4:    two small glasses, water, vegetable oil, salt, vaseline or solid butter/margarine, spoon

10-5:    plastic or paper cup and a water tap over a sink

### Chemistry in Action

10-1:    four aspirin tablets, three glasses, two coffee filters, cold and boiling water, measuring cup

10-2:    9-volt battery, aluminum foil, tape, salt, white vinegar, 3% hydrogen peroxide solution, tape, small shallow container

10-3:    zip-lock bag, milk, sugar, ice, vanilla flavoring, salt, four-cup bowl/pan, measuring cup

## CHAPTER 11    CHEMICAL REACTIONS REVEALED

### Activities

11-3:    two zip-lock bags, measuring cup, two cups/beakers, white vinegar, two Tums tablets, cold and boiling water

11-4:    wooden pencil, cotton wad/Q-tip, matches

11-5:    metric measuring cup, three small glasses, white vinegar, baking soda, teaspoon, water

11-15:    eight paper clips (same size)

11-17:    eight paper clips (same size)

## Chemistry in Action

11-1: 3% hydrogen peroxide solution, tablespoon, small glasses, shiny and dull pennies, leaf from green plant, household bleach, rusty object

11-2: water, vegetable oil, saturated salt solution, measuring cup, seven steel paper clips, three glasses, matches, samples of plastic materials

# CHAPTER 12    ACIDS AND BASES

## Activities

12-1: white vinegar, red and blue litmus paper, two shiny zinc–copper pennies

12-2: red and blue litmus paper, milk of magnesia, $Mg(OH)_2$, white vinegar, teaspoon, toothpick

12-7: white vinegar, baking soda, glass, eggshell (or calcium tablet or chalk)

## Chemistry in Action

12-1: milk of magnesia (regular strength only), white vinegar, two glasses, two teaspoons, knife, laxative tablet (phenolphthalein type)

12-2: red cabbage, nonaluminum pan (to heat cabbage), several glasses, teaspoon, saturated solution of baking soda, milk of magnesia, white vinegar, laundry detergent, seltzer water, aspirin or vitamin C tablets, sample of saliva, measuring cup, water

# CHAPTER 13    OXIDATION AND REDUCTION

## Activities

13-7: 9-volt battery, paper towel, salt, water, white paper, laxative tablet (phenolphthalein type)

## Chemistry in Action

13-1: small glass, shiny penny, 10 dull pennies, white vinegar, baking soda

13-2: iodized table salt, two glasses, water, measuring cup, 3% hydrogen peroxide solution, white vinegar, four aspirin tablets, zip-lock bag, mineral oil (or charcoal lighter or lighter fluid), two coffee filters (or napkins)

13-3: 9-volt alkaline battery, three shiny pennies, Magic tape, aluminum foil, teaspoon, tablespoon, calamine lotion (or zinc oxide cream), white vinegar, a Pepto-Bismol tablet, salt, mercurochrome, water

# CHAPTER 14    CHEMICAL EQUILIBRIUM

## Activities

14-4:  baking soda, water, two glasses

14-6:  aluminum foil, carbonated water, matches

## Chemistry in Action

14-1:  iodized salt, coffee filter (or napkins), 3% hydrogen peroxide solution, white vinegar, cornstarch (or flour), two zip-lock bags, three glasses, ice, hot water

14-2:  red cabbage, nonaluminum pan with lid, milk of magnesia, white vinegar, lemon juice, seven glasses, measuring cup, teaspoon, eyedropper, water

# CHAPTER 15    NUCLEAR CHEMISTRY

## Activities

15-1:    dime

15-6:    battery-operated smoke detector

15-10:  bubble-making solution, two bubble-making loops (wire or plastic)

## Chemistry in Action

15-1:  empty clear plastic 2-liter bottle with cap, steel paper clip, matches, pen/pencil, ruler, Magic tape, stopwatch or watch with second hand, water

15-2:  paper, scissors, metric rule

# CHAPTER 16    ORGANIC CHEMISTRY

## Activity

16-8:  sugar, dry baker's yeast, tablespoon, measuring cup, water, zip-lock bag, bowl/cup

## Chemistry in Action

16-1:  zip-lock bag, white vinegar, rubbing alcohol (70% isopropyl alcohol), bowl, boiling water, measuring cup, teaspoon

## CHAPTER 17    THE CHEMISTRY OF LIFE: BIOCHEMISTRY

### Activities

17-4:   bar of soap, glass, white vinegar, water, measuring cup, tablespoon

17-9:   4-inch twist tie, pencil

17-10:   six 4-inch twist ties

### Chemistry in Action

17-1:   iodized salt, several glasses, water, 3% hydrogen peroxide solution, white vinegar, mineral oil, vegetable oil, two coffee filters or napkins, zip-lock bag, teaspoon, eyedropper

17-2:   raw egg, five glasses, salt, rubbing alcohol (70% isopropyl alcohol), boiling water, aluminum foil, matches, eyedropper

17-3:   3% hydrogen peroxide solution, seven glasses, variety of fruits/vegetables/green leaves, dry baker's yeast, tablespoon

# GLOSSARY

**Absolute Zero:** The lowest possible temperature corresponding to zero on the Kelvin scale.

**Acid (Brønsted–Lowry):** A species that can lose a proton ($H^+$) in a chemical reaction. A proton donor.

$$H\text{—}A \longrightarrow H^+ + A^-$$

**Acid–Base Indicator:** A chemical species that changes color at the end point of an acid–base titration.

**Acid–Base Reaction:** A reaction in which a proton is transferred from an acid to a base to give the corresponding conjugate acid and base.

**Acid Constant or Acid Ionization Constant, $K_a$:** The equilibrium constant for a solution of a weak acid, HA; the equilibrium constant expression for a weak acid.

$$K_a = \frac{[H_3O^+][A^-]}{[HA]}$$

**Acidic Solution:** A solution that has a hydronium ion concentration greater than pure water. An acid dissolved in water forms an acidic solution.

**Acids:** Chemical compounds that display specific chemical properties called acidic properties. Normally, the formula of an acid begins with hydrogen (e.g., HF, $H_2S$).

**Activity Series of Metals:** A redox table of metals and metal ions that ranks the metals according to increasing strength as reducing agents.

**Addition Reaction:** A typical reaction involving alkenes in which the double bond is broken to form a single bond and other atoms become bonded to the carbons of the original double bond.

**Alcohols:** A series of organic compounds involving a hydroxy group bonded to an alkyl group, ROH (e.g., methyl alcohol, $CH_3OH$, and ethyl alcohol, $CH_3CH_2OH$).

**Aldehydes:** A series of organic compounds involving an aldehyde group bonded to an alkyl group,

(e.g., acetaldehyde, or $CH_3CHO$).

**Alkali Metals:** The group 1 or group IA elements excluding hydrogen.

**Alkaline Earth Metals:** The group 2 or group IIA elements.

**Alkanes:** A homologous series of saturated hydrocarbons that have the general formula $C_nH_{2n+2}$ ($CH_4$, $n = 1$; $C_2H_6$, $n = 2$; $C_3H_8$, $n = 3$; etc.).

**Alkenes, Unsaturated Hydrocarbons, or Olefins:** Hydrocarbons that have a double bond between any two carbons. The alkenes are a homologous series of hydrocarbons corresponding to the general formula of $C_nH_{2n}$ ($CH_2{=}CH_2$, $n = 2$; $CH_3CH{=}CH_2$, $n = 3$; etc.).

**Alkyl Groups:** A group of covalently bonded atoms that can be considered to be formed by removing one hydrogen from an alkane, leaving an open bonding position. Groups are hypothetical structural entities to which other atoms or groups can be bonded to form a variety of organic compounds. Alkyl groups are named by using the name of the alkane from which they are derived with the -ane ending changed to -yl (e.g., methyl, $CH_3$-; ethyl, $CH_3CH_2$-).

**Alpha Particle:** A nuclear particle containing two protons and two neutrons; it is equivalent to a helium-4 nucleus.

**Alpha-Particle Decay:** A mode of radioactive decay in which a nucleus decays by emitting a high-speed helium-4 nucleus, $_2^4He$, called an alpha particle. A new nucleus is formed with nucleon number four less and atomic number two less than the original nucleus.

**Amphiprotic:** The ability of a species to behave as an acid or a base.

**Animal Cells:** The microscopic units that make up animal organisms and that differ in shape and size depending on their functions. A typical cell is surrounded by a membrane containing the cytoplasm or cellular fluids. The membrane-enclosed nucleus of a cell contains chromosomes that dictate the function and heredity characteristics of the cell. The mitochondria of the cell are involved in energy production. The endoplasmic reticula are areas containing ribosomes involved in protein synthesis. The lysosomes contained in cells are involved in the breakdown and dissolution of a cell when the cell membrane is ruptured.

**Anion:** A negatively charged ion.

**Anode:** The positively charged electrode in an electrical cell at which oxidation occurs.

**Aqueous Solution:** A solution of a solute dissolved in water.

**Aromatic Hydrocarbon:** A hydrocarbon that contains the benzene ring in its structure.

**Atmosphere (atm):** A unit of pressure defined as 1 atm = 760 torr. Also the name used for the layer of gases that surrounds the earth.

**Atmospheric Pressure:** The pressure exerted by the gas particles in the air on all objects exposed to the atmosphere.

**Atom:** The smallest representative particle of an element.

**Atomic Mass Unit (u):** A special unit of mass used to express isotope masses and atomic weights. An atom of the isotope carbon-12 is defined as having a mass of exactly 12 u. The masses of all other isotopes of elements can be measured relative to carbon-12 and expressed in atomic mass units.

**Atomic Number:** The number of protons in an atomic nucleus. The number of protons in an atom of an element equals the number of electrons in the atom. Each element has its own unique atomic number and is listed in the periodic table according to this number.

**Atomic Theory:** (1) Elements are composed of tiny, fundamental particles called atoms. (2) Atoms of a particular element are the same but differ from atoms of all other elements. (3) Atoms enter into combinations to form compounds.

**Atomic Weight:** The average mass of an atom of an element determined using the contribution of each natural isotope.

**Avogadro's Law:** Equal volumes of gases at the same temperature and pressure contain the same number of moles.

**Avogadro's Number:** The number of atoms in one mole of an element or the number of formula units in a mole of a compound. It has the value $6.022 \times 10^{23}$.

**Barometer:** A pressure-measuring device consisting of a mercury-filled glass tube sealed on one end and inverted in a bowl of mercury. Barometric pressures are measured in terms of millimeters of mercury (mm Hg).

*Base (Brønsted-Lowry):*   A species that can gain a proton ($H^+$) in a chemical reaction. A proton acceptor.

$$H^+ + B \longrightarrow B-H^+$$

*Basic Solution:*   A solution that has a hydroxide ion concentration greater than pure water. A base dissolved in water forms a basic solution.

*Battery:*   A device in which oxidation and reduction are carried out in separate compartments so that useful electrical work is obtained from a redox reaction.

*Beta Particle:*   A nuclear particle that is a high-speed electron emanating from a nucleus.

*Beta-Particle Decay:*   A mode of radioactive decay in which a nucleus decays by emitting a high-speed electron, $_{-1}^{0}e$, called a beta particle. A new nucleus is formed with the same nucleon number as the original nucleus and atomic number one greater.

*Binary Solution:*   A solution consisting of two components, a solvent and a solute.

*Biochemistry:*   The study of compounds involved in biological processes, and the changes they undergo during life processes.

*Bohr's Atomic Model:*   The concept of the atom developed by Niels Bohr in which the atom is described as a nucleus around which an electron moves in circular orbit.

*Boiling Point:*   The temperature at which a liquid boils or rapidly changes from a liquid to a vapor.

*Boiling Point Elevation:*   The phenomenon in which the presence of a solute causes the boiling point of the solution to be higher than the boiling point of the pure solvent.

*Bonding Pairs:*   Pairs of valence electrons around the central atom of a molecule or polyatomic ion that are involved in bonding to other atoms.

*Boyle's Law:*   The volume of a gas is inversely proportional to the pressure at constant temperature. $V = k/P$.

*Brønsted–Lowry Acid:*   A species that can lose a proton ($H^+$) in a chemical reaction. A proton donor.

*Brønsted–Lowry Base:*   A species that can gain a proton ($H^+$) in a chemical reaction. A proton acceptor.

*Buffer Solution:*   A solution that resists a change in pH on the addition of small amounts of an acid or base. A buffer solution contains a weak acid in equilibrium with its conjugate base.

*Calorie (cal):*   The amount of heat needed to change the temperature of one gram of water by one degree Celsius. 1 cal = 4.184 J.

*Carbohydrates or Sugars:*   Compounds containing carbon, hydrogen, and oxygen. Produced mainly by plants, they range in size from simple sugars called monosaccharides to polymeric sugars called polysaccharides, and serve as food energy sources for humans.

*Carbon Chains:*   A term used to refer to the occurrence of sequences of covalently bonded carbon atoms in some organic compounds.

*Carboxylic Acids:*   A series of organic compounds involving a carboxyl group bonded to an alkyl group,

$$\begin{array}{c} O \\ \| \\ RC-OH \end{array} \quad \text{or } RCO_2H \text{ (e.g. acetic acid,} \quad \begin{array}{c} O \\ \| \\ CH_3C-OH \end{array}$$

or $CH_3CO_2H$).

*Catalyst:*   A chemical that increases the speed or rate of a chemical reaction without being chemically changed. A catalyst provides an alternate path for a reaction that is faster and gives the same products.

*Cathode:*   The negatively charged electrode in an electrical cell at which reduction occurs.

*Cathode-Ray Tube:*   A glass tube that has the air pumped out and metal electrodes sealed to opposite ends. When a high-voltage source is attached to the electrodes electricity flows through the tube. The electricity flow occurs as a cathode-ray beam, which is a beam of free electrons.

*Cation:*   A positively charged ion.

*Celsius °C:*   The unit of temperature used on the Celsius temperature scale.

*Cellulose:*   A polysaccharide that makes up the structural material of many plants. Cellulose consists of long chains of beta-glucose units.

*Chain or Sequential Nuclear Fission:*   A series of subsequent fission events that results when the neutrons of fission cause the fission of more nuclei. Such a sequential reaction is possible in a large collection of fissionable nuclei since a single fission event produces two or more neutrons that can enter into further fission events.

*Charles' Law:*   The volume of a gas is directly proportional to the temperature at constant pressure. $P = kT$.

**Chemical Bond:** A force that holds atoms in combinations.

**Chemical Equation:** The symbolic representation of a chemical reaction showing the formulas of the reactants and products separated by an arrow.

**Chemical Equilibrium:** A dynamic equilibrium between reactants and products in which the rate at which reactants form products equals the rate at which products form reactants. A state of chemical equilibrium can exist between reactants and products in a reversible chemical reaction.

**Chemical Formula:** A representation of a compound that gives the symbol of each element in the compound followed by a subscript indicating the relative number of combined atoms.

**Chemical Properties:** Properties that relate to the chemical behavior of matter. Such properties relate to how a chemical interacts with other chemicals. Does it burn in air? Does it rust or corrode? Does it react with water? And so on.

**Chemical Reaction:** A process in which one set of chemicals called reactants is converted into another set of chemicals called products.

**Chemical Symbol:** The symbol used to represent a chemical element.

**Chemistry:** The science of the properties, composition, and behavior of matter.

**Coefficient:** A number placed in front of a formula when balancing an equation.

**Colligative Properties:** Properties of solutions that depend only upon the presence of solute particles not on their identity.

**Columns:** The vertical (up and down) groupings of elements within the periodic table.

**Combined Gas Law:** $V_2 P_2 / T_2 = V_1 P_1 / T_1$.

**Combustion:** The reaction of a carbon-containing compound with oxygen to give carbon dioxide and water.

**Common or Trivial Names:** Names for compounds that have developed historically and give no information concerning structure.

**Compounds:** Pure chemicals made up of combinations of chemical elements. A given compound always contains the same elements in definite proportions by mass.

**Concentration:** An expression of the amount of solute per unit amount of solution.

**Condensation:** The change from a vapor to a liquid.

**Condensed Structural Formula:** A formula for a molecular compound that indicates the bonding sequence without showing all the bonds. A condensed structural formula for an organic compound (e.g., butane, $CH_3CH_2CH_2CH_3$) should be interpreted as indicating that the carbons are bonded to one another in sequence, and that each carbon is bonded to the hydrogens (or other atoms) that are next to it in the formula.

**Conjugate Acid–Base Pair:** An acid and a base related by the loss or gain of a proton.

**Conversion Factor or Unit Factor:** A factor relating units or properties that can be used to convert a measurement made in terms of one unit to another unit.

**Core Meltdown:** The melting of a reactor core resulting from the interruption of coolant flow and the subsequent dramatic increase in temperature from the energy of fission. Core meltdown could result in destruction of the reactor and radioactive contamination of the environment.

**Covalent Bond:** The force of attraction between atoms arising from the sharing of an electron pair.

**Covalent or Molecular Compound:** A compound consisting of a collection of molecules. Compounds involving two or more nonmetals or those that include metalloids and nonmetals are molecular compounds.

**Critical Mass:** The minimum amount of fissionable material in which a fission chain reaction can be self-sustaining and become supercritical.

**Crystal or Crystal Lattice:** The three-dimensional pattern or arrangement of the chemical particles making up a solid.

**Crystal Lattice Sites:** The specific positions or locations that particles occupy in the crystal lattice of a solid.

**Dalton's Law of Partial Pressures:** The total pressure of a mixture of gases is the sum of the partial pressures of the components of the mixture.

$$P_t = P_a + P_b + \ldots$$

**d Block:** The elements of the periodic table that correspond to the filling of the *d* sublevels.

**Decay Chains or Series:** The isotopes of the elements with atomic numbers greater than 82 belong to one of three series of elements that decay by alpha or beta emission. The elements in the series are formed in sequence according to the pattern of decay and the series ends in the formation of a stable isotope of lead.

**Density:** The mass per unit volume of a substance. $d = m/V$. A property of a substance determined by measuring the mass and volume of a sample then dividing the mass by the volume.

**Deoxyribose:** A monosaccharide related to ribose except that the hydroxy group of C-2 is replaced by a hydrogen.

**Deoxyribonucleic Acid (DNA):** Nucleic acids that structure the genes of cells. DNA is polymeric and composed of phosphoric acid, deoxyribose sugar, and the heterocyclic amines adenine, guanine, cytosine, and thymine.

**Diatomic Elements:** Elements that occur in the form of molecules composed of two atoms of the element. The seven diatomic elements are $H_2$, $O_2$, $N_2$, $F_2$, $Cl_2$, $Br_2$, and $I_2$.

**Dietetic Calorie:** A unit used for food energy equal to one kilocalorie and often given a capital C. 1 Cal = 1000 cal = 1 kcal.

**Dilution:** The process of preparing a solution of lower concentration by mixing more solvent with a solution of higher concentration.

**Dipole Molecular Attraction:** Attractions between polar molecules caused by the electrostatic forces of attraction between oppositely charged portions of molecules.

**Disaccharide:** A sugar composed of two monosaccharide units joined by mutual sharing of an oxygen.

**DNA:** See deoxyribonucleic acid.

**Double Bond or Double Covalent Bond:** The sharing of two pairs of electrons between atoms so that they are bonded by two covalent bonds. Carbon atoms are capable of double-bond formation in certain organic compounds.

**Double-Helix Structure of DNA:** A theory of the structure of DNA as it exists in genes first proposed by Francis Crick and James Watson. DNA exists as two strands intertwined in the form of two helices by hydrogen bonding between thymine and adenine units and between guanine and cytosine units on opposite strands.

**Ductile:** A property of metals corresponding to the ability to be stretched into wires.

**Dynamic Equilibrium:** A phenomenon in which the rates of two opposing processes are equal. For a solute dissolved in a solvent, a state of dynamic equilibrium exists when the rate at which the solute dissolves equals the rate at which the solute comes out of solution (undissolves). Once a state of dynamic equilibrium exists there is balance between the two processes and no net change occurs.

**Electrical Cell:** A device in which oxidation and reduction are carried out in separate compartments so that useful electrical work is obtained from a redox reaction.

**Electrolysis:** A process in which an external source of electricity is used to cause an oxidation–reduction reaction to occur.

**Electrons:** Negatively charged subatomic particles found in atoms and as particles of electricity.

**Electronic Configuration:** The pattern in which the electrons in an atom of a given element are distributed. The distribution of the electrons in an atom in terms of the sublevels occupied.

**Electron Dot Symbol:** A symbol for an element using the normal element symbol surrounded by dots that indicate the number of valence electrons.

**Electron Dot Structure:** See Lewis electron dot structure.

**Electron Orbitals:** Each energy sublevel is made up of one or more electron orbitals in which electrons can be located within an atom. Orbitals are visualized as three-dimensional electron clouds. The four kinds of orbitals are named s, p, d, and f.

**Electron Transfer Reaction:** A chemical reaction in which one species loses electrons and another species gains electrons. An oxidation–reduction reaction.

**Electrolyte:** A substance that forms an aqueous solution that conducts electricity. Electrolytes dissolve in water to form cations and anions.

**Electronegativity:** The tendency of an atom of an element to attract electrons in a chemical bond. Except for the noble gases the electronegativities of elements increase from bottom to top in columns and from left to right in rows. Fluorine is the most electronegative element followed by oxygen and chlorine.

**Elements:** Pure chemicals that cannot be separated into simpler chemicals. Fundamental forms of matter.

**Emergency Core Cooling:**  A reactor safety measure in which a reserve coolant is available to be used in case the primary coolant flow is interrupted.

**Empirical Formula:**  The simplest formula of a compound that is deduced from the experimentally determined composition of the compound.

**Endothermic Reaction:**  A chemical reaction in which heat is absorbed or required for the reaction to occur; also called an endergonic reaction.

**End Point:**  The point in a titration at which the indicator changes color.

**Energy:**  The capacity for doing work; it takes various forms, such as radiant energy, heat, potential energy, kinetic energy, and electrical energy.

**Energy Levels:**  Various energy positions in the atom in which electrons may be located. Sometimes called energy states or energy shells, they are referred to by the numbers 1, 2, 3, 4, 5, 6, and 7.

**Energy States:**  The possible quantized energy positions of electrons in an atom; sometimes called energy levels.

**Energy Sublevels:**  Each energy level is made up of one or more sublevels in which electrons of varying energies can reside. Sublevels are sometimes called energy subshells. An energy level can have $s$, $p$, $d$, and $f$ sublevels.

**Enzymes:**  Biological catalysts that are polymeric molecules of varying complexity. Some enzymes consist of protein only; some consist of a protein apoenzyme and a nonprotein cofactor. The body synthesizes thousands of enzymes to catalyze metabolic reactions.

**Enzyme and Protein Synthesis:**  A cellular process in which the DNA of the genes produces various RNAs that direct the synthesis of enzymes and proteins as they are needed.

**Ester:**  An organic compound corresponding to the formula

$$R-\overset{\overset{\displaystyle O}{\|}}{C}-O-R$$

Esters can be formed by reaction of an alcohol and a carboxylic acid.

**Equilibrium Constant Expression:**  A constant for an equilibrium reaction equal to the ratio of the molar concentrations of products to reactants. For the general reaction

$$a\,A + b\,B \rightleftharpoons c\,C + d\,D$$

the equilibrium constant expression is

$$K_{eq} = \frac{[C]^c[D]^d}{[A]^a[B]^b}$$

The numerical value of an equilibrium constant is determined by substituting experimental values of equilibrium concentrations into the equilibrium constant expression for an equilibrium system.

**Exothermic Reaction:**  A chemical reaction in which heat is released or produced; also called an exergonic reaction.

**Fats or Oils:**  Naturally occurring esters of glycerol. Fats are solid and are found mainly in animals. Oils are liquids and are found mainly in plants. Fats and oils are required foods for humans and are included in such common products as butter, lard, olive oil, peanut oil, corn oil, safflower oil, margarine, fish oils, and meat fats.

**Fatty Acids:**  A variety of long-chain carboxylic acids (normally containing from 12 to 20 carbons) obtained by chemical decomposition of fats and oils to form glycerol and the acids.

**f Block:**  The elements of the periodic table that correspond to the filling of the $f$ sublevels.

**Fission:**  A special kind of nuclear transmutation in which a neutron interacts with a nucleus, the nucleus splits into two new nuclei and a few neutrons, and energy is released.

**Fission Products (High-Level Radioactive Wastes):**  A variety of radioactive substances that accumulate in a nuclear reactor when they are formed by nuclear fission in fuel rods. Fission products have to be collected periodically and disposed of by special methods. Fission products include many isotopes. The most notable are plutonium-239, cesium-137, and strontium-90.

**Fissile Isotopes:**  The isotopes uranium-235 and plutonium-239 that can undergo thermal neutron fission.

**Force:**  An action that can cause some effect (e.g., gravity force, magnetic force, electrostatic force).

**Formula Unit:**  An amount of a compound that contains the number of atoms of each element given in the formula. The smallest representative unit of a compound.

**Freezing Point Depression:**  The phenomenon in which the presence of a solute causes the freezing point of the solution to be lower than the freezing point of the pure solvent.

**Fuel Cell:** A device in which oxidation and reduction are carried out in separate compartments to obtain useful electrical work. The reactants are continuously supplied and the products are continuously removed to give a continuous source of electricity.

**Fusion:** A special type of nuclear transmutation that occurs at thermonuclear temperatures in which certain lighter nuclei fuse to form heavier nuclei and release energy.

**Gamma-Ray Emission:** The daughter nuclei formed in alpha or beta decay are in some cases energetically excited. When excited nuclei fall to lower energy states, they emit electromagnetic radiation called gamma rays.

**Gas:** A physical state in which a sample of matter has no definite volume, no definite shape, and occupies all of its container.

**Gas Constant, R:** The proportionality constant in the ideal gas law that applies to any gas behaving according to the kinetic molecular theory. For volume expressed in liters, pressure in atmospheres, and temperature in kelvins

$$R = \frac{0.0821 \text{ L atm}}{\text{K mol}}$$

**Gas Discharge Tube:** A cathode-ray tube that contains a small amount of a particular gas. The flow of electricity through the tube results in the emission of colored light characteristic of the gas in the tube.

**Geiger–Müller Tube:** A gas-filled electronic tube designed to detect the presence of ionizing radiation. Radiation passing through the tube causes ionizing events in the gas, which are amplified into an electrical current detected and registered by built-in electronic circuits.

**Genes:** Heredity units of chromosomes contained in cell nuclei; they are composed of DNA.

**Genetic Code:** Information about the inherited characteristics of an organism contained in the genes as specific sequences of adenine, guanine, cytosine, or thymine coded on DNA strands.

**Germ Cells or Gamete Cells:** Cells located in the gonads that are capable of being formed into sperm or eggs. Human gamete contain 23 chromosomes.

**Glucose:** A very common monosaccharide that is chemically a polyhydroxy aldehyde having the molecular formula $C_6H_{12}O_6$. Glucose usually occurs as cyclic molecules.

**Glycogen:** A polysaccharide found in animals and called body starch in humans. Glycogen is very similar to amylopectin and consists of branched chains of α-glucose units.

**Groups or Families of Elements:** Vertical columns of elements in the periodic table. Elements within a group or family have similar properties.

**Half-Life:** A characteristic property of a radioactive substance defined as the time in which one-half of a sample of a radioactive substance decays.

**Heat:** A mode of energy transfer between objects. Heating an object can increase the heat content and the temperature, whereas cooling can decrease the heat content and temperature. Heat is a measure of the total energy content of a sample of matter.

**Heat of Reaction or Enthalpy Change:** The energy exchanged in a chemical reaction. Heats of reaction are usually expressed in units of kilocalories per mole or kilojoules per mole.

**Heat of Solution:** The energy exchange associated with the dissolving of a solute in water. Heats of solution are usually expressed in units of kilocalories or kilojoules per mole of solute.

**Homologous Series:** A series of organic compounds in which consecutive members differ in structure by a group consisting of a carbon and two hydrogens ($—CH_2—$). In the alkanes, for example, methane, $CH_4$, ethane, $CH_3CH_3$ ($CH_3—CH_2—H$), and propane, $CH_3CH_2CH_3$($CH_3—CH_2—CH_3$) differ in structure by a $—CH_2—$ unit and are the simplest members of an extensive homologous series.

**Human-Made Isotopes:** Nuclides formed by nuclear transmutations induced by nuclear scientists in the laboratory or a nuclear reactor.

**Hund's Rule:** Electrons in a set of orbitals within a sublevel tend to occupy empty orbitals and pair only when all orbitals have at least one electron.

**Hydrocarbons:** Organic compounds containing carbon and hydrogen.

**Hydrogen Bond:** Electrostatic attraction between hydrogen atoms of one polar molecule and a highly electronegative atom (F, O, N) of another polar molecule.

**Hydronium Ion:** The ion formed when water gains a proton, $H_3O^+$.

**Ideal Gas:** An idealized or theoretical gas that obeys the kinetic molecular theory and the ideal gas law.

**Ideal Gas Law:** For an ideal gas the product of its pressure, $P$, and its volume, $V$, equals the product of the number of moles, $n$, the gas constant, $R$, and its Kelvin temperature, $T$. $PV = nRT$.

**Inner Transition Elements:** Those elements comprising the $f$ block of the periodic table (the actinide series and the lanthanide series). They are listed in the lower portion of the periodic table.

**Intermolecular Attractions:** The attractive forces that exist between molecules; sometimes called van der Waals' forces.

**Ion:** A chemical particle that carries a negative or positive charge. Simple ions are charged atoms; polyatomic ions are groups of covalently bonded atoms that carry charges.

**Ion Combination Reaction:** A chemical reaction that occurs when solutions containing cations and anions of an insoluble substance are mixed. A precipitation reaction.

**Ionic Bond:** The electrostatic force of attraction between ions that binds ions in ionic compounds.

**Ionic Compound:** A compound consisting of a collection of ions held together by ionic bonds. Compounds of metals and nonmetals are typically ionic compounds. Ionic compounds are normally crystalline solids.

**Ionization Energy:** The amount of energy required to remove one outer-energy-level electron from an atom of an element.

**Ion Pairs:** Electrons and positive ions formed by ionization of molecules or atoms exposed to particles of radiation; a single high-energy particle can form numerous ion pairs.

**Ionizing Radiation:** Radiation that causes ion pair formation or ionization of molecules and atoms in the material through which it passes.

**Inhibitor:** A chemical that slows down or decreases the rate of a chemical reaction.

**Isomers or Structural Isomers:** Compounds with the same molecular formulas but different structural formulas.

**Isotopes:** Atoms of an element that have the same number of protons but different numbers of neutrons and different masses.

**Joule:** The metric or SI unit for energy. 1 cal = 4.184 J.

**$K_a$:** The equilibrium constant for a solution of a weak acid, HA; the acid ionization constant for a weak acid.

$$K_a = \frac{[H_3O^+][A^-]}{[HA]}$$

**Kinetic Energy:** Energy of motion. The kinetic energy of a moving object is given by the expression KE = $\frac{1}{2}mv^2$ where $m$ is the mass and $v$ is the speed.

**Kinetic Molecular Theory:** The view of a gas as a diffuse collection of molecules in rapid, random straight-line motion continually colliding with one another and any objects in their vicinity.

**Law of Conservation of Energy:** Energy cannot be created or destroyed in chemical processes.

**Law of Conservation of Matter:** Matter cannot be created or destroyed in chemical processes.

**Law of Constant Composition:** A compound always contains the same elements in definite proportions by mass.

**Le Châtelier's Principle:** When a factor affecting an equilibrium system is changed, the equilibrium will shift in a direction that tends to counteract the change.

**Lewis Electron Dot Structure:** A symbolic representation of a molecule showing the shared pairs of electrons between atoms and any other nonbonding pairs of outer-energy-level electrons associated with each atom.

**Limiting Reactant:** In a chemical reaction the reactant whose amount limits how much product can be formed.

**Liter, L:** A unit of volume equal to 1000 cubic centimeters, $1000 \text{ cm}^3$.

**Lipids:** Triglycerides or esters of fatty acids and the trihydroxy alcohol, glycerol ($CH_2OHCHOHCH_2OH$). Fats and oils are simple lipids.

**Liquefaction:** The conversion of a gas to a liquid.

**Liquid:** A physical state in which a sample of matter occupies a definite volume and takes on the shape of the portion of the container it occupies.

**Lock-and-Key Theory:** A specific substrate fits the three-dimensional structure of a specific enzyme. Once the enzyme and substrate form an aggregate, the substrate is exposed for the reaction. After the reaction occurs, the products move away and the enzyme is available for another substrate.

**Lone Pairs or Nonbonding Pairs:** Pairs of valence electrons around the central atom of a molecule or polyatomic ion that are not involved in bonding to other atoms.

**Low-Level Radioactive Wastes:** Impurities in the coolant of the reactor that become radioactive when exposed to the reactor core.

**Malleable:** A metallic property related to the ability to be bent, cast, and formed.

**Maltose:** A disaccharide composed of an $\alpha$-glucose unit bonded through C-1 to C-4 to another glucose unit.

**Mass:** A property of matter that determines its resistance to being set in motion or resistance to any change in motion. Mass relates to the amount of material in a sample of a substance.

**Mass Number or Nucleon Number:** The sum of the number of protons and neutrons in a nucleus of an atom. Each different isotope of an element has its own unique nucleon number.

**Mass Spectrometer:** An instrument used to measure the atomic masses of isotopes relative to carbon-12.

**Matter:** Anything that occupies space and has mass.

**Metabolism:** The numerous chemical reactions that occur in the body including digestion of food, utilization of food molecules for energy, and the breakdown and formation of new cells.

**Melting Point:** The temperature at which a solid melts.

**Metals:** Those elements that have metallic properties. They are good electrical and heat conductors, are flexible enough to be deformed, and possess metallic luster. Of the 109 elements, 84 are metals; they are located to the left of the metalloids in the periodic table.

**Metalloids:** Those elements (Si, Ge, As, Sb, Te, Po) that display both metallic and nonmetallic properties and are located between the metals and nonmetals in the periodic table.

**Metric System:** An international system of measurement in which defined units of mass, length, time, and temperature have been established.

**Molality (m):** The number of moles of solute per kilogram of solvent.

**Molar Mass of a Compound:** The number of grams per mole of a compound; found by multiplying the molar mass of each element by the corresponding subscript in the formula and adding up these products.

**Molar Mass or Number of Grams per Mole of an Element:** The number of grams of an element corresponding to one mole; it is numerically equal to atomic weight and has units of grams in the numerator and moles in the denominator.

**Molar Ratio:** A numerical ratio expressing the number of moles of one element to the number of moles of another element or the number of moles of an element per mole of a compound. The subscripts in the formula reveal the molar ratios. For example, in water, $H_2O$, there are

$$\frac{2\ mol\ H}{1\ mol\ O}\ , \quad \frac{2\ mol\ H}{1\ mol\ H_2O}\ , \quad \frac{1\ mol\ O}{1\ mol\ H_2O}$$

**Molar Volume:** The volume occupied by 1 mole of an ideal gas at STP. The value of the molar volume is

$$\left(\frac{22.4\ L}{1\ mol}\right)_{STP}$$

**Molarity (M):** The number of moles of solute per liter of solution. $M = n/V$

**Mole of a Compound:** The amount of a compound that contains Avogadro's number of formula units. The amount of a compound that contains the number of moles of each element given by the subscripts in the formula of the compound.

**Mole of an Element:** The amount of an element that contains as many atoms as there are carbon atoms in 12 grams of carbon-12.

**Molecular Formula:** A formula of a molecular compound that indicates the type and number of atoms comprising a molecule. A molecular formula is a conventional formula and does not convey the bonding sequence in the molecule.

**Molecule:** A group of two or more covalently bonded atoms.

**Multiple Bond:** The sharing of two pairs (double bond) or three pairs (triple bond) of electrons between two atoms.

*Nanometer:* A metric unit of length used to describe atomic, molecular, or ionic sizes. One nanometer equals $10^{-9}$ m.

*Net-Ionic Equation:* An equation for a reaction occurring in aqueous solution showing only the molecules and ions that react and that are produced.

*Neutral Solution:* A solution in which the concentrations of hydronium and hydroxide ions are equal as in pure water. A solution that is neither acidic nor basic.

*Neutralization Reaction:* A reaction between a solution of a strong acid and a solution of a strong base that produces a neutral solution or a reaction between an acid and a base that neutralizes their properties.

*Neutrons:* Subatomic particles that have no charge and are found in atomic nuclei.

*Noble Gases:* The elements in the far right column of the periodic table. They have completely filled *s* and *p* sublevels in the outer energy level and form few compounds. The noble gases are He, Ne, Ar, Kr, Xe, and Rn.

*Nonbonding Electrons:* See lone pairs.

*Nonelectrolyte:* A substance that forms an aqueous solution that does not conduct electricity.

*Nonmetals:* Those elements that do not possess metallic properties. They are gases or brittle solids, are poor or nonconductors, and have no metallic luster. All 19 nonmetals are located in the upper right of the periodic table.

*Nonpolar Molecular Attractions:* Attractions between nonpolar molecules caused by the attractions of electrons of some nonpolar molecules for the nuclei of other nonpolar molecules.

*Nonpolar Molecule:* A molecule with no net separation of positive and negative charge centers. A nonpolar molecule has either no polar bonds or polar bonds that are distributed in a symmetrical manner about the center of the molecule.

*Nuclear Fission:* See fission.

*Nuclear Fusion:* See fusion.

*Nuclear Reactor:* A device in which self-sustaining fission can be carried out under controlled conditions. The basic components of a nuclear reactor are (1) nuclear fuel rods that contain fissionable material, (2) a moderator used to slow down neutrons to induce fission, (3) control rods that absorb neutrons and can be manipulated to control or stop fission, (4) a coolant circulating about the fuel rods to carry away the energy of fission as heat, and (5) a thick shielding around the reactor to protect people from its radioactive core.

*Nuclear Transmutation:* A phenomenon in which a high-speed nuclear particle (projectile) collides with a nucleus (target), and a nuclear reaction occurs in which a new nucleus and particle are formed.

*Nucleic Acids:* Polymeric cellular substances made up of monomer units called nucleotides. Nucleotides can be chemically broken down into phosphoric acid, some heterocyclic amines, and either deoxyribose or ribose sugar.

*Nucleons:* Basic nuclear particles; protons and neutrons. See Mass Number.

*Nucleus:* The very small, highly dense, positively charged center of atoms made up of clusters of protons and neutrons.

*Nuclides:* A general term used to refer to the nuclei of isotopes of all the elements.

*Octet Rule:* Atoms generally lose, gain, or share electrons to attain a total of eight in the outer energy level.

*Orbital:* See electron orbitals.

*Organic Compounds:* A special group of carbon-containing compounds; most include carbon and hydrogen combined with oxygen, nitrogen, phosphorus, and/or sulfur.

*Organic Chemistry:* The field of chemistry devoted to the study of those compounds of carbon considered to be organic compounds.

*Osmosis:* The phenomenon in which water passes from a solution of lower concentration of solute to a solution of higher concentration of solute through a semipermeable membrane separating the two solutions.

*Outer-Energy-Level Electrons:* Those electrons in an atom that are located in the highest-numbered energy level occupied by electrons.

*Oxidation:* Loss of electrons; the increase in oxidation number of an element.

*Oxidation–Reduction Reaction:* An electron transfer reaction in which one species loses electrons (is oxidized) and another species gains electrons (is reduced).

***Oxidation Half-Reaction:*** The portion of an electron transfer reaction involving oxidation or loss of electrons.

***Oxidation Number:*** The charge an element has in a simple ion or the hypothetical charge it would have if the shared electrons in covalent bonds were assigned to the more electronegative elements.

***Oxidizing Agent:*** A species that causes another species to be oxidized and is reduced in the process; an electron acceptor.

***Oxoacid:*** An acid that contains hydrogen, oxygen, and some other element (e.g., $H_2SO_4$, $HNO_3$).

***Oxoanions:*** Negative polyatomic ions containing oxygen and another element. Some oxoanions are related to or derived from the common oxoacids (e.g., $SO_4^{2-}$, $NO_3^-$).

***Parts per Million (ppm):*** An expression of the number of parts of a given component per million parts of the whole.

***Pauli Exclusion Principle:*** An electron orbital can contain a maximum of two electrons and those two electrons must have opposite spins.

***p Block:*** The elements of the periodic table that correspond to the filling of the outer *p* sublevels.

***Peptide Bonds or Linkages:*** The amide linkages,

in peptides.

***Percent by Mass:*** The amount of a component in a solution or mixture expressed as a percent of the total mass of solution or mixture.

***Percent-by-Mass Composition of a Compound:*** The expression of the composition of a compound in terms of the mass of each element present with respect to the mass of the compound; the number of grams of each element in 100 grams of the compound.

***Period:*** A horizontal row in the periodic table.

***Periodic Table:*** A table in which the elements are listed according to increasing atomic number and are grouped in columns according to similarities in properties and electronic configurations.

***pH:*** A term used to express the relative hydronium ion concentration in a solution. pH is defined as the negative logarithm of the hydronium ion concentration.

$$pH = -\log[H_3O^+]$$

For typical solutions the pH values range from about 0 to 14. A solution of pH 7 is neutral. An acidic solution has a pH of less than 7, and a basic solution has a pH greater than 7.

***Photochemical Smog:*** See smog.

***Physical Properties:*** Properties that relate to the physical nature of matter. Examples are appearance, color, state, density, conductivity, melting point, and boiling point.

***Physical Law:*** A general statement concerning a consistent pattern in nature.

***Plastics:*** A group of synthetic organic compounds made by polymerization reactions. Plastics can be molded, laminated, and extruded into forms such as films, fibers, and bottles.

***Plutonium or Plutonium-239:*** The fissionable isotope of plutonium produced in nuclear reactors. In the future it may be the major fuel used in nuclear reactors; it can also be used to make nuclear weapons.

***pOH:*** A term used to express the relative hydroxide ion concentration in a solution. pOH is defined as the negative logarithm of the hydroxide ion concentration.

$$pOH = -\log[OH^-]$$

For typical solutions the pOH values range from about 0 to 14. A solution of pOH 7 is neutral. A basic solution has a pOH of less than 7 and an acidic solution has a pOH greater than 7.

***Polar Bond:*** A covalent bond in which the more electronegative atom attracts the electrons to a greater extent than the other atom, producing an uneven distribution of charge.

***Polar Molecular Attraction:*** Attractions between polar molecules caused by the electrostatic forces of attraction between oppositely charged portions of molecules.

***Polar Molecule:*** A molecule in which there is a net separation of positive and negative charge centers. A polar molecule has polar bonds and an unsymmetrical shape that results in a separation of charge centers.

***Polyatomic Ion:*** A group of two or more covalently bonded atoms that carries an electrical charge as a unit.

*Polymerization Reactions:* Chemical reactions involving specific compounds in which hundreds or thousands of the molecules add to one another to form relatively large molecules called polymers. The original molecule used to prepare a polymer is called the monomer. For example, the monomer ethylene can react in the presence of a catalyst to form the polymer polyethylene.

*Polymers:* See plastics.

*Polypeptides:* See proteins.

*Polysaccharides:* Polymeric carbohydrates made of hundreds or thousands of glucose units joined by mutual sharing of oxygens. Starch and cellulose are polysaccharides.

*Polyunsaturated Oils:* Vegetable oils that contain a high percentage of unsaturated or double-bond-containing fatty acids combined with glycerol as esters.

*Power Reactor:* A nuclear reactor in which the heat absorbed by the coolant is used to generate steam from which electricity is produced.

*Precipitate:* An insoluble solid formed in a precipitation or ion combination reaction.

*Precipitation Reaction:* See ion combination reaction.

*Prefix Nomenclature:* A system of nomenclature for binary nonmetal–nonmetal compounds in which the number of combined atoms of each element is given by a prefix in the name.

*Pressure:* Force per unit area. The pressure exerted by a gas results from the force of the rapidly moving particles of the gas per unit area of the container.

*Pressure–Temperature Law:* The pressure of a gas is directly proportional to the temperature at constant volume. $P = kT$

*Primary Air Pollutants:* Waste gases produced by activities of an industrial society. The five primary air pollutants as defined by the U.S. Environmental Protection Agency (EPA) are carbon monoxide, sulfur dioxide, nitrogen oxides, volatile organic compounds, and particulates.

*Primary Structure of Proteins:* The specific amino acid sequence that makes up the peptide chain of a specific protein.

*Products:* The chemicals produced in a chemical reaction.

*Proteins or Polypeptides:* Polymeric peptides consisting of chains of hundreds or thousands of amino acids. Proteins are synthesized in body cells and serve as structural units of cells, skin, muscles, bone interior, and nerves; as enzymes and hormones; and in many other important functions of the body. All body proteins are made from about 20 different amino acids. Proteins are an important food source for humans since they provide amino acids needed by the body to synthesize proteins.

*Protons:* Positively charged subatomic particles found in atomic nuclei.

*Proton:* In acid–base chemistry a proton is a hydrogen ion ($H^+$) formed by the breaking of a covalent bond involving hydrogen and some other element:
$$H—A \longrightarrow H^+ + A^-$$

*Proton-Transfer Reaction:* An acid–base reaction in which protons transfer from an acid to a base.

*Qualitative Analysis:* A chemical analysis in which only the identity or presence of one or more components in a mixture or a compound is determined.

*Quantitative Analysis:* A chemical analysis in which the percentage of one or more components in a mixture or a compound is determined.

*Quantum Jump:* A change in the energy level of an electron within an atom. Energy is required for a jump to a higher level, and energy is released in a jump to a lower level.

*Quantum Mechanical Model of Atom:* The concept of the atom in which the atom is described as a nucleus around which the electrons exist in the form of electron clouds called electron orbitals. The electrons are located in energy levels. These energy levels are made up of energy sublevels, which in turn are composed of electron orbitals in which the electrons reside.

*R:* See gas constant.

*Radiation:* The high-speed particles emitted in radioactive decay. The common forms of radiation are alpha ($\alpha$) radiation, beta ($\beta$) radiation, and gamma ($\gamma$) radiation.

*Radiation Sickness:* An illness resulting from malfunction of bodily processes caused by exposure to high levels of ionizing radiation. The illness is characterized by nausea, vomiting, weakness, weight loss, fever, diarrhea, internal bleeding, loss of hair, and, very often, death.

*Radioactivity:* Spontaneous decay of a nucleus to form another nucleus and a subatomic particle.

*Rate of Reaction:* A measure of how fast a chemical reaction occurs. Normally, the rate of a reaction is expressed in terms of how fast a reactant is used up or how fast a product is produced using units of moles per second or moles per minute.

*Reactants:* The initial or starting chemicals in a chemical reaction.

*Redox Reaction:* An electron transfer reaction in which one species loses electrons (is oxidized) and another species gains electrons (is reduced). An oxidation-reduction reaction.

*Reducing Agent:* A species that causes another species to be reduced and is oxidized in the process; an electron donor.

*Reduction:* Gain of electrons; the decrease in oxidation number of an element.

*Reduction Half-Reaction:* The portion of an electron transfer reaction involving reduction or gain of electrons.

*Rem or Roentgen Equivalent Human:* An absorbed dose unit used to express the amount of radiation absorbed by humans upon exposure to radiation.

*Representative Elements:* The elements comprising the *s* and *p* blocks of the periodic table that make up the following groups:

| | |
|---|---|
| Group 1 or IA | Alkali metals |
| Group 2 or IIA | Alkaline earth metals |
| Group 13 or IIIA | |
| Group 14 or IVA | |
| Group 15 or VA | |
| Group 16 or VIA | |
| Group 17 or VIIA | Halogens |
| Group 18 or VIIIA | Noble Gases |

*Reversible Reaction:* A chemical reaction in which the reactants form the products and the products form the reactants. A double arrow is used to represent a reversible reaction.

*Ribonucleic Acid (RNA):* Nucleic acid involved in protein synthesis in cells. RNA is polymeric and composed of phosphoric acid, ribose sugar, and the heterocyclic amines adenine, guanine, cytosine, and uracil.

*Ribose:* A common monosaccharide that is chemically a polyhydroxy aldehyde having the molecular formula $C_5H_{10}O_5$. Ribose usually occurs in a cyclic form.

*RNA:* See ribonucleic acid.

*Rows:* The horizontal (left to right) sequences or periods of elements in the periodic table.

*Salts:* A general term sometimes used to refer to ionic compounds.

*Saturated Hydrocarbons:* Hydrocarbons that have only single-bonded carbons with no double or triple bonds.

*Saturated Solution:* A solution in which the dissolved and undissolved solute are in dynamic equilibrium. A solution containing as much solute as possible at a specific temperature.

*s Block:* The elements of the periodic table that correspond to the filling of the outer *s* sublevels.

*Secondary Structure of Proteins:* The shape of a protein chain determined by hydrogen bonding and other forces of attraction between the amino acid groups along the chain. A common secondary structure for proteins is the α-helix in which the protein chain is coiled in a three-dimensional helical shape.

*Semimetals:* See metalloids.

*Shielding Effect:* The decrease in the attraction of the nucleus for outer-level electrons resulting from the presence of the inner core of the lower-energy-level electrons.

*SI Units:* An international system of measurement units essentially the same as the metric system.

*Significant Digits:* All the numerical digits that are part of a measurement. They indicate how accurately the measurement was made and do not include zeros used as placeholders.

*Simple Ion:* An atom carrying a positive or negative charge resulting from the loss or gain of electrons.

*Simple Lipids or Lipids:* Triglycerides or esters of fatty acids and the trihydroxy alcohol glycerol ($CH_2OH$-$CHOHCH_2OH$). Fats and oils are simple lipids.

*Single Bond:* A covalent bond involving the sharing of one pair of electrons.

*Smog:* The variety of chemicals that accumulate in the immobile air mass during a temperature inversion. These chemicals include primary air pollutants and numerous secondary pollutants. The secondary pollutants are formed by sunlight-induced (photochemical) reactions between primary pollutants and atmospheric gases.

**Solid:** A physical state in which a sample of matter occupies a definite volume and has a definite, rigid shape.

**Solidification, Crystallization, or Freezing:** The change from a liquid to a solid.

**Solubility:** The amount of a chemical that can be dissolved in a specific amount of solvent to form a saturated solution.

**Solute:** The component of a solution that has been dissolved by the solvent.

**Solution:** An intimate mixture of two or more substances in which the components intermingle on an atomic, molecular, or ionic basis; a homogeneous mixture of a solute in a solvent.

**Solvent:** The component of a solution that is the dissolver; the component of a solution that is of the same physical state as the solution or is present in the greatest amount if all components are of the same state.

**Somatic Cells:** Normal body cells that in humans carry 46 chromosomes.

**Spectator Ions:** Ions that are present in solution during a reaction but do not actually react.

**Standard Solution:** A solution of known concentration.

**Standard Temperature and Pressure (STP):** A commonly used set of reference conditions for gases defined as

$$0°C \quad \text{or} \quad 273 \text{ K}$$
$$760 \text{ torr} \quad \text{or} \quad 1 \text{ atm}$$

**Starch:** A polysaccharide found in seeds and roots of many plants, it has two forms: amylose consists of long chains of α-glucose units; amylopectin consists of branched chains of α-glucose units.

**States of Matter:** The physical forms, solid, liquid, or gas in which matter occurs.

**Steroids:** A special class of lipids most of which have the basic polycyclic ring structure.

**Stoichiometry:** Calculations involving mass relations in chemical reactions.

**STP:** See standard temperature and pressure.

**Strong Acids:** Those acids that completely react with water to form the hydronium ion ($H_3O^+$) and the conjugate base of the acid. Common strong acids are hydrochloric acid, sulfuric acid, and nitric acid.

**Strong Bases:** Substances that are soluble in water and dissolve to give the hydroxide ion and the metal cation. The most common strong bases are NaOH and KOH.

**Structural Formula:** A formula for a molecular compound that indicates the atoms present and the bonding sequence of the atoms. The covalent bonds between atoms are drawn as lines between the atom symbols.

**Subatomic Particles or Nuclear Particles:** Particles including protons, neutrons, electrons, and alpha particles that structure atoms and atomic nuclei.

**Sublimation:** The change from the solid state to the vapor or gaseous state.

**Sucrose:** A common disaccharide called table sugar that is composed of an α-glucose unit bonded through C-1 to C-1 of a β-fructose unit (fructose is a sugar similar to glucose).

**Supercritical Fission:** A sequential nuclear fission process in a sample of fissile isotope that can result in a nuclear explosion.

**Supersaturated Solution:** An unstable solution that contains more solute than a saturated solution at the same temperature.

**Substrate:** The reactant with which an enzyme interacts when it catalyzes a reaction.

**Synthetic Organic Compounds:** Substances synthesized or prepared from naturally occurring compounds (usually obtained from petroleum) by industrial chemical processes that convert the natural products to more useful compounds. Synthetics include consumer products such as plastics, plasticizers, paints, pesticides, preservatives, and pharmaceuticals.

**Synthesis:** A chemical processes by which a specific chemical is made from other chemicals.

**Systematic Names:** Names for compounds based on a system of nomenclature designed to indicate the chemical nature of the compounds.

**Temperature:** An expression of the degree of hotness or coldness of an object. Temperatures are measured relative to specific reference temperatures used to establish temperature scales. The Celsius and Kelvin scales are commonly used in science.

**Temperature Inversion:** The atmospheric situation that occurs when a cooler air mass moves in at lower altitude and underlies a warmer air mass. This traps a warm air mass between the cool air mass and the cooler air at high altitude. The cool–warm–cool layers of air differ from the normal warm–cool layers, and the condition is termed an inversion. A temperature inversion can result in an immobile air mass in which air pollutants accumulate. Inversions normally occur during warm, clear weather.

**Tertiary Structure of Proteins:** The folded and twisted three-dimensional form of a protein chain created by interaction between various portions of the chain.

**Tetravalent:** The tendency of the atoms of an element to form four covalent bonds in compounds; carbon is tetravalent.

**Theory or Model:** A concept that serves as a description of why objects behave as they do or why certain processes occur.

**Titration:** Adding a measured amount of a solution of known concentration to a solution of unknown concentration to determine the unknown concentration.

**Torr:** A unit of pressure defined as 1 torr = 1 mm Hg.

**Transition Metals or Elements:** Those elements comprising the $d$ block of the periodic table; they are located between the $s$ block and the $p$ block.

**Transmutation:** See nuclear transmutation.

**Transuranium Elements:** Those elements having atomic numbers above uranium that are formed by nuclear transmutation; they do not occur naturally.

**Triple Bond or Triple Covalent Bond:** The sharing of three pairs of electrons between atoms so that they are bonded by three covalent bonds. Carbon atoms are capable of triple-bond formation in certain organic compounds.

**Unit Factor:** A ratio of two equivalent terms used to convert from one unit to another; a type of conversion factor.

**Universal Gas Constant:** See gas constant.

**Valence Electrons:** The outer-energy-level electrons of an atom available for bond formation.

**Vapor Pressure:** The pressure of a vapor associated with a liquid.

**Voltaic Cell:** See electrical cell.

**Volume Ratio:** The ratio of the number of liters of one gas to the number of liters of another gas involved in a chemical reaction. These ratios are used in volume-to-volume stoichiometric computations.

**Weak Acids:** Those acids that react only slightly with water when dissolved.

**Weight:** The product of the mass of an object and the gravitational attraction of the earth.

**Zygote:** A new cell that is formed when an egg is united with a sperm; a fertilized egg.

# ANSWERS TO

## SELECTED

## QUESTIONS

**Chapter 1**

**Chapter 2**

6. **(a)** $6.9595 \times 10^5$ km   **(b)** 568,400,000,000,000,000,000,000,000 kg   **(c)** 29,980,000,000 cm/s
   **(d)** $1.672 \times 10^{-24}$ g   **(e)** 31,536,000 s   **(f)** $2.8841 \times 10^{-8}$ cm   **(g)** $1.252 \times 10^5$ cm/s   **(h)** 345,000 cm/s

9. 71 mm; 0.5 g

12. **(a)** (1 L/1000 mL)   **(b)** (1000 g/1 kg)   **(c)** (100 mL/1 dL)   **(d)** (1 m/100 cm)
    **(e)** (1000 mg/1 g)   **(f)** (1 cm/10 mm)   **(g)** (1000 ms/1 s)   **(h)** (1000 m/1 km)

13. **(a)** 0.0437 m   **(b)** 24,000 mg   **(c)** $2.8 \times 10^7$ nm   **(d)** $3.0 \times 10^4$ ms   **(e)** 0.537 m   **(f)** 0.750 L
    **(g)** $1.689 \times 10^{-2}$ kg   **(h)** 82.1 mL

15. **(a)** 2   **(b)** 4   **(c)** 2   **(d)** 1   **(e)** 5   **(f)** 4   **(g)** 3   **(h)** 6

17. **(a)** 3   **(b)** 2   **(c)** 3   **(d)** 2   **(e)** 4 (2 past decimal)   **(f)** 5 (2 past decimal)   **(g)** 4 (2 past decimal)   **(h)** 2

20. $1.00 \, \cancel{yd} \left( \dfrac{3 \, \cancel{ft}}{1 \, \cancel{yd}} \right) \left( \dfrac{12 \, \cancel{in.}}{1 \, \cancel{ft}} \right) \left( \dfrac{2.54 \text{ cm}}{1 \, \cancel{in.}} \right) = 91.4$ cm = 914 mm

22. $403 \text{ ft} \left( \dfrac{12 \text{ in.}}{1 \text{ ft}} \right) \left( \dfrac{25.4 \text{ mm}}{1 \text{ in.}} \right) \left( \dfrac{1 \text{ m}}{1000 \text{ mm}} \right) = 123$ m     27. $\left( \dfrac{55 \, \cancel{mi}}{1 \text{ hr}} \right) \left( \dfrac{1.609 \text{ km}}{1 \, \cancel{mi}} \right) = 88$ km/hr

29. $0.0018 \, \cancel{oz} \left( \dfrac{28.35 \, \cancel{g}}{1 \, \cancel{oz}} \right) \left( \dfrac{1000 \text{ mg}}{1 \, \cancel{g}} \right) = 51$ mg, 20,000 trips (two digits)

32. 0.264 gal     37. $34.0 \, \cancel{L} \left( \dfrac{1 \, \cancel{gal}}{3.785 \, \cancel{L}} \right) \left( \dfrac{41.5 \text{ mi}}{1 \, \cancel{gal}} \right) = 373$ mi

**40.** (a) $\left(\dfrac{34.2 \text{ g}}{34.1 \text{ mL}}\right) = 1.00 \text{ g/mL}$   **(b)** $\left(\dfrac{2568 \text{ g}}{133 \text{ mL}}\right) = 19.3 \text{ g/mL}$   **(c)** 0.92 g/mL   **(d)** 0.793 g/mL

(e) $2.70 \text{ g/cm}^3$   **(f)** 1.27 g/L

**43.** (a) $10{,}000 \text{ mL}\left(\dfrac{1.00 \text{ g}}{1 \text{ mL}}\right) = 1.00 \times 10^4 \text{ g}$   **(b)** $6.72 \text{ g}\left(\dfrac{1 \text{ cm}^3}{2.07 \text{ g}}\right) = 3.25 \text{ cm}^3$   **(c)** $11.6 \text{ cm}^3$   **(d)** 3.59 kg   **(e)** 377 g

(f) $5900 \text{ mL}\left(\dfrac{1 \text{ L}}{1000 \text{ mL}}\right)\left(\dfrac{1.29 \text{ g}}{1 \text{ L}}\right) = 7.6 \text{ g}$   **(g)** $15 \text{ gal}\left(\dfrac{3.785 \text{ L}}{1 \text{ gal}}\right)\left(\dfrac{1000 \text{ mL}}{1 \text{ L}}\right)\left(\dfrac{0.70 \text{ g}}{1 \text{ mL}}\right) = 4.0 \times 10^4$

$39{,}700 \text{ g}\left(\dfrac{1 \text{ lb}}{453.6 \text{ g}}\right) = 88 \text{ lb}$

**45.** $3619 \text{ cm}^3\left(\dfrac{19.3 \text{ g}}{1 \text{ cm}^3}\right) = 69{,}800 \text{ g}, \quad 154 \text{ lb}$   **47.** $\left(\dfrac{1.00 \text{ g}}{1 \text{ cm}^3}\right)\left(\dfrac{28{,}319 \text{ cm}^3}{1 \text{ ft}^3}\right)\left(\dfrac{1 \text{ lb}}{453.6 \text{ g}}\right) = 62.4 \text{ lb/ft}^3$

**53.** 19.9 g/mL. Thus it is not pure gold.   **56.** $\left(\dfrac{1.44 \times 10^6 \text{ kg}}{72 \text{ mL}}\right)\left(\dfrac{1000 \text{ g}}{1 \text{ kg}}\right) = 2.0 \times 10^7 \text{ g/mL}$

**59.** 0.951 g/mL, yes   **61.** $\dfrac{0.216 \text{ g}}{(1/6)(3.142)(0.3742 \text{ cm})^3} = 7.87 \text{ g/cm}^3$

# Chapter 3

**13.** (a) 14.3% C   **(b)** 11.1% H   **(c)** 42.0% C
**15.** (a) 40.0 g C/100 g vinegar   **(b)** 42.1 g C/100 g table sugar   **(c)** 33.3 g C/100 g baking soda
**29.** law of constant composition, the percent by mass of O and H in water is the same in many widely separated places in the world

**52.**

|  | *Protons* | *Neutrons* | *Electrons* |
|---|---|---|---|
| (a) carbon-14 | 6 | 8 | 6 |
| (b) chlorine-37 | 17 | 20 | 17 |
| (c) $^{234}$U | 92 | 142 | 92 |
| (d) $^{18}$O | 8 | 10 | 8 |
| (e) phosphorus-32 | 15 | 17 | 15 |
| (f) calcium-45 | 20 | 25 | 20 |
| (g) $^{90}$Sr | 38 | 52 | 38 |
| (h) iron-56 | 26 | 30 | 26 |
| (i) $^{125}$I | 53 | 72 | 53 |
| (j) $^{137}$Ba | 56 | 81 | 56 |

**63.** 28.09 u

# Chapter 4

**7.** (a) $24.8 \text{ g}\left(\dfrac{1 \text{ mol Ag}}{107.87 \text{ g}}\right) = 0.230 \text{ mol Ag}$   **(b)** $53.5 \text{ kg}\left(\dfrac{1000 \text{ g}}{1 \text{ kg}}\right)\left(\dfrac{1 \text{ mol Fe}}{55.847 \text{ g}}\right) = 958 \text{ mol Fe}$   **(c)** 0.209 mol S

(d) $3.80 \times 10^{-4}$ mol W    (e) $235 \text{ torr} \left( \dfrac{1000 \text{ kg}}{1 \text{ torr}} \right) \left( \dfrac{1000 \text{ g}}{1 \text{ kg}} \right) \left( \dfrac{1 \text{ mol Al}}{26.98154 \text{ g}} \right) = 8.71 \times 10^6$ mol Al

(f) $4.81 \text{ g} \left( \dfrac{1 \text{ mol He}}{4.003 \text{ g}} \right) = 1.20$ mol He    (g) $131 \text{ mg} \left( \dfrac{1 \text{ g}}{1000 \text{ mg}} \right) \left( \dfrac{1 \text{ mol Si}}{28.086 \text{ g}} \right) = 4.66 \times 10^{-3}$ mol Si

(h) $18.7$ mol Mg    (i) $43.4$ mol Cu    (j) $4.22$ mol Pb    (k) $3.3 \times 10^{-2}$ mol Be

9. (a) $124 \text{ mol Pt} \left( \dfrac{195.09 \text{ g}}{1 \text{ mol Pt}} \right) = 2.42 \times 10^4$ g    (b) $3.18 \times 10^{-6} \text{ mol Mo} \left( \dfrac{95.94 \text{ g}}{1 \text{ mol Mo}} \right) = 3.05 \times 10^{-4}$ g

(c) $1.70 \times 10^{-5}$ g    (d) $19.75 \times 10^6$ g    (e) $475$ g    (f) $5.28 \times 10^{-8}$ g

11. (a) $84.8 \text{ g} \left( \dfrac{1 \text{ mol Hg}}{200.59 \text{ g}} \right) \left( \dfrac{6.022 \times 10^{23} \text{ atoms Hg}}{1 \text{ mol Hg}} \right) = 2.55 \times 10^{23}$ atoms Hg

(b) $0.249 \text{ g} \left( \dfrac{1 \text{ mol Ti}}{47.90 \text{ g}} \right) \left( \dfrac{6.022 \times 10^{23} \text{ atoms Ti}}{1 \text{ mol Ti}} \right) = 3.13 \times 10^{21}$ atoms Ti

(c) $3.77 \times 10^{22}$ atoms Kr    (d) $3.805 \times 10^{25}$ atoms Ni

(e) $8.23 \text{ mg} \left( \dfrac{1 \text{ g}}{1000 \text{ mg}} \right) \left( \dfrac{1 \text{ mol Ag}}{107.868 \text{ g}} \right) \left( \dfrac{6.022 \times 10^{23} \text{ atoms Ag}}{1 \text{ mol Ag}} \right) = 4.59 \times 10^{19}$ atoms Ag

(f) $1.00 \text{ oz} \left( \dfrac{31.1 \text{ g}}{1 \text{ oz}} \right) \left( \dfrac{1 \text{ mol Au}}{196.9665 \text{ g}} \right) \left( \dfrac{6.022 \times 10^{23}}{1 \text{ mol Au}} \right) = 9.51 \times 10^{22}$ atoms Au

13. $1.674 \times 10^{-24}$ g/atom H; $2.657 \times 10^{-23}$ g/atom O

15. (a) lysine: $C_3H_7NO$    (b) potassium tartrate: $KH_5C_4O_6$    (c) calcium fluoride: $CaF_2$

(d) hydroquinone: $C_3H_3O$    (e) borax: $Na_2B_4O_{17}H_{20}$    (f) saccharin: $C_7H_5O_3NS$

(g) barium phosphate:

$$\left( \dfrac{0.4981 \text{ mol Ba}}{0.3325 \text{ mol P}} \right) = \left( \dfrac{1.5 \text{ mol Ba}}{1 \text{ mol P}} \right) \qquad \left( \dfrac{1.3313 \text{ mol O}}{0.3325 \text{ mol P}} \right) = \left( \dfrac{4 \text{ mol O}}{1 \text{ mol P}} \right)$$

Empirical formula $= Ba_{1.5}PO_4$; doubled, it is $Ba_3P_2O_8$    (h) vitamin C: $C_3H_4O_3$

17. $C_{14}H_{18}N_2O_5$    19. $Al_2O_3$

25. (a) potassium oxalate:

K  $2(39.0983) = 78.1966$

C  $2(12.011) = 24.022$

O  $4(15.9994) = \underline{63.9976}$

$166.2162 \quad 166.2$ g/mol $K_2C_2O_4$

(b) pentane:

C  $5(12.011) = 60.055$

H  $12(1.0079) = \underline{12.0948}$

$72.1498 \quad 72.15$ g/mol $C_4H_{10}$

(c) $94.94$ g/mol $CH_3Br$    (d) $58.12$ g/mol $C_4H_{10}$    (e) $78.07$ g/mol $CaF_2$

(f) acetylcholine:

$$
\begin{array}{llll}
C & 7(12.011) & = & 84.077 \\
H & 16(1.011) & = & 16.126 \\
N & 14.007 & = & 14.007 \\
O & 2(15.9994) & = & \underline{31.998} \\
& & 146.208 & 146.2 \text{ g/mol } C_7H_{16}NO_2
\end{array}
$$

(g) 227.1 g/mol $C_7H_5N_3O_6$   (h) 84.30 g/mol $MgCO_3$

27. (a) 48.1 mol $NH_3$ (b) 0.0418 mol $K_2Cr_2O_7$ (c) $2.216 \times 10^{-3}$ mol NaCN
    (d) 0.0827 mol $C_{12}H_{22}O_{11}$ (e) $3.75 \times 10^4$ mol NaCl (f) 0.245 mol $C_2H_2$
    (g) 0.344 mol $C_9H_{11}NO_2$

29. (a) 610 g (b) $1.89 \times 10^4$ g (c) 557 g
    (d) 55.0 g (e) 0.582 g (f) 31.4 g
    (g) $9.88 \times 10^7$ g (h) 550 g (i) 12.9 g

31. 18.8 mol $MgSO_4$    33. 6.4 mol $Na_2CO_3$    35. 0.0697 mol $C_{12}H_{22}O_{11}$

37. (a) empirical $C_{11}H_{12}N_2O_2$, molecular $C_{11}H_{12}N_2O_2$ (b) empirical $CCl_2F_2$, molecular $CCl_2F_2$
    (c) empirical $C_8H_8O_3$, molecular $C_8H_8O_3$ (d) empirical HgCl, molecular $Hg_2Cl_2$
    (e) empirical $CH_2O$, molecular $C_5H_{10}O_5$ (f) empirical $C_4H_9$, molecular $C_8H_{18}$

39. $Au_2S_3 = 80.38\%$ Au; $Au_2S = 92.47\%$ Au

41. (a) $\left(\dfrac{14.0067}{183.18}\right) 100 = 7.646\%$ N   (b) $\left(\dfrac{4(126.9045)}{776.874}\right) 100 = 65.34\%$ I

    (c) 40.05% S (d) 49.97% C (e) 27.37% Na

44. (a) 46.68% N, 53.32% O (b) 46.65% N, 6.71% H, 20.00% C, 26.64% O
    (c) 31.90% K, 28.93% Cl, 39.17% O (d) 81.71% C, 18.29% H
    (e) 28.83% Mg, 14.25% C, 56.93% O (f) 79.95% C, 9.69% H, 10.36% N

46. (a) LiF

56. (a) 2; 1; 2 (b) 4; 3; 2 (c) 2; 2; 3 (d) 4; 1; 2 (e) 2; 3; 2 (f) 2; 9; 6; 6 (g) 3; 1; 1 (h) 1; 3; 2 (i) 1; 3; 1; 3
    (j) 2; 13; 8; 10 (k) 6; 2; 3; 1 (l) 2; 5; 1 (m) 2; 4; 1; 3; 1

# Chapter 5

1. $0.055 \text{ mL} \left(\dfrac{13.6 \text{ g}}{1 \text{ mL}}\right)\left(\dfrac{1 \text{ mol Hg}}{200.6 \text{ g}}\right) = 3.7 \times 10^{-3}$ mol Hg

3. 1.65 mol $SiO_2$

5. $3.15 \mu g \left(\dfrac{1 \times 10^{-6} \text{ g}}{1 \mu g}\right)\left(\dfrac{1 \text{ mol I}}{776.8 \text{ g}}\right)\left(\dfrac{4 \text{ mol I}}{1 \text{ mol I}}\right) = 1.62 \times 10^{-8}$ mol I

7. 10.3 g chlorophyll

9. $2.0 \text{ tsp} \left(\dfrac{0.125 \text{ oz}}{1 \text{ tsp}}\right)\left(\dfrac{28.35 \text{ g}}{1 \text{ oz}}\right)\left(\dfrac{1 \text{ mol } NaHCO_3}{84.007 \text{ g}}\right)\left(\dfrac{6.022 \times 10^{23} NaHCO_3}{1 \text{ mol } NaHCO_3}\right) = 5.1 \times 10^{22} NaHCO_3$

11. $2.25 \text{ kg rock} \left(\dfrac{1000 \text{ g}}{1 \text{ kg}}\right)\left(\dfrac{12.5 \text{ g } Al_2O_3}{100 \text{ g rock}}\right)\left(\dfrac{1 \text{ mol } Al_2O_3}{101.96 \text{ g}}\right) = 2.76 \text{ mol } Al_2O_3, 8.28 \text{ mol } O^{2-}$

16. $2\,C_2H_2 + 5O_2 \longrightarrow 4CO_2 + 2H_2O$
2 mol of $C_2H_2$ and 5 mol of $O_2$ react to form 4 mol of $CO_2$ and 2 mol of $H_2O$

(a) $\left(\dfrac{2\text{ mol }C_2H_2}{5\text{ mol }O_2}\right)$  (b) $\left(\dfrac{2\text{ mol }C_2H_2}{4\text{ mol }CO_2}\right)$

(c) $\left(\dfrac{2\text{ mol }C_2H_2}{2\text{ mol }H_2O}\right)$  (d) $\left(\dfrac{5\text{ mol }O_2}{4\text{ mol }CO_2}\right)$

17. (a) $1.43$ mol $O_2$  (b) $0.200$ mol $C_2H_2$  (c) $17.6$ mol $H_2O$  (d) $3.11$ mol $CO_2$

20. $C_2H_6O + 3O_2 \longrightarrow 2CO_2 + 3H_2O$

(a) $52.6\,g\left(\dfrac{1\text{ mol }C_2H_5OH}{46.07\,g}\right)\left(\dfrac{3\text{ mol }O_2}{1\text{ mol }C_2H_5OH}\right) = 3.43$ mol $O_2$

(b) $52.6\,g\left(\dfrac{1\text{ mol }C_2H_5OH}{46.07\,g}\right)\left(\dfrac{3\text{ mol }O_2}{1\text{ mol }C_2H_5OH}\right)\left(\dfrac{31.9988\,g}{1\text{ mol }O_2}\right) = 110$ g

(c) $52.6\,g\left(\dfrac{1\text{ mol }C_2H_5OH}{46.07\,g}\right)\left(\dfrac{2\text{ mol }CO_2}{1\text{ mol }C_2H_5OH}\right)\left(\dfrac{44.01\,g}{1\text{ mol }CO_2}\right) = 100$ g

(d) $75.0\,g\left(\dfrac{1\text{ mol }O_2}{31.9988\,g}\right)\left(\dfrac{2\text{ mol }CO_2}{3\text{ mol }O_2}\right)\left(\dfrac{44.01\,g}{1\text{ mol }CO_2}\right) = 68.8$ g

Compared with part (c) $O_2$ is the limiting reactant.

22. (a) $500\,mg\left(\dfrac{1\,g}{1000\,mg}\right)\left(\dfrac{1\text{ mol }CaCO_3}{100.08\,g}\right)\left(\dfrac{1\text{ mol }CO_2}{1\text{ mol }CaCO_3}\right)\left(\dfrac{44.01\,g}{1\text{ mol }CO_2}\right) = 0.220$ g

(b) $1.00\,g\left(\dfrac{1\text{ mol }CaCO_3}{100.08\,g}\right)\left(\dfrac{2\text{ mol HCl}}{1\text{ mol }CaCO_3}\right) = 0.0200$ mol HCl

(c) $1200\,mg\left(\dfrac{1\,g}{1000\,mg}\right)\left(\dfrac{1\text{ mol }Ca^{2+}}{40.08\,g}\right)\left(\dfrac{1\text{ mol }CaCO_3}{1\text{ mol }Ca^{2+}}\right)\left(\dfrac{100.08\,g}{1\text{ mol }CaCO_3}\right) = 2.997$ g

(d) $9.45\,g\left(\dfrac{1\text{ mol HCl}}{36.46\,g}\right)\left(\dfrac{1\text{ mol }CO_2}{2\text{ mol HCl}}\right)\left(\dfrac{44.01\,g}{1\text{ mol }CO_2}\right) = 5.70$ g

$28.4\,g\left(\dfrac{1\text{ mol }CaCO_3}{100.08\,g}\right)\left(\dfrac{1\text{ mol }CO_2}{1\text{ mol }CaCO_3}\right)\left(\dfrac{44.01\,g}{1\text{ mol }CO_2}\right) = 12.5$ g

HCl is the limiting reactant since it gives less $CO_2$.

24. (a) $34.1$ g  (b) $992$ g

(c) $500\,kg\left(\dfrac{1000\,g}{1\,kg}\right)\left(\dfrac{1\text{ mol }NH_3}{17.03\,g}\right)\left(\dfrac{1\text{ mol }CaCN_2}{2\text{ mol }NH_3}\right)\left(\dfrac{80.10\,g}{1\text{ mol }CaCN_2}\right) = 1.18 \times 10^6$ g

(d) $500\,kg\left(\dfrac{1000\,g}{1\,kg}\right)\left(\dfrac{70\,g}{100\,g}\right)\left(\dfrac{1\text{ mol }CaCN_2}{80.1\,g}\right)\left(\dfrac{2\text{ mol }NH_3}{1\text{ mol }CaCN_2}\right)\left(\dfrac{17.03\,g}{1\text{ mol }NH_3}\right)\left(\dfrac{1\,kg}{1000\,g}\right) = 1.5 \times 10^2$ kg

26. (a) $11.6\,g\left(\dfrac{1\text{ mol Na}}{22.99\,g}\right)\left(\dfrac{1\text{ mol Ti}}{4\text{ mol Na}}\right) = 0.126$ mol Ti

(b) $385 \cancel{g} \left( \dfrac{1 \cancel{mol\ TiCl_4}}{189.7 \cancel{g}} \right) \left( \dfrac{1 \cancel{mol\ Ti}}{1 \cancel{mol\ TiCl_4}} \right) \left( \dfrac{47.90\ g}{1\ mol\ Ti} \right) = 97.2\ g$

(c) $9.6 \times 10^3$ g  (d) $1.05$ g  (e) $TiCl_4$ is the limiting reactant.

28. (a) $9.25 \times 10^{-5}$ g  (b) $2.44 \times 10^{-12}$ g  (c) $5.55 \times 10^{-5}$ mol $H_2O_2$  (d) $1.56 \times 10^{-8}$ g

30. (a) $525 \cancel{g} \left( \dfrac{1 \cancel{mol\ C_6H_{12}O_6}}{180.13 \cancel{g}} \right) \left( \dfrac{2 \cancel{mol\ C_2H_5OH}}{1 \cancel{mol\ C_6H_{12}O_6}} \right) \left( \dfrac{46.07\ g}{1 \cancel{mol\ C_2H_5OH}} \right) = 269\ g$

(b) $81.5 \cancel{mol\ C_6H_{12}O_6} \left( \dfrac{2\ mol\ CO_2}{1 \cancel{mol\ C_6H_{12}O_6}} \right) = 163$ mol $CO_2$

(c) $255 \cancel{kg} \left( \dfrac{1000 \cancel{g}}{1 \cancel{kg}} \right) \left( \dfrac{1 \cancel{mol\ C_2H_5OH}}{46.07 \cancel{g}} \right) \left( \dfrac{2 \cancel{mol\ CO_2}}{2 \cancel{mol\ C_2H_5OH}} \right) \left( \dfrac{44.01 \cancel{g}}{1 \cancel{mol\ CO_2}} \right) \left( \dfrac{1\ kg}{1000 \cancel{g}} \right) = 244\ kg$

33. (a) $25.9$ g of $CO_2$ formed.  (b) $7.2$ g of $HC_2H_3O_2$ remains.

35. (a) $286$ g of aspirin are formed.  (b) $48$ g of $C_4H_6O_3$ remains.

37. (a) $25.0 \cancel{g} \left( \dfrac{1 \cancel{mol\ CaO}}{56.08 \cancel{g}} \right) \left( \dfrac{1 \cancel{mol\ Ca(OH)_2}}{1 \cancel{mol\ CaO}} \right) \left( \dfrac{74.095\ g}{1 \cancel{mol\ Ca(OH)_2}} \right) = 33.0\ g$

$21.0 \cancel{g} \left( \dfrac{1 \cancel{mol\ H_2O}}{18.02 \cancel{g}} \right) \left( \dfrac{1 \cancel{mol\ Ca(OH)_2}}{1 \cancel{mol\ H_2O}} \right) \left( \dfrac{74.095\ g}{1 \cancel{mol\ Ca(OH)_2}} \right) = 86.3\ g$

(b) $H_2O$ is in excess. Grams remaining: $13.0$ g

$33.03 \cancel{g} \left( \dfrac{1 \cancel{mol\ Ca(OH)_2}}{74.095 \cancel{g}} \right) \left( \dfrac{1 \cancel{mol\ H_2O}}{1 \cancel{mol\ Ca(OH)_2}} \right) \left( \dfrac{18.02\ g}{1 \cancel{mol\ H_2O}} \right) = 8.03\ g$

$21.0$ g $- 8.03$ g $= 13.0$ g of $H_2O$ remaining

40. $91.4\%$ yield

46. (a) $1.00 \cancel{kcal} \left( \dfrac{4.184 \cancel{J}}{1 \cancel{cal}} \right) \left( \dfrac{1\ mol\ C_{12}H_{22}O_{11}}{5649 \cancel{kJ}} \right) \left( \dfrac{342.3\ g}{1\ mol\ C_{12}H_{22}O_{11}} \right) = 0.254\ g$

(b) $3630$ kJ; $868$ kcal or $868$ Cal

(c) $\left( \dfrac{5649 \cancel{kJ}}{1 \cancel{mol\ C_{12}H_{22}O_{11}}} \right) \left( \dfrac{1 \cancel{mol\ C_{12}H_{22}O_{11}}}{342.3\ g} \right) \left( \dfrac{1\ cal}{4.184 \cancel{J}} \right) = 3.944$ kcal/g

Sucrose has $3.944$ Cal/g or $4$ Cal/g.

48. $181$ kJ $+ N_2(g) + O_2(g) \longrightarrow 2NO(g)$

$175.2 \cancel{g} \left( \dfrac{1 \cancel{mol\ N_2}}{28.014 \cancel{g}} \right) \left( \dfrac{181\ kJ}{1 \cancel{mol\ N_2}} \right) = 1130$ kJ

50. $\left( \dfrac{803\ kJ}{1 \cancel{mol\ CH_4}} \right) \left( \dfrac{1 \cancel{mol\ CH_4}}{16.04\ g} \right) = \left( \dfrac{50.1\ kJ}{1\ g\ CH_4} \right)$

$\left( \dfrac{2046\ kJ}{1 \cancel{mol\ C_3H_8}} \right) \left( \dfrac{1 \cancel{mol\ C_3H_8}}{44.10\ g} \right) = \left( \dfrac{46.39\ kJ}{1\ g\ C_3H_8} \right)$

$$\left(\frac{5322 \text{ kJ}}{2 \text{ mol } C_4H_{10}}\right)\left(\frac{1 \text{ mol } C_4H_{10}}{58.12 \text{ g}}\right) = \left(\frac{45.78 \text{ kJ}}{1 \text{ g } C_4H_{10}}\right)$$

Methane gives the most kilojoules per gram.

**51.** (a) $1.00 \times 10^3 \text{ kJ} \left(\dfrac{1 \text{ g}}{50.1 \text{ kJ}}\right) = 20.0 \text{ g}$

(b) $65.0 \text{ g} \left(\dfrac{46.39 \text{ kJ}}{1 \text{ g}}\right)\left(\dfrac{1 \text{ Btu}}{1.05 \text{ kJ}}\right) = 2.87 \times 10^3 \text{ Btu}$

(c) $35.0 \text{ Btu} \left(\dfrac{1.05 \text{ kJ}}{1 \text{ Btu}}\right)\left(\dfrac{1 \text{ g}}{45.78 \text{ kJ}}\right) = 0.803 \text{ g}$

(d) $\left(\dfrac{30 \times 10^3 \text{ Btu}}{1 \text{ hr}}\right)\left(\dfrac{1.05 \text{ kJ}}{1 \text{ Btu}}\right)\left(\dfrac{1 \text{ g}}{50.1 \text{ kJ}}\right) = \left(\dfrac{6.3 \times 10^2 \text{ g CH}_4}{1 \text{ hr}}\right)$

## Chapter 6

**17.** $2d;\ 1d;\ 3f;\ 1p$

**22.** (a) H     $1s^1$

(b) Al     $1s^2 2s^2 2p^6 3s^2 3p^1$

(c) Ga     $1s^2 2s^2 2p^6 3s^2 3p^6 3d^{10} 4s^2 4p^1$

(d) Se     $1s^2 2s^2 2p^6 3s^2 3p^6 3d^{10} 4s^2 4p^4$

(e) Kr     $1s^2 2s^2 2p^6 3s^2 3p^6 3d^{10} 4s^2 4p^6$

(f) Zr     $1s^2 2s^2 2p^6 3s^2 3p^6 3d^{10} 4s^2 4p^6 4d^2 5s^2$

(g) Sn     $1s^2 2s^2 2p^6 3s^2 3p^6 3d^{10} 4s^2 4p^6 4d^{10} 5s^2 5p^2$

(h) O     $1s^2 2s^2 2p^4$

(i) Pt     $1s^2 2s^2 2p^6 3s^2 3p^6 3d^{10} 4s^2 4p^6 4d^{10} 4f^{14} 5s^2 5p^6 5d^9 6s$

(j) P     $1s^2 2s^2 2p^6 3s^2 3p^3$

(k) I     $1s^2 2s^2 2p^6 3s^2 3p^6 3d^{10} 4s^2 4p^6 4d^{10} 5s^2 5p^5$

**41.** (a) Sc transition metal      (f) Sb representative metalloid

(b) Ar noble gas      (g) Th inner transition metal

(c) Ra representative metal      (h) Cl representative nonmetal

(d) Rn noble gas      (i) Os transition metal

(e) Cd transition metal      (j) In representative metal

**43.** (a) $1s^1$    (b) $[\text{Ne}]3s^2 3p^1$    (c) $[\text{Ar}]3d^{10} 4s^2 4p^1$    (d) $[\text{Ar}]3d^{10} 4s^2 4p^4$    (e) $[\text{Ar}]3d^{10} 4s^2 4p^6$    (f) $[\text{Kr}]4d^2 5s^2$

(g) $[\text{Kr}]4d^{10} 5s^2 5p^2$    (h) $[\text{He}]2s^2 2p^4$    (i) $[\text{Xe}]4f^{14} 5d^9 6s^1$    (j) $[\text{Ne}]3s^2 3p^3$    (k) $[\text{Kr}]4d^{10} 5s^2 5p^5$

**52.** $8.19 \times 10^{-9}$ in.

**54.** $\left(\dfrac{0.154 \text{ pm}}{1 \text{ C atom}}\right)\left(\dfrac{1 \times 10^{-9} \text{ m}}{1 \text{ pm}}\right) 6.022 \times 10^{23} \text{ C atoms} = 9.27 \times 10^{13} \text{ m}$

**57.** $4 \times 10^2$ Au atoms

| 58. | (a) N | 7 | 14.00674 | $p$ | nonmetal | $[He]2s^2 2p^3$ | 5 |
| | (b) Na | 11 | 22.989768 | $s$ | metal | $[Ne]3s^1$ | 1 |
| | (c) Fe | 26 | 55.847 | $d$ | metal | $[Ar]4s^2 3d^6$ | 2 |
| | (d) Ni | 28 | 58.6934 | $d$ | metal | $[Ar]4s^2 3d^8$ | 2 |
| | (e) Si | 14 | 28.0855 | $p$ | metalloid | $[Ne]3s^2 3p^2$ | 4 |
| | (f) No | 102 | 259.1009 | $f$ | metal | $[Rn]7s^2 6d^1 5f^{13}$ | 2 |

## Chapter 7

**11.** (a) $H^+$  (b) $K^+$  (c) $Be^{2+}$  (d) $Sr^{2+}$  (e) $Ga^{3+}$  (f) $Fr^+$  (g) $Ca^{2+}$  (h) $Rb^{2+}$

**13.** (a) $S^{2-}$  (b) $I^-$  (c) $P^{3-}$  (d) $Br^-$  (e) $Se^{2-}$  (f) $As^{3-}$

**19.** (a) $Na_2S$  (b) LiBr  (c) CaO  (d) $MgF_2$  (e) $Al_2O_3$  (f) $AlBr_3$

**20.** (a) $K_2S$  (b) NaF  (c) $BaCl_2$  (d) BeO  (e) AlP  (f) $Al_2S_3$

**31.** (a) $K_2SO_3$  (b) $Al(NO_2)_3$  (c) $Mg_3(PO_4)_2$  (d) $KC_2H_3O_2$  (e) $Ca(HCO_3)_2$  (f) AgI  (g) $HgBr_2$  (h) $(NH_4)_2SO_4$

**38.** (a) $:\ddot{C}l—\ddot{C}l:$    (b) $H—\ddot{I}:$

(c) 
$$
\begin{array}{c}
H \\
| \\
:\ddot{C}l—C—\ddot{C}l: \\
| \\
:\ddot{C}l:
\end{array}
$$

(d)
$$
\begin{array}{c}
:\ddot{O}—H \\
| \\
H
\end{array}
$$

(e)
$$
\begin{array}{c}
:Te—H \\
| \\
H
\end{array}
$$

(f)
$$
\begin{array}{c}
:\ddot{O}—\ddot{C}l: \\
| \\
:\ddot{C}l:
\end{array}
$$

(g)
$$
\begin{array}{c}
:\ddot{O}: \\
| \\
:\ddot{O}—Cl—\ddot{O}—H \\
| \\
:\ddot{O}:
\end{array}
$$

(h)
$$
\begin{array}{c}
H \quad H \\
| \quad | \\
H—\ddot{N}—\ddot{N}—H
\end{array}
$$

(i)
$$
\begin{array}{c}
:\ddot{C}l—P—\ddot{C}l: \\
| \\
:\ddot{C}l:
\end{array}
$$

(j)
$$
\begin{array}{c}
:\ddot{F}—B—\ddot{F}: \\
| \\
:\ddot{F}:
\end{array}
$$

(k)
$$
\begin{array}{c}
H \quad H \\
| \quad | \\
H—C—C—H \\
| \quad | \\
H \quad H
\end{array}
$$

(l)
$$
\begin{array}{c}
H \quad H \\
| \quad | \\
H—C—C—H \\
| \quad | \\
H—C—C—H \\
| \quad | \\
H \quad H
\end{array}
$$

(m)
$$
\begin{array}{c}
:\ddot{C}l: \\
| \\
:\ddot{C}l—C—\ddot{C}l: \\
| \\
:\ddot{C}l:
\end{array}
$$

(n)
$$
\begin{array}{c}
\ddot{O} \quad H \\
\backslash \quad / \\
H \quad \ddot{O}
\end{array}
$$

(o)  [:Ö—H]⁻

(p)  [ H—N—H with H above and H below ]⁺

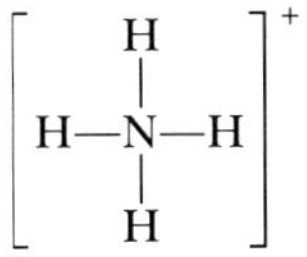

(q)  [ :Ö—P—Ö: with :Ö: above and :Ö: below ]³⁻

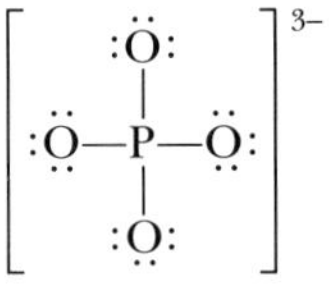

(r)  [ :O: above N, with :O. and .O: below ]⁻

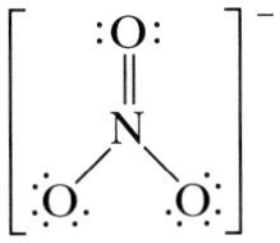

42.  (a)  :O: double bonded to C, with :Cl: and :Cl: below

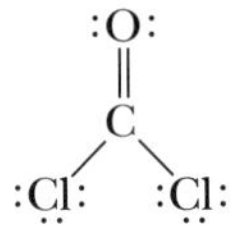

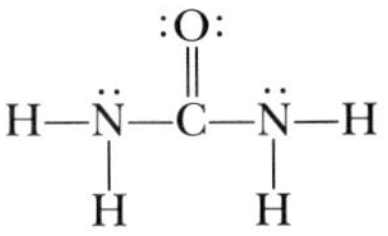

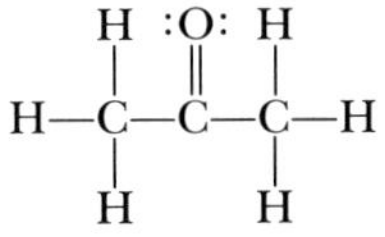

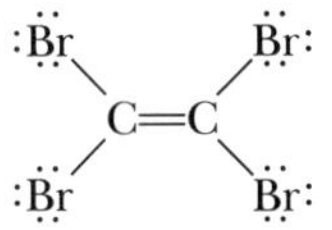

(b)  H—C≡C—H

(c)  :Ö=C=Ö:

(d)  H—C≡N:

(e)  H—N—C—N—H with :O: above C and H below each N

(f)  H—C—C—C—H with H above, :O: above middle C, H below each outer C

(g)  :Br:  and  Br: bonded to C=C with :Br: and Br: below

(h)  H and H bonded to C=C with H and Cl: below

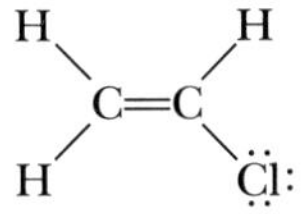

45.  (a)  Tetrahedral

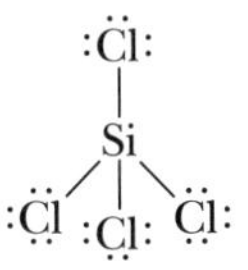

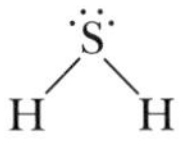

:Cl: above Si, with :Cl, :Cl:, Cl: below

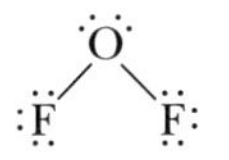

(e)  Tetrahedral

:Br: above C, with H H H below

(b)  Triangular pyramid

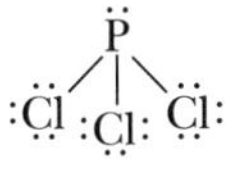

P above, :Cl, :Cl:, Cl: below

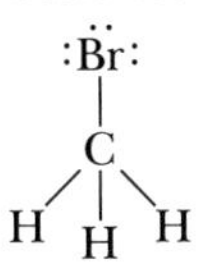

(f)  Bent

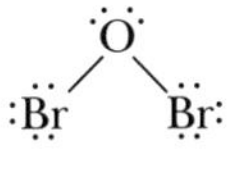

:O: above, :Br and Br: below

(c)  Bent

:S: above, H and H below

(g)  Tetrahedral

H above C, with :F, H, F: below

(d)  Bent

:O: above, :F and F: below

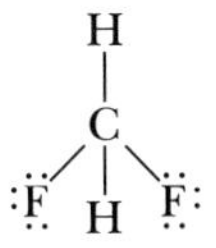

(h)  Triangular pyramidal

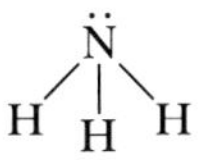

:N: above, H H H below

(i) Linear

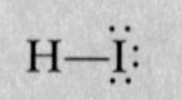

$$H—\overset{\displaystyle ..}{\underset{\displaystyle ..}{I}}:$$

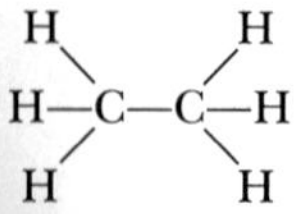

(j) Two angular

(k) Two tetrahedral

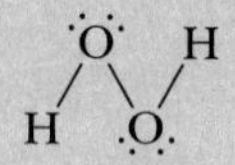

**47.** (b) linear  (c) tetrahedral  (d) bent  (e) bent  (f) bent  (i) triangular pyramid  (j) triangular
(m) tetrahedral  (o) linear  (p) tetrahedral  (r) triangular

## Chapter 8

m = metal, nm = nonmetal or metalloid, pi = polyatomic ion

**9.**

| | | | |
|---|---|---|---|
| (a) $PBr_5$ | nm–nm | phosphorus pentabromide |
| (b) $ClO_2$ | nm–nm | chlorine dioxide |
| (c) $Cu_2S$ | m–nm | copper(I) sulfide |
| (d) AlN | m–nm | aluminum nitride |
| (e) $N_2O$ | nm–nm | dinitrogen oxide |
| (f) $Ba(OH)_2$ | m–pi | barium hydroxide |
| (g) $Ni(C_2H_3O_2)_2$ | m–pi | nickel(II) acetate |
| (h) $SnCl_4$ | m–nm | tin(IV) chloride |
| (i) $BCl_3$ | nm–nm | boron trichloride |
| (j) $Na_2CrO_4$ | m–pi | sodium chromate |
| (k) $H_3PO_4$ | acid | phosphoric acid |
| (l) $HgCl_2$ | m–nm | mercury(II) chloride |
| (m) $Hg_2Br_2$ | m–nm | mercury(I) bromide |
| (n) $Cr_2O_3$ | m–nm | chromium(III) oxide |
| (o) $Fe(CN)_3$ | m–pi | iron(III) cyanide |
| (p) $Ca_3P_2$ | m–nm | calcium phosphide |
| (q) $KClO_3$ | m–pi | potassium chlorate |
| (r) HF(aq) | acid | hydrofluoric acid |
| (s) $Sn(NO_2)_2$ | m–pi | tin(II) nitrite |
| (t) $H_2SeO_4$ | acid | selenic acid |
| (u) NaCN | m–pi | sodium cyanide |
| (v) $N_2O_3$ | nm–nm | dinitrogen trioxide |
| (w) $SF_6$ | nm–nm | sulfur hexafluoride |
| (x) $HNO_3$ | acid | nitric acid |
| (y) CuO | m–nm | copper(II) oxide |
| (z) NaOH | m–pi | sodium hydroxide |
| (aa) $CCl_4$ | nm–nm | carbon tetrachloride |
| (bb) $I_2O_5$ | nm–nm | diiodine pentoxide |
| (cc) $H_2S$ | nm–nm | dihydrogen sulfide or hydrogen sulfide |
| (dd) $H_2Se$ | nm–nm | dihydrogen selenide or hydrogen selenide |
| (ee) $Na_2CO_3$ | m–pi | sodium carbonate |
| (ff) $SbBr_5$ | nm–nm | antimony pentabromide |
| (gg) $LiNO_2$ | m–pi | lithium nitrite |

11. 
| galena | PbS | lead(II) sulfide |
| anglesite | $PbSO_4$ | lead(II) sulfate |
| cerusite | $PbCO_3$ | lead(II) carbonate |
| litharge | PbO | lead(II) oxide |
| chrome yellow | $PbCrO_4$ | lead(II) chromate |

12. $CCl_4$ and $SbF_3$

15. (a) sodium nitrate, $NaNO_3$
    (b) tin(II) acetate, $Sn(C_2H_3O_2)_2$
    (c) manganese(II) sulfate, $MnSO_4$
    (d) ammonium acetate, $NH_4C_2H_3O_2$
    (e) chlorine trifluoride, $ClF_3$
    (f) diarsenic pentasulfide, $As_2S_5$
    (g) dinitrogen trioxide, $N_2O_3$
    (h) calcium flouride, $CaF_2$
    (i) iron(II) nitrate, $Fe(NO_3)_2$
    (j) magnesium carbonate, $MgCO_3$
    (k) methane, $CH_4$
    (l) copper(I) oxide, $Cu_2O$
    (m) silver phosphate, $Ag_3PO_4$
    (n) strontium cyanide, $Sr(CN)_2$
    (o) nitric acid, $HNO_3$
    (p) ammonium sulfide, $(NH_4)_2S$
    (q) phosphoric acid, $H_3PO_4$
    (r) magnesium oxalate, $MgC_2O_4$
    (s) zinc chromate, $ZnCrO_4$
    (t) mercury(II) iodide, $HgI_2$
    (u) potassium sulfite, $K_2SO_3$
    (v) sodium dichromate, $Na_2Cr_2O_7$
    (w) carbon disulfide, $CS_2$
    (x) dinitrogen pentoxide, $N_2O_5$
    (y) magnesium chloride, $MgCl_2$
    (z) hydrogen fluoride, HF

19. 
| | | | |
|---|---|---|---|
| (a) | $CCl_4$ | nm–nm | carbon tetrachloride |
| (b) | $FeSO_4$ | m–pi | iron(II) sulfate |
| (c) | MgO | m–nm | magnesium oxide |
| (d) | $HgBr_2$ | m–nm | mercury(II) bromide |
| (e) | KI | m–nm | potassium iodide |
| (f) | $Li_3As$ | m–nm | lithium arsenide |
| (g) | $NaC_2H_3O_2$ | m–pi | sodium acetate |
| (h) | $KMnO_4$ | m–pi | potassium permanganate |
| (i) | $PF_3$ | nm–nm | phosphorus trifluoride |
| (j) | HCl | nm–nm | hydrogen chloride |
| (k) | $UF_6$ | m–nm | uranium(VI) fluoride |
| (l) | $CoCl_2$ | m–nm | cobalt(II) chloride |
| (m) | $BaSO_3$ | m–pi | barium sulfite |
| (n) | $CaF_2$ | m–nm | calcium fluoride |
| (o) | $SnBr_2$ | m–nm | tin(II) bromide |
| (p) | $Zn(OH)_2$ | m–pi | zinc hydroxide |
| (q | NaBr | m–nm | sodium bromide |
| (r) | $Al_2S_3$ | m–nm | aluminum sulfide |
| (s) | AgCl | m–nm | silver chloride |
| (t) | $Ni(OH)_2$ | m–pi | nickel(II) hydroxide |
| (u) | $CrO_3$ | m–nm | chromium(VI) oxide |
| (v) | $HC_2H_3O_2$ | acid | acetic acid |
| (w | $NH_4I$ | pi–nm | ammonium iodide |
| (x) | $CaC_2O_4$ | m–pi | calcium oxalate |

22. 
| | | | | | |
|---|---|---|---|---|---|
| (a) | $HNO_3$ | nitric acid | (e) | $H_2SeO_4$ | selenic acid |
| (b) | $NO_2^-$ | nitrite ion | (f) | $AsO_4^{3-}$ | arsenate ion |
| (c) | $H_3PO_4$ | phosphoric acid | (g) | $H_2SO_3$ | sulfurous acid |
| (d) | $BrO_3^-$ | bromate ion | | | |

## Chapter 9

**15.** (a) 365 mL  (b) 231 mL  (c) 924 mL  (d) 23.1 mL

**17.** $109 \text{ L} \left( \dfrac{314 \,\cancel{K}}{263.1 \,\cancel{K}} \right) \left( \dfrac{0.931 \,\cancel{\text{atm}}}{1.08 \,\cancel{\text{atm}}} \right) = 112 \text{ L}$

**21.** (a) $625 \text{ mL} \left( \dfrac{329 \,\cancel{K}}{301 \,\cancel{K}} \right) = 683 \text{ mL}$  (b) $625 \text{ mL} \left( \dfrac{602 \,\cancel{K}}{301 \,\cancel{K}} \right) = 1.25 \times 10^3 \text{ mL}$

   (c) $625 \text{ mL} \left( \dfrac{100.3 \,\cancel{K}}{301 \text{ K}} \right) = 208 \text{ mL}$

**23.** (a) $1.08 \text{ atm} \left( \dfrac{369 \,\cancel{K}}{305 \,\cancel{K}} \right) = 1.31 \text{ atm}$

   (b) $1.08 \text{ atm} \left( \dfrac{915 \,\cancel{K}}{305 \,\cancel{K}} \right) = 3.24 \text{ atm}$

   (c) $1.08 \text{ atm} \left( \dfrac{101.7 \,\cancel{K}}{305 \,\cancel{K}} \right) = 0.360 \text{ atm}$

**25.** 2.94 atm    **26.** 770 torr

**29.** (a) $1.00 \text{ L} \left( \dfrac{1.00 \,\cancel{\text{atm}}}{0.756 \,\cancel{\text{atm}}} \right) \left( \dfrac{383 \,\cancel{K}}{273 \,\cancel{K}} \right) = 1.86 \text{ L}$  (b) 1.00 L  (c) 4.00 L

**31.** $8.96 \times 10^3 \text{ L}$    **36.** $(6.24 \times 10^4 \text{ mL torr/K mol})$

**38.** $n = \dfrac{PV}{RT} = \dfrac{(2.12 \,\cancel{\text{atm}})(50 \,\cancel{\text{gal}})(3.785 \,\cancel{L} /\cancel{\text{gal}})}{(0.0821 \,\cancel{L}\,\cancel{\text{atm}} /\cancel{K} \text{ mol})(295 \,\cancel{K})} = 17 \text{ mol C}_2\text{H}_2$

**41.** 4.55 atm

**43.** $\dfrac{(5.91 \times 10^6 \,\cancel{L})(763 \,\cancel{\text{torr}})(1 \,\cancel{\text{atm}}/760 \,\cancel{\text{torr}})(4.0026 \text{ g}/\cancel{\text{mol}})}{(0.0821 \,\cancel{L} \,\cancel{\text{atm}}/K \,\cancel{\text{mol}})(300 \,\cancel{K})} = 9.64 \times 10^5 \text{ g}$

**50.** 0.758 atm + 0.187 atm + 0.0511 atm = 0.996 atm

**52.** $n = \dfrac{PV}{RT} = \dfrac{1.07 \times 10^{-18} \,\cancel{\text{atm}})(1.00 \,\cancel{L})}{(0.0821 \,\cancel{L}\,\cancel{\text{atm}} /\cancel{K} \text{ mol})(291 \,\cancel{K})} = 4.48 \times 10^{-20} \text{ mol}$

   $4.48 \times 10^{-20} \,\cancel{\text{mol}} \left( \dfrac{6.022 \times 10^{23} \text{ molecules}}{1 \,\cancel{\text{mol}}} \right) = 2.70 \times 10^4 \text{ molecules}$

**53.** $P_{\text{oxygen}} = 756 \text{ torr} - 26.7 \text{ torr} = 729 \text{ torr}$

   $n = PV/RT = 0.0251 \text{ mol O}_2$

**55.** 197 g/mole

**57.** empirical formula $C_2H_5$, molecular formula $C_4H_{10}$

**61.** (a) 1000 mL  (b) 1.08 g  (c) 37.5 mL

**62.** **(a)** $n = \dfrac{PV}{RT} = \dfrac{(0.999 \text{ atm})(0.5000 \text{ L})}{(0.0821 \text{ L atm}/\text{K mol})(898 \text{ K})} = 6.775 \times 10^{-3} \text{ mol CO}_2$

$6.775 \times 10^{-3} \text{ mol CO}_2 \left( \dfrac{2 \text{ mol C}_2\text{H}_2}{4 \text{ mol CO}_2} \right) = 3.388 \times 10^{-3} \text{ mol C}_2\text{H}_2$

$3.388 \times 10^{-3} \text{ mol C}_2\text{H}_2 \left( \dfrac{22.4 \text{ L}}{1 \text{ mol C}_2\text{H}_2} \right) = 7.59 \times 10^{-2} \text{ L}$

**(b)** $\dfrac{(0.989 \text{ atm})(25.0 \text{ L})}{(0.0821 \text{ L atm}/\text{K mol})(296 \text{ K})} \left( \dfrac{5 \text{ mol O}_2}{2 \text{ mol C}_2\text{H}_2} \right) \left( \dfrac{32.0 \text{ g}}{1 \text{ mol O}_2} \right) = 81.4 \text{ g}$

**(c)** $286 \text{ L C}_2\text{H}_2$

**63.** **(a)** 2.38 atm

**(b)** The pressure of $CO_2$ builds up in the stomach as the gas is formed. When the pressure is great enough, the excess $CO_2$ escapes to the atmosphere, causing a burp.

**(c)** $1.19 \text{ g NaHCO}_3$

## Chapter 10

**25.** $\left( \dfrac{18.9 \text{ g}}{341 \text{ mL}} \right) \left( \dfrac{1 \text{ mol HC}_2\text{H}_3\text{O}_2}{60.05 \text{ g}} \right) \left( \dfrac{1000 \text{ mL}}{1 \text{ L}} \right) = 0.923 \, M \, \text{HC}_2\text{H}_3\text{O}_2$

**27.** $\left( \dfrac{0.088 \text{ g}}{0.237 \text{ L}} \right) \left( \dfrac{1 \text{ mol caffeine}}{194.193 \text{ g}} \right) = 1.9 \times 10^{-3} \, M \, \text{caffeine}$

**31.** $20.5 \text{ mL} \left( \dfrac{1.049 \text{ g}}{1 \text{ mL}} \right) \left( \dfrac{1 \text{ mol HC}_2\text{H}_3\text{O}_2}{60.05 \text{ g}} \right) \left( \dfrac{1}{0.115 \text{ L}} \right) = 3.11 \, M \, \text{HC}_2\text{H}_3\text{O}_2$

**33.** **(a)** $0.175 \text{ M SO}_4^{2-}$    **(b)** $0.350 \, M \, \text{Na}^+$

**35.** $0.0323 \text{ mol KNO}_3 \left( \dfrac{1000 \text{ mL}}{0.215 \text{ mol KNO}_3} \right) = 150 \text{ mL}$

**37.** 142 mL

**40.** $300 \text{ mL} \left( \dfrac{0.925 \text{ mol KOH}}{1000 \text{ mL}} \right) \left( \dfrac{56.11 \text{ g}}{1 \text{ mol KOH}} \right) = 15.6 \text{ g}$

**(b)** $520 \text{ mL} \left( \dfrac{0.487 \text{ mol NaNO}_3}{1000 \text{ mL}} \right) \left( \dfrac{84.99 \text{ g}}{1 \text{ mol NaNO}_3} \right) = 21.5 \text{ g}$

**(c)** 6.74 g    **(d)** 1.9 g    **(e)** 114 g

**42.** $5.85 \text{ L} \left( \dfrac{2.99 \times 10^{-4} \text{ mol aspirin}}{1 \text{ L}} \right) \left( \dfrac{180.16 \text{ g}}{1 \text{ mol aspirin}} \right) = 0.315 \text{ g}$

**44.** $1.00 \text{ qt} \left( \dfrac{0.946 \text{ L}}{1 \text{ qt}} \right) \left( \dfrac{9.75 \text{ mol NH}_3}{1 \text{ L}} \right) \left( \dfrac{17.03 \text{ g}}{1 \text{ mol NH}_3} \right) = 157 \text{ g NH}_3$

**46.** $25.0 \text{ mL} \left( \dfrac{1.37 \cancel{M}}{0.375 \cancel{M}} \right) = 91.3 \text{ mL}$

**48.** $100 \text{ mL} \left( \dfrac{0.250 \cancel{M}}{6.00 \cancel{M}} \right) = 4.17 \text{ mL}$

**50.** 1.25 mL      **52.**   10.0 mL

**54.** (a) $0.625\ M$   (b) $4.69\ M$   (c) $2.62\ M$

**56.** $300 \cancel{\text{ g tea}} \left( \dfrac{12.5 \text{ g sugar}}{100 \cancel{\text{ g tea}}} \right) = 37.5 \text{ g}$

**58.** 13.4 g

**60.** $\left( \dfrac{1.31 \text{ g Mg}^{2+}}{1000 \text{ g}} \right) \times 10^6 = 1.31 \times 10^3 \text{ ppm Mg}^{2+}$

**62.** $2.1 \times 10^{14} \text{ g Ag}$

**64.** The water is likely to be lethal to fish.

**66.** $0.91\ m$

**68.** 1670 mL

**70.** 5.38 L

**79.** (a) $Na^+(aq), SO_4^{2-}(aq)$    (b) $K^+(aq), Cr_2O_7^{2-}(aq)$
    (b) $Na^+(aq), F^-(aq)$    (d) $Li^+(aq), NO_3^-(aq)$
    (c) $NH_4^+(aq), Cl^-(aq)$    (f) $Ag^+(aq), NO_3^-(aq)$

**81.** (a) $H_3O^+(aq), NO_3^-(aq)$
    (b) $C_2H_5OH(aq)$
    (c) $H_2S(aq)$
    (d) $C_{12}H_{22}O_{11}(aq)$
    (e) $HF(aq)$
    (f) $Cu^{2+}(aq), SO_4^{2-}(aq)$
    (g) $H_2O_2(aq)$
    (h) $Ca^{2+}(aq), Cl^-(aq)$
    (i) $NH_3(aq)$

**87.** $-1.7°C$

**89.** $100.47°C$


## Chapter 11

**11.** (a) The finely divided starch particles from the flour may explode when ignited in air.
    (b) The increase in temperature in the internal combustion engine increases the rate of the reaction.
    (c) The increase in pressure of a reaction involving gases increases the rate.
    (d) The microorganisms can make ammonia at low temperature by use of enzymes that serve as catalysts.

(e) The increase in temperature increases the rate of internal metabolic reactions of the fish.

(f) The increased concentration of oxygen will increase the rate of combustion, resulting in a fire hazard.

(g) The wood shavings represent an increased state of subdivision of the wood.

(h) As the concentration of the oxygen decreases, the candle will stop burning.

**25.** (a) +5  (b) +2  (c) +4  (d) +7  (e) +6  (f) +4  (g) +6  (h) +6  (i) +3  (j) +5  (k) +4  (l) +4  (m) +3
(n) 0  (o) +1  (p) +3  (q) +5  (r) $-3$  (s) +4  (t) +3  (u) $-1$

**34.** (a) $Zn(s) + Cu^{2+}(aq) \longrightarrow Cu(s) + Zn^{2+}(aq)$
(c) $Zn(s) + 2H_3O^+(aq) \longrightarrow 2H_2O + Zn^{2+}(aq) + H_2(g)$
(e) $OH^-(aq) + HPO_4^{2-}(aq) \longrightarrow H_2O + PO_4^{3-}(aq)$

**39.** (a) $Hg_2^{2+} + 2Br^- \longrightarrow Hg_2Br_2$   (e) $Ca^{2+} + 2OH^- \longrightarrow Ca(OH)_2$
(b) $Ba^{2+} + 2F^- \longrightarrow BaF_2$   (f) $3Ba^{2+} + 2PO_4^{3-} \longrightarrow Ba_3(PO_4)_2$
(c) no reaction   (g) $Pb^{2+} + 2I^- \longrightarrow PbI_2$
(d) no reaction

**41.** (a) $Mg^{2+} + 2OH^- \longrightarrow Mg(OH)_2$   (g) $Fe^{3+} + 3OH^- \longrightarrow Fe(OH)_3$
(b) $Ca^{2+} + 2F^- \longrightarrow CaF_2$   (h) $Ba^{2+} + SO_4^{2-} \longrightarrow BaSO_4$
(c) $3Pb^{2+} + 2PO_4^{3-} \longrightarrow Pb_3(PO_4)_2$   $Cd^{2+} + 2OH^- \longrightarrow Cd(OH)_2$
(d) $Pb^{2+} + 2Cl^- \longrightarrow PbCl_2$   (i) no reaction
(e) no reaction   (j) $Pb^{2+} + CrO_4^{2-} \longrightarrow PbCrO_4$
(f) $Ag^+ + I^- \longrightarrow AgI$   (k) $Mg^{2+} + 2F^- \longrightarrow MgF_2$

**51.** $1.8 \times 10^3$ ethylene units

# Chapter 12

**8.** $NO_2^-$  $H_3O^+$  $SO_4^{2-}$  $NH_4^+$  $OH^-$

**32.** 1.09 *M* acetic acid   **33.** 0.928 *M* hydrochloric acid   **34.** 2.17 g $Na_2HPO_4$

**45.** (a) 8.05, basic   (e) 14.00, basic   (i) 3.64, acidic
(b) 7.57, basic   (f) 0.30, acidic   (j) 3.55, acidic
(c) 6.16, acidic   (g) 1.49, acidic   (k) 7.40, basic
(d 3.00, acidic   (h) 6.96, acidic

**47.** (a) 4.64, basic   (e) 0.00, basic   (i) 10.11, acidic
(b) 5.21, basic   (f) 13.70, acidic   (j) 9.68, acidic
(c) 5.16, basic   (g) 5.49, basic   (k) 6.22, basic
(d) 11.00, acidic   (h) 4.04, basic

# Chapter 13

**6.** O.A. = oxidizing agent   R.A. = reducing agent
(a) redox   $O_2$ = O.A.   $CH_4$ = R.A.   (c) redox   $I_2O_5$ = O.A.   CO = R.A.
(b) not redox   (d) not redox

(e) not redox          (h) not redox

(f) redox   $NO_2 = O.A.$   $SO_2 = R.A.$      (i) redox   $H^+ = O.A.$   $Ni = R.A.$

(g) redox   $Cl_2 = O.A.$   $CH_4 = R.A.$

7.   (a) C   $-4$ to $+4$     O    $0$ to $-2$     (c) I    $+5$ to $0$     C   $+2$ to $+4$

    (f)   S   $+4$ to $+6$     N   $+4$ to $+2$     (g) Cl    $0$ to $-1$     C   $-4$ to $0$

    (i)   Ni   $0$ to $+2$     H   $+1$ to $0$

12.   (a) $CH_4$ and $CO_2$; $H_2O$ and $H_2$    (b) $Fe_2O_3$ and Fe; CO and $CO_2$

    (c) Ca and $Ca(OH)_2$; $H_2O$ and $H_2$    (d) $Cl_2$ and $Cl^-$; $I^-$ and $I_2$

    (e) $Ag^+$ and Ag; Zn and $Zn^{2+}$

15.   (a) $Na^+ + Br^- + H_2O_2 + H^+$

        $H_2O_2 + 2H^+ + 2Br^- \longrightarrow Br_2 + 2H_2O$

    (b) $Na^+ + Cr_2O_7^{2-} + Na^+ + I^- + H^+$

        $Cr_2O_7^{2-} + 6I^- + 14H^+ \longrightarrow 2Cr^{3+} + 3I_2 + 7H_2O$

    (c) $K^+ + MnO_4^- + Fe^{2+} + Cl^- + H^+$

        $MnO_4^- + 5Fe^{2+} + 8H^+ \longrightarrow 4H_2O + Mn^{2+} + 5Fe^{3+}$

    (d) $H_2C_2O_4 + H_2O_2 + H^+$

        $H_2C_2O_4 + H_2O_2 \longrightarrow 2CO_2 + 2H_2O$

    (e) $Fe^{3+} + NO_3^- + K^+ + I^-$

        $2Fe^{3+} + 2I^- \longrightarrow 2Fe^{2+} + I_2$

20.   (a) $2K(s) + 2H_2O \longrightarrow 2K^+ + H_2 + 2OH^-$    (b) No reaction

    (c) $Cu(s) + Hg^{2+} \longrightarrow Hg(\ell) + Cu^{2+}$   (d) $Zn(s) + Pb^{2+} \longrightarrow Zn^{2+} + Pb(s)$    (e) no reaction

# C h a p t e r   1 4

4.   $1.8 \times 10^{-5}$

6.   (a) $K = \dfrac{[H_2][Cl_2]}{[HCl]^2}$   (b) $K = \dfrac{[SO_3]^2}{[SO_2]^2[O_2]}$   (c) $K = \dfrac{[O_3]^2}{[O_2]^3}$

    (d) $K = \dfrac{[H_3O^+][CN^-]}{[HCN]}$   (e) $K = \dfrac{[NH_4^+][OH^-]}{[NH_3]}$

9.   (a) $C_2H_4 + H_2O \rightleftharpoons C_2H_5OH$   (b) $CO + H_2O \rightleftharpoons H_2 + CO_2$

    (c) $CO_2 + H_2O \rightleftharpoons H_3O^+ + HCO_3^-$   (d) $HF + H_2O \rightleftharpoons H_3O^+ + F^-$

13.   (a) $K_a = \dfrac{[H_3O^+][HPO_4^{2-}]}{[H_2PO_4^-]}$   (b) $K_a = \dfrac{[H_3O^+][CO_3^{2-}]}{[HCO_3^-]}$

    (c) $K_a = \dfrac{[H_3O^+][NH_3]}{[NH_4^+]}$

14.   (a) $0.012$   (b) $5.7 \times 10^{-11}$   (c) $5.6 \times 10^{-10}$

20.   (a) Cooling causes a shift toward the $H_2$–$I_2$

    (b) Adding $H_2$ causes a shift toward the HI side.

    (c) A catalyst does not affect the equilibrium.

(d) A decrease in pressure has no effect since there are 2 moles of gas on each side.

(e) An increase in HI concentration causes a shift toward the $H_2$–$I_2$ side.

22. (a) Increasing the temperature causes a shift toward the $N_2$–$H_2$ side.

(b) Increasing the concentration of $H_2$ causes a shift toward the $NH_3$ side.

(c) Decreasing the concentration of $N_2$ causes a shift toward the $N_2$–$H_2$ side.

(d) Decreasing the temperature causes a shift toward the $NH_3$ side.

(e) Increasing the pressure causes a shift toward the $NH_3$ side since this side has fewer moles of gas.

24. (a) Cooling causes a shift toward the $H_2O(s)$ side.

(b) A catalyst does not affect the equilibrium.

(c) Adding extra $OH^-$ causes a shift toward the $Mg(OH)_2(s)$ side.

(d) Adding $H_3O^+$ causes a shift toward the $CO_2$–$H_2O$ side.

(e) Decreasing the pressure causes a shift toward the $CO_2(g)$ side.

(f) Both changes cause the equilibrium to shift toward the $H_2$–$O_2$ side.

(g) Adding $Cl^-$ causes a shift toward the $AgCl(s)$ side.

30. $pH = 1.5$

32. $[H_3O^+] = 7.5 \times 10^{-3}\ M$, $pH = 2.1$

36. $[H_3O^+] = 3.0 \times 10^{-5}\ M$

37. $[H_3O^+] = 4.9 \times 10^{-4}\ M$

38. $pH = 7.51$

39. $pH = 3.69$

## Chapter 15

12. (a) $^{239}_{94}Pu \longrightarrow ^{235}_{92}U + ^{4}_{2}He$    (b) $^{214}_{84}Po \longrightarrow ^{210}_{82}Pb + ^{4}_{2}He$

(c) $^{238}_{92}U \longrightarrow ^{234}_{90}Th + ^{4}_{2}He$    (d) $^{241}_{95}Am \longrightarrow ^{237}_{93}Np + ^{4}_{2}He$

(e) $^{207}_{84}Po \longrightarrow ^{203}_{82}Pb + ^{4}_{2}He$    (f) $^{210}_{82}Pb \longrightarrow ^{206}_{80}Hg + ^{4}_{2}He$

14. (a) $^{197}_{78}Pt \longrightarrow ^{197}_{79}Au + ^{0}_{-1}e$    (b) $^{14}_{6}C \longrightarrow ^{14}_{7}N + ^{0}_{-1}e$

(c) $^{82}_{35}Br \longrightarrow ^{82}_{36}Kr + ^{0}_{-1}e$    (d) $^{214}_{82}Pb \longrightarrow ^{214}_{83}Bi + ^{0}_{-1}e$

(e) $^{239}_{93}Np \longrightarrow ^{239}_{94}Pu + ^{0}_{-1}e$    (f) $^{60}_{27}Co \longrightarrow ^{60}_{28}Ni + ^{0}_{-1}e$

16. (a) $^{28}_{13}Al$ (b) $^{222}_{86}Rn$ (c) $^{0}_{-1}e$ (d) $^{238}_{92}U$

19. (a) $^{222}_{86}Rn \longrightarrow ^{218}_{84}Po + ^{4}_{2}He$

(b) Radon in the atmosphere can be breathed with air.

(c) 0.063 g after 3 half-lives.    (d) 977 atoms after 10 half-lives.

21. (b) $1.3 \times 10^{-6}$ g, $9.8 \times 10^{-9}$ g

35. (a) $^{0}_{-1}e$ (b) $^{254}_{102}No$ (c) $^{30}_{15}P$ (d) $^{32}_{16}S$ (e) $^{3}_{1}H$ (f) $^{6}_{3}Li$ (g) $^{13}_{6}C$

40. $1.67 \times 10^{12}$ J

## Chapter 16

19. (a) $H-\overset{\displaystyle O}{\overset{\|}{C}}-O-CH_3$    (b) $CH_3CH_2-\overset{\displaystyle O}{\overset{\|}{C}}-O-\overset{\displaystyle CH_3}{\underset{\displaystyle CH_3}{\overset{|}{\underset{|}{CH}}}}$

**20.** (a) $CH_3(CH_2)_6CH_2OH + HOCCH_3$   (b) $CH_3(CH_2)_3CH_2OH + CH_3CH_2CH_2COH$
$$\underset{\|}{O} \qquad\qquad\qquad\qquad \underset{\|}{O}$$

(c) $CH_3CH_2OH + CH_3COH$   (d) $CH_3CH_2OH + H_3PO_4$
$$\underset{\|}{O}$$

**24.** $NH_2CH_2CNHCHCOH$   **25.** $NH_2CHCNHCH_2COH$
$$\underset{\|}{O} \quad \underset{\|}{O} \qquad\qquad \underset{\|}{O} \quad \underset{\|}{O}$$
$\qquad\qquad CH_3 \qquad\qquad\qquad\qquad CH_3$

**30.** (a) alcohol  (b) carboxylic acid  (c) aromatic ester  (d) alkane  (e) aromatic carboxylic acid
(f) aromatic amide  (g) aromatic amino acid  (h) heterocyclic amine  (i) amine  (j) alkene
(k) aromatic  (l) aldehyde  (m) aromatic carboxylic acid  (n) amine  (o) amino acid  (p) cyclic alcohol

## Chapter 17

**20.** Thr-Val-Leu, Val-Leu-Thr, Leu-Val-Thr, Thr-Leu-Val, Leu-Thr-Val, Val-Thr-Leu
**26.** $2.8 \times 10^6$ amino acid units/day

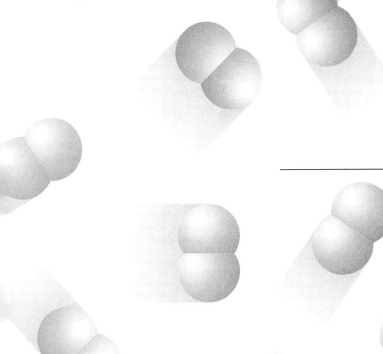

# INDEX

# Alphabetical List of Elements with Atomic Numbers and Atomic Weights*

| Element | Symbol | Atomic Number | Atomic Mass | Element | Symbol | Atomic Number | Atomic Mass |
|---|---|---|---|---|---|---|---|
| Actinium | Ac | 89 | 227.0278 | Mercury | Hg | 80 | 200.59(2) |
| Aluminum | Al | 13 | 26.981539(5) | Molybdenum | Mo | 42 | 95.94(1) |
| Americium | Am | 95 | 243.0614 | Neodymium | Nd | 60 | 144.24(3) |
| Antimony | Sb | 51 | 121.757(3) | Neon | Ne | 10 | 20.1797(6) |
| Argon | Ar | 18 | 39.948(1) | Neptunium | Np | 93 | 237.0482 |
| Arsenic | As | 33 | 74.92159(2) | Nickel | Ni | 28 | 58.6934(2) |
| Astatine | At | 85 | 209.9871 | Nielsborium* | Nr | 107 | 262.12 |
| Barium | Ba | 56 | 137.327(7) | Niobium | Nb | 41 | 92.90638(2) |
| Berkelium | Bk | 97 | 247.0703 | Nitrogen | N | 7 | 14.00674(7) |
| Beryllium | Be | 4 | 9.012182(3) | Nobelium | No | 102 | 259.1009 |
| Bismuth | Bi | 83 | 208.98037(3) | Osmium | Os | 76 | 190.2(1) |
| Boron | B | 5 | 10.811(5) | Oxygen | O | 8 | 15.9994(3) |
| Bromine | Br | 35 | 79.904(1) | Palladium | Pd | 46 | 106.42(1) |
| Cadmium | Cd | 48 | 112.411(8) | Phosphorus | P | 15 | 30.973762(4) |
| Calcium | Ca | 20 | 40.078(4) | Platinum | Pt | 78 | 195.08(3) |
| Californium | Cf | 98 | 251.0796 | Plutonium | Pu | 94 | 244.0642 |
| Carbon | C | 6 | 12.011(1) | Polonium | Po | 84 | 208.9824 |
| Cerium | Ce | 58 | 140.115(4) | Potassium | K | 19 | 39.0983(1) |
| Cesium | Cs | 55 | 132.90543(5) | Praseodymium | Pr | 59 | 140.90765(3) |
| Chlorine | Cl | 17 | 35.4527(9) | Promethium | Pm | 61 | 144.9127 |
| Chromium | Cr | 24 | 51.9961(6) | Protactinium | Pa | 91 | 231.03588(2) |
| Cobalt | Co | 27 | 58.93320(1) | Radium | Ra | 88 | 226.0254 |
| Copper | Cu | 29 | 63.546(3) | Radon | Rn | 86 | 222.0176 |
| Curium | Cm | 96 | 247.0703 | Rhenium | Re | 75 | 186.207(1) |
| Dysprosium | Dy | 66 | 162.50(3) | Rhodium | Rh | 45 | 102.90550(3) |
| Einsteinium | Es | 99 | 252.083 | Rubidium | Rb | 37 | 85.4678(3) |
| Erbium | Er | 68 | 167.26(3) | Ruthenium | Ru | 44 | 101.07(2) |
| Europium | Eu | 63 | 151.965(9) | Rutherfordium* | Rf | 104 | 261.11 |
| Fermium | Fm | 100 | 257.0951 | Samarium | Sm | 62 | 150.36(3) |
| Fluorine | F | 9 | 18.9984032(9) | Scandium | Sc | 21 | 44.955910(9) |
| Francium | Fr | 87 | 223.0197 | Seaborgium* | Sg | 106 | 263.118 |
| Gadolinium | Gd | 64 | 157.25(3) | Selenium | Se | 34 | 78.96(3) |
| Gallium | Ga | 31 | 69.723(4) | Silicon | Si | 14 | 28.0855(3) |
| Germanium | Ge | 32 | 72.61(2) | Silver | Ag | 47 | 107.8682(2) |
| Gold | Au | 79 | 196.96654(3) | Sodium | Na | 11 | 22.989768(6) |
| Hafnium | Hf | 72 | 178.49(2) | Strontium | Sr | 38 | 87.62(1) |
| Hahnium* | Ha | 105 | 262.114 | Sulfur | S | 16 | 32.066(6) |
| Hassium* | Hs | 108 | 265 | Tantalum | Ta | 73 | 180.9479(1) |
| Helium | He | 2 | 4.002602(2) | Technetium | Tc | 43 | 98.9072 |
| Holmium | Ho | 67 | 164.93032(3) | Tellurium | Te | 52 | 127.60(3) |
| Hydrogen | H | 1 | 1.00794(7) | Terbium | Tb | 65 | 158.92534(3) |
| Indium | In | 49 | 114.82(1) | Thallium | Tl | 81 | 204.3833(2) |
| Iodine | I | 53 | 126.90447(3) | Thorium | Th | 90 | 232.0381(1) |
| Iridium | Ir | 77 | 192.22(3) | Thulium | Tm | 69 | 168.93421(3) |
| Iron | Fe | 26 | 55.847(3) | Tin | Sn | 50 | 118.710(7) |
| Krypton | Kr | 36 | 83.80(1) | Titanium | Ti | 22 | 47.88(3) |
| Lanthanum | La | 57 | 138.9055(2) | Tungsten | W | 74 | 183.85(3) |
| Lawrencium | Lr | 103 | 262.11 | Uranium | U | 92 | 238.0289(1) |
| Lead | Pb | 82 | 207.2(1) | Vanadium | V | 23 | 50.9415(1) |
| Lithium | Li | 3 | 6.941(2) | Xenon | Xe | 54 | 131.29(2) |
| Lutetium | Lu | 71 | 174.967(1) | Ytterbium | Yb | 70 | 173.04(3) |
| Magnesium | Mg | 12 | 24.3050(6) | Yttrium | Y | 39 | 88.90585(2) |
| Manganese | Mn | 25 | 54.93805(1) | Zinc | Zn | 30 | 65.39(2) |
| Meitnerium* | Mt | 109 | 266 | Zirconium | Zr | 40 | 91.224(2) |
| Mendelevium | Md | 101 | 258.10 | | | | |

In some atomic weights the uncertainties in the last digit between ±1 and ±9 are given in parentheses.
*As recommended by the American Chemical Society. Not official.